MICHAEL Ralph

CW01669949

Fundamentals of Motor Vehicle Technology

Fourth Edition

V.A.W. Hillier IEng, MIRTE, FIMI

Formerly Principal Lecturer in Automobile Engineering at Croydon College

Stanley Thornes (Publishers) Ltd

First published by Hutchinson Education 1966
Second edition 1972
Reprinted 1975, 1977, 1978, 1979 and 1980
Third edition 1981
Reprinted 1982, 1984, 1986, 1987
ISBN 0 09 143161 1
Reprinted 1990 by:
Stanley Thornes (Publishers) Ltd
Old Station Drive
Leckhampton
CHELTENHAM GL53 0DN
England
ISBN 0 7487 0317 9

Fourth edition 1991
Reprinted 1992

British Library Cataloguing in Publication Data
Hillier, V. A. W.
Fundamentals of motor vehicle technology. 4th ed.
1. Transport engineering
I. Title
629.2

ISBN 0 7487 0516 3

Typeset by Tech-Set, Gateshead, Tyne & Wear.
Printed and bound in Great Britain at The Bath Press, Avon.

Contents

Preface v
Acknowledgements vii
Units and symbols viii

Part One: Vehicle Development and Layout

1 Vehicle development 1
2 Vehicle layout 4

Part Two: The Petrol Engine

3 Main parts and working principles 15
4 Practical applications of the cycle of operations 18
5 Crankshafts 21
6 Connecting rods 26
7 Pistons, rings and gudgeon pins 29
8 Valves and valve-operating gear 36
9 The cylinder block and crankcase 48
10 The four-stroke cycle in detail: valve and ignition timing 53
11 Combustion chambers 58
12 Multi-cylinder engines 64

Part Three: The Fuel System

13 The fuel supply system 75
14 The functions of the carburettor and fuel injection system 80
15 The simple carburettor 84
16 Constant choke carburettors 88
17 Constant-depression carburettors 100
18 Inlet manifolds 106
19 Petrol injection 112
20 Air cleaners and silencers 126
21 Emission control 131

Part Four: The Cooling System

22 The function of the cooling system 137
23 Main features of an air-cooled system 140
24 Main features of a liquid-cooled system 141

Part Five: Engine Lubrication

25 The functions and properties of lubricants 155
26 The lubrication system 158

Part Six: Bearings

27 Plain bearings 167
28 Ball and roller bearings 171

Part Seven: The Compression-ignition Engine

29 Cycle of operations 175
30 Engine construction: combustion chambers 178
31 The fuel system 183

Part Eight: Supercharging

32 Principles of supercharging 199
33 Types of supercharger 201

Part Nine: Alternative Types of Engine

34 Propulsion units 207

Part Ten: Clutches

35 The single-plate clutch 213
36 The multi-plate clutch 221

Part Eleven: Gearboxes

37 The gearbox 223
38 Sliding-mesh gearbox 231
39 Constant-mesh and synchro-mesh gearbox 237
40 Planetary gearing and unidirectional clutches 244
41 The automatic gearbox 250
42 The overdrive 277

Part Twelve: Propeller Shafts and Drive Arrangements

43 Propeller shafts and universal joints 279
44 Drive arrangements 288

Part Thirteen: Rear Axles

45 Final-drive gears 293
46 The differential 299
47 Rear axle construction 305
48 Four-wheel drive 308

Part Fourteen: Steering Systems

49 Directional control 313
50 Camber, castor and swivel-axes inclination 318
51 Steering components 324
52 Four-wheel steering 331

Part Fifteen: Vehicle Structure

53 General frame construction 337
54 Integral construction of frames 341
55 Heating, ventilating and air conditioning 346

Part Sixteen: Suspension Systems

56 Types and characteristics of metal springs 357
57 Rubber suspension 360
58 Air suspension 362
59 The damper 364
60 Independent front suspension (IFS) 367
61 Independent rear suspension (IRS) 371

Part Seventeen: Wheels and Tyres

62 Wheels 375
63 Tyres 378
64 Wheel balancing 383

Part Eighteen: Brakes

65 Breaking principles 385
66 Brake arrangements 391
67 Hydraulic operating systems 399
68 Servo operation 407
69 Anti-lock braking systems 412

Part Nineteen: Electrical Equipment

70 General layout and basic principles 419
71 The battery 424
72 The charging system 429
73 The lighting system 436
74 The ignition system 441
75 The starter system 458
76 Auxiliary equipment 464
77 Fault diagnosis 470

Index 476

Preface

The first edition of this book was written in response to a need for a textbook covering the construction of motor vehicles and their components in a manner simple enough to be understood by young apprentices beginning their training as mechanics, and detailed enough to serve as a solid foundation for later work.

However, even the brightest students will not find the use of this book sufficient in itself, but will require the help and guidance of a competent teacher who has had no little experience of practical work on motor vehicles, and students should themselves be working in a motor vehicle repair shop, handling the actual components. They will need also to develop the habit of examining the components in order to become familiar with the details of their design and construction, so that they can, if called upon to do so, describe and sketch these details from memory with reasonable accuracy.

The book is divided into parts, each of which covers a major topic or component. Each part is divided into chapters which deal with a particular aspect of the topic. This enables teachers to select appropriate chapters from various sections to suit their own scheme of work. I must emphasize that I expect the teacher to amplify the content of the book wherever necessary. The widest possible range of examples of each component, assembly, unit and layout should be handled and discussed by students.

Most of the drawings have been kept simple but, bearing in mind that usually a teacher has little time for more than a simple sketch on the blackboard, many of them are deliberately pictorial. Drawings have been preferred to photographs on account of their greater clarity of detail.

The book is intended particularly for students preparing for examinations set by the City and Guilds of London Institute and BTEC, but is flexible enough to be suitable for any course of a similar level.

Since the first edition the book has passed through two further editions, and nearly ten years have now elapsed since it was last revised. During this period motor vehicles have undergone many changes, with the result that many statements made and components covered in the third edition have now become obsolete.

The revision associated with the updating of material has enabled new topics to be introduced. In addition the use of modern reproduction methods has allowed the quality of the artwork to be improved.

The reasons for the changes in vehicle design are due to new emission legislation and the need for each manufacturer to offer a product that either matches or beats the competiton. New features and systems have been introduced to improve safety, power, economy and comfort of a vehicle. In addition special systems, which in the past were regarded as luxury items, are now used on the full range, including the cheaper models. Extra systems and refinements mean that there are more things to go wrong, so in order to perform an efficient repair, the personnel engaged in this activity must be competent, both in the theoretical and practical aspects. The aim of this book is to give the reader a basic knowledge of the various systems; this study should provide a foundation to enable a manufacturer or training establishment to develop the skill needed to enable a complex motor vehicle to be repaired in a safe, economic and efficient manner.

As vehicle construction becomes more complicated, it is difficult to cover the many principles and systems in one book, so in order to give a suitable depth of treatment, it has been found necessary to exclude some of the heavy vehicle topics that were covered in earlier editions.

Developments in electronics have resulted in the widespread use of solid-state control devices and these have replaced many of the mechanical systems. Since an effective repair of fully integrated electronic/ mechanical systems demands a good knowledge of both subjects, it is anticipated that this book will be used in conjunction with the companion book entitled *Fundamentals of Automotive Electronics*.

Sincere thanks are expressed to the staff of the companies listed for the help and advice given during the preparation of this edition. With their assistance it is possible to include new material such as engine structural features, engine balance, fuel systems, new lightweight diesel-type engines, automatic gearboxes, universal joints, visco drive units, four-wheel drive systems, four-wheel steering, air conditioning and anti-lock braking systems.

V.A.W.H.

Acknowledgements

I would like to thank the following firms for permission to make use of copyright and other material:

AE Group plc
Automotive Products plc
Borg Warner Ltd
British Standards Institution
Bosch, Robert, Ltd
Champion Sparking Plug Co. Ltd
Duckham (Alexander) & Co. Ltd
Dunlop Ltd
Eaton Ltd
Ford Motor Co. Ltd
GKN
Honda (UK) Ltd
Jaguar Cars
Jensen Cars
Johnson Matthey
Lucas Industries
Mazda Cars (UK) Ltd
Perkins Engines Ltd
Porsche Cars, Great Britain, Ltd
Renault UK Ltd
Renold Ltd
Rover Group plc
Saab (Great Britain) Ltd
Schrader Automotive Products Ltd
SKF Ltd
SU-Butec
Teves, Alfred, GmbH
Toyota (GB) Ltd
Volkswagen (UK) Ltd

Though many of the drawings are based on commercial components, they are mainly intended to illustrate principles of motor vehicle technology. For this reason, and because component design changes so rapidly, no drawing is claimed to be up to date. Students should refer to manufacturers' publications for the latest information.

Units and symbols

SI units are used throughout this book; the actual unit given is that used by the manufacturers of the component being considered.

In cases where it is felt that the reader might experience difficulty in comprehending the SI value, approximate Imperial equivalents are shown in brackets.

The following table gives a summary of the main units and symbols together with conversion factors for changing from one system to the other.

Quantity and symbol	*SI unit*		*Imperial unit*		*Conversion factor*
Length, l	millimetre	(mm)	inch	(in)	25.4
	metre	(m)	foot	(ft)	0.3048
	kilometre	(km)	mile	(mile)	1.609
Area, A	sq. centimetre	(cm^2)	sq. inch	(in^2)	6.452
	sq. metre	(m^2)	sq. inch	(in^2)	0.000 645
Volume, V	cu. centimetre	(cm^3)	cu. inch	(in^3)	16.387
	cu. metre	(m^3)	cu. foot	(ft^3)	0.028 317
Capacity, V	litre	(l)	cu. inch	(in^3)	0.016 39
			pint	(pt)	0.568 26
Mass, m	kilogram	(kg)	pound	(lb)	0.4536
	tonne	(t)	ton	(t)	1.016
	(1000 kg)				
Force, F	newton	(N)	pound force	(lbf)	4.448
	(1 kgf = 9.81 N)				
Torque, T	newton metre	(N m)	pound inch	(lbf in)	0.113
			pound foot	(lbf ft)	1.4
Pressure, p	newton/metre^2	(N/m^2)	pound/sq. inch	(lbf/in^2)	6895
	pascal	(Pa)	pound/sq. inch	(lbf/in^2)	6895
	kilopascal	(kPa)	pound/sq. inch	(lbf/in^2)	6.895
	bar	(bar)	atmosphere	(atm)	1

Quantity and symbol	*SI unit*		*Imperial unit*		*Conversion factor*
Stress, σ	newton/metre^2 kilopascal megapascal	(N/m^2) (kPa) (MPa)	pound/sq. inch pound/sq. inch ton/sq. inch	(lbf/in^2) (lbf/in^2) (ton/in^2)	6895 6.895 15.44
Power, p	kilowatt	(kW)	horsepower	(hp)	0.746
Velocity, v	metre/second kilometre/h	(m/s) (km/h) (or kph)	foot/second mile/hour	(ft/s) (mile/h) (or mph)	0.3048 1.609
Acceleration, a	metre/second^2	(m/s^2)	foot/second^2	(ft/s^2)	0.3048

To change:

Imperial to SI, multiply by conversion factor;
SI to Imperial, divide by conversion factor.

PART ONE

Vehicle development and layout

1 Vehicle development

At a very early stage in his descent from the apes, man must have realized that the means of transport with which nature had provided him left much to be desired. He was severely limited in the loads he could carry, the distance he could carry them and the speed at which he could travel, even without a load. Furthermore, it is a safe guess that the physical exertion involved was no more to his liking then than it is today.

The taming and training of suitable animals enabled heavier loads to be carried greater distances, often at greater speeds than man was capable of attaining, and there was the added advantage that most of the effort was provided by the animal, while the man could travel at his ease in comfort.

Heavy loads were dragged upon sledges until an early and unknown engineer invented the wheel. This made it possible to construct crude carts upon which even heavier loads could be carried more easily. The one drawback to the use of wheeled vehicles was – and still is – the necessity of providing a reasonably smooth and hard surface upon which the wheels could run. The development of wheeled vehicles is closely related to the development of roads.

As new materials and manufacturing methods were developed, it became possible to make improvements in vehicles, but so long as animals were the only form of motive power it was not possible to increase loads and speeds very much. The development of the steam engine during the eighteenth and nineteenth centuries led to its application to the driving of vehicles, and though some of the early attempts were crude and not very successful, several extremely promising carriages were produced which might have been developed into very practical vehicles had not restrictive legislation forced them off the roads.

The steam engine proved less suited to road vehicles than it did to the railway. It was the successful development of the light, high-speed internal combustion engine towards the end of the nineteenth century which really opened up the way to the power-driven road vehicle, and which made possible the development of the modern motor-car, truck, bus and coach.

Motor vehicles were developed from horse-drawn carriages – they were, in fact, called 'horseless carriages' – and naturally owed something of their general form to those carriages. For instance, the system of four wheels arranged one at each end of two transverse axles so that their points of contact with the ground are at the corners of a rectangle, as shown in Fig. 1.1a, has been used on carts and waggons from time immemorial and is still by far the commonest arrangement. While three wheels are sufficient to give stability, they do not provide so much 'useful space' for a given amount of road space taken up (compare Fig. 1.1a with Fig. 1.1b).

The horse was invariably put in front of the cart where he could see where he was going and where the driver could keep an eye on him. He steered the vehicle through shafts attached to a front axle which could pivot about its centre, and when it came to replacing the horse by an engine, it was natural that

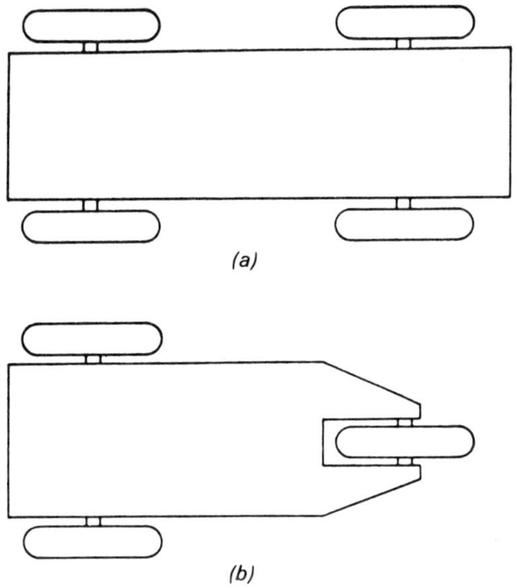

Fig. 1.1 Wheel arrangements for a simple vehicle

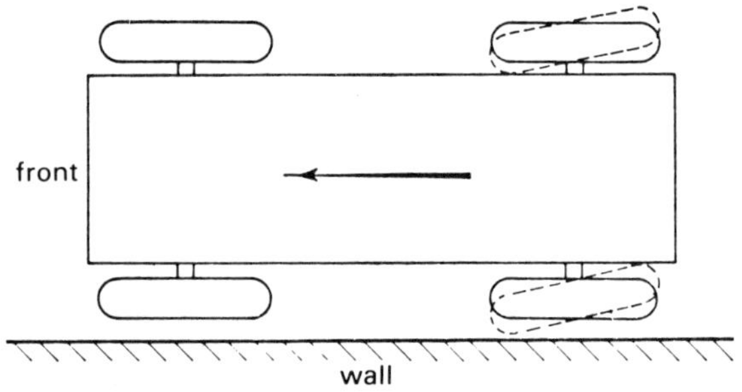

Fig. 1.2 One inconvenience of rear-wheel steering

front-wheel steering should be retained at least for a while. It was not long, however, before vehicles with rear-wheel steering were tried, and it was soon found that rear-wheel steering had disadvantages which ruled it out for general use. For example, a vehicle steered by its rear wheels would steer to the *right* by deflecting its rear end to the *left*, making it impossible to drive away *forwards* from a position close to a wall or kerb, as illustrated in Fig. 1.2. A car moving at any speed and coming alongside a wall, kerb, ditch or another vehicle could very easily find itself in a situation in which a collision could not be avoided.

The swivelling axle arrangement is not very satisfactory for powered vehicles, partly because a good deal of space must be left for the axle and wheels to swivel, and also because if one wheel strikes an obstruction, such as a large stone, it is extremely difficult, without the leverage of the long shafts, to prevent the axle swivelling about its pivot, causing the vehicle to swerve off the road. An alternative arrangement in which the wheels were carried on *stub* axles free to pivot at the ends of a fixed axle had already been used on some horse carriages, and this was soon adopted for motor vehicles.

When it came to using mechanical power to drive vehicles, it was natural that the power should be applied to the non-steerable wheels, since the problem of driving these is simpler than the one in which the driving wheels have also to be swivelled for steering purposes. This explains why, in the past, rear-wheel drive was universally adopted. The increased load on the rear wheels when climbing hills or accelerating gives a better grip so this makes rear-wheel drive attractive. Front-wheel-drive vehicles lose this advantage, so under these conditions the wheels spin more easily. This drawback, together with the extra complication of the drive arrangement to the steered wheels, is sacrificed to achieve either extra space for the vehicle's occupants or a shorter vehicle.

The most convenient source of power so far developed for driving road vehicles is the internal-combustion engine, which derives its power from the burning of fuel *inside the engine itself*. Alternative power units are the steam engine and the electric motor. The former requires a boiler to generate the steam, in addition to the engine itself, making the complete installation rather bulky. Also, there are heat losses in the boiler as well as in the engine, so that it is less efficient than the internal-combustion engine. During temporary halts, it continues to consume fuel in order to maintain steam pressure, whereas the internal-combustion engine consumes fuel only when it is actually running.

An electric motor needs a supply of electrical energy to operate it. If this is to be carried on the vehicle it must be in the form of *batteries* or *accumulators*, which are both heavy and bulky in relation to the amount of energy they can store, limiting the range, speed and load-carrying capacity of the vehicle. If an external supply of electrical energy is to be used, the vehicle must be connected to a distant power station by some means such as a system of wires, suspended at a safe height above the road, with which contact is made by suitable arms mounted on top of the vehicle and provided with some form of sliding connection to the wires. Such a system has obvious limitations.

Although both steam and electric vehicles have been used, the internal-combustion engine is now almost universal.

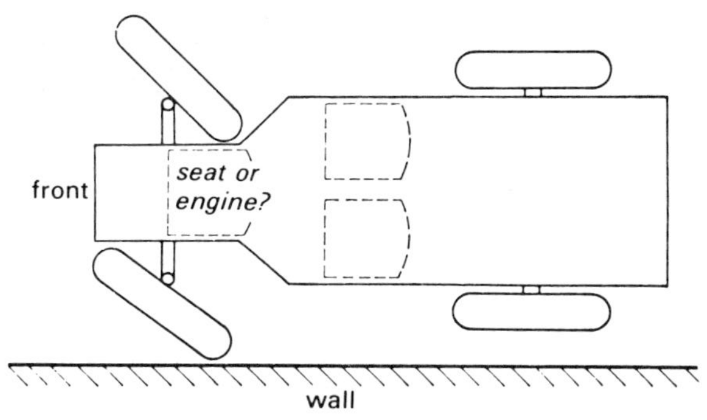

Fig. 1.3 One reason for placing the engine at the front

If you look at Fig. 1.3 you will see that the width between the front wheels is restricted by the necessity for the wheels to swivel for steering purposes, but there is a space which might provide room for one seat. However, people seem to prefer to sit side by side, and, though there may be much to be said in favour of placing the driver by himself in the extreme front of the vehicle, this arrangement is not popular. The engine can be made to fit into this space very conveniently, and although the earliest vehicles had their engines elsewhere, this position was adopted by almost all manufacturers from a very early stage.

The internal-combustion engine has two characteristics that necessitate certain arrangements in the mechanism connecting the engine to the driving wheels:

1 The engine cannot produce any driving effort when it is not running. When a steam engine is at rest, it is necessary merely to admit steam under pressure from the boiler, by opening a valve, and the engine will at once start to work. An electric motor needs only to be switched on. An internal-combustion engine, on the other hand, must be driven by some external means before it will begin to run under its own power. Once it is running, it has to be connected to the driving wheels of the vehicle by some arrangement which permits the running engine to be coupled smoothly and without shock to the initially stationary wheels.

   Most modern vehicles have a device called a *clutch*, which enables the running engine to start the vehicle smoothly.

2 The power developed by an engine depends upon the speed at which it runs. A small engine running at high speed can develop as much power as a larger engine running at low speed, and is to be preferred since it will be lighter and take up less space. Such small high-speed engines generally run at a speed some four or five times greater than the road wheels, so a speed-reducing gear has to be included in the driving mechanism connecting the engine to the driving wheels.

   There is, morever, a minimum speed below which the engine will not run. Most modern engines have a speed range from about 800 to about 6000 revolutions per minute (rev/min) or more, so that if the maximum speed of a car is, say, 150 km/h, its minimum speed will be about 20 km/h, unless the clutch is partly disconnected or *slipped*. In starting off from rest, the clutch cannot be fully engaged until the road speed has reached at least 13 km/h, during which time a great deal of heat will have been generated in the slipping clutch, which will damage it. This difficulty is overcome by providing additional gearing in the driving arrangements which can be brought into use – either by the driver or automatically – to enable the car to move slowly with the clutch fully engaged and the engine running faster than its minimum speed. Several such *speeds* or *gears* are usually provided in a mechanism called the *gearbox*.

   Another function of these gears is to allow the engine to run at high speeds when the car is moving slowly, so making its maximum power available for climbing steep hills or accelerating rapidly from low speeds.

2 Vehicle layout

Light vehicles

Vehicles classified in this category have a laden mass of less than three tonnes: this group is the main subject for this book. A wide range of different body shapes and sizes come into this light vehicle category; these range from two-seater cars to personnel carriers (mini-buses) and small trucks.

Layout of components

Fig. 2.1 shows a simplified layout of the various basic components used on a vehicle. The layout of this vehicle, which is based on a minibus, is a comparatively simple construction. This is chosen to enable the reader to see the location, and study the duty, of each of the main components. Once this foundation has been established it will be possible to appreciate the finer details of the more sophisticated systems in common use.

Wheels Most light vehicles run on four wheels fitted with hollow rubber tyres filled with air under sufficient pressure to support the load they have to carry. These

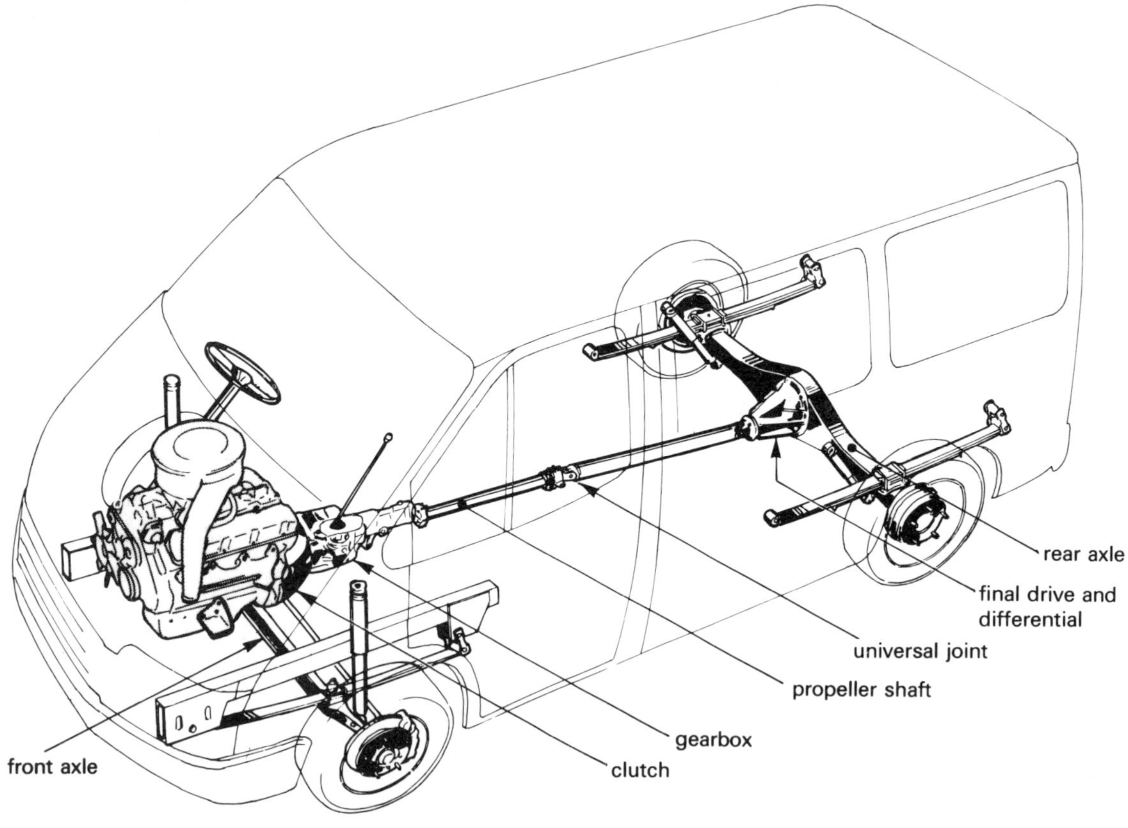

Fig. 2.1 Layout of main components

provide grip to the road and absorb shocks caused by small road irregularities. Larger shocks are taken by suspension springs and these allow the wheels to move vertically in relation to the rest of the vehicle.

Front axle This arrangement supports the front of the vehicle and is also used for steering. In Fig. 2.1 the axle is mounted on leaf-type springs to form the flexible suspension system needed to absorb the road shocks. Each wheel is carried on a *stub axle*; this is pivoted to the extremity of the front axle by a *king pin*. The two stub axles are linked together by *steering arms*, which are joined together by a *track rod* and connected to a linkage coupled to the driver's steering wheel.

On most modern light vehicles the one-piece axle beam has been superseded by a suspension arrangement that allows each front wheel to rise and fall with its own spring without affecting the other front wheel. This arrangement is called *independent front suspension*.

Rear axle This carries the wheels and supports the weight of the rear of the vehicle. The axle type shown is tubular in section and contains two axle *shafts* (*half shafts*), one for each side, to drive the roadwheels.

On the vehicle centre line the axle is enlarged to house the *final drive*. This pair of gears 'turns the drive through 90°' and reduces the driving speed to that which is suitable for the size of the road wheels.

When the vehicle is turning a corner, the inner and outer road wheels travel at different speeds. When the two road wheels are both rigidly connected to one axle shaft, the greater distance covered by the outer wheel causes one or both wheels to slip on the road. In addition to causing excessive tyre wear, this action makes the vehicle difficult to steer in a path other than the straight-ahead direction. These problems are overcome by using a *differential*. Fitted adjacent to the final drive gears, this unit ensures that each road wheel can rotate at a speed that suits the cornering conditions; in addition it arranges for each wheel to receive an equal driving effort irrespective of the speed of rotation of the wheel.

An alternative to the construction shown is to mount the final drive and differential casing to the vehicle frame and arrange each wheel to be independently sprung. This *independent rear suspension* arrangement uses a short *drive shaft* with a *universal joint* at each end to transmit the drive from the differential to the wheel.

Power unit The normal source of power is provided by an *internal combustion engine*. The petrol (gasoline) engine is the most popular for its superior performance, but if the vehicle is used extensively, the excellent fuel economy given by a diesel (compression ignition) engine makes this type attractive. The economy feature of the diesel engine offsets the higher initial cost and slightly reduced output.

Transmission system In the UK the transmission system covers the complete driveline between the engine and the road wheels. However, in many other countries the term 'transmission' refers to the gearbox unit and applies in particular to the type of arrangement used on a given vehicle, i.e. whether the selection of the gears has to be carried out manually by the driver or uses automatic means.

Gearbox consists of sets of gears that amplify the output given by the engine to enable the driving force at the road wheels to be increased sufficiently to overcome the resistance to movement of the vehicle. Also the gearbox enables the engine speed to be kept within its working limits irrespective of the speed of the vehicle. In addition it provides a neutral, where the engine can run without moving the vehicle, and a reverse, to drive the vehicle backwards.

Automatic transmission systems relieve the driver of the task of gear selection other than choosing the direction of travel and gear pattern needed to suit the general driving conditions. The system used allows the vehicle to be either moved off or brought to rest in a smooth manner without the driver having to do anything other than press the accelerator or brake pedal.

Clutch The gears in a manual gearbox have to be *changed*. This requires the driver to move a selector lever which, in turn, either engages or disengages a pair of gears. This action should only be performed when the gears are not under load, so a clutch is fitted to meet this need. This enables the driver to disconnect the engine from the gearbox.

A sudden connection of the drive from the engine to the transmission would cause a severe jolt and subsequent damage to the vehicle. The clutch is designed to avoid this; it allows the driver to accelerate the vehicle away from a stationary position gradually. Also it enables the driver to disconnect the engine from the wheels temporarily when road conditions dictate such a need.

Propeller shaft This is the long tubular shaft which links the gearbox to the final drive. Normally an *open-type* arrangement is used in which the shaft is exposed.

Universal joints are fitted to each end of the propeller shaft to enable the drive to be transmitted through a varying angle. This variation is due to the movement of the axle relative to the engine and gearbox.

Even when the final drive is fixed to the frame and the wheels are independently sprung, the propeller shaft still needs universal joints. This is to allow for the flexing of the frame structure that occurs when the vehicle is travelling over a bumpy surface.

Frame The component parts of a vehicle need a structure of some kind on to which they can be attached. Light trucks use a rectangular steel frame made up of two long *side members* which are linked together by a number of *cross members*. The assembled vehicle, without the body, is called a *chassis* and the frame alone the *chassis frame*.

Chassisless construction This *unitary* structure is much lighter in weight than the separate frame construction so it is normally used for motor cars. The system uses a specially designed body to withstand the various loads of the components and stresses that occur when the car is in use.

Brakes Vehicles are required by law to be fitted with an efficient braking system; this usually takes the form of brakes attached to each wheel.

Two types of brake are used: usually the front wheels and rear wheels are fitted with disc and drum brakes respectively.

A *disc brake* consists of an exposed disc which is attached to, and rotates with, the road wheel hub. A fixed caliper, mounted to the axle, prevents a friction pad on each side of the disc from rotating with the wheel when the disc is sandwiched by the pads during brake application. Braking is achieved by converting the energy of motion (*kinetic energy*) to heat energy, so exposure of the disc to the air stream is an important feature of this type of brake.

A *drum brake* is a system that has been in use for many years and is still popular for rear brakes. This is because it can also be used as a *parking brake (handbrake)* to keep the vehicle stationary when it is not in use.

This type has two brake shoes, which are lined with a friction material, and secured to a backplate fixed to the axle. The shoes are contained in a drum that is attached to the rotating hub. Retardation of the drum is achieved by moving the shoes outwards; an expansion which is normally produced by a piston in a hydraulic cylinder.

Light vehicles use a hydraulic system to operate the brakes and insertion of a *vacuum servo* between the driver's foot pedal and the hydraulic actuating system boosts the force applied to the brakes; this minimizes the effort that the driver has to apply.

Electrical equipment Vehicles are required by law to show certain lights during the hours of darkness, in poor visibility, for direction indication and during the application of the brakes.

An electric motor is used to start the engine, and the fuel in the engine is ignited by an electric spark produced by an *ignition system*.

Also various other items are operated electrically. These include: windscreen wipers and washers, horns, heaters, fuel injection arrangements, audio systems and various other aids for the comfort of the driver and passengers.

Electrical power to operate this equipment comes from a *generator* which is driven from the engine. Since certain items may be needed when the engine is not running, a *battery* (*accumulator*) is fitted. This battery is charged by the generator when the engine is running.

Cars

The layout shown in Fig. 2.1 was similar to that used on cars up to about 1950. Since that time various developments have taken place that have considerably improved the overall performance.

Many different layouts are used with each arrangement offering specific advantages. Variations occur in the location of the engine and the driving arrangement, i.e. the number and position of the driving wheels.

Engine position

Front engine Apart from tradition there are a number of reasons for siting the engine at the front of a car as shown in Fig. 2.2. The large mass of an engine at the front of the car gives the driver protection in the event of a head-on collision; also engine cooling is simpler to arrange. In addition, the cornering ability of

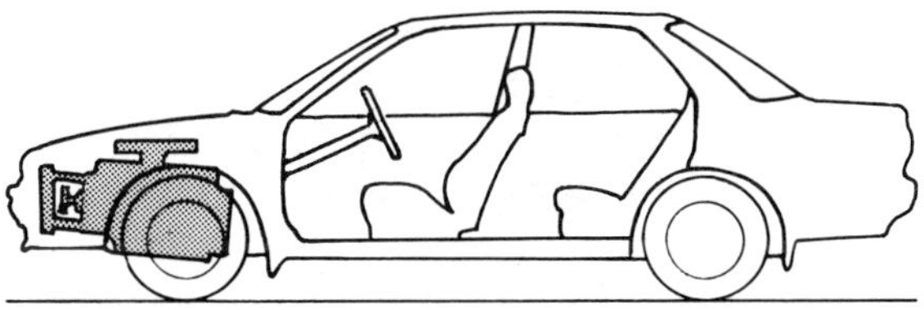

Fig. 2.2 Front-engined car

a vehicle is normally better if the weight is concentrated at the front.

Rear engine By placing the engine at the rear of the vehicle it can be made as a unit that incorporates the clutch, gearbox and final drive assembly. With such an arrangement it is necessary to use some form of independent rear suspension.

Most rear-engine layouts have been confined to comparatively small cars, because the heavy engine at the rear has an adverse effect on the 'handling' of the car by making it 'tail-heavy'. Also it takes up a good deal of space that would be used on a front-engined car for carrying luggage. Most of the space vacated by the engine at the front end can be used for luggage, but this space is usually less than that available at the rear. Although vertical engines of conventional form are occasionally used in rear-engine cars, a 'flat' engine, or a conventional engine laid on its side, offers the advantage of leaving some space for luggage above the engine.

One advantage attributed to a rear-engined layout is that it increases the load on the rear driving wheels, giving them better grip of the road.

Fig. 2.3 shows one of the most famous rear-engined cars - the Volkswagen Beetle. Notice that the front seats are nearer the front wheels than a front-engined car, and the floor is quite flat.

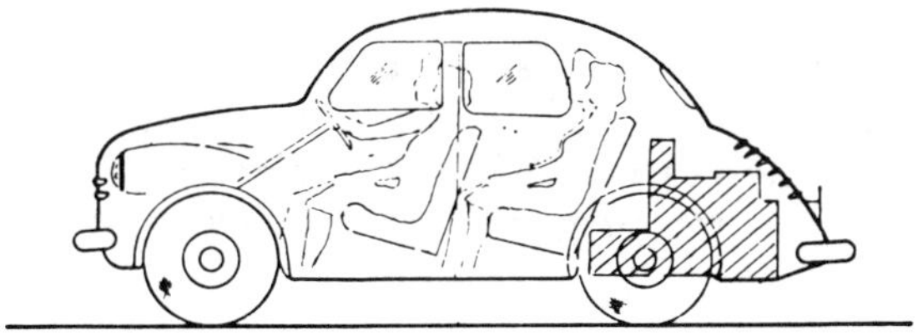

Fig. 2.3 Rear-engined car

Central and mid-engine These engine situations generally apply to sports cars because the engine siting gives a load distribution that achieves both good handling and maximum traction from the driving wheels.

These advantages, whilst of great importance for special cars, are outweighed in the case of everyday cars by the fact that the engine takes up space that would normally be occupied by passengers.

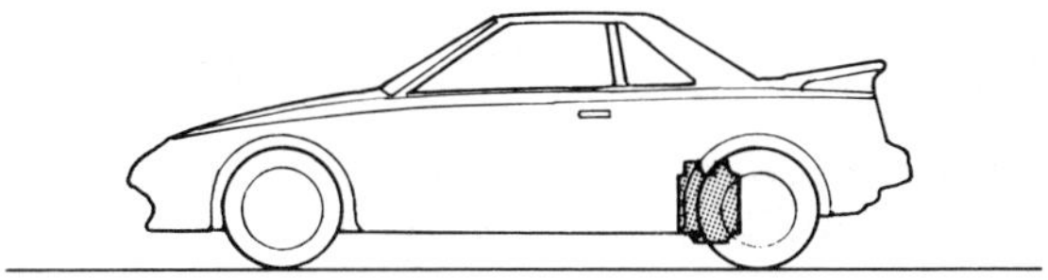

Fig. 2.4 Mid-engined car

The mid-engined layout shown in Fig. 2.4 combines the engine and transmission components in one unit. The term mid-engine is used because the engine is mounted in front of the rear axle line.

Drive arrangements

Rear-wheel drive The traditional layout shown in Fig. 2.5 has the engine situated with its output shaft set longitudinally. In this arrangement the rear wheels act as the driving wheels and the front wheels swivel to allow the vehicle to be steered. In the past rear-wheel drive was a natural choice because of the difficulty of transmitting a drive to a wheel that had to swivel for steering purposes.

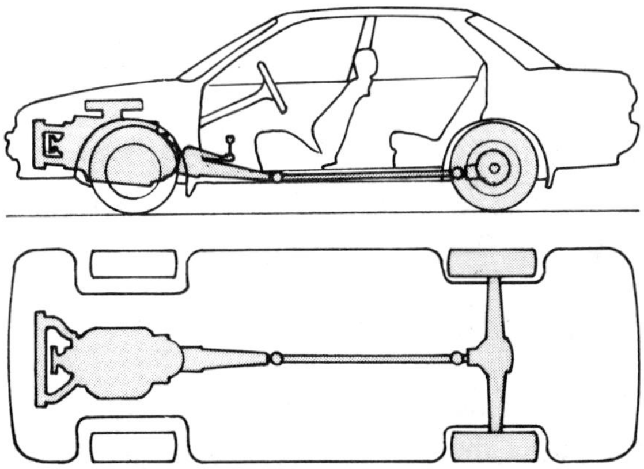

Fig. 2.5 Rear-wheel drive

Spacing out the main components in this layout makes each unit accessible but a drawback is the intrusion of the transmission components into the passenger compartment. These create a large bulge in the region of the gearbox and a raised long bulge, called a *tunnel*, down the centre of the car floor for the accommodation of the propeller shaft.

Using the rear wheels to propel the car utilizes the load transfer that takes place from the front to rear of the vehicle when the car is climbing a hill or accelerating. Good traction is obtained but when the wheels lose adhesion, the driving wheels move the rear of the car sideways. This motion causes the car to 'snake' and requires considerable movement of the steering wheel if a straight path is to be maintained.

Front-wheel drive The compactness of the layout shown in Fig. 2.6 has made it very popular for use on cars. Originally introduced for 'baby' cars, the advantage associated with the engine being placed across the vehicle, i.e. mounted transversely, has spread the use of front-wheel drive to many larger cars. For space reasons the length of the engine is the limiting factor, but the increased use of vee-type engines for larger power units has enabled many of these engines to be placed transversely. Accommodating all the main components under the bonnet (hood) in one compartment give maximum space within the car for the occupants; also the absence of floor bulges and tunnel provides more room for the rear passengers.

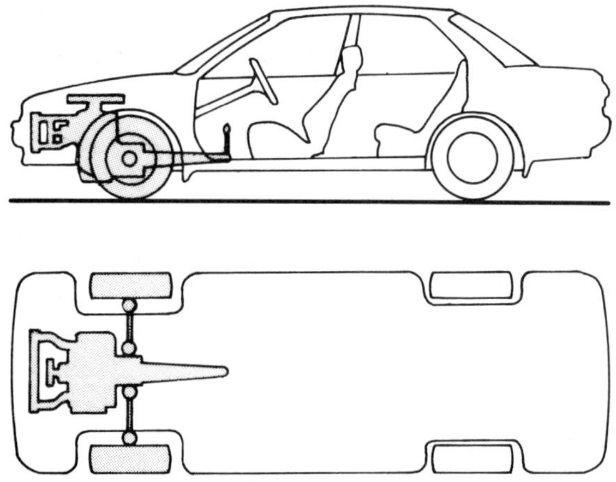

Fig. 2.6 Front-wheel drive

Transverse mounting of the engine simplifies the transmission, because the output shafts from the engine and gearbox move in a similar direction to the wheels. This avoids the need for a bevel-type final drive; instead a simple reduction gear, incorporating a differential, transmits the power by short drive shafts to the road wheels. Each drive shaft is fitted with an inner and outer universal joint. Since the outer joint must accommodate the steering action, it is specially designed to ensure that an even speed is obtained as it rotates; this avoids vibration even when the drive is being transmitted through a large angle.

Mounting the main units in one assembly sometimes makes it difficult to gain access to some parts. Nowadays this problem has largely been overcome; units such as the clutch can be easily removed for overhaul and by making removal of the engine assembly more simple, the cost of many major repair tasks has been minimized.

One criticism of front-wheel drive is that the driving wheels have less grip on the road when the vehicle is accelerating and hill-climbing. Although this characteristic can be partly corrected by placing the engine well forward to increase the load on the driving wheels, the car is then liable to become 'nose-heavy'. The effect of this is to make the steering of the car more arduous. In cases where the driver's steering effort is considered excessive, the car is often fitted with *power-assisted steering*; this system uses hydraulic pressure generated by the engine to help the driver turn the steering wheel.

Using the front wheels for steering allows the driving force to act in the same direction as the wheel is pointing. This feature, together with the fact that the vehicle is being 'dragged' behind the front driving wheels, improves vehicle handling especially in slippery conditions.

Four-wheel drive This arrangement, shown in Fig. 2.7, is safer because it distributes the drive to all four wheels. The sharing of the load between the four wheels during acceleration reduces the risks of wheel spin. Also the positive drive to each wheel during braking minimizes the possibility of wheel lock-up.

A further advantage of this layout is shown when the vehicle is driven on slippery surfaces such as

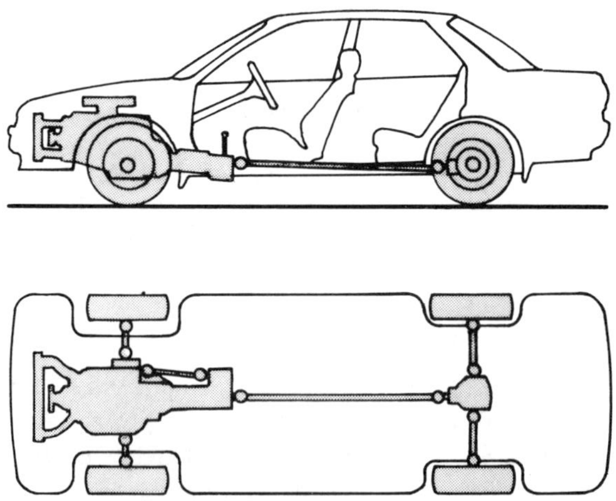

Fig. 2.7 Four-wheel drive

snow and mud. When on an icy road or driving cross-country (off-highway) a two-wheel-drive vehicle soon becomes undrivable because the loss of grip of one of the driving wheels causes the wheel to spin. Since a normal differential always transmits an equal share of the power to both wheels, the small driving force applied to the non-spinning wheel is insufficient to propel the vehicle.

Bodywork

The main purpose of the bodywork is to provide accommodation for the driver and passengers, with suitable protection against wind and weather. The degree of comfort provided will naturally depend upon the type of car and its cost.

The majority of mass-produced cars have a pressed steel body. Alternatively when the production is low in scale it is usual to construct the body by either hand-working aluminium alloy into shape or moulding glass reinforced plastics (g.r.p.). All steel surfaces must be treated and painted. Primarily this is to give protection against rust corrosion, and secondly to improve the appearance. External *trim*, made of stainless steel, chromium-plated brass or plastics, embellishes the body and appeals to the eye by providing a contrast with the plain coloured surface.

Fig. 2.8 shows some of the body styles used for cars.

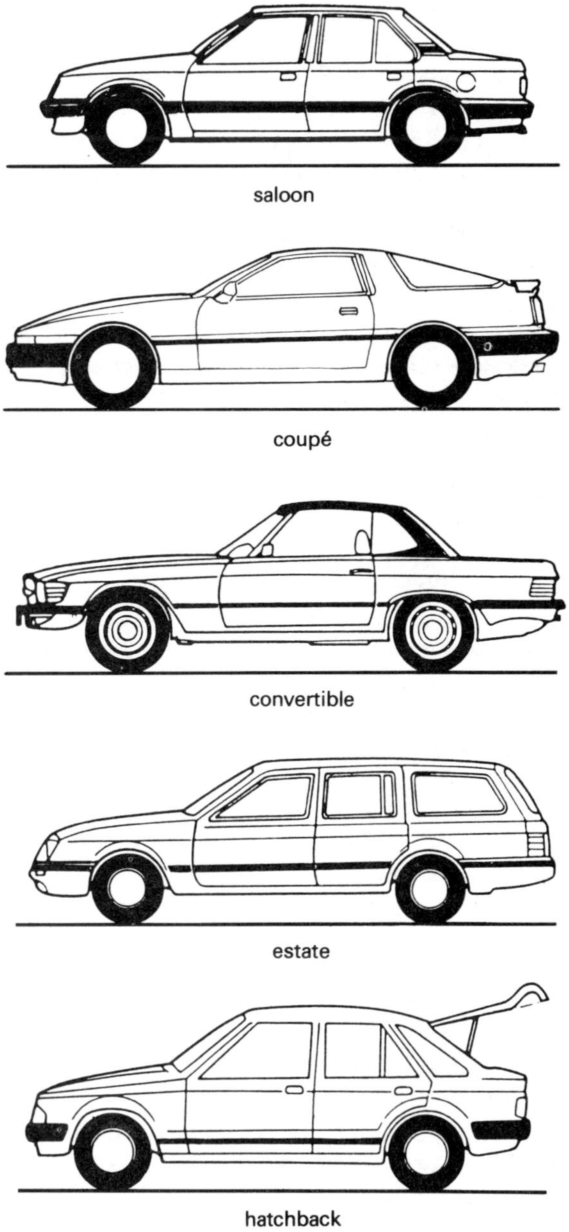

Fig. 2.8 Bodywork styles

Saloon A covered two- or four-doored car for four or more people. The common shape of *body shell* is based on three 'boxes'; the front box forms the engine compartment, the centre section the container for the occupants and the rear box a storage space, called a *boot* (*trunk*) for the luggage. The three boxes are blended together to give a pleasing appearance and are shaped to enable the car to move through the air with the minimum drag.

For safety purposes the passenger capsule must be suitably strengthened to keep its shape after the car has collided with some other object. Extra safety during front or rear impact can be obtained by arranging for the front and rear regions of the car to fold up in a concertina fashion. This absorbs the shock of impact and although damage to these regions is more extensive, the passenger capsule remains intact.

Coupé Generally this two-door type has a rigid roof and is intended for two people, a driver and passenger. Today some coupé models are called '2 + 2'; this generally suggests that the rear seat accommodation is rather cramped for adults but is suitable for children or for occasional adult use.

Convertible Also called *cabriolet* and *drop-head coupé*, this car can be changed into an open car by either removing a rigid roof or lowering a fabric collapsible roof. Prior to the introduction of air-conditioning this type was widely used in hot countries.

Estate Also known as a *shooting-brake* or *station-wagon*, this type has the roof extended to the rear to enlarge the internal capacity. Folding the rear seats down gives a large floor area for the carriage of luggage or goods. A *tail-gate* (door at the rear) enables bulky or long objects to be loaded easily. Stronger suspension springs are fitted at the rear to support the extra load.

Hatchback This type is half-way between a saloon and an estate car. Counting the tailgate as one door, this car is made in three- and five-door versions. As with the estate, the rear seats fold down to give a flat floor for the transportation of luggage or other objects.

Jacking points

It is sometimes necessary to raise the wheels of a vehicle clear of the ground: this is usually done by using a *jack*, of which there are several types. When jacking up a vehicle great care must be taken to see that the jack is both on a firm surface and is lifting the vehicle at a point which is sufficiently strong to withstand the concentrated load. Modern cars have special *jacking points* that are suitable for the jack supplied with the car. Reference to the handbook of the car will show the points at which the jack should be placed.

Jacks, even expensive hydraulic trolley types, should NOT be trusted when a person has to lie under the car or when the collapse of the jack can cause damage, human or mechanical. In these situations some form of *axle stand* or rigid prop must be used for safety purposes. The use of house bricks as an axle stand is highly dangerous.

Apart from securing the jack and stand, precautions should also be taken to ensure that the car cannot roll forwards or backwards; the handbrake should be applied and the wheels should be chocked.

Internal furnishings and fittings

Quality of the interior fittings and fabrics depends on the amount that the purchaser is prepared to pay for the car. Today the quality of the interior finish of cars is high so it is the level of luxury that separates the various models in a given range.

Most cars have individual *bucket-type* front seats. In the past, a single one-piece *bench-type* seat was used because, with this type, three persons could be seated. Besides giving greater comfort and safety, the space between the bucket seats allows for the fitting of the gear lever (or selector) and other controls. Each seat is carried on *rails* or *runners* securely fixed to the floor by an arrangement that allows the seats to be moved backwards and forwards, (and sometimes up and down) by manual or electrical means, to suit the occupants.

The rear seat for two or three passengers is normally a non-adjustable bench-type with a centre armrest to improve comfort when only two people are carried. Extra-large cars often have extra rear seats, called *occasional* seats, that fold into a division across the car behind the front seats.

Seat belts must be provided by law for the driver and all passengers to minimize the risk of severe injury if the car is involved in a collision. These belts must be anchored to a substantial part of the vehicle structure in such a way that the anchorages will not be torn out when an accident occurs. On most cars an *inertia reel* arrangement is used for the attachment of the lower end of the belt. The most popular type of belt used today has three mounting points which connect with a *diagonal shoulder strap* and *lap belt* to restrain the occupant. An automatic inertia-controlled belt reel attachment is generally preferred because this type allows the occupant to move freely in the seat when the car is stationary or moving normally. When a ball in the mechanism senses that the deceleration (slowing down) of the car is rapid, it rolls forward and locks up a toothed wheel attached to the belt; then forward movement of the occupant is resisted.

Carpets and interior trim Fitment of a carpet to the floor and covering the metal body panels enhances the appearance and improves the noise level and comfort. A fabric or plastic material is used as a *head lining* for the roof and for facing other interior surfaces. Panels are fastened to the interior surface of the doors to cover the mechanism for door locking and window winding.

Driving position Right-hand drive, whereby the driver sits on the right, is used in the UK. Driving on the left-hand side of the road evolved from the days when a horse rider, who normally carried his sword on the left-hand side of his body, mounted his steed on its left side for obvious reasons.

In the case of motor vehicles, where the traffic is keeping to the left side of the road the driver must sit on the right side in order to see that the road ahead is clear before overtaking.

Most countries now drive on the right, so cars sold in these areas have the steering wheel and driving

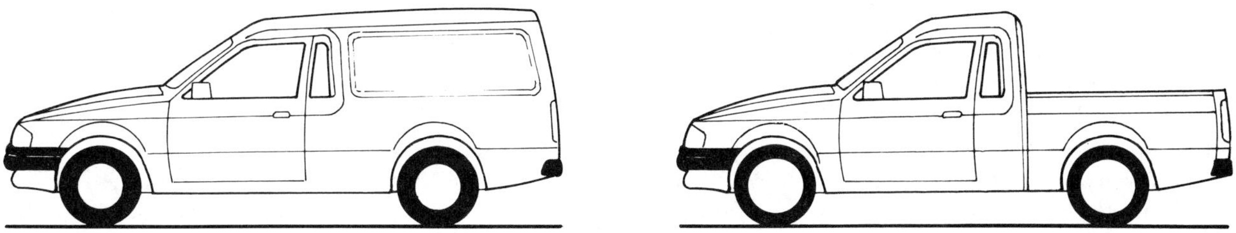

(a) Van

(b) Pick-up

Fig. 2.9 Light commercial vehicles

controls situated to give *left-hand drive*. Foot pedals in both right- and left-hand drive systems are normally arranged in the same order with the accelerator pedal on the right.

Controls and instrumentation Besides the main driving controls, various knobs and switches are provided within easy reach for the driver to operate the vehicle. A speedometer, together with other visual instruments or warning signals for the indication of their messages, are grouped in an instrument panel let into a *fascia panel* that runs across the front of the vehicle interior.

Light commercial vehicles

These small commercial vehicles are used for the conveyance of relatively light goods usually over short distances, e.g. the type of vehicle used by tradesmen to carry their equipment or by shopkeepers to make local deliveries to their customers.

At the light end of this range the vehicle is classed as a van (Fig. 2.9a). Often a van utilizes many of the components used by a similar sized car in the manufacturer's range. In these cases the term *car derivative* is used since the two vehicles share many of the major components. When fitting new parts it should be borne in mind that components such as clutches, braking systems and suspension units are uprated when they are used on commercial vehicles; this ensures that the parts can withstand the extra stresses. Car and van parts are similar in appearance so it is not always obvious to the mechanic that the van components have been strengthened to take account of the hard use associated with commercial use.

Manufacturers normally offer a number of different body shapes and sizes to suit the buyer's needs. When a vehicle is required for the transportation of bulky equipment, e.g. materials carried by builders and decorators, an open body is often preferred. A light vehicle having this partly enclosed body is called a *pick-up* and an example of this type is shown in Fig. 2.9b.

Today small vans and pick-ups have two seats for the driver and passenger and a standard of comfort comparable to the base model of the car from which the vehicles were derived.

Most vans have a load-carrying capacity of about 0.5 tonnes. Larger vehicles such as light trucks and minibuses need to carry heavier loads, so they require a stronger construction and have to be specially designed to suit the application. Since low-cost operation is an essential factor, most of these vehicles are fitted with a *diesel (compression-ignition) type engine*. In Fig. 2.10, which shows a typical construction for a four-wheeled light truck, twin wheels are used at the rear to support the load.

Heavy vehicles

Although this book concentrates mainly on light vehicles, a review of some heavier vehicles is included for comparison purposes.

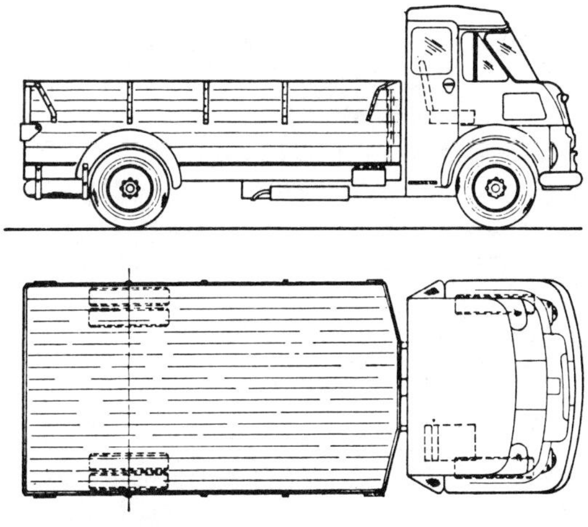

Fig. 2.10 A four-wheeled truck

Trucks

A heavy vehicle has a load-carrying capacity in excess of about 3 tonnes. For long-distance transportation of goods a truck is used and these come in various shapes and sizes.

A large flat platform is needed to carry the load: since this can only be provided above wheel height, no special effort is made to lower the frame, which usually consists of two straight and deep side-members joined by several cross-members. This frame supports the main components of the vehicle as well as the platform which forms the basis of the body, and the driver sits in a *cab* at the front of the vehicle.

The engine is usually a compression-ignition, or oil engine. (Briefly this means that the engine burns a light oil which is ignited by the high temperature produced by compressing air to a very high pressure in the engine cylinders.) It is fitted at the front, and the driver sits high up at one side of it, with his mate at the other side. Thus the whole of the space from the engine rearwards is available for carrying the load.

To carry heavy loads the rear wheels either have *twin tyres*, fitted side by side, or special wide-section single tyres. Vehicles exceeding a certain total loaded weight are required to have six wheels carried on three axles, whilst even heavier vehicles are required to have eight wheels, two on each of four axles. These legal requirements are laid down in *Construction and Use Regulations* drawn up by Parliament: since they are constantly liable to revision, readers seeking further details of maximum permitted weights and dimensions of vehicles are advised to consult the regulations currently in force. Fig. 2.10 shows a four-wheeled truck.

When six wheels are used the two extra wheels may be carried on an additional axle at the rear of the vehicle, as shown in Fig. 2.11. These extra wheels are usually, but not always, driven, but no provision is made for steering them. Fig. 2.12 shows an alternative arrangement in which the extra axle is placed at the front, in which case the wheels are steered but not driven. Eight-wheelers have two front axles - steered but not driven - and two rear axles - driven but not steered - as shown in Fig. 2.13.

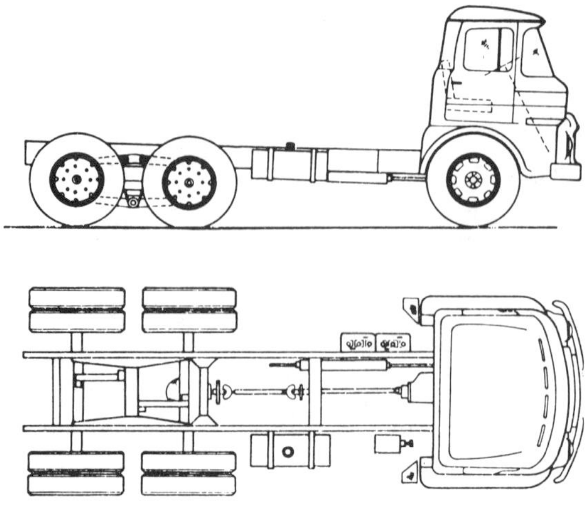

Fig. 2.11 A six-wheeled truck

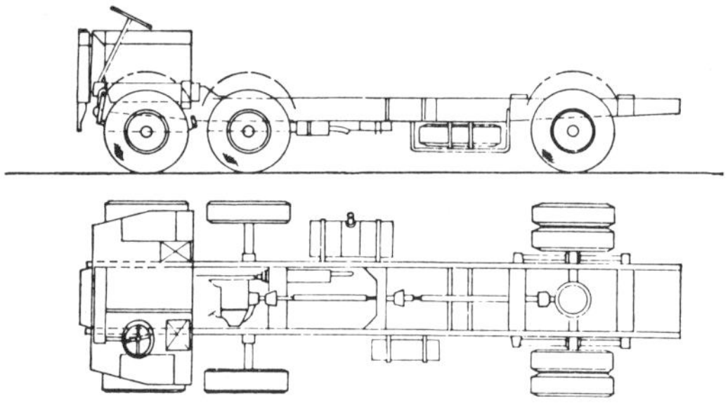

Fig. 2.12 Six-wheeled vehicle

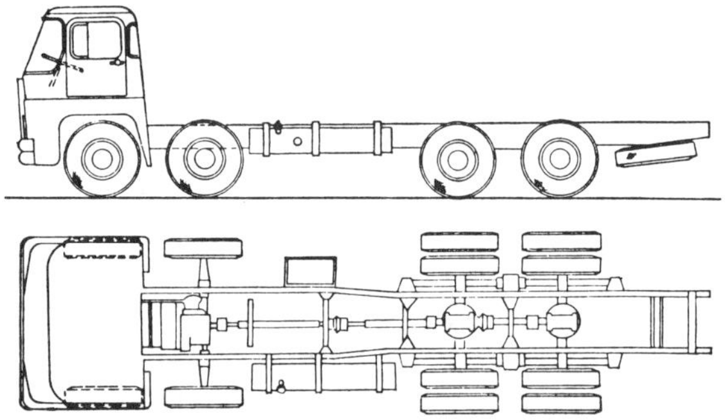

Fig. 2.13 Eight-wheeled vehicle

Trucks are allowed to tow a trailer on which an additional load can be carried. There is, in addition, a type of vehicle that consists of two parts: a four- or six-wheeled tractor (which does not itself carry any load) to which is attached, by a special turntable coupling, the front end of a semi-trailer having two or four wheels at its rear end. In this way some of the weight carried by the semi-trailer is supported by the tractor. The combination of tractor and semi-trailer is called an articulated vehicle. The trailer has retractable wheels on which its front end can be supported: this enables the tractor to be uncoupled and used with another trailer while its former trailer is being loaded or unloaded. An example of an articulated vehicle is shown in Fig. 2.14.

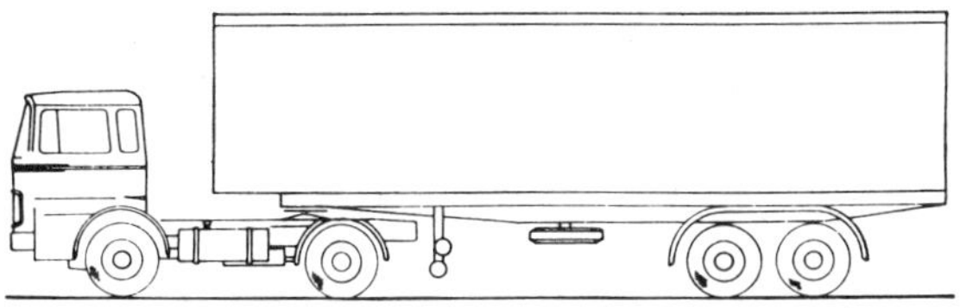

Fig. 2.14 Articulated vehicle

Buses (omnibuses)

These vehicles are designed to transport fairly large numbers of passengers. Up to about forty may be carried on a single floor or deck. Fig. 2.15 shows a single-decker bus which has its entrance to the front, where the driver can see passengers boarding or alighting from the vehicle. (Doors are not always fitted, but when they are they are often power-operated by the driver or conductor to prevent passengers getting on or off when the bus is moving.) In order to make room for the front entrance the engine is often laid on its side below the frame, as shown in the drawing of the chassis (Fig. 2.16).

Fig. 2.15 Single-decker bus

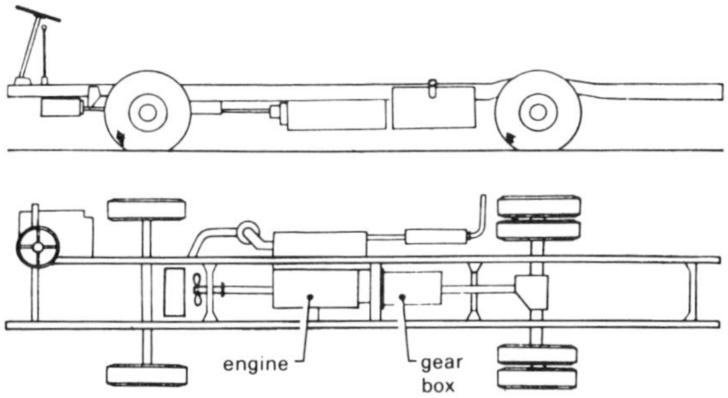

Fig. 2.16 Single-decker bus chassis

A double-decker bus capable of seating about sixty passengers is shown in Fig. 2.17. The *Construction and Use Regulations* stipulate a maximum height for the vehicle and a minimum ceiling height for both upper and lower decks. This means that the floor of the lower deck must be fairly close to the ground, at least along the central gangway between the seats: to permit this the final-drive gears are not at the centre of the rear axle, but are moved to one side, so that the gangway does not have to clear the highest point of the axle casing.

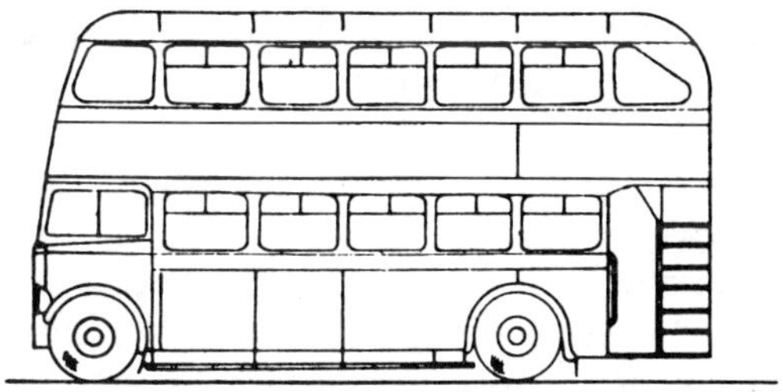

Fig. 2.17 Double-decker bus

The engine is placed well forward directly over the front axle, and slightly to the nearside to allow the driver to sit beside it, leaving all the space behind the front wheels free for the accommodation of passengers. The gearbox is separately mounted on the nearside some way behind the engine, where it can be low enough to be completely below floor level, and the two propeller shafts - one between engine and gearbox and the other between gearbox and rear axle - are clear of the gangway. In towns and cities there are usually short distances between stops, and speed of loading and unloading passengers is very important. At the rear of the bus is a large platform about 300 mm above ground level and an easy step up from the edge of the pavement. From this platform another step up leads into the lower deck and a staircase leads to the upper deck. The arrangement of chassis (Fig. 2.18) shows how the rear end of the frame is shaped to support this platform.

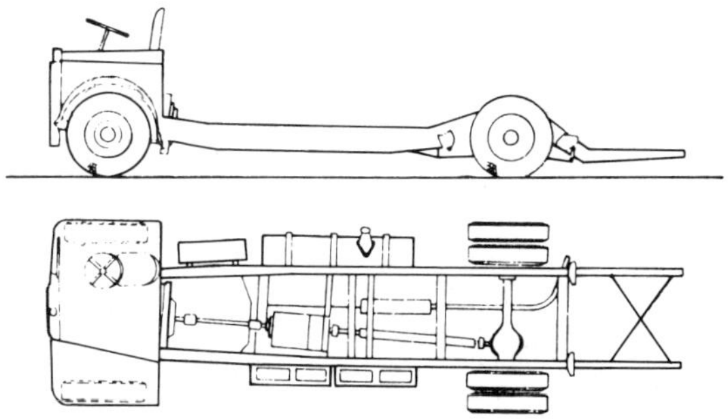

Fig. 2.18 Double-decker bus chassis

An objection to this arrangement is that the driver is unable to see passengers getting on and off except with the aid of mirrors, which do not give an adequate view of people on the platform or running to catch the bus. By placing the platform at the front the driver can have a direct view of passengers boarding and alighting, as in the case of the single-decker bus shown in Fig. 2.15. In the case of a double-decker the engine cannot be placed beneath the floor, so the forward platform necessitates placing the engine transversely at the extreme rear, and this layout is sometimes used.

Coaches

Vehicles of this type are designed to carry between thirty and forty passengers over fairly long distances in greater comfort than is provided in buses. Speed of loading and unloading is less important, but a fair amount of luggage accommodation is required. A chassis similar to that shown in Fig. 2.16 is generally used, but more comfortable seats are provided. The entrance is either at the front or behind the front wheels.

Some buses and coaches are made with the body structure forming the main frame. Though this is not very common it has the advantage of reducing total mass.

PART TWO

The petrol engine

3 Main parts and working principles

The main parts

Fig. 3.1 shows the main parts of an elementary engine:

1 *The cylinder* In its simplest form this is a tube of circular cross-section, closed at one end.
2 *The piston* fits closely inside the cylinder. Ideally it would be perfectly gas-tight yet perfectly free to move up and down inside the cylinder.
3 *The connecting rod* connects the piston to the crankshaft. At the piston end is a pin called the *gudgeon pin* which is fitted into holes in the piston and the connecting rod to couple them together.
4 *The crankshaft* is the main shaft of the engine and is carried in bearings in the *crankcase*. Offset from the main part of the shaft is the *crank pin* on which the connecting rod is fitted and is free to turn.

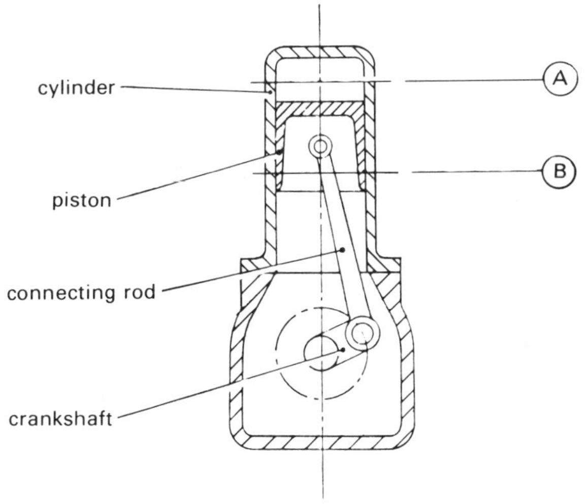

Fig. 3.1 The main parts of an engine

The arrangement is such that rotation of the crankshaft causes the piston to move up and down inside the cylinder: the lines A and B (Fig. 3.1) indicate the limits of travel of the top of the piston. As the piston moves upwards the space between its top surface and the closed end of the cylinder is reduced, i.e. the gas trapped in this space is *compressed*. As the piston moves downwards the space above it is increased, i.e. the gas in this space *expands*.

The crankshaft can be rotated by pushing the piston up and down in the cylinder. Starting with the position shown in Fig. 3.1, the crankshaft rotates clockwise as the piston is pushed downwards until the piston reaches the lowest point of its travel. At this point the crank pin will be directly under the centre of the crankshaft, and the centres of the gudgeon pin, crank pin and crankshaft will all lie in a straight line. In this position pressure on the piston will have no turning effect on the crankshaft, and this position is therefore called a *dead centre*. Another dead centre occurs when the piston is at the extreme top of its travel. These two dead centres, which are known as *bottom dead centre* (b.d.c.) and *top dead centre* (t.d.c) respectively, mark the extreme limits of the piston's travel. They are illustrated in Fig. 3.2. Movement of the piston from one dead centre to another is called a *stroke*, and there are two strokes of the piston to every revolution of the crankshaft.

Method of working

The piston is pushed down the cylinder by applying pressure to its upper surface. The method by which the pressure is produced is based upon the fact that if a gas is heated in a confined space its pressure will

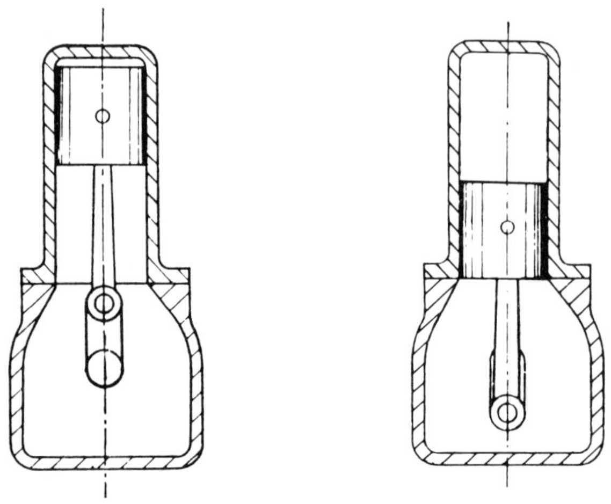

Fig. 3.2 The dead centres

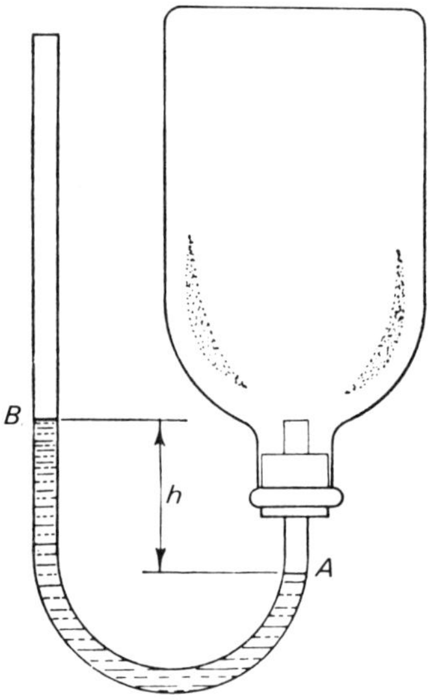

Fig. 3.3 Demonstration of increase in pressure of air with rise in temperature

increase. Fig. 3.3 illustrates a simple demonstration of this fact: the apparatus is very simple and you should try it for yourself. Bore a hole in the cork of a bottle and into this hole push one end of a glass tube bent into a U-shape. Partly fill this tube with water, and then push the cork into the bottle, holding the bottle upside-down as shown. The difference in pressure between the air in the bottle and the air outside is shown by the difference, h, in the levels of the water at A and B. Warming the air in the bottle - even by holding it in the hand - will increase the pressure of the air in the bottle which in turn will push down the level at A and raise the level at B.

In an engine the air can be heated to a very high temperature and a correspondingly high pressure created inside the cylinder, thus exerting a considerable force on the piston.

Pressure above the piston can only push it downwards, and at the end of a downward stroke the pressure must be released and the piston moved back to the top of the cylinder before it can be pushed downwards again. The pressure is released by opening a hole in the cylinder called the *exhaust port*, and the piston is returned up the cylinder by the rotation of a wheel with a heavy rim - called a *flywheel* - fitted to the crankshaft. Once this flywheel has been made to turn, it will continue to rotate for several revolutions.

The air inside the cylinder can be heated by playing a flame on the outside of the cylinder. To reach the air inside, however, the heat would have to pass through the wall of the cylinder, so that the air could be made no hotter than the cylinder. Much of the heat used would be lost by heating the cylinder and the air outside it.

This waste can be lessened by heating the air inside the cylinder directly: to do this a suitable fuel can be mixed with the air in the cylinder and burnt *inside the engine*. The cylinder will still, of course, absorb a good deal of heat, and arrangements must be made to prevent it getting too hot, but the air inside can be raised to a much higher temperature and so will reach a correspondingly higher pressure.

Any engine which uses the heat produced by burning a fuel to develop mechanical power is called a *heat engine*. One in which the fuel is burnt *inside* the engine is called an *internal combustion engine*. Internal combustion engines can use any one of a variety of fuels. Petrol is a liquid refined from crude petroleum, and is particularly suitable as a fuel for motor vehicles. It is clean to handle and relatively cheap. It is liquid at normal temperatures, and a vehicle can carry, in quite a small tank, enough to take it over 300 km (200 miles). It gives off a flammable vapour even at quite low temperatures, enabling the engine to be started from cold with little difficulty.

Cycle of operations

Before petrol can be burnt it must be vaporized and mixed with a suitable quantity of air. This mixture must then be introduced into the cylinder, and this is done through a hole called the *inlet port*. Once inside the cylinder the mixture is compressed before burning, since this greatly increases the pressure after burning. After compression the vaporized fuel is ignited by an electric spark, which, at the appropriate moment, jumps across a small gap on a *sparking plug*, screwed into the top of the cylinder.

After the fuel has burned and the resulting pressure has pushed the piston down the cylinder, the burned gases are released, as already mentioned, through the exhaust port. The inlet and exhaust ports are normally closed, but are opened at the correct times to allow the gases to pass through them (see Chapter 10).

The running of the engine involves the continuous repetition of *four operations* which make up what is called the *cycle of operations*. These operations are continuously repeated as long as the engine is running, in the following order:

1 The space in the cylinder above the piston is filled with a mixture of petrol vapour and air.
2 This mixture is compressed into the top end of the cylinder (called the combustion chamber).
3 The petrol vapour is burned, and the resulting pressure drives the piston down the cylinder.
4 The burned gases are expelled from the cylinder.

Notice that the gas helps the piston on its way during only one of these operations. The remaining operations give no direct help, but actually impede the piston's movement, particularly the compressing of the gas. Thus the work of the flywheel is not merely to push the piston back up the cylinder, but to keep the crankshaft turning as steadily as possible between one impulse on the piston and the next.

Engine size

The usual method of indicating the size of an engine is to state the volume of air and fuel taken into the engine for each cycle of operations. This is clearly the volume contained in that part of the cylinder between the t.d.c. and b.d.c. positions of the piston.

The volume of a cylinder is given by the formula

$$V = \pi r^2 h \quad \text{or} \quad V = \pi d^2 h/4$$

where V is the volume of the cylinder, r the radius, d the diameter and h the height or length.

The internal diameter of the engine cylinder is called the *bore*, whilst the distance the piston moves between t.d.c. and b.d.c. is called the *stroke*. Representing these by d and l, respectively, the above formula can be written:

$$V = \pi d^2 l/4$$

Since this is the volume *displaced* or *swept* by the piston, it is called the *displacement volume* or *swept volume* of the cylinder. If the engine has several cylinders, as most have, the total swept volume of the engine is swept volume per cylinder multiplied by the number of cylinders.

If the bore d and stroke l are measured in millimetres and if n represents the number of cylinders, then the total volume, V_t, is:

$$\begin{aligned} V_t &= \pi d^2 ln/4 \text{ mm}^3 \\ &= \pi d^2 ln/4000 \text{ cm}^3 \\ &= \pi d^2 ln/4\,000\,000 \text{ litres} \end{aligned}$$

(*Note:* $1000 \text{ cm}^3 = 1$ litre)

When the bore is equal to the stroke the engine is called a 'square engine'. In a similar way, when the bore is larger than the stroke the engine is called 'oversquare'.

Compression ratio

An important feature of the dimensions of an engine cylinder is the number of times the volume enclosed above the piston before compression is greater than that after compression. When the piston is at t.d.c. a space is left between its top and the end of the cylinder: the volume contained in this space is called the *clearance volume*. The volume enclosed above the piston at b.d.c. consists of the *clearance volume plus the swept volume* and is called the *total volume*.

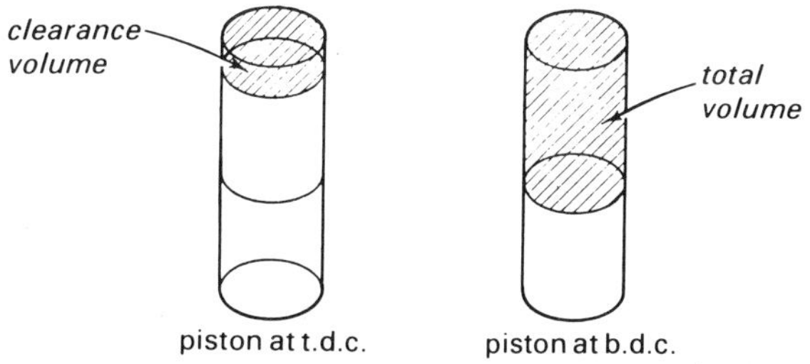

Fig. 3.4 Compression ratio

From Fig. 3.4:

$$\text{Compression ratio} = \frac{\text{Total volume}}{\text{Clearance volume}}$$

$$= \frac{\text{Swept volume} + \text{Clearance volume}}{\text{Clearance volume}}$$

$$= \frac{\text{Swept volume}}{\text{Clearance volume}} + 1$$

4 Practical applications of the cycle of operations

The four-stroke cycle

The first commercially successful internal combustion engine was made by a Frenchman, Etienne Lenoir, in 1860. It ran on coal gas, but worked on a cycle of operations which did not include compression of the gas before ignition: as a result it was not very efficient. In spite of this it was, in some respects, superior to small steam engines of the time, and a great many were sold and did useful work driving machinery in factories. In 1862 Lenoir made a horseless carriage powered by his engine and possibly drove it on the roads, but he lost interest in this venture and nothing came of it.

A method of carrying out the cycle of operations outlined in the previous chapter was described in a patent dated 16 January 1862 taken out by a French civil servant, M. Beau de Rochas. Since he did not have the means to develop the patent himself, the patentee offered it to Lenoir who, failing to realize its importance, turned it down.

In Germany Dr N. A. Otto started the manufacture of gas engines about 1866. His first engines were extremely noisy, though quite effective, but in about 1875 he took out a patent describing a method of carrying out the cycle of operations which was identical with that of Beau de Rochas' thirteen years earlier. (It is, however, most unlikely that Otto had heard of the Frenchman or his patent). Otto's new engine was an immediate success. It was much more efficient than Lenoir's and was very quiet, a characteristic which led to its being named 'Otto's silent gas engine'.

Lenoir, realizing his mistake, began to manufacture engines working on the same principle. Otto, of course, sued him for infringing his patent rights, but Lenoir had no difficulty in proving that his engines were made under the earlier patent of Beau de Rochas, which had by now lapsed. The court proceedings at last brought poor Beau de Rochas the fame he deserved, and he was awarded a sum of money by the Academy of Sciences in Paris in recognition of his invention. Even so, the method of operation which he was the first to describe, and which is the one used in most modern engines, was for many years (and sometimes still is) known as the *Otto cycle*.

In this method one complete stroke of the piston is used to carry out each of the four operations forming the cycle of operations. To complete the cycle, therefore, *four strokes of the piston*, occupying two revolutions of the crankshaft, are needed: consequently the method is generally known as the *four-stroke cycle*.

The inlet and exhaust ports are normally closed by valves (see Chapter 8) which are mechanically opened at the correct times. If the engine runs on petrol, the petrol and air are mixed in the correct proportions by a *carburettor*, fitted to the outer end of the inlet port.

Starting with the piston at t.d.c., as the crankshaft rotates the method of operation is as follows:

First stroke The piston moves *down* the cylinder with the *inlet port open* and *exhaust port closed*, filling the cylinder with a mixture of petrol vapour and air. This is the *induction stroke*.

Second stroke The piston moves *up* the cylinder with *both ports closed*, compressing the mixture into the combustion chamber at the top end of the cylinder. This is the *compression stroke*.

Third stroke At about the end of the compression stroke an electric spark is made to jump across a small gap on a *sparking plug* screwed into the end of the cylinder. This spark ignites the petrol vapour

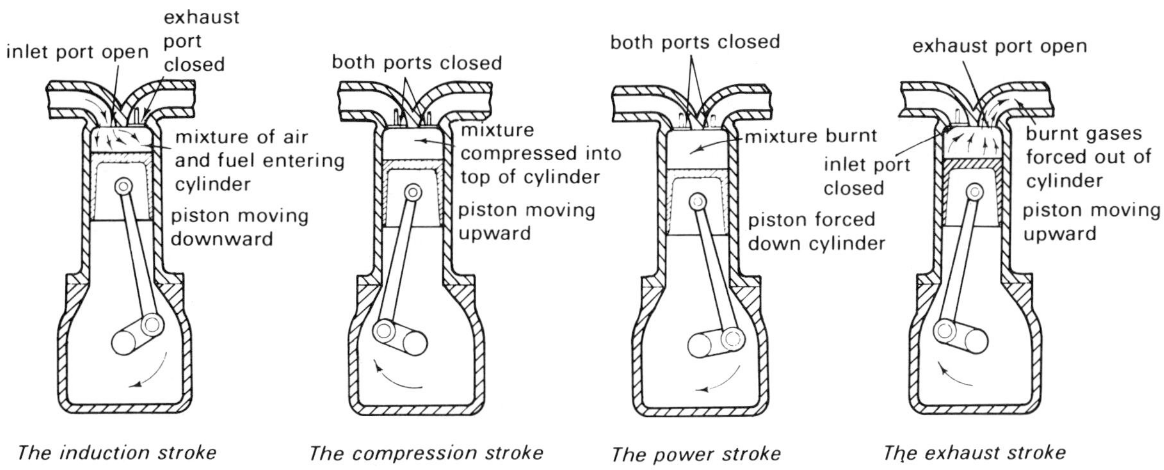

Fig. 4.1 The operation of an engine on the four-stroke cycle

which burns very rapidly, heating the gas in the cylinder to a high temperature and considerably increasing its pressure. The pressure forces the piston *down* the cylinder, both ports remaining closed. This is the *power stroke*.

Fourth stroke The piston returns *up* the cylinder with the *inlet port still closed* but the *exhaust port open*, expelling the burnt gases from the cylinder. This is the *exhaust stroke*.

At the end of this stroke the exhaust port closes and the inlet port reopens for the next induction stroke, which follows immediately.

Fig. 4.1 illustrates these four operations. Observe the movements of the piston and operation of the valves as the crankshaft of an actual engine is rotated.

The two-stroke cycle

It has always been considered a disadvantage of the four-stroke cycle that there are three 'idle strokes' to every 'working stroke'. Between 1878 and 1881 a Scotsman, Dugald Clerk, developed an engine in which the cycle of operations was completed in only *two strokes* of the piston, thus providing a power stroke for every revolution of the crankshaft. This engine used a second cylinder and piston to force the fresh mixture into the working cylinder.

In 1891 Joseph Day invented a modified form of Clerk's engine, in which he dispensed with the second cylinder by using the space in the crankcase *underneath* the piston. He also avoided the use of valves by using the piston itself to cover or uncover the ports.

The operation of this engine, which is illustrated in Fig. 4.2, is as follows.

Beginning with the piston about half-way on its upward stroke, all three ports are covered. The upward movement of the piston compresses a fresh charge of mixture in the *top* of the cylinder, and at the same time decreases the pressure *under* the piston below the pressure of the atmosphere (the crankcase being sealed). Near the top of the stroke the *lower* edge of the piston overruns the *inlet* port, allowing the pressure of the atmosphere to fill the lower part of the engine with fresh mixture from the carburettor.

At about the top of the stroke the mixture above the piston is ignited in the same way as in the four-stroke engine, and with the same result: the high pressure of the burned gases drives the piston down the cylinder. A little below t.d.c. the piston *covers the inlet port*, and further downward movement compresses the mixture in the crankcase. Near the bottom of the stroke the *top* edge of the piston overruns the *exhaust port*, allowing the burnt gases to begin to rush out of the cylinder under their own pressure.

Slightly further down the stroke the *transfer port* is uncovered, and the mixture compressed *below* the piston flows into the cylinder above the piston, where it is deflected upwards by the specially shaped piston. This prevents it shooting straight across the cylinder and out of the exhaust port.

As the piston rises on its next stroke the transfer and exhaust ports are covered and the cycle of operations begins again.

In modern engines of this type the deflector on the crown of the piston has been dispensed with, and the transfer port or ports (there are usually two) are

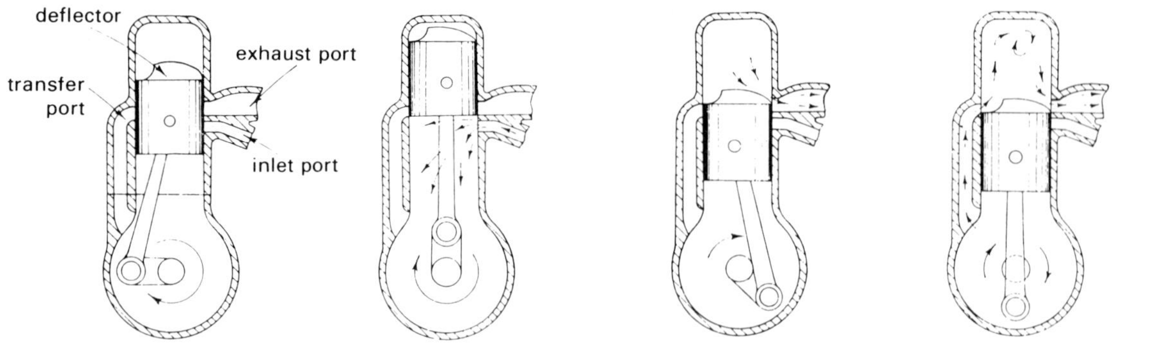

1 Piston rising, compressing mixture above, decreasing pressure below. All ports closed

2 Piston passing t.d.c., mixture above piston ignited, fresh mixture entering crankcase

3 Piston moving downward. The fresh mixture in the crankcase has been compressed. The piston has just uncovered the exhaust port, and is about to uncover the transfer port

4 Piston passing b.d.c. exhaust and transfer ports open. Fresh mixture entering cylinder via transfer port, directed to top of cylinder by deflection on piston, driving out burnt gas through exhaust port

Fig. 4.2 The operation of an engine on the two-stroke cycle

shaped and aimed so as to direct the fresh mixture towards the top end of the cylinder away from the exhaust port.

It might be expected that the two-stroke engine would develop twice the power of a four-stroke engine of the same size, but experience does not confirm this. The operations are less effectively carried out, and despite the deflector on the piston, mixing the fresh charge with burnt gas cannot be avoided: there is usually some loss of fresh mixture through the exhaust port and incomplete scavenging of burnt gas from the cylinder. The main advantages of the two-stroke engine, therefore, are its greater simplicity and smoothness. It is commonly used for the smaller sizes of motor-cycles and is seldom used for motor-cars.

5 Crankshafts

Main features

The crankshaft of an engine is formed of a number of 'sections' such as that illustrated in Fig. 5.1. The *main journals* rotate in the main bearings (a part of a shaft which rotates in a bearing is called a *journal*), and the *crank pin* or *crank journal*, to which the connecting rod is fitted, is offset from the main journals by a distance called the *crank radius*. The *webs* connect the main journals to the crank pin, and where the journals join the webs a *fillet* or *radius* is formed to avoid a sharp corner, which would be a source of weakness. This is of vital importance in crankshafts which are subjected to particularly heavy loads.

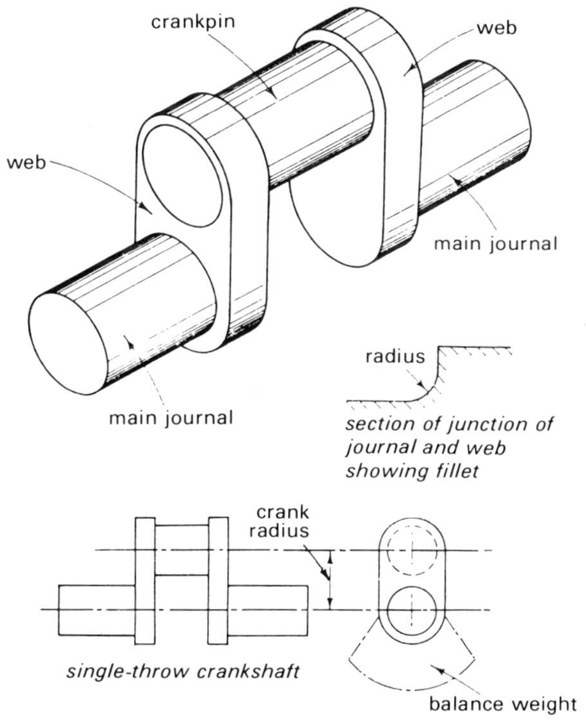

Fig. 5.1 Single-throw crankshaft

Balance masses

In some cases the webs are extended to form *balance masses* which are used in certain types of engine to ensure that the rotating parts are balanced as effectively as possible and will cause no vibration, or as little as possible, when the engine is running. It is not possible to balance completely the rotating parts of all types of engine; this is, for instance, one of the objections to the use of a single-cylinder engine, which is impossible to balance perfectly, though it can be made satisfactory for certain purposes.

In some engines in which the rotating parts are perfectly balanced, masses similar to balance masses may still be used: centrifugal force acting on the crank pins causes heavy loading of the adjacent main bearings at high speeds, and by extending the webs to form balance masses a counter-centrifugal force is applied in opposition to that on the crank pin, thus reducing the bearing loading. Masses used in this way are more properly known as *counter-balance masses*.

Throw

There appears to be some confusion as to the meaning of this term. It is often used as an alternative to *crank radius*, but old books on steam engines define the throw as the *diameter* of the crank pin circle, i.e. the throw is equal to the stroke of the piston. It might, therefore, be better to avoid the use of the word 'throw' in this sense and clarify what is intended by using the terms *crank radius* or *crank circle diameter* as appropriate.

The same word 'throw' is also used in a slightly different sense: it is the name given to a crank pin together with its adjacent webs and main journals. Thus Fig. 5.1 illustrates a single-throw crankshaft.

Crank pins do not always have a main journal on *both* sides. Fig. 5.2 illustrates a *two-throw* crankshaft

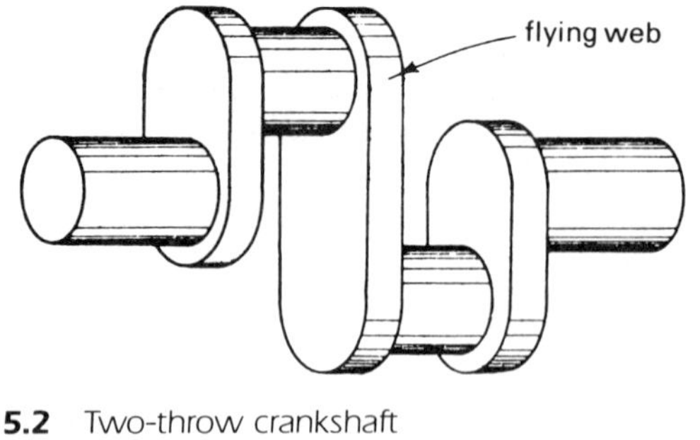

Fig. 5.2 Two-throw crankshaft

having only two main journals, the crank pins being connected together by a *flying web*.

Flywheel attachment

The flywheel is usually attached to the rear end of the crankshaft. The attachment must be perfectly secure, and it should be possible to assemble the flywheel in one position only, partly because the flywheel is often marked to indicate the position of the no. 1 crank pin, but more particularly because the crankshaft and flywheel, besides being balanced separately, are also balanced as an assembly, and if the flywheel is not fitted in the same position as that in which the assembly was balanced, some imperfection of balance may arise which may cause vibration. (It is possible that the crankshaft itself may be balanced within the limits permitted, but not perfectly; the flywheel may also have a small but 'tolerable' imbalance. Should the crankshaft and flywheel be assembled so that the imbalance of the crankshaft and of the flywheel are 'additive' the resulting vibration might be eliminated by turning the flywheel through 180° on the crankshaft, and this latter position would be the 'correct' one.)

Fig. 5.3 shows a more common method. A flange is formed on the end of the crankshaft which is a close fit into a counterbored hole in the centre of the flywheel. The flywheel is then secured by a number of screws, which pass through holes drilled axially through the flywheel face, and screwed into threaded holes in the crankshaft flange. One or more dowels are fitted to relieve the screws of sharing loads, and the dowels, or

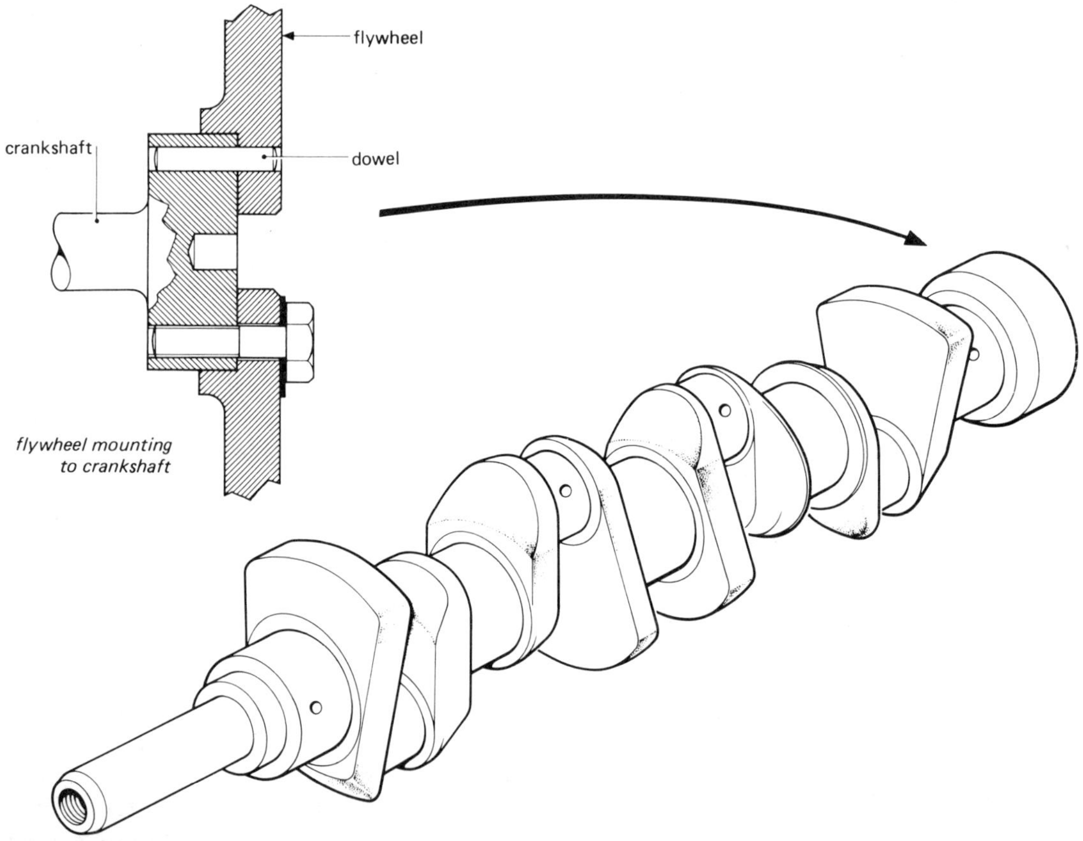

Fig. 5.3 Crankshaft – four-cylinder, five-bearing

screws, are often unevenly spaced to permit assembly in one position only. Bolts are sometimes used instead of screws; a suitable locking arrangement is provided for the nuts or screws.

Oilways

Engine lubrication systems are dealt with in Part 5, but it is necessary to mention here the method by which oil is supplied to the crank pins. Oil is delivered under pressure to the main bearings, where it is fed into a groove running around the bearing at about the middle of its length. A hole is drilled through the crankshaft (as shown in Fig. 5.4) running from the main journal, through the web to the surface of the crank pin. The main-journal end of the hole moves around the groove in the main bearing as the shaft rotates, allowing a continuous supply of oil to pass to the crank pin.

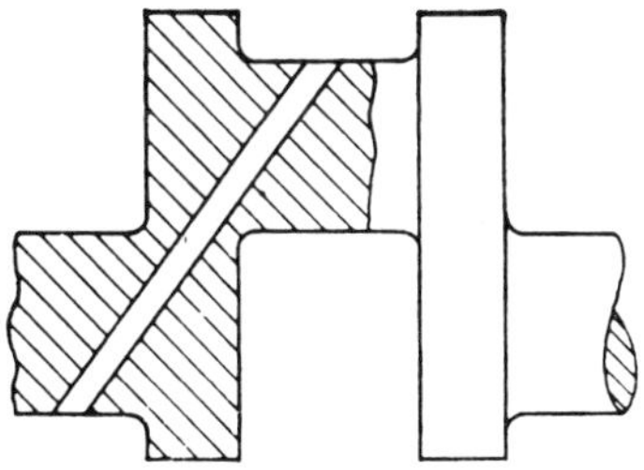

Fig. 5.4 Oilway drilled through crankshaft

Some large crankshafts have their crank pins and journal bored out to reduce the weight of the shaft. The ends of these hollow pins and journals are usually closed by caps or plugs held in place by bolts (see Fig. 5.5) to prevent the escape of oil.

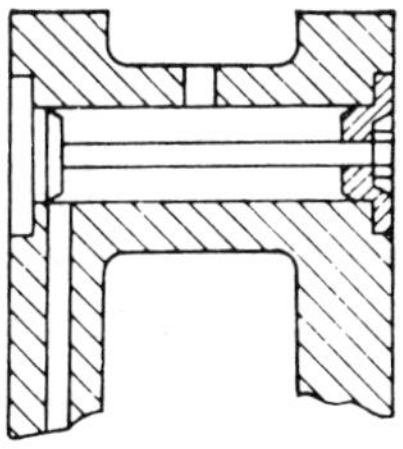

Fig. 5.5 Method of sealing hollow journals

Oil retainers

The crankshaft projects from the rear end of the crankcase, and some of the oil which is pumped into the rear bearing to lubricate it will escape from the

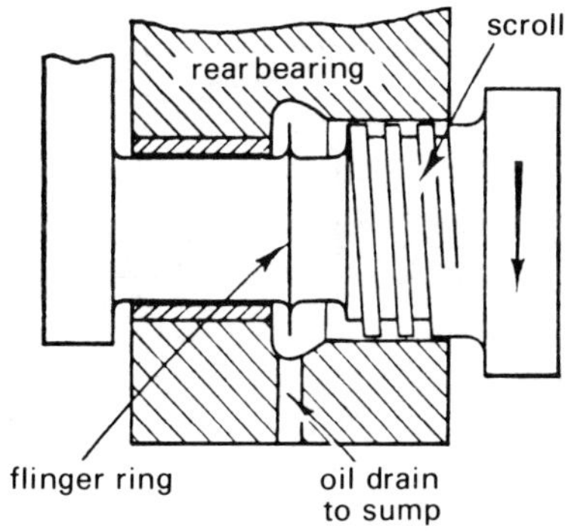

Fig. 5.6 'Scroll'-type oil retainer

rear end of the bearing to the outside of the crankcase and be wasted, besides making a mess of the engine and possibly getting into the clutch where it would be very undesirable. Fig. 5.6 shows two methods of preventing this escape of oil. Immediately behind the rear journal a thin ring or fin of metal is formed around the circumference of the shaft. Oil reaching this ring from the rear bearing is flung off by centrifugal force as the shaft rotates, and is caught in a cavity from which a drain hole leads the oil back inside the crankcase.

Behind this flinger ring a square-section helical groove or scroll, rather like a coarse screw thread, is machined in the surface of the shaft. The outer surface of the shaft has a small clearance inside an extension of the rear bearing housing, and any oil that reaches this part of the shaft will tend to be dragged round with the shaft but at the same time be held back by the stationary housing: as a result the oil will be drawn along the groove, the *hand* of which is such as to return it to the *inside* of the crankcase. In most cases both the methods illustrated and described are used on the crankshaft: sometimes only one may be considered sufficient.

Fig. 5.7 shows a section of a *lip-type* seal. Seals of this type have been used for many years in axles and gearboxes and are finding increasing use at both ends of the crankshaft. The seal consists of a specially shaped synthetic rubber ring supported by a steel shell and fitting into a recess in the crankcase or timing case. The ring has a shaped *lip* which is held in light contact with the shaft by a *garter spring*. This type of seal usually replaces both the scroll and the flinger ring.

Clutch shaft spigot bearing

The power developed by the engine is transmitted through the clutch to the gearbox, and the shaft on

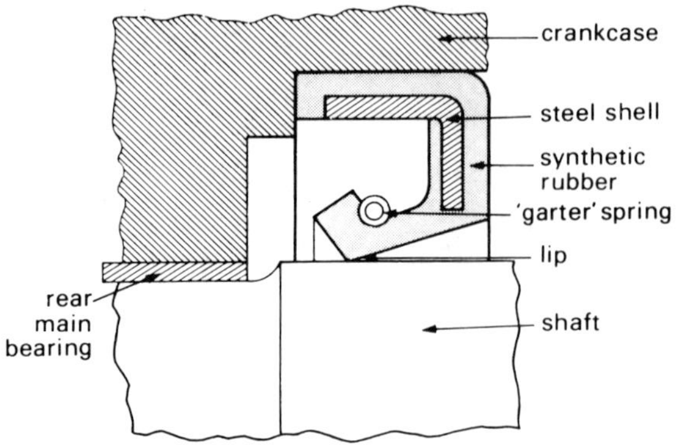

Fig. 5.7 Section of lip-type seal

which the driven part of the clutch is fitted is usually supported at its forward end in a bearing in the rear end of the crankshaft. This bearing, called a *spigot bearing*, usually consists simply of a bronze bush pressed into an axial hole in the rear end of the crankshaft (see Fig. 5.8), but sometimes a ball or roller bearing may be used.

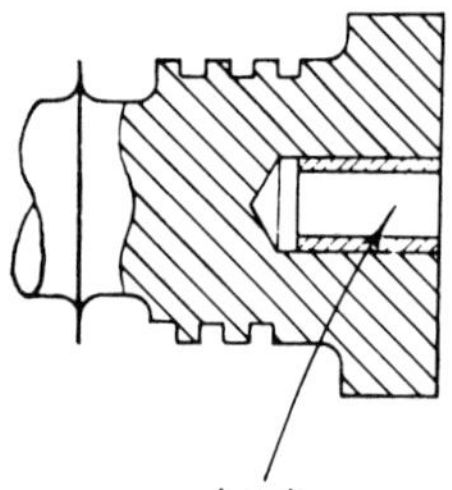

Fig. 5.8 Clutch shaft spigot bearing

Front end of shaft

The mechanism that operates the valves is driven from the crankshaft by gears or chains, and the drive is usually taken from the front end of the shaft, although in a few cases it is taken from the flywheel end. The front end of the shaft is extended beyond the front main journal, the extension being parallel or sometimes *stepped*, i.e. the forward part is of slightly smaller diameter than the rearward part. On to this extension is pressed the *timing gear* which drives the valve mechanism, and the pulley which drives the fan, water pump and dynamo: both are located on the shaft by *Woodruffe keys*. They are secured by a nut or screw threaded into the end of the shaft.

The timing gear is enclosed inside a cover called the *timing case*, but the pulley is outside this case. Oil-retaining arrangements, similar to those at the rear, are provided at this point, though the oil flinger is usually a separate part gripped between the timing gear and pulley bosses, and the scroll is formed on a rearward extension of the pulley boss.

Fig. 5.9 shows a lip-type oil seal retained in the timing cover. The lip should be smeared with oil before fitting the pulley on to the crankshaft.

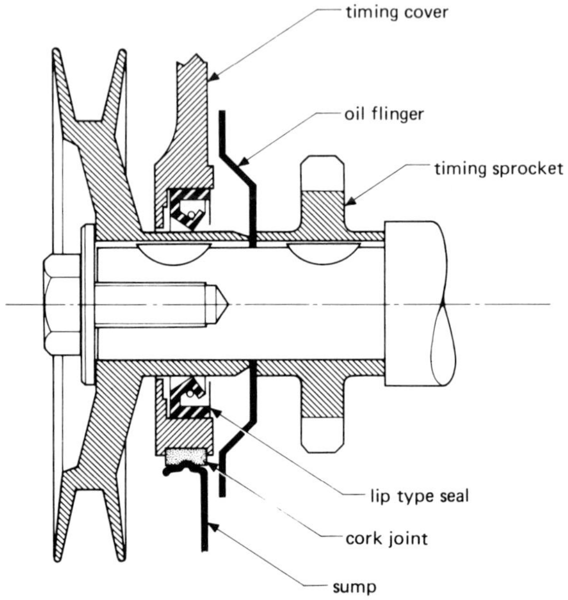

Fig. 5.9 Mounting of crankshaft pulley and timing chain sprocket

Crankshafts are made of high quality steel or iron. These materials are processed by either forging or casting.

Forging gives greater strength, but stiffness is of more importance than mere strength. Most modern shafts have large journals and a relatively short throw that allows the crankpins to overlap the journals when viewed from the end of the shaft. This feature allows the shaft to be cast; this gives a lighter shaft which is cheaper to manufacture because it requires less machining. In addition modern *nodular irons* are exceptionally strong and when ground to a fine finish and surface-hardened provide an excellent bearing surface. In this case the term 'nodular' applies to the inclusion of minute rounded lumps of graphite, i.e. carbon; these particles are not combined with the base material so their existence in a free state classifies the material as iron and not steel.

Following the initial forming, the shaft is ground to give a very smooth and accurately dimensioned surface of the journals and crankpins. Often the webs are not machined, but in the case of a high-performance engine the shaft would be machined all over to reduce oil drag.

Unless special precautions are taken, the repeated load applied to the shaft and the continual torsional flexing of the shaft, especially in the region of the rear journal fillet, causes the shaft to fracture. This *fatigue* can be minimized by *fillet rolling* (Fig. 5.10).

For long life there must be a large difference in hardness between the shaft and the bearing. Many modern engines use a comparatively hard bearing, so the journals and crankpins must be hardened. This is achieved by either *nitriding* or *induction hardening*. The former process involves heating the complete shaft to a temperature of about 500 °C in an atmosphere of ammonia gas for several hours. During this period the steel absorbs nitrogen from the ammonia and forms a hard surface of *iron nitride*. The induction hardening method uses an electrical process.

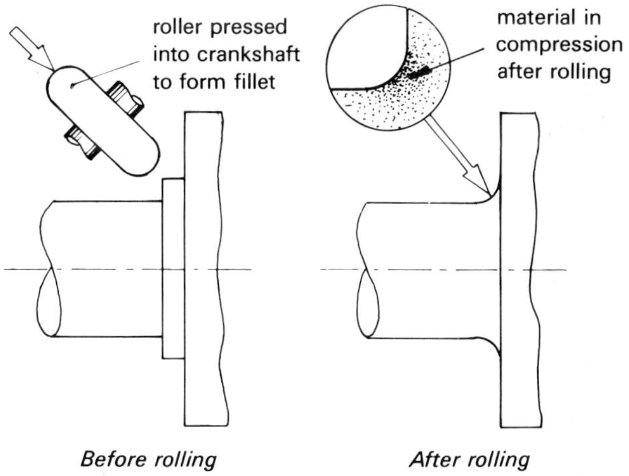

Fig. 5.10 Fillet rolling

6 Connecting rods

General design

Fig. 6.1 shows a simple connecting rod. Owing to limitations of space inside the piston, the end which fits on the gudgeon pin is smaller than that which fits on the crankshaft, and these ends are called the *small-end* and *big-end* respectively.

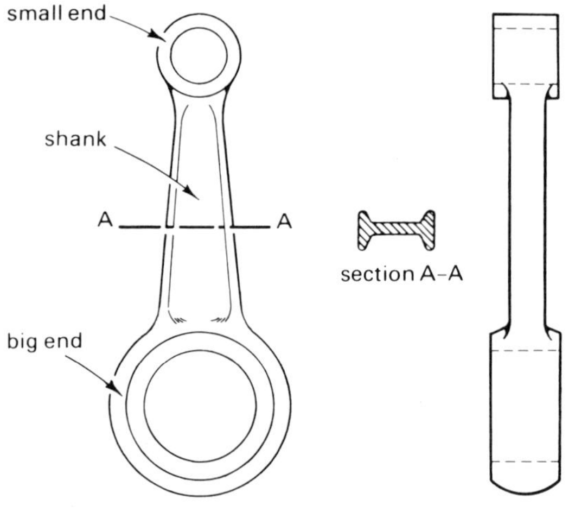

Fig. 6.1 A simple connecting rod

The load due to gas pressure on the piston has a buckling effect on the shank of the rod: in the direction of the crankshaft axis this tendency to buckle is resisted by the support provided by the gudgeon pin and the crankshaft, but no such support is provided to counteract buckling sideways. The shank of the rod is therefore generally made of H section, which gives the greatest possible resistance to sideways buckling, without excessive weight.

The big-end

The connecting rod shown in Fig. 6.1 has a one-piece or *solid-eye* big-end. This can only be used if the crankshaft is of *built-up* construction, i.e. if the shaft is made in sections assembled and suitably fixed together, allowing the big-end to be slipped over the end of the crank pin before the shaft is assembled. While this practice was once common in single-cylinder motor-cycle engines and has even been used in a few multi-cylinder car engines it is much heavier and more expensive than the usual one-piece crankshaft construction described in Chapter 5, and is seldom or never used in modern motor-vehicle engines.

It is therefore necessary to split the big-end across the centre of the crank pin and bolt the two halves together after assembly on to the crank pin. The detachable portion is called the *big-end cap*. The bolts used are of *high-tensile steel*: their heads are specially formed to prevent them rotating while the nuts are being tightened, and the nuts are usually locked by split pins, tab-washers, or some other arrangement.

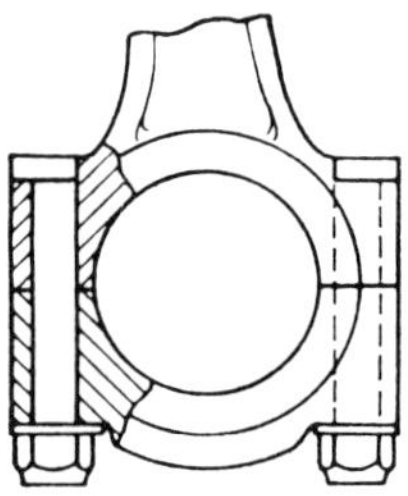

Fig. 6.2 Split big-end

Wherever possible the split is made at right angles to the length of the connecting rod (Fig. 6.2), as this gives the best combination of lightness and strength. The piston and connecting rod are usually assembled together and the assembly then placed in position in the engine, either by passing the connecting rod through the cylinder from the top end, or entering the piston into the cylinder from the lower end. The removal of the piston and connecting rod assembly is made in the reverse way. If the crankshaft is in position it is often impossible to get the piston past the crankshaft, especially if the shaft has balance weights; and though it would be quite practicable for the engine manufacturer to insert the pistons and connecting rods before fitting the crankshaft into its bearings, it is often necessary during repair work to remove one or more pistons; if the crankshaft also had to be removed just for this, a great deal of extra work would be needed.

On many modern high-speed engines the size of the crank pins required to give the crankshaft the necessary stiffness makes the big-end so large that it is impossible for the connecting rod to be assembled or withdrawn through the cylinder if the big-end is split at right angles to its length. In such cases the difficulty is overcome by splitting the connecting rod at an oblique angle and fixing the two parts together with screws (see Fig. 6.3).

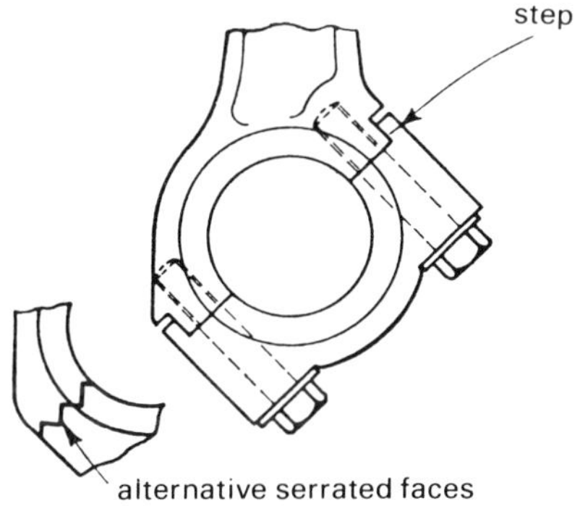

Fig. 6.3 Big-end split obliquely

At high engine speeds large inertia forces act on the piston and impose loads on the connecting rod, tensile around t.d.c. and compressive around b.d.c. When the big-end is split perpendicularly the inertia forces around t.d.c. impose high tensile loads on the big-end bolts. On the other hand, when the rod is split obliquely, the forces also tend to slide the cap sideways across the rod, and impose heavy shearing loads in addition to the tensile loads on the screws securing the cap. To relieve the screws of these shearing loads the joint faces of the rod and cap have steps machined in them (see Fig. 6.3). Alternatively the joint faces may be serrated.

Big-end bearing

To provide a suitable bearing surface to run on the steel crank pin, the inside surface of the big-end eye is lined with a thin coating of a special *bearing metal*. This is described in Part 6.

Small-end

The construction of the small-end of the connecting rod varies according to the method of securing the gudgeon pin, which may either be fixed in the small-end or free to move. In the latter case a solid-eye small-end is used, and this is generally lined by a bronze bush pressed in. One or more oil holes are drilled from the top of the rod to allow oil to reach the bearing surface: in some cases a hole is drilled, through a thickening of the web of the shank, to carry oil from the big-end to the small-end under pressure from the big-end. A bushed solid-eye small-end is shown in Fig. 6.4.

If the gudgeon pin is fixed in the small-end it is usually done in one of the two ways illustrated in Fig. 6.5, (a) being the more common. The screw is arranged so that a small groove must be made in the surface of the gudgeon pin to allow the screw to be fitted; this provides positive endwise location of the

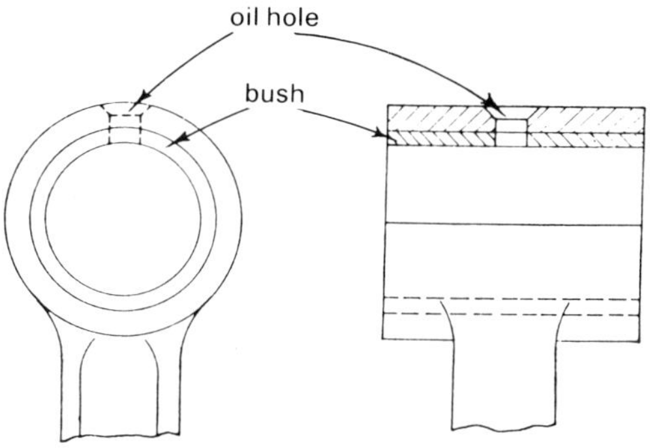

Fig. 6.4 Connecting rod small-end for fully floating gudgeon pin

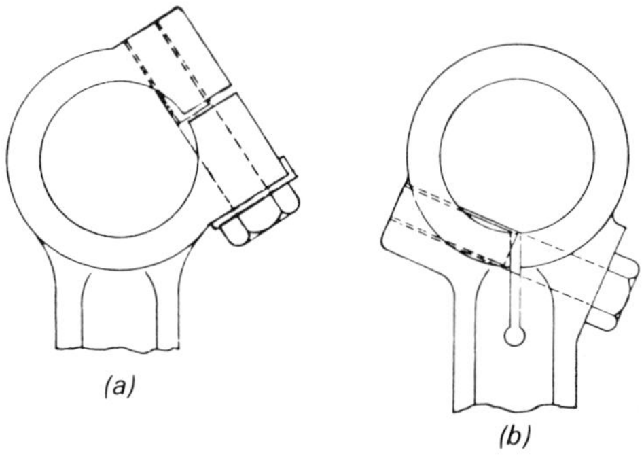

Fig. 6.5 Method of clamping gudgeon pin in small-end

pin independently of the clamping action of the screw, besides ensuring correct assembly. The screw is locked after tightening by some method such as the tab-washer illustrated in Fig. 6.5.

In many engines a third method of securing the semi-floating gudgeon pin is adopted. This consists of making the pin an interference fit in the unbushed solid-eye of the connecting rod small-end. Assembly is usually carried out by heating the small-end of the rod causing it to expand and allowing the gudgeon pin to be fitted easily. On cooling the rod contracts and grips the gudgeon pin securely. Special techniques are necessary for removing the pin.

Materials

Connecting rods are nearly always made from steel forgings, accurately machined where necessary. In high-performance engines they are often machined all over to reduce weight to the minimum, and polished to remove surface scratches which might lead to fatigue failure.

In a few cases an aluminium alloy is used for the rods. The main advantage sought is reduced weight, but a further advantage with aluminium is that it is a good material for bearings, and aluminium rods can be used without any bush in the small-end.

7 Pistons, rings and gudgeon pins

The piston: main features

Fig. 7.1 shows the main features of a piston. The *crown* forms the upper surface on which the gas pressure acts, and the force due to this pressure is equal to the cross-sectional area of the cylinder multiplied by the gas pressure. This force, which acts along the centre-line of the cylinder, is transmitted through the structure of the piston to the *gudgeon pin bosses* and thence through the *gudgeon pin* to the connecting rod.

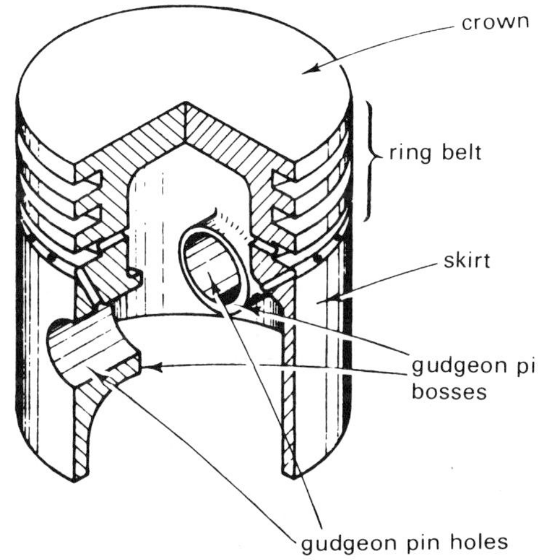

Fig. 7.1 Main features of a piston

During the greater part of the stroke the connecting rod operates at an angle to the centre-line of the cylinder. This causes a side force to be applied by the piston to the cylinder wall (Fig. 7.2), and it is necessary to provide bearing surfaces on the piston to carry this side force: these bearing surfaces are formed on the *skirt*.

To allow the piston to move freely in the cylinder it must have some clearance. This in turn allows gas to leak from the combustion chamber past the piston. Since the greatest leakage occurs when pressures are highest and the gas is hottest, much of the oil film lubricating the piston will be burnt away or carbonized. After combustion the gases contain water vapour, carbon dioxide and, probably, small amounts of sulphur dioxide which may contaminate the lubricating oil and lead to corrosion of the engine parts. To reduce the leakage as much as possible *piston rings* are fitted into grooves formed on the piston just below the crown.

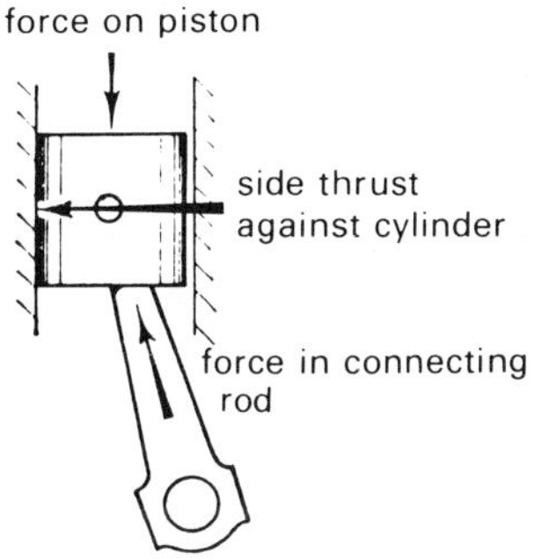

Fig. 7.2 How side thrust of piston against cylinder is caused

The crown of the piston is directly exposed to the full heat of the burning gases during combustion. These gases are still extremely hot during the power and exhaust strokes: the piston absorbs a great deal of heat from these hot gases and will reach a very high temperature unless heat is removed from the piston quickly enough to keep its temperature within reasonable limits. The piston can pass this heat on to the cylinder walls through the piston rings and skirt, and it can do this better if the metal of which the piston is made is a good conductor of heat.

Most metals expand with a rise in temperature, and since the piston gets hotter than the cylinder (which can be cooled more effectively), under running conditions it will expand more and the clearance between cylinder and piston will become smaller as the engine parts heat up. Thus, the clearance allowed when the engine is assembled must be large enough to allow for the decrease in clearance which occurs at running temperature, and the material used for the piston should preferably have a low coefficient of thermal expansion. Large clearances allow excessive side-to-side movement of the piston as the crank passes the dead centres and the connecting-rod angle changes, and this causes a noise called *piston slap*, which is objectionable.

Materials

Cast iron Early engines had pistons made of cast iron. This has good strength and hardness at operating temperatures and forms a good bearing surface against the cylinder wall. But it is relatively heavy, rather brittle and liable to develop cracks, and is not a particularly good conductor of heat. It has a coefficient of expansion comparable with that of the cylinder, allowing quite small clearances to be used when cold, thus avoiding piston slap.

Aluminium In its pure state aluminium has a high coefficient of thermal expansion, is relatively soft and low in strength. The addition of about 12 per cent silicon and small quantities of other elements makes a strong alloy which has a low coefficient of thermal expansion.

The main advantages resulting from the use of aluminium alloys for pistons are:

1 Improved thermal conductivity, which makes for lower piston-crown temperatures. This in turn causes less heating of the fresh mixture during the induction stroke, so that the mixture filling the cylinder is cooler and denser, leading to improved engine power output.
2 Reduced mass of the piston due to the lower density of aluminium, permitting the engine to run at a higher speed and so develop more power.
3 Because of its lower melting point it can be cast in steel moulds or *dies*. This gives greater accuracy of casting and reduced costs if the pistons are made in large numbers.
4 Aluminium alloys are usually softer than cast iron and are easier to machine.

The chief disadvantages are:

1 Greater coefficient of thermal expansion. This necessitates the use of larger clearances when cold, resulting in piston slap. This can be minimized by special construction of the piston skirt and by the use of alloys which have a relatively low coefficient of expansion.
2 Aluminium is not as strong as cast iron, particularly at high temperatures. It is thus necessary to use a greater thickness of metal to avoid distortion under load. Whilst this reduces the saving of mass resulting from its lower density, it does increase the ability of the piston to conduct heat.
3 Aluminium is also softer than cast iron, especially at high temperatures; this may lead to excessive wear of the ring grooves. Improved alloys have been developed and in some cases other methods have been adopted to overcome this difficulty.
4 Aluminium is more expensive than cast iron, though this is to some extent offset by its greater ease of casting and machining.

Composite construction Attempts to combine the low weight and good heat conductivity of aluminium with the small clearances when cold permissible with cast iron or steel resulted in a two-piece construction in which the crown and gudgeon pin bosses were made of aluminium and the skirt of cast iron or steel.

Piston details

The crown

In its simplest form the crown is flat and at right angles to the cylinder axis, and this involves only the simplest of machining operations.

In some cases the crown is slightly *dished* (Fig. 7.3a). One reason for this is to make possible the use of higher compression ratios, should this become desirable, simply by reducing or eliminating the 'dish'. It is sometimes done to provide a particular form of combustion chamber.

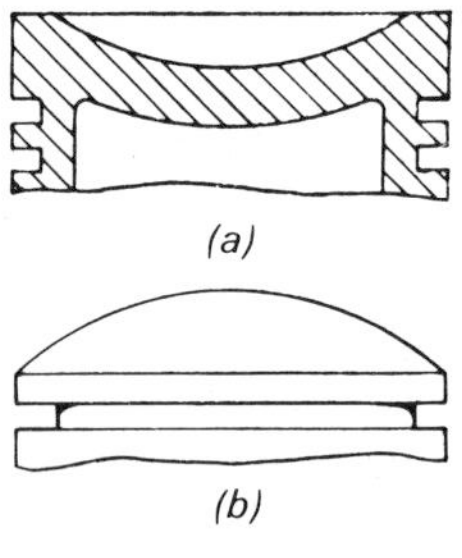

Fig. 7.3 (a) Concave piston crown **(b)** Domed piston crown

Certain designs of combustion chamber require the piston crown to be made a particular shape, such as the *domed* crown (Fig. 7.3b).

Note that, whatever the shape of the crown, the effective force on the piston due to gas pressure is always given by multiplying the pressure by the cross-sectional area of the cylinder bore.

The ring belt

It is most important that this part of the piston be accurately made. In particular, each ring groove should lie in a plane at right angles to the cylinder axis. This part of the piston is generally given more clearance in the cylinder than the skirt, and does not normally bear against the cylinder. The *ridges* which separate the grooves are called *lands*.

There are usually three or four of these grooves above the skirt, and sometimes one in the skirt below the gudgeon pin. These are discussed in greater detail later in this chapter in connection with piston rings.

Pistons that are prone to ring groove wear can be fitted with an iron ring carrier which is metallically bonded into the piston. This process provides good heat flow and ensures that the ring life is extended.

Gudgeon pin bosses

These must be connected to the piston crown in such a way that the loads which result from gas pressure are transmitted from crown to gudgeon pin without any possibility of distortion of the piston. The bosses are usually connected directly to the crown by substantial struts or thick webs.

It is important that the gudgeon-pin holes should be bored accurately and with an extremely fine surface finish.

The skirt

The simplest form of this is illustrated in Fig. 7.1, in which the skirt forms a downward tubular extension below the ring belt: this arrangement is called a *solid skirt*. It is the strongest form of skirt and is always used for engines where the loads on the piston are particularly great.

To permit the use of small clearances with aluminium alloy pistons when cold, special types of skirt construction are used which involve some degree of flexibility of the skirt. For example, the skirt may be almost completely separated from the ring belt except at the gudgeon-pin bosses, and split down one side. This allows it to fit in the cylinder with very little clearance when cold; expansion due to rise in temperature is taken up by reducing the width of the split. The split usually extends the full length of the skirt and is at a slight angle, to avoid leaving a ridge on the cylinder as wear occurs.

When assembling split-skirt pistons on to the connecting rod it is important that the split side of the skirt is on that side which bears against the cylinder during the compression stroke, so that the unsplit – and stronger – side takes the greater side thrust during the power strokes.

The high piston load of a modern engine has meant that the split skirt has been superseded by low-expansion, solid-skirt pistons such as the *thermal slot* or *link strut* types:

1 *Thermal slot* Cutting a lateral thermal slot at the bottom of the ring groove on the thrust and non-thrust sides of the piston directs the heat from the hot (250°C) piston crown to the gudgeon-pin boss regions. In some designs these slots are extended axially downwards into the skirt and terminate with a circular hole (Fig. 7.4). Thermal expansion of the crown and bosses, combined with the effect of the gas load, pulls in the thrust faces and allows the piston to be fitted with a small clearance. To compensate for this action the piston is oval when cold.

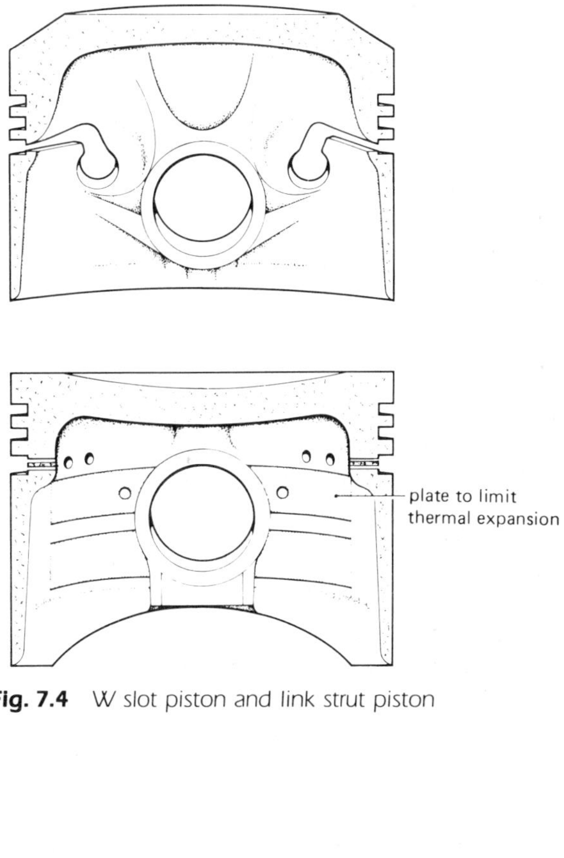

Fig. 7.4 W slot piston and link strut piston

2 *Link strut* pistons have special alloy steel (Invar) plates cast into the skirt. This material has an extremely low coefficient of thermal expansion, so the resistance given by these plates limits the expansion of the skirt.

Piston clearance

The clearance left between the cylinder and piston when the engine is assembled will depend on the piston material, piston design and the operating temperature of the engine. Too small a clearance is likely to cause seizure, whereas excessive clearance results in gas blow-by, noise and high oil consumption.

For a particular engine the manufacturer's manual should be consulted to obtain the actual clearances. In the past, a rough guide for piston skirt to cylinder clearance was taken as:

Cast iron	0.001 × bore diameter (mm)
Aluminium alloy, solid skirt	0.002 × bore diameter (mm)
Aluminium alloy, split skirt	0.001 × bore diameter (mm)

Design improvements and better materials have allowed the skirt clearance to be reduced. Many modern solid skirt, aluminium alloy pistons have a clearance as small as 0.0005 × bore diameter. For example, a certain piston of diameter 82 mm has a clearance of 0.041 mm; this is the distance between the cylinder and the thrust face of the piston.

The crown runs much hotter than the skirt, so the piston is slightly tapered to allow for this, the maximum clearance being given to the top. The taper is critical since too small a clearance at the top land causes the land to contact the cylinder and give scuffing and noise, whereas excessive clearance results in gas blow-by.

Piston rings

It was mentioned earlier that the purpose of these is to prevent gas leakage through the clearance which must be left between piston and cylinder to allow the piston to move freely.

Fig. 7.5 shows the main features of a piston ring: it is rectangular in section, and the *nominal diameter* is the diameter of the cylinder into which it is to fit. (This is measured on the ring with the gap closed, using a steel tape which is passed around the ring and pulled tight.) A radial cut is made in the ring so that it can be placed in position in the groove on the piston. When in position in the cylinder the gap between the ends of the ring, called the *working gap*, needs to be large enough to ensure that the ends of the ring do not actually touch, however hot the ring may get. When out of the cylinder the actual diameter of the ring is larger than its nominal diameter. The gap, now called the *free gap*, is also larger than the working gap: this ensures that when in position in the cylinder the ring will exert an outward pressure against the wall of the cylinder along its whole circumference.

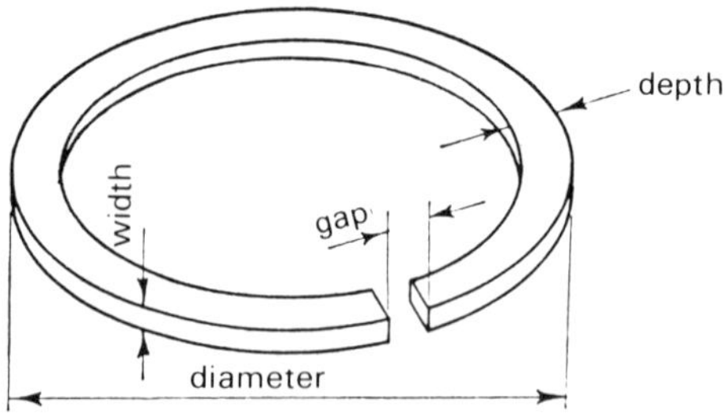

Fig. 7.5 Piston ring dimensions

Fig. 7.6 shows a section of the ring in position in its groove and filling the clearance between the cylinder wall and piston. The depth of the ring should be slightly less than the depth of the groove, to ensure that side forces acting on the piston are transferred to the cylinder via the bearing surfaces on the skirt and not through the piston rings, and also to ensure that the clearance between the inner surface of the ring and the bottom of the ring groove is not less than the correct piston clearance.

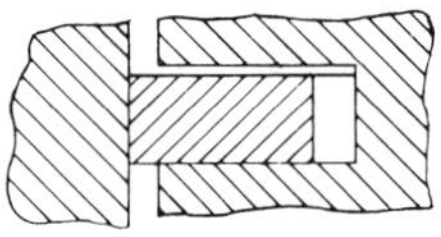

Fig. 7.6 Plain rectangular-section piston ring

The width of the ring must also be slightly less than the width of the groove, to ensure that the ring will be perfectly free to move about in its groove. If the ring is too close a fit in its groove it is liable to become securely stuck by carbonized oil after a period of use.

Fig. 7.7 shows that the ring has a rectangular cross-section. In some cases this is slightly modified in one of three ways:

1 The *outer surface* of the ring is given a slight taper towards the top (about 1°). The purpose of this is to speed up the 'bedding in' process when the ring is new (Fig. 7.7a).

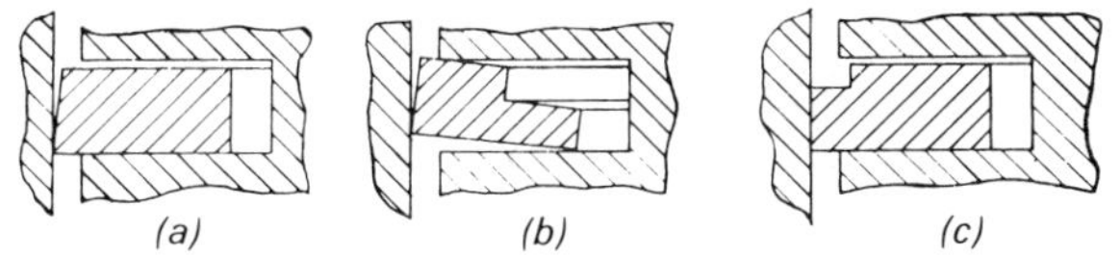

Fig. 7.7 Special types of compression ring

2. The *top inner corner* of the ring is cut away, leaving the cross-section of the ring like a thick letter L. This serves the same purpose as the taper-faced ring above (Fig. 7.7b).
3. The *top outer corner* of the ring is cut away (Fig. 7.7c). This ring should be used when new top rings are fitted to worn cylinders. The cut-away portion avoids contact with the ridge which develops at the upper limit of ring travel in a worn cylinder. Some manufacturers name this ring a 'ridge dodger'.

Oil-control or scraper rings

The piston rings so far described are fitted for the purpose of preventing the escape of gas through the clearance between cylinder and piston, and are called *compression rings*, *pressure rings* or *gas rings*. Piston rings also control the oil film on the cylinder wall, in order to permit adequate lubrication without excessive quantities of oil getting past the piston and into the combustion chamber. In the combustion chamber the oil would be decomposed and partly burnt, forming deposits of carbon on the combustion chamber walls and causing smoke in the exhaust gas. Special types of ring have been developed for this purpose which are intended to glide over the oil film as the piston moves upward, but to scrape off all but a very thin film of oil on the downward stroke. Provision is made, usually by holes at the back of the ring groove or by *chamfering* the land immediately below the groove and drilling oil drain holes from this chamfer to the inside of the piston, to return oil collected by the scraper ring to the underneath of the piston. Fig. 7.8 shows sections of several types of oil control rings and their grooves. The bevelled scraper is the least severe in its action, the grooved type the most severe.

Pistons are usually fitted with two or three compression rings and one scraper ring located in the bottom groove. In the past an additional scraper ring was fitted around the skirt, but the desire for low bonnet lines has resulted in a reduction in the engine stroke and piston length: this means that there is insufficient space available for the skirt scraper ring.

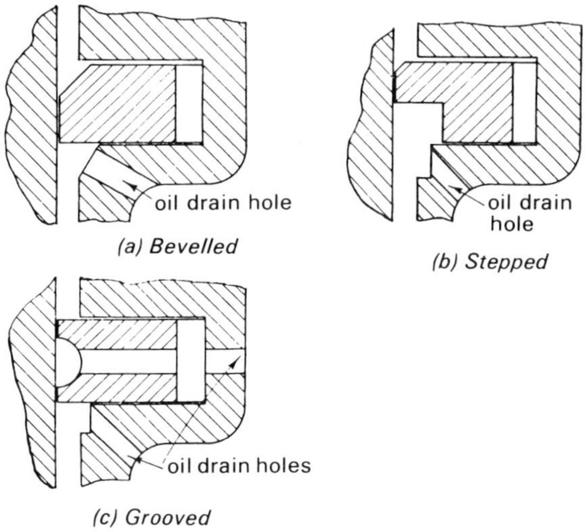

Fig. 7.8 Types of oil control ring

Furthermore the proximity of the piston to the oil being discharged from the bearings, combined with cylinder distortion, has demanded a scraper ring that exerts an outward pressure of about 3150 kPa (450 lbf/in^2): this is about twenty times as great as the standard iron types.

Fig. 7.9 shows a steel oil-control ring. It consists of two narrow steel rails separated from each other by a specially shaped spacer. Large outward pressure is obtained by fitting an expander spring behind the ring. To avoid incorrect assembly, some designs combine the various ring elements into one unit.

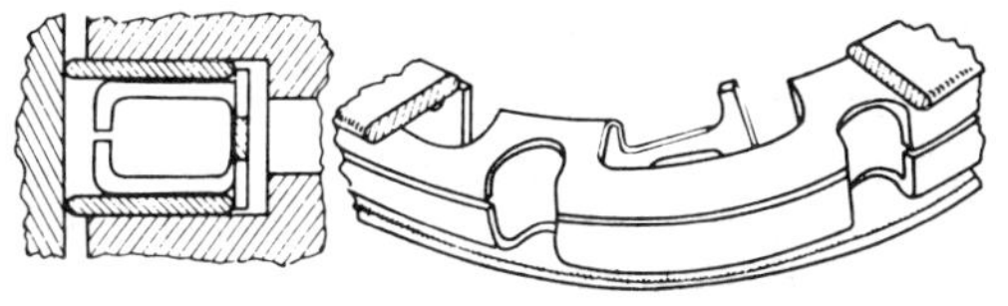

Fig. 7.9 Steel oil control ring

Fit of piston rings

When new piston rings are fitted to an engine there are two important points to check – apart from ensuring that the ring is of the correct size and type. These are (a) the gap, and (b) the side clearance in the groove. The correct figures for these are given in the engine manufacturer's service manual, but approximate values are given below for guidance.

Working gap
Water-cooled engines 0.003 × bore diameter
Air-cooled engines 0.004 × bore diameter
Side clearance in groove 0.04 mm (0.0015 in)

For example, a ring fitted to a water-cooled engine of bore 82 mm should have a gap clearance of 0.246 mm.

Measurement of working gap

This gap is measured by putting the ring, removed from the piston, into its cylinder and *squaring it up* by pushing it up against the end of the piston also placed inside the cylinder. The gap between the ends of the ring is then measured by feeler gauges. If the cylinder is worn, the ring should be in the *least worn* part of the bore when this measurement is made. Too small a gap can be rectified by carefully filing or grinding one end of the ring, preferably in a special jig which ensures squareness of the ends.

Piston ring materials

The most important properties of the material used for piston rings are:

1 Wear resistance, especially - in the case of the compression rings - under conditions of somewhat scanty lubrication.
2 Elasticity, to ensure that the ring will maintain sufficient outward pressure against the cylinder wall to maintain contact with the surface of the cylinder.
3 These properties must not be adversely affected by the temperatures to which the ring is subjected in use.

The material which most nearly fulfils these requirements is cast iron. In some cases the wear resistance is improved by chromium-plating the outer face of the ring by a special process, but rings of this type are normally only used for the top compression ring.

In some oil-control rings the material used is steel, generally chromium-plated on the edges which bear upon the cylinder wall.

Gudgeon pins

These are of two types, depending upon the method of locating them in position. To provide the strength necessary to carry the high loads imposed upon them by gas pressure on the piston, and a wear-resistant surface, they are made of steel, usually alloyed with 3–4 per cent of nickel to increase toughness, and case-hardened to obtain a wear-resistant surface. Clearly the ends of the hardened pin must be prevented from rubbing against the cylinder wall - otherwise it would soon score deep grooves in the cylinder surface.

The two types of gudgeon pin are:

1 Semi-floating These are securely fixed in either the piston or the connecting rod, usually the latter, as described in the previous chapter. Although it was at one time common practice to fix the gudgeon pin in the piston by means of a screw threaded into one of the gudgeon-pin bosses, this practice has been discarded in favour of fixing the pin in the connecting rod. Movement is therefore confined to oscillation of the pin in the piston.

2 Fully floating This type is free to turn in both piston and connecting rod, and is generally used in engines in which the loads are particularly high. The pin is shorter than the cylinder diameter and is prevented from contacting the cylinder by one of the methods illustrated in Fig. 7.10. The method shown at (a) consists of a spring steel ring called a *circlip*, which is sprung into a groove inside the gudgeon pin hole after the gudgeon pin has been fitted - one at each end of the pin. The circlip is sometimes made of round-section steel wire and may have its ends bent inwards as shown at (b) to make it easier to fit and remove with pointed-nose pliers. Some circlips do not have the bent-in ends: in this case the ends of the gudgeon pin are slightly chamfered as shown at (a) (this helps to prevent the circlip being forced out of its groove by endwise forces), and small notches are cut in the piston so that a pointed instrument can be inserted behind the circlip to remove it. Another type of circlip often used is the *Seeger circlip* (c) which is stamped from spring steel sheet. Special circlip pliers are used for fitting and removing this type.

Gudgeon pins are usually made tubular to reduce weight, and when this is so an alternative method of preventing scoring of the cylinder is to fit mushroom-shaped pads of soft metal such as aluminium or

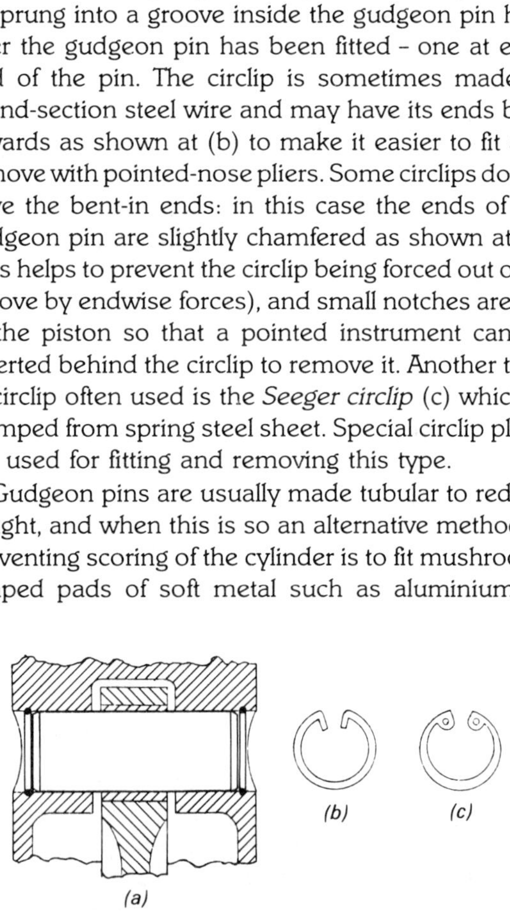

Fig. 7.10 Location of gudgeon pin by circlips

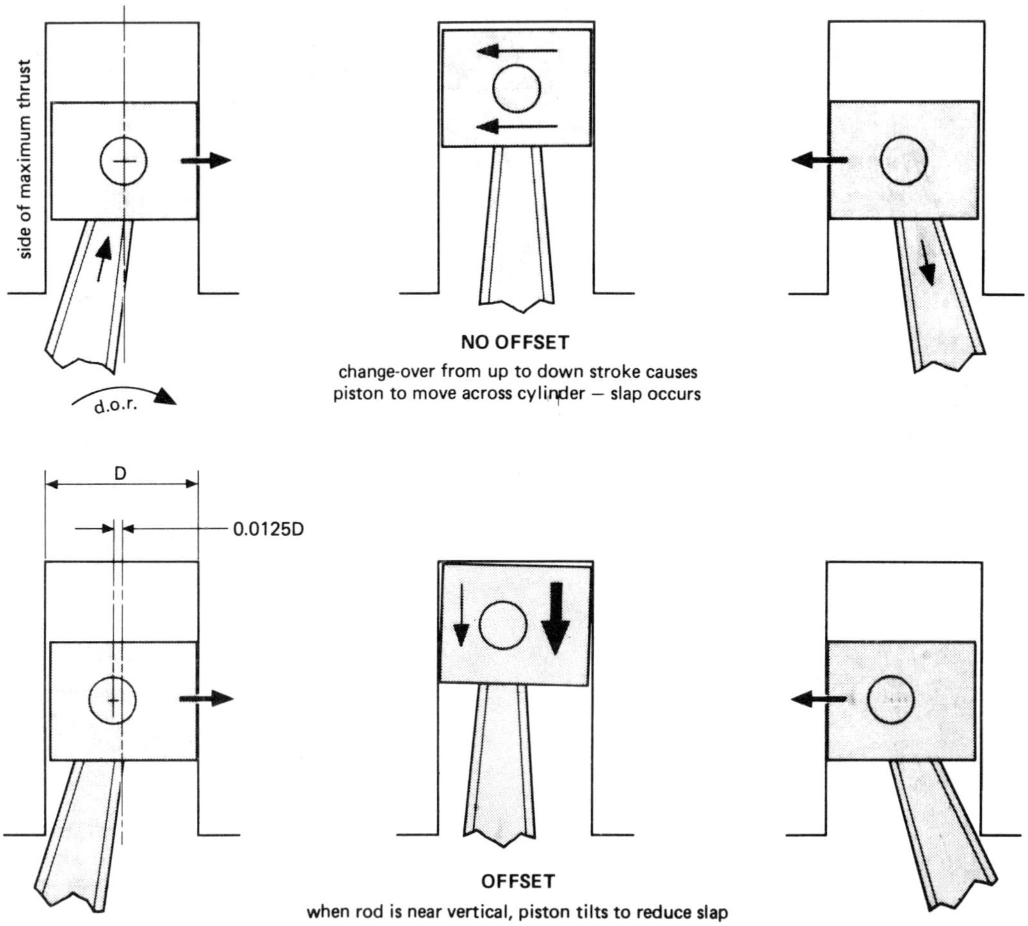

Fig. 7.11 Effect of offset of gudgeon pin

brass. The 'stem' of the mushroom is a press fit in the bore of the gudgeon pin. This method is seldom used in modern engines.

Gudgeon pin fit

The gudgeon pin is usually made a tight push fit in the piston when cold: this eases off slightly when the piston reaches normal running temperature. You must not use force to remove a tightly fitting pin unless you do it very carefully, and great care must be taken to ensure that no force whatever is applied to the connecting rod, to avoid the risk of bending it. If possible, heat the piston in hot water or oil: if the piston cannot be immersed in oil or water, you can wrap rags soaked in hot water round the piston, after which the gudgeon pin can be pushed out easily without risk of damage.

A recent development is the use of an interference fit of the gudgeon pin in the small end of the connecting rod. This is in effect a semi-floating pin in which the interference fit is relied upon to hold the gudgeon pin in place (see page 27).

Offset gudgeon pin

The skirt clearance needed for a solid skirt alloy piston has increased the occurrence of piston slap when the engine is cold. This noise can be reduced by offsetting the gudgeon pin axis towards the side of maximum piston thrust (see Fig. 7.11).

At t.d.c. the slight tilting action eases the transfer of piston thrust from one side of the cylinder to the other. Since the offset is only about 0.0125*d*, where *d* is the piston diameter, then care must be taken to observe the marking 'front' when the piston is fitted.

8 Valves and valve-operating gear

The function of valves

The gases enter and leave the working cylinder through ports, but these ports must be open *only* during that part of the cycle of operations when gas is required to pass through them. At all other times they must not only be closed, but closed sufficiently tightly to prevent the passage of gas at considerable pressure. *It is the function of the valves to close the ports except when they are required to be open*. (Note that in the case of two-stroke engines the ports can be closed by covering them with the piston, but a little thought will show that this method cannot be successfully used on an engine working on the four-stroke cycle.)

In performing this function the valves must

1 make a completely gas-tight seal in the ports when closed
2 offer no opposition to the flow of gases when open
3 require the simplest possible mechanism to operate them
4 operate with the minimum of friction.

Types of valve

During the development of internal combustion engines, several types of valve have been used. The chief of these are:

1 *The poppet valve* This was already in use on steam engines before successful internal combustion engines were developed, and was used on the early types of such engines. It has proved so successful that its use has continued and it is almost the only type to be used in modern engines.
2 *The slide valve* This too was very commonly used on some early internal combustion engines, but proved less satisfactory than the poppet valve and is now obsolete.
3 *The sleeve valve* Several versions of this type of valve have been used. The two most successful were the Knight and the Burt–McCullum.

The *Knight* type originated in about 1905 and consisted of a sleeve free to slide inside the cylinder with a second sleeve free to slide inside it, the piston moving within the inner sleeve. The sleeves were moved up and down inside the cylinder, and ports cut in the sleeves were arranged to uncover the cylinder ports at the correct times.

This valve was used by the Daimler Company from about 1909 until about 1933: it was also used by a number of Continental manufacturers such as Panhard, Minerva, Voisin, Peugeot and on a few Mercedes models.

The *Burt–McCullum* sleeve valve consisted of a single sleeve which was given a combination of up-and-down motion and part-rotary motion inside the cylinder, ports in the sleeve uncovering the cylinder ports at the appropriate times. This was developed in about 1909 and was first used in the Argyll car in 1911. It has also been used in a few other cars, but its greatest success has been in aircraft engines, chiefly the large engines made by the Bristol Aeroplane Co. beginning about 1935. It was also used in the famous Napier 'Sabre' engine, a twenty-four-cylinder engine developing about 3500 hp, and in the last of the Rolls-Royce piston-type aero-engines, the 'Eagle', also a twenty-four-cylinder engine developing over 3000 hp.

Although these types of valve seemed to have many advantages over the cruder poppet valve they have not survived in motor-vehicle engines. Their main drawback was a relatively high oil consumption and a smoky exhaust.

4 *The rotary valve* Several types of rotary valve have been developed and some have given very good results. They consist of rotating 'plugs' fitted across the ports, having holes which, at the correct times, uncover the ports and allow gas to pass.

The only type of valve at present in use in motor-vehicle engines is the poppet valve and this is the only type which will be described in detail.

The poppet valve

A poppet valve, with its immediate attachments, is shown in Fig. 8.1. The valve itself consists of a disc-shaped *head* having a *stem* extending from its centre at one side. The edge of the head on the side nearest the stem is accurately ground at an angle - usually 45° but sometimes 30° - to form the *face*. When the valve is closed, the face is pressed into contact with a similarly ground *seating* at the inner end of the port. Beyond the seating the port curves smoothly away clear of the valve.

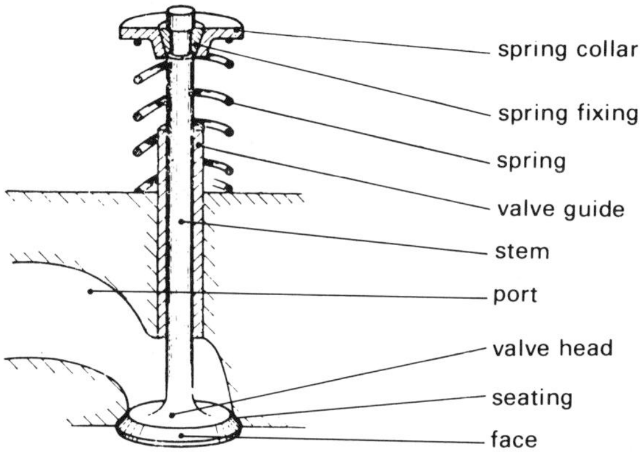

Fig. 8.1 A poppet valve

The condition of the face and seating is of vital importance in ensuring the gas-tightness of the valve. They are ground on special machines and the valve face is sometimes *lapped in* to its seating with a very fine abrasive. Whether this is necessary or even advisable if the grinding is sufficiently accurate is often in dispute.

When, after a considerable period of use, the condition of face and seating deteriorates and allows leakage, they must be reground. Special grinding machines have been developed for this purpose.

Fig. 8.2 shows the face and seating in more detail. The seating is narrower than the face, to reduce the risk of trapping particles of carbon between face and seating, and to ensure a surface pressure great enough to provide a gas-tight seal. The thickness of the head indicated by the arrows is most important: a sharp edge, particularly in the case of the exhaust valve, is liable to become excessively hot and may

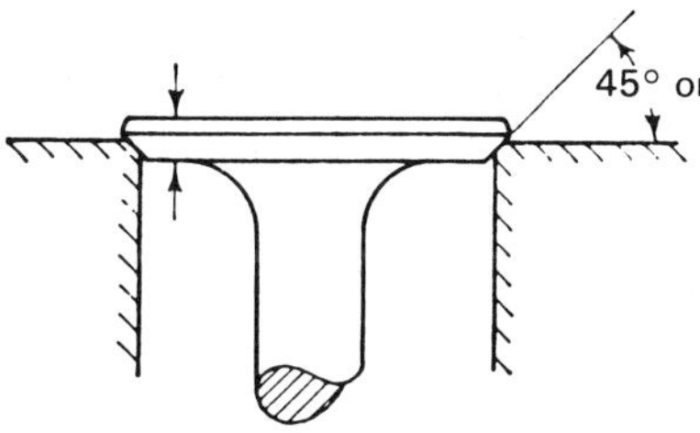

Fig. 8.2 Valve face angles

cause *'pre-ignition'* (ignition of the mixture inside the cylinder *before* the spark occurs).

In a few cases the exhaust valve face is ground to an angle about $\frac{1}{2}$° less than the seat angle, as shown in Fig. 8.3. There are three reasons for this:

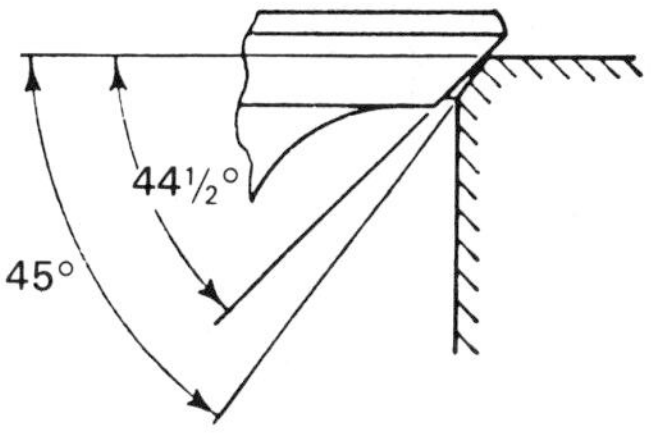

Fig. 8.3 Valve face angle different from seat angle (exaggerated)

1 The hottest part of the valve, under running conditions, is the stem side of the head, and the additional expansion of this side makes the face and seat angles equal at running temperatures.
2 The exhaust valve gets very hot (often red hot when the engine is running at full throttle) and is then less strong. Under these conditions the spring load tends to cause the head to dish slightly, which can lift the inner edge of the face (nearest the combustion chamber) clear of the seating if the angles are the same when cold.
3 It reduces the risk of trapping carbon between face and seating. In this case the face and seating cannot be lapped in.

Valve guide

The valve slides in a hole in the cylinder or cylinder head called the *valve guide*. This must be perfectly true with the seating, and a small clearance is allowed between stem and guide to prevent any possibility of the valve sticking. Excessive clearance may lead to the seating being worn oval, and in the case of the inlet valve will allow air to leak into the mixture entering the cylinder.

The guide is usually made in the form of a detachable sleeve to permit easy renewal when worn, but in some cases it is integral with the cylinder or head, since this assists the transfer of heat from the hot exhaust valve to the cooling system. In this case wear is rectified by boring or reaming out the guide slightly and fitting a valve with an 'oversize' stem.

When the guides are detachable they sometimes have a shoulder formed on them. They are an interference fit in the cylinder or head and are pressed into the shoulder. Guides without shoulders are easier and cheaper to make: when fitting, press these in to leave a definite length projecting from the head, as shown in Fig. 8.4b.

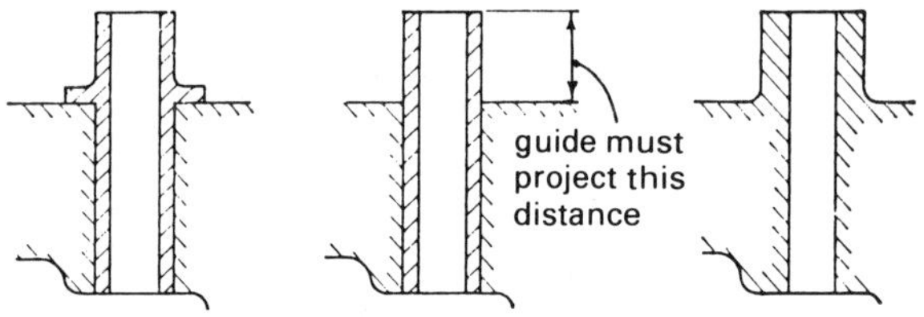

Fig. 8.4 Types of valve guide

The valve spring

The spring returns the valve to its seat after it has been opened, and holds it there (assisted by gas pressure during the compression and power strokes) until it is next opened.

The spring must be strong enough to close the valve quickly at maximum engine speeds. The spring strength needed depends upon the weight of the valve and any other part of its operating gear which it has to move, and upon the maximum speed at which the engine is required to run. If the force exerted by the spring is insufficient the valve will be *late* in closing at high rev/min and so limit maximum engine speeds.

Valve springs are usually of the *helical coil* type, as shown in Fig. 8.1 and 8.5a. Such springs are liable to *surging* at certain engine speeds, when the centre coils vibrate in a direction parallel to the valve stem. In extreme cases this can lead to breakage of the springs, and can also allow the valves to *bounce* off their seatings after closing. This may be overcome in several ways:

1 The springs are so designed that when the valve is fully open they are compressed until adjacent coils almost touch, so preventing the building up of excessive surging.
2 The coils are spaced closer together at one end than the other (Fig. 8.5b). Springs of this kind must be fitted with the 'close' coils nearest the cylinder head.
3 Two (or occasionally three) springs are used, one inside the other. This does not prevent surging, but the two springs will not both surge at the same engine speed, and 'valve bounce' will be prevented. Also, if one spring breaks the second one will maintain pressure on the spring attachment and prevent the valve dropping into the cylinder.

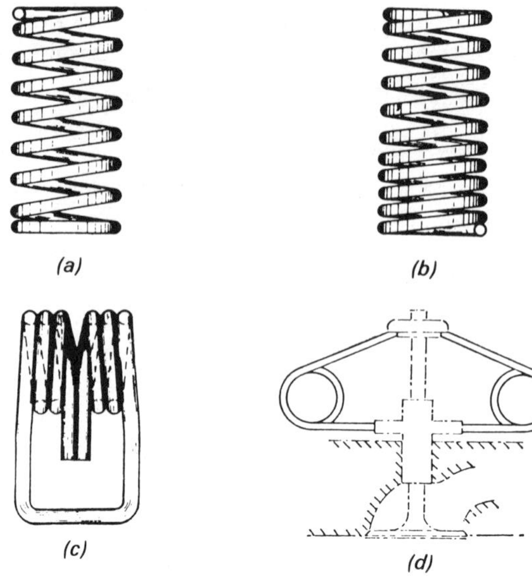

Fig. 8.5 Types of valve spring

Another type of spring occasionally used is the *hairpin* spring shown in Figs. 8.5c and 8.5d. This is not liable to surge, but is more difficult to fit into the space available.

Valve spring fixing

Several different methods have been used to attach the spring to the end of the valve stem. Some commonly used methods are shown in Fig. 8.6.

Materials

The exhaust valve is exposed to the full heat of the burning gas during combustion, and when the hot gases are released from the cylinder they sweep past the exhaust valve head. The exhaust valve is the hottest part of the engine, and it is estimated that under full power conditions it reaches a temperature

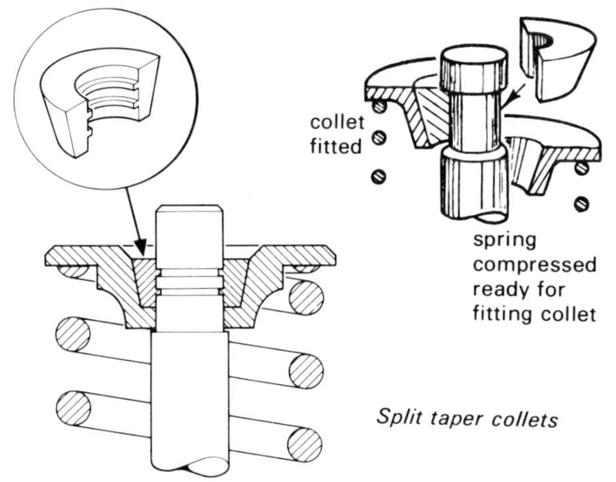

Fig. 8.6 Methods of fixing valve spring to valve stem

of around 650 °C. A rather special steel is required to operate satisfactorily for long periods, and alloys have been developed for this purpose containing varying amounts of manganese, silicon, nickel and chromium.

The exhaust valve is cooled by passing on its heat in two directions: (1) along the stem and through the guide into the cylinder head, and (2) directly into the cylinder head from the face to the seating when it is closed. For extreme operating conditions the valve stem is sometimes made hollow and partly filled with sodium, which is a very soft metal having a melting point of 98 °C. Under running conditions it is molten, and in splashing from end to end of the valve stem it assists the transfer of heat from the hot head. This is an expensive form of construction and is only used when absolutely necessary.

Some valves are coated with aluminium to improve the heat transfer from the valve to the engine block. 'Aluminized valves' cannot be reground in the normal way because the surface coating is comparatively thin.

Detachable valve guides are usually made of cast iron, though bronze is sometimes used, particularly for exhaust valves, because of its better heat-conducting properties.

Valve springs are made of *spring steel*. This contains about 0.6 to 0.7 per cent carbon and usually small amounts of silicon, manganese and vanadium.

Valve position

The position of the valve in the cylinder head depends upon the design of the combustion chamber which is discussed on page 58. It is sufficient for the present to list the possible positions of the valves, as follows:

1 *Side valve or L-head* Both valves are at one side of the cylinder, heads uppermost with the stems approximately parallel to the cylinder axis. To accommodate the valves in this position the combustion chamber is extended sideways from the top of the cylinder forming a shape similar to an inverted letter L (Fig. 8.7). This arrangement is not used on modern engines.

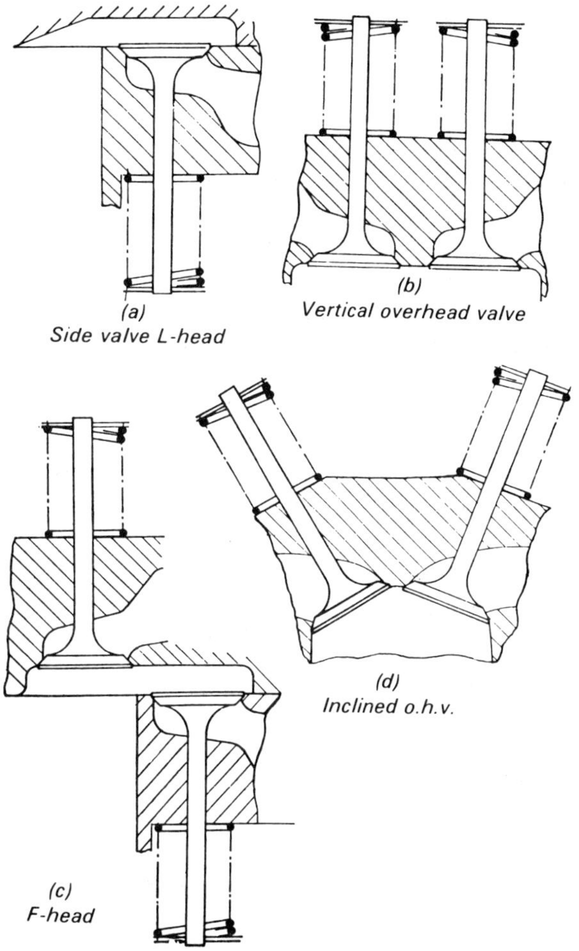

Fig. 8.7 Alternative valve arrangements

2 *Side valve or T-head* This is like the above except that inlet and exhaust valves are fitted at opposite sides of the cylinder. This arrangement, though common on early engines, has been obsolete for many years.

3 *Vertical overhead valves* Both valves are fitted over the top of the cylinder with their stems approximately vertical and parallel (Fig. 8.7b), usually in a single row.

4 *Overhead inlet and side exhaust or F-head* The inlet valve is arranged as in 3 above and the exhaust as in 1 (Fig. 8.7c).
5 *Inclined overhead valves* Both valves are fitted over the top of the cylinder head, but in two rows inclined at an angle to one another (Fig. 8.7d).

Valve-operating gear

It has already been explained that the valves are moved to, and held in, their closed position by their springs. They are opened by *cams* carried upon one or more *camshafts*, which are driven by suitable gearing from the crankshaft. The cams do not usually act directly upon the valves since their rotary motion would impose a side force upon the end of the valve stem: to combat this side force *cam followers* are interposed between the cam and the valve. These followers may take one of two forms:

1 *tappets* which rest upon the cams and are guided so as to move in a straight line, and
2 *rockers* or levers which rock upon fixed pivots. If the cams are placed close to the ends of the valves no additional mechanism may be needed, but in the case of overhead valves this would involve a long and possibly complicated driving gear between camshaft and crankshaft. In many engines it is considered preferable to place the camshaft reasonably near the crankshaft, so that a short and simple driving gear can be used. In such cases the motion of the tappets is conveyed to the valves through *push rods* and overhead rockers.

Valve clearance

A valve is closed when its face is pressed into contact with the seating by the full force of the spring.

Fig. 8.8 shows a simplified arrangement of valve, tappet and cam, and at (a) the valve is shown in the fully open position. As the cam rotates, the valve moves under the action of its spring (not shown) towards its closed position. At (b) a condition is shown in which the tappet has reached the *base* of the cam before the face of the valve has come into contact with the seating. Since the tappet cannot move downwards any further *the valve cannot completely close*.

At (c) the tappet is shown in the same position on the cam, but the valve face has reached the seating *before* the tappet has reached the base of the cam, leaving a gap or clearance between the end of the valve and the tappet. This is called *valve clearance* or *tappet clearance*, and it is clear that in this case the tappet cannot prevent the valve from closing properly.

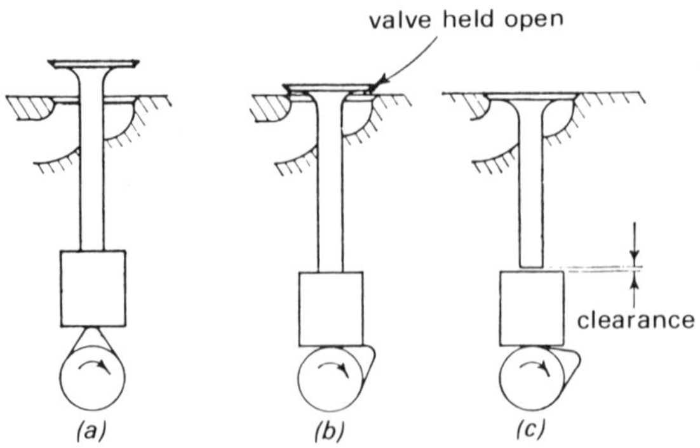

Fig. 8.8 The need for valve clearance

There must always be a small clearance in the valve-operating gear when the valve is in its closed position *to ensure that the valve will close completely*. When the clearance is checked, allowance must be made for the fact that under different running conditions the amount of clearance will be altered by different temperatures - and therefore different amounts of expansion - of the valves, cylinder block and valve-operating gear. The valve clearance and the engine conditions under which it should be measured, i.e. engine hot or cold, are specified by the engine manufacturer, and it is most important that the clearance should be correct under the specified conditions. Excessive clearance will result in noisy operation and excessive wear. Too little clearance may result in the valve not fully closing under certain conditions. This causes loss of pressure in the cylinder and burnt exhaust valves (since one of the paths for the escape of heat is cut off if the valve does not close properly, and gas at a very high temperature will be forced at high speed through the gap between valve face and seating).

There is always provision for adjusting the valve clearance during assembly, usually (but not always) by means of an adjusting screw. In most cases the clearance should be checked at intervals to ensure that it has not been altered by wear.

It is often assumed that the valve clearance is decreased as the engine becomes hotter, but this is by no means always the case. It is suggested that students should themselves check the valve clearances of a number of engines both hot and cold, and note the results.

It should be noted that incorrect clearance will

affect the angle of the cam at the instant the valve leaves or returns to its seat. Excessive clearance will cause the valve to leave its seat *later* and return it *earlier* in terms of cam rotation: i.e. the angle through which the cam – and therefore the crankshaft also – rotates while the valve is open will be *shortened*. Insufficient clearance will have the reverse effect.

Cams and camshafts

A cam is a component so shaped that by its rotation it causes another part in contact with it to move in a different manner. There are very many types of cams in use in all manner of machines.

The type of cam used to operate the valves of a motor-vehicle engine is a relatively simple one such as is illustrated in Fig. 8.9. The cam follower remains stationary during that part of the cam's rotation when it rests upon the *base circle*, and its 'lift' begins at the point where the *opening flank* joins the base circle. At the *peak* of the cam the lift is maximum, and the follower falls as the *closing flank* passes beneath it. By the time it is again in contact with the base circle at the end of the closing flank it will have returned to its original position.

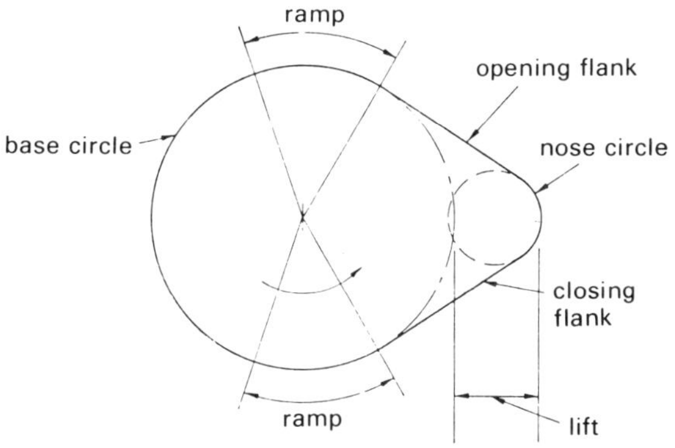

Fig. 8.9 Details of cam

In some engines there may be fairly large variations of clearance under different running temperatures, and to ensure quiet operation even when the clearances are large the flanks are joined to the base circle by *ramps* which extend over about 30° of cam rotation. When checking valve clearances it is most important that the cam follower should be resting on the base circle of the cam. One way of ensuring this is to turn the engine to the position where the valve whose clearance is to be adjusted is *at full lift*, and then turn the engine *one complete revolution*. A simpler method is to note that in most multi-cylinder engines (see Chapter 12) the pistons are arranged to move up and down together in pairs, so that if one of a pair is moving, say, upwards on its exhaust stroke, the other will also be moving upwards but on its compression stroke. Thus, to get any valve in the correct position for adjusting its clearance, *turn the engine to the position where the corresponding valve in the other cylinder of the pair is at maximum lift*.

All the cams for an engine are usually formed on a single shaft called the *camshaft*, although in some cases separate shafts may be used for the inlet and exhaust valves.

The shafts are made of steel, either by forging or casting, with subsequent machining. Forged camshafts usually have the cams case-hardened, whilst the cams of cast shafts are generally hardened by *chilling* during casting.

The camshafts are carried in suitable bearings. These are usually plain bearings, but in some cases ball or roller bearings may be used. In most engines the shaft is moved into position from the front end of the engine, and to enable this to be done the bearings are larger than the cams. The shaft is located endwise by some such arrangement as that shown in Fig. 8.10, which also shows one of the methods used to secure the camshaft driving gear.

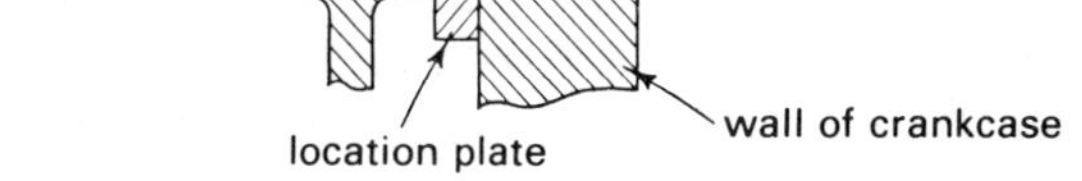

Fig. 8.10 Attachment of camshaft timing gear

Camshaft drives

In an engine working on the four-stroke cycle each valve must open once every two revolutions of the crankshaft, which makes it necessary for the camshaft to rotate at half-crankshaft speed, i.e. the speed ratio of crankshaft to camshaft is 2 : 1.

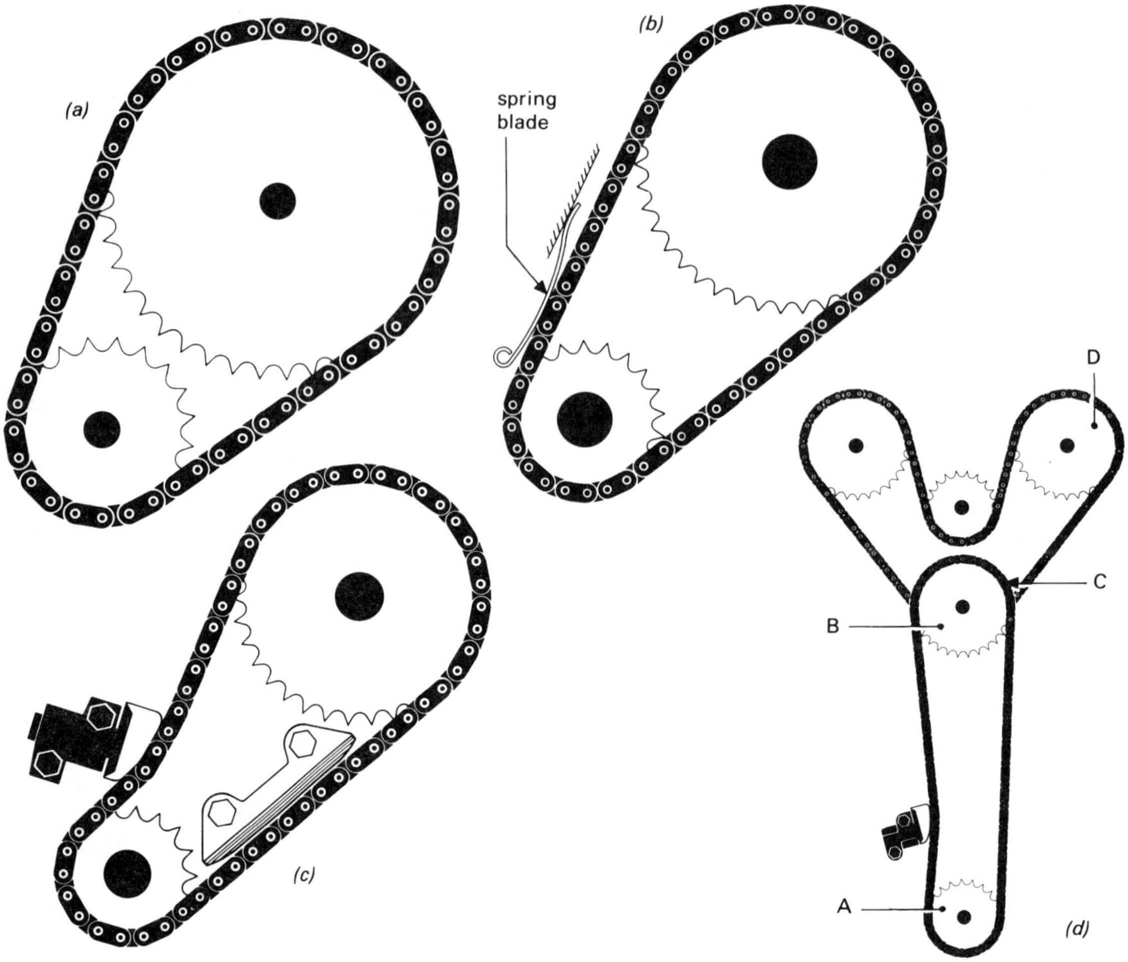

Fig. 8.11 Chain drives

Chain drives

When the camshaft is close to the crankshaft, a short chain connects the crankshaft sprocket to the camshaft sprocket, the latter having twice as many teeth as the former (Fig. 8.11a). The chain is either a single row type or a duplex with a double row of rollers. Unless special provision is made, wear will cause the chain to lengthen and a rattling noise will be heard as the irregular rotation of the camshaft slaps the chain against the timing cover. Compensation for wear is achieved by fitting an automatic adjuster to rub on the non-driving (slack side) of the chain. Fig. 8.11b shows a spring blade adjuster and in Fig. 8.11c a hydraulic type of automatic adjuster is used in conjunction with a fixed slipper made of a plastics material; the slipper damps chain 'flutter'.

The camshaft is sometimes carried on the cylinder head and this means that a longer and more complex driving arrangement is necessary. The system used should operate with minimum noise and also must allow for thermal expansion of the block and head. Fig. 8.11d shows a two-stage chain layout suitable for driving twin camshafts. In this case, the sprockets A, B, C and D give a speed ratio of $B/A \times D/C = 2:1$.

Automatic adjuster

The construction of the Renold SCD (Short Centre Distance) auto adjuster is shown in Fig. 8.12. The plunger in this compact type of adjuster is forced against the chain by hydraulic pressure to compensate for chain wear, but return travel of the plunger is resisted by a ratchet. This non-return feature overcomes the need for undue pressure on the chain which would cause power loss and extra wear.

The rubber slipper head is bonded to a hollow plunger which has a peg that engages in a helical slot

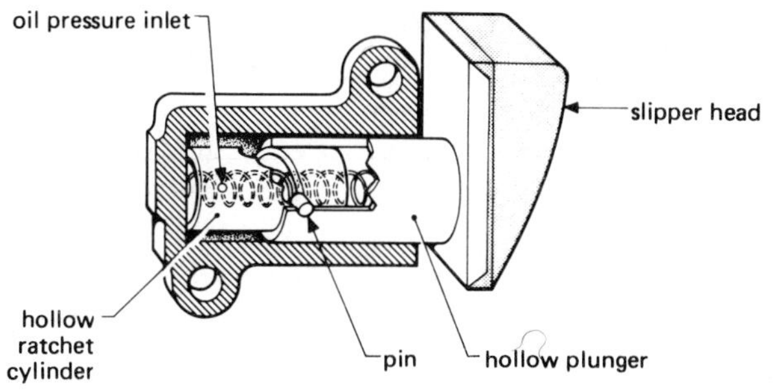

Fig. 8.12 Renold SCD hydraulic adjuster

in the ratchet cylinder. Oil pressure, supplied from the lubrication system, acts on the plunger to assist the spring in tensioning the chain. When chain wear occurs the movement of the plunger causes the peg to push on the smooth face of the helical slot in the hollow ratchet and rotate the ratchet a small amount. Oil, trapped in the adjuster, acts as a damper and seepage past the plunger lubricates the chain. Return of the plunger in excess of that required to allow for thermal expansion is resisted by the peg engaging in one of the notches in the groove of the hollow ratchet.

Provision is made for the mechanic to lock the plunger to the ratchet for engine assembly purposes.

Gear and belt drives

When the camshaft is fitted close to the crankshaft a gear layout similar to that of Fig. 8.13a can be used. Spur gears are noisy, so helical teeth are generally used. Sometimes the large camshaft gearwheel is made of plastic material and in cases where the distance is large an idler gear is fitted.

Overhead camshafts are often driven by a rubber belt. The belt is notched to form teeth to maintain the correct valve timing (Fig. 8.13b). This type of drive is a cheap, quiet and efficient way of driving a camshaft that is mounted far away from the crankshaft. Also

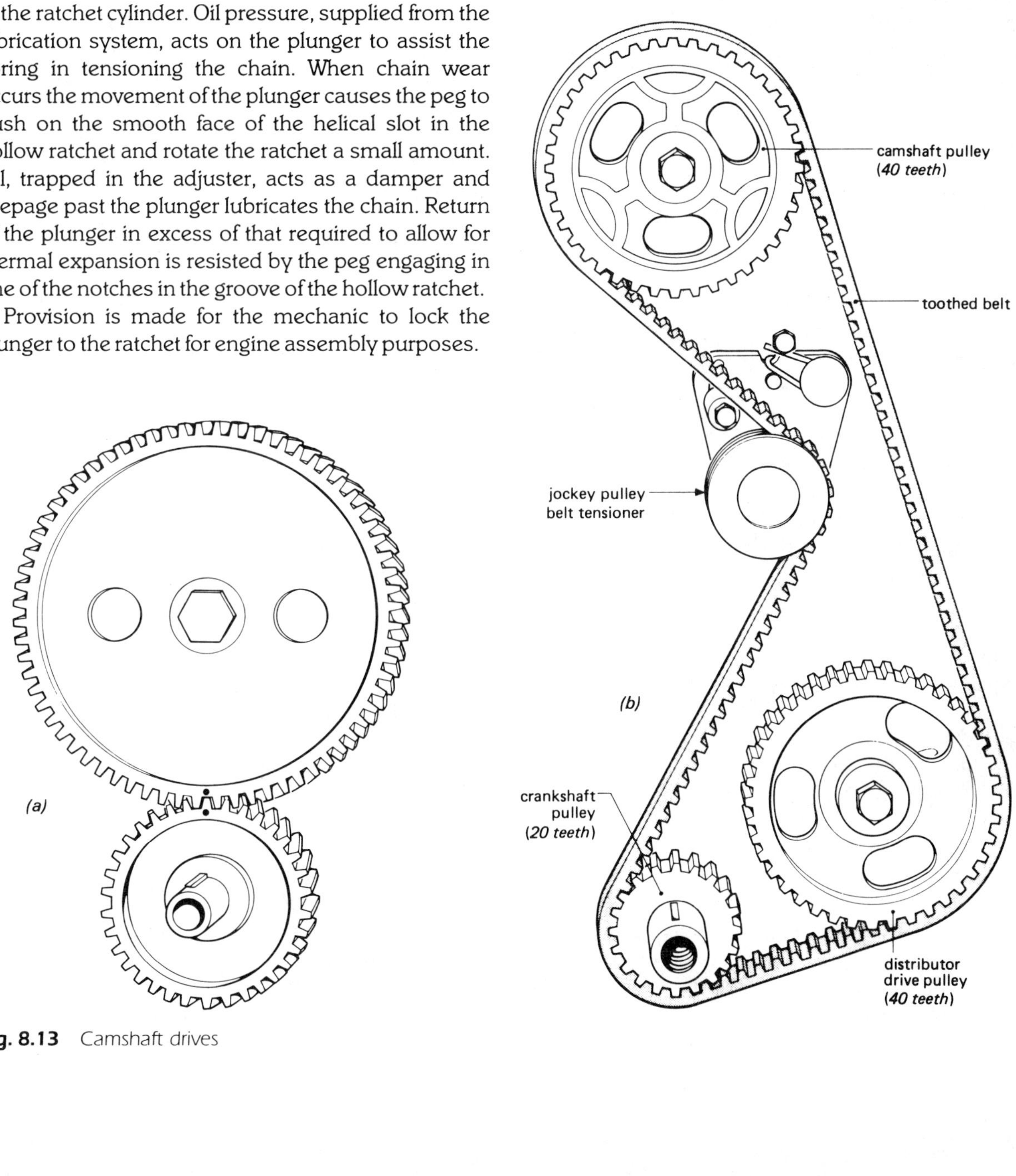

Fig. 8.13 Camshaft drives

this method enables the drive to be easily disconnected when the cylinder head has to be removed. Breakage of the belt can cause a problem especially if the piston strikes an open valve, but reinforced belt materials make belt breakage a rare occurrence.

Valve timing

Whenever the camshaft drive is disconnected, the valves must be retimed to ensure that they open and close at the correct time in relation to the piston position. Timing gears and sprockets are normally marked, but where not, as in the case of a belt drive, the method is to set the camshaft and the crankshaft to the recommended positions before the drive is connected. See also page 56.

Tappets

As mentioned on page 40 tappets are one form of cam follower, and some common types of tappet are shown in Fig. 8.14; (a) and (b) show *barrel* type tappets, which are made of cast iron, the bottom surface being *chilled* in the casting process to provide a hard surface bearing against the cam. They are hollow and usually have 'windows' which help to reduce weight and assist lubrication. The tappet shown at (a) is suitable for a side valve and has a hardened adjusting screw to adjust the valve clearance. That shown at (b) is suitable for an o.h.v. engine and operates a push rod, the lower end of which is located in the small hollow formed inside the base of the tappet. Both these types of tappet operate directly in the cylinder block.

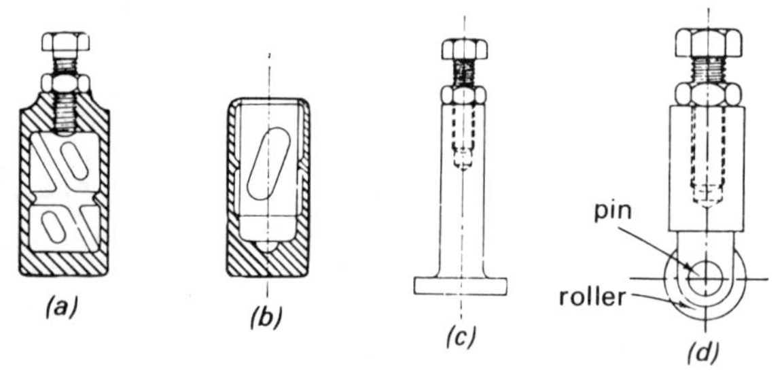

Fig. 8.14 Types of tappet

Fig. 8.14c shows an older type of tappet for a side-valve engine. This type is usually made of steel and is carried in a detachable tappet block attached to the cylinder block by screws.

These three types of tappet, and all tappets having a flat base, are used with cams which have convex curved flanks. Some cams have straight flanks, and these operate on followers having curved ends. In some cases the tappet has a roller rotating on a pin, as shown in Fig. 8.14d. This type is now seldom used, since for high-speed engines the roller must revolve much more rapidly round the flanks and nose than it does around the base circle, and at high engine speeds the angular accelerations of the roller involved are greater than can be achieved, and a considerable amount of skidding occurs. This being so, the more expensive roller construction might just as well be dispensed with and a plain rounded tappet base substituted. An additional weakness of this roller type is the pin, which is subjected to high shearing loads: increasing the pin diameter to make it stronger raises the rubbing speed between pin and roller, and little is gained compared with a plain tappet. For low-speed engines, however, a roller tappet may have advantages.

Flat-based tappets are usually allowed to rotate in their guides; indeed they are encouraged to do so since this reduces the rubbing speed between cam and tappet and spreads the wear over the whole of the tappet base. Rotation of the tappet is encouraged by offsetting the tappet from the centre-line of the cam or by grinding the cams with a very slight taper (about 1°) in which case the foot of the tappet is ground very slightly convex, forming part of a sphere of large radius. Roller tappets and curved-ended tappets must not, of course, rotate on their axes, and some means of positively preventing rotation must be provided.

Hydraulic tappets

Engine temperature depends on the operating conditions; as a result the change in valve clearance is considerable. Setting of a clearance to suit an engine during its warm-up period will alter drastically when the engine is subjected to a period of high-speed or high-load use. This clearance variation will result in excessive noise and the change in valve timing will alter the power output.

Fig. 8.15 shows the layout and operating principle of one type of hydraulic tappet. The tappet body (1) contains a plunger (2), which is formed into two chambers: a feed chamber (3) and a pressure chamber (4). Oil from the main lubrication system is passed to the feed chamber and then to the pressure chamber via a one-way ball valve (5). The oil flow from the pressure chamber is controlled by the amount of clearance between the tappet body and the plunger. By accurately setting this clearance, a given amount of

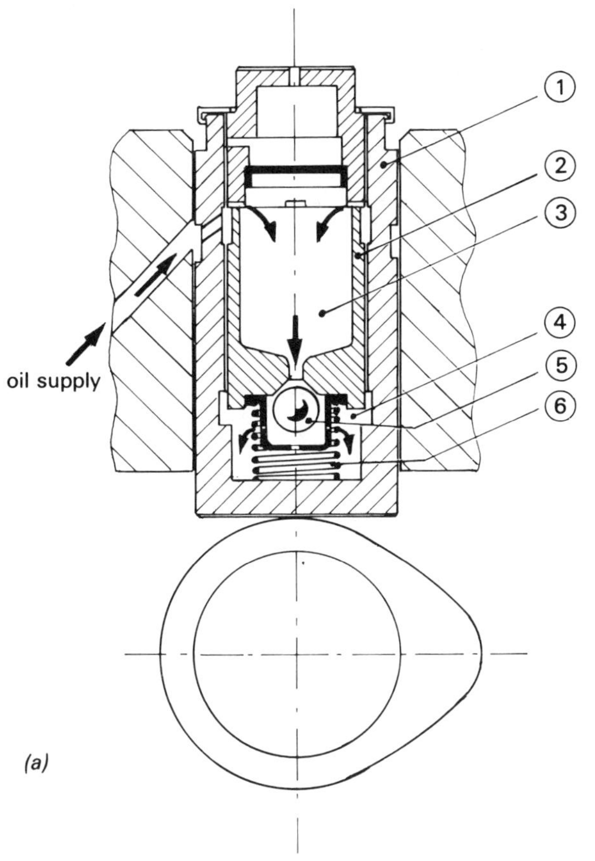

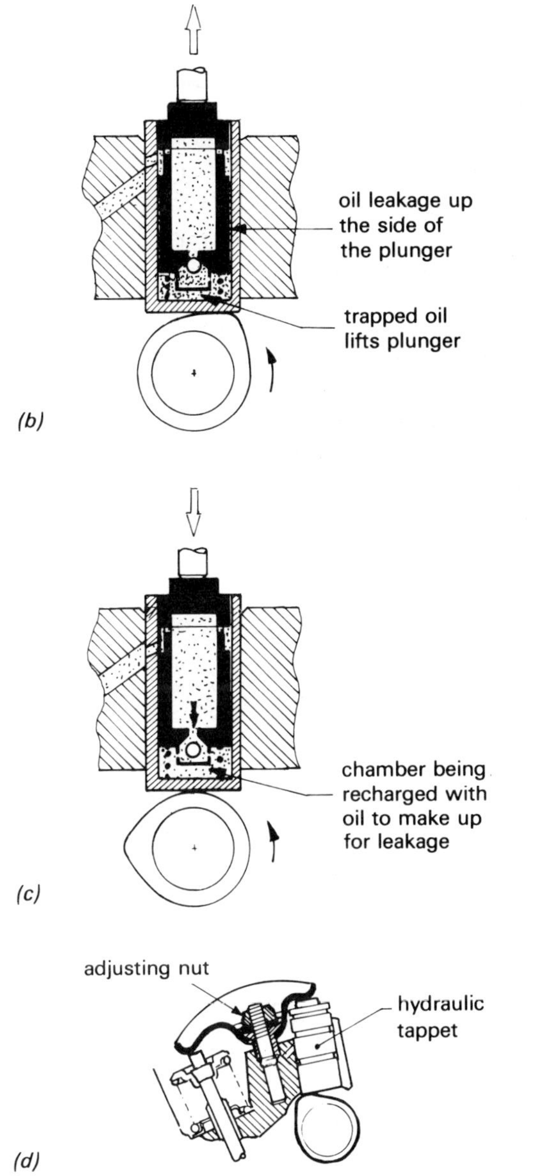

Fig. 8.15 Hydraulic tappet

oil is allowed to escape up the side of the plunger each time the tappet is operated. A tappet spring (6) tends to force the plunger out of the tappet body, but although this spring reduces the valve clearance to zero, it does have the strength to operate the engine valve.

When the valve is being opened (Fig. 8.15b), oil in the pressure chamber is trapped, since the ball valve will not allow it to return back to the feed chamber. As a result, the upward movement of the tappet body pressurizes the oil and causes the plunger to be moved a similar amount. During this stage the tappet behaves like a solid cam follower.

After the engine valve has closed (Fig. 8.15c) the small leakage up the side of the plunger during the time that the oil was pressurized will mean that the pressure chamber requires more oil to make up for the oil loss. This is provided by allowing fresh oil from the feed chamber to pass the ball valve until the chamber is filled. During engine warm-up the engine valve expands and this decreases the volume of oil in the pressure chamber needed to take up the play between the cam and the engine valve. This occurs over a period of time and is compensated for by the slight oil loss from the pressure chamber that occurs every time the tappet is operated. Consequently zero valve clearance and correct seating of the valve is achieved irrespective of the temperature of the engine.

Slight oil loss from the tappet occurs when the engine is unused for a period and this causes the tappets to rattle for a short time after the engine is started. If this noise persists for more than about ten seconds the faulty tappet should be replaced.

Rocker and push-rods

Rockers are sometimes used instead of tappets between camshaft and valve; this arrangement can amplify the movement given by the cam. Valve clearance is adjusted by the screw at the pivot, or by mounting the rocker pivot (or the bush) eccentrically. Fig. 8.16 illustrates three types. In Fig. 8.16a an adjusting screw is shown as the means of adjusting the valve clearance. In the arrangement shown in Fig. 8.16b the eccentric bush is a close fit in the rocker, but can be rotated in the rocker by a hexagonal flange at one end. The other end of the bush is threaded, and the bush is locked in the rocker by tightening a nut on this threaded end. To adjust the clearance, slacken the nut, rotate the bush to give the correct clearance, and re-tighten the nut.

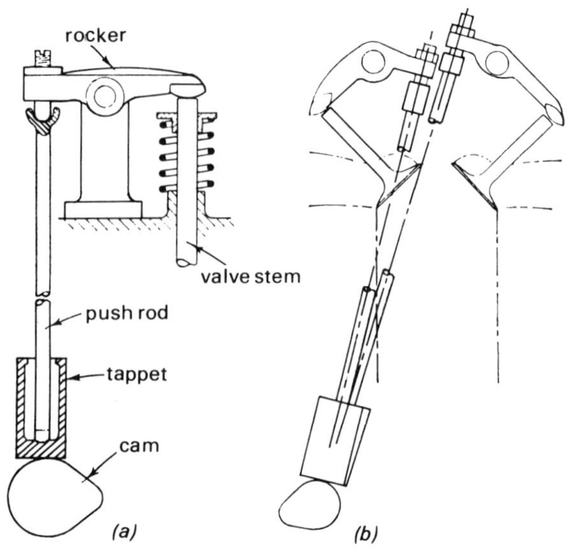

Fig. 8.17 Push-rod operation of overhead valves

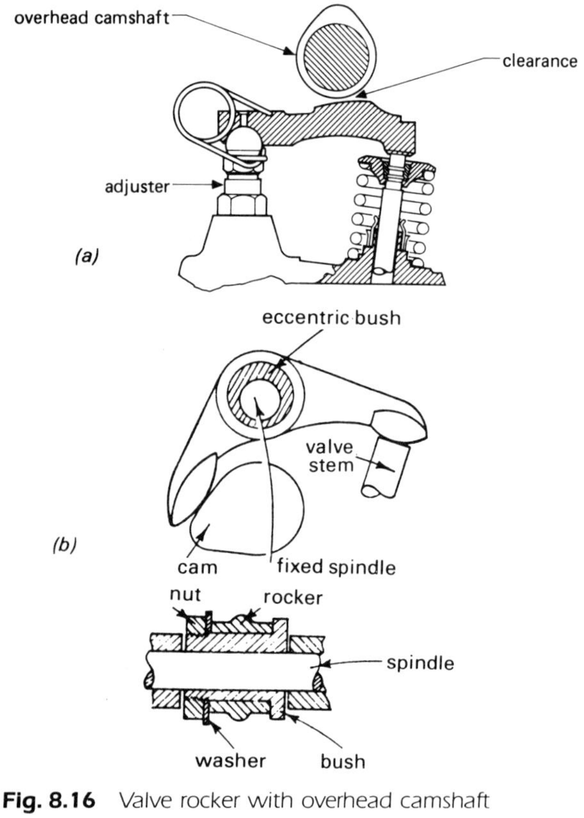

Fig. 8.16 Valve rocker with overhead camshaft

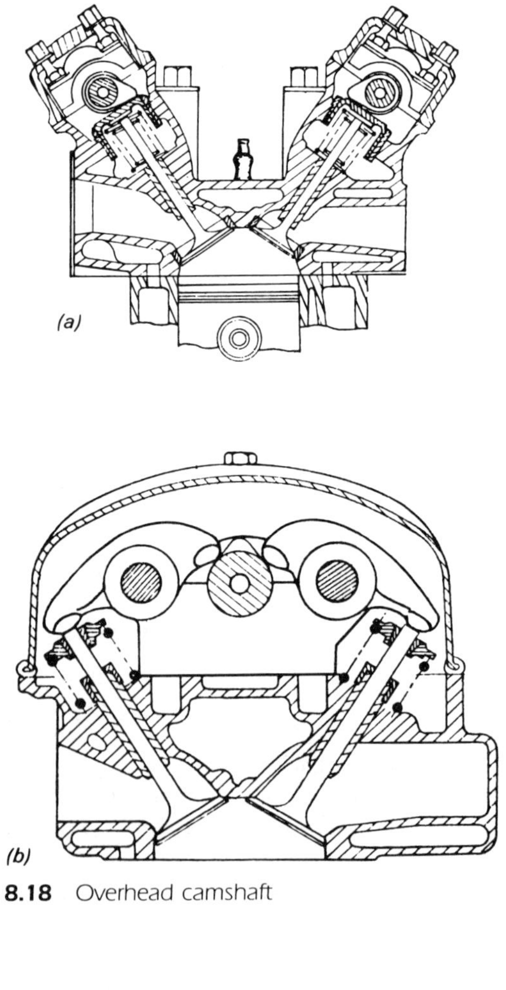

Fig. 8.18 Overhead camshaft

Fig. 8.15d shows a layout which incorporates a hydraulic tappet to operate a lightweight pressed steel rocker arm. No provision is made for manual adjustment because the hydraulic tappet automatically takes up any clearance in the valve train. Each layout must contain some system of lubrication to reduce wear of the rubbing surfaces. Seals fitted to the valve stems ensure that oil used to lubricate the rocker assembly does not pass down the valve guides (see page 164).

Many engines have overhead valves operated by a camshaft in the crankcase, close to the crankshaft. This arrangement necessitates the use of push-rods and overhead rockers to transmit the motion from the tappets to the valves, the usual arrangement being as illustrated in Fig. 8.17a. The rockers pivot on a common rocker spindle supported in pillars attached to the cylinder head, and the rockers are spaced apart on the spindle by distance pieces, or springs, or both. Valve clearance is adjusted by a hardened ball-ended screw in the end of the rocker, the screw fitting into a small cup formed in the top end of the push-rod; the cup retains oil to ensure lubrication at this point. The lower end of the push-rod is ball-ended and is located in a cup formed in the tappet. The push rod may be made of a solid steel rod, the two ends being suitably 'formed' from the rod itself. Alternatively the push-rod may be tubular, of either steel or aluminium alloy, with separate hardened steel ends pressed on.

The rocker is usually made from a steel forging, but is sometimes formed from a steel pressing, and some are made in a special type of cast steel.

When the valves are arranged as shown in Fig. 8.7d, it is generally preferred to operate them by a camshaft immediately over each row of valves, as illustrated in Fig. 8.18a, or by a single overhead camshaft and rockers as shown in Fig. 8.18b. Sometimes, however, a designer may prefer to retain the advantages of a camshaft mounted in the crankcase close to the crankshaft, in which case the valves may be operated by a more complicated system of push-rods and rockers such as that shown in Fig. 8.17b.

9 The cylinder block and crankcase

Once the early motor vehicle engines had developed beyond the initial experimental stage, design settled down to a type that had up to four or even more separate cylinders mounted upon a separate crankcase. This arrangement was probably dictated by the limitations imposed by the foundry practice and machine tools then available, which made complicated castings difficult to produce and to machine. Furthermore, bearing in mind the unreliability of early engines, it was no doubt thought desirable to be able to get at any individual piston without having to disturb too much of the engine. The cylinders, which had to resist wear from the pistons and valves, were made of cast iron, whilst the crankcase was generally made of aluminium to save weight. (It should be remembered that at the beginning of this century aluminium was a relatively new metal and the art of alloying it with other metals to improve its properties had not been developed to anything like its present state.)

As manufacturing techniques improved, it became possible to cast and machine cylinders in *blocks* of two, three or four, and it eventually became the practice to cast up to six or even eight cylinders in one block, but still separate from the crankcase.

This arrangement of having all the cylinders of an engine in one casting was known as *monobloc* construction: its chief merits compared with the previous practice of having separate cylinders were: (1) a reduction in the number of machining operations, (2) simplification of assembly, and (3) (most important), increased rigidity of the engine structure.

Carried to its logical conclusion, this development brings us to the arrangement, now almost universal, in which the cylinders and crankcase are contained in a single casting which forms the main structure of the engine. Fig. 9.1 illustrates the cylinder block and crankcase for a four-cylinder engine. The chief advantages gained by this construction are: (1) it provides the greatest possible rigidity of the engine structure, (2) it reduces machining and assembly operations to the minimum, and (3) it reduces cost, for the reasons given in (2).

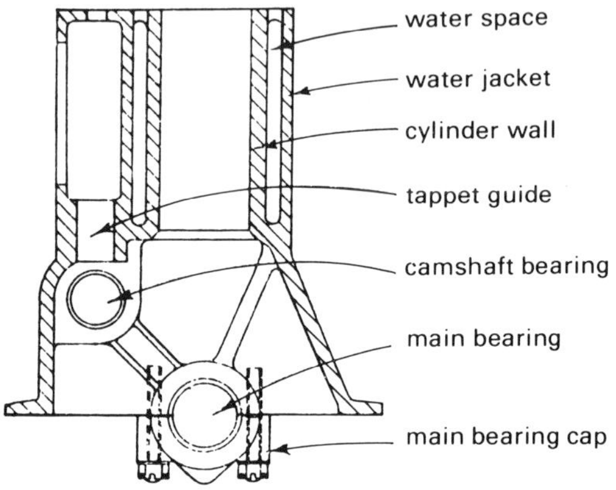

Fig. 9.1 Section of a cylinder block and crankcase

The arrangement has, of course, some disadvantages, the chief being that it makes a large and heavy component, particularly on large engines, which may necessitate lifting-tackle to handle it in the factory and workshop.

Early engines usually had the cylinders made with integral cylinder heads, but when the monobloc construction was adopted there were advantages, both in manufacture and in maintenance, in making all the cylinder heads in a single separate casting secured to the cylinder block by screws or studs and nuts. In manufacture, for instance, it is easier to bore the cylinders if they are open at both ends, and assembly of the engine is simplified. In maintenance, the combustion chambers, piston crowns and valves are more easily accessible for decarbonizing. In spite of these advantages, however, *fixed* or non-detachable cylinder heads continued to be used in a few engines.

There are difficulties in obtaining a gas- and water-tight joint between head and block. Sufficient thickness of metal has to be left in the region of the joint to accommodate the screws or studs by which the head is secured to the block, and this impedes the circulation of cooling water in a region where free flow is very important. Unless the bosses into which the studs are screwed are substantial and very carefully positioned, the upper surface of the block will be distorted by the tightening of the nuts and a gas-tight joint will be impossible to maintain. The use of fixed cylinder heads simplifies the problem of water flow around the ports and sparking plugs and avoids distortion and gas- and water-leaks.

The adoption of the unit cylinder block and crankcase construction makes the detachable cylinder head a necessity, and this is now the universal practice.

The crankshaft main bearings are carried in webs which lie transversely across the crankcase. One of these webs forms the front wall of the crankcase and another the rear wall, whilst the others stiffen the structure besides acting as supports for the main bearings. These bearings are usually split across their diameters to permit assembly of the crankshaft. The machining of the block is simplified if its lower face is level with the centre of the main bearings.

In some engines the crankcase walls are extended below the centre of the main bearings. This stiffens the structure by making it deeper, but more machining operations are involved: the bearing housings must be machined flat across a diameter, and the detachable half of the bearing housing - called the *cap* - must be securely attached by studs and nuts: thus the lower face of the crankcase comes at a different level, complicating the machining operation. If the front and rear main bearing caps are machined flush with the crankcase joint face, the attachment of the *sump* which closes the bottom of the crankshaft is simplified.

The sump, which usually contains the supply of oil used for lubricating the engine, is attached to the lower face of the crankcase by a number of screws or bolts and nuts. If the joint is at the level of the main bearing centres, the sump and the front and rear bearing caps must be specially shaped to permit an oil-tight seal. A cork gasket is generally fitted between sump and crankcase to assist in making the joint oil-tight. The sump is usually made as a steel pressing, though in a few cases it may be an aluminium casting.

Cylinder liners

The wearing surface of the cylinder must be made of a relatively hard material which is resistant to wear. Cast iron is most commonly used because it is reasonably hard and can be cast into complicated shapes. There are, however, many different varieties of cast iron, and those which have the best casting properties are not necessarily the best for wear resistance. Furthermore, wear-resistant cast iron may be not only more difficult to cast but also more difficult to machine. This difficulty can be overcome by making the main cylinder block casting of a type of iron having good casting and machining properties, and making the wearing surfaces of the cylinders in the form of *sleeves* or *liners* of a harder material. These cylinder liners are of two main types:

1 Dry liners The cylinder block is made in the normal way but the cylinders are lined with thin sleeves of harder material - usually a high-quality cast iron but sometimes steel. The liners are made an interference fit in the block and are inserted by means of a press, the operation usually being assisted by heating the block to expand the bores or by freezing the liners in liquid air to contract them. Fig. 9.2 shows a liner of this type fitted in position in its cylinder bore.

Owing to the interference fit the liners shrink slightly in fitting. It is thus necessary for them to be accurately bored to the correct diameter after they

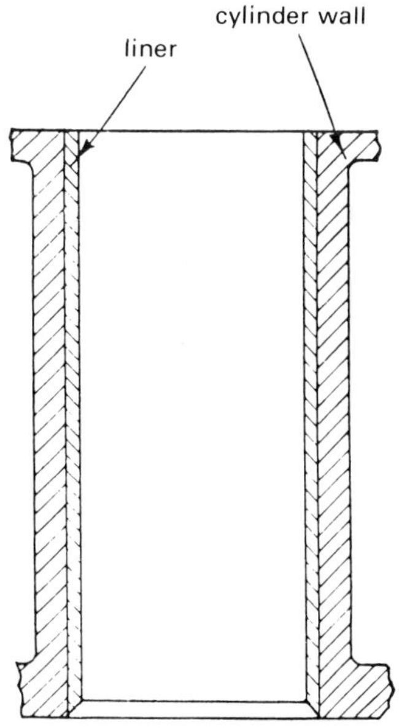

Fig. 9.2 Dry cylinder liner

have been fitted. To eliminate the necessity for this operation and to simplify fitting and removing the liners, some engines have liners which have no interference fit and which can be pushed in with only a slight pressure - in some cases *hand pressure*. Such liners are called *slip* liners, and these can be *pre-finished* to the correct size before fitting. They are prevented from moving when the engine is running by a flange at the upper ends which is clamped by the cylinder head into a recess, machined at the top of the cylinder block.

2 Wet liners When this type of liner is used the cylinder bores are omitted from the cylinder block casting. Each bore is made separately in the form of a relatively thick liner, which is fitted into the block in a suitable manner - Fig. 9.3 shows two methods in common use. At (a) a liner is shown clamped between the cylinder head and the bottom of the water jacket, in which holes are bored to allow the lower ends of the liners to pass through. A shoulder near the bottom of the liner seats on a gasket or sealing ring to make a water-tight joint. At the top the liner stands slightly proud of the top face of the block, so that it is clamped securely in position and a gas-tight joint with the head is ensured.

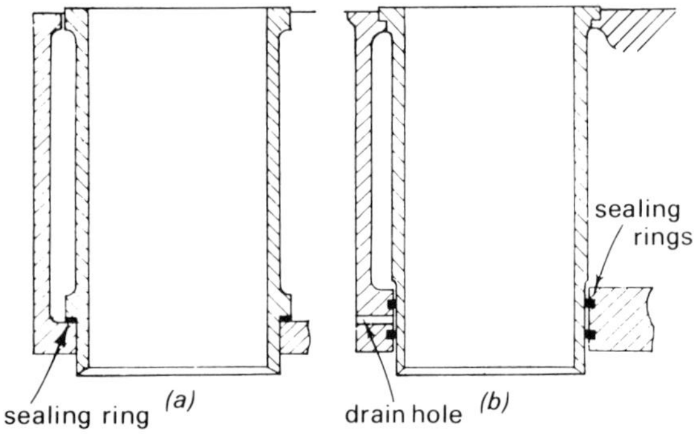

Fig. 9.3 Types of wet liner

In the arrangement shown at (b) the top of the liner is a light press fit in a recess in the top face of the block: the top face of the liner stands slightly proud of the block face to hold the liner firmly and ensure a sound joint. The lower end of the liner passes through a hole bored in the lower part of the block, and a seal is made by two rubber rings fitted in grooves which may be either in the block (as shown) or in the liner. A drain hole leads from the space between the sealing rings, to reduce the risk of water entering the crankcase should the upper seal leak, and to provide visible evidence of such leakage.

Comparison of dry and wet liners

The main advantages of wet liners over dry are:

1 They can be pre-finished to correct size.
2 They are easily fitted and removed.
3 Their outer surfaces are in direct contact with the cooling water.

There are two objections to wet liners: the most obvious is the risk of water leakage into the crankcase, and the other is the reduced rigidity of the main engine casting due to the omission of the cast-in cylinder walls.

Aluminium cylinder blocks

When cylinder liners are used the cylinder block and crankcase may be made of aluminium alloy, thus reducing weight. Advantage is taken of this only where low weight is important, for aluminium is more expensive than cast iron, is less strong and is softer. If dry liners are used in an aluminium block they may be cast in position.

Air-cooled engines

The cylinder blocks so far described have all incorporated water cooling of the cylinders. Air-cooled cylinders have *fins* formed on their outer surfaces, and it is necessary that the cylinders should be made separately from the crankcase and often separate from one another, or in blocks of two.

Cylinder heads

In the case of *in-line* engines (those in which the cylinders are arranged in a single row or *bank* with their axes parallel) the cylinder head is generally made in a single casting containing the upper surfaces of all the combustion chambers.

If the engine has *side valves* the cylinder head is merely a cover incorporating the water jackets (or fins if the engine is air-cooled) with suitably placed holes into which the sparking plugs are screwed. In the case of *overhead-valve* engines the head casting will also have to incorporate the ports and valve seatings with suitable provision for the valve-operating gear. Whenever possible the top face of the cylinder block is machined flat and square to the cylinder axis, although there are some engine designs in which this surface has to be at some other angle. The lower

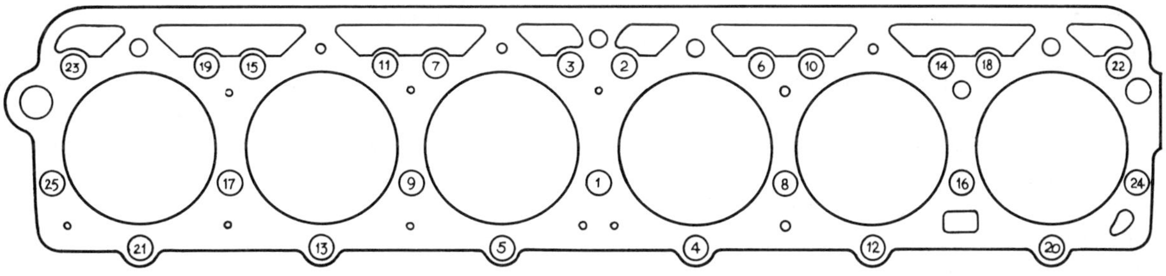

Fig. 9.4 Cylinder head tightening sequence

surface of the cylinder head is also machined flat, and the head is securely fastened to the block by a number of screws or studs and nuts.

Materials

As with the cylinder block, the head is usually made of cast iron, but in some cases aluminium alloy is preferred. This gives the advantages of reduced weight and better heat conductivity, the latter resulting in lower and more uniform temperatures of the combustion chamber walls, thereby permitting the use of a slightly higher compression ratio than is possible with a cast iron head.

When aluminium is used for the cylinder head it is generally necessary to fit *inserts* of harder material - usually cast iron or bronze - for the valves to seat on and for the sparking plugs to be screwed into.

Gaskets

To ensure a gas- and water-tight joint it is necessary to fit a *gasket* between the head and the block. For many years the type of gasket used consisted of a thin sheet of asbestos protected on either side by thinner sheets of copper, with suitable holes to clear the combustion chambers, water holes and studs. This type is still in use, but a later type which seems to be replacing the older one consists of a thin sheet of metal - such as stainless steel - with corrugations formed around the holes to assist in making a good seal.

Some metal-faced gaskets are varnished to prevent seepage of coolant fluid. Also the surface is often precoated with a lubricant which allows slight gasket movement: this occurs because of unequal thermal expansion as the engine heats up and also when the cylinder is being tightened down. This relative movement is the reason why it is inadvisable to use a solution that hardens on a new gasket. Any gasket not precoated can be treated with a thin film of oil, or a non-hardening lubricant.

Correct tightening down of the cylinder head is necessary to avoid the risk of a 'cylinder gasket blowing'; this is a general term used to describe a leaking gasket. Failure to follow the correct procedure results in early failure which can take many forms. The gasket should seal the cylinders, cooling system and lubrication system, so leakage between any two of these can produce many symptoms.

All cylinder head bolts must be tightened evenly if cylinder head distortion is to be avoided. The method used is to start at the centre and work outwards in the manner shown in Fig. 9.4. Most manufacturers use one of the following methods:

Standard method Tighten all the bolts or nuts fingertight and then use a wrench to 'pull-down' the head a bit at a time in the sequence shown in Fig. 9.4. The final tightness should be established by using a *torque wrench* set to the figure specified by the manufacturer. After the engine has run the equivalent of about 800 km the cylinder head should be retightened to the recommended torque. The engine temperature at which this is carried out depends on the material of the cylinder head: hot for cast iron and cold for aluminium alloy.

Torque to yield method When a steel bolt is gradually tightened, a stage is reached when the bolt stretches without further increase in the torque applied. This limit of elasticity is called the *yield point* and if metals are loaded beyond this limit the bolt will not return to its original length after the load has been released. However, taking a bolt beyond its yield point allows it to exert a greater load on the components it is clamping. Some manufacturers use this technique for highly stressed bolts such as those used to secure cylinder heads, main bearings and some suspension components. These *stretch bolts* can only be used once, so it is essential that they are scrapped after they have been removed from a vehicle.

The method of tightening is to follow the sequence shown in Fig. 9.4 and initially 'torque down the bolts' to the recommended setting. When this tightness has been achieved, the bolt heads are then turned through a given angle. For example, one manufacturer recommends that the cylinder head bolts are tightened to a torque of 54 N m and then rotated exactly 90° to take the bolt past its yield point. This angular movement of the bolt head can be determined either by marking the bolt head with chalk or by using a home-made protractor.

Special care should be exercised when removing a cylinder head if distortion is to be avoided. The head bolts should be slackened using a sequence similar to that shown in Fig. 9.4. Aluminium alloy heads should be allowed to cool before they are removed and all types of cylinder head should be placed only on flat surfaces.

10 The four-stroke cycle in detail: valve and ignition timing

In Chapter 4 the operation of an engine on the four-stroke cycle was described. We must now consider this more closely, particularly in connection with the opening and closing of the valves.

The first point to be appreciated is the fact that it is not possible for a valve to move from closed to fully open or vice versa *instantaneously*. In fact, the opening and the closing movements will each be spread over a considerable angle of crankshaft rotation. Thus, if a valve is required to be *effectively* open at the *beginning* of a stroke it must begin to open *before* the dead centre: similarly, if it is required to be *effectively* open at the *end* of a stroke it must not close *completely* until *after* the dead centre.

Even if the valves *could* be opened and closed instantaneously, the dead centres would not be the best points at which they should open and close except at very low engine speeds. The engine will give its best performance when the greatest mass of air and fuel is passed through the combustion chamber and burnt effectively. Consider an engine running reasonably fast. The most suitable point at which to open the inlet port depends upon conditions at the end of the exhaust stroke, and we will begin our study with the piston moving down the cylinder on the induction stroke and the inlet port already wide open (Fig. 10.1). The downward movement of the piston reduces the pressure inside the cylinder so that the pressure of the atmosphere forces air through the carburettor and along the inlet pipe, i.e. the flow of air along the pipe is *accelerated*. The *inertia* of this air (its resistance to change of speed) causes it to lag behind the piston movement, so that by the time the piston has reached its maximum speed (about half-way down the stroke) the pressure inside the cylinder is well below atmospheric pressure, but during the second half of the stroke, when the piston speed is decreasing, the air flow is able to catch up with the piston and the pressure inside the cylinder rises towards atmospheric pressure.

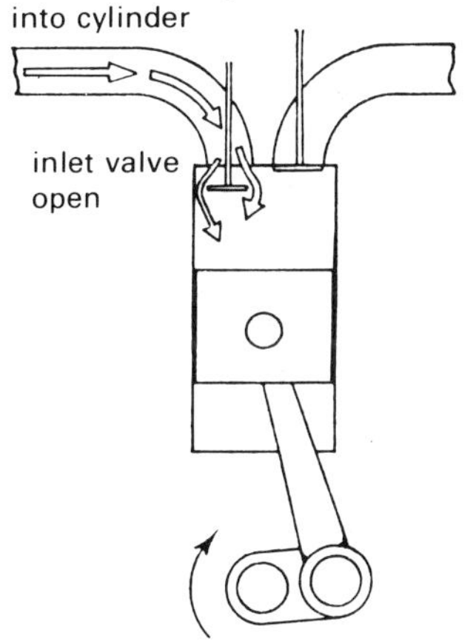

Air/fuel mixture entering cylinder during induction stroke

Fig. 10.1 Induction stroke

At b.d.c. the direction of piston movement reverses, but the momentum of the air flow along the inlet pipe *towards* the cylinder causes it to continue entering the cylinder until the piston has moved some way up the next stroke – provided the valve remains open (Fig. 10.2). At some point during this stroke the upward movement of the piston increases the pressure inside the cylinder sufficiently to stop the inward flow of air and to begin to force air *out* of the cylinder (Fig. 10.3), and the greatest possible amount of air will be trapped inside the cylinder if the inlet valve closes at this point. This will always be *after* b.d.c. How much after will depend upon engine speed, upon the length and diameter of the inlet pipe, and throttle opening.

At about the end of the compression stroke the fuel is ignited and burnt, the gas pressure increases and

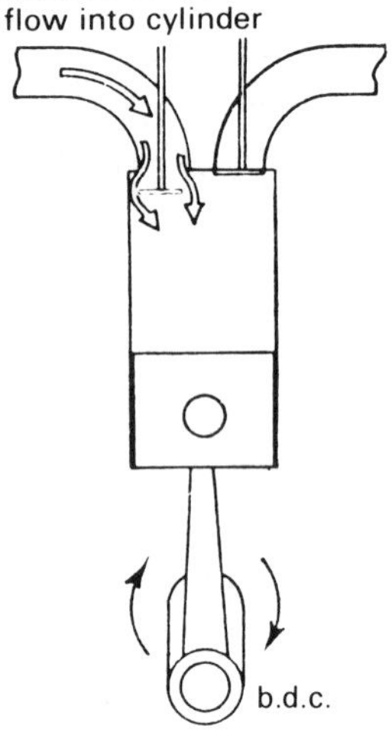

Momentum of mixture maintains flow into cylinder at b.d.c.

Fig. 10.2 End of induction stroke

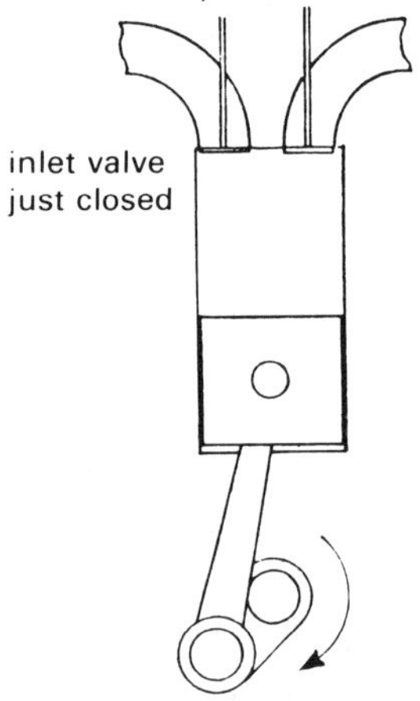

Inlet valve closing point

Fig. 10.3 Compression stroke

the piston is forced down the cylinder on its power stroke. During the downward movement of the piston, the increasing volume lowers the pressure of the gas, but if the exhaust valve remains closed the pressure will still be well above atmospheric when the piston arrives at b.d.c. This pressure will offer a good deal of opposition to the upward movement of the piston during the exhaust stroke, even though the exhaust valve may now be open. Moreover, as the piston approaches b.d.c., the leverage exerted by the connecting rod on the crankshaft decreases rapidly and the pressure on the piston thus has a rapidly decreasing turning effect on the crankshaft. By opening the exhaust valve *early*, i.e. *before* the piston reaches b.d.c., a good deal of the burnt gas is allowed to escape before the piston begins it upward exhaust stroke, and the pressure during this stroke - and consequently the opposition to the piston's upward movement - is very much reduced (see Fig. 10.4). The point at which the exhaust valve opens must be carefully chosen so as to sacrifice as little as possible of the effect of the pressure in driving the piston downwards during the power stroke, whilst ensuring that as much gas as possible will escape before b.d.c. to reduce to a minimum the opposition to the piston's upward movement during the exhaust stroke.

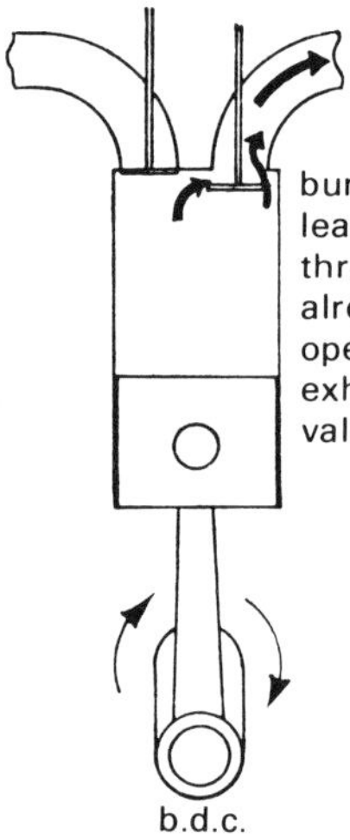

Burnt gas already escaping from cylinder at b.d.c.

Fig. 10.4 Start of exhaust stroke

Thus the exhaust gas is already moving out of the cylinder and along the exhaust pipe before the piston starts its exhaust stroke. During the stroke the piston increases the gas velocity along the exhaust pipe. Thus, as the piston approaches t.d.c. and its upward velocity decreases, the momentum of the gas rushing along the exhaust pipe creates a partial vacuum inside the cylinder. As soon as the pressure inside the cylinder falls below the pressure in the inlet manifold, fresh mixture will begin moving into the cylinder if the inlet valve is opened. This can usually occur *before* the piston has reached t.d.c. - i.e. by opening the inlet port before the exhaust port has closed, the momentum of the exhaust gas can be used to start

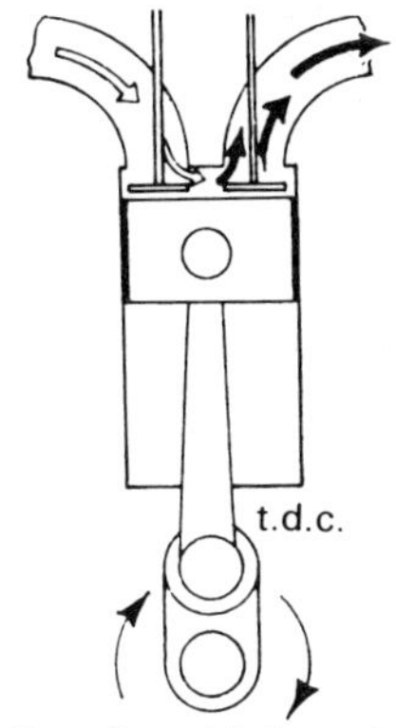

Fig. 10.5 End of exhaust stroke

the flow of fresh gas into the cylinder without any help from the piston (Fig. 10.5).

The fresh gas entering the cylinder will, of course, be drawn towards the exhaust port, displacing burnt gas. Placing the valves on opposite sides of the combustion chamber makes the fresh gas sweep right across the chamber on its way from inlet port to exhaust port. This ensures that the combustion chamber is thoroughly *scavenged*, or cleared of exhaust gas and filled with fresh gas. If the ports are very close together it is possible for fresh gas to reach the exhaust port without displacing *all* the burnt gas from the cylinder. The correct moment to close the exhaust port is the moment the fresh gas reaches it.

Thus the opening point of the inlet valve and the closing point of the exhaust valve depend upon:

1 The velocity of the flow of exhaust gases along the exhaust pipe. This in turn depends upon engine speed, throttle opening, the length and diameter of the exhaust pipe and the restricting effect of the silencer.
2 The pressure in the inlet manifold, which is dependent upon engine speed and throttle opening.
3 The position of the ports in relation to the combustion chamber and to each other.

The inlet valve usually opens a little *before* t.d.c. and the exhaust valve remains open a little *after* t.d.c. Thus, for some angle of crankshaft rotation around t.d.c., *both valves are open at the same time*. The opening of the inlet valve *overlaps* the closing of the exhaust valve: this is the period of *valve overlap*.

Whilst a large valve overlap may be beneficial at high engine speeds if the ports are arranged on opposite sides of the combustion chamber, it will make it impossible for the engine to *idle* or run slowly at light load. At low engine speeds and small throttle openings the pressure in the inlet manifold is well below atmospheric pressure (as a vacuum gauge connected to the manifold will show). The quantity of gas exhausted from the cylinder is small and moves slowly, having very little momentum, so that it does not reduce the pressure in the cylinder appreciably below atmospheric pressure. Under these conditions, opening the inlet valve before the exhaust valve has closed will result in a rush of exhaust gas into the inlet manifold. Hence, on the following induction stroke *the cylinder will be filled with a lot of exhaust gas and very little fresh mixture*, which will almost certainly fail to be ignited when the spark occurs. If the engine will never be required to run slowly, e.g. a racing engine, this may not matter, but all engines used in normal motor vehicles are frequently called upon to idle smoothly and reliably, and for such engines the amount of valve overlap must be strictly limited.

To sum up, therefore:

1 The inlet valve remains open *after* the piston has passed b.d.c. at the end of the induction stroke. This is called the *inlet valve closing lag*.
2 The exhaust valve opens *before* the piston reaches b.d.c. at the end of the power stroke (*exhaust valve opening lead*).
3 The exhaust valve usually remains open *after* t.d.c. at the end of the exhaust stroke (*exhaust valve closing lag*).
4 The inlet valve usually opens *before* the exhaust valve has closed (valve overlap) and usually *before* t.d.c. (*inlet valve opening lead*).
5 The amount of lead or lag of valve opening or closing points and the amount of overlap will depend upon the design of the engine - particularly the port arrangement and inlet and exhaust systems - and upon the performance characteristics the engine is required to have.

The opening and closing points of the valves in relation to piston and crankshaft position are called the *valve timing*, and the correct timing for an engine - that selected by the engine manufacturers - is given either in the form of a table or by means of a *valve timing diagram*. Some examples of valve timing tables and diagrams are given in Fig. 10.6, together with brief details of the engine to which they refer: students should make similar tables and diagrams for other engines.

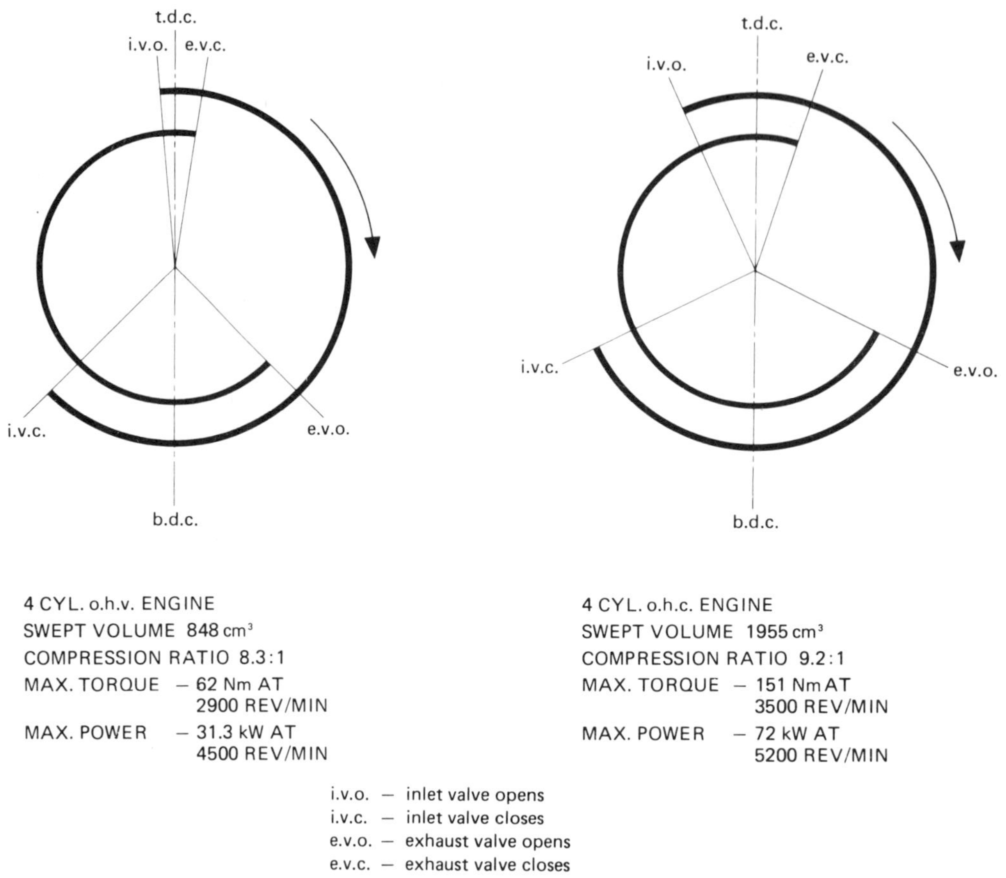

Fig. 10.6 *Valve timing diagrams*

Timing the valves of an engine

During the process of assembling an engine the drive between the crankshaft and the camshaft must be connected up so that the valves will open and close at the correct times in relation to the crankshaft and piston. This operation is known as *timing the valves*. There are three main stages to this operation:

1 Set the crankshaft in the position in which one of the valves should open or close. (It is usual to work on the opening point of the inlet valve, but any other point can be used.)
2 Set the camshaft in the position in which it is about to open the inlet valve (or whatever point is chosen).
3 Connect up the drive to the camshaft.

If the engine has more than one camshaft each one will have to be timed individually.

Suitable parts of the engine are usually marked, to assist in timing the valves. The timing gear or sprocket is keyed to the crankshaft and can be fitted in one position only: the camshaft gear or sprocket is similarly fixed to the camshaft in such a way that it can be attached in one position only. By lining up marked teeth on these gears the crankshaft and camshaft are placed in the correct positions for connecting up the camshaft drive. These gears are hidden inside the *timing cover* when the engine is assembled. To enable the timing to be checked without removing the cover, the pulley is usually marked in a position corresponding to t.d.c. in one of the cylinders. If the pulley is not marked in this way, the t.d.c. position can be found by any one of several methods, of which the following is an example.

On most modern engines the piston can be felt by inserting a rod through a sparking-plug hole. Turn the engine until the piston is approximately 10 mm below its highest position, and file a notch on the rod exactly opposite the top of the plug hole when the end of the rod is resting on the piston. Mark the crankshaft pulley opposite some suitable fixed reference point. Now turn the engine crankshaft so that the piston passes t.d.c. and begins to move down the cylinder: stop when the mark on the rod reaches the top of the plug hole, and put a second mark on the pulley opposite the fixed reference mark. Mark the mid-point of the distance between the two marks on the pulley. The piston is at t.d.c. when this last mark is opposite the fixed reference point.

Valve timing marks can also be made on the flywheel. If the circumference of the flywheel is divided by 360, a figure is obtained which represents the distance measured round the flywheel rim equivalent to one degree. For example if the circumference of the flywheel is 800 mm and the inlet valve opening point is 9° before t.d.c., the distance to be measured round the flywheel rim is:

$$800 \times (9/360) = 20 \text{ mm}$$

There are several ways by which the opening point of the valve can be found. First of all the valve clearance must be correct, since variation of clearance causes a considerable variation in timing, as a little experimenting will show. (Some manufacturers stipulate a different clearance for valve timing from that used for running.) If the valves are push-rod operated and the end of the push rod is accessible, it can be spun easily between finger and thumb while the valve is on its seat, but becomes stiff to turn as soon as the valve begins to open. If this method is not practicable, a feeler gauge inserted in the clearance between the rocker (or tappet) and the valve stem will be nipped at the moment the valve begins to open: in this case *the valve clearance must be increased by an amount equal to the thickness of the feeler used.*

Ignition timing

In describing the cycle of operations (Chapter 4), it was stated that the mixture was ignited 'at about the end of the compression stroke', and that after ignition the mixture 'burns very rapidly'. It is clear that the rate at which the mixture burns must be very high, because if it is not, the piston will have moved most of the way down the cylinder before maximum pressure is developed. To have its maximum effect the pressure must be developed as early as possible after t.d.c. Since, however, the mixture does take some time – however short – to burn and develop maximum pressure, the spark which starts combustion must be timed to occur *before* the piston reaches t.d.c., i.e. the timing of the spark must be *advanced*.

The amount of advance necessary to give best results varies with engine speed and throttle opening, and it is the usual practice to fit automatic devices to regulate the ignition timing according to conditions (see page 448), but it is necessary to 'time the ignition' correctly to set the automatic controls in their correct basic setting.

11 Combustion chambers

Combustion

The process of combustion in an engine is considered in Chapter 14. For the purpose of this chapter it is sufficient for the reader to understand that:

1 The rate of burning must be rapid, but not excessively so.
2 The rate is influenced by several factors, of which temperature, pressure and *turbulence* are the only ones which concern us here.
3 The higher the compression ratio of the engine the higher is the thermal efficiency.
4 *Combustion knock*, commonly called *detonation*, is a form of extremely rapid combustion which is liable to occur under certain conditions. It is harmful and must be avoided. It occurs if the mixture is heated and compressed to an excessive extend during combustion. It is also influenced by the quality of the fuel.

General principles of combustion chamber design

The power which can be obtained from an engine depends upon:

1 The amount of air which can be passed through the engine per minute, i.e. the *volumetric efficiency*.
2 The efficiency with which the air can be mixed with fuel and burnt to release heat, i.e. the *thermal efficiency*.
3 The efficiency with which the energy released by the burning fuel can be converted into mechanical work at the driving end of the crankshaft, i.e. the *mechanical efficiency*.

Whilst the design of the combustion chamber can have little direct influence on the mechanical efficiency it has a great deal to do with both the volumetric and thermal efficiencies. It has, however, an indirect influence on the mechanical efficiency.

Requirements for high volumetric efficiency

The chief requirements for high volumetric efficiency are:

1 Large and unobstructed ports offering the minimum obstruction to the flow of gases.
2 The most suitable valve timing for the engine speeds and throttle openings at which the engine will most be required to operate.
3 A mixture temperature as low as possible at the moment the inlet valve closes. (Because of thermal expansion the cylinder will hold a smaller *mass* of mixture if it is hot.)

Requirements for high thermal efficiency

The chief of these are:

1 The highest possible compression ratio which can be used without *knocking* (detonation).
2 The minimum loss of heat to the combustion chamber walls. This depends upon the temperature difference between the hot gases and the walls, and upon the surface area of the walls. Since a high temperature is necessary to produce high pressure it is important to reduce the surface area - i.e. to make the combustion chamber as compact as possible. This compactness has a further benefit - it reduces the time taken by the gas to burn, by reducing the distance the flame has to travel. It can be assisted by placing the sparking plug as near the centre of the combustion chamber as possible.

Possibly the most important consideration in the use of a high compression ratio is the *anti-knock rating* (octane number) of the fuel used. Since *high-octane* fuels are expensive, a compression ratio must be chosen which balances thermal efficiency against fuel cost, bearing in mind the purpose for which the engine is to be used. (Whilst a high compression ratio will result in a lower fuel consumption for a given amount of work done, any gain may be entirely offset if the extra cost of the high-octane fuel necessitated by so high a ratio results in an increase in overall running costs. If running cost is of no importance, the fuel with the highest octane rating available may be used with a correspondingly high compression ratio.)

The design of the combustion chamber can also contribute to the suppression of knocking. The risk of detonation is greatest in the last gas to burn - what is called the *end* gas - and this risk can be reduced in two ways:

1 By getting the gas burnt as quickly as possible.
2 By arranging that the end gas is burnt in the coolest part of the combustion chamber: it is easier to do this by placing the sparking plug nearer to the exhaust valve (which gets very hot) than to the inlet valve (which keeps relatively cool).

Turbulence

The rate at which the flame travels through stagnant gas is much too low to be of use in a high-speed engine, and it was realized many years ago that a violent agitation or *turbulence* of the gas is necessary to help spread the flame rapidly throughout the mixture of air and fuel. There are two ways of creating turbulence:

1 By arranging the inlet port so that the mixture enters the cylinder obliquely and at high velocity, setting up a swirling motion which persists during the compression stroke. This is known as *induction turbulence*, and, though helpful, it is not by itself sufficient.
2 By designing the combustion chamber so that at t.d.c. the piston approaches *part* of the cylinder head very closely, displacing mixture from this region with some violence. This action has been given the very descriptive name of *squish*, and turbulence created during the compression stroke is known as *compression turbulence*.

Mechanical efficiency

The positioning of the valves is one consideration in the design of the combustion chamber, and depending upon this the valve-operating gear will be more or less complicated. The greater the complication of the valve gear the more friction will be absorbed in operating the valves. This will also influence the cost of manufacturing the engine, and the maintenance it is likely to need.

Exhaust emission

Many countries have regulations which specify the maximum amount of undesirable exhaust products that may be discharged. When considering combustion chamber shape, the exhaust pollutants which influence the design are:

1 *Carbon monoxide (CO) and hydrogen (H)* These gases are produced if the fuel mixture is not completely burnt.
2 *Oxides of nitrogen (NO_x)* Gases which are formed when the combustion temperature exceeds a given value.
3 *Lead (Pb)* The anti-knock quality of a fuel can be improved by adding a lead compound, but this solid is passed through the engine to the exhaust.

In order to achieve a 'clean' engine, the designer must either produce a combustion chamber that does not generate pollutants or he must incorporate in the exhaust system a special catalytic converter to remove the problem gases. It takes a time for the manufacturer to achieve the former, so in the changeover period, some vehicles have exhaust devices in order for them to operate in countries where emission limits are very strict.

Unfortunately the engine modifications necessary to meet the regulations normally reduce the overall road performance and often increase the fuel consumption. When fuel conservation is necessary owing to restricted supplies, the emission requirements need careful attention by both the designer and the mechanic.

In general the effects of the shape of the combustion chamber on the pollutants are:

Unburnt gas The chamber shape should be as compact as possible. Long narrow chambers such as those used in the past on some o.h.v., over-square engines gave a 'dirty' exhaust due to the difficulty of completely burning all of the charge in the short time available. This effect was very pronounced when the idling speed was low or under conditions of over-run when the ignition timing was unsuitable.

NO_x Many high performance engines of the past used a high compression ratio in conjunction with a very compact chamber. This gave a very high flame speed which produced an excessively high temperature. Although the power output was exceptional, the percentage of NO_x was far above the safe limit. Flame speed is the controlling factor, so the chamber shape must achieve a compromise in flame speed - too fast causes NO_x and too slow results in some of the charge being unburnt.

A further difficulty arises when the effect of engine load is considered. An engine set to give a 'clean' exhaust when idling or cruising often exceeds the NO_x limit when full throttle is applied. To overcome this problem, some engines have an arrangement which recycles some of the exhaust gas back into the induction manifold: the system is called *exhaust gas recirculation* (EGR). The burnt gas mixing with the new charge slows down the rate of burning and so reduces the NO_x to an acceptable level. Valves controlling the flow of exhaust gas should only allow recirculation under full load conditions, i.e. at times when NO_x would be formed.

Lead Limiting the amount of lead in the petrol means that high anti-knock (octane) rating cannot be obtained economically, so this means that the compression ratio of the engine must not be so high as to cause combustion knock. For a given fuel the ratio which causes the onset of this fault depends on the chamber's resistance to knock, together with many other factors such as timing, mixture quality and temperature. The chamber shape selected for an engine must enable the engine to use a low octane fuel and produce a high power output together with good fuel economy - a difficult task for a designer.

A catalytic converter fitted in the exhaust system to remove the pollutants operates with chemicals which are contaminated by lead. Engines using this system must use only 'unleaded' fuels. (A catalyst is an agent that aids a chemical change without undergoing change itself.)

With the foregoing in mind, some of the commoner types of combustion chamber can be considered.

The L-head combustion chamber

Volumetric efficiency is not particularly high in this design (Fig. 11.1). The arrangement of the valves at one side of the cylinder restricts valve size to about half cylinder diameter or less, and there are sharp bends through the ports and into the cylinder.

The chamber is not very compact, but there is a good deal of freedom in the choice of plug position to obtain the best results. The compression ratio is limited to about 7 : 1, as it is necessary to leave sufficient space above the valves to allow them a reasonable lift.

By bringing the surface of the head down close over the piston at t.d.c., leaving a suitable passage between the offset combustion chamber and the cylinder, both induction and compression turbulence are improved: it is, in fact, easy to get too much turbulence causing rough running through excessively rapid combustion.

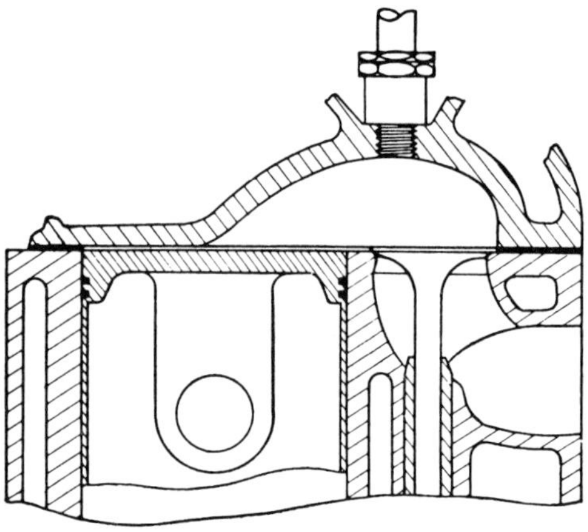

Fig. 11.1 Typical side-valve design

The position of the valves makes possible the use of the very simplest valve-operating gear, and mechanical losses are the least possible.

Though popular at one time, this combustion chamber is now obsolete in motor-vehicle engines on account of its low volumetric efficiency and restriction on compression ratio.

The 'bath-tub' combustion chamber

The valves are placed vertically over the top of the cylinder (Fig. 11.2), giving a better port shape leading directly into the cylinder. Thus the gas flow - and volumetric efficiency - are much better than in the L-head design. The valves are arranged in line, slightly offset from the cylinder centre, and the sides of the combustion chamber form a shape like an inverted bath (from which the name derives). Almost the only position for the sparking plug is the one shown, but this happens to fit in well with requirements, and the plug can be slightly nearer the exhaust valve than the inlet.

The chamber is reasonably compact, and in some cases the bath-tub shape is somewhat bent to make the chamber roughly heart-shaped when seen from below.

There is little induction turbulence, but this is made up for by the very good compression turbulence produced by the squish effect. The *squish area*, as it is

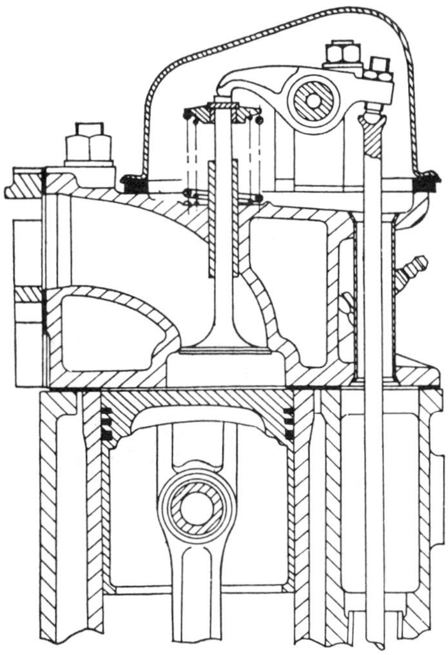

Fig. 11.2 A 'bath-tub' combustion chamber with vertical valve

called, is placed on the side of the combustion chamber opposite to the sparking plug, to ensure a thorough sweep of fresh gas over the sparking plug as the piston approaches t.d.c.

The valves are often inclined to the port side of the cylinder head, to give a slightly reduced bend to the ports. This also gives a wedge-shaped combustion chamber, which makes for more effective control of flame speed. The sparking plug is fitted at the thick end on the wedge, thus giving a large area of flame front during the early stages of combustion and speeding up the first part of this process. As combustion proceeds the flame area is reduced, thus reducing somewhat the speed of the final stages of combustion and minimizing the risk of detonation.

The hemispherical combustion chamber

It is commonly believed that the ideal shape for a combustion chamber is a perfect sphere. This belief is based upon the fact that the sphere has the smallest surface area per unit volume and thus the heat loss to the combustion chamber walls is reduced to the minimum. In most other ways, however, this shape is most unsuitable. For instance, if the bore and stroke of the engine are equal, and the spherical combustion chamber is formed half in the piston and half in the cylinder head, the compression ratio is only $2\frac{1}{2} : 1$, which is absurdly low. For a compression ratio of 8 : 1 and a similar form of combustion chamber the stroke would need to be $4\frac{2}{3}$ times the cylinder bore. This would severely limit engine rev/min and place an unacceptable restriction on valve size.

However, a combustion chamber, the upper surface of which consists of a hemisphere formed in the cylinder head, has several important benefits to offer. By inclining the valves at about 45° on opposite sides of the centre-line (Fig. 11.3) the valve diameter is much increased and an excellent port shape is obtained, giving excellent volumetric efficiency. This advantage is so important that where high power output is the main consideration this type is excellent, but high NO_x emission is a great disadvantage.

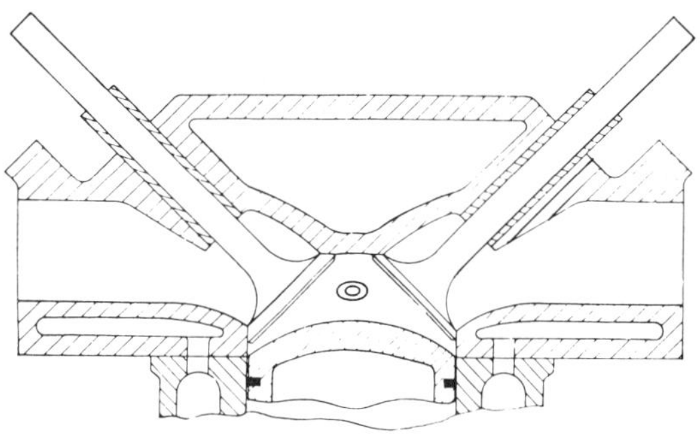

Fig. 11.3 Hemispherical combustion chamber

To obtain a high compression ratio the piston crown must be domed, giving the combustion chamber a shape like the peel of half an orange having a high surface area per unit of volume, quite the opposite of the 'ideal' spherical shape. However undesirable this may be, it is of little importance compared with advantages gained.

Induction turbulence can be arranged by having the inlet port at a slight angle to a cylinder diameter, and compression turbulence can be obtained by a suitable shape of piston crown.

The chamber is of a shape easy to machine all over, giving accurate control of compression ratio. Although the sparking plug cannot be fitted in the ideal position, it can be placed near the centre of the cylinder head, where good results are obtained.

The greatest drawback to this type of combustion chamber is undoubtedly the complication necessary in the valve-operating gear. The best method is twin overhead camshafts (Fig. 8.18a), one directly over each row of valves, but this is also the most

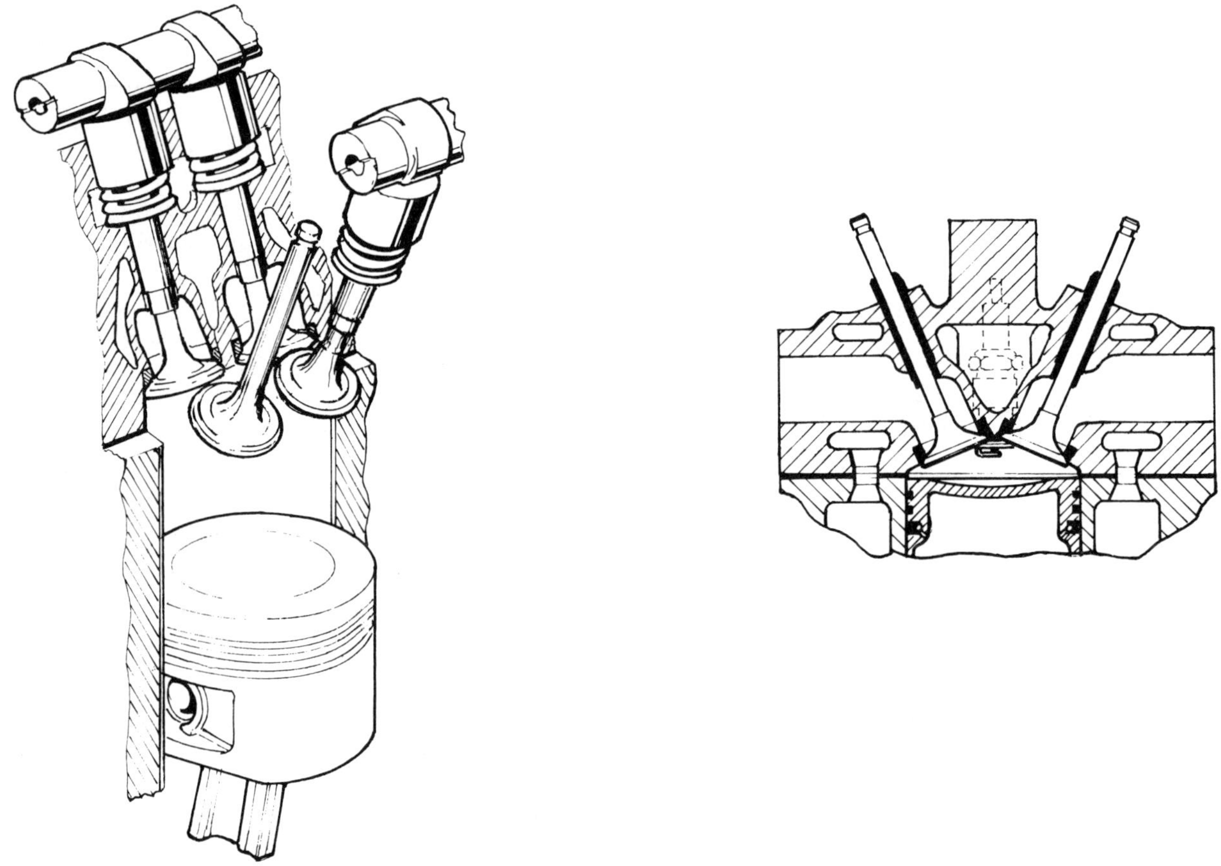

Fig. 11.4 Pent roof type combustion chamber with four valves

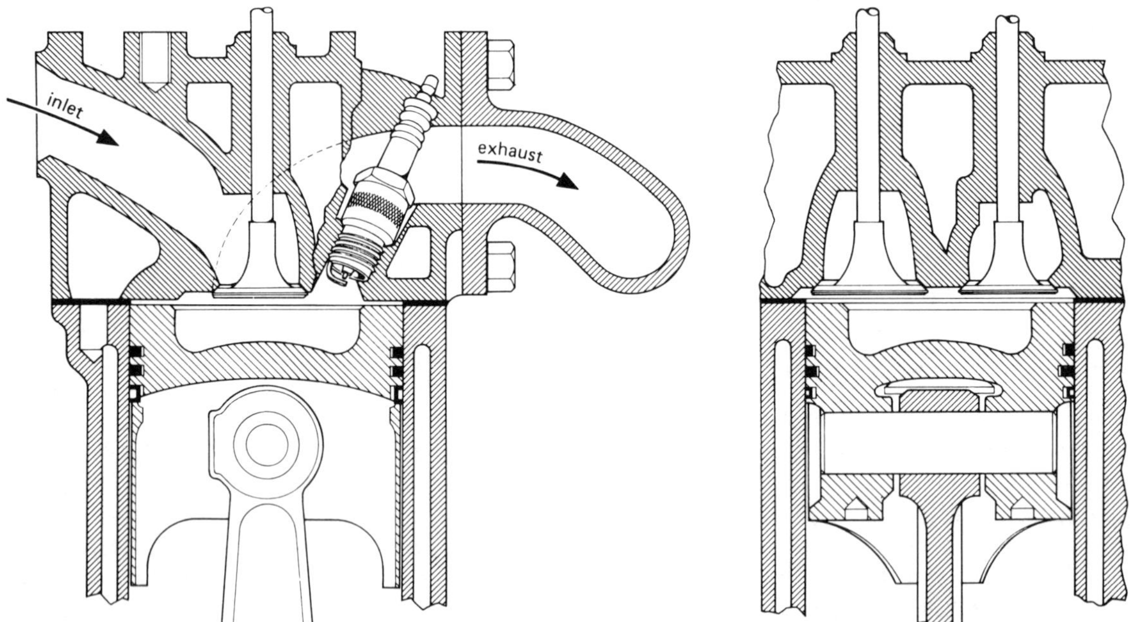

Fig. 11.5 Bowl-in-piston cross-flow head

complicated and expensive. Alternative arrangements are:

1 Single overhead camshaft with rockers (Fig. 8.18b).
2 Double side camshaft, push rods and rockers.
3 Single side camshaft, push rods and rockers (Fig. 8.17b).

With improved designs, materials and lubrication the objections to the use of the hemispherical head are becoming less important, and the increasing emphasis on engine power output, even for 'production' cars, is resulting in the use of the twin overhead camshaft layout (which has been used for practically all racing-car engines since about 1912).

The four-valve combustion chamber

A large-diameter valve is needed if good volumetric efficiency is to be achieved, but it is difficult to blend a large valve into a compact combustion chamber. This problem is minimized by using four valves, two inlet and two exhaust.

Fig. 11.4 shows a typical four-valve chamber. Since the total port area of two inlet valves is greater than that given by one valve, it is possible to combine good breathing and economy with low exhaust emission. Inclining the valves to form a *pent roof* type combustion chamber allows the sparking plug to be situated in the centre of the roof of the chamber. Four-valve engines can achieve more stable combustion over a wide fuel mixture range, so are suited for lean-burn designs (see page 134).

Cross-flow, bowl-in-piston

In this design (Fig. 11.5) the underside of the cylinder head is flat and a bowl in the piston forms the combustion chamber. The effect of having a machined chamber in the piston gives close control of the shape and capacity, so variation between cylinders of the compression ratio is minimized. A high ratio could be used if emission restrictions did not apply.

The flat head allows the large valves to open direct into the cylinder so good breathing and scavenging is possible: this can be improved still further by increasing the bore/stroke to oversquare.

Squish action around the chamber is obtained by the small piston-to-head clearance around the piston crown. In some engines, recesses are cut in the piston to prevent contact between piston and valve.

The term *cross-flow* applies to the port arrangement: the inlet and exhaust manifolds are fitted on opposite sides of the engine. This features gives space to use better manifold shapes, which further improves volumetric efficiency and exhaust scavenging.

See also 'lean-burn engines', p. 134.

12 Multi-cylinder engines

Limitations of the single-cylinder engine

Practically all motor-vehicle engines have more than one cylinder. To understand the reasons for this it is necessary to consider the limitations of the single-cylinder engine and the ways in which the use of a number of cylinders overcomes these limitations.

Irregular torque output

The torque exerted upon the crankshaft of a single-cylinder engine by the gas pressure acting on the piston fluctuates widely during each cycle of operations. As shown in Fig. 12.1, the torque reaches its maximum value during the power stroke, and during the exhaust, induction and compression strokes some effort has to be applied to the crankshaft to keep it turning. The energy necessary to do this is taken from the energy stored in the flywheel during the power stroke. Thus, although the *maximum* value of the torque developed may be high, its *average* or *mean* value is very much lower. (The *mean torque* is that steady value of torque which would do the same amount of work per minute as the actual fluctuating torque.)

The driving effort developed by the engine is, therefore, rather jerky, especially at low speeds: a heavy flywheel is necessary to give acceptable smoothness. Moreover the transmission system must be capable of transmitting the *maximum* torque developed, and is therefore undesirably heavy in relation to the *mean torque*. Fig. 12.2 shows that if *four* cylinders are used, working in succession, the mean torque is little less than the maximum torque, and this increased mean torque can be transmitted by the same transmission system as was needed for the single-cylinder engine.

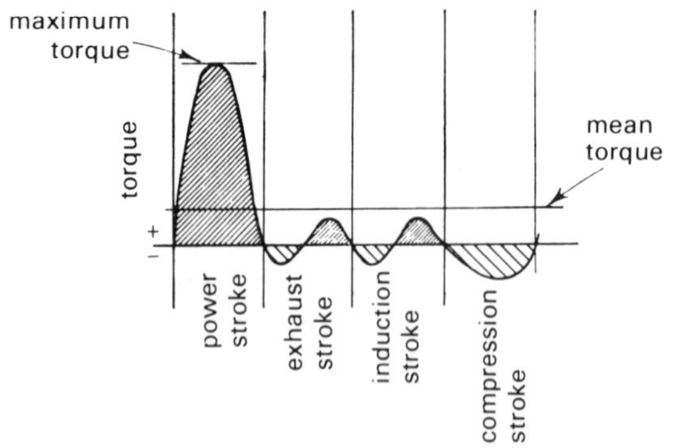

Fig. 12.1 Torque fluctuations in single-cylinder engine

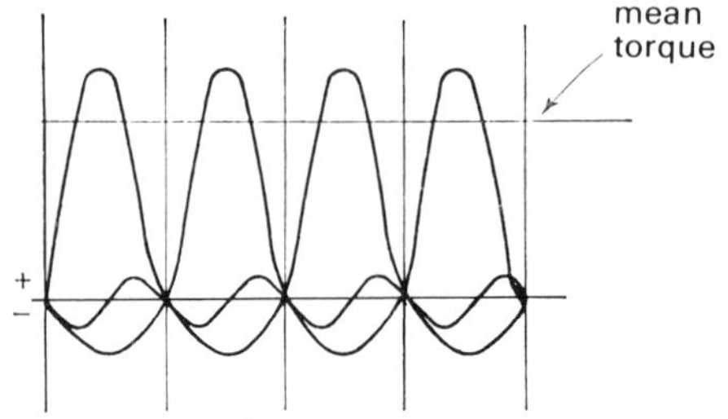

Fig. 12.2 Torque fluctuations in four-cylinder engine

Balance

The inertia forces which are caused by the acceleration of the reciprocating parts of the engine tend to cause vibration. To balance these forces and neutralize their effects, equal and opposite forces must be introduced. This can only be done by additional reciprocating parts making similar movements but in the opposite direction. While a single-cylinder engine cannot be perfectly balanced, it should be possible to arrange that, in multi-cylinder engines, the vibratory forces are neutralized.

Power output

The power which an engine can develop is determined by the amount of air it can consume per minute, i.e. upon its swept volume and the speed at which it runs.

The maximum speed is limited by the inertia forces on the reciprocating parts of the engine, and these forces depend upon the mass of these parts and their acceleration: this in turn depends upon piston stroke and the *square* of the engine speed.

For a given swept volume the bore and stroke of a multi-cylinder engine are smaller than those of a single-cylinder. The size and mass of each piston will be smaller and the stroke will be shorter, and so at the same engine speed the inertia forces on the reciprocating parts of a multi-cylinder engine will be considerably less than those in a single-cylinder: the multi-cylinder can safely run at higher speed, develop greater power and have longer life.

Shape of engine

The shape of a single-cylinder engine cannot be altered very much, and its size will depend upon its swept volume. This does not leave the designer much scope for making the engine compact so as to fit it into a smaller space in the vehicle. When several cylinders are used the designer has a choice of several different cylinder arrangements, by means of which the engine shape can be varied to suit the space into which it must fit.

Limitations on cylinder numbers

The advantages resulting from the use of a number of cylinders are not gained without some sacrifices. One of the chief objections to the multi-cylinder engine is the greater cost and complication due to the increased number of parts which have to be made and assembled. Other difficulties arise in connection with carburation and ignition systems. For these reasons the majority of motor vehicles use engines having four, six or eight cylinders. A few small vehicles have two-cylinder engines, whilst some large cars have used twelve or even sixteen cylinders, but these are exceptional cases.

Cylinder arrangements

There are four arrangements which may be used for the cylinders of an engine.

1. *In line* The cylinders are arranged in a single row, side by side and parallel to one another. They may be vertical, horizontal or inclined at any convenient angle.
2. *Vee* The cylinders are arranged in two rows at an angle to one another. For two-, four- and eight-cylinder engines, the V-angle is usually 90°, and for six- and twelve-cylinder engines it is usually 60° or occasionally 90° and 120°. There are, however, exceptions to this: for instance most V4 engines use a V-angle of 60°. In each case the crankpins are suitably offset in order to equalize the firing interval impulses.
3. *Opposed* This could be regarded as a V-type engine in which the V-angle is 180°. The cylinders are usually placed horizontally.

Cylinder numbering

The cylinders of an engine are identified by giving them numbers. When there is only one row of cylinders they are usually numbered from one end towards the other, but in the case of other arrangements there is considerable variation in practice. A method put forward by the British Standards Institution (BS 1599 : 1949) may be summarized as follows: The cylinders of *in-line* engines should be numbered 1, 2, 3, etc., commencing from the 'free' or non-driving end. In the case of Vee or opposed engines, each group of cylinders should be numbered in the same manner, but groups of cylinders should be identified by letters A, B, etc. The order of the lettering should be clockwise, looking at the 'free' or non-driving end of the engine, beginning at '9 o'clock' or the first after. This method is seldom used by motor-vehicle manufacturers: in V-type and opposed-type engines the two groups of cylinders are sometimes identified as Left (L) or Right (R): sometimes one group of cylinders contains even numbers and the other odd; sometimes one group of cylinders is numbered consecutively, followed by the cylinders in the other group.

Firing order

The cylinders of a multi-cylinder engine are arranged to have their power strokes in succession, and the order in which the cylinders work is called the *firing order* of the engine. For smoothest running the power strokes should be spaced at equal intervals, each interval being equal to the number of degrees per cycle of operations (720° for a four-stroke engine) divided by the number of cylinders. For a four-cylinder engine the interval will be 720°/4 = 180° and for a six-cylinder, 720°/6 = 120°.

The firing order of an engine is determined by *two* things:

1 The disposition of the cylinders and the cranks on the crankshaft. (This determines the *possible* firing orders.)
2 The arrangement of the cams on the camshaft. (This must be in accordance with *one* of the possible orders.)

Firing orders for some of the common types of engine are discussed in a little more detail in the following pages.

Four-cylinder in-line engines

Fig. 12.3 shows in the form of a line diagram the arrangement of cylinders and cranks in an engine of this type. Power strokes occur at intervals of 180° and the pistons move in pairs, 1 and 4 forming one pair and 2 and 3 the other. Suppose piston 1 is commencing its power stroke. Then piston 4 will move *down* its cylinder on its induction stroke. Piston 2 will move *upwards* on either its exhaust or compression stroke, and piston 3 on either its compression or exhaust stroke. We can now make up a table showing the four strokes made by each piston during one cycle of operations.

From this table we see that there are two possible firing orders for a four-cylinder engine. These are 1243 and 1342. Both are in common use.

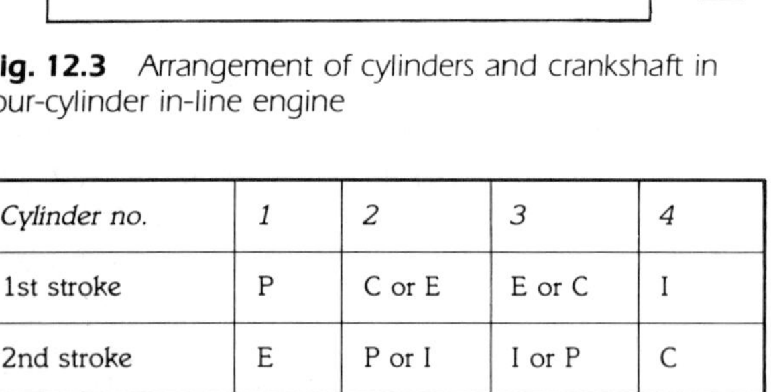

Fig. 12.3 Arrangement of cylinders and crankshaft in four-cylinder in-line engine

Cylinder no.	*1*	*2*	*3*	*4*
1st stroke	P	C or E	E or C	I
2nd stroke	E	P or I	I or P	C
3rd stroke	I	E or C	C or E	P
4th stroke	C	I or P	P or I	E

P — power stroke
I — induction stroke
E — exhaust stroke
C — compression stroke

In addition to taking the firing impulses, the crankshaft also has to resist the forces caused when the pistons change their direction of movement. The arrows in Fig. 12.3 indicate these inertia forces and the circular arrows show the effect on the crankshaft. By arranging the crank throws so that the two circular arrows oppose each other, then it is possible to achieve reasonable balance of the engine as a whole.

The high speed and power produced by a modern engine makes the load on the crankshaft and centre bearing very high. Extra main bearings are provided between each crank throw to stiffen the construction.

Fig. 5.3 shows a typical shaft for a four-cylinder engine.

Six-cylinder in-line engines

Fig. 12.4 is a line diagram of the cylinder and crank arrangement: (a) and (b) show alternative crankshaft arrangements as viewed from the front. In arrangement (a), commencing with pistons 1 and 6 at t.d.c., the next pair to come to t.d.c. will be 3 and 4, followed by 2 and 5, and then 1 and 6 again. If piston 1 is commencing its power stroke, the next power stroke will occur in either of the cylinders 3 or 4, and the next in either of the cylinders 2 or 5. There are thus *four* possible firing orders:

132645 . . . , 135642 . . . , 145632 . . . , 142635 . . .

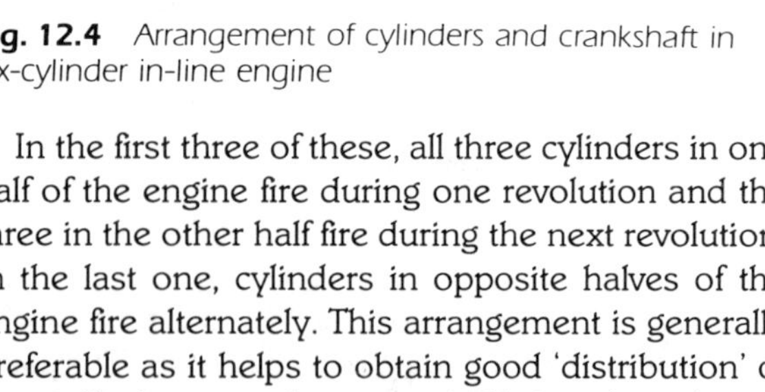

Fig. 12.4 Arrangement of cylinders and crankshaft in six-cylinder in-line engine

In the first three of these, all three cylinders in one half of the engine fire during one revolution and the three in the other half fire during the next revolution. In the last one, cylinders in opposite halves of the engine fire alternately. This arrangement is generally preferable as it helps to obtain good 'distribution' of the air/fuel mixture (see page 108). It is the only one in common use, though others have been used in rare cases.

If the cranks are arranged as shown at (b) the

pistons come to t.d.c. in the order 1 and 6, 2 and 5, 3 and 4, followed by 1 and 6 again. In this case the four possible firing orders are:

123654 . . . , 124653 . . . , 154623 . . . , 153624 . . .

Once again it is only in the last one that cylinders in opposite halves of the engine fire alternately, and this is the one in common use.

Both alternative crank arrangements are used, and there are thus *two* alternative firing orders for a six-cylinder engine in common use - 142635 and 153624.

Five-cylinder in-line engines

There are obvious manufacturing advantages of using similar-sized parts in the various engines offered by a vehicle producer. When this policy is adopted, there are occasions when a given vehicle requires an engine which has a power output, and smoothness factor, mid-way between that given by four- and six-cylinder units; this is achieved by using a five-cylinder layout. Similarly there are cases when a six-cylinder in-line unit is too long and a V6 is too heavy.

Fig. 12.5 shows the throw layout of a five-cylinder crankshaft. The angle between the throws is 72°, the firing interval is 144° and a typical firing order is 12453.

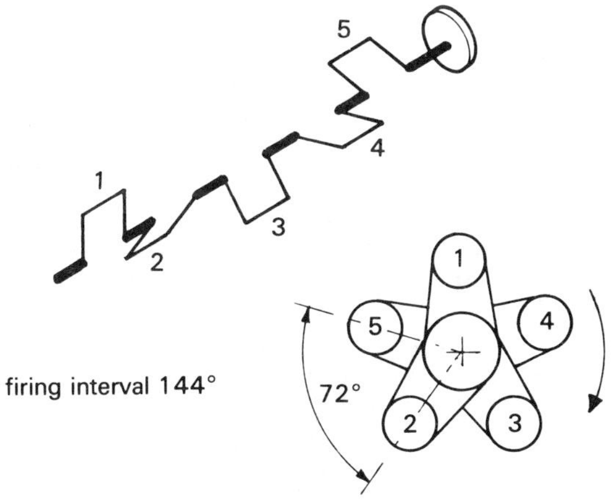

Fig. 12.5 Five-cylinder crankshaft for firing order of 12453

Torsional oscillation

A six-cylinder crankshaft is long and slender. One end is attached to a flywheel, which rotates at a near-constant speed, whereas at the front end, the firing impulses cause the shaft to wind-up and then unwind.

If the shaft is made in the form shown in Fig. 12.4, the rate at which the shaft would vibrate would be very high. As the firing impulses occur at a slower rate, no problem would arise from this vibration, but the shaft would *whip* due to the bending action caused by the centrifugal forces on each crank throw. Adding counter-balance masses (Fig. 12.6) will reduce the whip, but these heavy masses will now cause the shaft to vibrate in a rotational direction, i.e. to speed-up and slow down at a rate similar to that produced by the firing impulses.

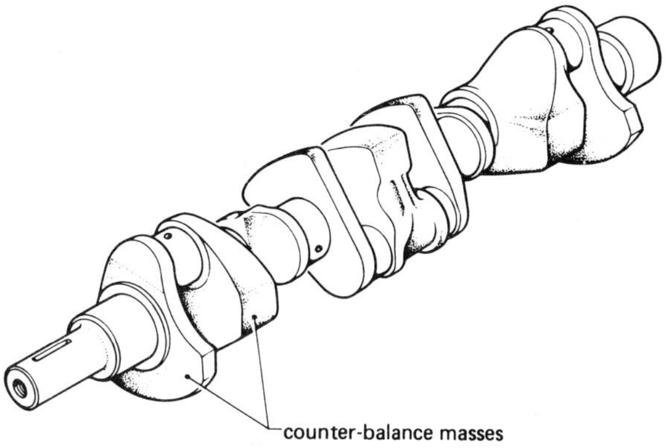

Fig. 12.6 Six cylinder crankshaft

At speeds where this occurs, the oscillation becomes severe and at these times the irregular movement of the front end of the shaft can cause problems with valve timing gears, general vibration of the engine and in severe cases, breakage of the shaft due to fatigue.

A *torsional oscillation damper* is fitted to prevent build-up of vibration at the speeds at which vibration is severe. The damper shown (Fig. 12.7) consists of a small flywheel member bonded by rubber to a hub attached to the front end of the crankshaft.

At times when oscillation occurs, the constant speed of the flywheel member opposes the winding up and unwinding of the shaft. The extent of the vibration movement is reduced by means of the energy-absorbing rubber.

Torsional oscillation dampers are not required on engines having short, stiff shafts such as those used on four-cylinder engines.

V6 engines

The diagrammatic sketch (Fig. 12.8) shows two banks of three cylinders set at an angle of 60° with a six-throw crankshaft also set at 60° between throws.

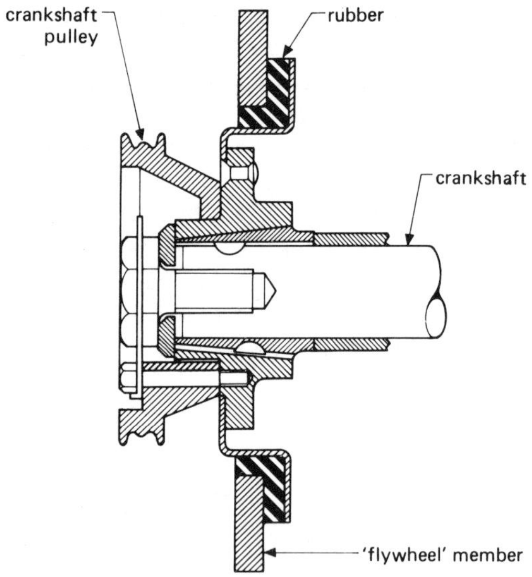

Fig. 12.7 Torsional oscillation damper

Numbering of the cylinders is that used by a large manufacturer who produces V6 engines instead of the longer in-line units.

The firing interval is 120° and the banks are fired in the order RLRLRL. The 240° angle moved by the crank between adjacent cylinders in any one bank being fired gives good distribution. It will be seen that the cylinders in each bank fire in the order: front, centre, rear.

V8 engines

Fig. 12.9 shows the arrangement of cylinders in this type of engine: the cylinders are shown numbered in accordance with BS 1599. The earliest engines of this type used a crankshaft similar to that used in the in-line four-cylinder engine, as shown at (a). Later the crank arrangement shown at (b) was adopted because it gives better balance. Power strokes occur at intervals of 90° in both cases, and the 'pairs' in which the pistons move are, for arrangement (a), 1A and 4A, 1B and 4B, 2A and 3A, 2B and 3B, giving *eight* possible firing orders. In the case of arrangement (b) the 'pairs' are 1A and 2B, 1B and 3A, 3B and 4A, 4B and 2A, also giving eight possible firing orders. It is suggested that you work out these sixteen possible firing orders for yourself.

Determination of firing order

It is necessary to know the firing order of an engine in order to be able to connect up the sparking-plug leads correctly. There are several methods by which it may be found. It is sometimes marked on the engine, or on a plate on or near the engine. It is invariably given in the workshop manual. If the engine is not marked and the workshop manual is not available the following method can be used:

1 Remove the valve gear covers and note which are the inlet valves and which the exhaust.
2 Rotate the engine in the direction in which it runs and watch the order in which *one* set of valves, inlet *or* exhaust, operate. This will give the order in which the inlet strokes or exhaust strokes occur, and the power strokes occur in the same order.

An alternative method is for the mechanic to position his thumbs and fingers over the plug holes (or fit corks in the plug holes *suitably secured* to each other): rotate the engine and note the order of 'compressions'.

Engine balance

Severe vibration occurs if the engine is in a poor state of balance. This vibration can be caused by design factors or may result from poor maintenance of the engine.

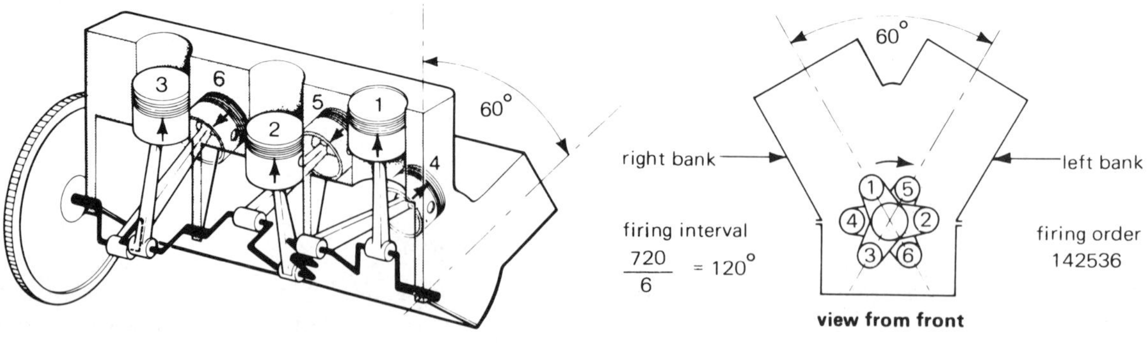

Fig. 12.8 Cylinder and crank arrangement for V6 engine

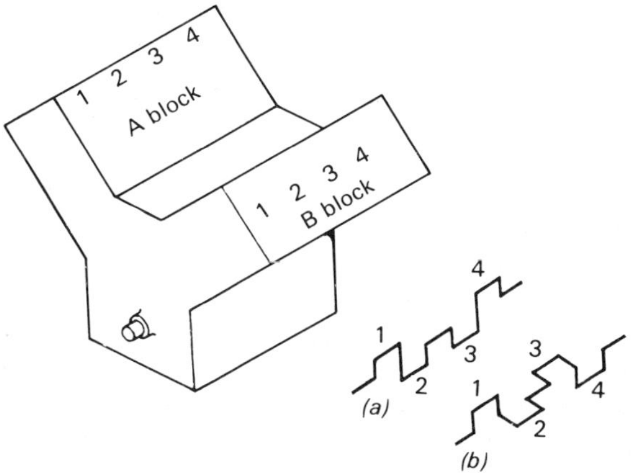

Fig. 12.9 Cylinder and crank arrangement for V8 engine

If vibration is to be minimized, attention must be given to the following:

- Primary balance
- Component balance
- Firing interval
- Secondary balance.

Primary balance

When a piston passes through t.d.c. and b.d.c., the change of direction produces an *inertia force* which attempts to keep the piston moving in the direction in which it was going before the change. This force, called the *primary force*, increases as the engine speed is raised, and unless counteracted it will produce a severe oscillation in the vertical plane, i.e. in line with the cylinder axis.

Single-cylinder Fig. 12.10 shows the primary inertia forces produced in a single-cylinder engine; the graph

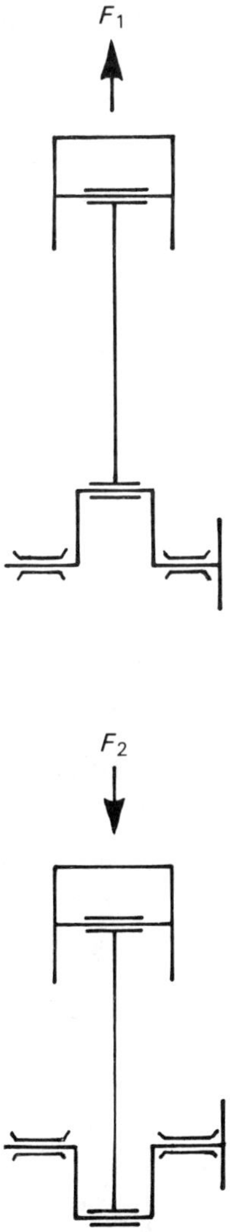

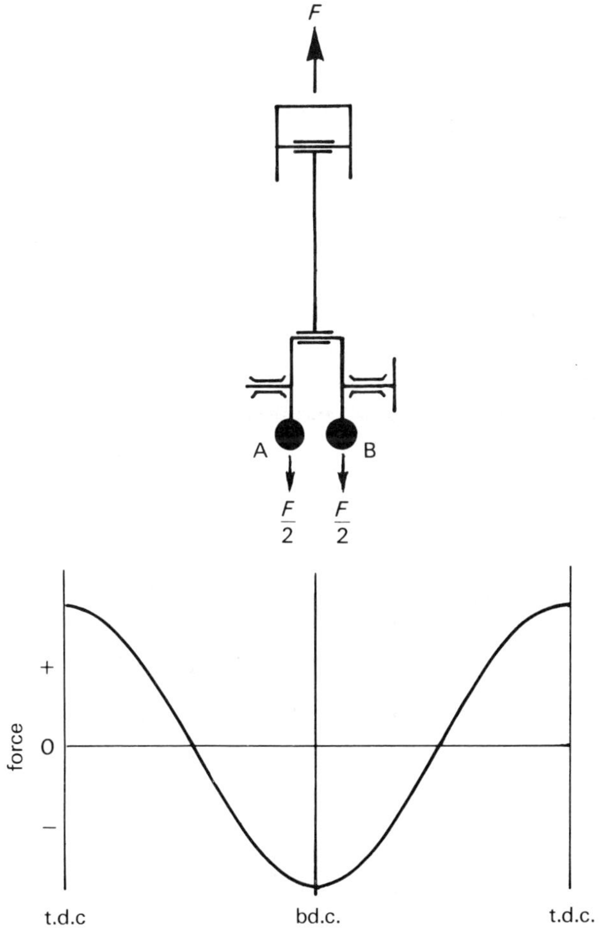

Fig. 12.10 Direction of primary force – single cylinder

indicates the direction and magnitude of the force for one revolution of the crankshaft, taking the upward direction as positive. This shows that at t.d.c., the deceleration of the reciprocating masses (piston assembly and one-third of the connecting rod) generates an upward force on the engine.

At b.d.c. a similar force is produced but in this position the direction of the force is downwards. Consideration of the effect of these two forces will show that when the engine is running it will oscillate up and down at a frequency equal to the engine speed.

This vibration can be reduced by adding at A and B counter-balance masses that exert an outward force as the crank is rotated. By varying the size of the two masses, the outward force can be made to equalize the inertia forces F_1 and F_2, but in positions other than the dead centres the counter-balance masses would themselves produce an out-of-balance force. This would be unsatisfactory because it would only shift the plane of vibration from the vertical to the horizontal. Because of this, the counter-balance mass used on a single-cylinder engine is set to balance only half the reciprocating mass. As a result, vibration in the vertical and horizontal planes is expected; this means that all nuts and bolts used on vehicles propelled by single-cylinder engines should be securely locked.

Four-cylinder Fig. 12.11 shows the crank throw layout on a four-cylinder in-line engine and the direction of the primary forces. Primary balance is achieved, because the forces on the two pistons at t.d.c. equal the forces on the pistons at b.d.c.

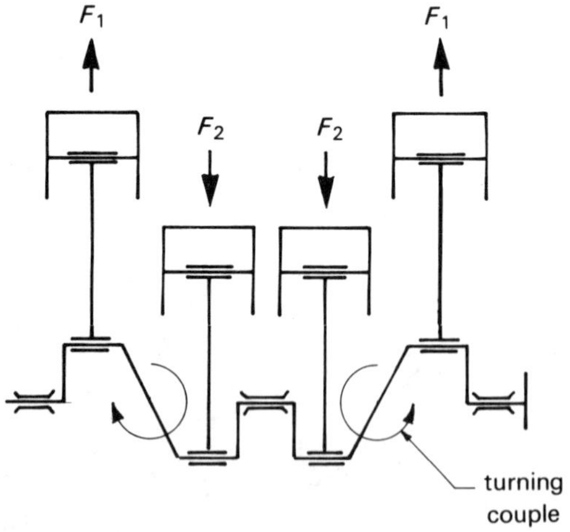

Fig. 12.11 Primary forces – four cylinders

It will be seen in Fig. 12.11 that the crankshaft throws are arranged so that forces acting on pistons 1 and 2 produce the opposite turning moment (called a *couple*) on the shaft axis to that given by the forces on pistons 3 and 4. The opposing couples introduced by this crank arrangement prevent a rocking action of the engine and, as a result, minimize fore and aft vibration of the engine.

The bending action on the shaft produced by the couples, and the high load on the centre main bearing, can be reduced by adding counter-balance masses to the crankshaft. Also, by using five main bearings to support the shaft, instead of the three commonly used in the past, a stiffer construction is achieved which is essential on modern engines because of the high operating speeds.

Three-cylinder Study of a three-cylinder in-line unit is useful because not only is it used as a 'straight' in-line engine also it forms the basic unit for both the in-line six and V6 engines.

Fig. 12.12 shows the crank layout and the primary forces when piston 1 is at t.d.c. Since the crank throws are set at 120°, the large force at each of the dead centres is balanced by the two smaller forces on the other two pistons. These smaller forces are the result of piston acceleration or deceleration as it approaches or leaves the end of the stroke.

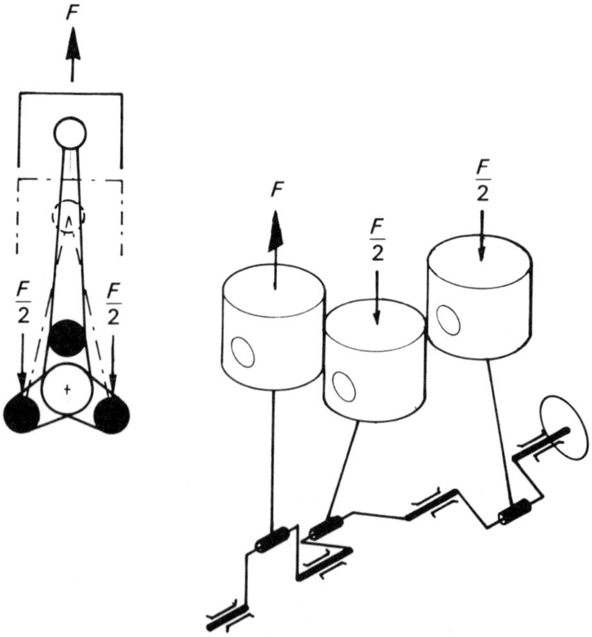

Fig. 12.12 Primary forces – three cylinders

Component balance

To minimize vibration, all components that have to rotate at high speed must be balanced. This is particularly important with large heavy components such as a flywheel and clutch assembly. Although these two parts are balanced separately to within given limits, the mating of each part so that they 'run true' with the crankshaft axis is very important. This is achieved by using various location devices such as spigots, registers and dowels.

Ideally the balancing of the crankshaft and flywheel assembly as one unit is desirable since it overcomes the 'build-up of tolerances'. (Vibration due to this occurs when 'heavy spots' of each part are positioned so that they all act in the same direction.) Cost, both at the manufacture and repair stages, generally rules out the use of this one-piece balancing method on mass-produced vehicles.

Reciprocating masses should also be balanced to achieve good primary balance. All parts that move in this manner should be selected so that they are all nearly equal in weight.

Balance of components generally covers two conditions: static balance and dynamic balance. The former can be carried out by placing the shaft and/or component on two horizontal 'knife-edges'; when released the heaviest part moves to the bottom. Dynamic balance requires expensive apparatus, which rotates the part at high speed and indicates the extent and location of the heavy spots.

Imbalance is normally corrected by removing metal by drilling one or more holes in the component at the heavy point.

Firing interval

The angle turned by the crankshaft between power strokes of a multi-cylinder engine should be regular if maximum smoothness is to be achieved. Also the more cylinders that are fired during the 720° period of the four-stroke cycle the lower the variation in the output torque, and the smoother the flow of power to the road wheels (see page 64).

Secondary balance

The inertia forces considered during the study of primary balance are based on a piston movement which is called *simple harmonic motion (SHM)*. This type of reciprocating movement is shown in Fig. 12.13a. Consider a point P to be travelling at a constant speed around a circle of diameter AB, and another point N moving in a straight path from A to B. The point N is said to move in simple harmonic motion if it always keeps at the foot of the perpendicular NP.

Point N varies its velocity as it travels across AB and this is shown by the graph, Fig. 12.13b.

Comparing the movement of an engine piston with SHM shows that during the first 90° rotation of the crank from t.d.c., the piston covers a greater distance; whereas within the range 90–180° a smaller distance is covered in the given time (Fig. 12.13c). This produces the following effects:

1 The piston travels more than half the stroke when the crank is rotated from t.d.c. to the 90° position.
2 Starting at t.d.c., the relative piston velocity for each 90° of crank movement is: fast, slow, slow, fast.
3 The *piston dwell* (the angular period where piston movement is small in relation to crankshaft motion) at b.d.c. is far greater than at t.d.c.
4 The inertia force at t.d.c. is far greater than at b.d.c.

This last point shows the reason why an engine designer needs to make a deeper study of engine balance if vibration is to be reduced. This study involves the analysis of secondary balance, which takes into account the difference between actual piston movement and the ideal SHM.

Fig. 12.14 shows the primary force produced by SHM, and the secondary force which must be added or subtracted to correspond to the actual motion. It will be seen that the frequency of the secondary force is twice the speed of the crank. By applying the information given by this graph, the direction of the secondary force can be determined; this has been added to the diagram of the engine shown in Fig. 12.15. The result shows that a four-cylinder in-line engine has very good primary balance but has poor secondary balance.

The effect of this imbalance is to produce a vibration in the vertical plane at a frequency twice the speed of the crank. In the past this vibration has been tolerated and, by using soft rubber engine mountings, the engine vibration is prevented from being transmitted to the remainder of the vehicle.

In the three- and six-cylinder in-line units, and V6, the secondary forces balance out, so this is one reason why the six-cylinder in-line engine was used extensively in the past. Nowadays the preference is for four-cylinder in-line units for engines up to about 2-litre capacity. This type has a high potential in respect of economy resulting from lower frictional losses. When it is combined with the use of simpler engine management systems, a higher power/weight

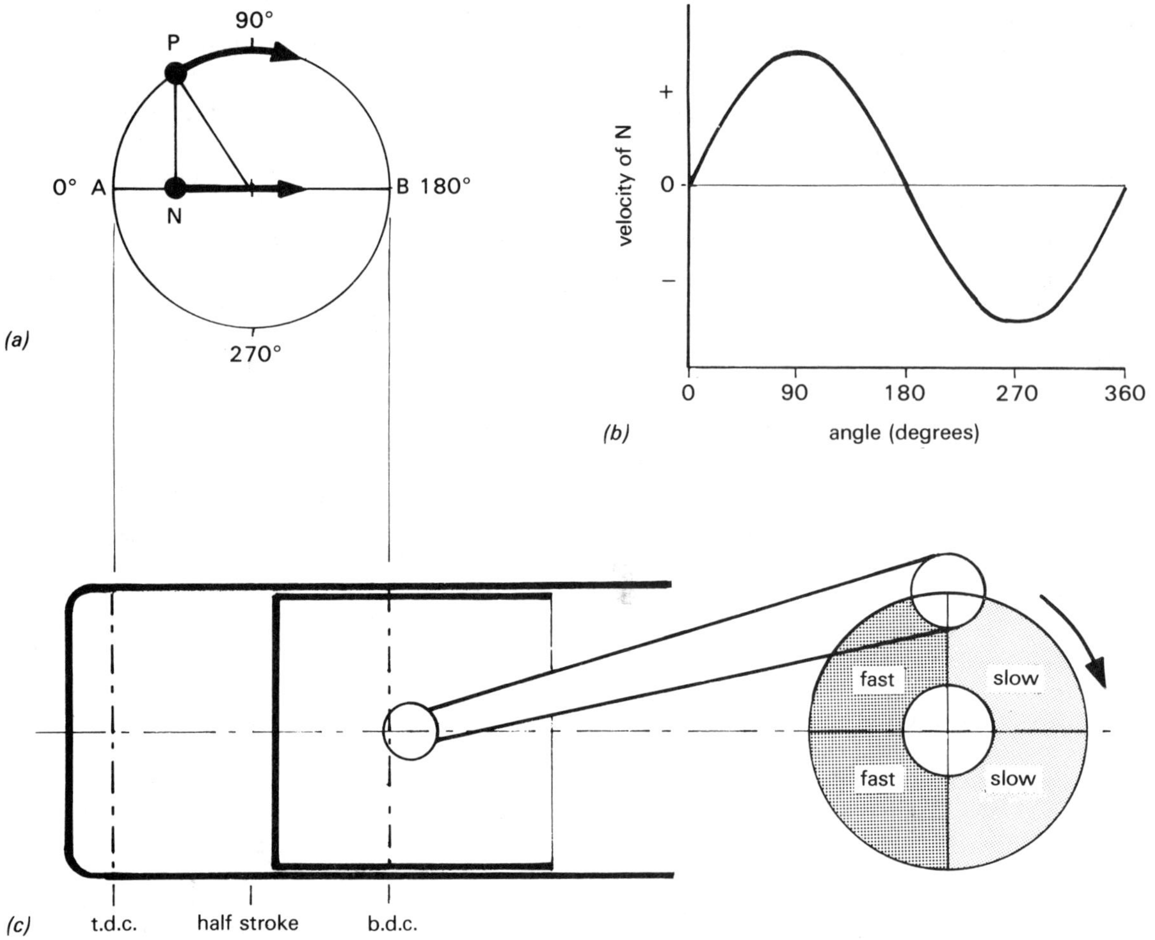

Fig. 12.13 SHM and actual piston motion

ratio can be obtained. In addition, the short, stubby crankshaft used on a four-cylinder unit does not suffer the severe torsional oscillation problems associated with longer shafts (see page 67).

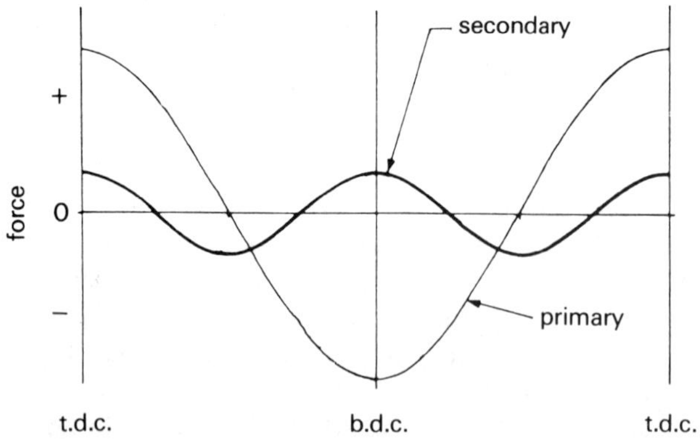

Fig. 12.14 Graph of secondary force

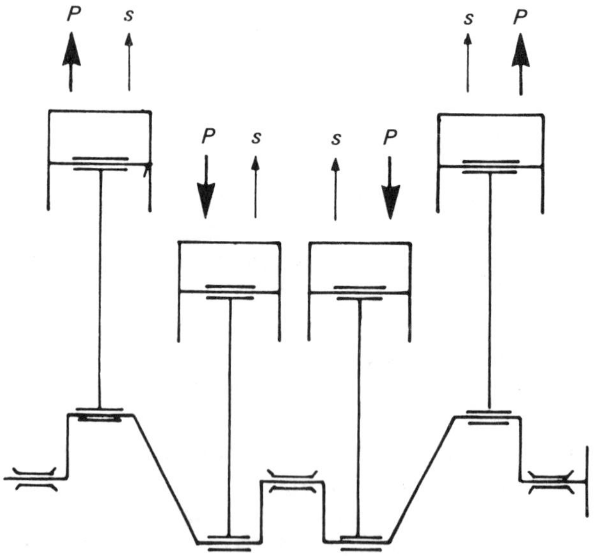

Fig. 12.15 Direction of primary and secondary forces

An effective way of eliminating secondary forces is to use a secondary harmonic balancer.

Secondary harmonic balancer This was first used in 1911 by Frederick Lanchester to balance four-cylinder engines. Although this device was very effective, for many years soft rubber mountings have been used instead of a damper for cost reasons. In 1975 Mitsubishi of Japan produced a secondary balancer; in many ways this is similar in principle to the Lanchester type. This is now being made, under licence, by a number of companies and engines using this arrangement are much smoother in operation.

Fig. 12.16 shows the principle of a secondary balancer. Two counter-balance shafts having offset masses are driven by the crankshaft at twice crankshaft speed. One counter-balance shaft is rotated clockwise and the other anticlockwise. Both shafts are 'timed' to the crankshaft so that when the piston is at t.d.c., both masses are set to exert a downward force.

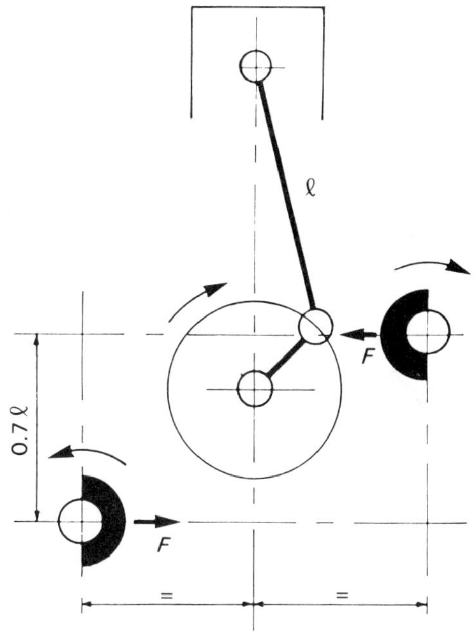

Fig. 12.17 Secondary balancer (Mitsubishi Motors)

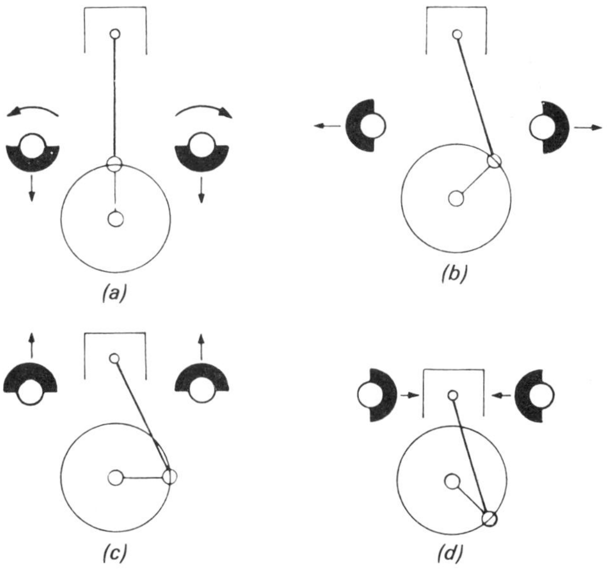

Fig. 12.16 Principle of secondary balancer

To counteract the secondary force on the engine, the balancer must exert an opposing force only at the time when the force is needed. For four-cylinder in-line engines this is a maximum when the crank is at 0°, 90°, 180° and 270°.

In Fig. 12.16a and c, this balancing force is downwards and upwards respectively, whereas in Fig. 12.16b and d, the two masses of the balancer oppose each other and produce a neutral effect; in these neutral positions the engine is already in a state of balance.

Mitsubishi Motors' *Silent Shafts* arrangement uses twin counter-balancing shafts with one shaft higher up the engine than the other (Fig. 12.17). In addition to damping the vertical vibration, this shaft arrangement also damps the secondary rolling couple which is produced when the crankshaft is rotated by the force of combustion.

The upper shaft turns in the same direction as the crankshaft and the vertical spacing of the shafts is 0.7 times the length of the connecting rod. Arranged in this way the counter-balance masses set up a couple which opposes the rolling couple. Balance of the rolling couple cannot be achieved throughout the complete engine load range, so a shaft position is selected which will minimize the unbalanced couple during the most frequently encountered road load conditions. With this arrangement the rolling couple of a balanced four-cylinder engine is better than that of a six-cylinder unit.

The Porsche 944 engine (Fig. 12.18) uses a double-sided toothed belt to drive the counter-balance shafts. It is claimed that the balancer system on this engine reduces the noise level by 20 dB.

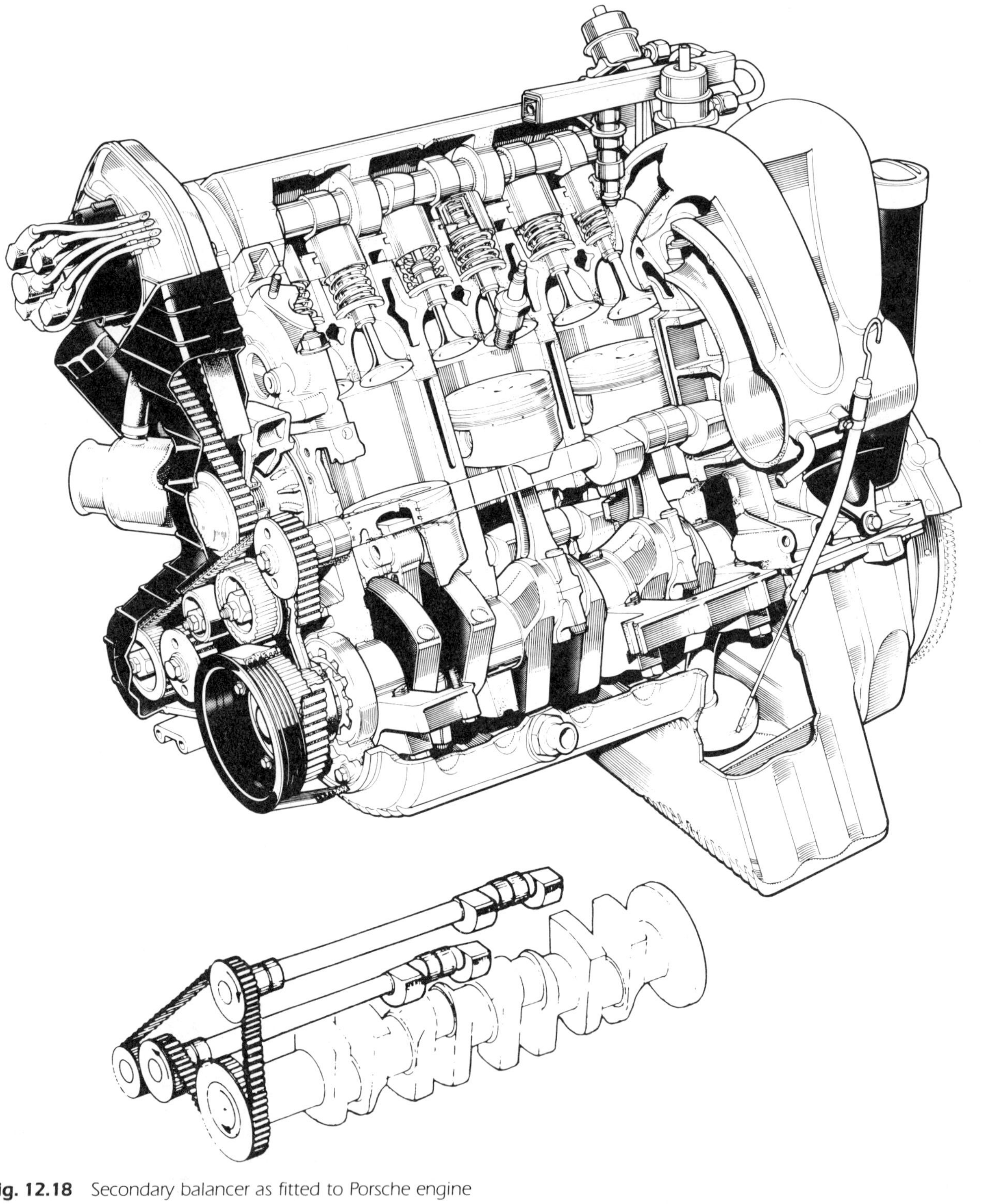

Fig. 12.18 Secondary balancer as fitted to Porsche engine

Minimizing the secondary vibration, especially at high engine speed, gives a reduction of the 'boom' which is felt and heard in the passenger compartment. In addition, a decrease in secondary vibration lengthens the life of engine auxiliaries such as emission control equipment, electrical and fuel supply components and management systems which incorporate electronic devices.

The fuel system

13 The fuel supply system

The fuel supply system consists of the fuel tank and the means for delivering fuel to the carburettor.

The tank

Most tanks are constructed by soldering, welding, or by moulding plastic.

1 *Soldered* The walls of the tank are made from tinned-steel sheets cut and bent to shape. The seams are either rolled or riveted before soldering, and the tank walls are internally supported at intervals by stiffeners. These consist of sheets which divide the tank into compartments but are pierced with large holes, so that although they allow fuel to pass from one compartment to another they prevent the fuel 'surging' from side to side of the tank.
2 *Welded* This construction uses steel pressings for walls and stiffeners, with joints and seams welded. The tank is often coated inside, and sometimes outside also, with tin or lead.
3 *Plastic* Materials such as high-density polyethylene can be easily moulded into irregular-shaped tanks to fit into spaces situated away from accident impact zones. Extra protection against impact and exhaust heat can be given by using steel plates.

Location of tank

It was at one time common practice on small and some medium-sized cars to fit the tank as high as possible in the *scuttle* (between the engine bulkhead and the base of the windscreen). With the carburettor mounted fairly low at the side of the engine the fuel

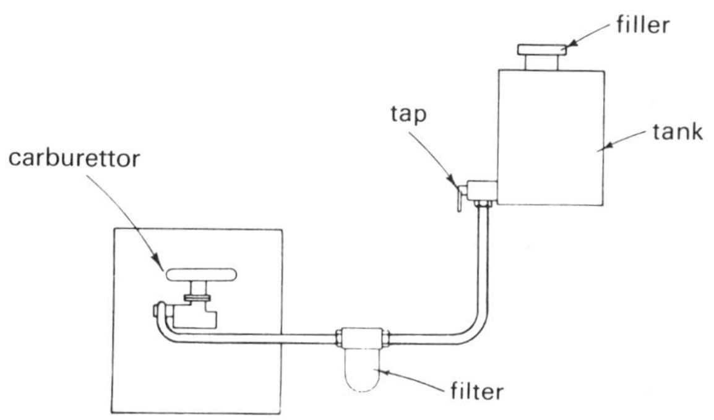

Fig. 13.1 Gravity-feed fuel system

could be allowed to flow to the carburettor by gravity, providing a simple and reliable sysem (see Fig. 13.1).

In large cars and commercial vehicles, a tank of sufficient size could not be conveniently accommodated in the scuttle. Furthermore, any leakage would allow fuel to drip into the interior of the driving compartment, with consequent risk of fire, and in case of accident a split tank would allow petrol to leak on to the exhaust pipe at a point where the pipe might be hot enough to ignite the fuel. Various alternative positions for the tank were tried, the most popular for cars being at the rear of the vehicle.

In this position the tank is below the carburettor level and it is necessary to provide some means of lifting fuel from tank to carburettor. This usually takes the form of a pump in which the pumping element is a flexible diaphragm: it may be operated either mechanically or electrically.

When mechanically operated, it is mounted on the engine and worked by a special cam on the engine camshaft. As the engine is carried on flexible mountings, it is necessary to include a length of

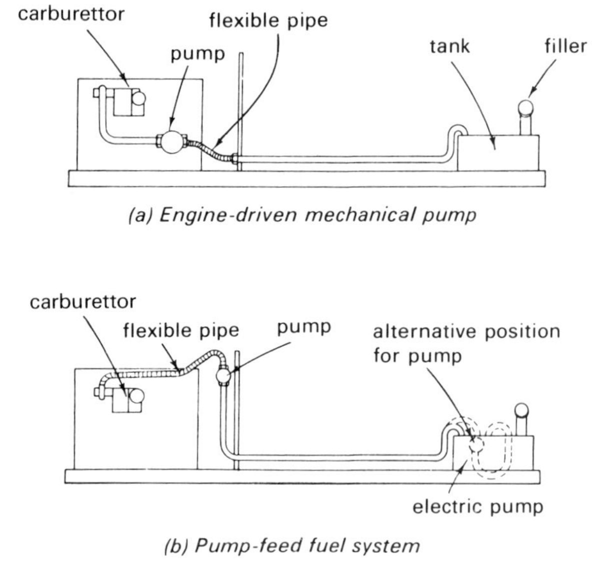

(a) Engine-driven mechanical pump

(b) Pump-feed fuel system

Fig. 13.2 Layout of fuel system

flexible piping in the pipeline connecting the pump to the tank, as shown in Fig. 13.2a.

When electrically operated, the pump may be mounted in one of two positions. The first of these is on the engine bulkhead: the pump is connected to the tank by a rigid pipeline, but a flexible pipe connects the pump to the carburettor to allow for engine movement on its flexible mountings. With the pump in this position, one difficulty which is sometimes met is the formation of bubbles of fuel vapour in the pipe between tank and pump. There are two main causes for this - heat and reduced pressure. If the fuel pipe is heated by passing near the exhaust pipe it should be rearranged, but if the heating is merely due to hot weather little can be done. The trouble is aggravated by the fact that the pipeline between tank and pump, being on the suction side of the pump, is operating under a pressure less than atmospheric, a condition which encourages the formation of vapour. Thus the alternative position for an electric fuel pump is close to the tank so that the main length of the pipeline is under pressure and the formation of vapour is discouraged. Both these positions are shown in Fig. 13.2b.

Engines which use an injection system instead of a carburettor generally have an electric pump fitted in, or close to, the fuel tank. This rotary type pump not only 'lifts' the fuel from the tank to the engine but also generates the pressure to inject the fuel into the engine manifold.

Mechanical fuel pump

The constructional features of a pump of this type are shown in Fig. 13.3. The upper part of the body (16) contains the inlet connection (1), the inlet valve (2), the outlet valve (3) and the outlet connection (4). The circular diaphragm (5) (made of fabric impregnated with synthetic rubber) is clamped around its edge between the upper (16) and lower (15) parts of the body. It is also clamped at its centre between two dished circular plates attached to the upper end of the diaphragm pull rod (6). A spring (7) is fitted between the lower diaphragm and the lower body pushes the diaphragm upwards. The pump is bolted to the wall of the crankcase. The lever (8) passes through an aperture to bear against an eccentric (9) on the engine camshaft, the end of the lever being pressed lightly against the eccentric by the spring (12). The lever pivots on a pin (10) on which is also pivoted a ⊓ section link (13), the other end of which engages the lower end of the diaphragm pull rod.

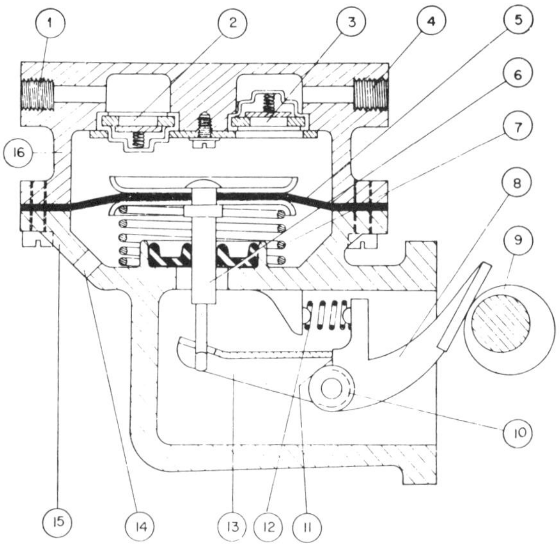

Fig. 13.3 Mechanically operated diaphragm fuel pump

When the engine is running, the lever (8) is rocked on its pivot. Movement of the lever *towards* the pump causes the step (11) on the lever to engage the link (13), which it moves about the pivot (10), so pulling the diaphragm and compressing the diaphragm spring (7). Movement of the lever (8) *away* from the pump allows the diaphragm to be moved upwards by the spring (7).

Downward movement of the diaphragm increases the volume of the pumping chamber above the diaphragm, holding the outlet valve (3) on its seat, but opening the inlet valve (2) and drawing fuel from the tank through the inlet connection into the pumping chamber. On the upward stroke of the diaphragm, pressure is applied to the fuel in the pumping chamber by the spring (7), closing the inlet valve but forcing the outlet valve off its seat and delivering fuel through the outlet connection (4) and a pipe to the carburettor.

When the carburettor is full of fuel a valve in the carbutettor closes (see page 84), preventing the entry of any more fuel, and consequently fuel will now be unable to leave the pump. The diaphragm will thus be held in its lowest position by the fuel in the pumping chamber under the loading of the spring (7), although the lever (8) will maintain contact with the eccentric (9) through the influence of the spring (12). The pressure at which fuel is delivered to the carburettor is determined by the force exerted by the spring (7). This force in turn depends upon the dimensions of the spring and the extent to which it is compressed when the diaphragm is in its lowest position. Within limits this may be varied by selecting a suitable thickness of packing to place between the pump mounting-flange and the engine crankcase, a greater thickness reducing the pressure slightly and a lesser thickness increasing it.

Should the diaphragm develop a leak, fuel can escape through a number of drain holes (14) drilled through the wall of the pump body.

Maximum fuel pressure at inlet and outlet points depends on the pump model: typical values are:

Inlet – vacuum of about 150 mm (6 in)
Outlet – 21-35 kPa (3-5 lbf/in^2)

Prior to about 1975, pumps were provided with a drain hole (14). Unless a seal is fitted around the rod (6), the hole would allow crankcase fumes to escape and so contravene emission regulations.

Electric fuel pump

Although electric fuel pumps have been extensively used in the past, the mechanical pump is now preferred because it is cheaper. Also the mechanical pump operates only when the engine is running, so the fire risk is reduced when the vehicle is involved in a crash.

The electric pump (Fig. 13.4) consists of two parts: the pumping section, which is similar to the corresponding section of a mechanical pump, and the electrical unit, which operates the diaphragm.

Movement of the diaphragm (9) is effected by energizing an electromagnet and attracting an iron armature (8). At the end of the stroke during which the fuel is entering the pump, the push rod (7) moves the toggle-spring assembly and clicks open the tungsten contacts (1): this interrupts the electrical supply to the magnet coil (2), allows the spring (3) to return the diaphragm and forces the fuel through the outlet valve to the carburettor. When the diaphragm reaches the end of its travel, the contacts close and the cycle is repeated.

This action will continue until the carburettor is full of fuel, when the diaphragm will be unable to make its return stroke, holding the contacts open until the carburettor requires more fuel. The pressure at which the fuel is delivered to the carburettor is governed by the strength of the spring (3).

Reference was made to the fire risk following a crash. To minimize this hazard, some vehicles have an *inertia-actuated cut-off* switch connected in the electrical supply to the fuel pump. A deceleration greater than 5 g (49 m/s^2) causes the switch to open.

Filters

All carburettors have a number of passages and small holes through which petrol must flow, and tiny pieces of dirt or blobs of water can obstruct the flow and prevent the correct functioning of the carburettor.

The petrol companies take a great deal of trouble to ensure that the petrol delivered by their pumps at filling stations is clean, but it is possible for dust to enter the tank of the car any time the filler cap is open, even for short periods. As fuel is taken out of the tank and delivered to the carburettor, air must be allowed to enter the tank. It does this through a vent hole or pipe, and it may take in dust or moisture with it, the moisture subsequently condensing inside the tank. A small sump is usually formed at the lowest point of the fuel tank, with a drain plug to enable any water collected in the sump to be drained. The fuel supply to the engine is taken from the above sump level to minimize the possibility of dirt and water getting into the pipelines and eventually into the carburettor.

It is, however, a sensible precaution to fit one or more filters in the fuel system to trap dirt and water, preventing them entering the fuel pump - where they might prevent a valve from closing properly - or the

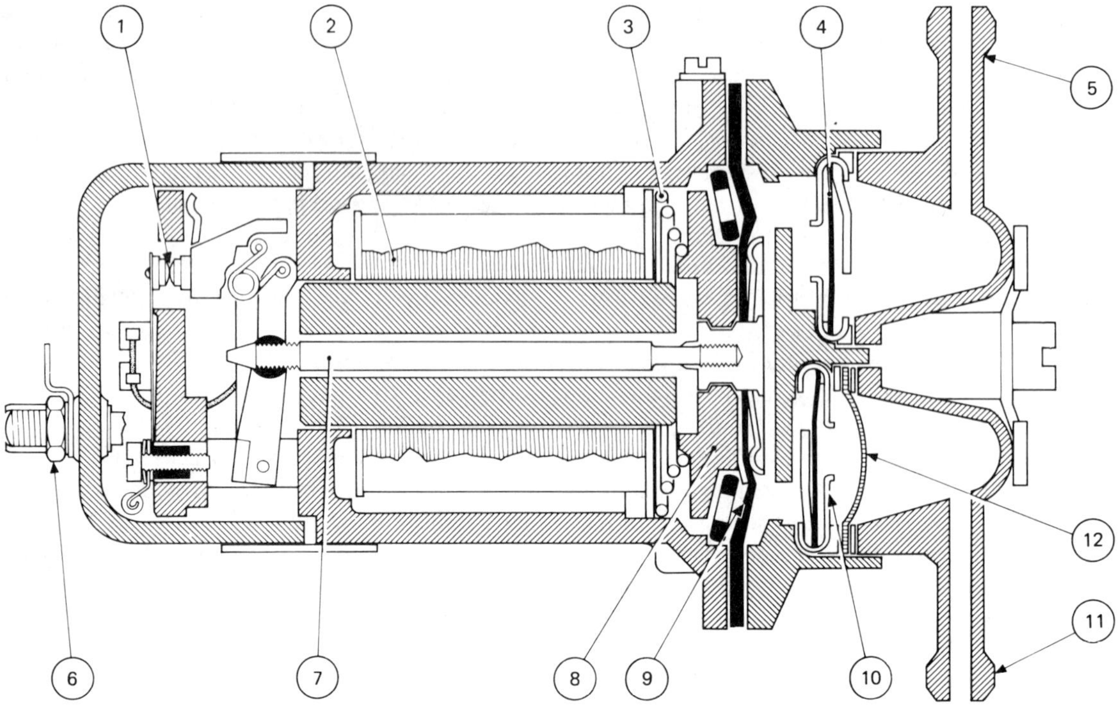

1 contact breaker points
2 magnet coil
3 return spring
4 outlet valve
5 outlet connection
6 electrical supply connection
7 push-rod
8 magnet armature
9 diaphragm assembly
10 inlet valve
11 inlet connection
12 filter

Fig. 13.4 Electric fuel pump

carburettor. Filters may be fitted in one or more of the following positions:

1 Around the outlet from the fuel tank.
2 At the fuel pump inlet.
3 At the carburettor inlet.
4 At any convenient point in the pipeline connecting tank to carburettor.

Fuel filters usually consist of brass wire gauze of a mesh sufficiently fine to trap those particles of dirt which are large enough to cause trouble in fuel pumps and carburettors, and of an area large enough to pass more than the maximum fuel flow. If previously damped with petrol such filters will also trap water.

Fig. 13.5 illustrates a type of filter suitable for connecting in a fuel pipeline. A cast body (4) has a drilled lug (1) by which it can be secured to a suitable position on the vehicle. A cylinder of fine-mesh brass gauze (7), closed at its lower end, is soldered at its top edge to a fitting (6) which screws into the underside of the body. A hole drilled through this fitting communicates with the outlet connection (2). A glass bowl (8) is clamped below the body by a stirrup (9) and a clamping nut (10), and seats on a cork washer (5) to prevent leakage.

Fuel enters at the inlet connection (3) directly into the bowl and passes through the gauze to the outlet connection, leaving dirt and water in the bowl. The filter will need cleaning at suitable intervals. There are several alternatives to brass gauze as a filtering element.

Pipelines

For many years the pipes used to convey the fuel from tank to carburettor were made of copper, a metal of

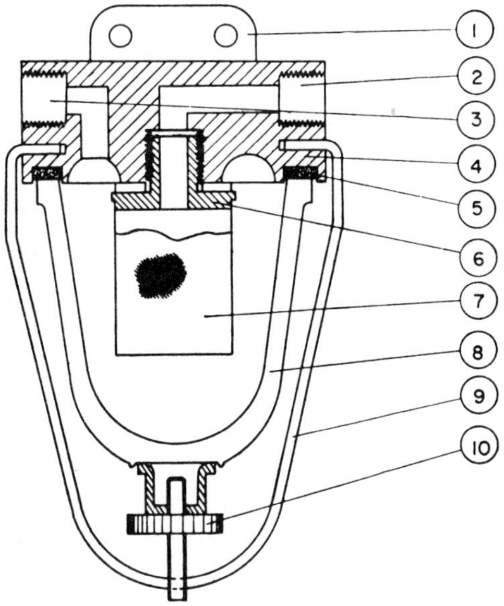

Fig. 13.5 Fuel filter

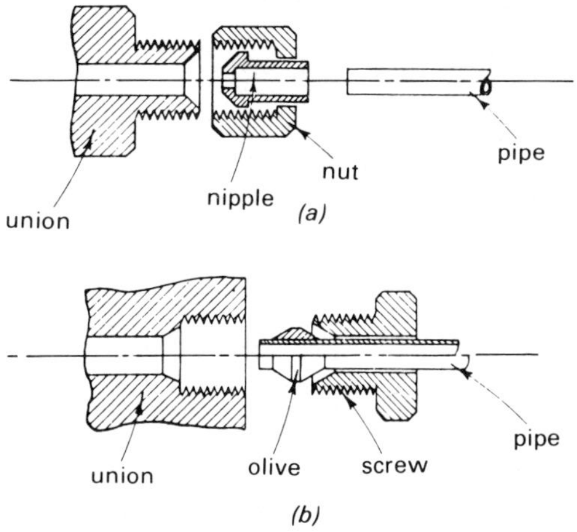

Fig. 13.6 Types of pipe connection

great ductility, which enables it to be easily bent to shape, and possessing good resistance to corrosion. Copper, however, is relatively expensive and has been almost entirely replaced on modern vehicles by tinned steel or plastic.

Modern vehicles have their engines supported on flexible mountings, which permit fairly extensive movement of the engine relative to the frame, to prevent the transmission of vibration to the bodywork and passengers. It is therefore necessary to include a length of flexible piping in the pipeline. When a mechanical pump is used, this is fitted on the engine, and a length of flexible pipe connects the pump with the rigid part of the pipe which is secured to the frame.

Plastic petrol piping, even when reinforced with braiding to improve strength, becomes soft and melts when exposed to excessive heat. In view of this, the pipe should be kept clear of the exhaust system.

Pipe connections

Where pipes are connected to tanks, pumps, filters and carburettors, some form of secure but easily detachable connection is required. This usually takes the form of screwed connectors such as those shown in Fig. 13.6. That shown at (a) involves soldering the nipple on to the end of the pipe, but the type shown at (b) requires no soldering, the olive being compressed between the coned surfaces of the union and the screw as the latter is tightened. Since the olive is made of a relatively soft brass it contracts on to the pipe making a fuel-tight connection. In a third type, not shown, the end of the pipe is flared out to mate with a suitable conical seating on the union.

With the increasing use of flexible plastic pipes a very simple connection is being used, in which the end of the plastic pipe is simply pushed on to a tightly-fitting tubular extension on the union (such as that shown in Fig. 13.4), sometimes with the use of a petrol-proof adhesive.

14 The functions of the carburettor and fuel injection system

It has been seen in Chapter 3 that the driving force in an internal combustion engine is the pressure resulting from the heating of a gas by burning a fuel inside the engine cylinder. During one cycle of operations the pressure acting on the piston varies widely, but it has an 'average' value which is called the *mean effective pressure*. The effort exerted by the engine is clearly proportional to the mean effective pressure (m.e.p.), which depends upon engine design, compression ratio, and the temperature to which the gas is heated by combustion.

Combustion of petrol

Petrol is a fuel which is particularly suitable for use in motor vehicles. It is a clean liquid which is easily stored (although reasonable precautions against fire are necessary) and which flows freely. It gives off a flammable vapour at quite low temperatures and when burnt liberates quite a large amount of heat.

Before petrol can be burnt it must be vaporized and mixed with a suitable quantity of air, since combustion is a process involving the chemical combination of a fuel with oxygen.

Crude petroleum, from which petrol is obtained, is a mixture of various compounds of hydrogen and carbon (called *hydrocarbons*) and petrol consists of these constituents of crude petroleum which have boiling points between the temperatures of, roughly, 30 and 200 °C.

Calorific value

The amount of heat liberated by complete combustion of unit mass of a fuel is called the *calorific value* of the fuel. For an average sample of petrol the calorific value is about 44 MJ/kg.

Mixture strength

The mass of air per kilogram of fuel in a mixture of air and fuel gives the air/fuel ratio of the mixture and, as already noted, the air/fuel ratio for complete combustion – called the *chemically correct* mixture – is about 15 in the case of petrol.

A mixture having a greater proportion of fuel, i.e. a *lower* air/fuel ratio than 15, is called a *rich* mixture, whilst one having a greater proportion of air, i.e. a *higher* air/fuel ratio than 15, is a *weak* mixture. Within limits both rich and weak mixtures will burn, but will produce different results.

Influence of air/fuel ratio

Mixtures having air/fuel ratios of less than about 8 or more than about 22 cannot normally be ignited in petrol engine cylinders, and within this range variation of air/fuel ratio has a considerable influence on the engine's performance. (It should be emphasized that the mixtures referred to are those in which the fuel is completely vaporized and thoroughly mixed with the air so that the proportions of air and fuel are the same throughout the whole bulk of the mixture. This is rarely the case in an engine cylinder under running conditions.)

It might be expected that the chemically correct mixture ought to give the best results, but this is not necessarily true. As indicated in the preceding paragraph, for instance, it is impossible, under normal running conditions, to ensure that the fuel is evenly distributed throughout the air, and under some conditions the fuel is not all completely vaporized when combustion occurs.

The main effects of variation of air/fuel ratio are upon:

1 *The mean effective pressure* This is found to reach maximum values with a slightly rich mixture.

The upper curve of Fig. 14.1 gives some idea of the way in which m.e.p. varies with air/fuel ratio.

2 *The rate of combustion* The speed of travel of the flame through the mixture varies in a manner roughly similar to the variation of m.e.p., being fastest with slightly rich mixtures and slowing down markedly as the mixture is weakened and to a lesser extent as it is enriched. One result of this is a tendency for the engine to run hotter if the mixture is weak.

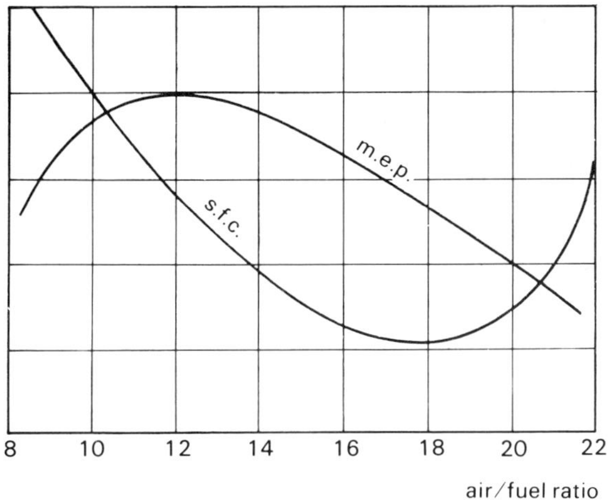

Fig. 14.1 Effects of variation of air/fuel ratio on mean effective pressure and specific fuel consumption

3 *The rate of fuel consumption* This is called the *specific fuel consumption (s.f.c.)* and it is found to be lowest with mixtures which are slightly weaker than the chemically correct mixture as shown by the lower curve of Fig. 14.1.

4 *Composition of exhaust gas* If an engine is supplied with the chemically correct mixture the exhaust gas should consist of a mixture of carbon dioxide, water vapour and nitrogen. Owing to imperfect mixing of the air and fuel, however, it is found that there are in fact small amounts of both oxygen and carbon monoxide in addition to the other gases. If the mixture is made richer the proportion of carbon dioxide in the exhaust gas decreases whilst the amount of carbon monoxide increases, and a small amount of unburnt hydrogen is also found. As the limit of combustibility is approached some unburnt carbon appears as *sooty black smoke* in the exhaust gas, and this is always a sign of excessive richness.

If the mixture is weakened from the chemically correct ratio the proportion of carbon dioxide in the exhaust gas again decreases. Carbon monoxide is absent once the mixture has been weakened slightly, but the proportion of oxygen increases.

These changes in the composition of the exhaust gases can be used to estimate the air/fuel ratio on which the engine is running. A sample of exhaust gas can be analysed chemically: though this is impracticable in a repair shop. An alternative is available in the form of an instrument called an *exhaust gas analyser*: this detects changes in the proportions of the gases by electrical means and indicates the air/fuel ratio on a scale.

Vaporization

Since it is petrol *vapour*, and not the liquid, which burns, it is important that all the petrol supplied to the engine should be vaporized before combustion occurs. As liquids change into vapours they absorb heat (called *latent heat*) which they obtain from their surroundings. If petrol were to be vaporized in the carburettor, the latter would be chilled by the heat taken from it to vaporize the petrol, and moisture in the air passing through the carburettor would be condensed, deposited on the throttle and choke tube, and frozen. In this way the passage through the carburettor would become blocked by ice and the engine would stop.

This could, of course, be prevented by arranging for the carburettor to be supplied with hot air drawn from the region around the exhaust manifold. Overheating the air should be avoided because:

1 It reduces power output. Since petrol vapour occupies a considerably greater volume than the equivalent quantity of liquid, a much smaller mass of petrol and air can be got into the cylinder if the petrol is in vapour form than liquid. If the air is heated any more than is absolutely necessary to prevent the formation of ice there is a further loss of power due to the density of the air (and hence the mass of air in the cylinder) being reduced by the rise in temperature. It is not easy to regulate the extent to which the air is heated.

2 If the petrol can be vaporized *inside the cylinder*, the heat absorbed during vaporization will have a beneficial cooling effect on the inside of the combustion chamber, particularly when the engine is developing full power.

Thus best results will be obtained if the fuel can be introduced into the cylinder in liquid form *provided it can all be vaporized before combustion*. Since vapour is given off from the *surface* of a liquid, the rate of evaporation can be increased by breaking up the liquid into a fine spray of minute droplets. (A millimetre cube has a volume of 1 cubic millimetre and a surface area of 6 square millimetres. The cube can be cut up into eight smaller cubes each having sides $\frac{1}{2}$ millimetre long and a surface area of $1\frac{1}{2}$ square millimetres, giving a total area of 12 square milli-

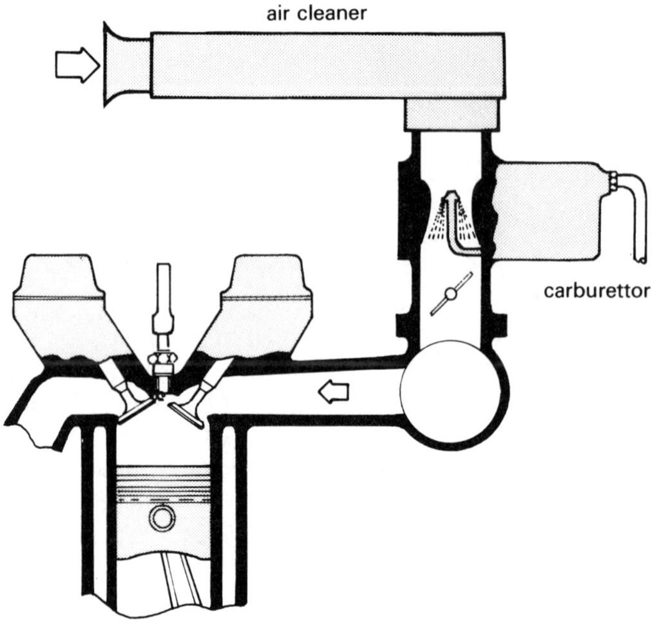

carburettor system

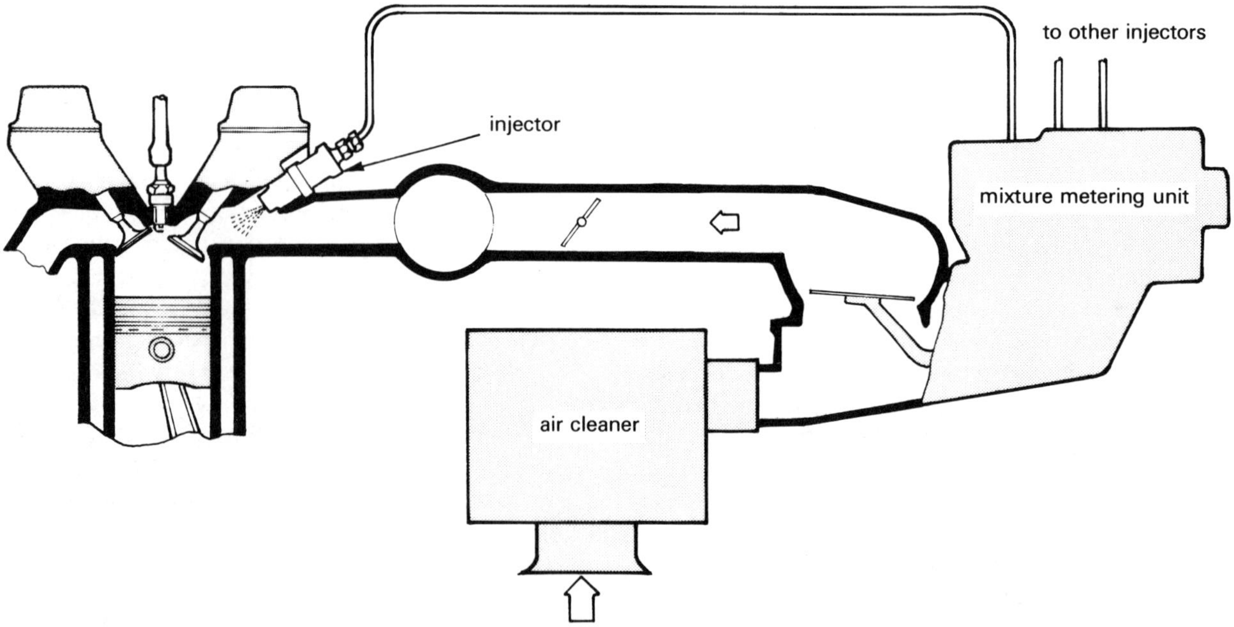

fuel injection system

Fig. 14.2 Carburation systems

metres for the same total volume. Hence, the smaller the size of the droplets forming a spray of fuel the greater the surface area for a given volume.)

Atomization

Generally the time is too short to vaporize the fuel completely before combustion, so additional means must be provided to break up the fuel mechanically into small particles by a process called *atomization*. This breaks up the fuel either by subjecting it to a turbulent air flow, or by pumping the fuel into the air stream through small holes in an injector. This action not only aids the production of a fast-burning mixture but also helps to mix the air and fuel equally, i.e. it aims to produce a *homogeneous* mixture.

Carburation systems

In the past the bringing together of the petrol and air was performed by a *carburettor*. This device mixes the air and petrol in suitable proportions to obtain the desired performance from the engine; it also atomizes the petrol to give good combustion.

Nowadays many engines in the higher performance category use a fuel injection system instead of a carburettor. In addition to providing a selling feature, a good injection system should return a higher performance from the engine, i.e. the engine will:

- develop more power to give a higher speed and to accelerate at a greater rate
- be more economical in its use of fuel.

Fig. 14.2 shows, in simple form, two systems of carburation in use today.

15 The simple carburettor

The term *simple carburettor* refers to the basic type of carburettor. Whilst it might be suitable for certain types of engine, it is not satisfactory for modern motor vehicle engines, and modifications to this basic type are necessary. These modifications will be described in later chapters.

A carburettor consists essentially of two parts:

1 The means for regulating the entry of fuel to the carburettor according to the rate at which the fuel is used. This usually takes the form of a *float chamber*.

2 The means for atomizing the fuel and mixing it with the necessary amount of air. This part is known as the *mixing chamber*.

These two parts are not necessarily separate from one another.

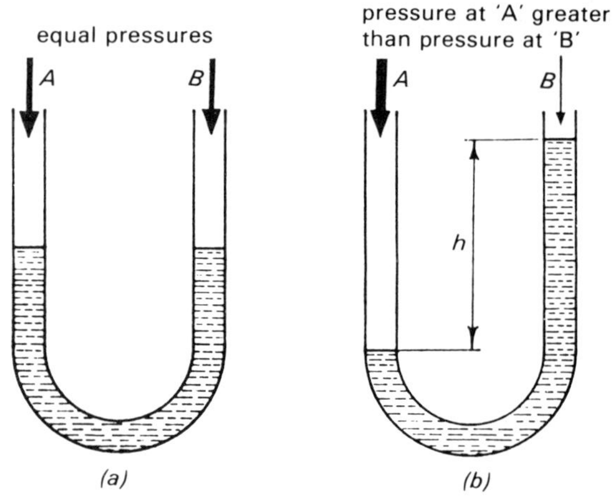

Fig. 15.1 A simple U-tube

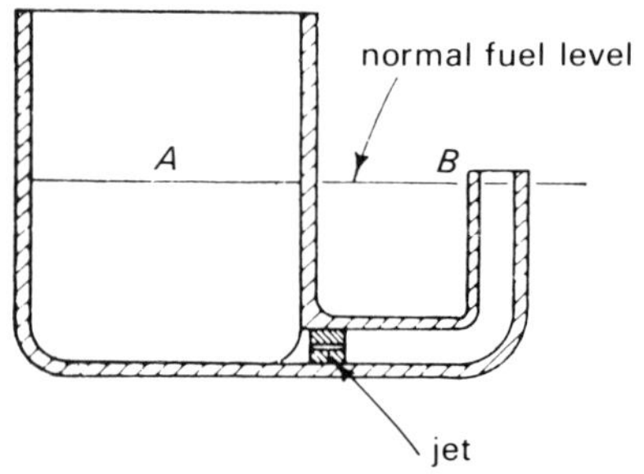

Fig. 15.2 Modified U-tube used in a carburettor

The float chamber

The action of this is based upon the principle of the simple U-tube, which is illustrated in Fig. 15.1. Diagram (a) shows a U-tube partly filled with liquid: so long as the pressure acting at A and B are equal, the levels of liquid in the two limbs of the tube will be at the same height, but if the pressure at A exceeds that at B, the level of liquid in limb A will be forced down and that in B raised. Thus there will be a difference in the height of the two levels, *h*, proportional to the difference in the two pressures, as illustrated in diagram (b). Note that it is not the *actual* pressures which matter, but their *difference*, and the same effect will be produced by either an *increase* of pressure at A or a *decrease* at B, or by a combination of both. Fig. 15.2 shows the U-tube modified for use in the carburettor, the modifications consisting of an enlargement of the size of limb A and a shortening of limb B. (It can easily be demonstrated that this does not affect the principle in any way whatsoever.) Under normal conditions the limb or chamber A is open to the pressure of the atmosphere, and is kept full of fuel to the level of the top of the limb B, or slightly below. If this level can be kept constant and limb B subjected to a pressure lower than that of the atmosphere, fuel will be drawn out from B at a rate which depends upon the difference in pressures and the opposition which the liquid encounters in flowing from A to B. It is a simple matter to provide the necessary opposition to flow by fitting, at any convenient point in the tube B, a plug with a small hole, and different sizes of hole will clearly regulate the rate of flow for any given pressure difference.

These drilled plugs are known as *jets*, and they are carefully calibrated and numbered by carburettor manufacturers to indicate the rate at which fuel flows through them under standardized conditions. The jets are externally threaded to screw into position and are easily interchangeable.

Fig. 15.3 illustrates the manner in which the level in the chamber A is kept constant. The top of the chamber is closed by a lid in which there is a small hole or air vent, by means of which atmospheric pressure is maintained on the liquid in the chamber. The lid also has a connection for the pipe from the fuel pump, and this connection leads to a hole entering the top of the chamber. Inside the chamber is a float which may be made of hollow brass pressings, soldered together, or a suitable plastic material. Attached to the top of the float is a needle, the pointed end of which enters the hole by which the fuel comes into the chamber.

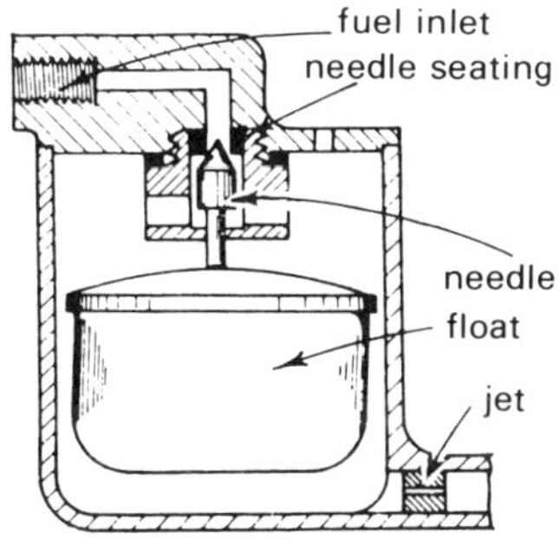

Fig. 15.3 Float mechanism

When the chamber is empty the float lies on the bottom of the chamber and the fuel inlet is open. Operation of the fuel pump delivers fuel to the chamber; as this fills, the float rises on the fuel level, bringing the end of the needle into contact with a seating at the end of the fuel inlet hole and cutting off any further entry of fuel until some fuel is drawn away through the jet, when the needle valve reopens as the fuel level falls to admit more fuel. The height of fuel in the chamber is arranged to be slightly below the top of the discharge nozzle.

The venturi

This device consists simply of a tube of which the bore diminishes smoothly to a *throat* and then smoothly enlarges to its original size, as illustrated in Fig. 15.4.

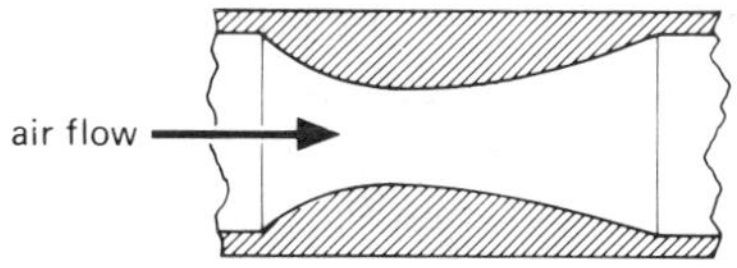

Fig. 15.4 Venturi

A fluid, i.e. a liquid or a gas, flowing along this tube will have to move faster as it passes through the throat than at other points. The result of this is that the pressure of the fluid will be reduced as its velocity increases, and will rise again as the velocity falls.

This can be demonstrated by the apparatus shown in Fig. 15.5. This consists of a series of glass tubes which dip into a trough of coloured water. A blower (like a vacuum cleaner) is used to pass air through the venturi and the level of water in the tubes is noted when the blower is switched on. When there is no air flowing through the venturi, the level in all tubes is equal, but when air is moving through the venturi, the level in tube B will be higher than that shown in the other tubes. This indicates that the air pressure in the throat of the venturi is *less* than that which exists on each side of the restriction. Increasing the air speed through the venturi causes a greater drop in air pressure; in a similar way if the throat of the venturi is made smaller, the drop in pressure for a given air speed will be greater. Whenever the air pressure is lowered below atmospheric pressure, the term *depression* can be used, so in this case *a venturi increases the air speed and forms a depression*.

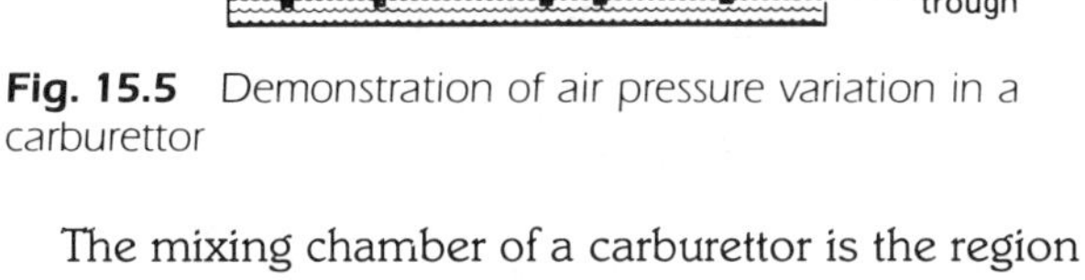

Fig. 15.5 Demonstration of air pressure variation in a carburettor

The mixing chamber of a carburettor is the region centred around the venturi; it is situated in the induction pipe that supplies the engine with air. Fitted at the throat of the venturi, which is also known as the *choke tube*, is an outlet from a pipe or drilling that conveys petrol from the float chamber.

The combination of the depression created by the

venturi and the petrol supply from the float chamber allows the petrol to mix with the air. Since the depression intensifies as the quantity of air passing through the induction pipe increases, the amount of petrol supplied to mix with this air will also increase.

The complete carburettor

Fig. 15.6 shows a simple carburettor. The only component not so far mentioned is the *throttle valve*, the purpose of which is to regulate the engine power by controlling the air flow to the engine. The action of the carburettor is as follows.

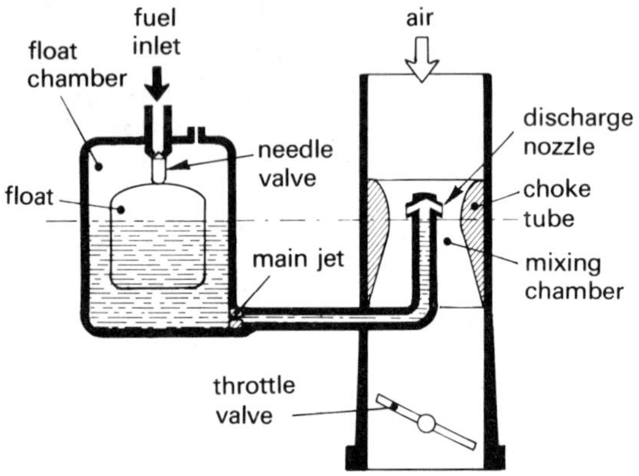

Fig. 15.6 A simple carburettor

When the fuel supply is turned on, fuel enters the float chamber through the inlet connection, and rises in both the float chamber and also the fuel discharge nozzle. As the level rises, the float also rises, and lifts the needle into the needle seating. This cuts off the entry of fuel when the level is just below the top of the discharge nozzle. If the engine is rotated, air is drawn in through the air intake and choke tube, producing a pressure drop or depression inside the choke tube. This depression draws fuel from the float chamber via the passage and jet into the mixing chamber. The rush of air through the choke tube will, if its velocity is great enough, atomize the fuel as it issues from the discharge nozzle. The mixture of air and atomized fuel will be drawn into the engine at a rate which depends upon engine speed and the extent of opening of the throttle valve. The driver controls the throttle by the accelerator pedal.

As the rate of air flow increases, because of either an increase in engine speed or a wider throttle opening or both, the depression in the choke tube also increases, thus drawing more fuel from the jet.

The size of the choke tube is selected so that the air velocity through it is sufficient to atomize the fuel at the lowest speed at which the engine is required to run. The desired mixture can be obtained by using a size of jet which allows the correct amount of fuel to flow and join the air stream.

Attitude of choke tube

In Fig. 15.6 the axis of the choke tube is vertical. It does not matter whether the choke tube is horizontal, vertical, or at some intermediate angle (so long as the float chamber remains vertical) and carburettors are made with the choke tube in any of these attitudes, as shown in Fig. 15.7.

The downdraught arrangement is most common on modern engines, for two main reasons: (1) the flow of mixture into the manifold is assisted by gravity, and (2) it enables the carburettor to be fitted in a very accessible position on top of the engine.

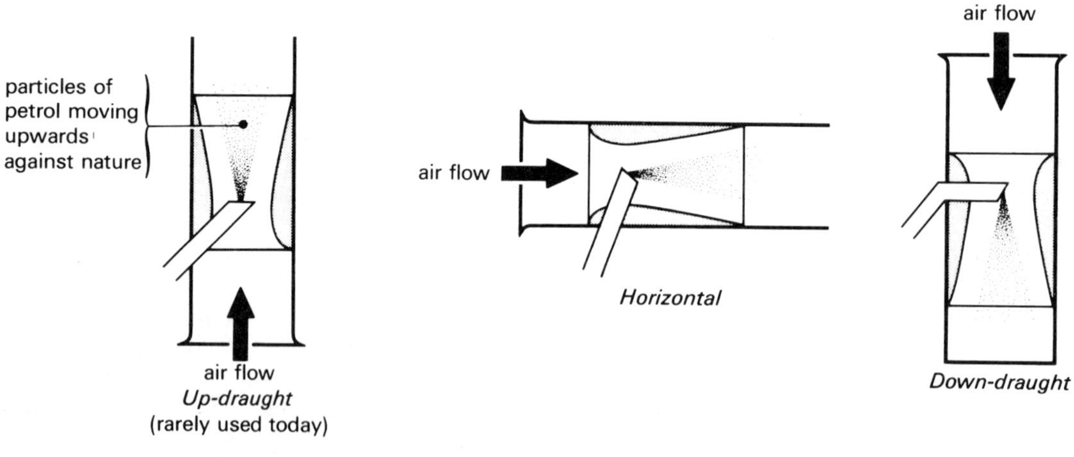

Fig. 15.7 Attitude of choke tube

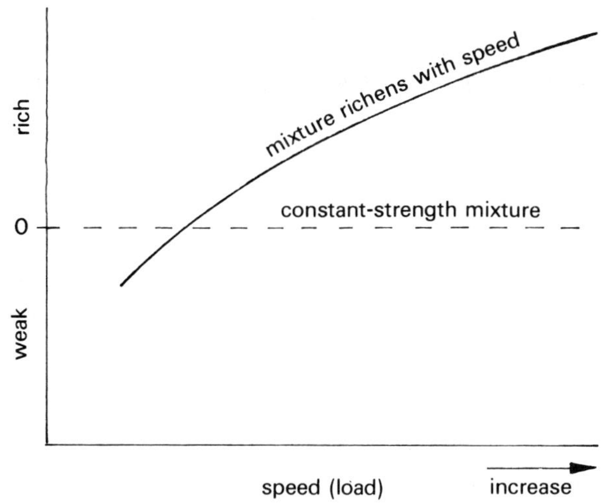

Fig. 15.8 Mixture variation with simple carburettor

Variation of air/fuel ratio

The simple carburettor shown in Fig. 15.6 would be suitable for a fixed-speed engine that operated against a constant load, but these conditions are quite different from those experienced by an engine fitted to a motor vehicle. When the load and/or speed of an engine fitted with a simple carburettor is increased, the mixture strength supplied by the carburettor becomes richer, i.e. the air/fuel ratio decreases as the speed or load is increased (Fig. 15.8). This problem is overcome either by incorporating a *compensation system* or by using a carburettor design which uses the *constant depression* principle. The difference in construction that is used to overcome the mixture variation problem enables carburettors to be classified as:

- *constant-choke*
- *constant-depression*.

As these names suggest, the former has a fixed size of venturi and incorporates a compensation system to maintain a near-constant air/fuel ratio. A constant-depression type has a variable choke which enlarges as the engine speed and load increases. Externally the shape is quite different, so identification is simple.

16 Constant-choke carburettors

This type of carburettor is also known as a *fixed-choke*, or *variable-depression* type because the depression in the fixed size venturi varies when the engine speed changes. This variation must be balanced by the fuel flow if weakness or richness of the mixture is to be avoided.

Need for mixture compensation

The change of mixture strength supplied by a simple carburettor is due to the fact that the carburettor is attempting to meter two different fluids, air and petrol. Air is a gas which flows very easy, whereas petrol is a liquid that opposes flow. In addition, petrol has a tendency for its particles to stick together as well as to the walls of the passage through which it flows.

By selecting suitable sizes for the choke tube and jet, it is possible to obtain a chemically correct mixture strength to suit a set engine speed, but when this speed is increased, the subsequent increase in the depression would promote an excessive petrol flow. This enrichment with increase in speed and/or load is prevented by incorporating a compensation system in a carburettor.

Compensation systems

A *compensation* or *correction* system is necessary to maintain a near-constant air/fuel ratio. Careful matching of the jets to the system is necessary to avoid *under-correction* or *over-correction*; the former implies that the carburettor becomes richer, and the latter weaker as the speed or load increases.

Air bleed compensation

This method of correction is the most common system in use today. The constructional details of the system vary between different makes, but the principle remains the same.

The arrangement shown in Fig. 16.1 consists of an air jet that bleeds air into an *emulsion*, or *diffuser tube*, in the side of which are drilled a number of small holes. Petrol is supplied to the well via a main jet and the outlet from the well is situated in the venturi.

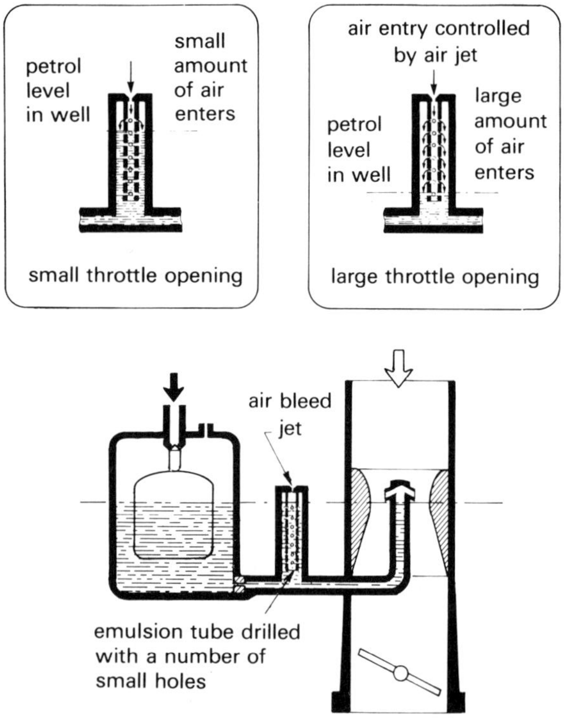

Fig. 16.1 Air bleed correction

When the main jet is not in use, the petrol levels in the well and float chamber are similar, but when the venturi depression is sufficient to cause fuel to discharge from the outlet, the level in the well falls. This drop in level increases as the throttle is opened and when full throttle is reached the fuel level in the well is very low.

This opening of the throttle would cause a simple

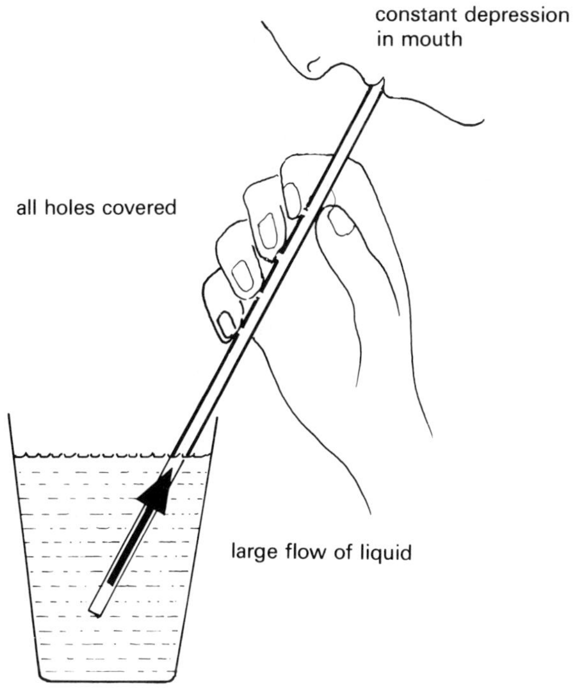

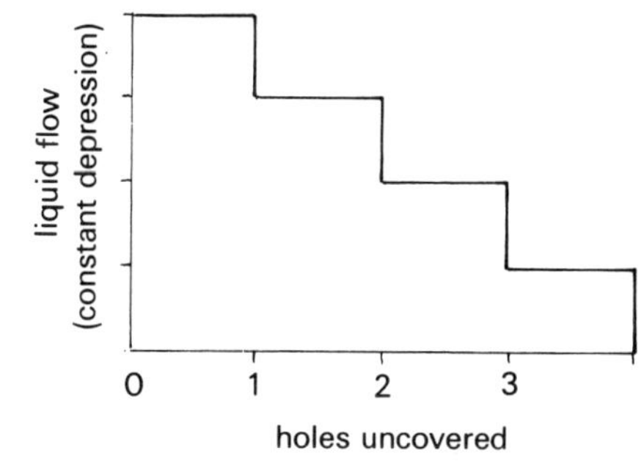

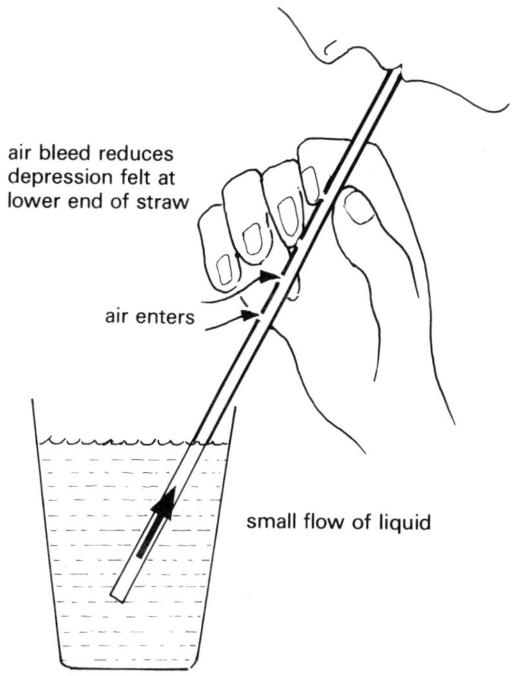

Fig. 16.2 Effect of air bleed

carburettor to supply a mixture which becomes richer and richer as the speed increases. A correction system overcomes this problem by using the drop in fuel level in the well to expose air holes in the emulsion tube. When the mixture strength shows a sign of becoming rich, a hole is uncovered and air is bled into the system. This reduces the depression which is felt on the main jet and, as a result, restores the correct mixture strength.

The action of this type of compensation system is similar to the glass and straw experiment shown in Fig. 16.2. With all bleed holes covered, the creation of a slight suction (depression) in the mouth causes fluid to move up the straw. Intensifying the depression increases the fluid flow. At this point if one hole is uncovered the flow will decrease; the flow will decrease even further if more holes are uncovered. It is possible to adjust the flow of fluid, by varying the number of holes that are uncovered.

Tuning the correction system to give a near-constant air/fuel ratio is achieved by altering the hole size and spacing in the emulsion tube to suit a given size of main jet. Some systems have an air bleed jet (sometimes named the air correction jet) which is detachable to enable its size to be changed. This air jet controls the maximum entry of air and is only effective at large throttle openings, because at smaller throttle openings the exposed holes in the emulsion tube control the entry of bleed air.

Slow-running systems

A decreased air flow into the engine results in a reduction in the air velocity at the choke tube, and consequently the pressure at this point rises nearer to atmospheric. If the carburettor is properly corrected this will not alter the mixture strength, but it will result in a progressive coarsening of the spray owing to the reduced velocity of the air flow. Also, as the air flow is reduced, the velocity will eventually become insufficient to maintain the fuel droplets in suspension, owing both to their increased size and the low air velocity in the manifold.

There will, therefore, be a speed below which the engine will not run, and this will depend primarily upon the size of the choke tube and, to a lesser extent, upon the diameter and length of the inlet manifold. The conditions necessary for good low-speed operation are quite unsuitable for developing reasonably high power at high speeds, and there is obviously a limit to the extent to which power at high rev/min can be sacrificed to obtain a good low-speed operation.

The extreme condition of low air velocity through the choke tube occurs when the engine is running but not driving the car, and therefore has merely to develop sufficient power to overcome its own internal friction losses. This is the case whenever the vehicle is stationary but may be required to move at any moment, as, for instance, when halted in traffic or at traffic lights. For economy and comfort, the engine should run slowly and quietly, but respond instantly to the opening of the throttle when the time comes for it to drive the vehicle.

This condition of engine operation is known as *slow running, ticking over* or *idling*. The air flow through the choke tube is then not only too slow to atomize the fuel, but there is insufficient depression in the choke tube to draw fuel from the jets.

To pass the amount of mixture needed to keep the engine running at idling speed, the throttle is barely open, and the air velocity is greatest where it passes through the very small gap around the edge of the throttle. A vacuum gauge will show that the depression in the manifold is at its highest (about 380–450 mm or 15–18 in of mercury below atmospheric).

An arrangement similar to that shown in Fig. 16.3 is used to provide a suitable mixture for slow running. A passage connects the float chamber, via the main jet, to an outlet positioned in the region of the throttle valve. Fuel flow is regulated by a slow-running jet or idling jet, and to ensure that fuel does not flow continually the top of the passage is taken above the fuel level in the float chamber. Siphoning is prevented by having an air bleed at the top of the inverted U of the passage and an adjusting screw adjacent to the throttle enables the volume of emulsified fuel entering the engine to be regulated to suit the condition of the engine. The speed at which the engine runs is determined primarily by the extent of the throttle opening; normally this is set by a *throttle stop adjusting screw*.

When the engine is idling a strong depression exists on the engine side of the throttle, causing fuel to be drawn from the idling system. This fuel mixes with the air spilling past the edge of the throttle.

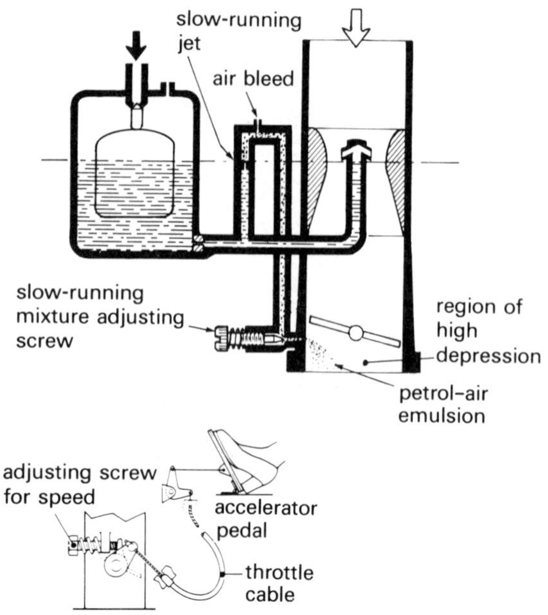

Fig. 16.3 Slow-running system

Mixture strength for slow running

At idling speed only a whiff of mixture is needed to drive the engine over, and under these conditions the cylinder contains a comparatively large quantity of exhaust gas. This dilutes the new charge, so in order to provide an ignitable mixture, a charge that is richer than normal is provided during slow running. Today this enrichment of the mixture is strictly limited because any unburnt, or partially burnt, fuel causes the exhaust to pollute the atmosphere. In view of this, many engines today will not slow-run at a speed less than about 900 rev/min.

Pick-up from idling

Opening the throttle from the idling position causes more air to pass the throttle. This lowers the depression (raises the pressure) and reduces the fuel discharge from the slow-running outlet. At the same time, the air flow through the choke tube causes the main jet system to come gradually into operation. This changeover does not always occur smoothly, with the result that any deficiency in fuel supply is accompanied by a sudden drop in engine power.

To avoid this *transfer flat-spot* the arrangement shown in Fig. 16.4 is used. At a throttle position where the 'flat-spot' would otherwise occur, one or more

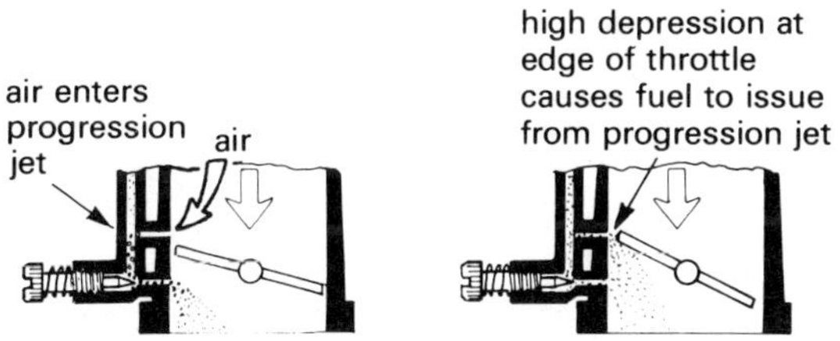

Fig. 16.4 Two-hole idling system

holes are drilled adjacent to the edge of the throttle valve to provide additional fuel discharge outlets. Fuel discharge from the upper hole is controlled by an interchangeable jet called a *progression jet*.

When the engine is idling a high depression on the engine side of the choke causes petrol to discharge from the lower drilling. During this phase the progression hole will be above the throttle, so air will enter this hole, pass through the progression jet and bleed into the slow-running mixture to assist in the emulsifying process. As the throttle is opened, the rush of air past the edge of the throttle will create a depression in this region and will cause fuel to discharge from the progression outlet.

Slow-running adjustment

To meet current regulations, final adjustment of the slow running requires a tachometer to set the engine speed accurately and an exhaust gas analyser to measure the quantity of carbon monoxide (CO) present in the exhaust gas.

The adjustment of the slow running varies with different engines but the main points are as follows:

1 Check that ignition system is serviceable and that the timing is correct.
2 Warm up the engine to its normal running temperature.
3 With the engine running adjust the throttle stop screw to obtain the recommended slow-running speed.
4 Adjust the volume control screw until the highest and smoothest speed is found. Moving the screw outwards enrichens the mixture and when it is moved too far the engine speed surges in a manner called *hunting*.
5 Weaken the mixture from the point where it runs the fastest until the speed drops by the small amount recommended. (Often this is about 100 rev/min.)
6 Reset the throttle stop screw to make the engine slow-run at the recommended speed.

Emission control

To meet the regulations, the slow-running system of a modern carburettor is designed to take into account the following:

1 *Tamper-proof adjusting screws* Screws provided for mixture alteration are arranged so that unqualified persons are deterred from altering the mixture. Adjusting screws are normally hidden in a recessed hole which is sealed by a metal or plastic plug.

2 *Carbon monoxide (CO) content in the exhaust* The mixture for slow running is set weaker than that used in the past. An exhaust gas CO test, taken after the slow running has been set, should show that the CO content is less than the value specified by the manufacturer - a typical maximum value is 1 per cent.

The weaker mixture needed to obtain this low value necessitates other carburettor alterations. These are:

a *Higher slow-running speed* The dilution effect of the residual exhaust gas on a weak slow-running mixture causes the engine to stall at a speed of 500–600 rev/min. To give smooth, consistent engine operation, a slow-running speed in the range 800–900 rev/min is used.
b *Better mixing of the petrol and air* in the slow-running circuit, together with improvements to distribute this emulsified fuel throughout the air which is flowing along the main air tract in the carburettor. The aim is to overcome a 'patchy' mixture. Fuel-rich regions in the combustion chamber give high CO emission, whereas weak mixtures in the vicinity of the sparking plug cause misfiring; this means that the complete charge for that stroke is exhausted and high hydrocarbon (CH) pollution results.
c *Slow-running throttle by-pass* A reduction in the richness of the slow-running mixture makes the engine very sensitive to slight alterations in the throttle position. A small amount of wear in the throttle spindle causes the butterfly valve to vary its closed position. Consequently, the engine may either race or stall when the throttle is released.

This problem can be overcome by making the valve close fully at 'slow running' and providing a separate throttle by-pass duct (Fig. 16.5). Slow-running speed is controlled by a screw which governs the *quantity* of mixture being supplied to the engine.

Unless carefully designed, this system is prone to throttle seizure when the carburettor cools after strenuous use of the engine.

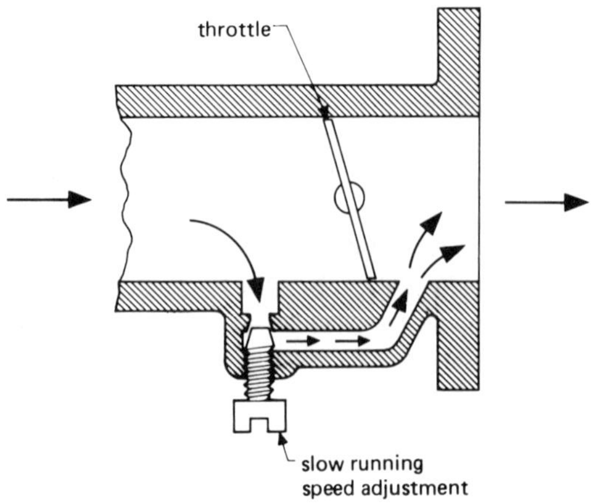

Fig. 16.5 Throttle by-pass

3 *Running on* When the ignition is switched off the driver expects the engine to stop, but often the engine continues to run very erratically for a considerable time or until it is stalled by the driver. During this *running-on* period, combustion is initiated by some part of the combustion chamber, such as a valve or sparking plug electrode, that is hotter than normal. This extraordinary high temperature is often produced when the engine is operated on a weak mixture as needed to meet the emission regulations.

Various arrangements are used to overcome this problem; the method shown in Fig. 16.6 is an electric solenoid which cuts off the slow-running mixture when the ignition is switched-off. This system is called an *anti-run-on* valve or *anti-dieselling valve*.

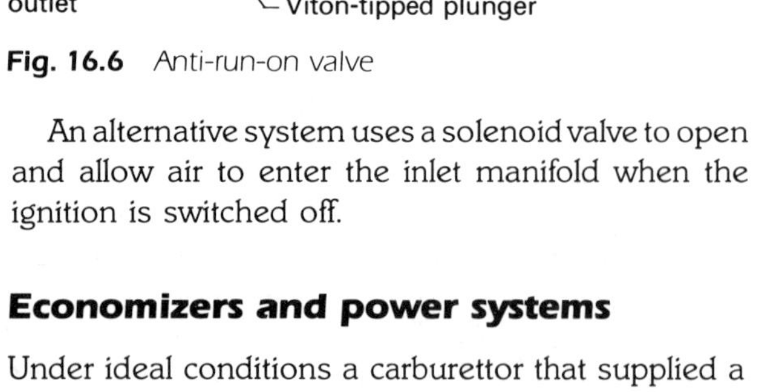

Fig. 16.6 Anti-run-on valve

An alternative system uses a solenoid valve to open and allow air to enter the inlet manifold when the ignition is switched off.

Economizers and power systems

Under ideal conditions a carburettor that supplied a chemically correct mixture (15 : 1) would cause the engine to produce maximum power and economy at all speeds and loads. In practice the physical problems associated with distributing the correct amount of fully vaporized fuel to each cylinder makes it impossible to achieve this ideal. Even if it could be achieved, the difficulty of bringing each particle of fuel into intimate contact with the correct amount of oxygen needed to burn it completely to carbon dioxide (CO_2) and water (H_2O) makes the ideal remote from reality.

Varying the mixture strength supplied to an engine that is operating under load at a fixed speed causes the engine performance to change as shown in Fig. 16.7. This graph plots the 'fuel used per unit of power' against the 'torque output'.

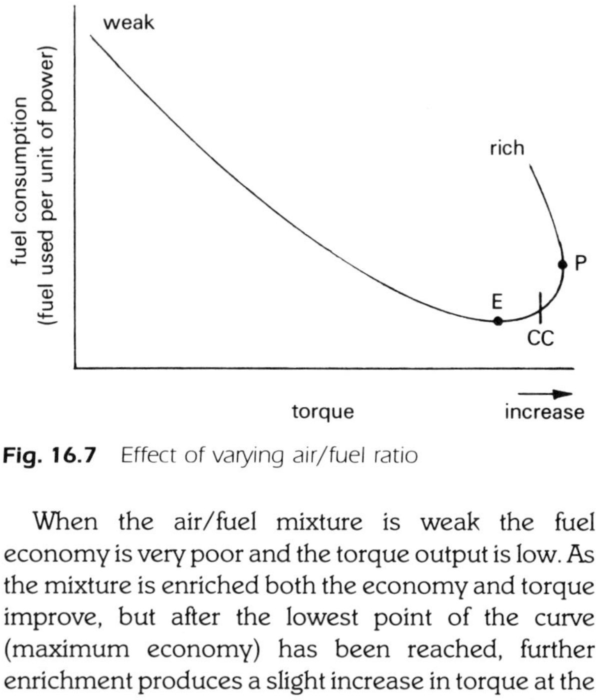

Fig. 16.7 Effect of varying air/fuel ratio

When the air/fuel mixture is weak the fuel economy is very poor and the torque output is low. As the mixture is enriched both the economy and torque improve, but after the lowest point of the curve (maximum economy) has been reached, further enrichment produces a slight increase in torque at the expense of economy. Continuing to enrich the mixture past the maximum torque point produces a drop in torque and a considerable rise in fuel consumption. Three points on this graph should be noted:

1 The chemically correct (CC) mixture strength gives neither maximum economy nor maximum power.
2 Maximum power (P) requires a mixture that is slightly rich.
3 Maximum economy (E) requires a mixture that is slightly weak.

In order to achieve both maximum power and greatest economy the mixture strength supplied by the carburettor must be varied to suit the operating conditions, i.e. it must provide a slightly rich mixture for power and slightly weak mixture for economy. Emission regulations normally prevent the use of mixtures richer than 15 : 1, because they do not fully burn during combustion, so the engine power output at 15 : 1 is the maximum and the operating mixture range is on the weak side of this value.

The choke cannot distinguish between the high load (power) and cruising (economy) conditions, so an alternative sensing source must be used. This is either the movement of the throttle or the pressure in the inlet manifold. Since the cruising and high-load conditions give a manifold depression that is high and low respectively, the variation in the manifold pressure is used by many systems to sense the load acting against the engine.

The devices used in a carburettor to vary the mixture strength to suit either the power or cruising conditions are called *economizers* or *power systems*.

In all cases the weak cruising mixture is obtained by reducing the petrol flow through the main jet system. This can be achieved either by using a smaller jet or by reducing the pressure difference across the jet.

Restoration of the air/fuel ratio to 15 : 1 for power is achieved either by providing an additional path from the float chamber to the choke, or by exposing the main jet to the normal pressure difference.

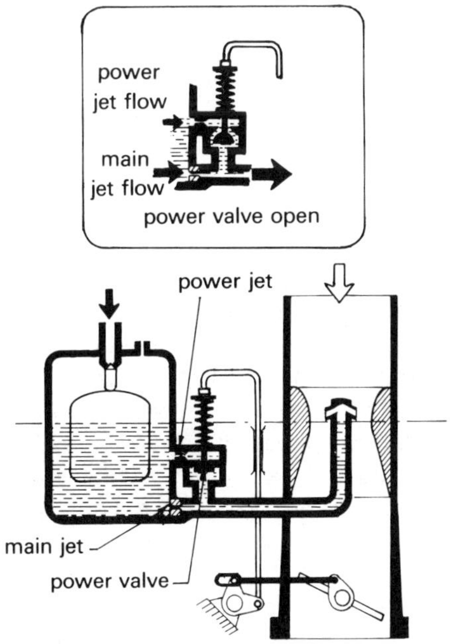

Fig. 16.8 Power valve – mechanical

Fig. 16.8 shows a mechanically operated power valve. The smaller-than-normal main jet provides the cruising mixture. The throttle-controlled power valve, which comes into operation when the throttle is opened in excess of about 75 per cent, supplies extra fuel for power. In the high-load throttle range, the petrol flow through both the power and main jets gives a suitable mixture to enable the engine to develop its maximum power.

Vacuum control of the power valve can be achieved by using a vacuum chamber diaphragm to act on a rod which, in turn, controls the valve. Alternatively the system can be arranged as shown in Fig. 16.9. The diaphragm in this arrangement acts directly on the power valve which can be made to control either the air bled into the system or the petrol flow as shown in Fig. 16.9.

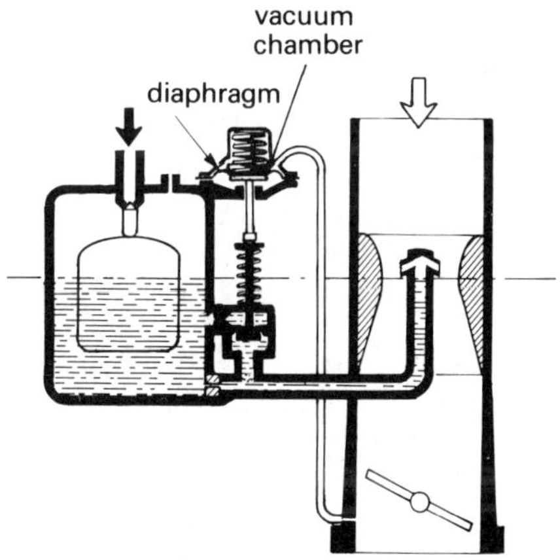

Fig. 16.9 Power valve – vacuum-operated

Failure of a power system is generally due to the diaphragm becoming torn. This prevents the mixture from becoming weak during cruising, so high fuel consumption results. Also the torn diaphragm allows air to enter the manifold via the sensing passage; this causes unstable slow running.

Acceleration pumps

Whenever the throttle is plunged open, there is a weakening of the mixture. One of the reasons for this is the delay caused by the reluctance of the fuel to respond compared with the fast action of the air. If the carburettor is already delivering a slightly weakened mixture to give maximum economy, the further weakening that occurs when the throttle is opened causes a delay before the engine responds; in some cases it will cause the engine to misfire.

Whereas a carburettor without an economy system has small reserves of fuel on the outlet side of the main jet to allow for sudden increases in speed or load, an economy type carburettor generally needs an *acceleration pump*. This delivers fuel into the mixing chamber when the throttle is plunged open.

There are two main types of acceleration pump: mechanical and vacuum-operated.

Mechanical type

The mechanical pump shown in Fig. 16.10 is controlled by a linkage connected to the throttle. This linkage acts on a diaphragm which draws fuel via a one-way valve from the float chamber and pumps it past a delivery valve fitted in the outlet passage. A separate jet, or restriction in the passageway, regulates the flow of fuel delivered by the pump.

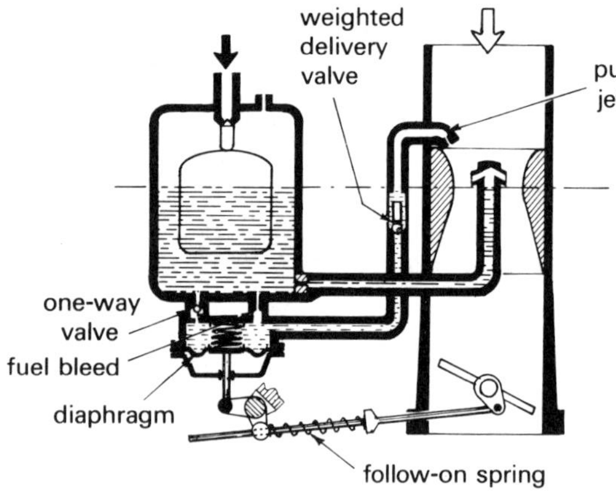

Fig. 16.10 Acceleration pump – mechanical

When the fuel outlet pipe terminates in a region where the pressure is lower than atmospheric, there is a risk that fuel will discharge continuously, so a device such as a weighted valve, or air bleed into the outlet passage, is used to prevent this flow.

The pump must not discharge when the throttle is moved slowly. In Fig. 16.10 this is achieved by having a bleed to allow fuel to by-pass the inlet valve of the acceleration pump when the throttle is opened slowly.

Most pump designs use a spring in the pump-operating linkage to increase the time that the fuel in the pump chamber is being discharged. This *follow-on* action also improves exhaust emission and economy.

Vacuum type

The pump shown in Fig. 16.11 operates by using the sudden change in manifold pressure that occurs when the engine is accelerated. The pumping action is produced by the diaphragm spring; this is compressed when a high depression acts in the vacuum chamber. A sudden collapse of this depression allows the spring to discharge fuel into the mixing chamber.

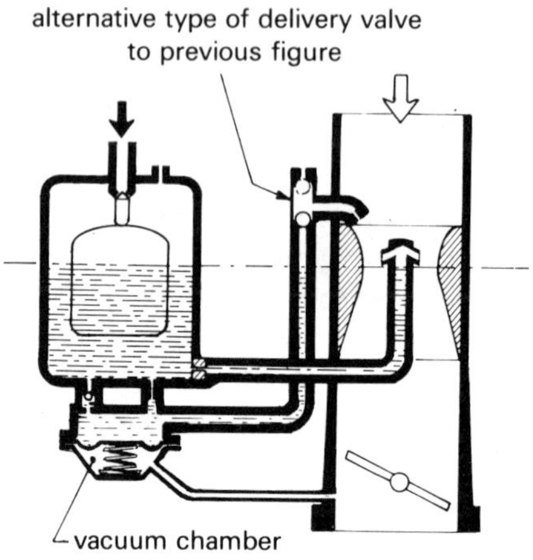

Fig. 16.11 Acceleration pump – vacuum-operated

Cold starting

Special provisions are necessary to start a cold engine. This is because the cranking speed of the engine is insufficient to allow the venturi to create an adequate depression. In addition, the slow-moving air entering the engine causes a large amount of fuel to be deposited on the walls of the inlet manifold. Since the manifold and cylinders are cold very little vaporization of the fuel takes place, so ignition of the charge under these conditions is very difficult.

Petrol consists of many different hydrocarbons (CH) and the fractions (groups of CH) that make up the fuel have different boiling points in the range 85 to 220 °C. A light-fraction fuel boils (vaporizes) at a low temperature whereas a heavy-fraction fuel does not boil until the temperature approaches 220 °C.

The provision of extra fuel during cold starting ensures there is an adequate supply of light-fraction fuels that are able to vaporize in the cold engine. If the quantity of gas formed by this vaporization is sufficient, it is possible to ignite the gas and produce enough heat to drive the engine over, even though the cold oil is causing a large drag.

The petrol supplier helps to overcome the cold-starting problem by providing a fuel for the winter period that incorporates a higher proportion of light fractions. This blend has to be changed in the summer because highly volatile fuels cause vapour locks in the fuel lines when exposed to summer heat.

Cold-starting systems used today can be divided into two main types:

- *Strangler*
- *Separate starting carburettor.*

Strangler

Fig. 16.12 shows the principle of a strangler. It consists of a flap valve which is positioned at the air entry point of the carburettor. When the valve is closed, the strangulation of the air supply intensifies the depression felt at the venturi. As a result, extra fuel is supplied to provide a very rich mixture in the region of 8 : 1.

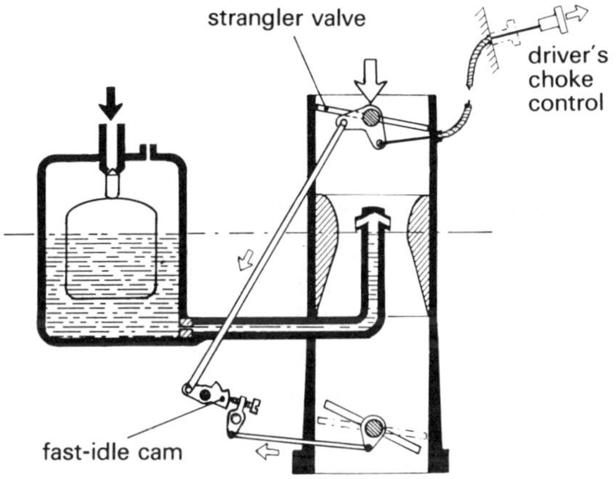

Fig. 16.12 Simple strangler system

Once the engine has started, the richness of the mixture must be reduced to a point where the engine runs smoothly. This reduces the risk of rapid cylinder wear caused by neat fuel washing the oil film from the top part of the cylinders: also excessive fuel 'wets' or 'fouls' the sparking plug and stops the plug functioning in the normal manner.

For historical reasons the strangler is sometimes called the 'choke'. That is the reason why the name is occasionally marked on the driver's control knob. This control connects with a cable that acts on the strangler valve; the first movement of the 'choke' control opens the throttle a small amount to give a *fast idle* to allow for the extra drag on the engine when it is cold.

The driver must not 'over-choke' the engine, because this action 'floods' the engine with fuel vapour and starves it of air. Combustion cannot occur under these conditions and this becomes apparent when a driver attempts to start a hot engine after it has not been used for a few minutes. During this time the fuel in the manifold vaporizes and drives out the air, therefore when the starter is operated, the engine will not fire until sufficient air has been induced. If the driver misreads the temperature condition and pulls out the 'choke', the result will be that the engine will be 'flooded' and most probably the plugs will be fouled. In order to clear the overchoked condition, the driver should push in the 'choke' and then keep the throttle fully depressed as the starter is operated. This action will allow the engine to start after a few seconds unless the sparking plugs have been 'flooded' with petrol.

To minimize the problems associated with 'over-choking', the strangler normally incorporates some arrangement to weaken the mixture as soon as the engine starts. Fig. 16.13 shows an offset strangler valve that allows the moving air to act on the valve and partially open it after the engine has started.

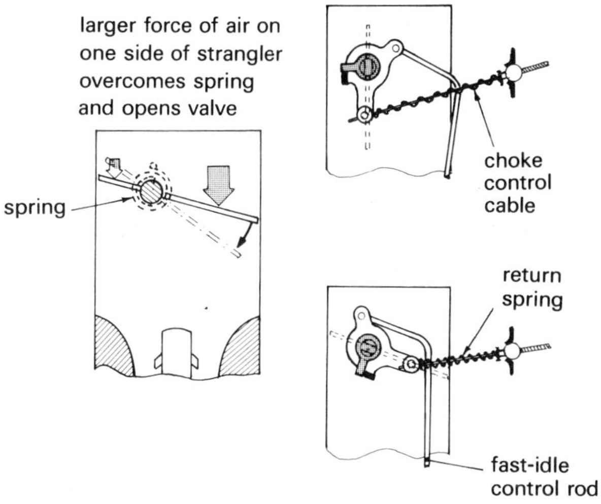

Fig. 16.13 Offset strangler

Automatic strangler Incorrect use of the 'choke' by the driver, especially the delay in returning it to the 'off' position, causes emission problems as well as those drawbacks previously outlined. To overcome these problems, the automatic choke was introduced.

Fig. 16.14 shows the principle of one system; this uses a bimetallic strip to sense the temperature of the coolant. When the engine temperature is less than

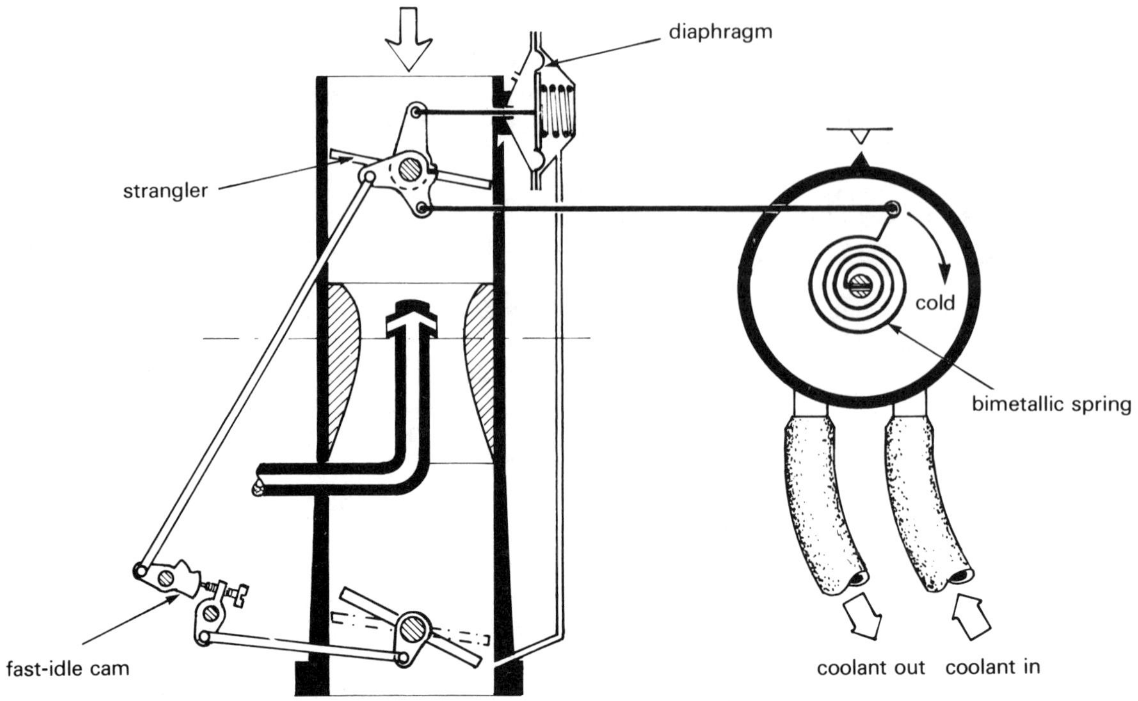

Fig. 16.14 Automatic choke

about 15 °C the bimetal strip pulls the strangler to the closed position.

In the type shown, the driver should depress and then release the throttle prior to starting the engine. This allows the strangler and fast-idle cam to take up a position appropriate to the engine temperature.

After the engine has started, the strangler is partly opened by the diaphragm in the vacuum chamber; this is activated by the high depression created in the inlet manifold.

As the engine warms up, the rise in coolant temperature causes the bimetallic strip gradually to reduce its pull on the strangler. This takes place over a period of time, so both the richness and the fast-idle action are slowly changed to suit the engine temperature.

Normally two adjustments are provided:

1 The wind-up of the bimetallic strip governs the temperature at which the system comes into operation. Often marks are provided on the housing to suit the engine and climatic conditions under which it operates.
2 A fast-idle adjusting screw controls the engine speed during 'choke' operation. This screw should be clear of the cam when the engine is warm.

Separate starting carburettor

This system may be regarded as a separate carburettor attached to the main carburettor, therefore it incorporates its own jets and mixing chamber. This enables the cold-starting air/fuel ratio to be accurately controlled and the fuel and air to be well mixed to suit the conditions. The system may be operated by manual or automatic means.

Fig. 16.15 shows the layout of a starting carburettor that is brought into action by means of a 'choke' cable which rotates a disc (1). When the cable is pulled, the disc rotates and aligns a small orifice in the disc with a fuel passage connected to the float chamber. This movement also allows a large enlongated hole at the bottom of the disc to uncover an outlet passage that leads to the main air intake on the manifold side of the throttle.

Providing the throttle is kept closed when the 'choke' is operated, the rotation of the engine by the starter motor causes the high depression in the manifold to be felt in the chamber (2). This depression causes air to enter the chamber by the bleed (3) and fuel to be drawn from the passage (6) after the fuel has been metered by the starter jet (4). When the system is not in use, a small quantity of fuel is held in the

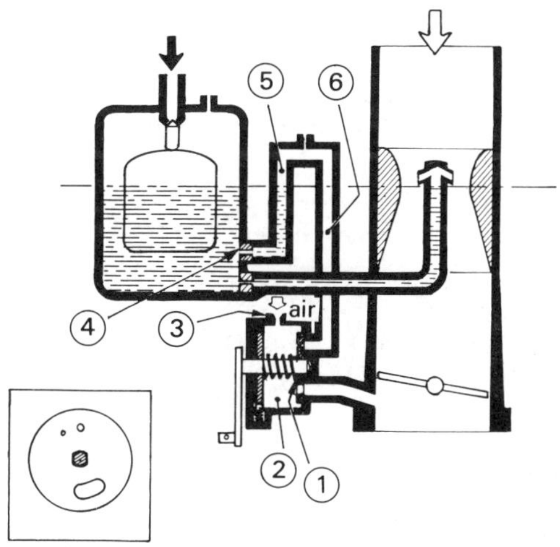

control disc 1

Fig. 16.15 Starting carburettor

reserve well (5), so this fuel gives extra richness when the system is initially brought into use.

After the engine has started, an opening of the throttle by the driver reduces the depression in the chamber (2); this decreases the quantity of rich mixture supplied by the starting carburettor. As the engine warms up, the mixture has to be weakened. This is achieved by rotating the disc a small amount to align a smaller orifice in the disc with the fuel supply passage.

When a rich mixture is no longer required, the system is put out of use by pushing in the 'choke' to the 'off' position. This action rotates the disc and blanks off the fuel and outlet ports.

There are many variations of this system. Some designs use a cylindrical valve instead of a disc valve; in this configuration it is often operated automatically.

Modern carburettor constructional features

The importance of emission control has meant that a modern carburettor must supply air and finely atomized fuel evenly distributed in the proportion needed to suit the engine operating conditions. These requirements can be met fairly easily if the engine operates at one constant speed, but wide speed limits coupled with other variable factors such as engine load, engine temperature, changes in atmospheric temperature, pressure and humidity, etc., make the carburettor's task very difficult.

Single fixed-choke carburettor

This type has been described in detail in the previous sections. Alterations during the past few years include:

Inside air An air vent to atmosphere (Fig. 16.16) allows dirt to enter and vapour to escape from the float chamber, so a transfer passage (1) between the 'engine side of the filter' to the float chamber is provided. Any slight restriction of air flow through the filter does not affect the air/fuel ratio provided that the joint (2) makes a good seal.

Auxiliary venturi Placing a small venturi in a position such that its outlet is in the waist of the main venturi causes the fuel to mix with the air in two stages. Petrol joining the air flowing through the auxiliary venturi gives a petrol-rich initial mix, but on meeting the main air flow at the auxiliary venturi outlet the fuel is distributed evenly throughout the air mass and a constant air/fuel ratio is obtained.

Twin-choke carburettor

Engines such as the V6 and V8 normally require two carburettors. Cost and space are saved when the two carburettors are joined together to share a common float chamber.

Fig. 16.17 shows a diagram of a dual-venturi down-draught unit. Normal practice is to arrange the manifold so that one venturi system supplies one set of cylinders and the other venturi the remaining cylinders. This layout shows why each venturi needs to be provided with its own system for cold starting, slow running, cruising and power.

The two throttle valves are interconnected and should be synchronized to give the same opening at all engine speeds.

Twin-barrel sequential carburettor

To obtain fuel economy, good atomization of the fuel is needed. In a fixed-choke carburettor this can be achieved by using a comparatively small-diameter venturi. Unfortunately the restriction to air flow of a small venturi gives a poor volumetric efficiency and results in low maximum power.

When both economy and high maximum power are required, a multi-barrel sequential type carburettor is often fitted.

Fig. 16.18 shows a diagram of this type. The two throttle valves are coupled together in such a way that the secondary throttle remains closed until the primary throttle has opened about two-thirds. When

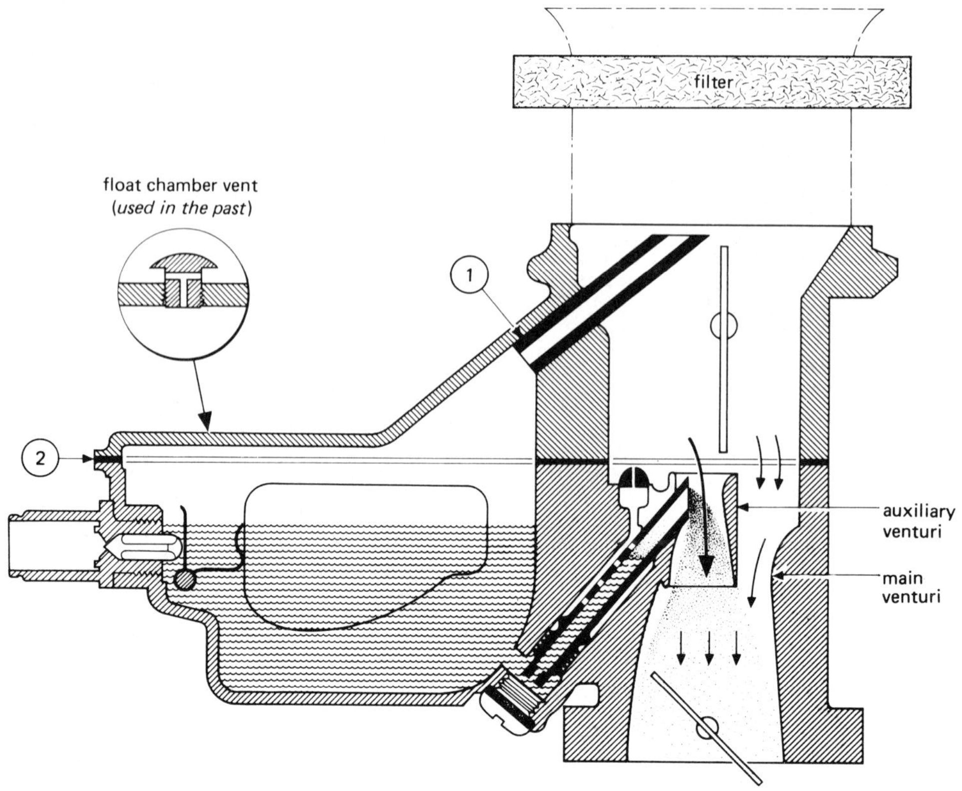

Fig. 16.16 'Inside-air' feature and auxiliary venturi

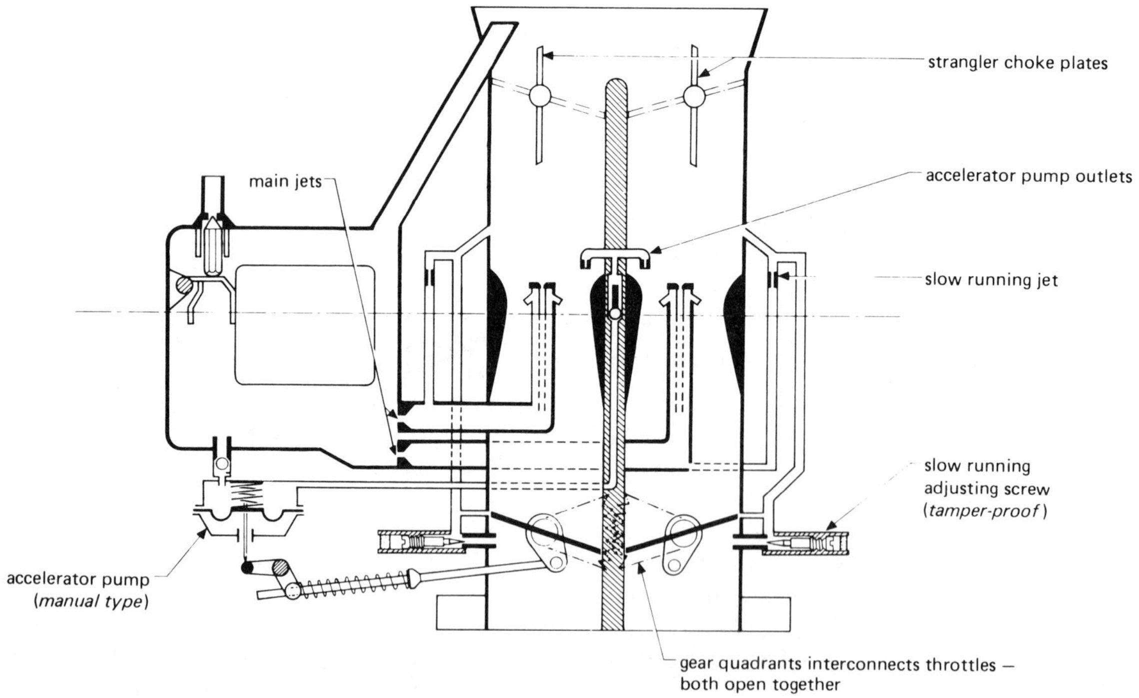

Fig. 16.17 Twin-choke carburettor

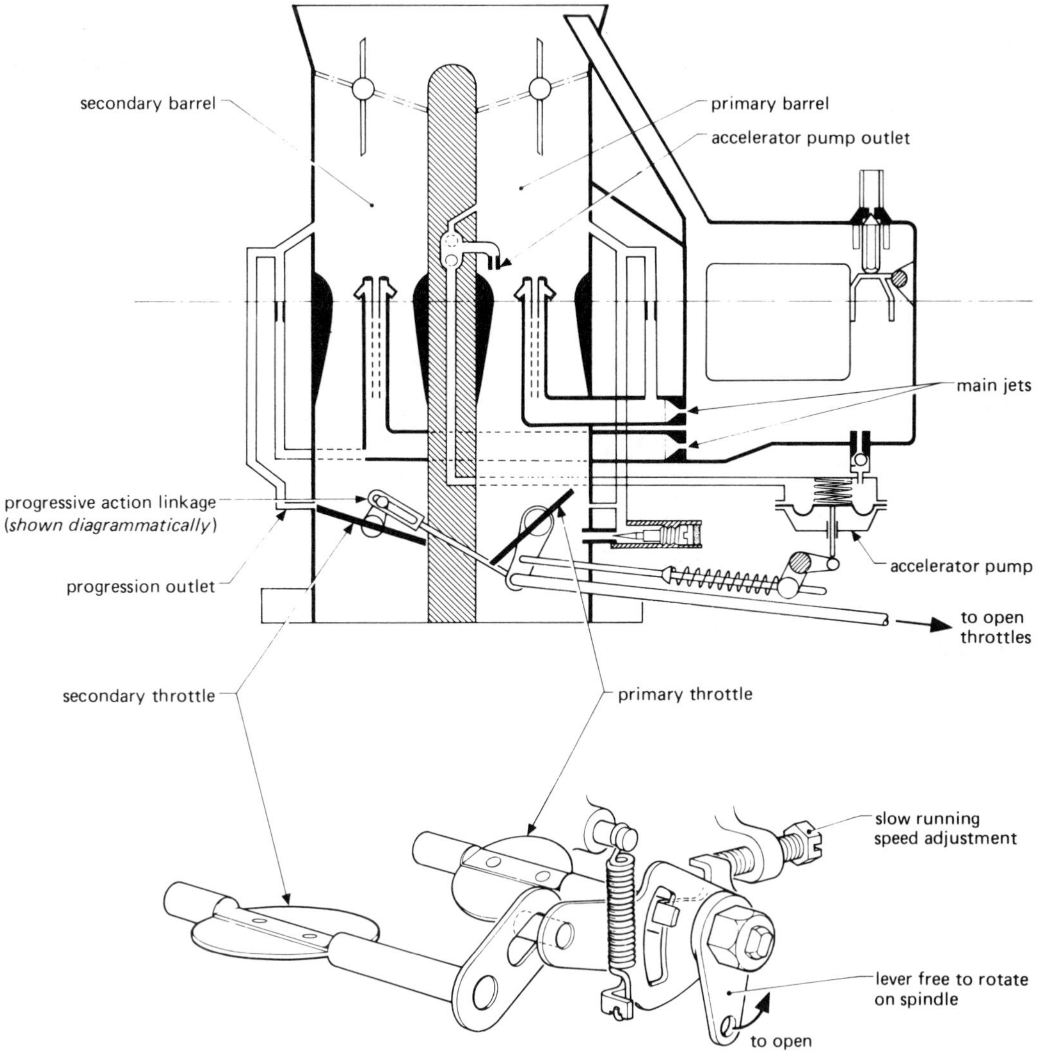

Fig. 16.18 Sequential twin-barrel carburettor

the throttle opens past this point, the linkage moves the secondary throttle so that both throttles reach the fully open position together. This sequential movement allows the primary barrel to supply all cylinders with a slightly weak, well-atomized mixture for cruising conditions. When power is required, the secondary barrel provides extra fuel and a good supply of air to meet power demands.

On some carburettors an auxiliary butterfly valve is fitted in the secondary barrel. Operated automatically, this valve closes when both throttles are open and the air speed through the venturi is low.

17 Constant-depression carburettors

Constant-choke limitations

The varying depression which acts on the jet of a constant-choke carburettor makes it necessary to fit some form of compensating device to prevent mixture enrichment with increase in engine speed. Also the size of the venturi is a compromise that gives neither maximum economy nor maximum power. Carburettors working on a constant-depression principle do not suffer these drawbacks because they have a variable choke area which alters to keep the *choke air speed constant*. This allows the air speed to be set to that required for good atomization over the full engine speed range. Also the constant depression over the jet overcomes the need for a compensating system.

Principle of the constant-depression carburettor

Fig. 17.1 shows the basic construction of a constant-depression carburettor which is also known as *constant vacuum, variable choke* and *variable venturi*.

The venturi or choke is formed by a movable piston that alters the choke area to suit the quantity of air being supplied to the engine. An air vent maintains atmospheric pressure in the space below the piston and a communicating passage transfers the depression from the mixing chamber (the space between the choke and throttle) to the space above the piston.

When air is flowing through the carburettor there is a difference in pressure between the air intake and the mixing chamber. This difference in pressure acts on the piston and will give an upward force to oppose the downward force caused by the piston weight and light spring. When the upward force due to air pressure difference is increased by opening the throttle, the piston will rise and the choke area will enlarge. Similarly, when the throttle closes, the mixing chamber depression reduces, the piston falls and the choke area decreases.

Whatever the air flow, the piston assembly always takes up a position that maintains a constant air speed through the choke to ensure that the petrol jet is acted upon by a constant depression - hence the name given to this type of carburettor.

As a tapered needle is attached to the piston, the rise or fall of the piston will vary the effective area of the petrol jet. By altering the taper of the needle it is possible for the manufacturer to vary the fuel flow to suit the quantity of air being supplied at any speed.

SU type

Fig. 17.2 shows a section through an SU constant-vacuum type set in the cruising position. The main features of its operation are:

Slow running The piston will be lifted very slightly from the lowered position and the depression caused by the air rushing over the petrol jet will give a small fuel supply to suit the conditions. A separate slow-running jet system is unnecessary unless emission restrictions are strict.

Throttle opening As the throttle is opened, the mixing chamber depression is increased; this causes the piston to rise to a point where the mixing chamber depression is just sufficient to support the piston. The higher the piston moves, the larger the area of the jet and the greater will be the flow of petrol.

Cold starting A lever, operated by a cable control, lowers the fuel jet in relation to the needle. This enlarges the jet opening and increases the amount of fuel which is mixed with the air. The initial movement of the control cable slightly opens the throttle to give 'fast idle' action.

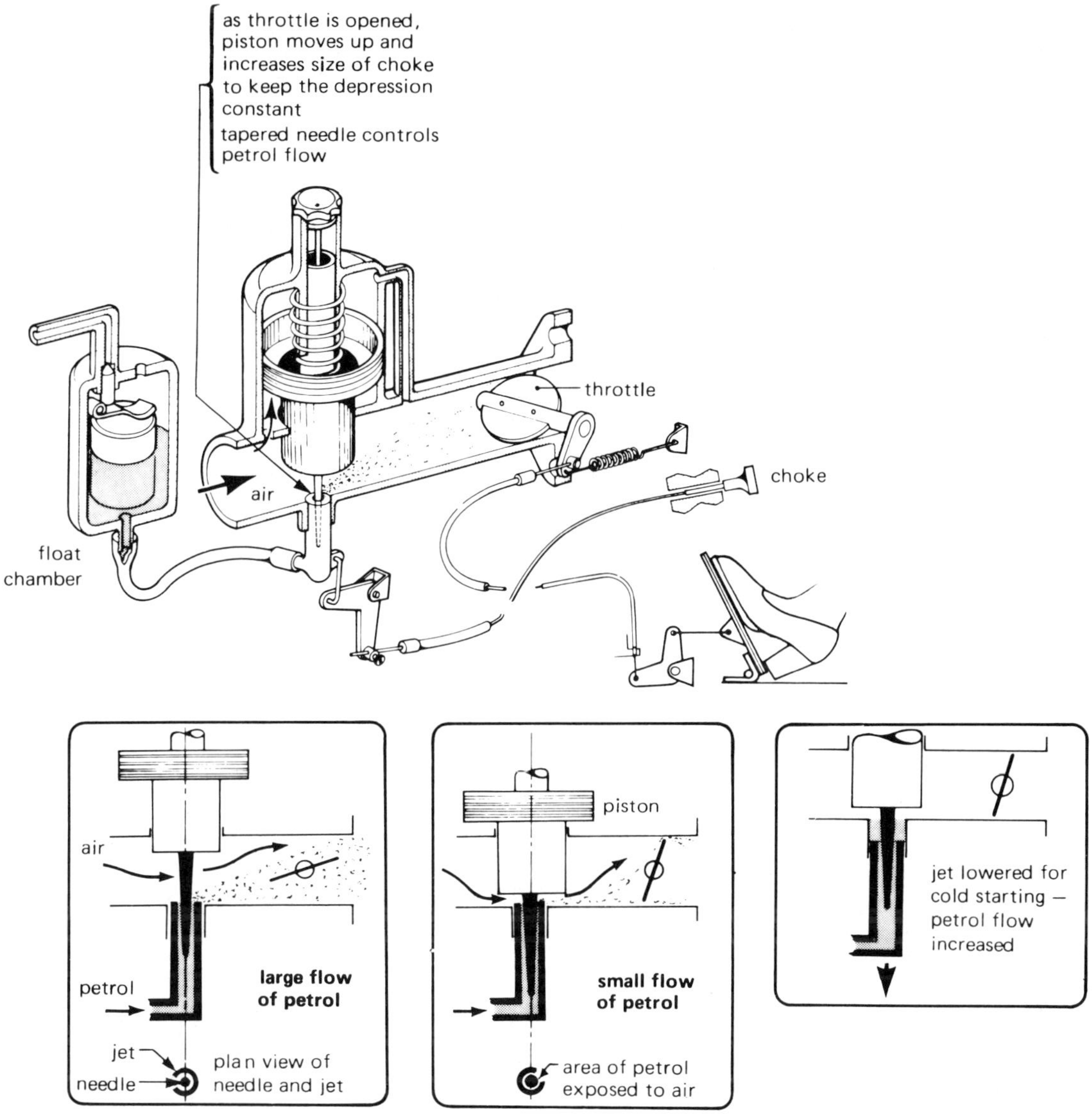

Fig. 17.1 Principle of the constant-depression carburettor

Mixture adjustment An adjusting nut acts as a stop to limit the upward movement of the jet. Unscrewing the nut lowers the jet and richens the mixture throughout the entire speed and load range of the engine.

Hydraulic damper restricts the rate of upward movement of the piston during acceleration and so gives a slight enrichment of the mixture. The damper also reduces piston flutter caused by the irregular flow of air through the induction systems.

Over-run limiting valve is a spring-loaded valve mounted in the throttle butterfly. When the engine over-runs (i.e. road wheels drive the engine) with the throttle closed, the valve opens to supply a fuel mixture which reduces the exhaust pollution.

Zenith–Stromberg constant-depression (CD) type

Fig. 17.3 shows the construction of this type. Whereas the SU type uses a piston, this carburettor has a synthetic rubber diaphragm.

Ford variable-venturi (VV) type

This addition to the range of variable-venturi carburettors has many interesting features which are

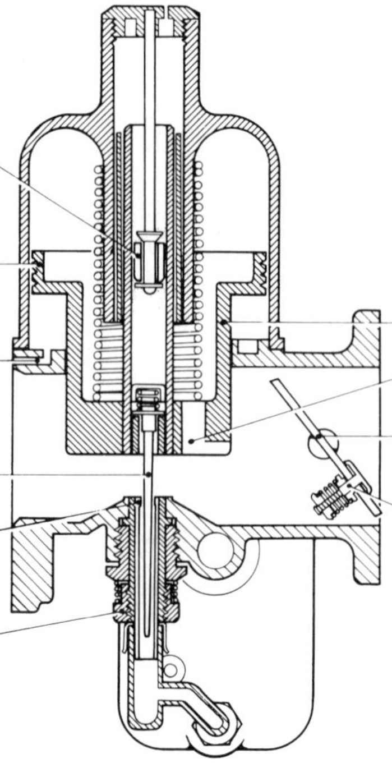

Fig. 17.2 SU carburettor

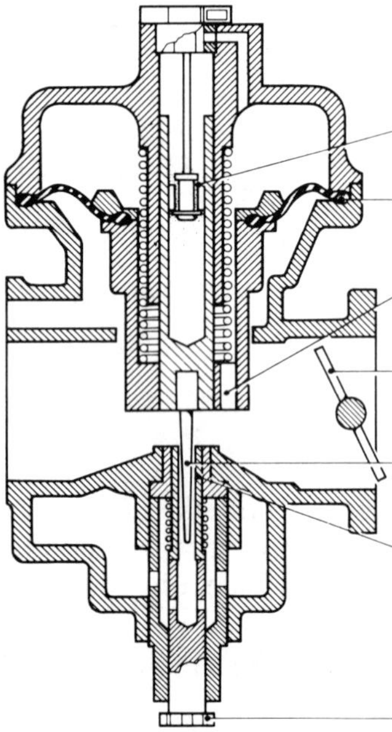

Fig. 17.3 Zenith–Stromberg carburettor

claimed to reduce harmful exhaust emissions. It also overcomes the two main disadvantages of the fixed-venturi type: the need for mixture correction devices and the choke air speed problem which gives poor atomization at low speeds and breathing restriction at high speeds.

Fig. 17.4 shows a diagram of this type of carburettor. An air valve, operated by a diaphragm, is used to vary the area of a large, fixed venturi. A tapered needle, attached to the air valve, fits into a main jet to control the petrol flow.

Air valve operation The operating principle is similar to other variable-venturi carburettors. In Fig. 17.4 the carburettor is shown in the cruising position. A depression in the mixing chamber (A) is communicated through the air valve chamber (B) to the diaphragm (C). When engine load is steady, the difference in air pressure on the diaphragm balances the force given by the diaphragm spring.

When the throttle is opened a given amount, the following events take place in quick succession:

1 Depression will become more intense in the mixing chamber (A) because insufficient air passes the air valve. This depression is transferred to the diaphragm chamber (C).
2 Atmospheric pressure will now push the diaphragm against the spring. As the diaphragm deflects, the linkage will move the air valve and increase the venturi opening. This will prevent the venturi air speed from increasing and will restore the

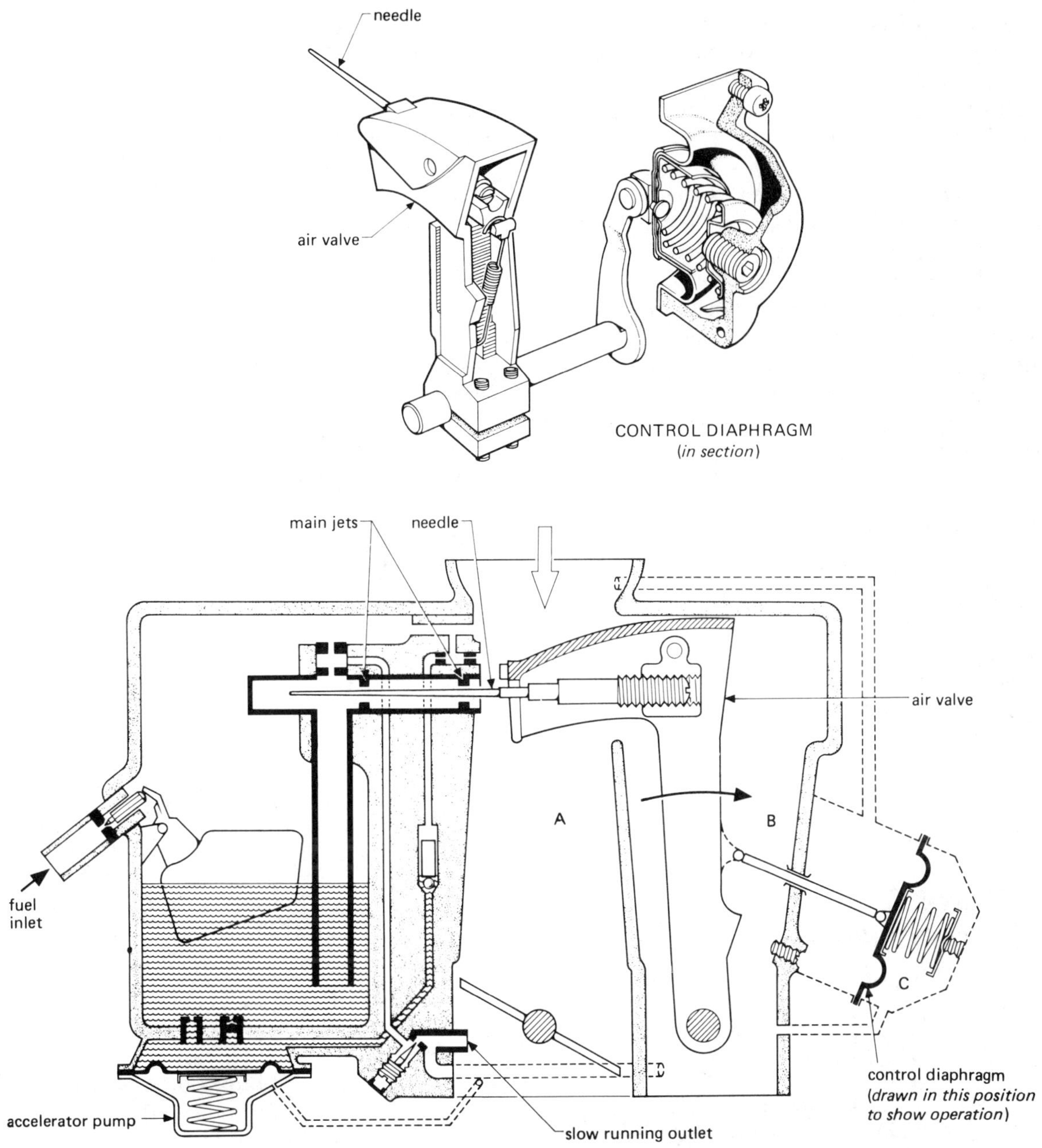

Fig. 17.4 Ford variable-venturi carburettor (shown in diagrammatic form)

depression in the mixing chamber (A) to the original value. At this point the various forces on the diaphragm will be balanced.

Closing the throttle gives a 'reverse' action – the depression in regions (A) and (C) will diminish and the spring will close the air valve to prevent a decrease in the venturi air speed. Operating in this manner, the air valve will keep the venturi air speed constant irrespective of throttle opening.

Slow running As emission requirements become more strict, so the modern carburettor must provide a better atomized and distributed slow-running mixture. The VV design achieves this by using a separate slow-running system. This is similar to that used in modern constant-choke units except that in the VV carburettor the slow-running system only supplies about 70 per cent of the total mixture; the remainder is provided by the main jet.

Acceleration A slightly rich, well-atomized mixture is obtained by using a vacuum-operated accelerator pump. The mixture provided by this pump compensates for the sudden drop in venturi air speed caused by the rapid opening of the air valve.

Cold starting When the engine is cold an automatic choke in the form of a miniature auxiliary carburettor supplements the mixture being provided by the main system. Automatic control is given by a bimetallic spring which is heated by the coolant from the engine system. The heat-sensitive bimetal spring acts on a needle in a fuel jet; this opens when the engine is cold. Fuel from this jet is mixed with air and the resultant mixture is discharged into the main system at a point beneath the throttle valve.

Electronic carburettors

Every year sees the introduction of more and more stringent anti-pollution regulations; also there is a constant need for improvements that give an engine better fuel economy.

These two requirements can only be met by a fuel system that is able to sense the engine's operating conditions accurately and then use the information gained to provide a near-ideal mixture. Such an arrangement must be very sensitive and quick in operation, so electronic control systems have recently come into prominence to meet these requirements.

The features covered at this stage apply to a constant-depression carburettor, but many aspects considered also apply to other types of carburettor.

Fig. 17.5 shows the layout of a typical electronic control system fitted to a constant-depression carburettor. In this system four sensing devices are used to measure engine conditions and ambient (surrounding) changes that affect the operation of the carburettor. Electrical signals from these sensors are passed to a computer called an *electronic control unit* (ECU), the brain of the system. During its manufacture the ECU is programmed to show the action it should take in response to a given set of conditions, so this feature enables the carburettor to operate efficiently over a wide speed and load range.

The electronic system can be used to control:

- Mixture for cold-starting
- Slow-running speed
- Fuel cut-off when the vehicle is on over-run or the ignition is switched-off.

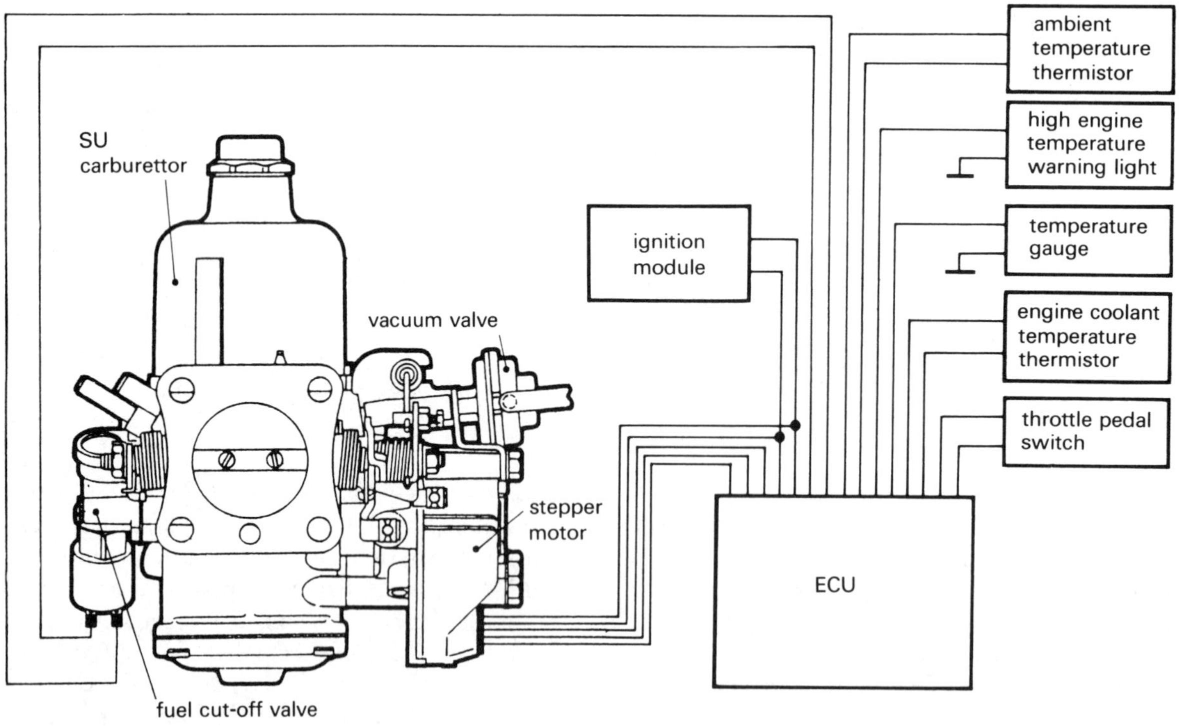

Fig. 17.5 SU carburettor with electronic control

Cold starting Accurate measurement of ambient and engine conditions by electronic sensing ensures that the mixture supplied during cold starting is set to suit the engine temperature.

Fig. 17.6 shows an auxiliary starting carburettor that is brought into operation when the cylindrical 'choke' is rotated by an electrical stepper motor. This type of rotary motor has a range of about three revolutions and is capable of moving in either direction through small angles such as 7.5°. The motor responds to an electrical pulse, so when a series of pulses is applied, it moves through a larger angle.

The diagram shows the 'choke' in operation and supplying an extra-rich mixture to augment that delivered by the main system. Air that enters the starting carburettor flows around the rotary choke and mixes with fuel coming from the float chamber. The quantity of air–fuel mixture is determined by the position of the rotary choke, which in turn is controlled by the stepper motor. As the engine warms up the associated movement of the rotary 'choke' gradually reduces the mixture supplied by the system until eventually the fuel and air ports are cut off and the system goes out of action.

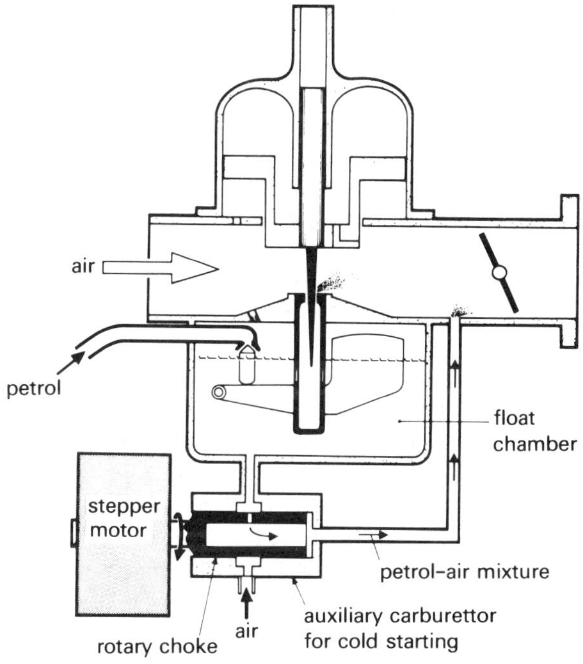

Fig. 17.6 Carburettor with auxiliary cold-starting system (simplified)

Slow running speed control Many modern carburettors use a stepper motor to control the slow-running speed. This motor acts on a throttle stop and 'jacks' opens the throttle when the ECU detects that the engine speed is too low. Engines using this system can be set to slow-run at a speed lower than normal without stalling.

In the SU type shown, the first movement of the stepper motor is arranged to control the throttle in a manner similar to that used on non-electronic carburettors.

Fuel cut-off Economy can be improved by cutting off the fuel supply when power is not required. This can be achieved by using a solenoid valve to reduce the air pressure in the float chamber.

Fluid flow through any orifice is governed by the pressure difference that exists between the inlet and outlet sides of the orifice. In this case petrol flow through the jet will cease when the air pressure in the float chamber equals the pressure on the output side of the jet.

Mixture control

Elaborate compensation systems as required on constant-choke carburettors can be eliminated by using an electronic system to control the mixture strength. This type of system sets the air/fuel ratio to suit operating conditions such as speed, load, temperature and throttle position of the engine, as well as the ambient temperature.

18 Inlet manifolds

Function of the inlet manifold

Put as simply as possible, the function of the inlet manifold is *to distribute the mixture of air and fuel from a single carburettor to the individual cylinders of a multi-cylinder engine*. There are, however, several difficulties to be overcome in designing a satisfactory induction manifold, difficulties so complex that the designer of a very famous engine has remarked that 'a certain amount of what can only be termed *witchcraft* is apparent' in the development of the induction manifold. These difficulties will be considered in two stages.

Air flow

If a single, straight, open-ended pipe is connected to the inlet port of a single-cylinder engine, and the engine driven by, say, an electric motor, the engine will draw air through the pipe during its induction stroke. It was explained in Chapter 10 that towards the end of the induction stroke the momentum of the gas in the inlet pipe causes the gas to continue to flow towards the cylinder even after the piston has passed b.d.c. This inward flow of gas, combined with the upward movement of the piston at the beginning of the next stroke, causes a rise in pressure of the gas in the cylinder, which stops the inward flow of gas. This is the point at which the inlet valve should close. If the valve timing is 'correct', therefore, the pressure at the port end of the inlet pipe will be higher than the pressure at the open end, and this will cause the air to *rebound* away from the port, setting up an air flow along the pipe *away* from the closed valve. If the valve closes too late this *reverse flow* will already have started before the valve closes, while if it closes too early the momentum of the gas will cause it to build up pressure at the valve end of the port still further, producing the same result - a reverse flow. During this reverse flow the gas will acquire momentum which will cause the flow to continue until the pressure at the port end of the pipe falls below atmospheric, and this will in turn cause a second rebound of gas *into* the port. These reversals of air flow in the pipe will continue for some time if there is little or no opposition to flow, the time for one complete oscillation depending upon the length of the pipe.

By selecting a suitable length of pipe it is possible to arrange that, at the moment the inlet valve opens, the gas in the pipe is moving *towards* the cylinder. The result is that more gas will get into the cylinder than would be the case if the gas were moving away from the port at the moment the valve opened, and a useful gain in power will be obtained. This, however, will only occur at one engine speed, and at other speeds the reverse effect will be produced. Whilst this can be very usefully applied to engines required to deliver high power outputs over a very narrow speed range, such as racing engines, it is impractical for engines such as those used in ordinary motor vehicles which are expected to give good results over a wide range of speeds. To achieve this object, the inlet manifold should not be too long, and the length of pipe between the carburettor and each cylinder should be as nearly as possible the same.

Finally, to obtain the best flow of gas into the cylinder, the pipe should be of large diameter and all bends should be gentle.

Behaviour of fuel spray

It is the presence of the fuel spray in the air passing along the manifold that is responsible for most of the difficulties. Since liquid fuel is much denser than air it has a much greater disinclination to change either its speed or direction, so that where changes of speed or direction occur in the manifold the air responds to these changes much more quickly than the fuel does.

Unless the speed of the flow of mixture along the

manifold is kept at a fairly high level the fuel spray tends to settle out on the floor and walls of the manifold, and fuel will then not reach the cylinders in a form in which it can easily be vaporized.

It is thus clear that the cross-sectional area of the manifold must be small enough to ensure that the flow of mixture through it is sufficiently rapid to maintain the fuel spray in suspension in the air, even at low engine speeds: this is contrary to what is required to obtain the best filling of the cylinders.

There are three other factors that influence the extent to which the fuel remains in suspension in the air stream. These are:

1 The fineness of atomization of the fuel.
2 The pressure (or depression) in the manifold.
3 The temperature of the mixture in the manifold.

The finer the atomization the less likely are the fuel particles to separate out from the air, but the greater the amount of evaporation which takes place in the manifold. Evaporation within the manifold is undesirable because it reduces the weight of air and fuel which can be got into the cylinder, and the internal cooling due to vaporization of the fuel inside the cylinder, as explained in Chapter 14, is lost. Since the fineness of the spray depends upon the speed of air flow through the choke tube, the spray is coarsest when the speed of the mixture through the manifold is lowest, i.e. both conditions are at their worst at low speeds, and any attempt to improve either or both at low speeds merely reduces the possible power output at high speeds.

The lower the pressure - or the higher the depression - in the manifold the more rapidly does the fuel evaporate in the manifold. Here again, conditions are worst at low speeds and wide throttle openings.

Heating of the mixture encourages evaporation in the manifold and so helps to keep the fuel in suspension in the air, but it also reduces the possible power output, for the reason stated above and because it reduces the density of the air. In this case, however, conditions are rather better, provided the heating is applied via the walls of the manifold, because the faster the mixture travels through the manifold the less time it has to absorb heat, and the mixture will pick up more heat under those conditions when it will be most helpful, i.e. at low speeds.

Features of manifold design

Enough has been said to indicate that manifold design is so complex that it is not possible to design one that gives satisfaction under all conditions, and it is to be expected that a great deal of variation in manifold design will be found on motor vehicle engines.

For what may be called the normal types of motor vehicle engines, smoothness and economy are of more importance than sheer power. Thus the manifold will be designed to provide the best possible distribution of mixture, both as to quantity and equality of air/fuel ratio, among the individual cylinders. Features that contribute to this are:

1 Not too large a size of pipe to maintain reasonably high mixture speeds at low engine speeds.
2 Sharp corners rather than gentle bends where changes in direction occur. The turbulence created at these sharp corners assists in keeping the fuel in suspension in the air.
3 A fairly short manifold to avoid marked gain in power at one part of the speed range at the expense of a marked loss at some other speed.
4 Some means of applying heat to the manifold walls.

Manifold heating

Applying heat to the manifold and using this heat to *vaporize* the fuel is essential if the fuel mixture is to be evenly distributed to all cylinders, easily ignited and completely burnt. Moderate heating is required because excessive heat, as used in the past when the exhaust gases were made to come into contact with the inlet manifold at a *hot spot*, lowers the volumetric efficiency, i.e. it reduces the ratio of the volume of air induced per cylinder per cycle to the swept volume of the cylinder.

Heat is transferred to the manifold in two ways: conduction from the cylinder block through the manifold gasket and by conduction from the coolant circulated through a jacket that surrounds the inlet manifold.

This conducted heat is absent until the engine warms up, so in addition to being provided with an extra-rich mixture with its associated pollution problems, many engines today are fitted with electrically powered heating devices to warm the air during this period.

Fig. 18.1 shows a water-heated manifold. Water from the coolant system is piped to a jacket adjacent to the carburettor and then returned by another pipe to the coolant system. An alternative design incorporates most of the inlet manifold in the water jacket of the engine block.

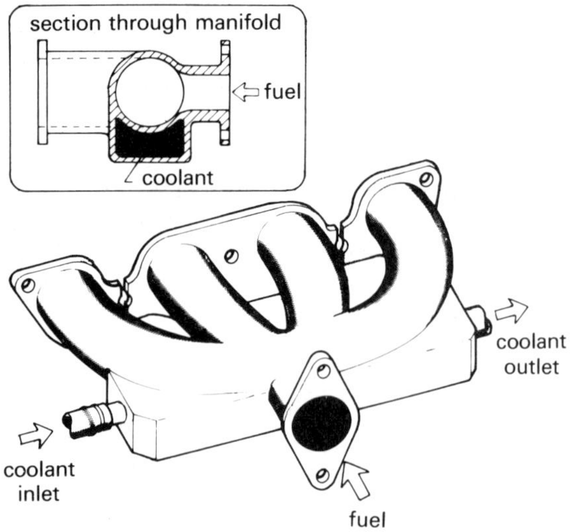

Fig. 18.1 *Water-heated manifold*

Typical manifolds

Various aspects of manifold design can be illustrated by studying some of the layouts used during recent years.

Four-cylinder in-line

The layouts shown in Fig. 18.2 indicate various manifold arrangements that can be used for a four-cylinder engine.

Fig. 18.2a is the cheapest arrangement and it was commonly used in the past for the more basic types of engine. Since the inlet valves for cylinders 1 and 2 and cylinders 3 and 4 are adjacent to each other the inlet ports in the cylinder head can be joined together, or *siamesed*. This simplifies the head casting and allows a 2-branch manifold to be used. For the majority of inlet manifolds, aluminium alloy is used because the excellent thermal conductivity of this material allows the manifold to warm-up quickly; also it is a lightweight metal. However, the siamesed arrangement has two serious limitations:

1 Poor volumetric efficiency due to the restriction of the gas flow caused by the large number of bends.
2 Severe induction robbery due to the overlap of induction strokes that occur in adjacent cylinders.

Power produced by an engine depends on the quantity of gas that can be charged into the cylinder during the induction stroke. For this reason a high volumetric efficiency (VE) is a prime objective when designing an inlet manifold. Since sharp bends in the tract reduce gas flow, an ideal manifold from a VE aspect should have gas paths as straight as possible.

Induction robbery, as the name suggests, means that some cylinders do not get their full share of new gas. Consequently the total power output from the engine is low and the variation in power from the cylinders reduces the smooth running of the engine.

One of the two robbery periods can be seen by considering an engine with siamesed ports and a firing order of 1342. Bearing in mind that each inlet valve opens early and closes very late, there is a large crank angle when the inlet valves of cylinders 3 and 4 are open simultaneously. Towards the end of the induction stroke of cylinder 3, at high engine speeds the gas will be flowing freely into the cylinder at a very fast rate. Opening the valve in cylinder 4 will not induce the gas to change its direction, so cylinder 4 will be starved of fuel until the valve in cylinder 3 has closed. This action is repeated for cylinders 2 and 1, so robbery occurs in two crank positions of the cycle. Both firing orders used on this design of manifold give similar robbery effects.

The carburettor used on a multi-branch manifold should have its throttle spindle parallel with the engine axis to avoid one branch being starved of mixture during part-throttle operation.

Fig. 18.2b uses twin carburettors and owing to the straighter air passages, the VE is raised, especially at high engine speeds. Robbery is still a problem here, and the irregular air flow through each carburettor requires a slightly richer mixture than before. The two separate manifolds are normally interconnected by a *balance pipe*, which gives smoother idling by equalizing the depression in each manifold.

The cylinder head layout used in Fig. 18.2c can be fitted with many different manifold arrangements. These options are often used by manufacturers to provide a range of engines, each with different power characteristics, to suit cars from the base model to the top-of-the-range carburettor version. The elimination of the siamesed ports overcomes the problem of robbery provided the individual tracts of the manifold are of adequate length.

Ideally each cylinder should receive the same amount of air, the ratio of petrol and air should be identical and the additives contained in the fuel should not vary in quantity between the cylinders. The term *distribution* is used to describe the ability of a manifold to meet these requirements, so the layout shown in Fig. 18.2c gives a better distribution than the first layout. Although the layout is an improvement, the unequal length tracts means that the distribution

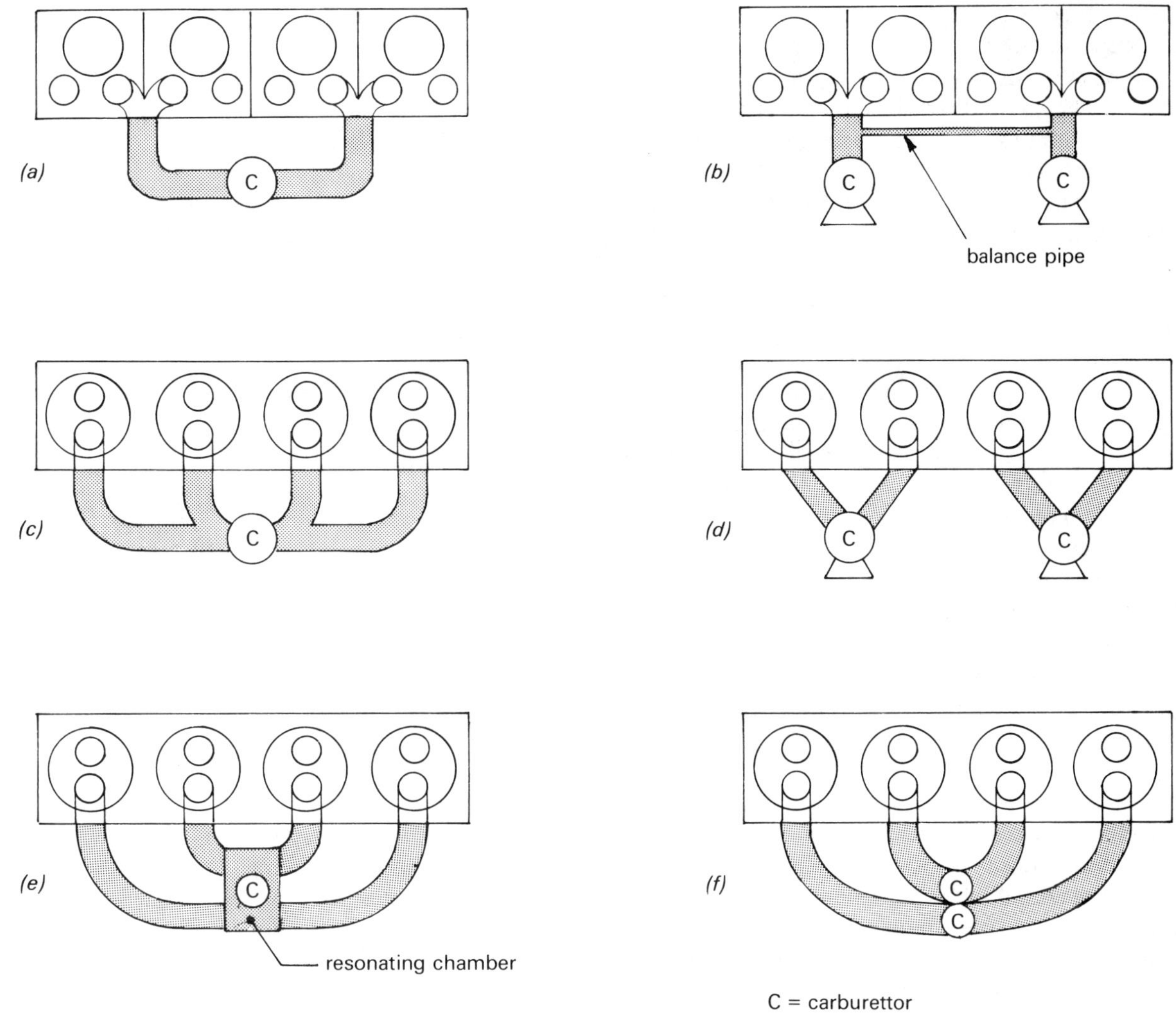

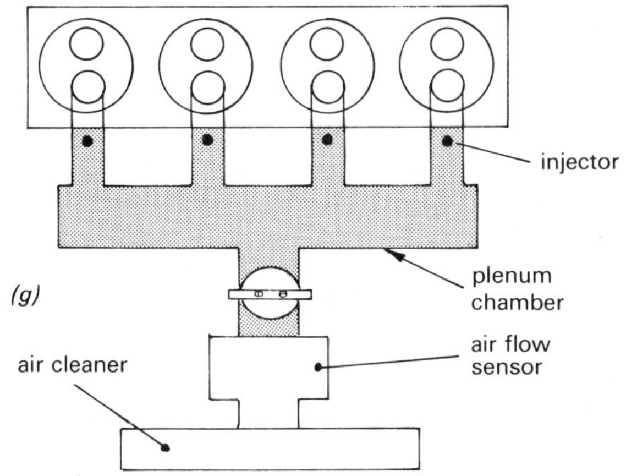

Layout	*(a)*	*(b)*	*(c)*	*(d)*	*(e)*	*(f)*	*(g)*
Volumetric efficiency	Poor	Good	Fair	Very good	Good	Good	Very good
Induction robbery	Poor	Poor	Good	Good	Ex	Ex	Ex
Distribution	Poor	Fair	Fair	Very good	Good	Very good	Ex
Economy	Poor	Poor	Very good	Poor	Very good	Good	Ex

Fig. 18.2 Manifold layouts – four cylinders

is still far from ideal. This is because the cylinders closest to the carburettor receive both a slightly richer and a larger quantity of mixture.

High power output at high speeds is the aim of the layout shown in Fig. 18.2d. As with most applications of twin carburettors on four-cylinder engines, the irregular induction impulses through the carburettor generally result in a higher fuel consumption than the single-carburettor version, but distribution is very good.

Fig. 18.2e is a layout that uses only one carburettor to achieve good VE and distribution; as a result a lively engine with a good fuel consumption is obtained.

Sometimes this design has a *resonating chamber* between the carburettor and the individual tracts to amplify the pressure pulses in the manifold. Assuming the manifold tract is of a suitable length, these pulses build up a positive pressure in the region of the manifold side of the inlet valve at the instant the inlet valve opens.

The layout shown in Fig. 18.2f has a twin-choke carburettor with each choke supplying two cylinders in order to obtain a high power. The regular induction impulses through each choke improve the mixture quality, and any interference to flow due to robbery is eliminated.

In the past the way to achieve good VE and ensure that each cylinder received its full share of mixture was to fit one carburettor to each cylinder. This was a very expensive and temperamental system which required constant *balancing* in order to keep the output of each carburettor equal to the others. The objectives of this layout can be equalled by using a multi-point fuel injection system as shown in Fig. 18.2g.

This layout passes air from a cleaner to a *plenum chamber*, which acts as an air reservoir and dampens the pressure pulsations. A chamber size of 0.8–1.2 of the engine capacity ensures a smooth flow of air through the air flow sensor so that it can accurately regulate the quantity of petrol to be injected.

Six cylinder in-line engines

The problems associated with four-cylinder manifolds are accentuated with long manifolds as fitted to six-cylinder engines, so the power output and economy of this type of engine can be changed dramatically by altering the manifold layout. This point is illustrated by the two manifolds shown in Fig. 18.3.

The siamesed port arrangement shown in Fig. 18.3a uses a single carburettor and a three-branch manifold.

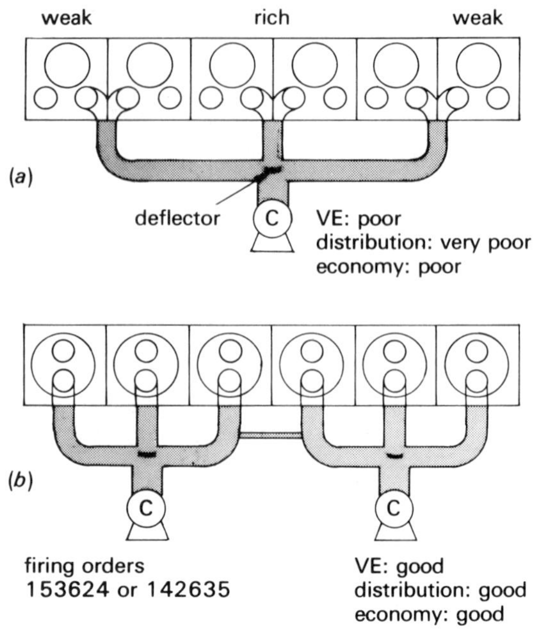

Fig. 18.3 Manifold layouts – six cylinders

This layout gives poor distribution as can be verified by examining the sparking plugs after the engine has been in use: the two centre plugs normally have a sooty black deposit that indicates a rich mixture; the end plugs have a white deposit, which shows a weak mixture. Cylinders 1 and 6 fire erratically if the mixture is too weak, so to avoid this the carburettor mixture is enriched, with the result that cylinders 3 and 4 run very rich. Fitting of a deflector improves distribution, but the layout shown will not meet current emission regulations.

The rear cylinder is farthest away from the coolant intake to the cylinder block, so when this is combined with poor distribution, the combined effect is to make this cylinder prone to combustion problems. Also this cylinder is at risk when the engine is supplied with low-grade fuel and/or is overheated.

To avoid induction robbery the firing order used must not fire two adjacent cylinders one after the other, i.e. if the engine is considered to be divided into two halves, the cylinders fire in the order: front, rear, front, rear, . . . This is one of the reasons why the firing orders used are 153624 or 142635.

The use of twin carburettors and a manifold layout as shown in Fig. 18.3b gives a more costly arrangement but with much better distribution and a higher VE. In this case the use of twin carburettors improves the economy because the air flow through the carburettors is regular and the mixture does not have to be set rich to satisfy the end cylinders.

In cases where a superior distribution and improved VE are demanded, fuel injection has to be used.

Vee engines

Compared with in-line engines, the vee arrangement gives a much shorter manifold, which reduces the mixture variation between cylinders.

In V6 and V8 engines, more than one cylinder will be on induction at the same time, so mixture robbery will take place if a single-manifold system is used. To overcome this problem, vee engines are normally fitted with a manifold cast as one piece but formed into two separate tracts.

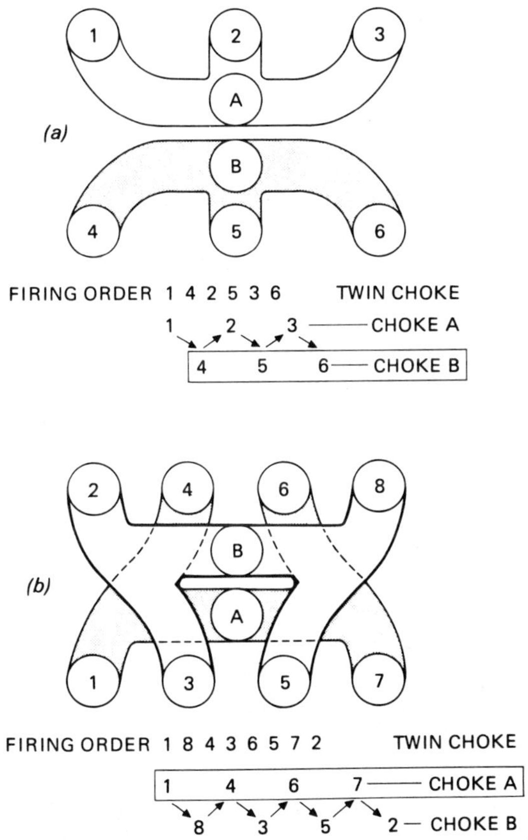

Fig. 18.4 Vee engine inlet manifolds

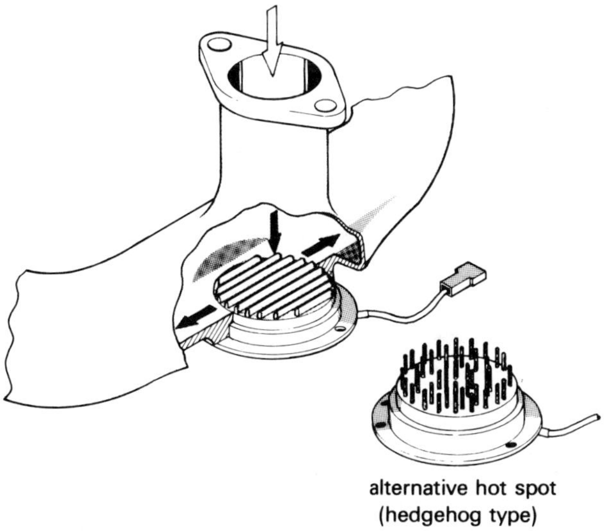

Fig. 18.5 Manifold heaters

Fig. 18.4 shows a V6 layout with two tracts. Applying the firing order to the cylinder numbering system used shows that the banks fire: right, left, right, left, right, left. The interval between firing strokes in a given bank means that there is a similar interval between induction strokes.

A V8 manifold is shown in Fig. 18.4b and once again the firing order and manifold shape are arranged to use each tract in alternate order so as to minimize the robbery caused by overlapping induction strokes.

The restricted space between the two banks makes the twin-choke carburettor most suitable for vee engines.

Manifold heaters

Fuel vaporization is very poor during both cold starting and the warm-up period; this fact results in low power output, poor economy and high exhaust emissions.

To overcome the problem, many engines today are fitted with some form of electric heater to preheat the air and fuel as it passes through the inlet manifold. Fig. 18.5 shows two forms of heater which come into operation when the engine is below a set temperature and remain in use until the engine has warmed up.

19 Petrol injection

The term *petrol injection* is used to describe any system in which the fuel is pumped out in an atomized state to mix with the air supply.

Most modern systems are controlled electronically because this form of control enables the fuel quantity and quality to be accurately set to suit the engine operating conditions. Strict emission control regulations have demanded precision in the metering of the fuel, and although petrol injection systems are more expensive, these systems are now commonly used on engines designed to output high power.

A petrol injection system that delivers the correct quantity of highly atomized fuel at each inlet valve of the engine gives the following advantages, when compared with a normal carburettor and manifold layout:

- Lower exhaust pollution
- Lower fuel consumption
- Higher power output
- Smoother engine operation due to an even power output from each cylinder
- Even supply of anti-knock additives to each cylinder
- Automatic adjustment of the air/fuel ratio to suit operating conditions.

Injector positions

Injection systems can be divided into two main groups: *multi-point* and *single-point*. This classification is based on the number and positions of injectors:

1 *Multi-point injection* - separate injectors for each cylinder, each injector positioned close to the inlet valve of the engine
2 *Single-point injection* - one injector only that discharges fuel into the air stream at the point used by a carburettor.

Fig. 19.1 shows the general layout of both arrangements.

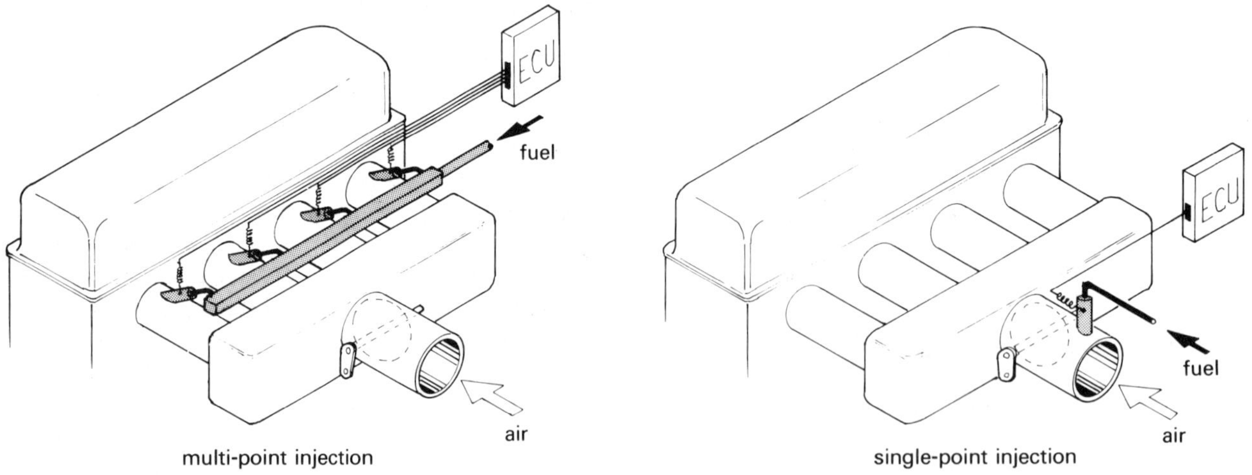

Fig. 19.1 Injection systems

Multi-point injection

During the development period of petrol injection, which saw many uses on two-stroke engines and aircraft, the well-tried system used on compression-ignition (diesel) engines was applied to spark-ignition engines. On most of these applications the injector sprayed fuel into the combustion chamber, but this *direct injection* system required high fuel pressures, and unless the injection commenced in the early part of the compression stroke, vaporization was poor.

Today, *indirect injection* is preferred (Fig. 19.2). In this arrangement an injector, operating at a low pressure, e.g. 2 bar (30 lbf/in^2), is fitted to spray the fuel *downstream* into the air as it passes the inlet valve of the engine.

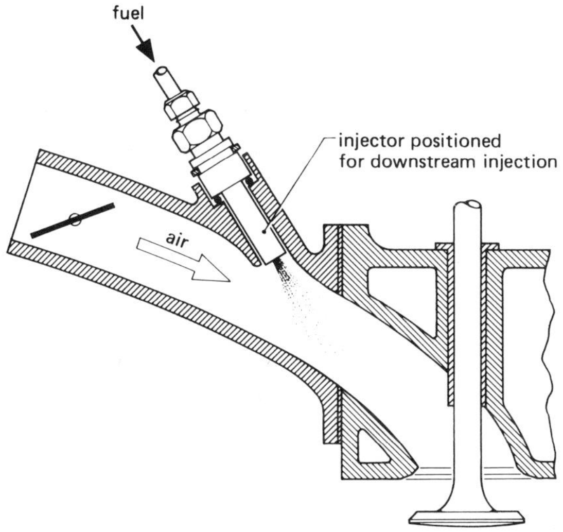

Fig. 19.2 Indirect petrol injection

As the fuel spray needs only a comparatively low pressure, a simple injector construction can be used. It consists of a body, screwed into the manifold, which encloses a spring-loaded conical valve which chatters to break up the spray. Fuel supply to the injector is carried by a pipe made of plastic material.

Injection of petrol can be *timed* or *continuous*. In the former, fuel is sprayed into the air stream only when induction is taking place.

Although a manifold is not essential, the need for an air cleaner and throttle valve is the reason why a manifold is still used.

Metering methods

An ideal injection system must supply to the engine a quantity of atomized fuel which is correct to suit the engine speed and load. Also, the system must vary this quantity to allow for alteration in engine temperature and changes in air density caused by ambient temperature and altitude variations. The initial cost of the equipment generally indicates the number of special refinements used to cater for these variables.

Systems can be separated into two main types: mechanical or electronic.

Mechanical system

The Bosch K-Jetronic is a moderately priced mechanical system using the principle of *continuous-flow fuel injection* with *air-flow sensing* to meter the fuel.

The operating principle is considered here by building up the system step by step. Fig. 19.3 shows the basic system. A permanent magnet electric motor drives a roller-type pump which pressurizes the fuel sufficient to give the required supply to operate the injectors. Maximum pressure in the primary fuel circuit is limited to 4.7 bar (about 70 lbf/in^2) by a pressure regulator valve, and excess fuel from this valve is returned to the tank. The spring-loaded diaphragm forming an accumulator:

1 Holds the pressure in the line to aid hot starting.
2 Slows-down the build-up of pressure on starting.
3 Smooths out the pulsations and reduces pump noise.

Fuel distributor and air-flow sensor This controls the quantity of fuel delivered to the injectors. The air sensor (Fig. 19.4) consists of a venturi (1) in the induction pipe into which is fitted a movable disc (2). Attached to the disc is a counter-balanced lever (3) which acts on a control plunger (4). Mounted around the plunger are a number of diaphragm valves (5), one for each injector. The purpose of these valves is to provide a fuel supply to the injectors that is proportional to the movement of the control plunger.

When the engine is stationary, the disc (2) is resting in a closed position in the venturi. The lever (3) and control plunger will be in the lowered position and no fuel will pass from the waist of the plunger to the chamber above the diaphragm, so no injection will take place.

On starting the engine, the air movement through the venturi will lift the disc – the extent of the movement will be governed by the quantity of air flowing through the venturi. When the disc has lifted a set amount, the lever will raise the control plunger (4) and open a slit to allow fuel to the chamber. Although

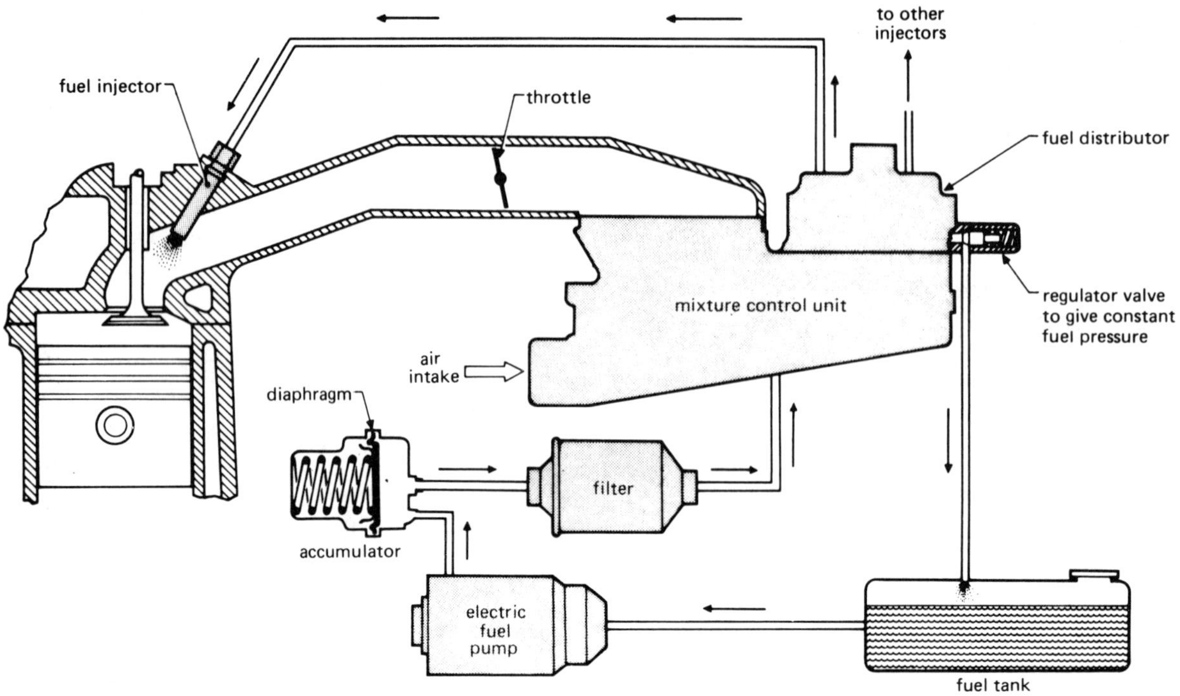

Fig. 19.3 Basic layout of a petrol injection system

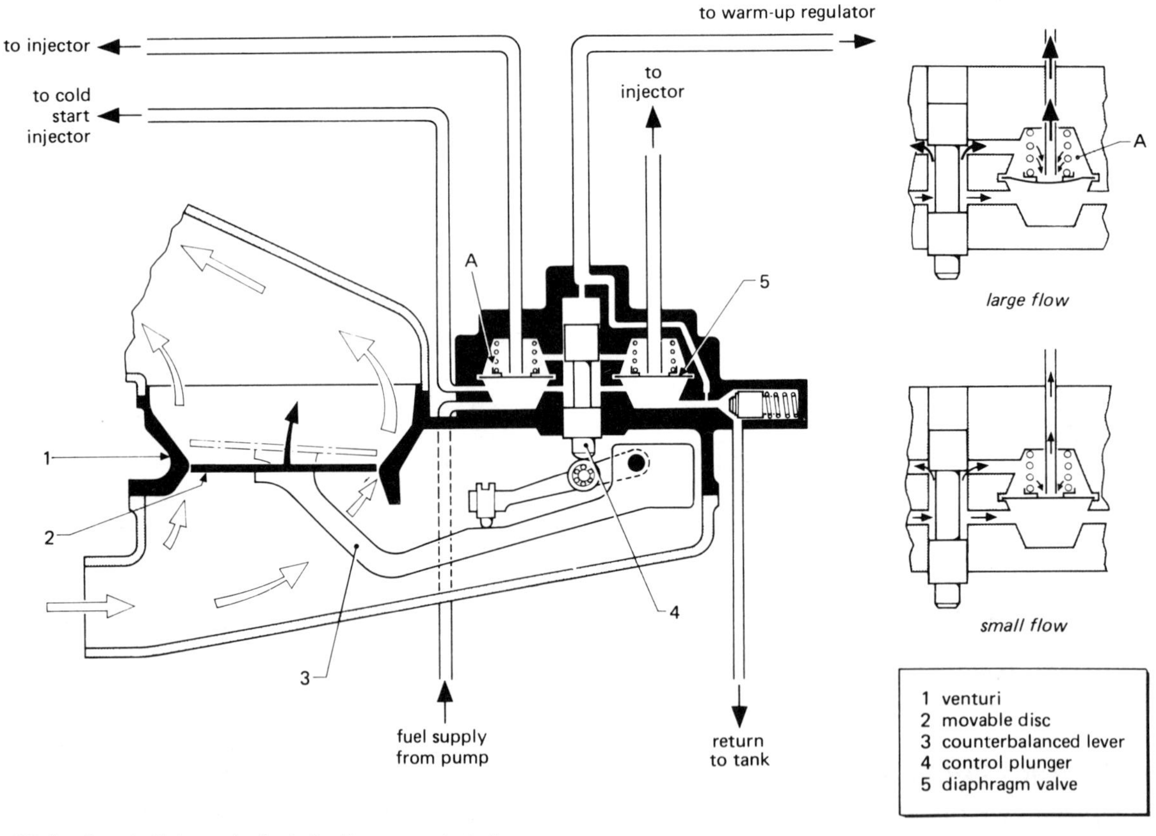

Fig. 19.4 Bosch K-Jetronic fuel distributor and air flow sensor

Although fuel pressure now acts on both sides of the diaphragm, the force of the spring moves the diaphragm downwards and opens the valve at the centre of the diaphragm to release the fuel to the injector. The flow of fuel from the chamber slightly lowers the downward force on the diaphragm, so a valve-open position is reached which corresponds to the amount that the control plunger is moved.

Opening the throttle will give a greater air flow, so the disc (2) will rise further, which, in turn, will raise the control plunger (4) and allow more fuel to pass to the chamber. This extra fuel flow will more than make up for the fuel that is 'escaping' past the valve to the injector, so the extra downward force produced by the fuel pressure lowers the diaphragm, opens the valve further and increases the quantity of fuel going to the injector.

From the description of the operation of the diaphragm unit it will be seen that it acts as a differential pressure valve. It regulates the pressure drop on each side of the diaphragm to a constant value of 0.15 bar (about 2 lbf/in^2) irrespective of supply pump pressure, nozzle opening pressure or fuel flow.

The air/fuel ratio given by this system can be altered by varying the control pressure which acts on the end of the control plunger. Reducing this pressure allows the sensor plate to move a greater amount for a given air flow and results in a richer mixture.

Warm-up regulator (Fig. 19.5) This controls the mixture enrichment during this operating phase. It consists of a valve, operated by a bimetallic strip, which lowers the control pressure acting on the end of the control plunger when the engine is cold.

Cold starting The layout of the system (Fig. 19.6) includes a cold-start injector positioned downstream of the throttle. This electrically operated valve, together with the auxiliary air device, comes into operation when the thermo-time switch senses that the engine is cold.

Mechanical system with electronic control Closer matching of the mixture to meet changing loads and operating conditions, and the introduction of stricter emission control regulations have induced the manufacturers to modify the K-Jetronic system. The modified system is called Bosch KE-Jetronic; the letter E in the title signifies 'electronic'.

Control of the mechanical system by electronic means allows the mixture to be set more accurately, and by virtue of the inherent 'intelligence' of the electronic control unit (ECU), the modified system is able to vary the mixture to suit a wider range of operating conditions.

Comparing the KE system shown in Fig. 19.7 with the K system, the following are some of the differences:

1 *System-pressure regulator* – more precise control of the system pressure is obtained by using a modified regulator valve which takes into account manifold depression.
2 *Electro-hydraulic pressure actuator* – functions as a fine mixture adjustment control to vary the air/fuel ratio slightly to suit all engine conditions,

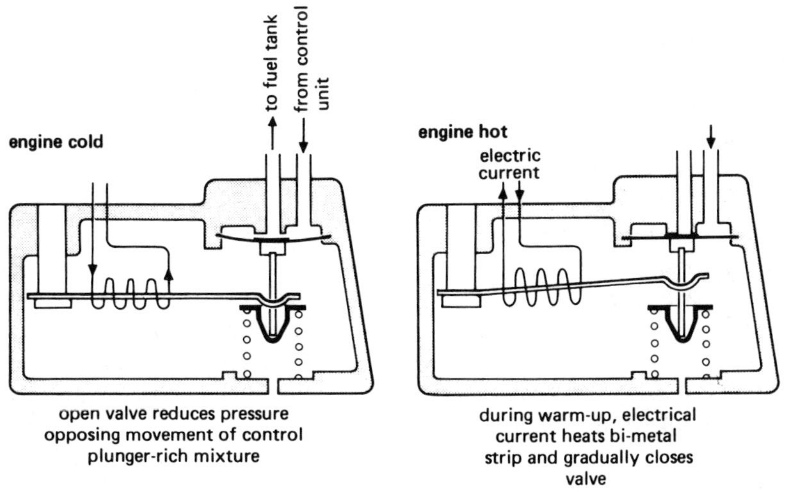

Fig. 19.5 Warm-up regulator

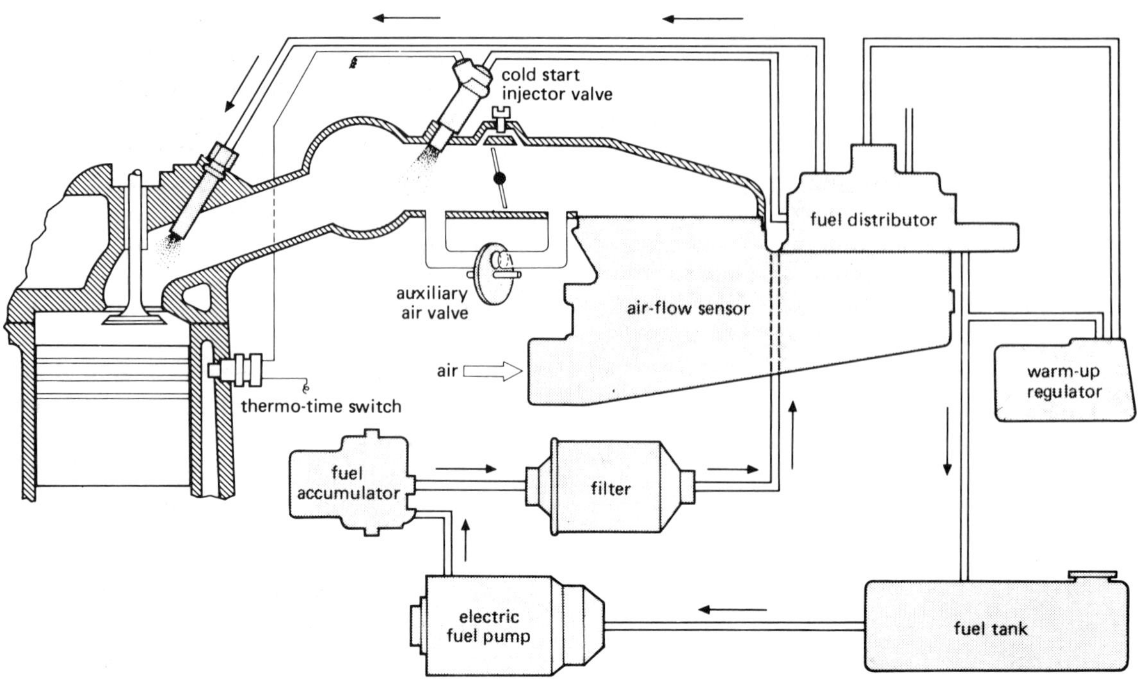

Fig. 19.6 Bosch K-Jetronic continuous-flow fuel injection with air flow measurement

especially during warm-up, acceleration and full-load operation.

3 *Electronic control unit* - acts as the brain of the system. It analyses the signals it receives from the *throttle valve switch, air flow potentiometer* and *engine speed and temperature sensors,* and then sends output signals to the actuators, which control the fuel mixture.

In addition, other features are included to allow fuel cut-off during over-run and, in countries where special emission laws are in force, controls the mixture by utilizing the variation in the oxygen content in the exhaust gas (Lambda closed-loop system).

Electronic system

The first full electronic system was introduced by Bendix in the USA in 1950. Seventeen years later a similar unit was made by Bosch and was fitted to a Volkswagen model. Since that time *Electronic Fuel Injection (EFI)* has been a common system for both sports type and luxury category cars.

The full electronic system uses a solenoid-operated fuel injector. This opens at set times in the engine cycle, hence the term *intermittent,* and is held open for a period of time proportional to the quantity of fuel that is required.

A number of different types of fully electronic system are in use. The main difference between them is the way in which the air flow is measured; the two main systems are:

1 *Indirect* or *pressure-sensed air flow.*
2 *Direct air flow measurement.*

1 a Pressure-sensed air flow This system uses a *manifold absolute pressure (MAP)* sensor to measure the manifold depression. Signals from the MAP sensor are passed to the ECU and, after taking into account the data received from the other sensors, the ECU signals the injector to open for a set time, proportional to the quantity of air that the engine is receiving.

b Bosch D-Jetronic This is a good example of a pressure-sensed system (the 'D' stands for *Druck*, which is the German word for *pressure*). Fig. 19.8 shows the basic layout of this type of system.

In this layout the quantity of air induced into the engine for a given throttle opening depends on the manifold pressure. This is measured by a MAP sensor which sends an electrical signal to the ECU to indicate the quantity of air entering the engine. This signal is processed by the ECU and is used to determine the period of time that the injectors should remain open.

Commencement of injection is triggered by

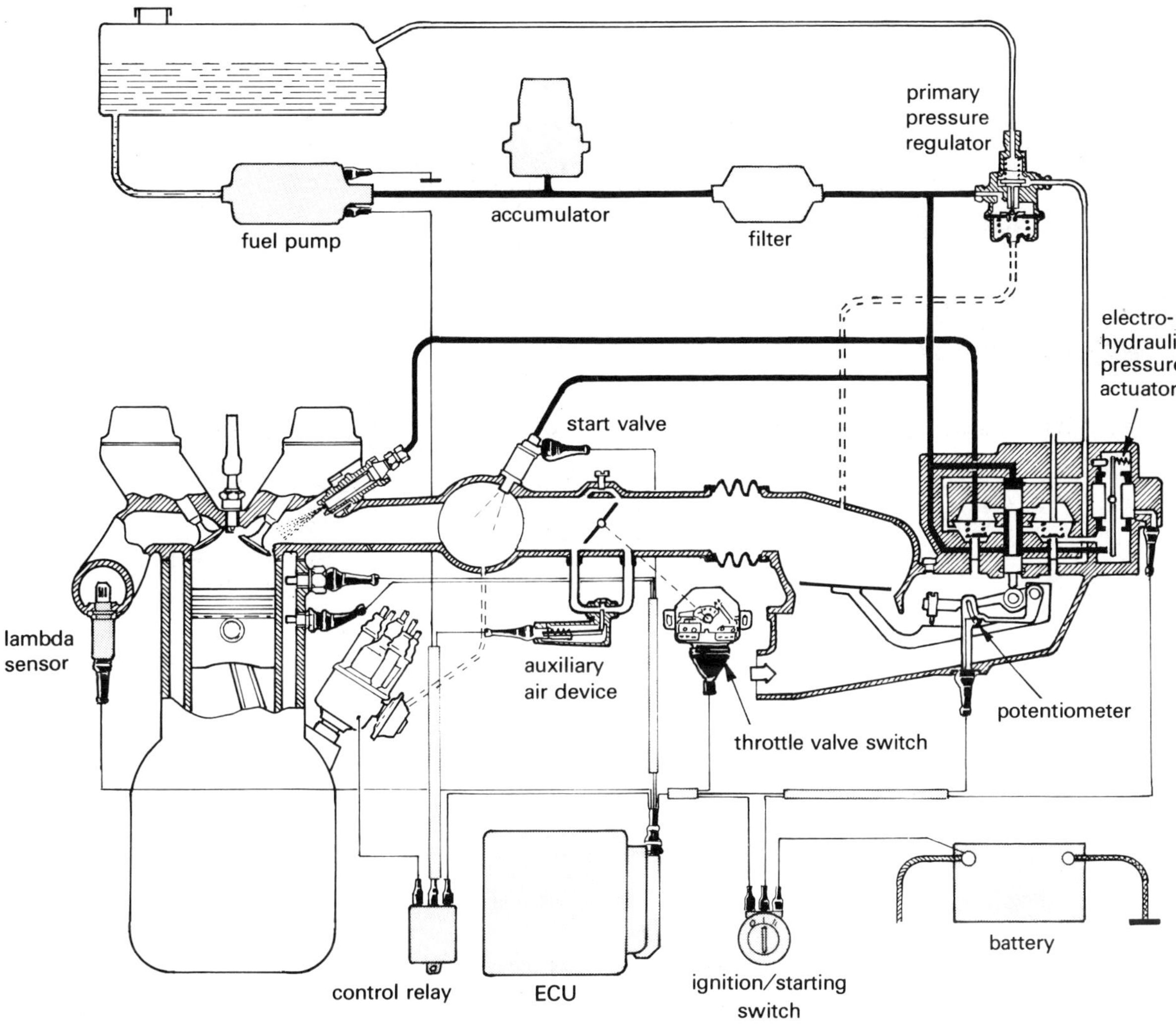

Fig. 19.7 Bosch KE-Jetronic system

either a switch in the ignition distributor or a sensor situated adjacent to the flywheel or crankshaft pulley. For six-cylinder engines, the injectors are operated in sets of three, i.e. three injectors spray at the same time.

Extra fuel provision for cold starting is similar to that used in the K-Jetronic.

The brain of the electronic system is the ECU; this receives messages from a number of sources (Fig. 19.9), and after comparing the data with the instructions programmed into its built-in memory, the ECU signals the injectors to operate accordingly.

Fig. 19.10 shows some of the other items that are used to control the system. The throttle position switch works in conjunction with the MAP sensor for measurement of air flow.

2 Direct air flow systems This type of system measures the air flowing into the engine by:
a vane or flap
b hot wire.

Both of these sensors produce a voltage output which increases as the rate of air flow increases. This voltage signal is passed to the ECU for computation of the fuel requirement; this is controlled by the time that the injector is kept open.

Both types of air flow sensor take account of the actual density of the air entering the engine, so it is possible to maintain the required air/fuel ratio to

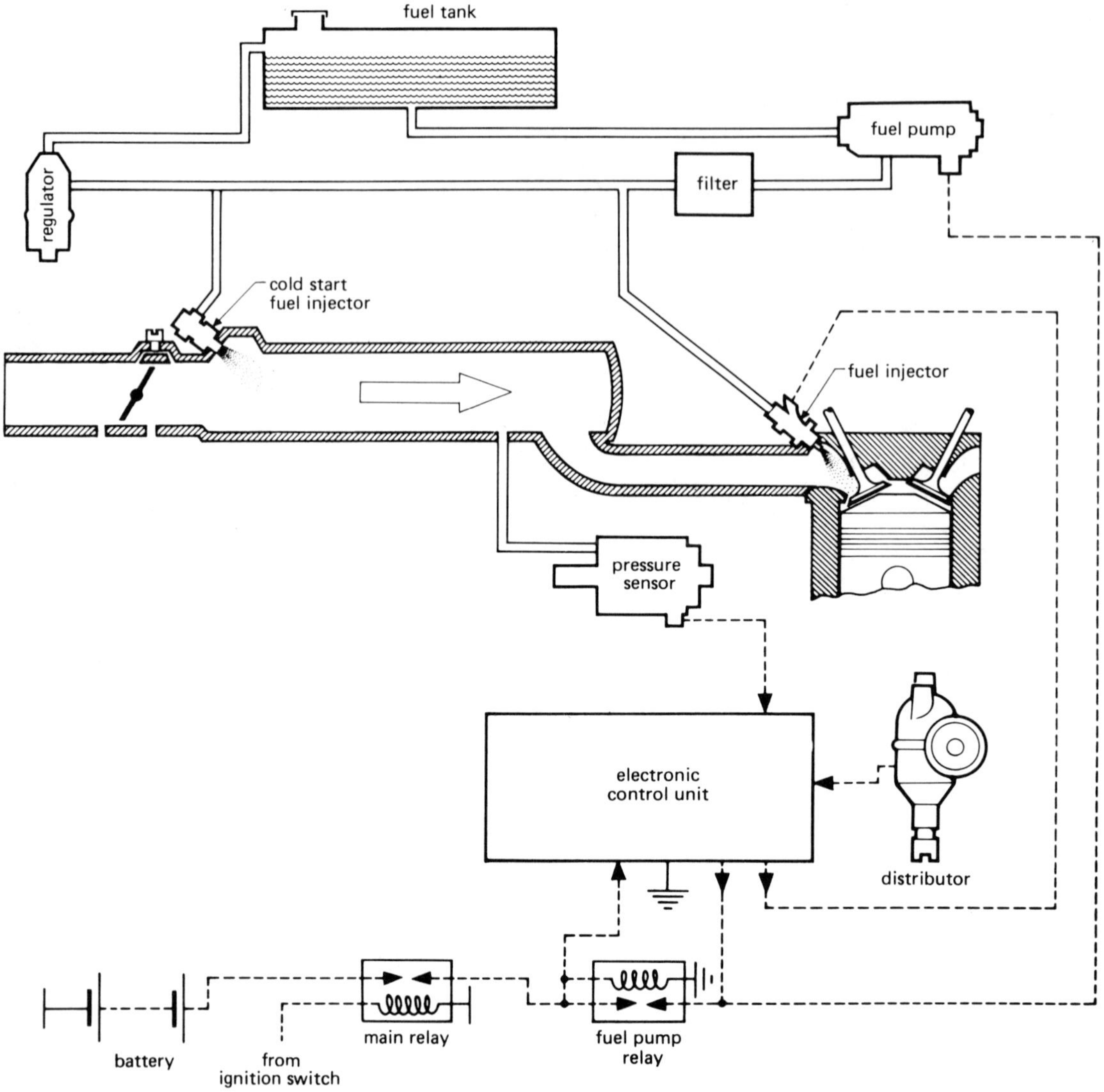

Fig. 19.8 Basic layout of electronic system of petrol injection

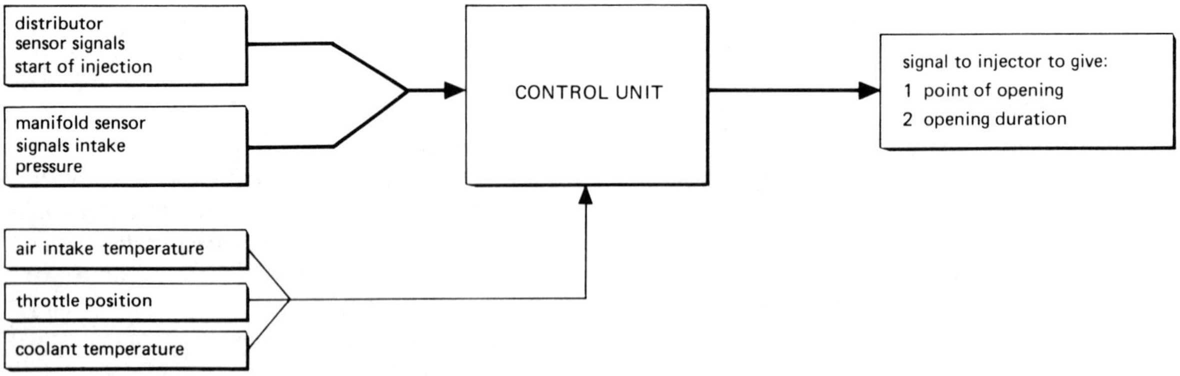

Fig. 19.9 Input and output signals of electronic system

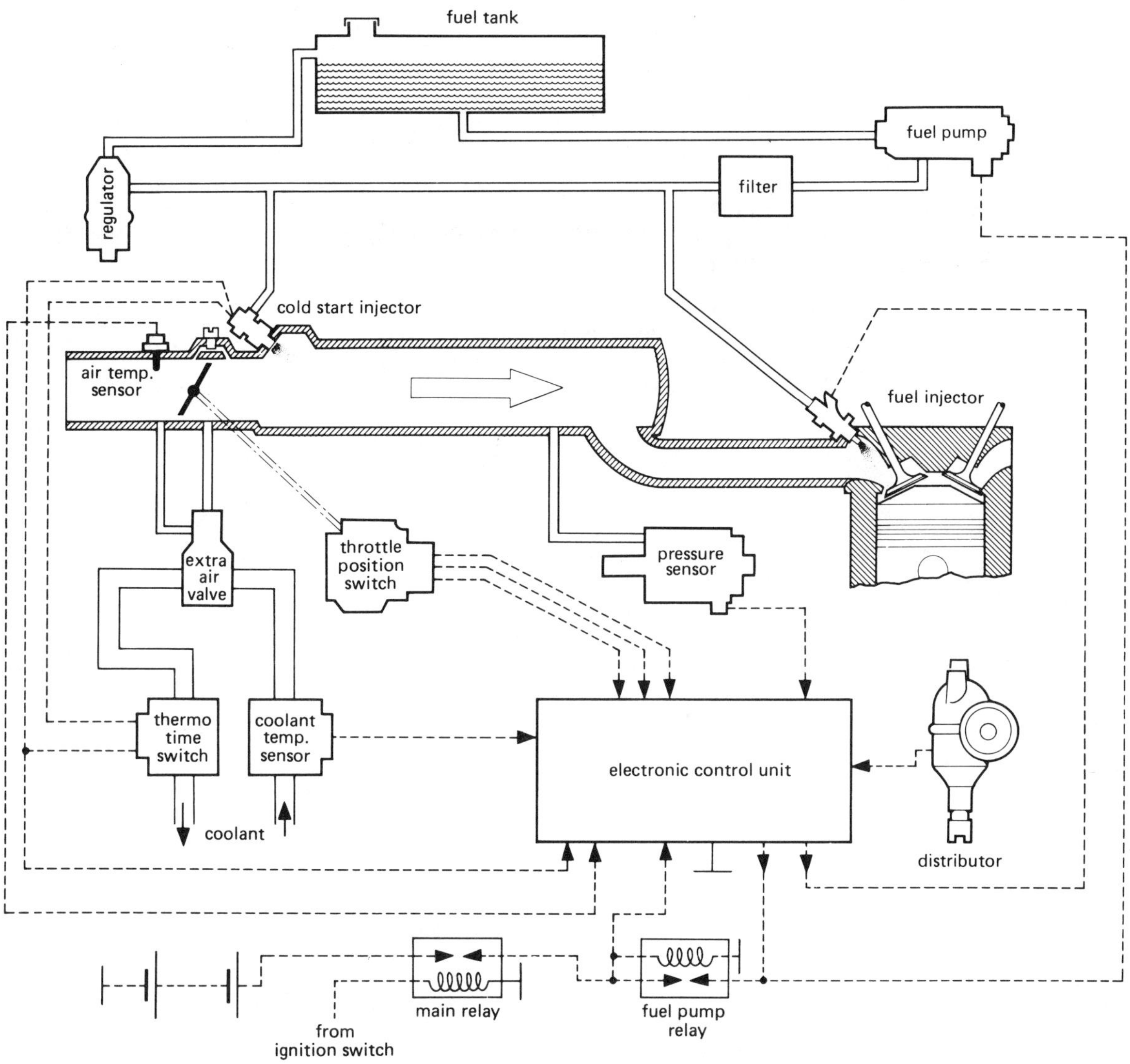

Fig. 19.10 Bosch D-Jetronic electronic system of fuel injection

compensate for variations in atmospheric temperature and pressure. This feature is especially significant when the vehicle has to operate at different altitudes.

a Vane or flap metering Fig. 19.11 shows the layout of a system similar to a Bosch L-Jetronic. (The 'L' stands for *Luft*, the German word for 'air.')

The action of the flap sensor is similar to a spring-loaded rigid flag which is exposed to a movement of air. As the air speed (or density) is increased, the force on the flag will also increase and will move it through an angle to a point where the force of the air equals the reaction of the spring. In the case of the vane, the position it occupies for a given air flow is registered by a potentiometer.

Other than the arrangement for sensing the air flow, the basic system is similar to those previously described. Fuel pressure, produced by an electrically driven fuel pump, is maintained constant at 2 bar by a pressure regulator. Injector valves are actuated once per crankshaft revolution and fuel quantity is governed by the period that the injector is held open by the solenoid.

b Hot-wire metering Both the Lucas EFI and the Bosch LH-Jetronic use this method of air flow measurement.

The principle of the system shown in Fig. 19.12

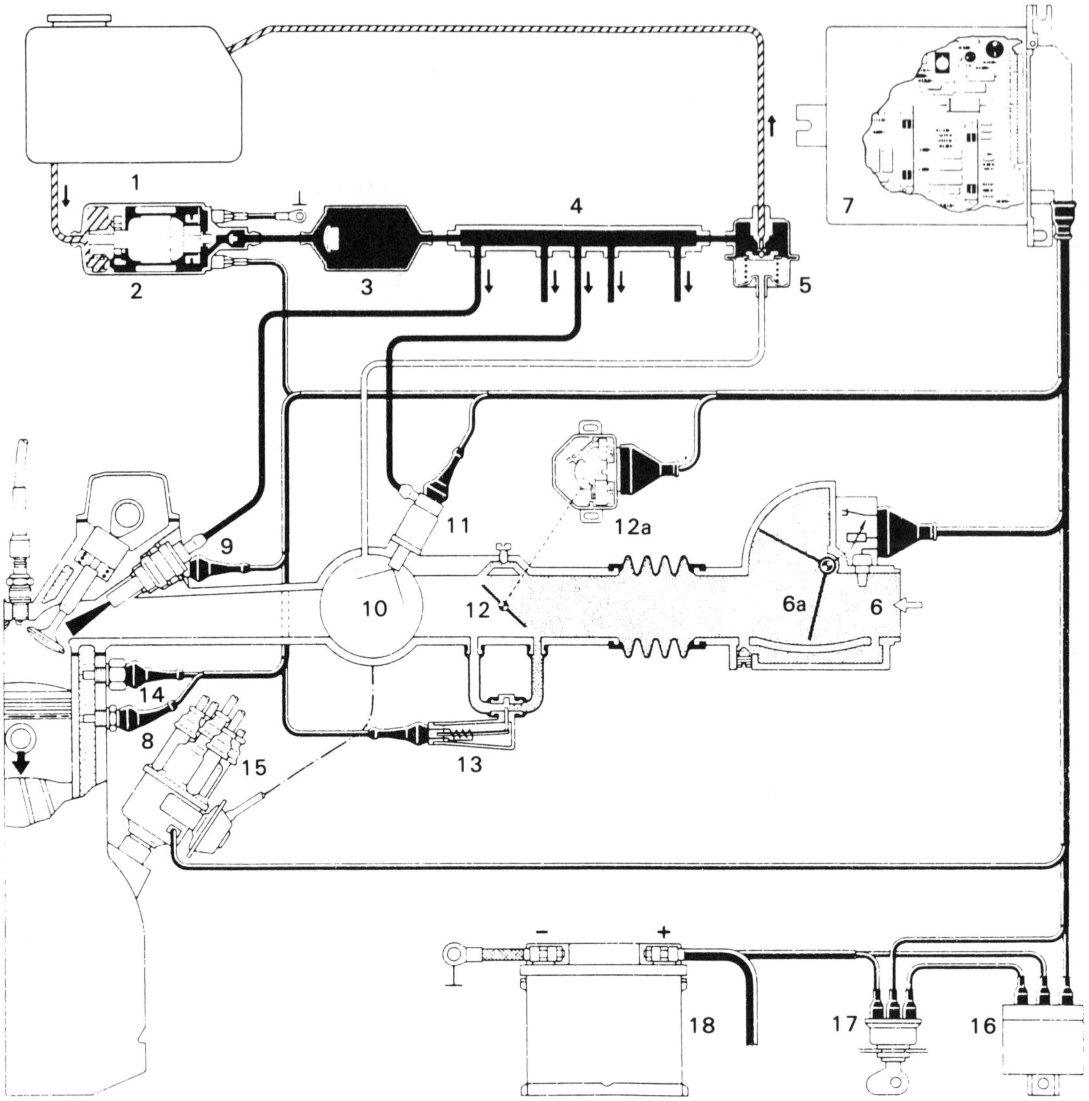

system pressure

'suction' line/return

atmospheric pressure

intake manifold pressure

1 fuel tank
2 electric fuel pump
3 fine filter
4 fuel distribution pipe
5 pressure regulator
6 air flow sensor with sensor flap (6a)
7 control unit
8 temperature sensor
9 injection valve
10 intake manifold
11 start valve
12 throttle valve with switch (12a)
13 auxiliary air device
14 thermo-time switch
15 ignition distributor
16 relay set
17 ignition and starting switch
18 battery

Fig. 19.11 Vane or flap metering (Bosch LE-Jetronic)

is based on the cooling action of air blown over a hot surface. Rate of heat transfer depends on the rate of air flow – a scientific fact easily demonstrated when you blow on a hot substance to cool it, e.g. a cup of hot tea.

The sensor has a hot wire element which is heated by an electric current sufficient to maintain it at a constant temperature. When the air flow is increased, the cooling action is greater, so a larger current has to be passed through the wire filament

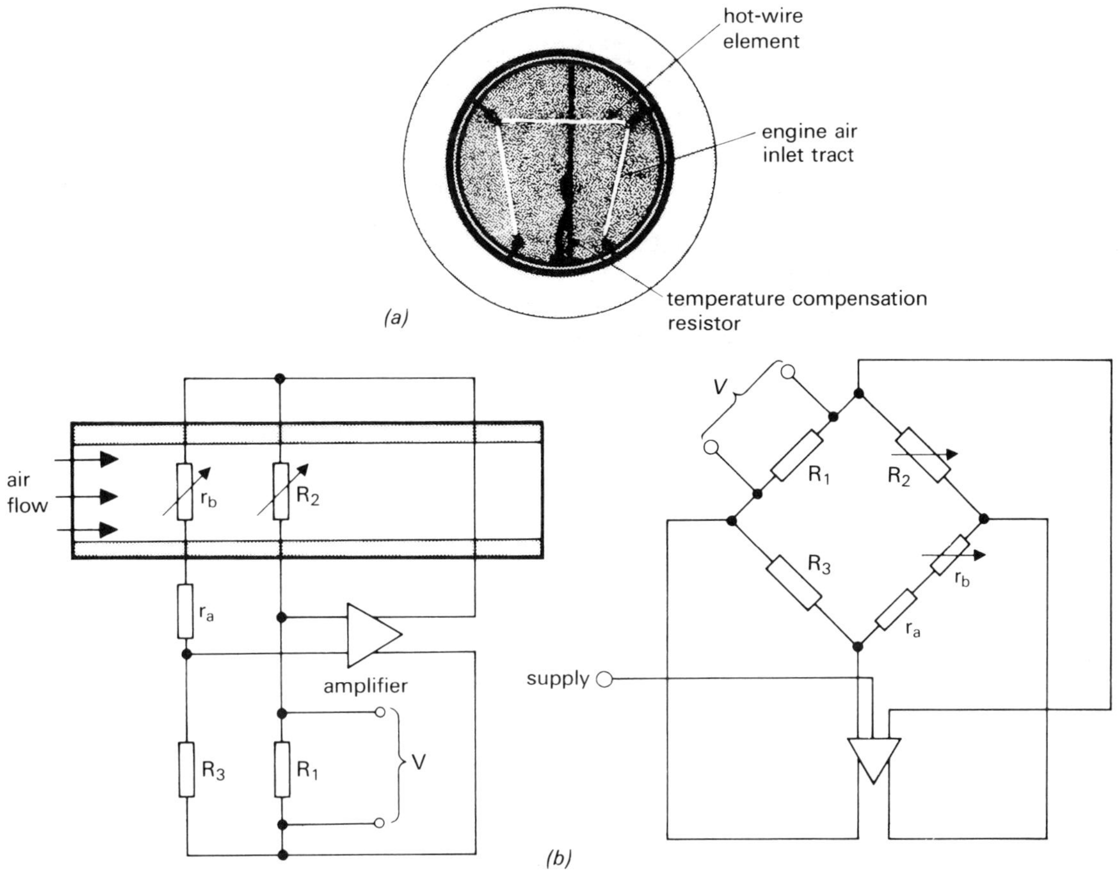

Fig. 19.12 Hot-wire sensing

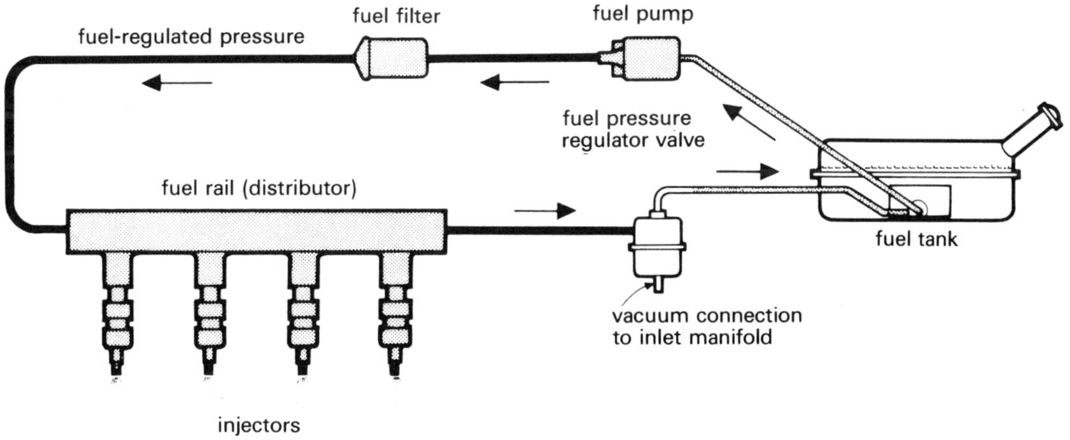

Fig. 19.13 Fuel system

to prevent a drop in its temperature. The voltage required to provide this current gives an indication of the inducted air mass. This voltage is signalled to the ECU to enable the ECU to compute initially the basic quantity of fuel. The quantity is then slightly altered to take account of the signals it receives from other engine sensors.

Air temperature affects the accuracy of the system as described, so to prevent temperature errors a compensating resistor is placed in the air

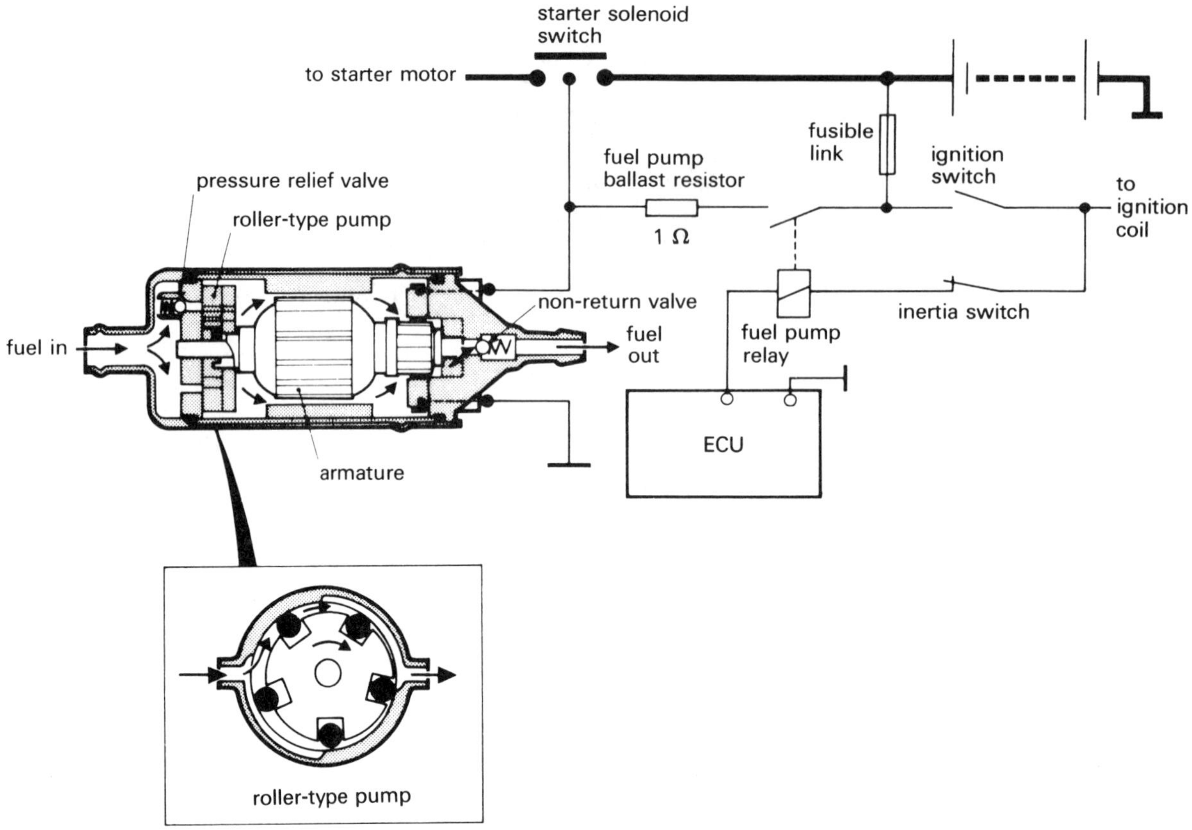

Fig. 19.14 Fuel pump and electrical supply circuit

stream. A change in air temperature causes this to alter its resistance and this compensates for the effect of a change of air temperature.

Today many cars use air flow sensing because this system allows the air/fuel ratio to be controlled within the limits required to meet emission regulations in force in various countries.

Although there are many different electronic systems of control in use throughout the world, the basic fuel supply system and the supporting electrical circuitry are often similar. The following description of a typical system is intended to show the basic principle of the main hydraulic components.

Fuel system The fuel supply system provides the injectors with adequate fuel at a pressure sufficient to allow the injectors to give good atomization. The quantity of fuel injected depends on the fuel pressure, so this must not vary.

Fig. 19.13 shows the main components of the fuel system; it includes the following:

Pump Normally a roller-type rotary pump driven by a permanent-magnet electric motor (Fig. 19.14). Rotation of the pump moves the rollers outwards and seals the spaces between the rotor and casing. As the fuel is carried around with the rotor, the combination of the rotor movement and the decrease in volume of the pumping chamber causes an increase in pressure.

Fuel from the pump passes through the motor; this aids cooling. More fuel is supplied than is needed; excess fuel is recirculated back to the tank and this reduces the risk of vapour-lock problems.

Two ball valves are fitted in the pump: a non-return valve at the outlet and a pressure relief valve limit the maximum pressure.

The pump is controlled by an ECU via a fuel pump relay, and the current supplied to the pump is passed through a ballast resistor to drop the voltage to 7 V. This resistor is shorted-out when the engine is cranked to compensate for the lower battery voltage.

On switching on the ignition, the pump motor runs for a short time to pressurize the system fully. After this initial period the pump is stopped until the engine is cranked.

For safety reasons an *inertia switch* is fitted in the electrical supply line to the pump relay. This switch opens if it is jolted, so, in the event of a collision, the pump ceases to operate. The switch can be reset by pushing down a protruding plunger.

Fuel pressure regulator This controls the operating pressure of the system and is set to maintain a constant pressure difference (e.g. 2.5 bar or 36 lbf/in^2) between the fuel line and the inlet manifold irrespective of throttle opening. It consists of a spring-loaded diaphragm and ball valve (Fig. 19.15).

Manifold depression depends on throttle opening, e.g. engine load, so when the opening is small the high depression encourages more fuel to leave the injector. To compensate for this, the fuel system operating pressure is lowered when the manifold depression is high. This is achieved by connecting one side of the regulator to the induction manifold. At times when the engine is operating under a light load, the regulator valve is slightly opened and the pressure is reduced.

The pressure controlled by the regulator is as shown in Table 19.1.

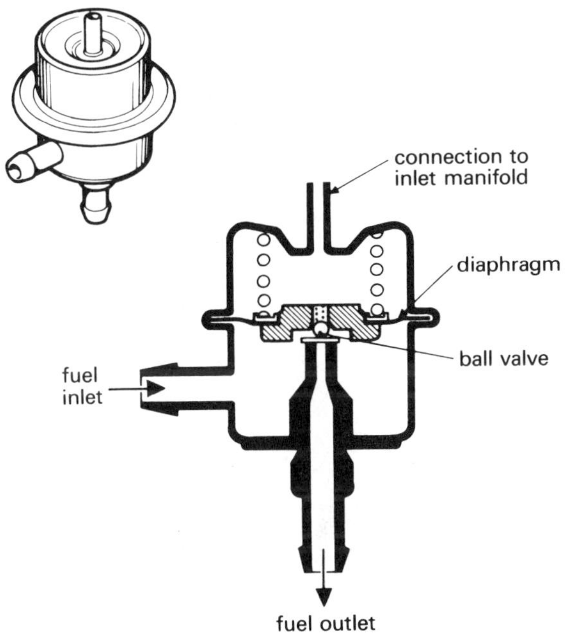

Fig. 19.15 Pressure regulator

Table 19.1 Regulator control pressure

Engine condition	*Manifold depression*	*Typical operating pressures*
Idling	Very high	1.8 bar (26 lbf/in^2)
Full throttle	Very low	2.5 bar (36 lbf/in^2)

Injectors The duty of an injector is to deliver a finely atomized spray downstream into the throat of the inlet port. In addition, the injector must vary the quantity of fuel to suit the engine operating conditions; this is achieved by varying the time that the injector is open.

The required conical spray pattern is obtained by pumping the fuel through a pintle-type nozzle. Fuel flow takes place when the nozzle valve is opened by a solenoid (Fig. 19.16). Movement of the valve is limited to about 0.15 mm (0.006 in) and the period of time that the valve is open varies from about 1.5 to 10 milliseconds (0.0015 to 0.0100 second).

This variation in opening time alters the amount of fuel that is supplied to each cylinder per cycle. Open time depends mainly on the rate of air flow and engine speed, but engine temperature and fuel temperature also have a bearing on the amount that needs to be delivered.

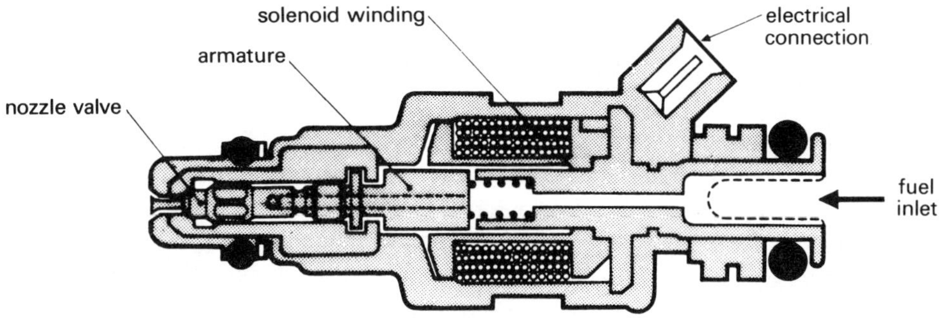

Fig. 19.16 Fuel injector

All injectors are electrically connected in parallel, so for four- and six-cylinder engines they all open and close at the same time. When the engine is warm, this opening normally occurs once every other revolution.

When the engine is cold-started extra fuel must be injected; this is provided by increasing the frequency of injections.

Injector opening and closing are by a solenoid and spring respectively.

Single-point injection

This system is also called *throttle body* injection because the single injector sprays fuel at a point on the air intake side of the throttle, i.e. the injection is made at the point where the carburettor would normally be situated.

This system loses some of the merits of 'full' fuel injection, especially the advantages associated with good fuel distribution, but its value should be assessed against a carburettor layout.

The call for better atomization of the fuel, especially at low engine speeds, more precise control of the air/fuel ratio to meet current emission regulations, and additional demands to meet modern requirements for cold starting and warm-up, has forced many manufacturers to abandon the low-cost carburettor. Whereas the high cost of multi-point injection often does not warrant its use on some models of car, the use of a single-point system is a compromise. Also this system is easier to control electronically than a carburettor, and because electronic systems are more sensitive in operation, the single-point system becomes an attractive proposition.

Fig. 19.17 shows a typical throttle body injection layout that is similar to a Bosch Mono-Jetronic system. This shows a single solenoid-operated injector situated centrally in the air intake. It is supplied by a pressurized fuel system similar to that used in a multi-point layout.

Injected fuel is directed into a venturi-shaped region around the throttle, so the increased air speed

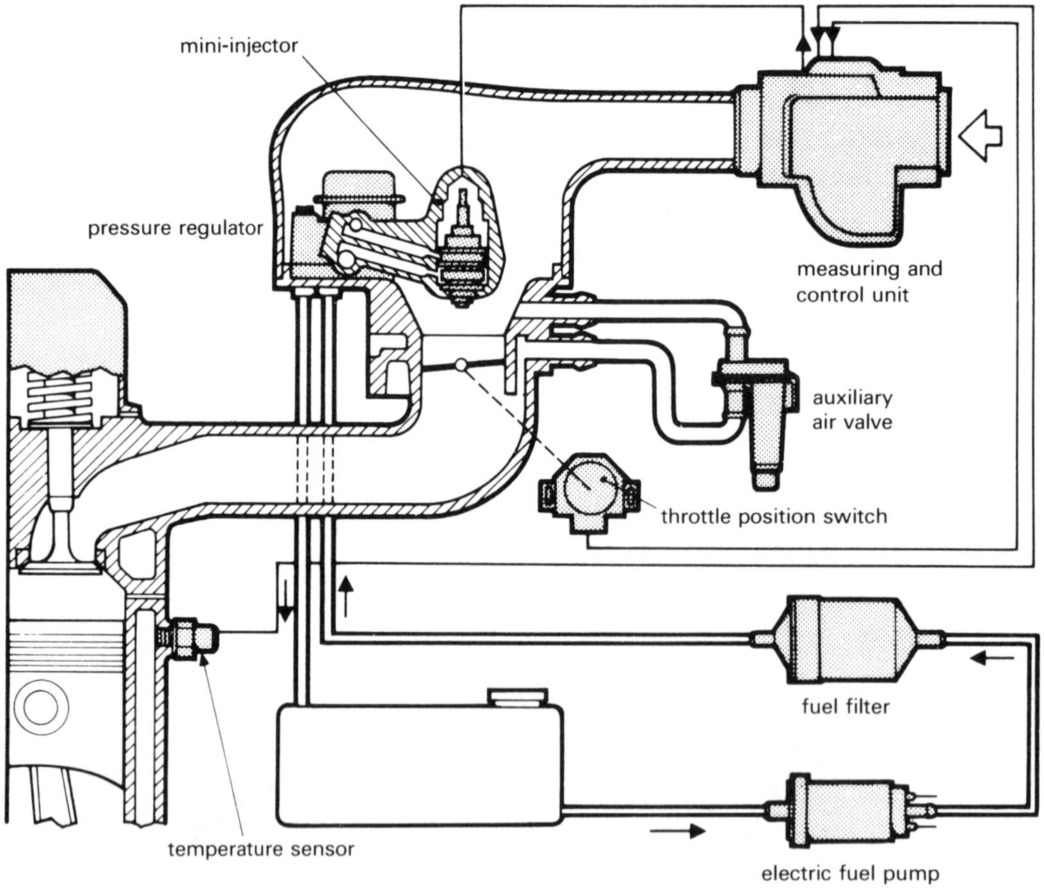

Fig. 19.17 Throttle body injection

at this point is used to break up the fuel further.

Air flow is measured by a flap or hot-wire sensor. On this housing is mounted the ECU, so a compact layout is obtained. Since the ECU contains a micro-computer it can easily process data from sensors other than the throttle and temperature units shown.

Working in conjunction with these sensors the ECU varies the fuel flow to give deceleration cut-off and enrichment during cold start, warm-up, acceleration and full-load operation.

When the engine is cold, fast idle of the engine is provided by an auxiliary air valve.

20 Air cleaners and silencers

Cleaning systems

In the course of only one day's running, an engine breathes in a considerable volume of air, which may contain a large proportion of dust. This dust will find its way into the engine cylinders and mix with the lubricating oil to form an abrasive compound which will cause wear of pistons, cylinders and bearings. In an attempt to reduce such wear, filters are generally fitted to the carburettor air intake to remove the dust particles.

These filters are often combined with silencers which eliminate or reduce the noise created by the air flow through the carburettor. This noise varies from an intense hiss at small throttle openings to a roar at full throttle.

Fig. 20.1a illustrates the construction of one type of air cleaner and silencer. Air first passes through the filtering element (1) which collects as much as possible of the dust, and then passes down the central tube (3), and into the carburettor through the connection (4). The sound waves produced in the carburettor are absorbed in the resonance chambers (2).

The filtering element shown in this diagram consists of a somewhat tangled mass of fine wire which is wetted with oil. As air threads its way through the wire mesh, dust sticks to the oil and the cleaned air passes on. Periodically (more often in very dusty conditions), the element should be thoroughly swilled in a can of paraffin (kerosine) to remove the trapped dirt. After drying, it should be dipped in clean oil and drained: this will leave a thin film of oil adhering to the wire mesh to collect more dirt.

An improved filtering element is composed of a special grade of paper through which the air passing to the engine is drawn. A considerable area of paper surface is necessary to avoid restriction of the air flow, and the necessary area is provided within a filter of reasonable size by making the paper into a large number of fine pleats as shown at (2) in Fig. 20.1b. This paper filtering element is surrounded by a pressed-steel frame and fitted into the lower part of the filter casing (4), and is enclosed by the cover (3) which is held in place by clips or screws. Air enters through the tube (6), passes through the paper element (2) and thence to the carburettor through the connection (5). A silencing action is produced partly by the form of the tube (6) and also by the absorption of sound waves from the carburettor by the paper

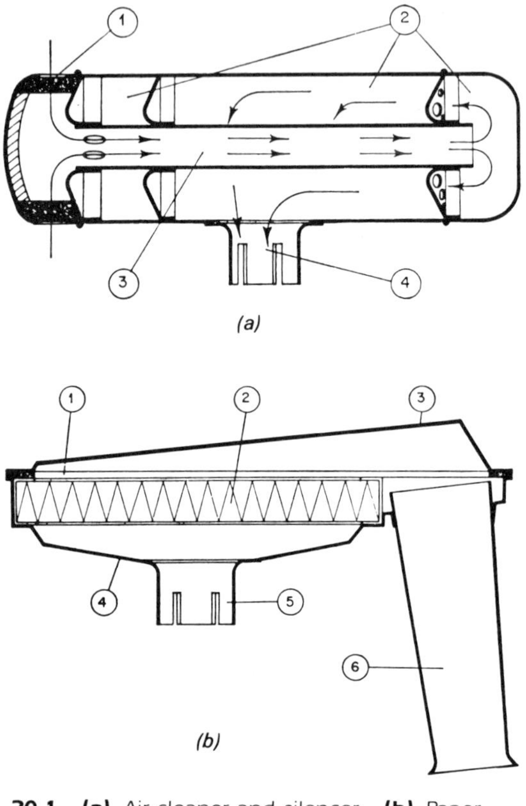

Fig. 20.1 **(a)** Air cleaner and silencer **(b)** Paper element air cleaner

element itself. An additional advantage of this type of filter is its small overall height.

Other material used for filtering elements are cloth and felt. These, like paper, cannot normally be cleaned and should be replaced at intervals as specified by the engine manufacturer.

Air cleaner intake temperature

Some air cleaners have an intake tube, (6) in Fig. 20.1b, that is adjustable to suit summer and winter conditions.

Hot air helps fuel vaporization and also prevents ice forming in the carburettor venturi, so the intake would be set to take in air from the region of the exhaust manifold during the winter season.

In summer the air is already hot, so further heating would result in low power, due to poor volumetric efficiency and overheating of the carburettor. To avoid these problems, the intake tube entrance should be sited in a cold-air region.

Automatic hot air intake system

Good vaporization of the fuel is necessary if the vehicle is to conform to emission control regulations, so an automatic control is often fitted to ensure that the air intake temperature is correct.

The air cleaner has two intake tubes; one sited to induce cold air and the other tube set to receive hot air preheated by the exhaust manifold (Fig. 20.2). Selection of the proportion of hot and cold air is by an air control flap valve, which is moved by a vacuum-operated diaphragm connected to the induction manifold. A thermo-valve mounted in the air stream in the air cleaner senses the temperature of the air entering the engine. When a given temperature is reached, say 25 °C, the thermo-valve opens and bleeds in air into the vacuum pipe. This destroys the depression in the diaphragm chamber and moves the valve towards the 'cool' position.

During slow-running and cruising conditions, the high manifold depression, e.g. above 100 mm of

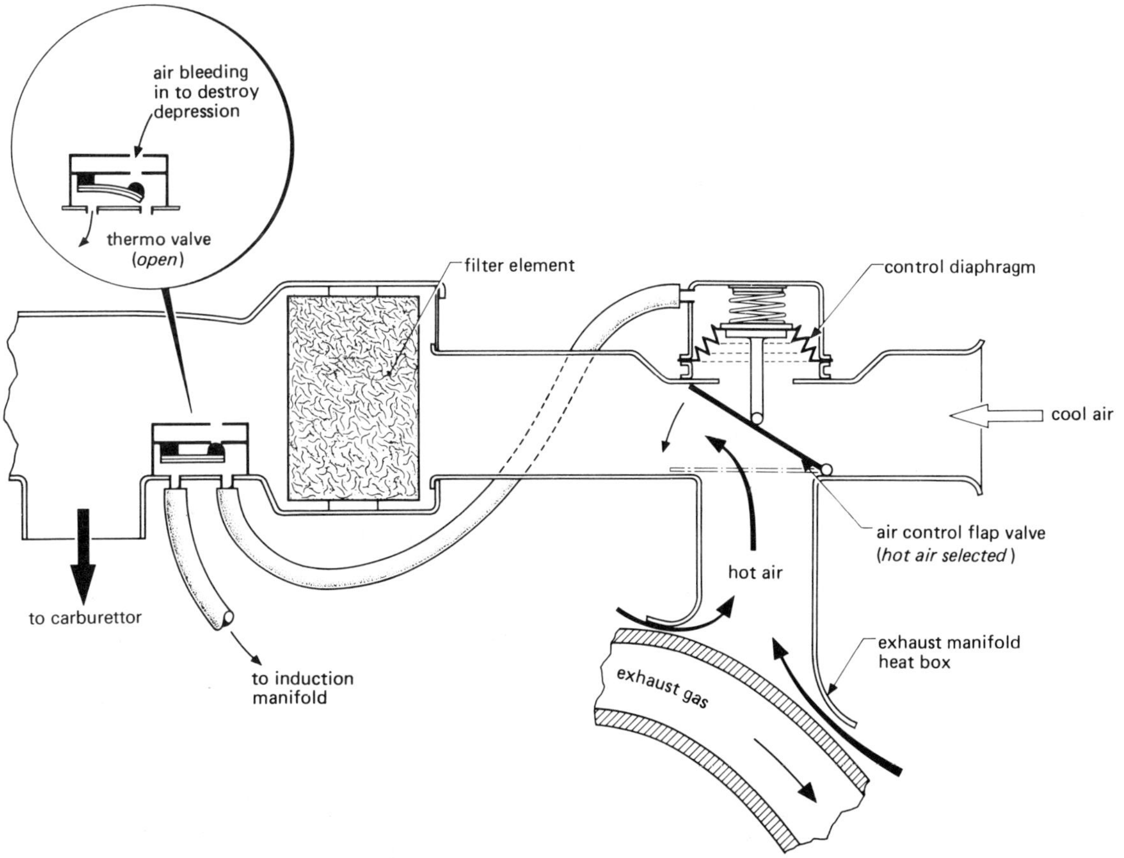

Fig. 20.2 Automatic hot air intake system

mercury, will lift the diaphragm against the spring and move the air control vent to select hot air.

Exhaust system

The exhaust system consists of three main parts:

1 The exhaust manifold, which collects the exhaust gases from a number of cylinders and leads them to a single pipe.
2 The exhaust pipe, which conducts the exhaust gases to a suitable point before discharging them into the air. This point is usually at the rear of the vehicle, but certain special types of vehicle (such as petrol tankers) may be required by law to discharge their exhaust gases at some other point.
3 The silencer.

Exhaust manifolds

When an engine is running at full throttle, the temperature of the exhaust gases may reach 800 °C or more. To withstand such temperatures, exhaust manifolds are generally made of iron castings or fabricated from steel pipes and plate by welding.

The chief problem in the layout of an exhaust manifold is possible interference between cylinders. Consider the manifold shown in Fig. 20.3, which is fitted to a four-cylinder engine having the firing order 1243. At the moment when piston 1 is approaching t.d.c. at the end of its exhaust stroke and the momentum of the gas moving along the exhaust pipe should be helping to scavenge the burnt gas from the cylinder, the exhaust valve of cylinder 2 opens and discharges into the manifold a cylinderful of burnt gas at high pressure. The shape of the manifold will clearly result in some of this gas discharged from cylinder 2 entering cylinder 1, which will thus retain an unduly large proportion of burnt gas when its exhaust valve closes.

The manifold shown in Fig. 20.4 is an improvement on that of Fig. 20.3 since the entry of each port into the manifold is curved so as to maintain an outward flow

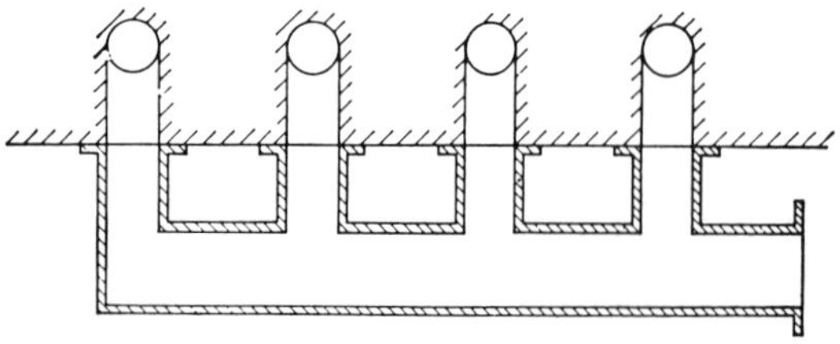

Fig. 20.3 Simple exhaust manifold

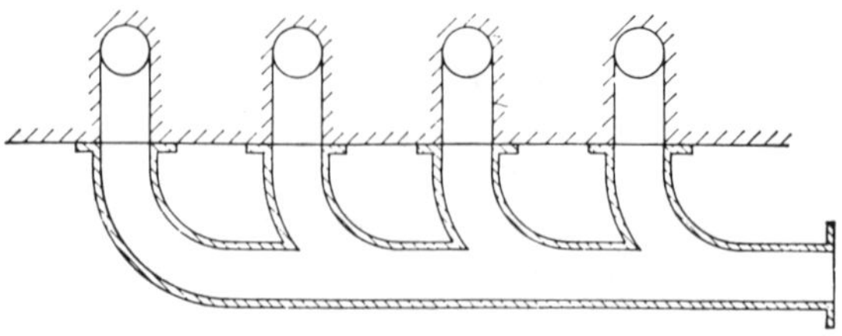

Fig. 20.4 Improved exhaust manifold

from each port rather than permit one port to blow back into another.

Exhaust ports are sometimes siamesed to simplify and cheapen the cylinder head casting, a common arrangement being that shown in Fig. 20.5. Note the shaping of the end branches of the manifold to avoid interference between cylinders, and note also that this cannot occur between the cylinders served by the siamesed ports since, on a four-cylinder engine, exhaust strokes occur in these cylinders one revolution apart.

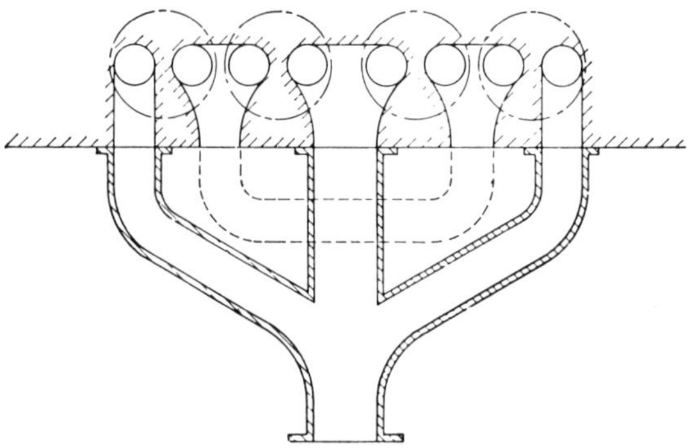

Fig. 20.5 Exhaust manifold for four-cylinder engine with one pair of exhaust ports siamesed

Interference can be completely avoided by having a separate pipe for each cylinder, but for ordinary cars and commercial vehicles this arrangement is too bulky and expensive. Furthermore, it is possible to use the gas pulsations from one cylinder to assist the flow of gas from another at certain engine speeds. This is done by joining separate pipes from each cylinder, first in pairs, making two pipes which are then joined into one. The lengths of the various sections of pipes are important, depending upon the engine speed at which the greatest benefit is desired; in a four-cylinder engine the pairs of pipes to be joined are 1 and 4, and 2 and 3. Thus, when performance is an important consideration, a pipe layout similar to that shown diagrammatically in Fig. 20.6 is used for a four-cylinder engine.

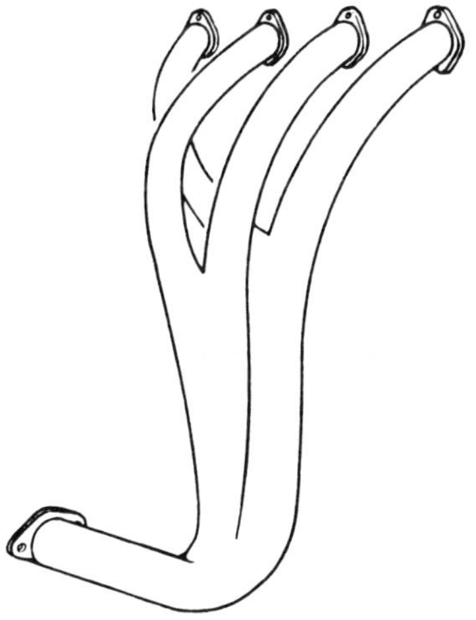

Fig. 20.6 High-efficiency exhaust manifold for four-cylinder engine

For six-cylinder engines the cylinders which should be *paired* in this manner are 1 and 6, 2 and 5, and 3 and 4, but this is much more difficult to arrange compactly and neatly than in the case of a four-cylinder engine. It is, however, easy and convenient to use two manifolds, one for cylinders 1, 2 and 3, and the other for cylinders 4, 5 and 6, each of which groups have equally spaced firing intervals. The pipes from these two manifolds usually join at some distance from the manifolds, but in a few cases two completely independent exhaust systems are used, one for each group of three cylinders.

Exhaust pipes

These pipes are made of steel tubing which is, in some cases, protected against corrosion and oxidation by being coated with aluminium by a metal-spraying process.

Since engines are usually carried on flexible mountings, provision must be made for their movement relative to the frame. In some cases, there is a length of flexible exhaust piping between the pipes or manifolds bolted to the ports and the rear part of the pipe which is attached to the frame by rigid brackets. Alternatively, the whole exhaust system is made up of rigid pipes and connections but is supported from the frame by flexible attachments, usually incorporating bonded rubber blocks.

Silencers (mufflers)

The exhaust noise of an 'unsilenced' engine is caused by the release of hot gas into the atmosphere in a series of high-pressure pulsations. This noise must be reduced to a level below the legal limit specified in the regulations.

Analysis of exhaust noise shows that although it is made up of a number of different sounds or notes, it is possible to group the majority of the sounds into two main frequency bands:

1 *Low frequency* - caused by the air pumping action of the engine. This sound is also produced when the engine is motored over without combustion taking place.
2 *High frequency* - caused by the release and rapid expansion of very hot gas at a temperature of 800–1000 °C.

To damp these frequencies two silencers are needed: a *capacity type* for the low frequencies and an *absorption type* for the high frequencies.

Capacity type Fig. 20.7 shows this type; it consists of a large chamber in which the regular gas pulsations can be damped (smoothed out) by the bulk of gas contained in the chamber. Ideally this expansion chamber should be large, but because the available space often restricts the size, more than one capacity-type silencer is sometimes fitted. The damping of the low-frequency waves is achieved by fitting baffles to create a turbulent zone to break up the flow of gas in the silencer.

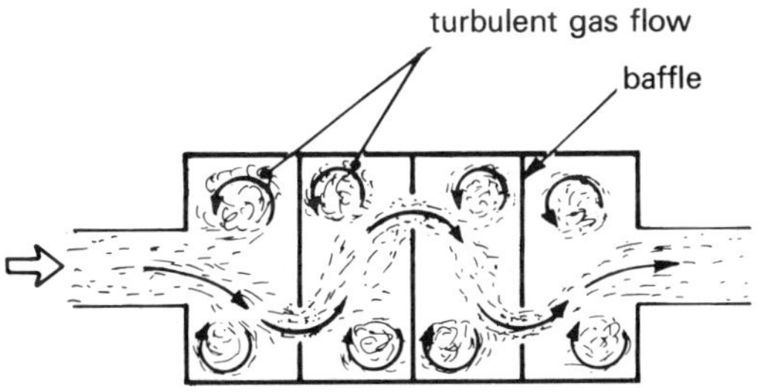

Fig. 20.7 Capacity silencer

Baffles reduce the free flow of gas, so a bad design could cause a build-up of pressure in the exhaust system. This *back pressure* reduces engine power because any exhaust gas that is allowed to remain in the cylinder (exhaust gas residual) will occupy space that would otherwise be filled with the fresh charge, i.e. a poor exhaust system can cause a reduction in the volumetric efficiency. The engine power suffers both from the loss of cylinder volume occupied by the exhaust gas residual, and from the heat of the burnt gas; the presence of this gas with its very high temperature heats the new charge as it enters, lowering its density and reducing its mass.

A well-designed, or finely tuned, exhaust system can improve engine power by using the negative and positive pressure pulses which travel back and forth along the front exhaust pipe to start the new charge flowing into the cylinder during the valve overlap period. An ideal pipe length gives a negative pressure pulse in the cylinder at the instant the inlet valve opens. The assistance given by the exhaust is only achieved at certain engine speeds, so the design should be tuned to give its support at the time when extra engine power is required.

Absorption type The construction of this type, also called a *Burgess silencer*, is shown in Fig. 20.8. It consists of a straight-through perforated tube which is surrounded by a sound-absorbing material such as glass fibre. Damping of the high-frequency sound waves is achieved by the material, which causes a slight retardation in the wave. When the wave returns from the outer chamber it is out of phase with the vibration of the gas passing through the perforated tube, so the opposing action of the two waves smooths out the high-frequency peaks.

Composite type This type of silencer, shown in Fig. 20.9, is a combination of the capacity and absorption types. This combination is attractive, assuming space or ground clearance is sufficient for its installation.

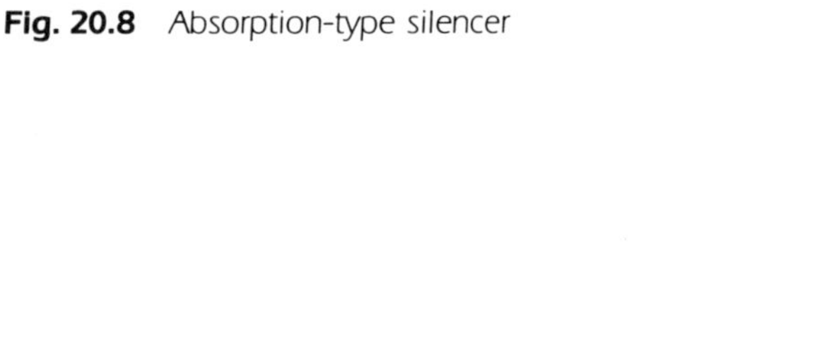

Fig. 20.8 Absorption-type silencer

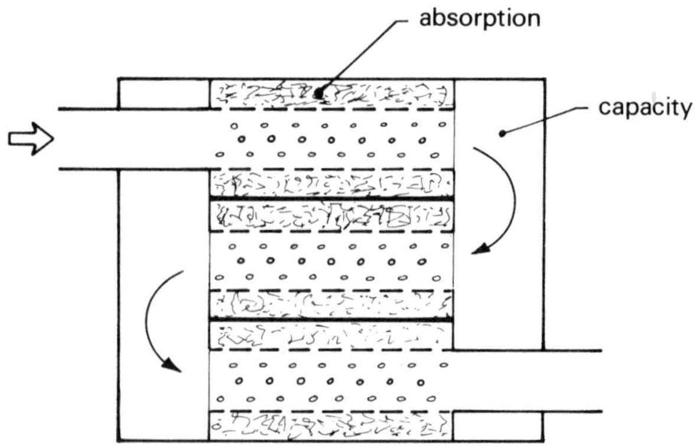

Fig. 20.9 Composite-type silencer

Some silencers have a number of tubes and/or chambers to produce out-of-phase sound waves that vibrate in sympathy with the main frequencies of the gas passing into the silencer; hence the name *resonator* that is sometimes used to describe this type.

Silencer material

Steam is one of the products of combustion. When the exhaust system is cold, the steam condenses to form water that either drips from the tailpipe or collects in the silencer until the system heats up. The combined effects of water and acids from combustion, high-temperature gas, and water from the road, produces a highly corrosive action which rapidly attacks the normal low-carbon steel exhaust system. This corrosion can be resisted by coating the steel with aluminium or by using a more expensive material such as stainless steel.

Catalytic silencers are described on page 133.

21 Emission control

Environmental considerations have forced many countries to introduce regulations to limit the pollution caused by motor-vehicles. It is claimed that emissions from motor-vehicles damage human health, plant life and the environment. This problem is particularly severe in areas where the geographic and climatic conditions create an atmospheric envelope which traps the pollutants.

Areas that first introduced stringent controls were Australia, Japan, Sweden and the USA, in particular the state of California. In recent years, as these countries lowered the maximum allowable emission limits, other countries including those of the EEC, introduced their own standards. Each year sees a tightening of the standards, so manufacturers have to update or modify their vehicles continually to meet the particular requirements laid down for the country in which the vehicle is to be sold.

Table 21.1 shows the main pollutants; the sources of these are:

1 *Exhaust gas* - can contain unburnt fuel (HC), partially burnt fuel (CO), dangerous nitrogen oxides (NO_x) from combustion, and lead (Pb) from petrol additives.
2 *Crankcase* - vaporized lubrication oil (HC) and corrosive acid compounds.
3 *Fuel system including fuel tank* - petrol vapour (HC).

Devices used to 'clean up' a vehicle are costly and their use often results in a lower power output and higher fuel consumption. Consequently, manufacturers often only fit the emission control devices that are needed to meet the local regulations.

Some emission controls are now used as standard and mention of some of these is made in this book under the appropriate topic heading.

Table 21.1 Main pollutants produced by motor vehicles

Pollutant	*Origin*	*Effects*
CO (Carbon monoxide)	Partially burnt fuel	Poisonous to human beings when inhaled. CO adheres to haemoglobin in the blood and prevents oxygen being carried to body cells
HC (hydrocarbon)	Unburnt fuel, vaporized fuel escaping from fuel system	Irritates eyes and nose. Cancer risk. Odour
C (carbon)	Partially burnt fuel	Smoke - restriction in visibility. Can carry carcinogenic (cancer) constituents. Odour
NO_x (oxides of nitrogen - NO and NO_2)	Very high combustion temperatures cause nitrogen to combine with oxygen	Toxic to humans. NO_x combines with water to form nitrous acid, which causes lung disorders. It combines with other exhaust products to give eye and nose irritants; also affects nervous system. Smog
Pb (lead)	Added to petrol to raise octane rating	Toxic to humans, causing blood poisoning and nervous disorders

Table 21.2 Emission limits for petrol engine, based on EEC Directive 70/220/EEC (89/458/EEC)

Engine capacity (litres)	*Mass of CO (g/test)*	*Combined mass of HC and NO_x (g/test)*	*Mass of NO_x (g/test)*
over 2.0	25	6.5	3.5
1.4 to 2.0	30	8.0	
under 1.4	19	5.0	

Emission regulations

Before any new vehicle is sold in EEC countries, a number of vehicles must be submitted to the regulating body of a Member State to enable tests to be carried out to confirm that the vehicle type meets current EEC standards. This *EEC type approval* procedure includes engine tests on a minimum of five vehicles to ensure that emission levels are within current limits.

Emission limits laid down by the Council of European Communities are gradually being tightened so that by 1992 they will equal the standards for vehicle emissions in force in the USA. These are:

HC 0.41 CO 3.4 NO_x 1.0

These values were introduced in 1983 and are expressed in grams per mile. The limits must be maintained for 80 000 km (50 000 miles).

New petrol-engined vehicles sold in EEC countries at an agreed introduction date for the standard after 1992 must have emission levels lower than the values shown in Table 21.2. These levels apply to engines which are bench tested in accordance with a special European programme; this includes cold/hot start conditions, urban driving and simulated motorway driving at various speeds up to 120 km/h.

Engines submitted for test must be designed to run on unleaded petrol; this is to reduce and stop the use of lead-based additives in fuels. In addition the petrol filler pipe must be so designed that it prevents the tank from being filled from a petrol pump delivery nozzle having an external diameter of 23.6 mm or greater. This regulation means that nozzles of pumps supplying lead-free petrol have to be smaller than those used on pumps dispensing leaded fuel.

The method adopted by an engine designer to meet the emission limits depends on the technology available to a manufacturer. The cost of fitting a catalyst, and its accompanying fuel mixture control system, on small and medium-sized cars, forced many manufacturers to develop lean-burn engines. Tests show that this type of engine gives a good fuel economy and a much lower emission level than other conventional engines. It was expected that the lean-burn type would meet future emission requirements, but the introduction in 1989 of a more stringent standard based on a new European Extra-urban Driving Cycle (EUDC) has meant that to attain the new limits the lean-burn engine needs an exhaust catalyst.

Diesel engines must also meet emission standards and, as in the case of petrol engines, the limits laid down by the EEC are gradually being reduced to give a cleaner environment. From 1 July 1988 the emission levels for new designs of diesel engines are shown in Table 21.3. **N.B.** The mass of gas measured in grams per hour is divided by the engine power to give the unit 'g/kW h'.

Table 21.3 Emission limits for diesel engines, based on EEC Directive 70/156/EEC (88/77/EEC)

Mass of CO (g/kW h)	*Mass of HC (g/kW h)*	*Mass of NO_x (g/kW h)*
11.2	2.4	14.4

Control of engine emissions

Various design modifications and extra fitments are necessary to control engine emissions. The following is a summary of the main features:

Carbon monoxide (CO)

This is produced when a cylinder receives either a rich mixture or a poor quality (patchy) mixture. Methods of improvement are:

1 *Closer control of mixture strength* supplied by the carburettor, especially under slow-running and cold-starting conditions.
2 *Better distribution*, both in the carburettor and manifold. Fuel injection is one way of overcoming this problem.
3 *More precise engine tuning* by mechanics. The CO content should be measured by a meter and all

adjusters should be sealed with tamper-proof plugs.

4 *More compact combustion chamber* Long narrow chambers associated with an oversquare engine often give a high CO content.

5 *Throttle positioner system* – to open the throttle slightly when slow-running or when decelerating. (This also reduces HC emission.)

6 *Precise ignition timing* to ensure that the spark occurs at the correct time and remains correct during the period between servicing. Transistorized ignition is sometimes necessary to achieve this requirement.

Hydrocarbons (HC)

These are produced when fuel vaporizes and escapes to the atmosphere from the fuel system or when the fuel is unburnt in the cylinder. Methods of improvement are:

1 *Closed crankcase ventilation system* Unburnt fuel passing the pistons and entering the crankcase is prevented from escaping to the atmosphere by a *positive crankcase ventilation (PCV)* system. The unburnt fuel is returned to the induction system.

2 *Sealed fuel system* A *fuel evaporative emission control (EVAP)* system seals the fuel tank and float chamber, collects the vaporized fuel, passes it through a charcoal-filled canister and delivers it to the induction manifold for combustion in the engine cylinders.

3 *Altering the vacuum advance of the ignition* A system such as a dual-diaphragm vacuum advance unit retards the ignition during the periods when the engine is slow-running or decelerating.

4 *Mixture adjustment during deceleration* The carburettor is provided with a device which on some engines cuts off the fuel and on other engines delivers extra fuel to ensure that the mixture burns efficiently when the engine is decelerated.

5 *Reburning fuel in exhaust system* *Air injection (AI)* into the exhaust port allows the unburnt and partially burnt fuel to be completely combusted in the exhaust system. This system is sometimes called an *after-burner* and it operates when the engine is slow-running or cruising.

6 *Exhaust gas oxidation* Many engines cannot be 'cleaned up' internally to meet the emission limits, so a catalytic converter has to be fitted in the exhaust system. This looks similar to a large conventional silencer but is much heavier and more costly.

The term 'catalyst' is the name given to a material that produces a chemical action without undergoing any change itself.

In *two-way* converters the catalyst material changes the gas by oxidizing the hydrocarbon and carbon monoxide emissions to form water and carbon dioxide. Later designs of converter necessary to meet tighter emission limits also have to remove oxygen from the nitrogen oxide compounds; these converters are called *three-way*.

Converter construction

There are two basic types of converter, *monolith* and *pellet (bead)*. In the monolith type the large catalyst surface area needed for exposure to the gas is achieved by using a honeycomb construction, whereas the pellet type has a number of aluminium-coated wire baskets to support the catalyst covering.

At present, the precious metals platinum, palladium and rhodium are the most suitable catalyst agents; the actual metal used, or mixed combination, governs the type of converter. A two-way oxidizing type uses platinum and palladium whereas the three-way oxidizing/reducing converter uses platinum and rhodium. Since each converter has a weight of about 4 grams of precious metal, cheaper converters using base metals such as copper and chromium are being developed. At present these metals are more prone to being 'poisoned' by small quantities of sulphur present in the fuel.

All converters become ineffective if the catalyst comes into contact with lead. Failure to use a lead-free fuel results in lead being deposited on the catalyst surface; this causes the catalyst to be isolated from the gas.

Monolith type The construction of this 'single block' type converter is shown in Fig. 21.1. In this type the aluminized ceramic or steel honeycomb forms a stable and secure base for the catalytic deposit of precious metals besides providing a gas exposure area equivalent to several football pitches.

When used with a lean-burn engine, a two-way catalyst uses the excess oxygen exhausted from the engine to react with the hydrocarbons. In the case of a three-way catalyst fitted to a normal engine, the nitrogen oxide content in exhaust gas acts as the

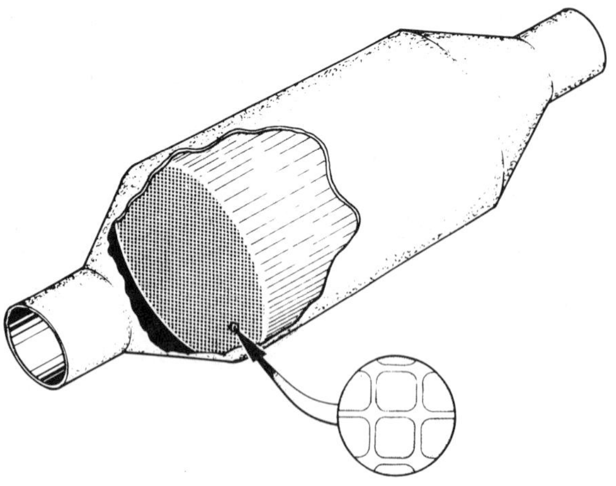

Fig. 21.1 Monolith-type catalytic converter

oxidizing agent to promote combustion of carbon monoxide and hydrocarbons to carbon dioxide and water while the nitrogen oxides, being stripped of their oxygen, are reduced to nitrogen.

The three-way catalyst only operates effectively when used with a fuel control system such as a 'closed-loop' electronic engine management system; this ensures that the air/fuel ratio is maintained at the precise *stoichiometric point* (chemically correct ratio).

Pellet type Fig. 21.2 shows a diagram of a dual-bed, three-way converter similar to that used for many years on large engines in the USA.

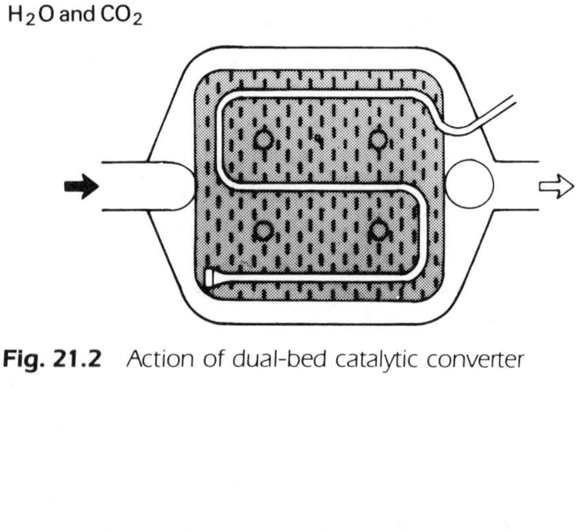

Fig. 21.2 Action of dual-bed catalytic converter

In the layout shown the first bed contains a three-way catalyst and the second bed an oxidizing catalyst. The first bed reduces the NO_x to basic nitrogen and then uses the released oxygen to oxidize some of the HC and CO. Secondary air, introduced by a pump to a chamber between the beds, mixes with the remaining HC and CO and allows the gas to be fully oxidized after it passes through the second bed. During this process a considerable amount of heat is generated.

For a converter to operate efficiently, and to avoid a fire hazard in the area of the converter, the engine must NOT:

- use leaded fuel
- be slow-run for long periods
- be tested by shorting-out the sparking plugs to earth
- be compression tested for long periods
- operate in a condition that delivers neat fuel to the exhaust system.

Lean-burn engine During recent years some engines have been designed to operate efficiently on lean mixtures up to about 22 : 1. When compared with conventional engines, the lean-burn arrangement gains in fuel economy and has a lower level of emission. For engines of capacity less than 2 litres the lower emissions levels, the retention of a high compression ratio and the ability of the engine to use low-octane unleaded fuels make the lean-burn design attractive.

Two major sources of unburnt HC in a conventional engine are:

1 Crevices in the combustion chamber that trap the unburnt mixture, especially those crevices in the regions of the piston ring and head gasket.
2 Absorption and desorption of HC by the oil film and combustion chamber deposits.

Attention to these factors, together with features that give high turbulence at low loads without sacrificing volumetric efficiency for full-load operation, enables the engine to operate satisfactorily on lean mixtures. At steady cruising speeds this gives a reduction in the HC emission, and because the combustion temperature is lower, the NO_x emission is also reduced. Lean-burn engines have lower power losses from pumping, heat transfer and dissociation (fuel breakdown at high temperature into its separate chemical 'elements' without being burnt), so these factors contribute to the higher fuel economy.

Most designs of lean-burn engine use a four-valve combustion chamber and have specially shaped inlet ports which speed up and direct the incoming

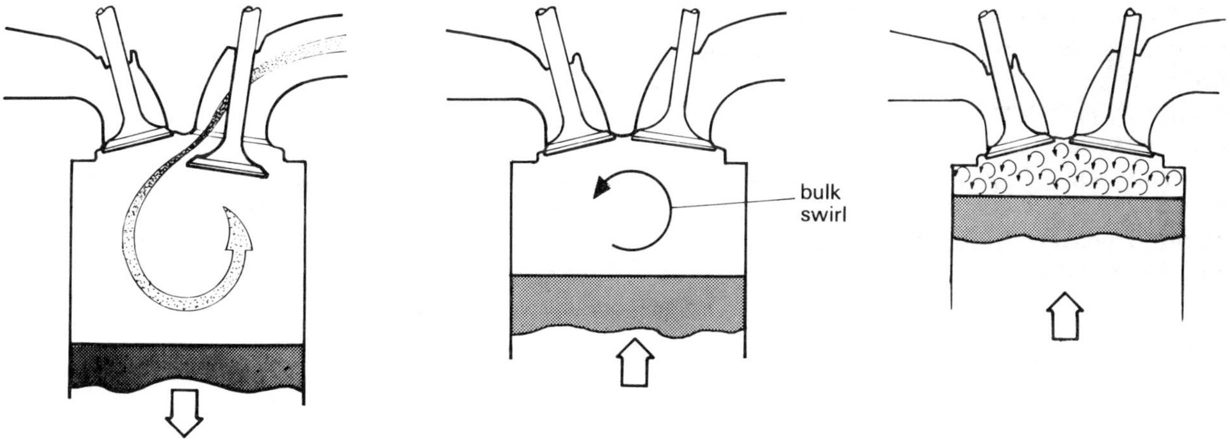

(a) Barrel swirl *(b) Micro-turbulence*

Fig. 21.3 Turbulence in a lean-burn engine

mixture to give *barrel swirl*, a gas movement shown in Fig. 21.3a.

The tumbling motion of the gas continues until the latter part of the compression stroke; at this stage the main swirl motion is broken up into small swirl patterns as shown in Fig. 21.3b. It is this 'micro-turbulence' action, generated just prior to the point of ignition, that allows the engine to operate on lean-burn mixtures.

Oxides of nitrogen (NO_x)

These are produced when the combustion temperature exceeds a given value. Methods of improvement are:

1 *Decrease in flame speed. Exhaust gas recirculation (EGR)* system directs some exhaust gas back to the induction manifold to slow down the combustion when the engine is under high load. Naturally this system reduces the maximum power of the engine.

2 *Combustion chamber alteration* The shape is changed to increase the flame travel. Used in conjunction with lower compression ratios, these chambers reduce the NO_x content, but normally only at the expense of fuel economy.

3 *Increase in air/fuel ratio* The highest flame speed and NO_x content occurs when the mixture is about 12 per cent richer than the chemically correct value. An engine designed to operate on a weak mixture improves the emission problem, but the vehicle driveability suffers unless very precise tuning is carried out. (*Driveability* is the manner in which a vehicle responds during acceleration, and the smoothness of engine operation throughout the speed range especially during the warm-up period.) Some engines, such as the *stratified charge* type, are designed to operate on ultra-weak mixtures with air/fuel ratios higher than the operating limit of a conventional engine. The stratified charge engine uses a layer of rich mixture in the region of the sparking plug to ignite the very weak mixture in the main part of the cylinder.

4 *Ignition timing alteration* A *spark control (SC)* valve in the vacuum pipe to the distributor delays the vacuum advance for a given time when the throttle is plunged open.

Lead – present in the exhaust when tetra-methyl-lead and/or tetra-ethyl-lead (TML/TEL) is used as a cheap additive to improve the octane rating of a fuel, i.e. to improve the fuel's resistance to combustion knock. With an octane rating of 98 RON (Research Octane Number), high compression ratios in the region of 9 : 1 were used that resulted in a high thermal efficiency, hence good fuel consumption.

To comply with EEC and UK regulations the lead additive in fuels has gradually been reduced from 0.84 g/litre in 1972 to 0.15 g/litre in 1989. This has made it difficult for many petrol companies to maintain the 97 RON minimum for the 4 star grade of fuel with the result that knock-free operation of high-compression engines is not always possible.

In addition to the 'leaded' petrols the introduction of a 'lead-free' fuel, with a target lead content similar to the USA content of 0.005 g/litre, has been introduced for engines having catalytic converters and for drivers who are conscious of the ill effects of lead.

Lead-free fuel, with its lower octane rating (95 RON), may be used on about 60 per cent of engines produced between 1985 and 1988, assuming the engine is tuned to accept this fuel. Compared with past designs these engines have a lower compression ratio, a less advanced ignition setting and a valve and seat material that allows a low-octane fuel to be used without damage.

By the early 1990s, the EEC, in company with many other countries, will only allow the manufacture of engines that are suitable for operation on lead-free fuel. Legislation will then follow which will ban the sale of leaded fuel.

PART FOUR

The cooling system

22 The function of the cooling system

During combustion, and when the engine is operating at full throttle, the maximum temperature reached by the burning gases may be as high as 1500–2000 °C. The expansion of the gases during the power stroke lowers their temperature considerably, but during the exhaust stroke the gas temperature may still be not far short of 800 °C (Fig. 22.1). All the engine parts with which these hot gases come in contact will absorb heat from them in proportion to the gas temperature, the area of surface exposed and the duration of the exposure; this heat will raise the temperature of the engine components. The temperature of even the exhaust gases is above red heat and much above the melting point of such metals as aluminium. Unless steps are taken to limit the temperature of the engine parts, a number of more or less serious troubles will arise:

1 The combustion chamber walls, piston crown, the upper end of the cylinder and the region of the exhaust port are exposed to the hottest gases and will reach the highest temperatures. The resulting thermal expansion of these parts will distort them from their correct shape, causing gas leakage, loss of power, valve burning, and possibly even cracking of cylinder or head.
2 The oil film which should lubricate the piston and cylinder walls will be burnt or carbonized, causing excessive wear and even seizure of the piston.
3 Power will be reduced by the heating of the fresh gas entering the cylinder, so reducing its density. The increased temperature of the fresh gas will also increase the liability to detonation, thus making a reduction in compression ratio necessary.
4 Some part of the surface of the combustion chamber may become hot enough to ignite the fresh gas before the spark occurs: this is called *pre-ignition*, and will result in serious damage to the engine if it persists.

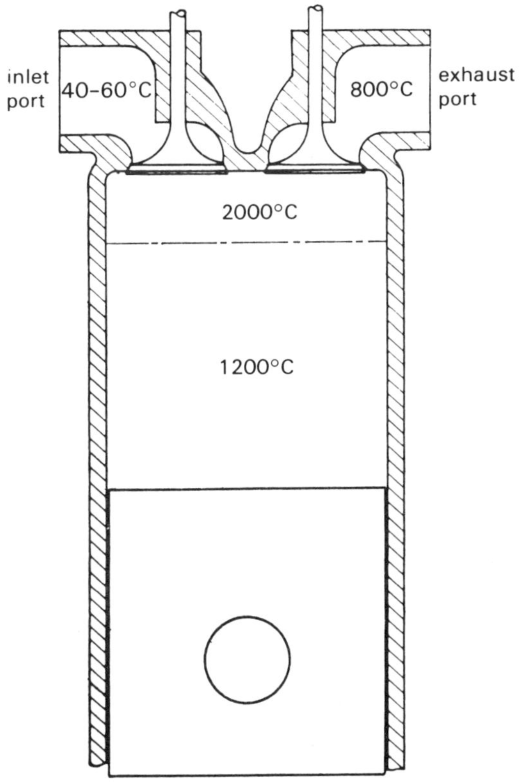

Fig. 22.1 Approximate gas temperatures in engine

The function of the cooling system is to remove heat from the engine parts at a high enough rate to keep their temperature within safe limits and so avoid these troubles.

It is, however, important not to overcool the engine, or other troubles will be encountered:

1 Heat is necessary to assist in vaporizing the fuel inside the cylinder during the compression stroke. Unvaporized fuel will be deposited on cold cylinder walls, and besides being wasted it will dilute the lubricating oil and destroy its lubricating properties.
2 Water vapour formed during combustion will condense on the cold cylinder walls, forming a sludge with the lubricating oil and corroding engine parts. Thus the rate of wear is considerably greater when the engine is cold than when hot.

Experience suggests that the temperature of the cylinder head must be kept below about 200–250 °C if overheating is to be avoided.

There are two main systems in use, both of which dissipate heat removed from the cylinder into the surrounding air:

Direct air cooling

In this system, more simply called *air cooling*, heat is radiated from the cylinder and head directly into the surrounding air. The rate at which heat is radiated from an object depends upon:

1 The difference in temperature between the object and the surrounding air.
2 The surface area from which heat is radiated.

Since (1) must be limited, the surface area of the cylinder and head exposed to the air must be increased, and this is done by forming *fins* on their external surfaces (Fig. 22.2). It is also necessary to remove the heated air from around the cylinder and deliver a constant supply of cool air around and between the fins. This means that the cylinders must be sufficiently widely spaced to permit a suitable depth of finning all around them, and the engine must be placed in an exposed position where the motion of the vehicle can provide the necessary supply of cool air. If it is necessary or desirable to enclose the engine, for the sake of protection or appearance, a fan must be used to supply the air, with suitable cowls to direct the air where it is needed.

Liquid cooling

In this system the outer surfaces of the cylinder and head are enclosed in a casing or *jacket*, leaving a space between cylinder and jacket through which a suitable liquid is circulated. The liquid generally used is water, which is in many ways the most suitable for this purpose, though it has certain drawbacks. During its passage through the jacket, the water absorbs heat from the cylinder and head, and it is cooled by being passed through a *radiator* before being returned to the jacket.

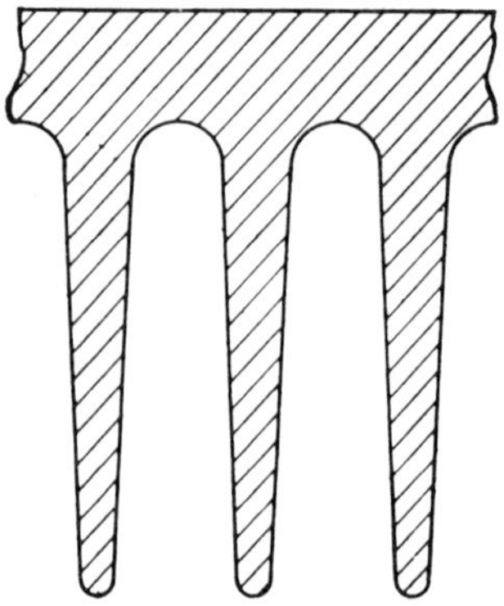

Fig. 22.2 Section of fins for air-cooled cylinders

Comparison of air- and water-cooled systems

Air cooling has several points in its favour:

1 An air-cooled engine should generally be lighter than the equivalent water-cooled engine.
2 The engine warms up to its normal running temperature very quickly.
3 The engine can operate at a higher temperature than a water-cooled engine.
4 The system is free from leakage problems and requires no maintenance.
5 There is no risk of damage due to freezing of the coolant in cold weather.

It has, however, a number of disadvantages:

1 A fan and suitable cowls are necessary to provide and direct the air flow. The fan is noisy and absorbs a fairly large amount of power. The cowling makes it difficult to get at certain parts of the engine.
2 The engine is more liable to overheating under arduous conditions than a water-cooled engine.
3 Mechanical engine noises tend to be amplified by the fins.
4 The cylinders usually have to be made separately to ensure proper formation of the fins. This makes the engine more costly to manufacture.
5 Cylinders must be spaced well apart to allow sufficient depth of fins.
6 It is more difficult to arrange a satisfactory car heating system.

The chief points in favour of *water cooling* are:

1 Temperatures throughout the engine are more uniform, thus distortion is minimized.
2 Cylinders can be placed close together and the engine made more compact.
3 Although a fan is usually fitted to force air through the radiator, it is smaller than that required in an air-cooled system and is thus quieter and absorbs less power.
4 There is no cowling to obstruct access to the engine.
5 The water and jackets deaden mechanical noise.
6 The engine is better able to operate under arduous conditions without overheating.

Its main disadvantages are:

1 Weight – not only of the radiator and connections but also of the water: the whole engine installation is likely to be heavier than an equivalent air-cooled engine.
2 Because the water has to be heated, the engine takes longer to warm up after starting from cold.
3 If water is used, the maximum temperature is limited to about 85–90 °C to avoid the risk of boiling away the water.
4 If the engine is left standing in very cold weather, precautions must be taken to prevent the water freezing in the cylinder jackets and cracking them.
5 There is a constant risk of leakage developing.
6 A certain amount of maintenance is needed, e.g. checking water level, anti-frost precautions, cleaning out deposits, etc.

23 Main features of an air-cooled system

The volume of air required to conduct a given amount of heat away from the cylinders of an air-cooled engine is about 2000 times the volume of water necessary to remove the same amount of heat, and in order to provide the necessary air flow around the cylinders of an enclosed engine, a powerful fan is essential. Fig. 23.1a shows a fan of the simple curved-blade type, known as an *axial-flow* fan because the direction of air flow is parallel to the axis of the fan spindle. This type of fan is sometimes used, but the *radial-flow* or centrifugal type, shown in Fig. 23.1b, is more often used since it is more effective and a fan of smaller diameter can be used for a given air flow. This type of fan consists of a number of curved radial vanes mounted between two discs, one or both having a large central hole. When the fan is rotated, air between the vanes rotates with it and is thrown outwards by centrifugal force.

Fig. 23.2 shows a simple air-cooled system for a four-cylinder in-line engine. A centrifugal fan (4), driven at approximately twice crankshaft speed, is mounted at the front of the engine and takes in air through a central opening (5) in the fan casing (6). From the fan the air is delivered into a cowl (3), from which it passes over the fins on the engine cylinders (1). Baffles (2) direct the air flow between the fins, from which the air picks up heat, so cooling the cylinders.

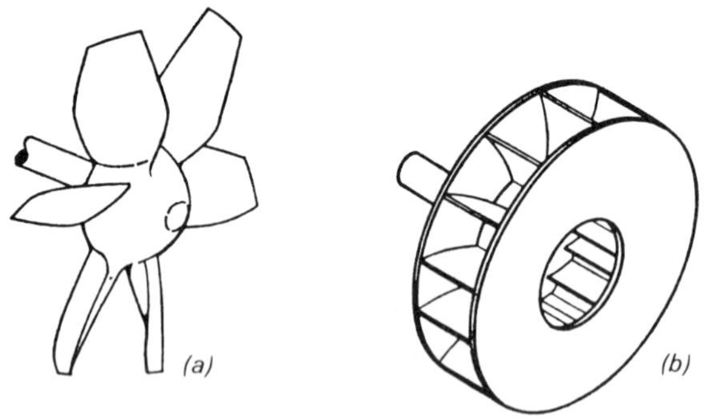

Fig. 23.1 Types of fan

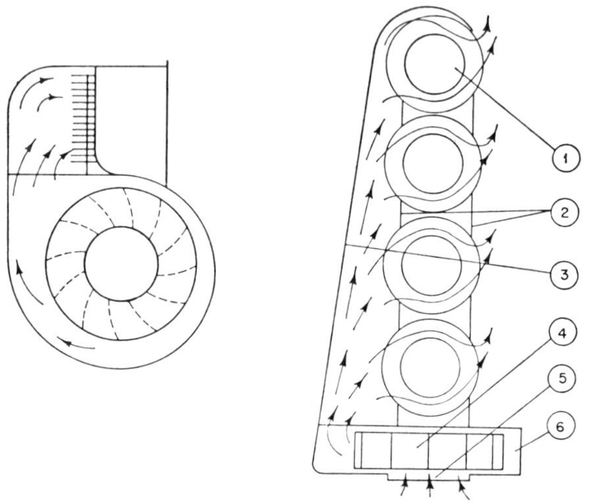

Fig. 23.2 Air-cooled system for four-cylinder in-line engine

The in-line type of engine shown is not the most suitable for air-cooling, since the cylinders have to be placed further apart than would otherwise be necessary, in order to allow enough air to flow between the cylinders. Vee-type engines, or horizontally opposed types, are better in this respect since the cylinders have to be spaced far enough apart to leave room for the crankshaft bearings: this allows a good air flow between the cylinders while keeping the total engine length short.

24 Main features of a liquid-cooled system

A brief outline of this system has already been given in Chapter 22 and we must now consider it in more detail.

Coolant

The liquid used to carry heat away from the cylinders and heads is almost invariably water, to which may be added certain chemicals. The chief advantages of water are:

1 It has a high specific heat, i.e. a given amount of water, heated through a given temperature range, absorbs more heat than almost any other substance.
2 It is readily available in *most* parts of the world at little or no cost.

It has, however, certain disadvantages:

1 It boils at a temperature somewhat lower than is desirable for best possible engine performance, especially at heights much above sea level.
2 It freezes at temperatures often encountered in winter in many parts of the world. Since ice expands with fall in temperature, there is a considerable risk of cracking the cylinder blocks if freezing is allowed to occur.
3 There may be some trouble with corrosion of the metals with which it comes in contact.

These objections can be wholly or partly overcome in the following ways:

1 The temperature at which the coolant boils can be raised by operating the system under pressure. The way this is done will be described later, but the boiling temperature is raised by about 1°C for every 4 kN/m^2 (0.6 lbf/in^2) that the pressure is increased.
2 The freezing temperature of the coolant can be lowered by adding an *anti-freeze* chemical. The one most commonly used is ethylene glycol, and a solution containing about 20 per cent of this in water is sufficient to give complete protection in Britain. The strength of the solution can be checked by means of a suitable hydrometer.
3 The risk of corrosion can be reduced or eliminated by the addition of suitable chemicals. These are usually included in the anti-freeze solution.

Thermo-syphon system

The coolant must be made to circulate around the cooling system so that heat absorbed by the coolant around the cylinders can be dissipated in the radiator. The simplest method of producing this circulation relies upon the convection currents in the coolant. These result from the reduction in density caused by expansion of the coolant with increase in temperature. A system using this method of circulation is known as a *natural circulation* system, or a *thermo-syphon* system, and is illustrated in Fig. 24.1.

The basic system consists of a water jacket which is connected by a synthetic rubber hose to a header tank. This tank forms part of the radiator, a heat exchanger made by connecting two tanks, header and lower, with a number of finned tubes to provide a large surface area for the disposal of unwanted heat. Air flow over the tubes and fins carries away the heat radiated from the hot coolant and so lowers the temperature of the coolant as it passes down the tubes. A bottom hose connects the lower tank to the jacket to provide a return path to the engine for the coolant.

When the engine is running and the vehicle is either stationary or travelling slowly, the air flow through the radiator is insufficient to give adequate cooling. At these times overheating is avoided by using a *fan* to act as an air pump. Energy to drive the fan is supplied by a vee belt from the crankshaft, or alternatively the fan can be driven direct by an electric motor.

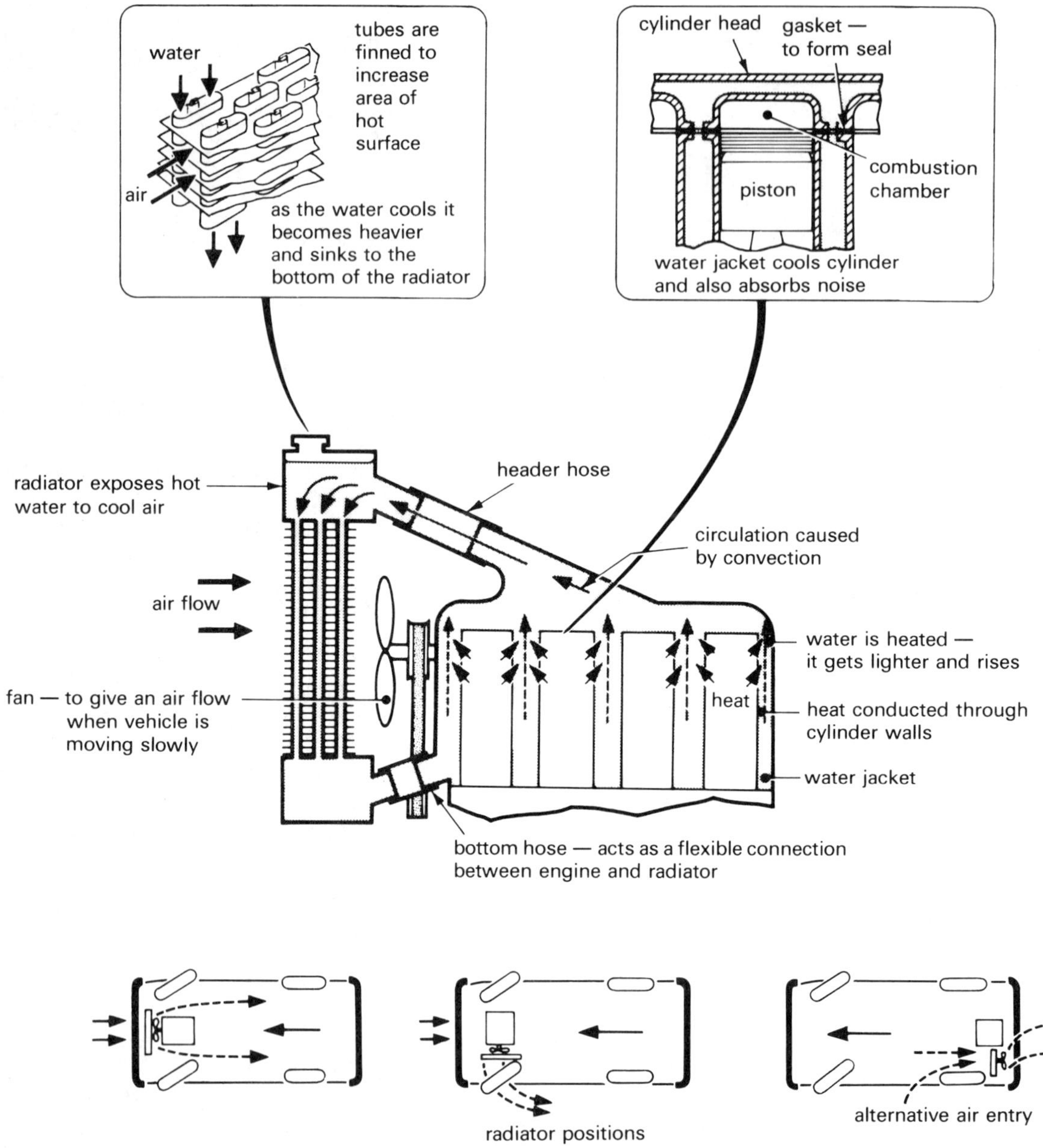

Fig. 24.1 Liquid cooling system

The principle of operation is based on three methods of heat transfer: *conduction, convection and radiation* - cooling of the engine takes place in this sequence.

Conduction
Heat flows from a hot substance to a cold substance. The hot internal parts of the cylinder pass the heat from one metal particle to the next through the walls to the comparatively cool outer surface. Coolant in contact with the metal surface receives the heat and carries it away.

Convection
This is based on the principle that at a temperature higher than 4 °C water gets lighter in weight as it is heated, i.e. the density decreases as the temperature increases.

Coolant in the water jacket becomes heated, and since it becomes 'lighter' it will rise to the top of the jacket and flows through the header hose to the radiator. The hot water in the jacket will be replaced by colder water fed in from the lower tank of the radiator. When the flow is combined with the upward flow in the jacket, a natural circulation of the coolant is achieved which is called *convection*.

Circulation due to the thermo-syphon action can only take place if the coolant level in the header tank is above the level of the header hose.

Radiation

Heat travels through air in a wave form similar to light. Radiated heat can be felt if your hand is placed close to a hot surface.

The purpose of a radiator is to transfer the heat from the coolant to the air. The air must move because, if it is stationary, the air temperature will soon become equal to the coolant temperature and heat transfer will cease. To avoid this problem, cold air is continuously supplied to the radiator and the heated air is carried away.

Modern vehicles do not rely on the thermo-syphon system for engine cooling. This is because the body line of the vehicle prevents the use of a tall radiator and the large quantity of coolant required is too heavy to be carried around. When these disadvantages are combined with the slow cooling action and long delay before the engine reaches its optimum running temperature after cold starting, it is seen why a more positive and compact cooling system, as given by a pump, is now used universally.

Pump circulation

Fig. 24.2 shows the main features of a pump-assisted cooling system. The layout is based on the thermo-syphon system and uses a pump, generally belt driven, to promote circulation of the coolant. The increased rate of flow of coolant given by a pump system allows a smaller radiator, and where necessary this can be mounted lower relative to the engine. Mounting the pump directly on to the cylinder block is common practice; this overcomes the need for an outlet hose between the pump and engine. The alternative is to make the pump a separate unit, which is convenient when the pumping capacity has to be increased to suit hot climates.

Unless special attention is given to the outlet flow of coolant from the pump, the cylinders at the end of the engine farthest from the pump will operate at a much higher temperature than those at the front. This is because the coolant will gain heat as it flows to the rear. Besides causing the rear cylinders to wear at a higher rate, the reduced heat flow to the coolant increases the risk of detonation especially when these cylinders are operating on a weaker air–fuel mixture than the remaining cylinders.

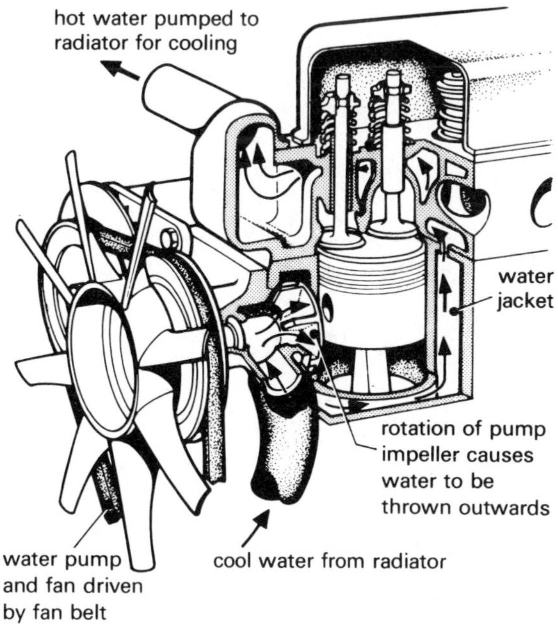

Fig. 24.2 Pump circulation

One method of minimizing this problem is to incorporate in the cylinder head a tube that runs from front to rear of the engine (Fig. 24.3). Holes (2) formed in the tube (1) direct jets of 'cold' coolant around the outside of the exhaust valve seats and any other parts of the combustion chamber that need extra cooling. Holes in the cylinder head gasket allow convection currents to carry away the heat from the regions where the heat is less intense.

The cooling pump

Fig. 24.4 shows a typical 'water pump'. The body of the pump is generally a cast iron or light alloy casting which carried a double-row ball bearing prepacked with grease and sealed for life. The bearing must be robust because it has to withstand the radial load caused by the driving belt. Mounted on the spindle, and retained in position by an interference fit, is the *impeller*. This has either radial or curved vanes as shown in Fig. 24.5, which act as the means either to pressurize the coolant or to agitate the fluid sufficiently to produce a given flow. Various materials can be

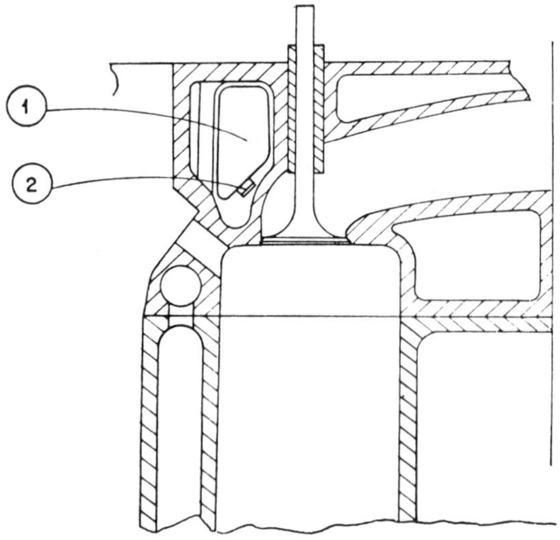

Fig. 24.3 Water distribution tube in cylinder head

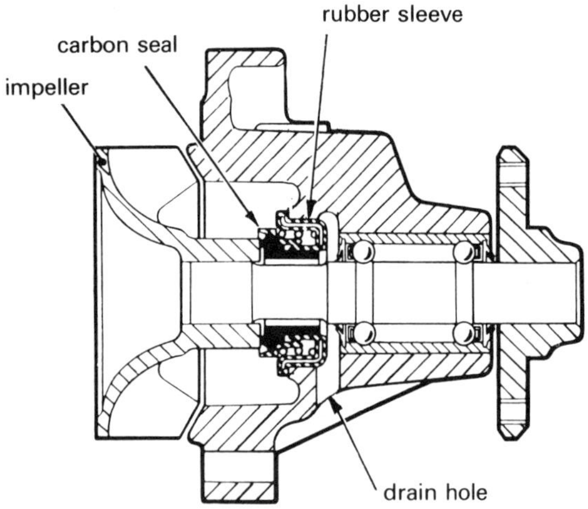

Fig. 24.4 Cooling pump

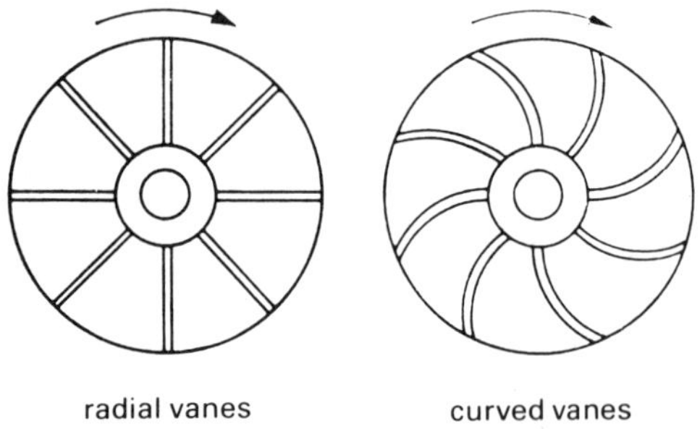

Fig. 24.5 Face of water pump impellers

used for an impeller; it can be cast iron, stamped in steel or moulded in plastic. The stamped impeller is very cheap but has a low pumping efficiency; as a result the energy consumed lowers the fuel economy of the engine. A moulded impeller is more efficient but since this type suffers from bearing rotation and creep at the fixings, the cast impeller, shrouded in a manner to give a flow about three times greater than the stamped type, is favoured.

One of the most important parts of the pump is the seal between the impeller and casing. This is intended to prevent coolant entering the bearing, which would cause noise and rapid wear. The seal shown is retained in the casing and is a spring-loaded carbon type; this is bonded to a synthetic rubber sleeve fitted to allow for the difference in expansion between the carbon and the casing. An annular groove in the casing, on the bearing side of the seal, collects any coolant that might pass the seal; a drain hole at the bottom allows the coolant to drip from the pump - a sign that the pump seal is defective and the pump needs replacing.

Normally the pump is situated at the 'front' of the engine and is driven at about crankshaft speed by a vee or notched belt.

Radiators

The radiator is a liquid-to-air heat exchanger that consists of two tanks and a *core* or *matrix* from which heat is radiated from the coolant to the air. To be effective the hot coolant contact area exposed to air moving through the core should be as large as possible. Also it should offer minimum resistance to the flow of air to ensure that the hot core is exposed to the coldest air.

The core can be constructed in various ways; two common forms are:

- *Tube and fin*
- *Cellular* or *film-type*.

Tube and fin (Fig. 24.6)

The tubes through which the coolant flows are of flattened oval section and made of brass about 0.125 mm thick. They are arranged in three or four rows and pass through a series of thin copper fins. The whole assembly is dipped in solder to bond the tubes and fins together.

A brass tank, fitted at each end of the core, connects with the tubes and provides a connection point for the synthetic rubber hoses that supply and receive the engine coolant.

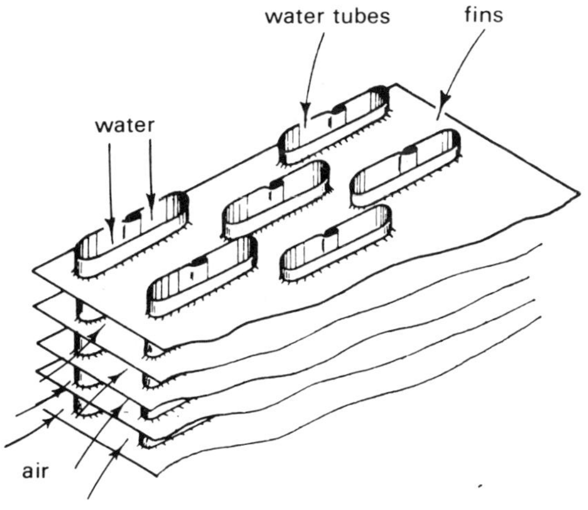

Fig. 24.6 Tube and fin radiator construction

Cellular or film-type (Fig. 24.7)

Two forms of strip are used in this construction.

The waterway strip is a shallow trough with corrugated flanges. Two of these strips are placed together to form a water space similar in shape to a well-flattened tube.

The airway strip is deeply corrugated and is placed between the water tubes to increase the hot surface area.

After arranging the strips to form a matrix and fitting a tank to each end, the assembly is dipped in a solder bath to bond the strips and seal the waterways.

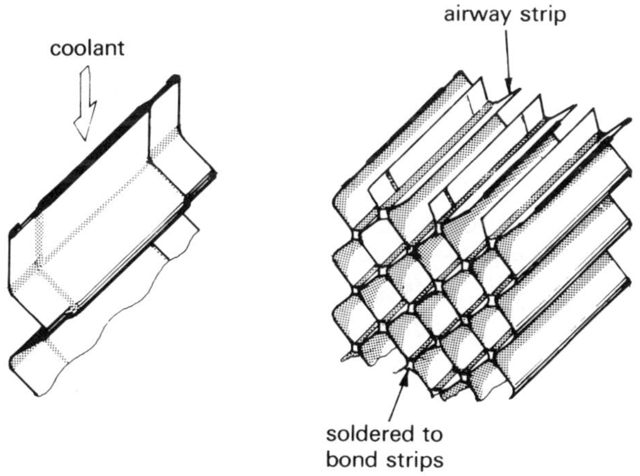

Fig. 24.7 Cellular film-type radiator construction

Flow direction through radiator

The coolant flow through a radiator can be from either top to bottom or side to side; the radiators giving this flow are called *vertical-flow* and *cross-flow* respectively.

Vertical-flow

This is the oldest type of radiator and is designed to use the downward flow caused by natural convection as the water cools on passing through the radiator. The increased speed of circulation given by the pump overcomes the dependence on natural circulation, but the direction flow is unchanged.

Modern radiators are much wider and shorter than earlier designs, the actual shape being governed by the available space. To ensure the system is completely filled with coolant, a filler cap, combined with a vent that also acts as an overflow, is fitted at the highest point of the radiator.

On some cars having automatic transmission, an oil cooler is used. This can be incorporated in the main radiator by mounting the oil cooler under the engine-cooling radiator.

Cross-flow

Lower bonnet lines used in the last few years have meant that a wide short radiator, often placed lower than the engine, is needed; this requirement is achieved with a cross-flow type. Used in conjunction with a pump, the system cools the water as it flows across the radiator. The two main engine hoses are connected to the end tanks; normally the 'hot' inlet to the radiator is situated higher than the outlet.

Fig. 24.8 shows the layout of a typical cross-flow arrangement. This uses a separate reservoir that is often made of a transparent material to allow the coolant level to be checked. This reservoir also performs two other duties: it acts as an expansion chamber to allow for the change of coolant volume with change of temperature, and a *degas tank* to condense the steam that collects at the highest point of the engine, the region under the filler cap, A.

Normal topping up of the system is made at the reservoir but whenever the engine is drained, the system must be refilled at point A to avoid an air-lock.

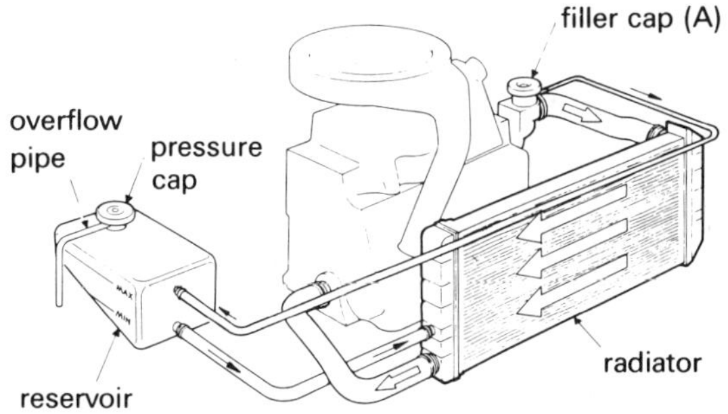

Fig. 24.8 Cross-flow cooling layout

Temperature control of the cooling system

The cooling system must be capable of keeping the engine temperature within safe limits under the most arduous conditions, such as when climbing long, steep hills at full throttle in hot climates. Since these conditions represent a very small proportion of the running time of most vehicles, it is clear that the cooling system will overcool the engine most of the time unless some method of reducing the effectiveness of the system when necessary is adopted.

There are two ways in which this may be done:

1 By controlling the circulation of the water. This is the commonest method and is done by a temperature-sensitive valve called a *thermostat*.
2 By controlling the air flow through the radiator by shutters or blinds.

Thermostats

There are two types in common use:

1 The bellows type (Fig. 24.9)

The operating element is a sealed, flexible metal bellows (8) partly filled with a liquid which has a boiling point somewhat lower than the boiling point of water (such as alcohol, ether or acetone). Air is excluded from the bellows, which consequently contains the liquid and its vapour only, so that the pressure inside the bellows is due to the vapour pressure of the liquid. This varies with temperature, being equal to atmospheric pressure at the boiling temperature of the liquid: it is less at lower temperatures and greater at higher temperatures.

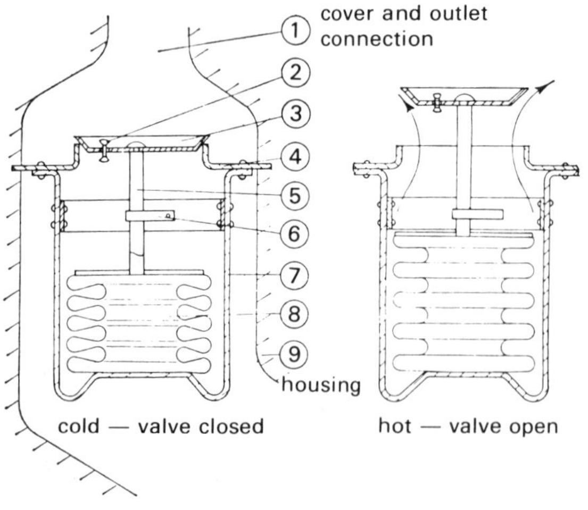

Fig. 24.9 Bellows-type thermostat

The lower end of the bellows is fixed to a frame (7) which is attached at its upper end to a circular flange (4), by which the thermostat is supported in its housing (9) which is usually formed in the outlet from the cylinder head to the radiator. A poppet type valve (3) attached by a stem (5) to the top of the bellows controls a circular opening in the flange to regulate the flow of water. A flat spring blade (6) holds the stem in light contact with a V-shaped groove in a cross-member of the frame which supports the stem, and provides sufficient friction to prevent flutter of the valve.

At low temperatures the pressure inside the bellows is lower than atmospheric pressure outside it and the bellows is contracted, holding the valve closed and preventing water circulating through the radiator.

As the water temperature increases, the hot water will collect around the thermostat which is usually in the engine outlet connection, heating the liquid in the bellows and so increasing the pressure inside the bellows. At about the boiling temperature of the liquid in the bellows, the internal and external pressures will be equal, and the bellows will begin to extend and open the valve. This occurs at a temperature of about 70–80 °C and by the time the temperature has reached about 85–90 °C the internal pressure in the bellows will have extended it sufficiently to open the valve fully.

The thermostat will, therefore, perform two important functions: (a) it will shorten the time required to get the engine warmed up after a cold start, and (b) it will prevent the temperature falling below about 70 °C when the engine is running at light loads.

A small hole (2) drilled in the valve acts as a vent to prevent air being trapped underneath the valve when the system is being filled, and a loosely fitting *jiggle-pin* prevents this hole becoming clogged.

The 'free length' of the bellows is such that when internal and external pressures are equal, the valve is open: thus, should the bellows develop a leak, the valve will remain open, i.e. the thermostat will *fail-safe*.

2 The wax element type (Fig. 24.10)

This depends for its operation upon the considerable change in volume which occurs in certain types of wax at around melting point. The operating element is a substantial metal cylinder or capsule (7) filled with wax into which is inserted a thrust pin (2). A flexible

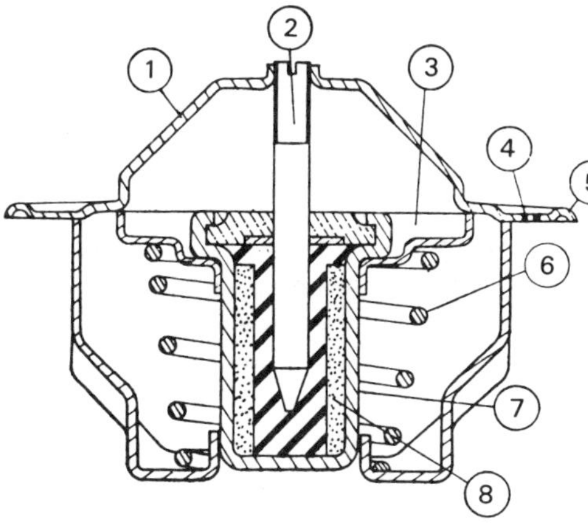

Cold – valve closed

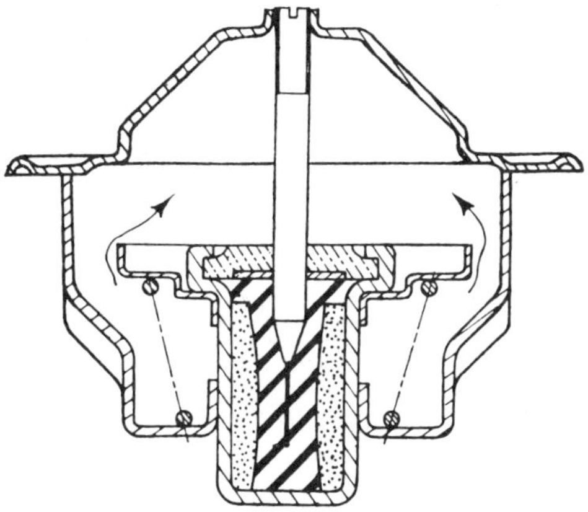

Hot – valve open

Fig. 24.10 Wax element thermostat

rubber sleeve (8) surrounds the pin and is sealed into the top of the capsule to prevent the escape of wax.

The thermostat is supported in its housing by the flange (5) in a similar manner to the bellows type, and the thrust pin (2) is attached to a bridge (1) spanning the flange. The valve (3) is attached to the capsule (7) and closes against an opening in the flange, being held in the closed position by the spring (6) when cold. Expansion of the wax during melting forces the thrust pin out of the capsule, so opening the valve.

The useful life of this thermostat is claimed to be over 100 000 km; it is limited by a tendency for its opening temperature to increase because of deterioration of the rubber sleeve. It is much more robust than the bellows type so that sudden and complete failure is extremely unlikely. This is just as well since failure would generally cause the valve to remain closed, resulting in overheating of the engine, but if a leak develops in the rubber sleeve (8) below the thrust pin, the valve will stick open.

The hole (4) serves the same purpose as the hole (2) in the bellows type (Fig. 24.9), and usually has a jiggle-pin also.

The opening temperature of wax element type thermostats is unaffected by coolant pressure, so this feature makes it suitable for use in most modern cooling systems.

Radiator blinds

A very simple arrangement for controlling the air flow through the radiator is illustrated in Fig. 24.11. A spring-loaded roller blind (4) is carried at the lower end of a rectangular channel-section frame secured to the front of the radiator. A cable control enables the blind to be raised to blank off as much of the radiator as may be necessary to maintain the required temperature. The end of the cable is taken into the driving compartment of the vehicle and incorporates some means of fixing the blind at a suitable height. It is desirable that a temperature gauge should be fitted to indicate to the driver the temperature of the cooling water.

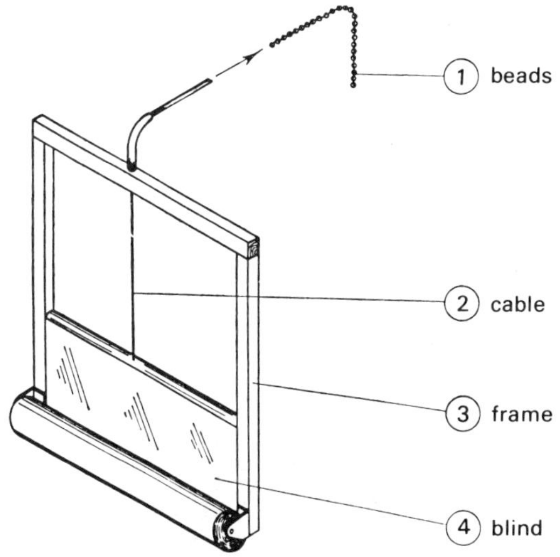

Fig. 24.11 Radiator blind

Radiator shutters

This rather more complex arrangement for regulating the air flow through the radiator is illustrated in Fig. 24.12. A rectangular frame fixed to the front of the radiator supports about twelve horizontal rods, each of which carries a metal strip about 40 mm ($1\frac{1}{2}$ in)

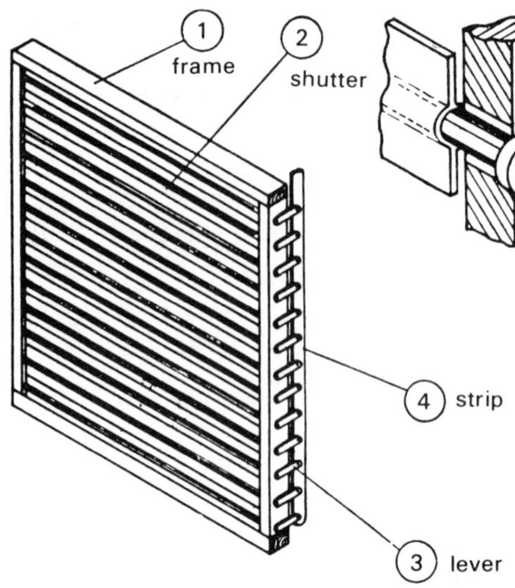

Fig. 24.12 Radiator shutter

wide. When these strips are all vertical they blank off the air flow through the radiator, but by turning them through 90° a free passage for air is provided.

Small levers fitted to one end of each spindle are connected to a vertical strip which may be raised or lowered to open or close the shutters. This may be operated manually as in the case of the blind, but the short movement necessary makes possible the use of a thermostat device. Shutters such as these were once fitted to some of the more expensive types of car and operated by a thermostat immersed in the radiator header tank. The arrangement used is normally fail-safe - any defect in the operating system results in the shutters being set in the fully open position.

Shutters are rarely used on cars but are quite common on heavy commercial vehicles.

Fans

In the simplest type of water-cooled system, the forward motion of the vehicle is alone relied upon to force sufficient air through the radiator. However, a positively driven fan gives a bigger air flow so that a smaller radiator can dissipate the required amount of heat. The simplest method of driving the fan is by belt, usually the same belt that drives the alternator and the water pump: the fan is, in fact, usually mounted on an extension of the water-pump spindle.

The fan is only needed when the air speed through the radiator is insufficient to control the temperature of the coolant. Other than at times when the vehicle speed is low, as in heavy traffic or hill climbing, the energy required to drive the fan continuously robs the engine of power and increases fuel consumption.

Although an energy saving is achieved by aero-dynamic shaping of the fan blades compared with pressed steel blades as used in the past, the modern vehicle achieves a better economy by using a drive system that 'cuts out' when fan operation is not needed. Two common systems are:

- *Electric*
- *Viscous*.

Electric fan (Fig. 24.13)

The fan is driven by a separate electric motor which is only switched on when the cooling water reaches a predetermined temperature; e.g. 90 °C. The energy for the motor is controlled by either a thermostatic switch, usually a bimetallic type, fitted in the region of the header hose, or by a relay energized by a signal from the electronic control unit (ECU).

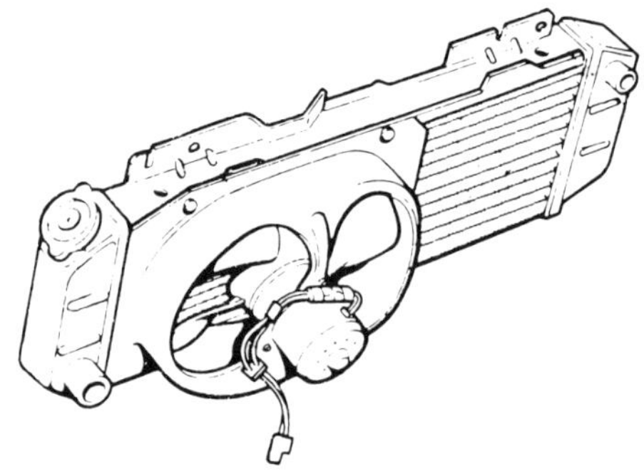

Fig. 24.13 Electric cooling fan

Viscous fan

This type of drive has a disc-shaped clutch plate that is placed in a container of silicone fluid. The viscous drag of the fluid, caused by its resistance to shear, provides a non-positive drive that is designed to slip at an increasing rate as the engine speed rises.

Viscous drives for fans are made in two forms:

- *Torque-limiting*
- *Air temperature sensing*.

Torque-limiting fan drive As the name suggests this type is capable of transmitting to the fan a maximum torque that depends on the viscosity of the fluid.

Fig. 24.14 shows a typical construction. A clutch disc is sandwiched between the two halves of the casing which is fitted with fluid seals to prevent leakage of the silicone fluid. Aluminium alloy is normally used as a material for the disc and casing.

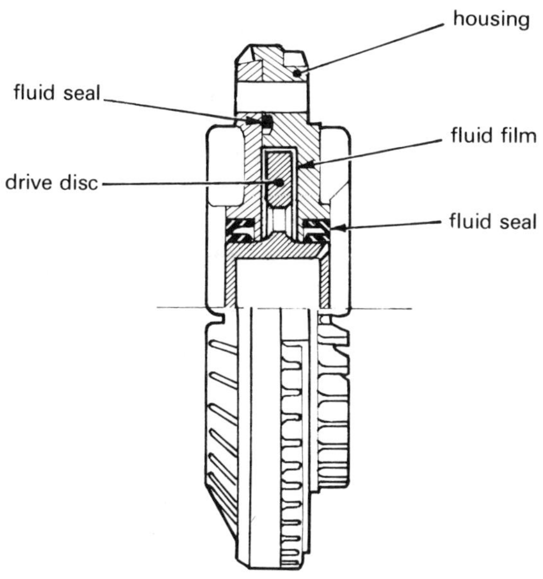

Fig. 24.14 Viscous fan drive

This lightweight material is chosen because it has good thermal conductivity for dissipating the heat generated by the shearing action of the fluid: fins on the outside of the casing also aid the heat transfer from the coupling to the air.

The two graphs shown in Fig. 24.15 show the performance characteristics of this type of drive.

Air-sensing fan drive As well as torque limiting, this type varies the fan speed to suit the temperature of the air that has passed through the radiator by controlling the amount of fluid in contact with the drive plate. When the temperature sensor detects that the coolant is below a temperature of about 75 °C, the fluid is evacuated from the drive chamber and the fan drive is disengaged. This saves more power than the previous type and also reduces fan noise.

Fig. 24.16 shows an air-sensing fan drive. At the front of the unit, and situated so that it is fully exposed to the air coming from the radiator, is fitted a spiral-shaped bimetal sensor. This sensor controls a valve arm which is made to rotate through a small angle as the temperature changes and either cover or uncover an inlet port in the pump plate.

The fan is not needed at low radiator temperatures, so the sensor causes the valve plate to close the inlet port; this action stops the flow of fluid from the reservoir to the drive chamber. Since the existing fluid in this chamber is being pumped back to the reservoir continuously by the combination of centrifugal action and scoop movement, and no fluid is entering the chamber to take its place, the chamber soon empties and the drive disengages.

At high radiator temperatures the thermostatic coil moves the valve arm and uncovers the inlet port; as a result fluid flows from the reservoir to restore circulation between the reservoir and drive chamber and the drive to the fan occurs.

Fig. 24.17 shows the performance characteristics of an air-sensing fan of the type fitted to light vehicles.

Use of antifreeze mixtures

One of the disadvantages of water as a coolant is the fact that in many parts of the world temperatures may drop to freezing during the winter months.

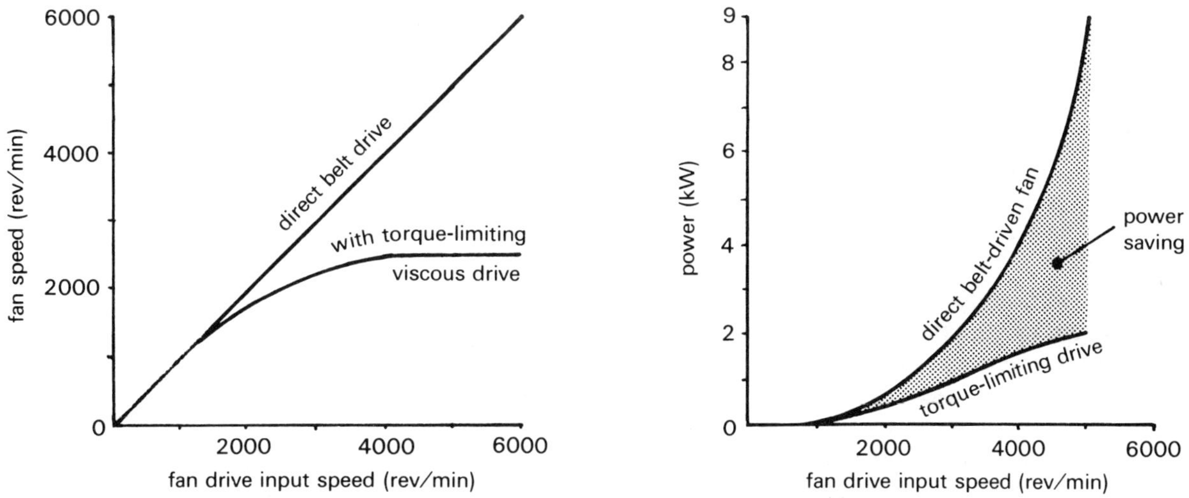

Fig. 24.15 Viscous fan performance characteristics

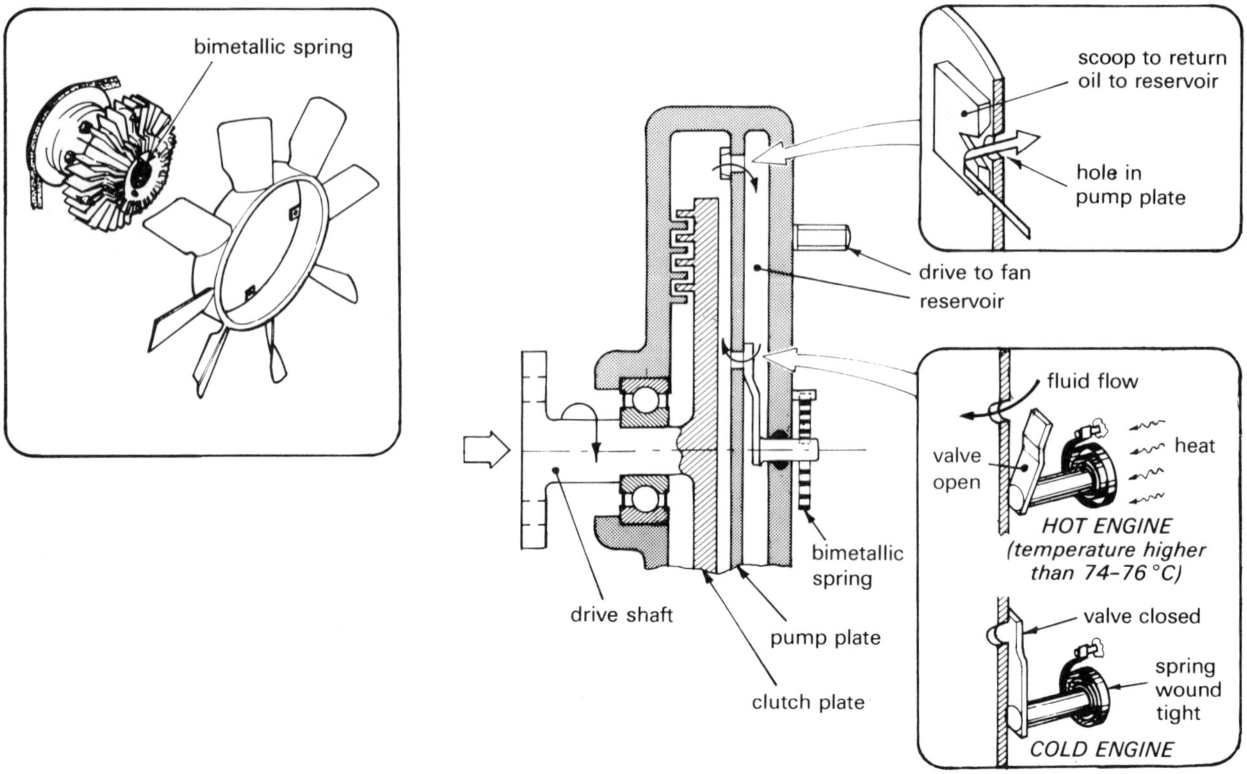

Fig. 24.16 Air-sensing fan drive

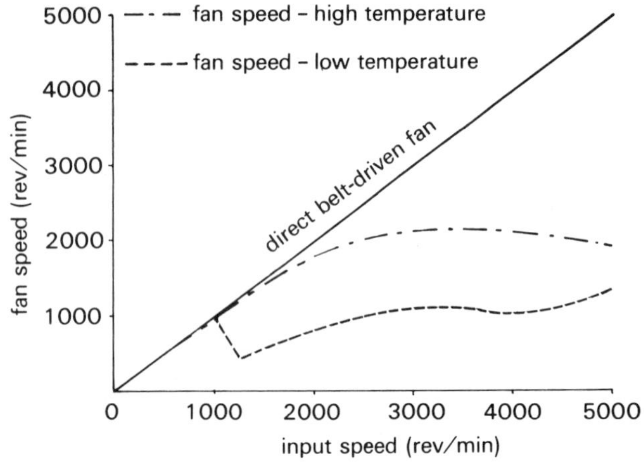

Fig. 24.17 Characteristics of air-sensing fan

Water is peculiar in that its maximum density occurs at a temperature of 4 °C and from this point it expands with either rise or fall in temperature. So long as it remains liquid this has no serious consequences, but solid ice expands on cooling and will transfer its expansion to its container. If this happens to be the brittle cast iron cylinder block, it is likely to crack and this results in a very expensive repair. A less drastic effect occurs when the expansion of the ice forces out the cylinder block *core plugs* (*Welch plugs*).

These plugs, often made of brass and about 2 mm thick, blank off the core holes in the cylinder block. The holes are the result of the manufacturing process and are the places where the sand core, used to form the water jacket, is supported in the casting mould.

Prior to fitting, the plugs are shaped so that they are slightly convex. This allows the plug to be placed in the block recess and expanded by tapping the centre with a hammer so as to allow the increase in its diameter to form a tight fit (Fig. 24.18).

The freezing point of water can be lowered by

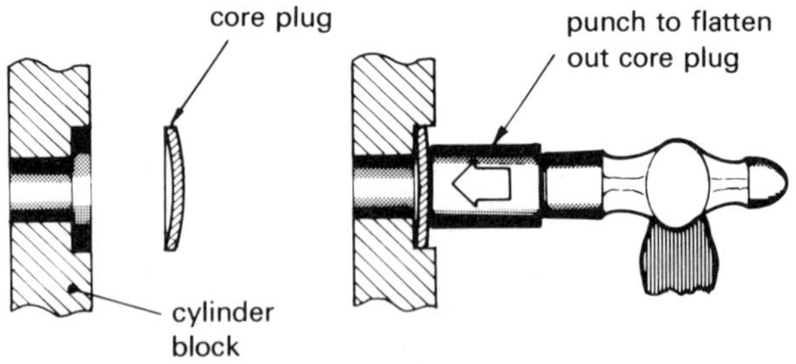

Fig. 24.18 Core plug fitting

dissolving some other substance in it. The one most commonly used is ethylene glycol. The effect of this liquid in lowering the freezing point of the coolant is shown in Fig. 24.19. This graph shows that the coolant passes through a 'mushy ice' stage before it reaches the solid state that causes the structural damage. When partial freezing occurs while the vehicle is in use, the mushy ice that forms in the coldest part of the radiator tank interrupts the circulation and causes the engine to overheat.

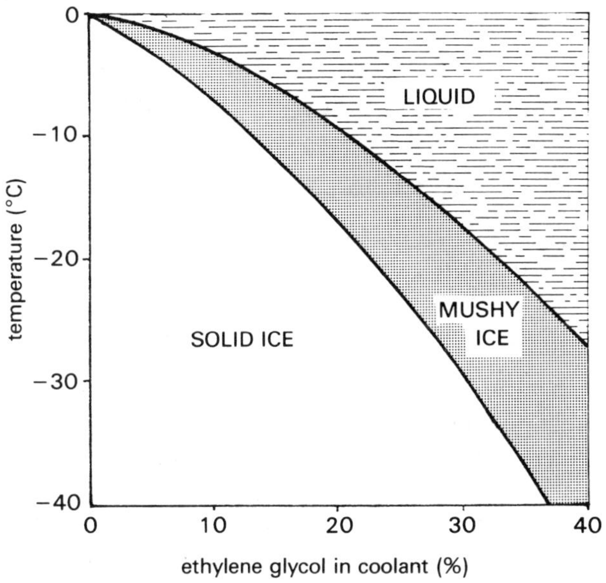

Fig. 24.19 Effect of ethylene glycol on freezing point of coolant

Ethylene glycol has the disadvantage of decomposing in use and forming acids that causes corrosion of parts, especially those made of aluminium alloy. To give protection to the many metals that come into contact with the coolant, suitable chemicals called *corrosion inhibitors* are added to the antifreeze solutions. Of recent years it has become common practice to leave the antifreeze in the cooling system for the whole year rather than drain it at the end of the winter. In view of this, a reputable make of antifreeze should be used in order to give the engine all-year protection from corrosion.

In the UK a coolant mixture containing 25 per cent antifreeze, i.e. 1 part antifreeze to 3 parts water, is often recommended to suit the conditions experienced during a normal winter. This proportion gives protection against ice damage down to −25 °C and gives coolant circulation down to about −12 °C.

Incidently, the use of ethylene glycol in the coolant raises the boiling point; a 25 per cent solution raises it to about 103 °C.

Since the specific gravity of ethylene glycol differs from that of water, it is possible to check the proportion of antifreeze in the engine coolant by using a special hydrometer.

Pressurized cooling systems

Another disadvantage of water as a coolant is the fact that its boiling point is lower than the most efficient engine-operating temperature, and to prevent boiling and the formation of steam pockets around exhaust ports and sparking plug bosses, it is necessary to keep the temperature of the water leaving the cylinder head below 85–90 °C.

The temperature at which a liquid boils rises as the pressure on it increases. In the case of water, the variation of boiling point with pressure is shown in the graph of Fig. 24.20.

Pressure can easily be imposed on the water in a cooling system by sealing it off from the atmosphere, so that any steam formed by the boiling water will raise the pressure and suppress further boiling until the temperature has risen still further. There is obviously a limit to the pressure that a radiator and rubber hoses can stand, and a pressure relief valve, such as that illustrated in Fig. 24.21, is necessary. The one illustrated is incorporated in the header-tank filler cap.

Starting from cold, the header tank will contain some air above the water at a pressure approximately that of the atmosphere. As the water is heated it expands, compressing the air, and if the temperature rises sufficiently for boiling to begin, the steam formed will raise the pressure still further. This suppresses the boiling until the pressure in the header tank rises sufficiently to lift the pressure valve (8) against the loading of the spring (3), whereupon air and steam will be able to escape through the vent pipe (5). By this means the system may be operated without boiling at a temperature slightly below that corresponding to the pressure needed to lift the valve.

The operating pressure that may be used with a cooling system is governed by the strength of the radiator and hoses. When this has been decided by the manufacturer the appropriate radiator pressure cap is fitted. For information purposes the 'blow off' pressure is stamped on the cap; typical values range from 0.5 bar (7 lbf/in^2) to 1.0 bar (15 lbf/in^2).

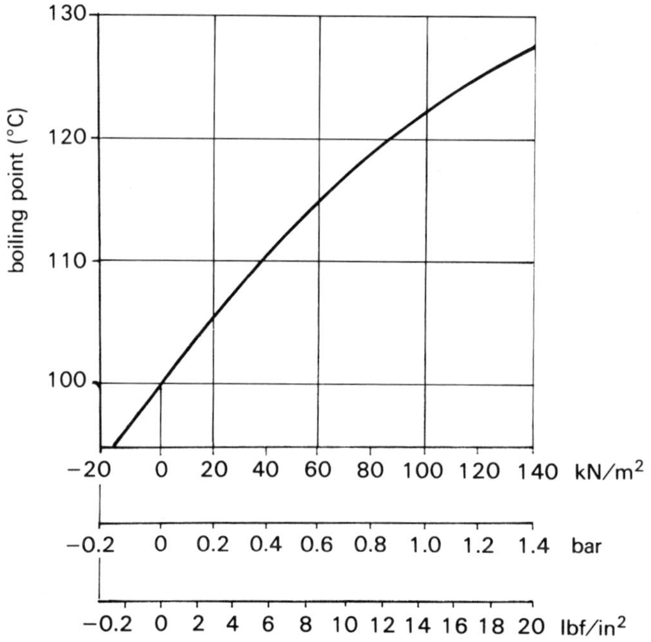

Fig. 24.20 Variation of boiling point of water with pressure

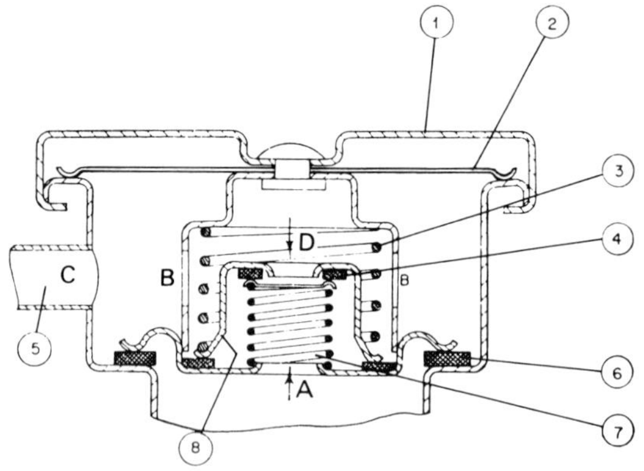

Fig. 24.21 Pressure cap for coolant system

When the engine is stopped and allowed to cool down, condensation of vapour and contraction of the water will reduce the pressure in the header tank. Should this pressure fall appreciably below atmospheric, there is a risk of the hoses and even of the header tank collapsing. To prevent this, a vacuum valve (4) is fitted in the centre of the pressure valve and acting in the opposite direction – i.e. it is closed by a spring (7) assisted by positive pressure inside the header tank, but opens against the spring loading if header-tank pressure falls about 7 kN/m^2 (1 lbf/in^2) below atmospheric.

The advantages to be gained by using a pressurized cooling system are as follows:

1 Elimination of coolant loss by surging of the coolant during heavy braking.
2 Prevention of boiling during long hill climbs, particularly, for example, in regions much above sea level.
3 Raising the working temperature improves engine efficiency.
4 It allows a smaller radiator to dissipate as much heat as a larger one operating at a lower temperature.

Never remove the cap when the coolant temperature is above 100 °C, since the release of pressure will allow the water to boil violently: the resulting jet of steam and water from the open filler can cause serious scalds. At temperatures below this, the cap should be released *slowly*: it is designed so that the spring disc (2) (which is riveted between the cover (1) and the frame which contains the valves) remains seated on the top of the filler neck until *after* the seal (6) has lifted, so allowing pressure to escape through the vent pipe (5) before it can escape from the main opening.

Since bellows type thermostats are sensitive to pressure changes, they are unsuitable for use in pressurized cooling systems. The wax element type does not have this disadvantage.

Sealed cooling systems

A further refinement of the cooling system consists of an arrangement whereby the system is kept completely full of coolant, expansion of the coolant being accommodated by providing an *expansion tank* into which the displaced coolant can pass and from which it can return to the system as the coolant in the system contracts on cooling. There are several variations of the arrangement, of which two are briefly described.

Fig. 24.22 shows an addition to the pressurized system already described, the modification consisting of leading the vent pipe from the filler neck to the bottom of an expansion tank. A vent pipe is fitted to the top of the expansion tank, which may - though not necessarily - have a drain tap and a filler cap.

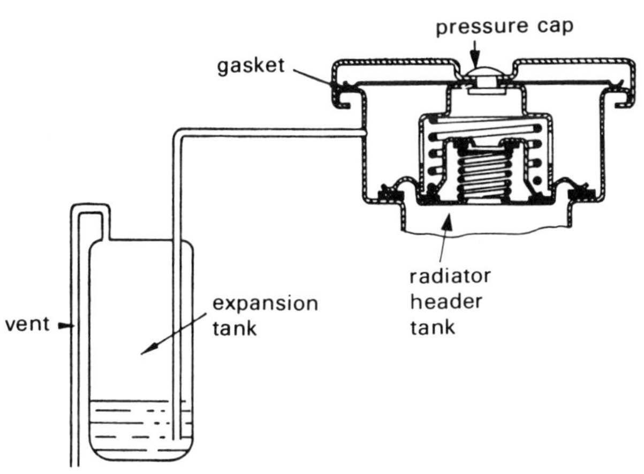

Fig. 24.22 Radiator expansion tank

The pressure cap differs from that shown in Fig. 24.21 only in having a sealing gasket fitted between the filler neck and the spring disc (2).

The system is completely filled with coolant up to the top of the filler neck and the cap fitted. As the engine warms up the coolant expands, lifting the pressure valve off its seating, and some coolant passes into the expansion tank. Air displaced from the expansion tank escapes through the vent pipe. The expansion tank is seldom more than about half full under normal conditions.

As the engine cools down after stopping, the coolant in the system contracts, withdrawing coolant from the expansion tank back into the system through the filler cap vacuum valve - hence the necessity for the gasket under the spring disc.

A variation of this system uses a plain, airtight filler cap on the header tank, the pressure and vacuum valves being contained in a small housing permanently attached to the header tank.

Fig. 24.23 shows an alternative system in which the pressure cap is fitted on the expansion tank which is connected to the top of the header tank by a small pipe. The filler neck on the header tank is sealed by a plain cap. The header tank is completely filled with coolant and a small amount of coolant is also put into the expansion tank. As the system warms up coolant expands into expansion tank which in this case is under pressure. As the system cools down coolant from the expansion tank is drawn back into the header tank.

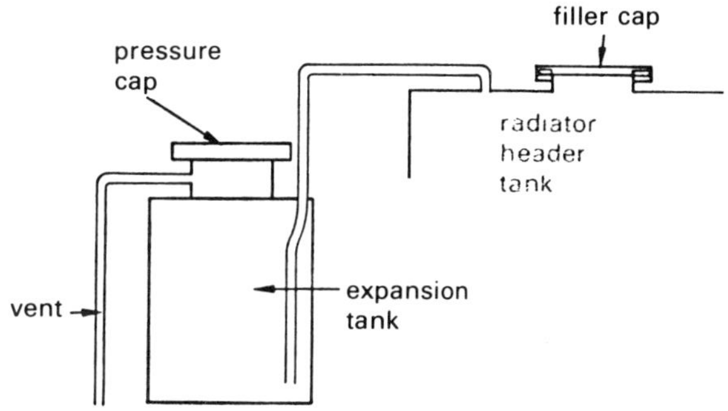

Fig. 24.23 Radiator expansion tank

The advantages claimed for a sealed system are:

1 It eliminates coolant loss by expansion.
2 It eliminates the need for periodic topping up, and prevents possible damage by neglecting to top up.
3 By excluding air from the main system it considerably reduces corrosion of the components in the cooling system and deterioration of the antifreeze additives.

PART FIVE

Engine lubrication

25 The functions and properties of lubricants

Friction

When two surfaces are in contact, there is an opposition to relative movement between them, which is called *friction*. If the surfaces are clean and dry, the force needed to overcome friction depends on:

1 The materials of which the surfaces are made.
2 The surface finish, i.e. whether rough, smooth or polished.
3 The load pressing the surfaces together.

For any one pair of clean dry surfaces it can be shown by simple experiments that the ratio

$$\frac{\text{Resistance to movement}}{\text{Load pressing surfaces together}}$$

is a constant number: it is called the *coefficient of friction* for those surfaces.

When friction is overcome and movement between two surfaces occurs, work is done against friction and an equivalent amount of heat is generated at the surfaces. In a continuously running bearing, this head must be dissipated into the surrounding air to keep the bearing temperature within reasonable limits.

Another result of movement between dry surfaces is wear of the surfaces. The rate of wear depends on the materials and also varies with the load and speed, but in a high-speed machine it is likely to be so rapid as to render the mechanism useless within a very short time.

If the surfaces can be kept apart neither friction nor wear can occur, and the primary function of a lubricant is to separate the moving surfaces. Lubricants may be solid, liquid or gaseous, but liquids are by far the commonest and are almost universally used in motor vehicles.

Viscosity

When moving surfaces are *completely separated* by a film of liquid lubricant, the only resistance to motion is due to the *viscosity* of the lubricant. Viscosity is a property of fluids by which they resist flow: the greater the viscosity the greater the resistance and vice versa. Thus the 'friction' in a lubricated bearing depends upon the viscosity of the lubricant. But the viscosity also influences the rate at which the lubricant is squeezed out from between two surfaces when load is applied, i.e. the greater the viscosity, the greater the lubricant's ability to withstand load. Hence the lubricant should have a high enough viscosity to withstand the maximum load to be carried without causing excessive resistance to movement.

The rating introduced by the Society of Automotive Engineers (SAE) is universally adopted to classify oils according to their viscosities. This method uses a number to represent the viscosity, e.g. SAE 20. The higher the SAE number, the higher the viscosity, or to use a more common expression, 'the thicker the oil'.

Whereas engines in the past used a comparatively thick oil, such as SAE 50, the need to reduce both fuel consumption and cold-cranking loads has brought about the common use of thin oils, e.g. SAE 10. An even better economy can be achieved with the very thin oil, SAE 5, but this can only be used on engines having extra-close fitting bearings; this suitability can only be determined by seeking manufacturers' advice.

Viscosity index

The viscosity of lubricants decreases with a rise in temperature; the extent of this change is measured by the *viscosity index*. A high index indicates a relatively small change in viscosity whilst a low index indicates a large change. The lubricant used should have a suitable viscosity at its normal operating temperature in the engine. This means that when the engine is cold, the viscosity will be unnecessarily high, leading to poor circulation of the lubricant and excessive friction (oil drag), possibly even to the extent of making the engine difficult to start.

Multi-grade oils

In the past the high viscosity oil used in an engine during the summer made engine cranking difficult in the winter, so different grades were specified for the two seasons. Nowadays special additives that reduce the change in an oil's viscosity with temperature are often used, and this has meant that the same grade can be used throughout the year. These oils are called *multi-grade* or *cross-grade*, or given trade names that suggest the viscosity remains constant.

These oils can be recognized by the special SAE rating; this has two numbers separated by the letter 'W'. A typical oil is SAE 10W 40: in this case the oil is equivalent to SAE 10 when tested at a sub-zero temperature, (hence the 'W' to indicate winter conditions) and has a viscosity of SAE 40 at the normal rated temperature.

Oiliness

This property is the ability of an oil to 'cling' or be attracted to a metal surface. The effect of this property is seen when a spot of oil is applied to a clean piece of metal; the oil film spreads out over the surface and resists being removed when wiped with a cloth.

The degree of oiliness varies with the type of oil; vegetable-based oils are excellent in this respect.

Oil

The lubricant used for motor vehicle engines - and most other components - is oil. Oils are obtained from three main sources:

1 *Animal* Purified and suitably treated animal fats, such as tallow and whale oil, are used for certain purposes, but decompose too readily to be suitable lubricants in modern motor vehicle engines.
2 *Vegetable* These also decompose too readily to be satisfactory, though one example, castor oil, was used quite extensively at one time. Its chief merit is ability to lubricate under arduous conditions, but after a fairly short time treacle-like deposits are formed, and it is now very seldom used.
3 *Mineral* Oils of this type are refined from crude petroleum and are far more stable than other types. They form the basis of practically all modern lubricants, and though by no means perfect, they can be improved by the addition of certain chemicals known by the general name of *additives*.

Additives

Among the most important of these for use in engines are:

Oxidation inhibitors At high temperatures mineral oils tend to *oxidize*, forming hard deposits on the hottest parts with which they come in contact, e.g. the underside of the piston crown, and varnish-like deposits on parts not quite so hot, e.g. the piston skirt. Other products of oxidation may be carried in the oil and deposited in other parts of the engine. If they settle in oil passages, they may eventually reduce the oil flow to a dangerous extent. Nowadays, oxidation inhibitors (or anti-oxidants) are added to the oil to reduce oxidation.

Detergents In use the oil becomes contaminated with oxidation products and with burnt or partly burnt products of combustion which escape past the pistons. These usually consists of extremely small and relatively soft particles which will not harm the bearings, but which tend to settle out and block up oil passages. Around piston rings they become baked hard and restrict the free movement of the rings, eventually sticking them completely in their grooves.

The function of detergent additives is to keep these oxidation products in suspension in the oil so that they are not deposited inside the engine. The oxidation products are removed from the engine with the dirty oil when the oil is changed.

Viscosity index improvers Certain chemicals have the property of reducing the change in the viscosity of mineral oil caused by change in temperature.

Anti-foam agents Some engines suffer from the formation of foam or froth in the oil, and suitable additives are used to reduce this tendency.

Methods of lubrication

Boundary lubrication

This form of lubrication relies on the oiliness property of an oil to coat the surfaces and fill the cavities of low-speed rubbing components to ensure that metal-to-metal contact and subsequent wear is avoided.

In an engine, boundary lubrication is used for all sliding engine components other than the highly loaded bearings that require a pressure feed. The supply of oil for the boundary lubrication film is provided by *splash*; this comes from the oil thrown out from the crankshaft bearings.

Hydrodynamic lubrication

The bearings of a modern engine must withstand great loads and high rubbing speeds. If the surfaces are to be adequately held apart when subjected to these arduous conditions, the quantity of oil supplied must be sufficient both to fill the space between the shaft and bearing and to make up for the oil squeezed out from the bearing as it rotates. To ensure these requirements are met, the bearing is force-fed with oil at a pressure sufficient to maintain the supply at all times the engine is in use.

Highly loaded bearings rely on *hydrodynamic lubrication* to separate the surfaces. This is achieved by using the natural movement of the oil (hence the term hydrodynamic) to create an 'oil wedge' to lift and centre the shaft in the bearing.

Fig. 25.1 shows the principle of this method of lubrication. The bearing is made larger than the shaft to the extent governed by the type of oil to be used and the expected thermal expansion.

When the engine is stationary the shaft rests on the bottom of the bearing and is supported only by a thin oil film (Fig. 25.1a). This boundary film provides lubrication when the engine is started.

Initial rotation causes the boundary layer of oil on the shaft to move around with the shaft owing to its oiliness property. This movement, together with the inherent resistance of the oil particles to shear (viscosity), forces the oil between the surfaces to create a 'wedge' (Fig. 25.1b). This wedge lifts and centres the shaft in the bearing (Fig. 25.1c). This diagram also shows an indication of the pressure variation and its relationship to the oil wedge that is generated by the hydrodynamic oil film.

To achieve effective lubrication in the bearing, an adequate supply of good-quality oil at a suitable pressure must be provided to maintain the oil wedge.

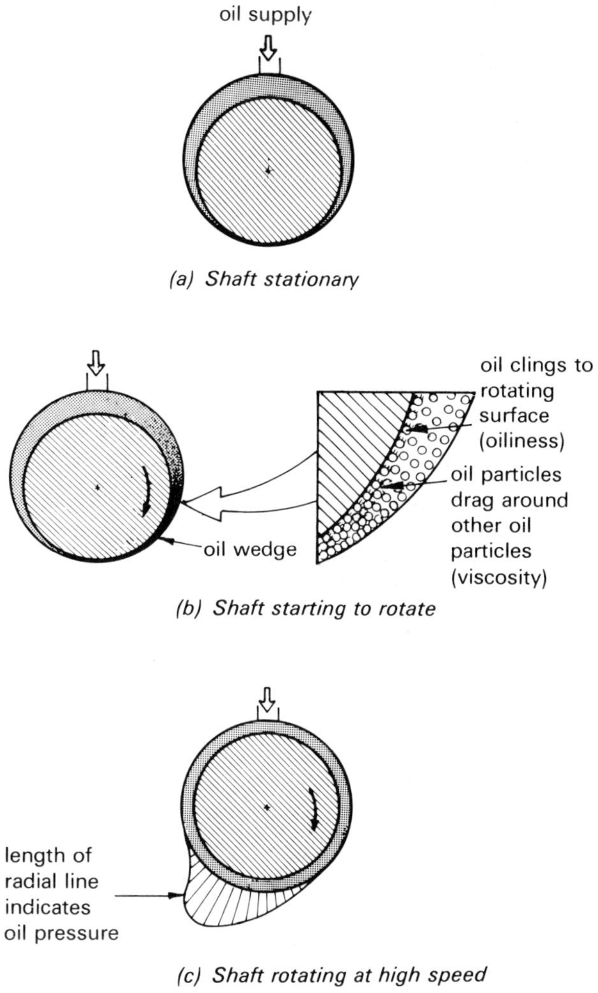

Fig. 25.1 Bearing lubrication

Oil as a coolant

In modern engines, the oil helps to cool such parts as pistons and bearings. In its passage around the engine, the oil picks up heat from the hot parts with which it comes in contact, and to prevent excessive oxidation and loss of viscosity, the oil itself must be cooled. In most normal types of engine the circulation of air around the engine and the oil reservoir is sufficient for this purpose, but on high-performance cars and some commercial vehicles it is necessary to fit an oil cooler.

26 The lubrication system

The function of the lubrication system of an engine is to distribute the lubricant to all the surfaces needing lubrication. In the very early motor vehicles, simple - even crude - lubrication systems were considered quite satisfactory.

As early as 1905, Lanchester used a system in which a pump forced oil under pressure into the crankshaft bearings, and this has developed into the typical modern system, of which an example is shown in Fig. 26.1. Fig. 26.2 shows the same system in the form of a diagram, omitting all components not included in the lubrication system.

The oil is carried in the sump (16) (Fig. 26.1), in which the level must be high enough to cover the pump inlet but not so high that the crankshaft dips into the oil. A dip-stick (6) is a simple means of checking the level of oil in the sump, though other means, such as a float indicator or an electric gauge, may be used. A mechanically driven pump (15) draws oil from the sump and delivers it via the main filter (18) to the main oil gallery (7). A rather coarse strainer (17) is fitted over the pump inlet to protect the pump from small hard objects which would damage it.

From the main oil gallery, oil is supplied to the main crankshaft bearings and the camshaft bearings through passages drilled through the crankcase walls. A feed is also taken through an external pipe (3) and a drilled oilway (2) to the hollow rocker shaft on the cylinder head. Radial holes in the rocker shaft convey oil to the rocker bearings, and oil seeping through the bearings is splashed about the valve chamber to lubricate valve stems, push rods, etc. This oil eventually drains back to the sump via the push-rod enclosure, lubricating the tappets on its way. It is usual to have a restrictor in the supply to the valve gear, and oil seals are generally fitted to inlet-valve systems to prevent oil being drawn into the combustion chambers through the inlet ports.

Holes (10) drilled through the crankshaft convey oil from a groove (12) round the main bearing to the big end: the groove is supplied from the main gallery (7) via the oilway (11), so that there is an uninterrupted supply to each big end.

A small hole (9) drilled in a suitable position in the big end bearing allows an intermittent jet of oil to spray onto the cylinder, and in some engines a hole (shown in dotted lines) is drilled through the shank of the connecting rod to take an intermittent supply to the small end bearing.

Oil splashed off the crankshaft lubricates the remaining parts, and eventually drains back to the sump.

The quantity of oil delivered into the system by the pump depends upon the pump capacity and the speed at which it is driven. Exactly the same quantity must escape from the system, and this can normally happen only through the bearing clearances. As engine speed increases, the pressure in the system increases in order to force a greater quantity of oil through the constant bearing clearances. The pressure must be sufficient to ensure an adequate flow through the bearings at relatively low speeds, so that at higher speeds the pressure will be unnecessarily high, with risk of burst pipes or joints and excessive power loss in driving the pump. The pressure in the system is therefore limited by using a pressure relief valve (14).

Oil pumps

Several types of pump are used:

Gear type For many years this type was almost universal, and it is still in common use. As shown in Fig. 26.3, it consists of a pair of gear wheels meshing together in a casing (1) which fits closely around the tips of the teeth and the ends of the gears. One gear (2) is fixed to the driving spindle (3) and drives the other gear (7), which rotates idly on a fixed spindle

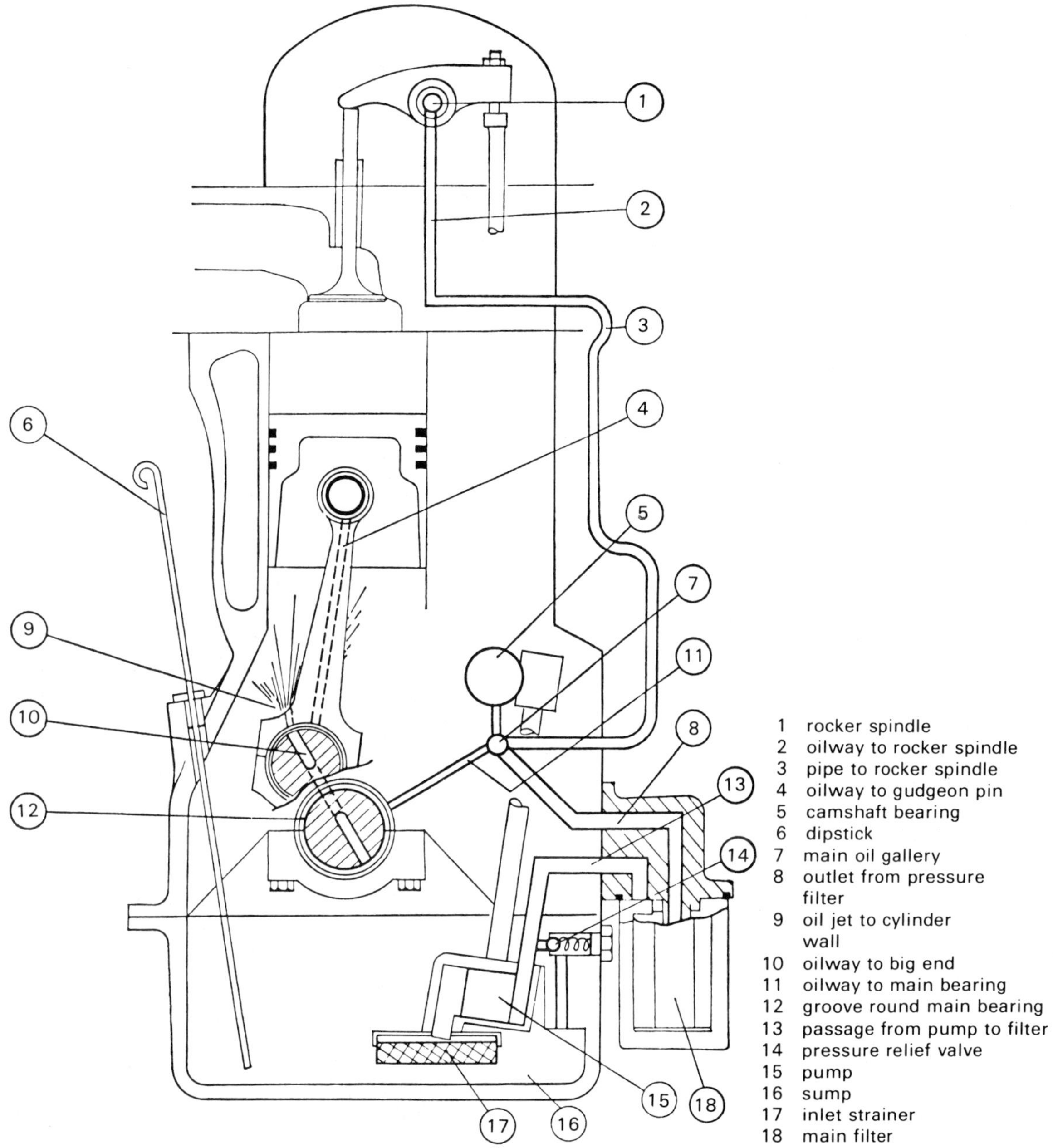

Fig. 26.1 Lubrication system

(6). Inlet (4) and outlet (5) ports are cut in the casing on either side of the meshing point of the gears. When the gears rotate they carry oil from inlet to outlet in the spaces between the teeth. As a tooth of one gear moves out of mesh with a space between two teeth of the other gear, oil flows in through the inlet to fill the void left. On the outlet side, oil is displaced through the outlet as a tooth of one gear moves into the space between two teeth of the other. Note particularly the direction of the oil flow and the direction of rotation of the gears, shown by arrows.

Eccentric rotor type (Fig. 26.4) The casing (1) has a cylindrical bore in which is fitted the outer rotor (2). The outer surface of this is cylindrical, but a number of lobes are formed on its inner surface. The inner rotor (3) has lobes formed on its outer surface, one fewer than the number on the outer rotor. It is fixed on the

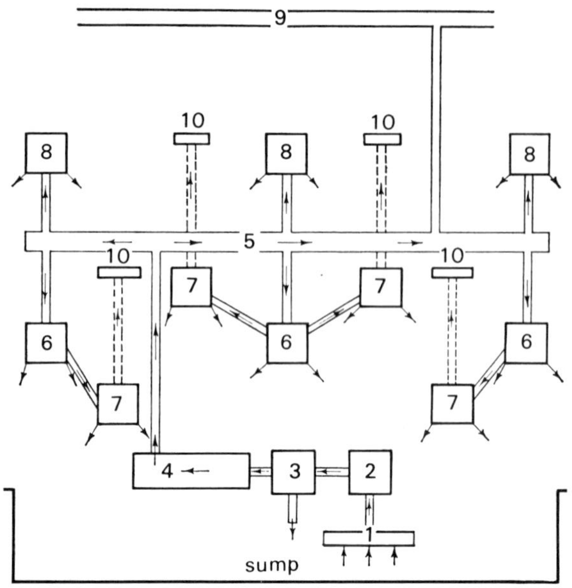

1	suction strainer	6	main bearings
2	pump	7	big-end bearings
3	pressure relief valve	8	camshaft bearing
4	main filter	9	rocker shaft
5	main oil gallery	10	gudgeon pins

Fig. 26.2 Block diagram of lubrication system

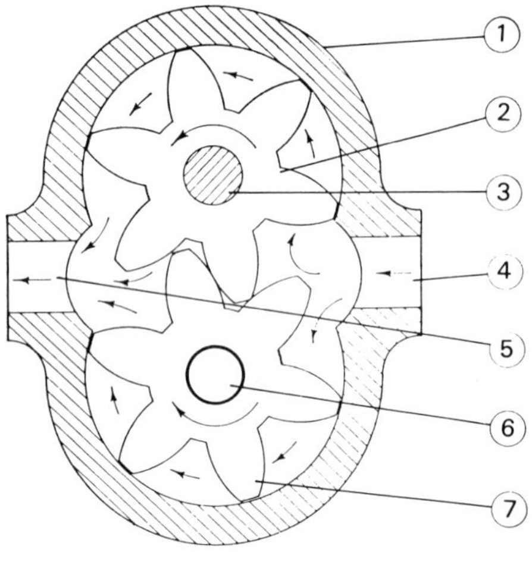

Fig. 26.3 Gear-type oil pump

driving spindle (4) and mounted eccentrically in the casing so that each of its lobes makes contact with the inner surface of the outer rotor, dividing the space between the rotors into a number of separate compartments of varying size, i.e. the size of each compartment varies as the rotors turn. Inlet and outlet ports are cut in the end plate of the pump, and positioned so that the pumping compartments sweep

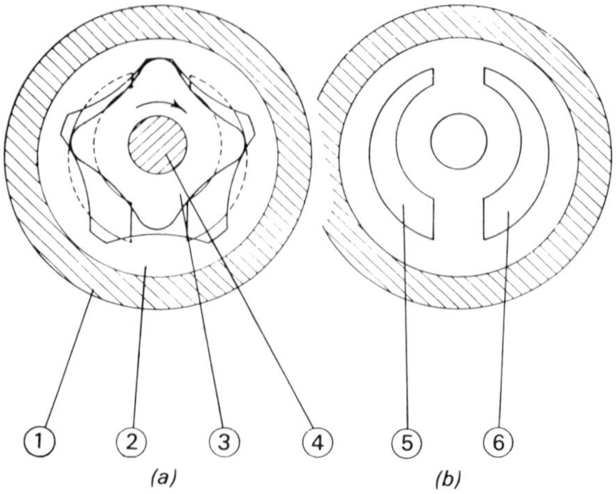

Fig. 26.4 Eccentric rotor pump

over the inlet port (6) as they increase in size and over the outlet ports (5) as they decrease. The pump is shown assembled at (a): sketch (b) shows the rotors removed to reveal the ports more clearly.

Eccentric vane type (Fig. 26.5) The earlier form of this pump is shown in Fig. 26.5a. The casing (1) has a cylindrical bore in which is fitted the rotor (2) mounted on a shaft eccentric to the casing bore and touching it at one place. Spring-loaded vanes (4) are a close-sliding fit in a slot cut diametrically through the rotor, their outer edges being kept in contact with the casing bore by the spring (3). Oil is carried from the inlet (5) to the outlet (6) as the rotor turns.

A later version of this pump used on some modern engines is shown in Fig. 26.5b. This has two one-piece vanes, each fitting in a diametral slot at right angles to one another. Each vane is cut away as

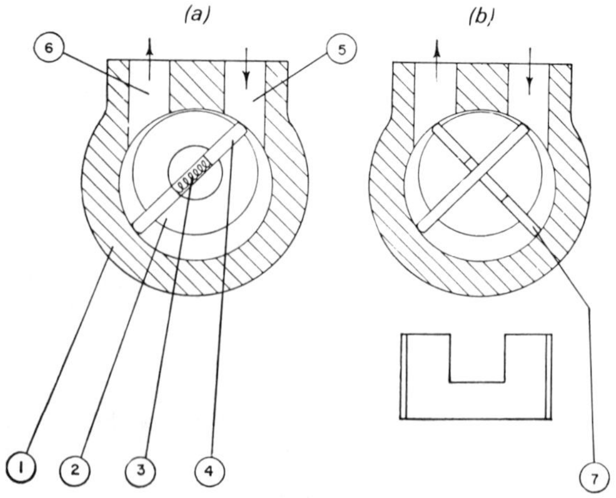

Fig. 26.5 Eccentric vane pump

shown at (7) and the bore of the casing is not truly cylindrical but shaped so that each vane touches the bore at both ends in all positions.

Internal–external type (Fig. 26.6) This type was introduced on automatic transmission units. Its compact size and ease in which it can be driven without the need for a separate drive shaft, has made the pump popular for use in engines.

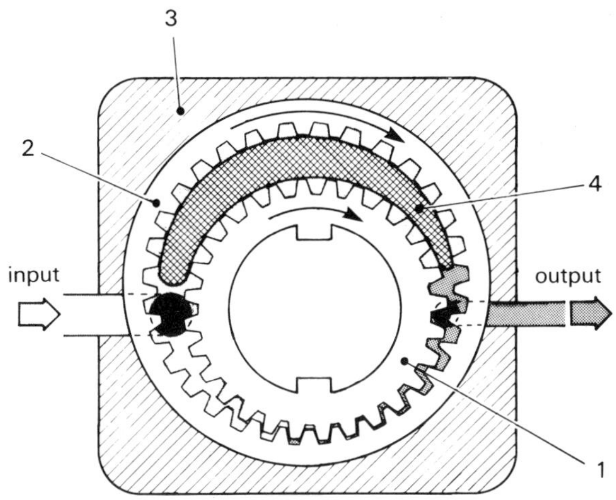

Fig. 26.6 Internal–external oil pump

Normally two flats on the 'front' end of the crankshaft are used to drive the inner gear (1); this meshes with and drives the outer gear (2), which is positioned off-centre. The outer gear is supported in a casing (3) and the wide region between the gears is filled with a crescent-shaped spacer (4); this projects from the casing.

At one end of the pump, two ports are formed in the casing to allow oil to flow to and from the pump. On the front face of the pump a flat plate is fitted to blank off the gears and ensure that the only path for the oil is via the gear teeth.

Rotation of the crankshaft causes *both* gears to revolve in the same direction. This motion carries oil from the inlet to outlet side in the tooth spaces on both sides of the spacer. Since the teeth mesh together to give a drive, the oil cannot return back to the inlet side. As a result, there is a build-up of oil pressure.

As with most pumps, any wear that takes place will allow oil to escape back to the inlet and, as a result, will lower the pumping capacity and prevent the pump from building up its full pressure. Since the siting of an internal–external pump makes it more difficult to remove, often it is recommended that the pump be inspected for wear when other work makes the pump accessible.

Internal–external pumps are more troublesome than eccentric rotor pumps if a car owner uses cheap oil that has no anti-foam agents. The air trapped in the system can pass to the hydraulic tappets and make them noisy.

Pump priming

Pumps submerged in the sump oil are self-priming. This means that the level causes the oil to flow into the pump so as to wet the gears and bridge the clearances.

When a pump is mounted above the sump level, the pump must be primed by filling it with oil from a convenient external point. This operation is essential if, for any reason, the pump becomes dry, a condition sometimes created when the engine is assembled or when the engine has been allowed to stand unused for a considerable time.

To avoid *drain-back* of oil from the pump, some systems incorporate a valve between the intake filter screen and the pump.

Oil pressure relief valves

Fig. 26.7 illustrates a simple type of relief valve, which consists of a ball held by spring pressure over a hole drilled into the main oil channel leading from the pump to the bearings. The pressure of oil in this channel exerts a force on the ball (5), tending to lift it off its seat against the load of the spring (3). The spring load is adjustable by screwing the cap (1) in or out, and locking it in the correct position by the locknut (2). When oil pressure is great enough to lift the ball off its seat, oil is allowed to escape from the main oil channel, thus relieving the pressure and preventing further rise. Oil escaping past the valve returns either to the sump or the pump inlet valve via the passage (4).

Fig. 26.8 shows an alternative type of valve in which a plunger (4) replaces the ball, and an alternative method of adjustment is shown. In this case the spring load is adjusted by adding or removing shims (2) above the spring.

Filters

The purpose of the filters in a lubrication system is to remove from the oil abrasive particles which would

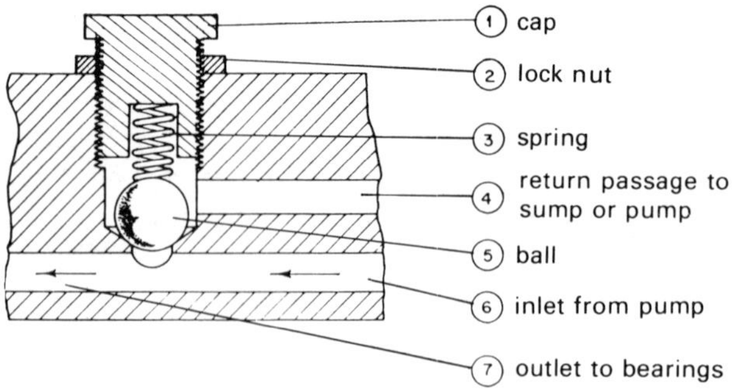

Fig. 26.7 Ball-type oil pressure relief valve

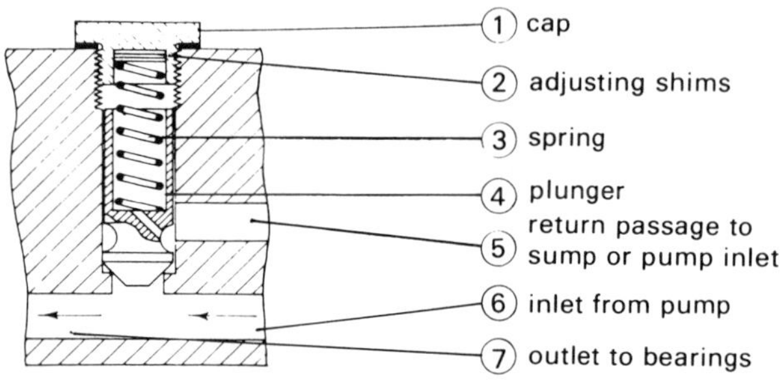

Fig. 26.8 Plunger-type oil pressure relief valve

cause rapid wear of the bearings. If a fluid containing solid particles is passed through a porous material the solid particles will either be too large to enter the pores, or will become lodged in the tortuous passages of the pores, or will pass completely through, depending upon the relative sizes of the pores and the particles. Clearly the finer the pores the smaller the particles the filter will remove from the fluid, but its resistance to the flow of fluid will be correspondingly greater. Thus a very fine filter will need to have a very large surface area if the flow of oil to the engine bearings is not to be restricted.

A rather coarse wire mesh strainer is usually fitted at the pump intake to protect the pump against odd nuts, etc., lying about in the engine, or hard objects large enough to damage the pump. On the pressure side of the system a much finer filter is used. Nowadays the replaceable filter consists of about 4 metres of resin-impregnated paper; this is pleated to expose a large external surface area to the oil and is retained in a cylindrical metal canister. The porosity of the paper is designed to trap nearly all particles of over 25 microns (0.025 mm), a proportion of smaller particles and any sludge.

The position of the filter in the lubrication 'circuit' governs the name used to describe the type; the two common types are:

- *Full-flow*
- *By-pass.*

Fig. 26.9 shows the two arrangements. The full-flow type filters all the oil delivered to the bearings *provided the filter is clean and the oil is not excessively viscous.*

In the case of the by-pass type, only a proportion of the oil delivered by the pump is cleaned, so although this type filters-out finer particles than the full-flow type as the oil returns to the sump, all the oil passing to the bearings is unfiltered. For this reason most engines today use full-flow filters.

Full-flow filters

In addition to the filter material, this type incorporates a by-pass valve that opens when the pressure drop across the filter exceeds about 1 bar (15 lbf/in^2). This valve opens when the filter is clogged, or when the oil is cold, to avoid oil starvation of the bearings.

Although this arrangement gives limited protection to the engine, the continued use of a filter that is not changed at the recommended time will, after a time,

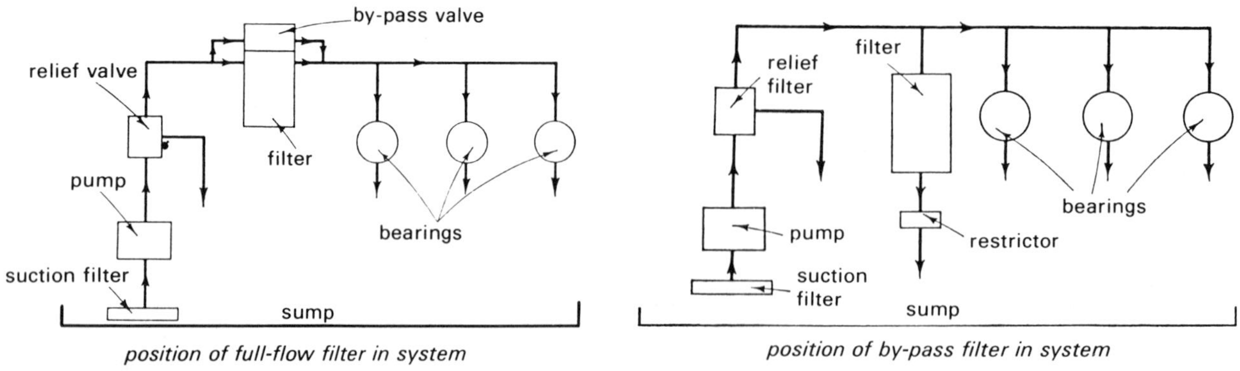

Fig. 26.9 Position of filter in system

cause damage to the bearings owing to the supply of unfiltered oil.

The replaceable full-flow filter is made in two forms:

- *Element-type*
- *Cartridge-type*.

Element-type Fig. 26.10 shows the construction of this type of filter. The assembly is mounted directly on the side of the crankcase by a pad which overcomes the need for troublesome external piping. Today most filters use a paper element but in some cases a felt element, carried on a wire mesh frame and convoluted to give a large surface area, is occasionally fitted to suit world conditions.

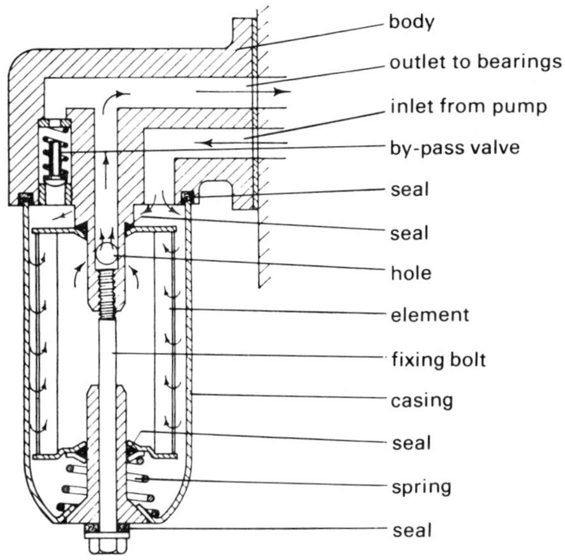

Fig. 26.10 Full-flow filter

The filter should be changed at the recommended intervals and this is carried out by removing the bolt at the base of the filter casing. This type has a number of seals and these should be renewed when the filter is changed. After renewing the element it takes a few seconds for the engine to fill the filter, so the engine should not be accelerated during this time.

Cartridge-type This type, shown in Fig. 26.11, is also called a *spin-on* filter. It is a 'throw-away' filter designed to simplify the task of removing and refitting.

The canister unit, often screwed directly on to the side of the engine crankcase and sealed with a synthetic rubber ring, houses a paper filter element and by-pass valve.

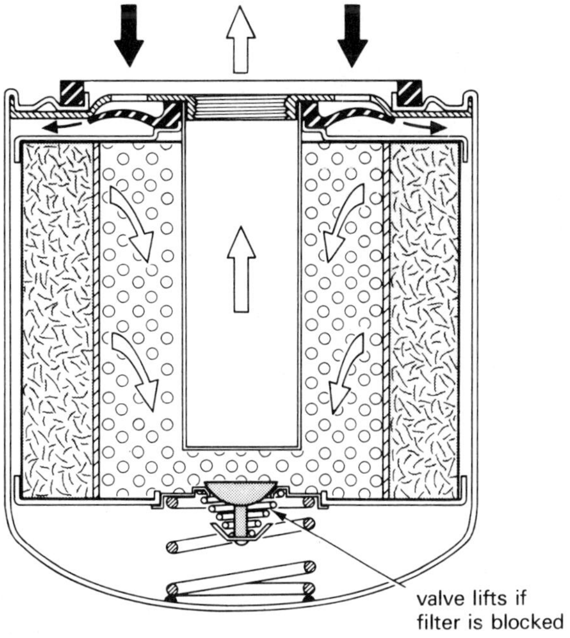

Fig. 26.11 Cartridge-type filter

Valve stem sealing

An overhead valve mechanism, using either rockers or an overhead camshaft, needs a good supply of oil. This requires the valve stems to be fitted with effective seals to stop the oil trickling down the valve guides by gravity and, in the case of the inlet valve, being 'drawn in' by the manifold depression. Although this oil leakage is beneficial as regards stem wear, the oil consumption and exhaust smoke make sealing necessary.

Ineffective valve seals and worn guides can normally be diagnosed by allowing a warm engine to stand for about ten minutes: on restarting, a puff of blue smoke is emitted from the exhaust.

Fig. 26.12 shows arrangements for sealing a valve system. In some cases, special care is necessary when removing or refitting the valve to avoid seal damage.

Crankcase ventilation

If ideal conditions prevail within the engine, the air spaces in the crankcase, rocker cover and timing cover would be filled with a fine oil mist that coats all working parts. Unfortunately this is not so, especially when the vehicle is used for short journeys. The main problems are:

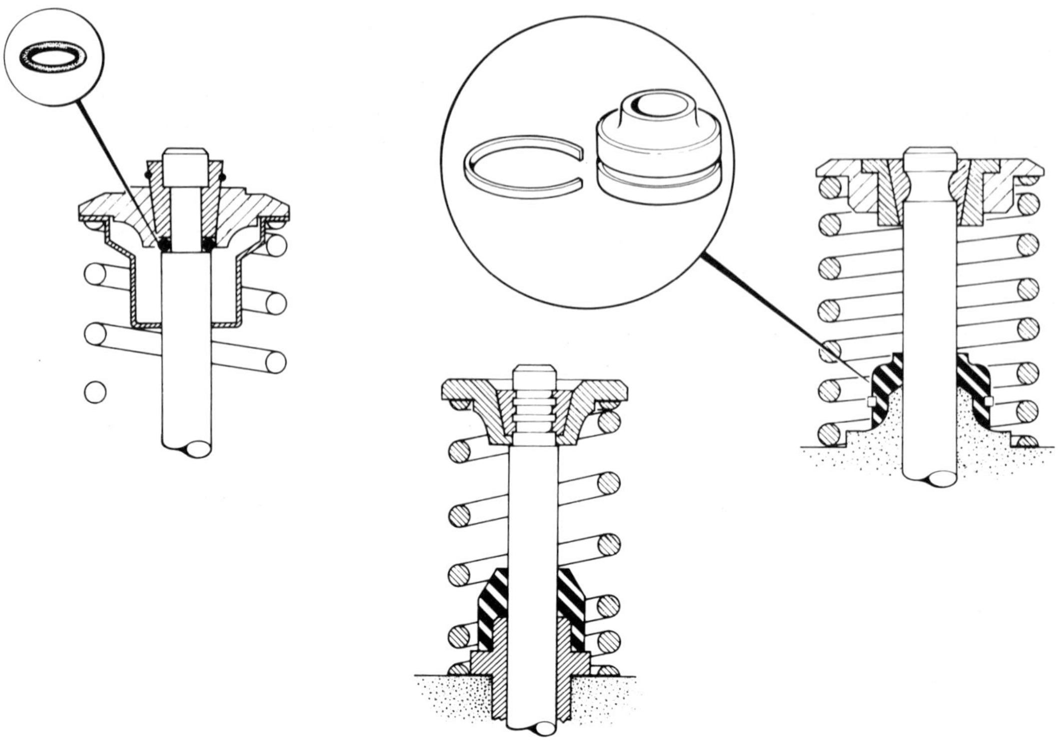

Fig. 26.12 Valve oil-sealing arrangements

1 Blow-by of the pistons carries unburnt fuel, corrosive gases and steam into the crankcase.
2 Steam coming into contact with cold surfaces, such as the covers exposed to the cold air stream through the radiator, causes condensation. Water produced in this way mixes with the oil to form an emulsion called *cold sludge* - a dirty, black, smelly mess which causes corrosion and obstructs oil flow.

The ill effects of these problems can be minimized if the air can be urged to circulate around the engine. In the past this was achieved by providing a *crankcase breather* for clean air to enter the crankcase and a vent pipe to expel the dirty fumes. Sometimes the end of the vent pipe was fitted with an air flow detector to lower the pressure and 'draw out' the fumes. Systems such as this were effective, but caused air pollution, especially if the engine was worn.

Positive crankcase ventilation (PCV) is necessary to conform with emission control regulations. This system uses a *closed crankcase ventilation* arrangement to prevent partially burnt fuel being discharged to the atmosphere. Instead the fumes are returned to the induction system for burning in the combustion chamber.

Fig. 26.13 shows the layout of a typical closed system. In this arrangement the oil filler cap contains a calibrated passage and regulator valve that allows air to enter the crankcase only when the throttle is less than half open. At full throttle the two hoses connected to the air cleaner and manifold convey the fumes to the induction system for burning in the engine.

To allow the engine to function correctly, especially during slow running, air must not be allowed to enter the system at points other than those designed for entry. This means that the oil dipstick must make an effective seal.

Pressure indication

Mechanical type pressure gauge

An indication of the oil pressure can be signalled to the driver by either a pressure gauge or a low pressure warning light.

In the past a pressure gauge system was used that

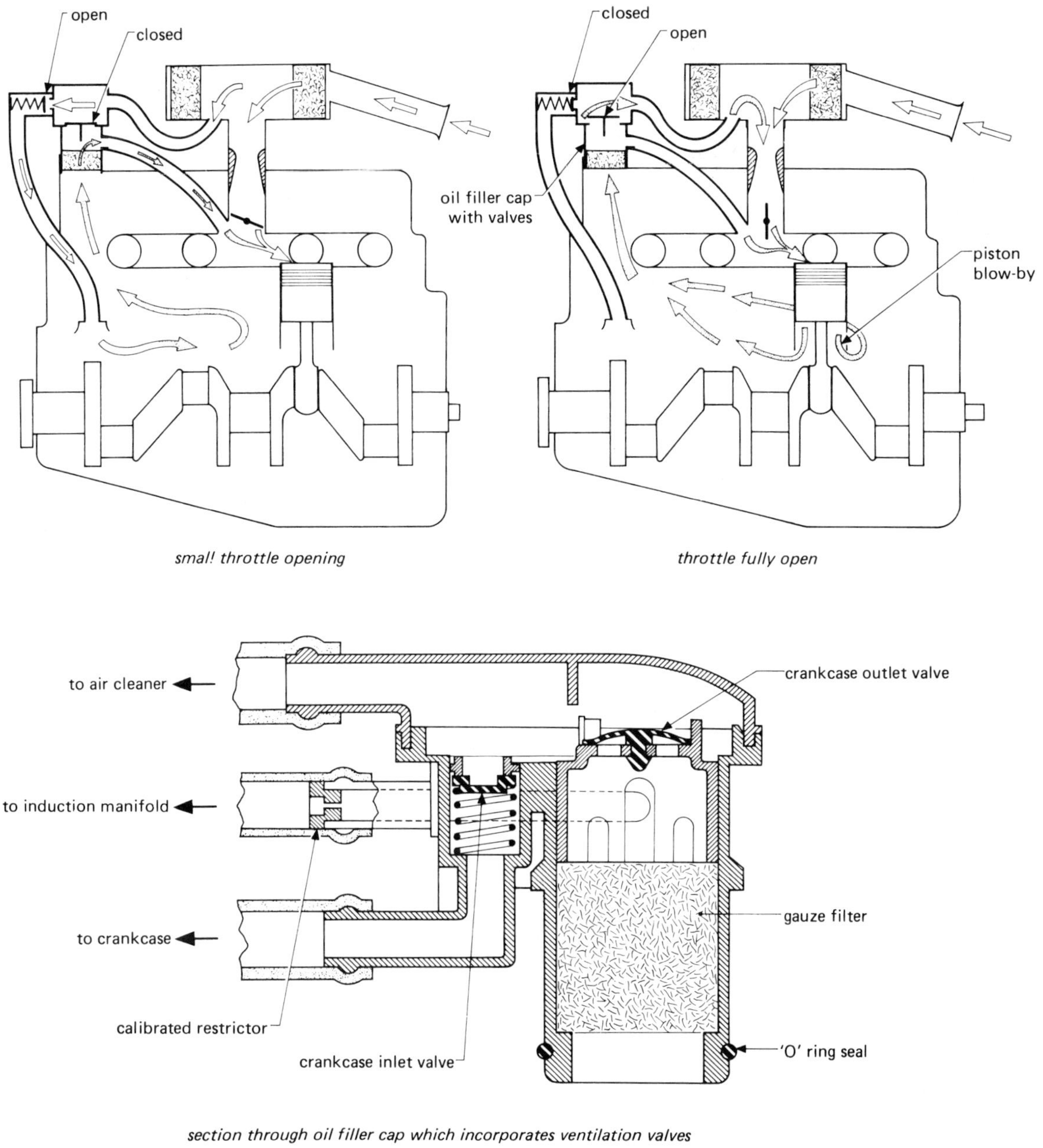

Fig. 26.13 Closed engine ventilation system – positive crankcase ventilation

was based on the Bourdon tube principle. The gauge of this system contains a Bourdon tube, which is a curved tube as shown in Fig. 26.14. When pressure is applied to the tube via a small-bore copper or plastic tube from the engine, the tube tends to straighten out and this moves a needle across the scale on a gauge.

This principle is still used today to operate a vacuum gauge, an instrument used for tuning an engine and for fault diagnosis.

Electrical type pressure indicator

Modern vehicles use electrical/electronic systems to indicate to the driver when the pressure drops to a 'danger' point or uses an illuminated display to show the actual pressure.

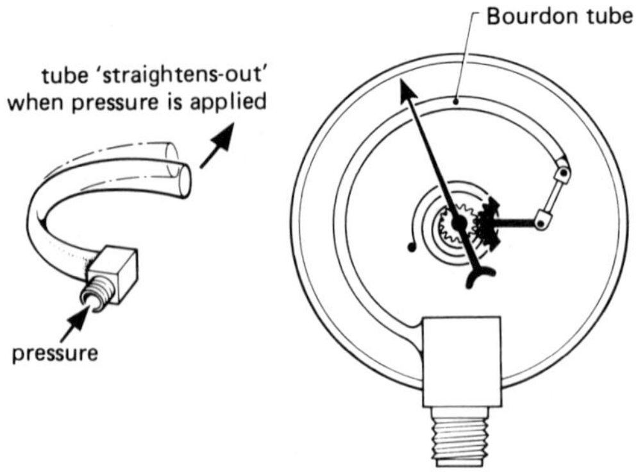

Fig. 26.14 Pressure gauge – Bourdon type

Fig. 26.15 shows a low-pressure warning light system. The pressure switch, fitted on the side of the engine and connected to the main oil gallery senses when the oil pressure is satisfactory. At this point a pair of electrical contacts are separated and this extinguishes a light on the display panel.

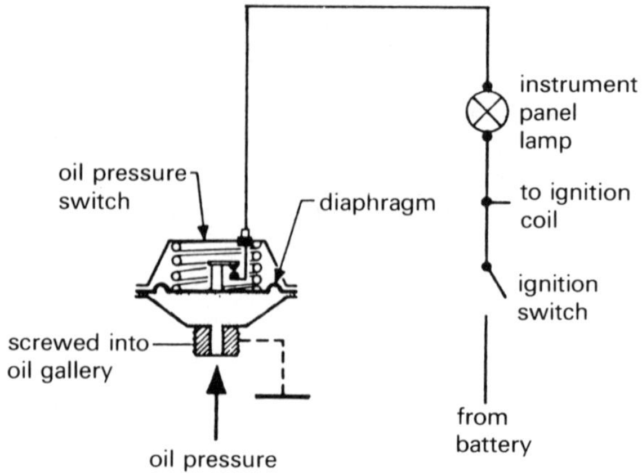

Fig. 26.15 Low oil pressure warning circuit

In addition to pressure sensing some vehicles use a 'hot wire' dipstick to indicate the level of oil in the sump; this electrical arrangement signals to the driver when the oil level is low.

PART SIX

Bearings

27 Plain bearings

A bearing is a part of a mechanism that supports and guides a moving part. By far the commonest type of bearing is that used to support a rotating shaft.

Plain bearings are those in which the moving surfaces are in sliding contact with one another, apart from a film of lubricant and nearly all engine bearings are of this type.

Bushes

The simplest type of bearing consists of a plain hole in the supporting component in which the shaft rotates. In order to provide a suitable combination of materials to minimize friction and wear, and to provide a simple and inexpensive way of repairing a worn or damaged bearing, a sleeve - called a *bush* - is fitted in the hole. Bushes are usually made an interference fit in the hole so that all movement and wear takes place between the bore of the bush and the shaft. There is, however, a type of bush which has a clearance fit in its housing as well as on the shaft; when correctly fitted, this bush rotates at about half the speed of the shaft, thus reducing the rubbing speed of the surfaces. This type is not often used.

Fig. 27.1 shows three types of bush. Type (a) is a plain bush and is used to support a shaft where radial loads only, i.e. loads acting at right angles to the shaft, are carried.

Type (b) has a flange at one end which, in conjunction with a collar or shoulder on the shaft, will resist an end force - or axial load - on the shaft in one direction. It also has one end closed or *blind*, to prevent lubricant escaping or dirt getting in.

Type (c) shows a popular and inexpensive type of bush made from metal strip bent into cylindrical shape. It is called a *wrapped* bush.

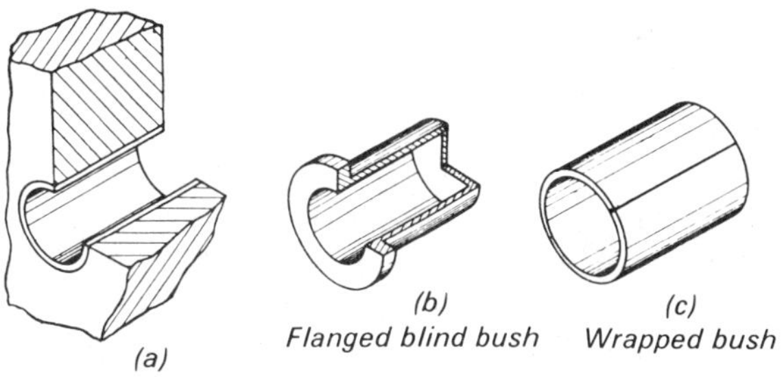

Fig. 27.1 Types of bush

Split bearings

A bush can only be used when the part of the shaft which runs in it can be inserted from one end. If the shaft is of such a shape that this is not possible, e.g. an engine crankshaft, the bearing must be split as illustrated in Fig. 27.2, the two halves of the bearing being held together by bolts, screws or studs and nuts. In this case the equivalent of the bush is known as a *pair of bearing shells, or half bearings*.

The lower part (4) is the housing in which the bottom shell (3) is fitted. The top shell (2) fits in the cap (1). As in the case of a bush, the shells are prevented from turning in the housing by an interference fit. When a shell is pressed down into its housing, its edges should stand slightly proud of the housing faces, and similarly for the shell fitted in the cap. If the cap is now fitted in place and the two nuts screwed down *finger-tight*, there will be a small clearance between the faces of the cap and housing, although the edges of the shells are touching, as

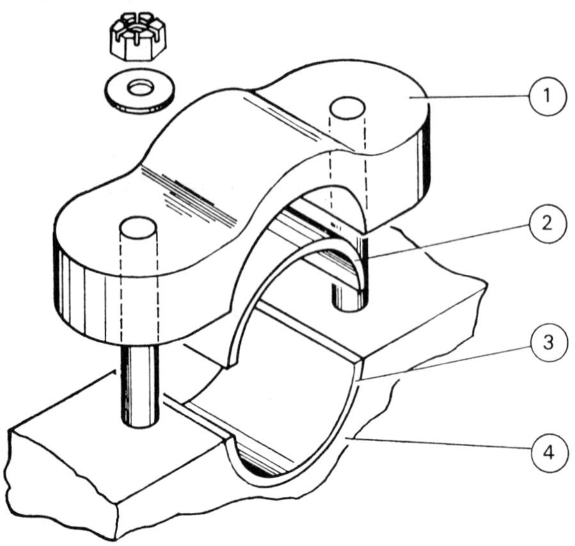

Fig. 27.2 Split bearing with detachable shells

shown (much exaggerated) in Fig. 27.3. The gap - or *nip* - is actually only a few hundredths of a millimetre, the exact amount depending upon the size of the bearing and the materials of which the housing and cap are made. When the nuts are correctly tightened down, the nip disappears and the shells are tightly held.

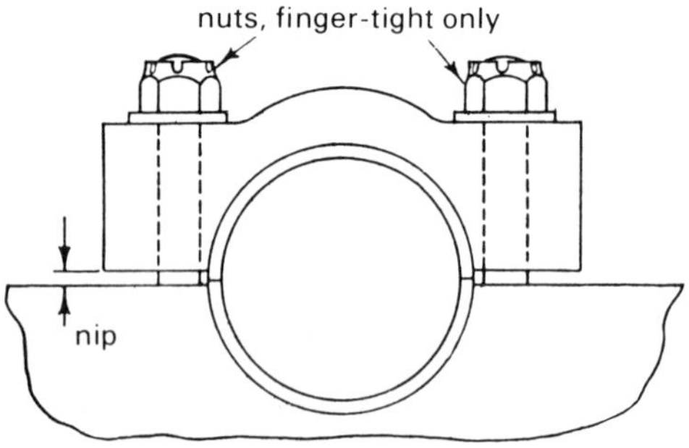

Fig. 27.3 Bearing nip (greatly exaggerated)

Location devices

These are provided to ensure correct positioning of the shells during assembly. Two common types are illustrated in Fig. 27.4. A *dowel* (a) is a short peg, one end of which is made a tight fit in a hole in the housing, the other (protruding) end being a clearance fit in a hole in the shell.

Lugs (b) are pressed out at *one* joint face of each shell and engage with corresponding notches cut in the housing and cap. One lug per shell is used, and the lugs of both shells of a pair are normally arranged

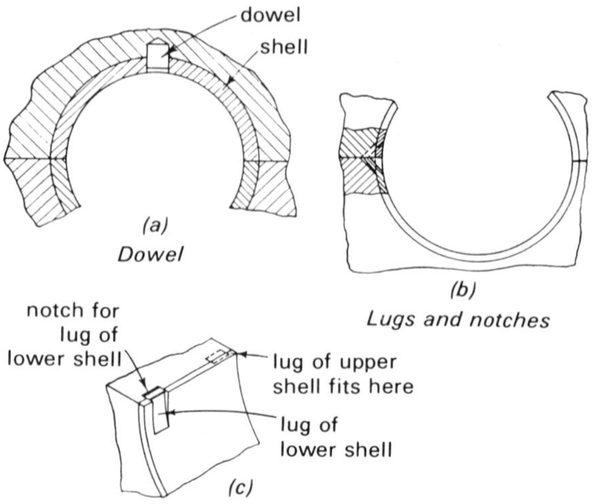

Fig. 27.4 Methods of locating shells

at opposite ends of *the same* joint face, as shown in Fig. 27.4c. It is most important that there should be a clearance between the back of the lug and the notch to prevent the shell being pushed inwards and causing local metal-to-metal contact at this point. You can check this by pushing down on the joint face of the shell on the lug side. It should be possible to move it about 1 mm lower than the housing face before the lug touches the back of the notch.

It should be emphasized that the shells must be prevented from moving in the housing and cap by their interference fit alone. The dowels or lugs are purely for correct location during assembly and they must not be relied upon to prevent rotation of the shells, though they may contribute to it.

Oil holes and grooves

In most bearings, the oil for lubrication is supplied through a hole drilled through the housing, and there must be a corresponding hole in one of the shells. This hole should be placed at that part of the bearing where radial load is least, which is usually the centre of the upper shell. From this hole, oil is carried round by shaft rotation, and pressure causes the oil to spread towards the ends of the bearing. Oil grooves are unnecessary in most pressure-fed bearings and are undesirable in the areas supporting heavy loads, since not only do they reduce the surface area of the bearing but they also let the oil escape easily from the region where it is most important to retain it.

Certain bearings, such as the big-end bearing, are carried on moving parts and cannot conveniently be supplied with oil directly and independently. In such cases the oil is supplied through a drilling in the shaft, and therefore there is no need for an oil supply hole in the shells. To ensure a continuous supply of oil to the big ends, a circumferential groove is formed around the main bearing shells, into which the oil is fed from the main oil gallery (Fig. 27.5). The main journal end of the hole through the crankshaft runs around this groove so that the big end is in constant communication with the main oil gallery. The width and depth of this groove around the main bearing shells should be no more than is necessary to carry enough to the big ends.

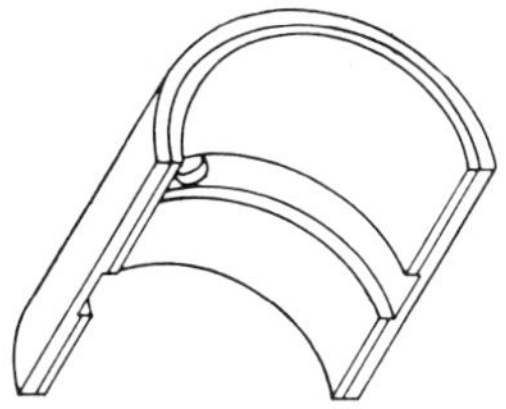

Fig. 27.5 Bearing shell with transfer channel oil groove

Bearing materials

Shafts are almost invariably made of steel or iron, and the surface of the bearing should make a satisfactory combination with the shaft. Some materials used for the housing may be satisfactory in this respect: for instance both cast iron and aluminium are quite satisfactory for many conditions of operation and could be used without bushes or shells. But for operation at high speeds, or heavy loads, or both, some other material is generally preferred. A few are mentioned below.

Bronze

Bronzes are primarily alloys of copper and tin, the exact properties depending upon the proportions and also upon what other elements, if any, are included. A type commonly used contains a small amount of phosphorus and is known as *phosphor bronze*.

Bronzes are fairly hard - though softer than steel - and have good load-carrying capacity, but are less suited than some other materials to high rubbing speeds.

In cases where the supply of lubricant is difficult, then the plain bearing could be either pre-lubricated or self-lubricated.

Pre-lubricated Powdered lead bronze can be compressed into a mould and heated to partially fuse the metal particles together; this process is called *sintering*. The porous composition has the ability to absorb and retain a lubricant, which is applied either by soaking the bearing before fitting, or by occasional re-oiling. Some bushes are made with the bronze sintered to a steel backing and coated with a layer of acetal resin polymer. This surface gives a good bearing performance even when there is only a trace of lubricant. The bearing surface is often indented to give pockets to hold the oil or grease.

Self-lubricated As the name suggests, this bearing is intended to operate in a dry state. One type in common use is made up of three layers - a steel backing, a sintered bronze interlayer and a coating, or *overlay* as it is called, of lead mixed with PTFE (polytetrafluoroethylene). This synthetic polymer (long-chain chemical compound) has an exceptionally low frictional resistance and complete immunity to most solvents including water, so it has many uses for rubbing surfaces on a vehicle.

White metal, or Babbitt alloys

First introduced by Babbitt over a century ago, this is an alloy of tin and small amounts of copper and antimony. It has very good anti-friction properties and when used with a comparatively hard shaft, the soft material allows abrasive particles to embed below the surface and so reduce wear.

Softness is good for embeddability, but is bad when heavy loads have to be carried. Repeated heavy loads cause *fatigue* (break-up and flaking of the surface), so in order to reduce this problem the *thin-wall bearing* is used. This type has a steel liner, about 2 mm thick, coated with a thin layer of about 0.1 mm of bearing metal. White metal was used for a number of years for main and big-end bearings but the large inertia load associated with modern engines is too great for this material. Camshaft bearings and thrust washers are the present-day uses of white metal.

An ideal material for big-end and main bearings should have the following properties:

1 High mechanical strength - to withstand heavy loads and resist fatigue.
2 High melting point - to resist damage due to high oil temperature.
3 Resistance to corrosion.
4 Conformability - to yield easily when slight misalignment occurs.
5 Good surface properties - to prevent seizure or pick-up if the oil film breaks down momentarily. This property is also called compatability.

In practice no material meets the ideal requirements of all of these properties, so the alloy selected must match the properties that are special to the engine type, e.g. a compression-ignition engine requires a material of high strength, whereas in a spark-ignition engine the material has to withstand high oil temperatures (over 120 °C), extremely thin oil films and high rubbing speeds.

Aluminium-based alloys

A popular alloy used for car crankshaft and big-end bearings is 20 per cent tin–aluminium. This material, formed in a thin-wall bearing, combines high fatigue strength with good surface properties and can be used without an overlay. When bonded together, the tin and aluminium form a network structure, so this material is sometimes called *reticular tin–aluminium*.

Where higher loadings prevail, either 6% tin–aluminium or aluminium–silicon alloys may be used. The former has a high fatigue strength and the latter a very high strength, but both materials require an overlay to improve compatibility and conformability.

Copper–lead and lead–bronze

These alloys have an exceptionally high fatigue strength and are used for heavily loaded bearings. Being harder and less able to embed abrasive particles, they generally require a harder shaft.

The materials have poor corrosion resistance: the lead is dissolved by acidic oil and this destroys the boundary and wear-resisting properties. To overcome this problem a very thin overlay (0.013–0.038 mm) of lead–tin or lead–indium is used. This overlay is soft so embeddability and conformability are also improved.

The life of these bearings is linked to the time it takes for abrasive particles in the oil to cut away the overlay. When the overlay is worn off the wear is extremely rapid, so attention to oil filtration is essential.

Thrust washers

Shafts usually need to be located endways, usually against a certain amount of end thrust. Bushes may be flanged (see Fig. 27.1) in conjunction with shoulders or collars on the shaft, and shells may be flanged in a similar manner. In the case of wrapped bushes and thin-wall shells it is more usual to use separate thrust washers or half washers. These have steel backs faced with white metal, copper–lead or lead–bronze, and are located in a recess in the bearing housing (see Fig. 27.6).

In many cases only one half washer is used, and it is prevented from rotating by letting its ends abut the unrecessed edge of the housing half which does not have a half washer. When a pair of half washers is used, one has a lug which engages a notch in the housing to prevent rotation.

Shafts are usually located endways by one bearing only, thrust washers being fitted at both ends of this bearing but omitted from all others.

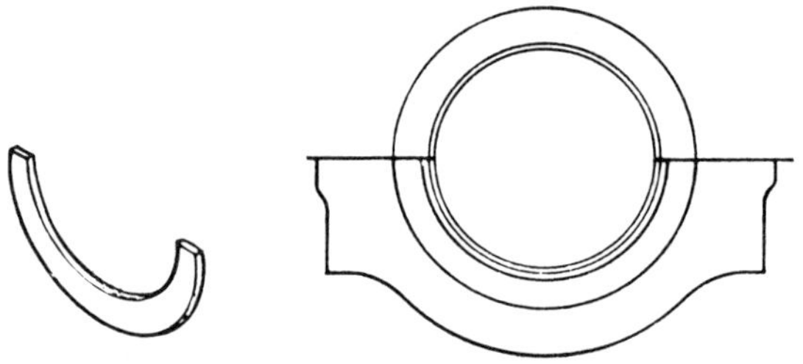

(a) Single half-washer located in cap

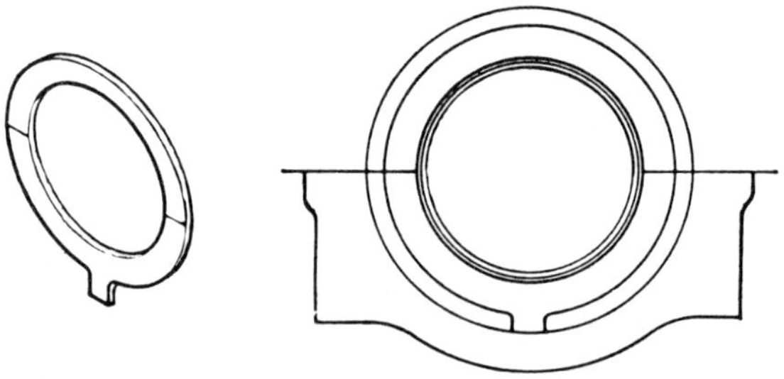

(b) Two half-washers, lower one located by lug

Fig. 27.6 Thrust washers

28 Ball and roller bearings

In this class of bearing, rolling motion is substituted for sliding by interposing rolling elements between the shaft and its housing. By this means friction is reduced, especially when the speed of rotation is low and no continuous supply of oil under pressure is available.

A complete bearing consists of four components:

1 An inner race which is fixed on the shaft. (In a few cases the shaft itself may form the inner race if it is made of suitable material, but replacement may then be more costly when excessively worn.)
2 An outer race which is fixed in the housing. (As with the inner race, this may sometimes be formed by the housing itself.)
3 A suitable number of balls or rollers.
4 A cage to prevent adjacent balls or rollers from rubbing against one another.

The races and balls or rollers are made from a special quality steel, suitably hardened. There may be slight variations in the composition and heat treatment of the steel between different manufacturers and for different types of bearing. Cages may be made of bronze, aluminium or mild steel.

There are four main types of bearing, classified by the shape of the rolling elements as ball, cylindrical roller, spherical roller and taper roller.

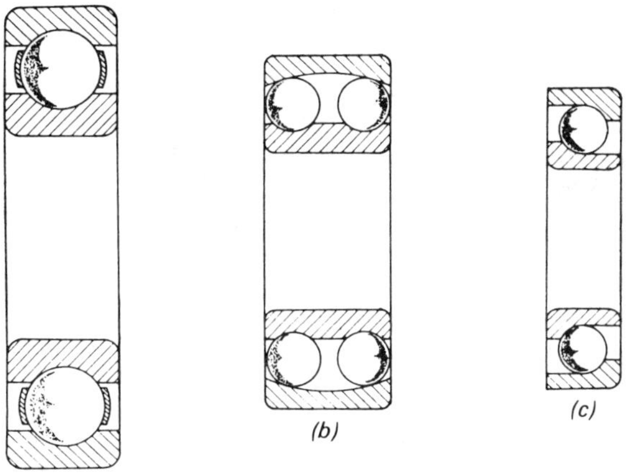

Fig. 28.1 Types of ball bearing

Ball bearings

Fig. 28.1 shows cross-sections of three examples of ball bearing:

1 *Single-row* bearing with the balls running in grooves in the races. This type is intended for carrying mainly radial loads, but can also support some axial load in both directions. Double-row bearings of this type are also made and are capable of carrying heavier loads.
2 *Self-aligning* bearing. The inner race has two ball tracks and both rows of balls run in an outer race whose inner surface forms a section of a sphere having its centre at the shaft centre. This allows the shaft axis to run at a small angle to the axis of the housing, and bearings of this type are used for applications where precise alignment of shaft and housing cannot be maintained.
3 *Angular contact* ball bearing which is capable of taking axial loads comparable with the radial loads, but in one direction only. Bearings of this type are generally used in pairs one at each end of a shaft, and care must be taken when fitting the bearings to ensure that they are fitted the correct way round.

Cylindrical roller bearings

An example of this type is shown in Fig. 28.2. These are capable of carrying greater radial loads than ball bearings, *but no axial load*. The guiding flanges may be either on the inner or outer races.

Where heavy radial loads have to be carried at low

Fig. 28.2 Cylindrical roller bearing

speeds, or where the motion is of an intermittent or oscillating nature, specially long rollers of small diameter – called *needle rollers* – may be used.

Spherical roller bearings

As shown in Fig. 28.3, this is the roller bearing equivalent of the self-aligning ball bearing. It is constructed on similar principles and used for similar applications, but is capable of carrying greater loads.

Fig. 28.3 Spherical roller bearing

Taper roller bearings

An example of this important and useful type is shown in Fig. 28.4. The working surfaces of both races and rollers are conical, the taper being such that the cones, of which races and rollers form a part, each have their apex at a common point on the axis of the shaft.

They are always used in pairs facing opposite

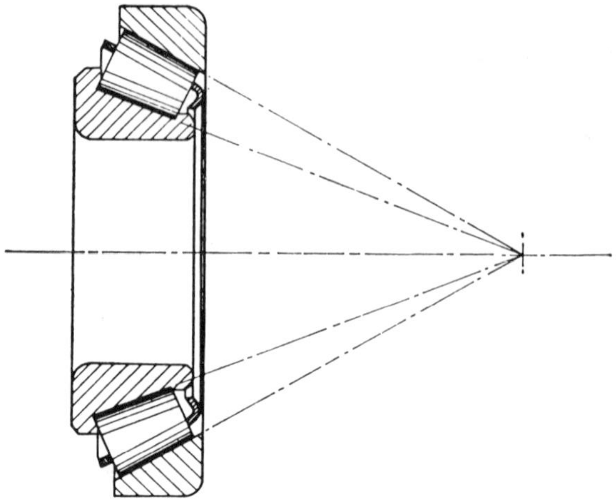

Fig. 28.4 Taper roller bearing

ways, and are capable of dealing with considerable axial loads as well as radial loads. Different taper angles may be used, depending upon the amount of axial load to be carried. They are provided with some means of axial adjustment to control the amount of end play or pre-load allowed.

A single-row bearing as illustrated can only deal with thrust in one direction (hence the need for using two) but double-row types, each row having its taper facing opposite ways, are also made and are capable of taking thrust in *both* directions.

Fitting bearings

Cleanliness is vitally important. Do not unwrap new bearings from their packages until the last possible moment. The bearings are coated with preservative and this need not be removed.

In most cases the races are an interference fit both in housings and on shafts. The best way to fit the bearings is by pressing them into place with a press, using a mild steel tubular drift of correct diameter for each race. *Under no circumstances should pressure be applied to one race in order to press the other into place*, i.e. the pressure must be applied directly to the race being fitted. Where a press is not available, strike the drifts carefully with a suitable hammer.

Removing bearings

Unless the bearing is to be scrapped, the same care must be exercised in removing as in fitting it. Using a suitable extractor, drift and press or hammer, force must be applied directly to the race being removed. In many cases, special extractors are available for this purpose.

Bearing adjustment

Where adjustment is provided it may be made by nuts, threaded rings or by distance pieces and shims. For many applications the ideal adjustment is that there should be no end float and no pre-load (or over-tightening). For certain applications, however, such as rear-axle pinion shafts, correct meshing of the gears demands complete elimination of play, and the bearings are generally *pre-loaded*, i.e. they are tightened slightly beyond the point of elimination of play. Specific instructions are given in workshop manuals on this point.

Lubrication

The bearings are lubricated with oil or grease depending upon the component into which they are fitted. In engines, gearboxes and final-drive gears, oil is used and the design ensures a constant supply to each bearing. In other parts of the vehicle, such as wheel hubs, bearings are grease-lubricated by packing the bearings and the inside of the hub cavity with a suitable grease. Details as to the type of grease to be used, and the amount to be put into the housing, are given in the service manual for the vehicles.

PART SEVEN

The compression-ignition engine

29 Cycle of operations

History

This type of engine is often called a *diesel* engine; the name is derived from the German engineer, Dr Rudolf Diesel, who in 1892 took out a patent (no. 7241) on an engine which relied on the heat generated during compression to ignite a fuel of coal dust. This fuel was blasted into the cylinder by air pressure at the end of the stroke. The aim of the design, which was successfully applied to an engine five years later, was to achieve a higher thermal efficiency or improved fuel consumption by using a compression ratio higher than that employed on petrol engines. In those early days, pre-ignition (ignition before the spark) occurred in a petrol engine if the compression ratio exceeded a given value.

Many authorities do not recognise Diesel as the inventor of the engine which was the forerunner of the modern compression-ignition engine. They state that the patent (No. 7146) taken out in 1890 by a British engineer, Herbert Ackroyd-Stuart, and put into commercial production two years later, contained all the fundamental features of the modern unit. This patent, which was the result of practical development work on low-compression oil engines, included the induction and compression of air, and the timed injection of a liquid fuel by means of a pump.

To avoid taking sides in this controversy, the terms *compression-ignition (CI)* or *oil engine* are often used.

Four-stroke operation

In many ways, this operation is similar to the four-stroke petrol cycle; both occupy 720 degrees, or two revolutions of the crank, to complete the four strokes - induction, compression, power and exhaust.

Fig. 29.1 shows the sequence of operations, and also the main constructional differences between petrol and CI engines, i.e. the elimination of the carburettor and the substitution of a fuel injector in place of the sparking plug. A difference which is not apparent is the compression ratio, which is much higher in the case of the CI engine. Some engines have ratios as low as 11 : 1, or as high as 26 : 1, whereas petrol engines seldom use a ratio greater than 10 : 1.

The sequence is as follows:

Induction The descending piston increases the cylinder volume and decreases the pressure. Atmospheric pressure forces *air* through the open inlet port into the cylinder.

Compression Both valves are closed and so the ascending piston compresses and raises the temperature of the air. For a compression ratio of 14 : 1, the final temperature and pressure will be 650 °C and 3.5 MN/m^2 (500 lbf/in^2) respectively.

Power Just before t.d.c., fuel oil, having a self-

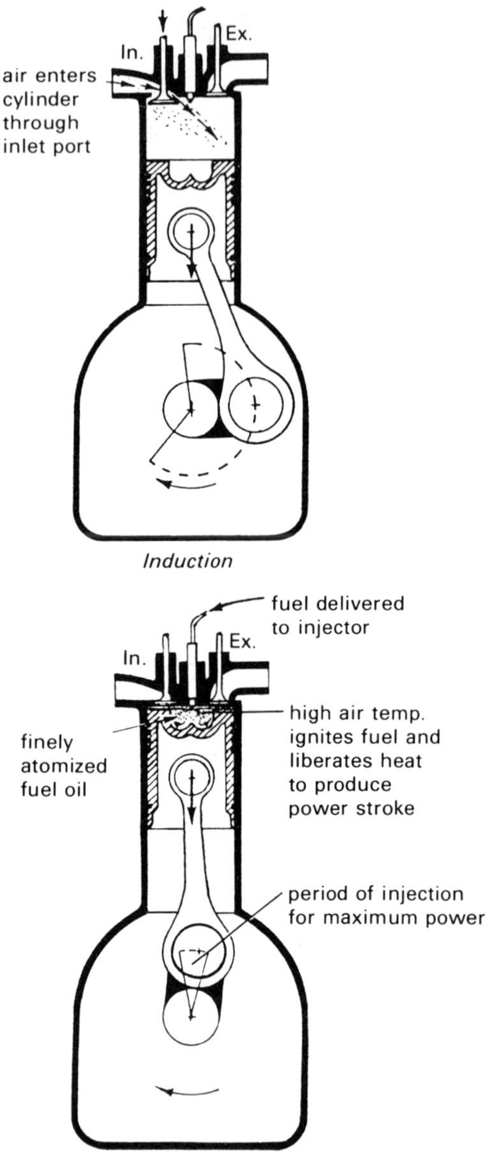

Fig. 29.1 Four-stroke compression-ignition engine

ignition temperature of 400 °C, is injected into the cylinder at a high pressure, e.g. 175 bar or 17.2 MN/m^2 (175 atmospheres), by means of a jerk-type pump. After a short delay, the fuel begins to burn and liberates heat which raises the pressure to 6.2 MN/m^2 (900 lbf/in^2), providing the thrust necessary for the power stroke. The amount of power is controlled by the period of injection, i.e. the quantity of fuel injected.

Exhaust As the piston nears the end of the power stroke, the exhaust port is opened. The ascending piston pumps out the burnt gas in readiness for the new cycle.

(The temperatures and pressures quoted are approximate and are only intended as a guide, since engine design and other factors can alter conditions considerably.)

Two-stroke operation

Although the two-stroke CI engine is not so popular as the four-stroke unit, some manufacturers regard the two-stroke engine's smoother torque, simpler construction and smaller unit as supreme advantages. Originally, this type was restricted to low-speed industrial and marine applications, but nowadays high-speed units are fitted to commercial vehicles. The great disadvantage of the two-stroke petrol engine - the loss of fuel to the exhaust when both ports are open - does not apply to the CI engine, since the cylinder only contains air: therefore it may be stated that the two-stroke cycle is most suitable for CI operation.

Fig. 29.2 shows the sequence of operations for a simple two-port valveless form of engine. Fig. 29.2a shows the air entering the cylinder through the inlet port, and 'burnt' gas flowing to the exhaust system.

As the piston ascends (Fig. 29.2b) inlet and exhaust ports are closed to provide the compression stroke. After the air has been compressed to a c.r. of 12–16 : 1, the high temperature will ignite the fuel injected into the cylinder and produce the power stroke (Fig. 29.2c).

Vehicle applications normally employ a blower to pressure-charge the cylinder with air, but it is possible to operate the engine by utilizing the pressure waves or pulses in the exhaust system to induce the 'new' air into the cylinder.

Fig. 29.3 shows a uniflow type of two-stroke engine which is fitted with an exhaust valve and pressure-charged with a Roots-type blower. This arrangement gives excellent cylinder evacuation and recharging, which is claimed to give double the power of a four-stroke engine of the same capacity.

Opening the exhaust valve before uncovering the circumferential air ports allows the remaining gas pressure to start pumping out the exhaust gas. This is followed by an air charge approximately 30 per cent greater in volume than the cylinder capacity, to cool and scavenge the cylinder effectively.

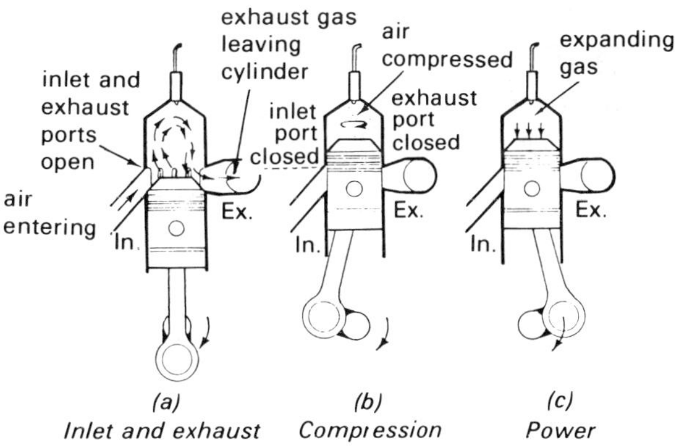

Fig. 29.2 Two-stroke cycle: valveless form

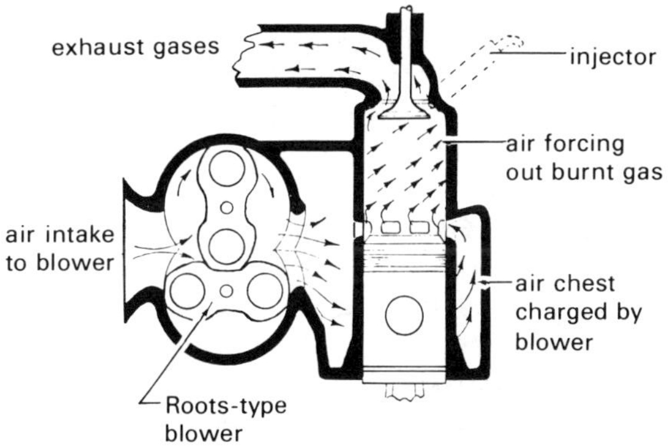

Fig. 29.3 'Uniflow'-type two-stroke engine

30 Engine construction: combustion chambers

Combustion in the chamber of a petrol engine originates at the sparking plug and then progresses throughout the cylinder. In the case of the CI engine, combustion of the fuel is started by the heat of the air in the chamber. As the fuel 'droplet' passes through the air, it absorbs heat, and, if the temperature is high enough, the fuel will vaporize and ignite. Wide distribution of the fuel occurs during the heating phase and so combustion will begin at many points in the chamber. In the case of one type of direct injection engine, once ignition has taken place, most of the burning will tend to concentrate in zones fairly close to the injector. These zones must be fed with air in order to sweep away the *burnt* gases and supply the oxygen necessary for complete combustion: lack of oxygen in the combustion region leads to black smoke in the exhaust. Since power is governed by the quantity of fuel injected, and this, in turn, is limited by the point at which smoke is emitted, some system must be used to introduce an orderly supply of air to the fuel. The air flow is called *swirl*, a term used in preference to *turbulence*, since this expression implies a disorderly movement.

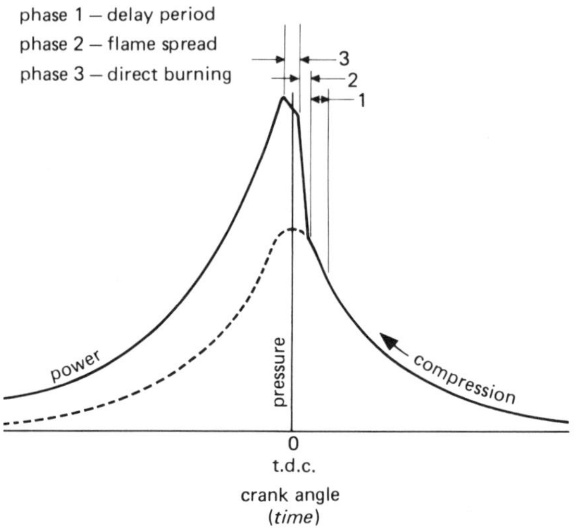

Fig. 30.1 Phases of combustion

Phases of combustion

Fig. 30.1 shows a time-base indicator taken from a CI engine that is operating at full-load. Ricardo indicated that the combustion process which gave this form of diagram may be separated into three phases.

Phase 1: ignition delay period This is the time taken (or angle turned by the crank) between the start of injection to the commencement of the pressure rise. During this important period, the injected fuel particles are being heated by the hot air to the temperature required for the fuel to self-ignite.

Phase 2: flame spread causes a sharp pressure rise due to the sudden combustion of the fuel that was injected during the first phase. The rate of pressure rise governs the extent of the combustion knock. This is commonly called *diesel knock* and is considered to be the main disadvantage of the CI engine.

Phase 3: direct burning of the fuel as it enters the chamber gives a more gradual pressure rise. When the engine is operating at less than full load, this phase does not exist.

Consideration of the diagram shows that if the delay period is shortened the quantity of fuel in the chamber is less so the engine is quieter and smoother. Design factors which increase swirl, atomization and temperature give a shorter delay period, but perhaps one of the most important factors is the ignition quality of the fuel. Unlike petrols, CI fuels must be capable of igniting in a heated air mass without the aid of an electrical spark. The ability of a fuel to ignite, or its 'ignition quality', is expressed as a *cetane rating*. Classification is obtained by comparing the fuel with a good fuel, cetane, rated at 100, and a poor fuel, alpha-methylnaphthalene, which is rated as 0. Most of the fuels which are commercially available have a cetane rating of about 50.

Types of combustion chamber

There are many good designs of CI combustion chamber, each arranged so as to provide an effective swirl pattern. These designs can be divided broadly into two main classes:

- *Direct injection*
- *Indirect injection*.

In the former, fuel is injected directly into the closed end of the cylinder, whereas in the latter type, fuel is sprayed into a separate small chamber, which is connected to the cylinder by a small passage or throat.

Direct injection

Fig. 30.2 shows the essentials of an *open type* combustion chamber. For many years the direct injection (DI) type shown has been used on heavy vehicles and, in a slightly modified form, is currently becoming popular for car engines in the 2-litre category.

A deep cavity, machined in the piston, contains most of the air, since at t.d.c. the piston is very close to the flat cylinder head. To obtain the necessary compression ratio, overhead valves are essential. Shallow recesses in the piston crown provide clearance for the valve heads. (Inaccurate setting of the valve timing allows the valves to strike the piston). A multi-hole injector allows finely atomized fuel under high pressure (175 bar) to penetrate the fast-moving air and just enter the cavity of the piston.

Swirl is produced in two planes, vertical and horizontal. The ascending piston causes the air to be directed into the cavity and move in the manner shown in the diagram. As the piston approaches t.d.c., this motion will be speeded up because of the *squish* action of the air between piston and head. Horizontal or rotary swirl can be obtained by inclining the inlet port tangentially to the cylinder, or masking the inlet valve. Fig. 30.2a shows the latter arrangement, which is the most popular. Combining both swirl movements gives a vortex air flow in the cavity, and ensures a good supply of oxygen to the combustion region.

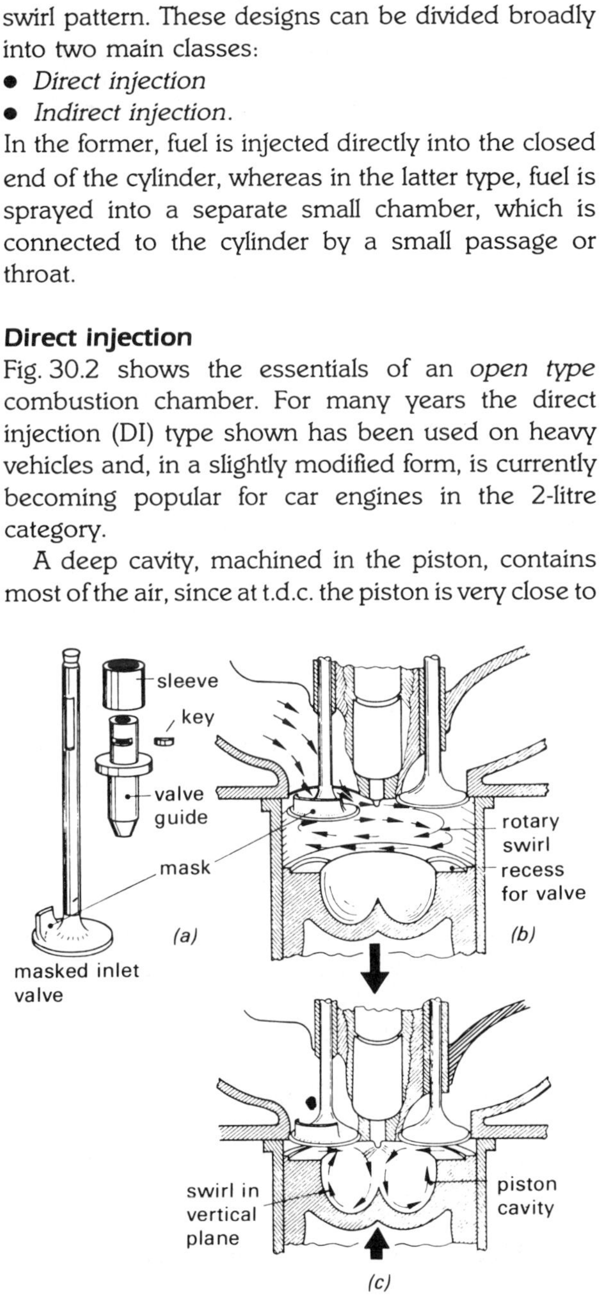

Fig. 30.2 CI combustion chamber – direct injection

Indirect injection

Up until the mid-1980s the indirect injection (IDI) type was commonly used on small CI engines as fitted to light vehicles. Compared with the traditional heavy vehicle DI engine, the IDI ran more smoothly and since the IDI type used lower injection pressures, the engine was more able to operate over a large speed range.

Many IDI combustion chambers have been based on the Ricardo Comet design shown in Fig. 30.3. In this arrangement a *swirl chamber* is connected to the main chamber by a throat, which is allowed to operate at a higher temperature than the surrounding metal.

Air is pumped through the hot throat into the chamber during compression, so by the end of this stroke the ante-chamber contains very hot air in a

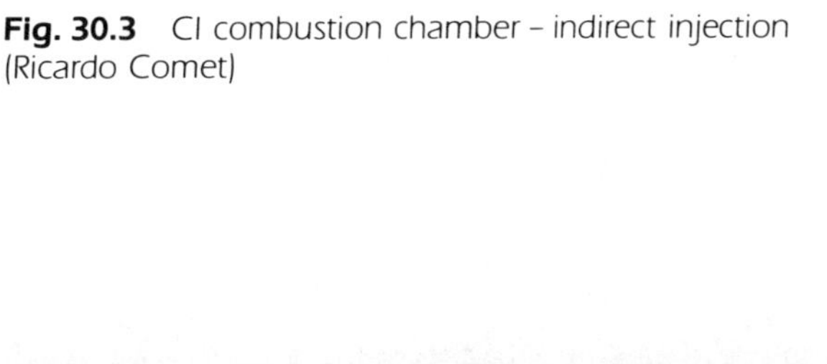

Fig. 30.3 CI combustion chamber – indirect injection (Ricardo Comet)

high state of swirl. Fuel injected into this fast moving air mass is quickly atomized into a very fine state. This atomization is very effective even though the fuel is injected in the form of a 'soft' spray by a pintle or single-hole nozzle set at a comparatively low pressure (e.g. 100 bar).

After combustion has been initiated in the swirl chamber, burning fuel, together with the unburnt and partially burnt fuel, is carried into the piston cavity in the main chamber. When the injection period is increased to produce a higher engine power, most of the fuel that is injected towards the end of the spraying period does not ignite until it mixes with the air in the main chamber. This ensures that during the combustion period the burning can continue for a relatively long time until it finally reaches a stage at which the fuel cannot find sufficient oxygen. Beyond this point black smoke starts to be emitted from the exhaust; this smoke point dictates the maximum amount of fuel that can be injected without sacrificing economy and also represents the maximum power that can be legally obtained from the engine.

In an IDI engine the combination of hot air and excellent atomization gives a short ignition delay. Compared with the DI type, the intensity of diesel knock is lower, the engine runs more smoothly and the cetane rating of the fuel that is required can be lower with IDI engines.

All CI engines require some special provision for cold starting. The injection of a larger quantity of fuel, and the greater amount of easily ignitable fractions contained in the injected charge, are generally sufficient to start a cold DI engine, but the greater heat losses of IDI units require these engines to have extra cold-start provisions. Compared with ratios of about 16 : 1 that are used with DI, the IDI engines use higher ratios of the order of 22 : 1; in some cases a ratio as high as 30 : 1 is used.

In addition to the cold-starting requirement, a high ratio is also used to raise the thermal efficiency, i.e. economy, to that of the DI engine. This feature tends to counteract the greater heat loss due to the larger surface area of an IDI combustion chamber.

When cold-starting an IDI engine, one or more of the following are used:

1 *Heater plug* – a hot bulb, often called a *glow plug* that is fitted in the swirl chamber. The air in the chamber is heated by electrical means for a few seconds prior to starting a cold engine. Today these plugs are often controlled automatically.
2 *Manifold heaters* – an electrical unit fitted to pre-heat the air as it passes through the inlet manifold to the cylinder.
3 *Pintaux injector* – a pintle-type injector which has an auxiliary hole to direct fuel down the throat of the chamber during the cranking period (see page 197).

Modern power units used on light vehicles

The small modern CI engine, with a fuel consumption up to 40 per cent better than a petrol engine of equivalent power, has made the CI engine very attractive for use in cars. This advantage becomes even more attractive when the vehicle is used more than normal to use the saving in fuel cost to offset the higher initial cost of the engine.

This benefit, together with the rising demand for this type of engine, has induced most car manufacturers to offer at least one small 'diesel' engine alternative in their range.

In the past, small CI engines were noisy and they were unresponsive compared with a spark-ignition unit, but considerable advancements in these areas have been made during recent years. Combustion improvements and sound damping have reduced the noise levels, and by fitting an engine of slightly larger capacity the performance gap has been narrowed.

Perkins Prima

One example of an advanced design of CI engine suitable for light vehicles is the Perkins Prima. Besides achieving above average performance and economy, this 2-litre DI power unit also meets the EEC emission regulations recently introduced.

Traditionally IDI is the system normally associated with small CI units. Perkins departed from the norm because they found that the IDI was less efficient because of its high energy consumption; this was caused by the high compression ratio and large pumping losses that resulted from the rapid movement of air between the main chamber and the swirl chamber.

Fig. 30.4 shows the layout of the combustion chamber. The *bowl-in-piston* combustion cavity, together with an inclined (helically shaped) inlet port to achieve high-speed rotary swirl, gives high power and excellent economy up to a rated engine speed of about 4500 rev/min; a speed range higher than the majority of other units in current use.

Automatic glow plugs are used for cold starting;

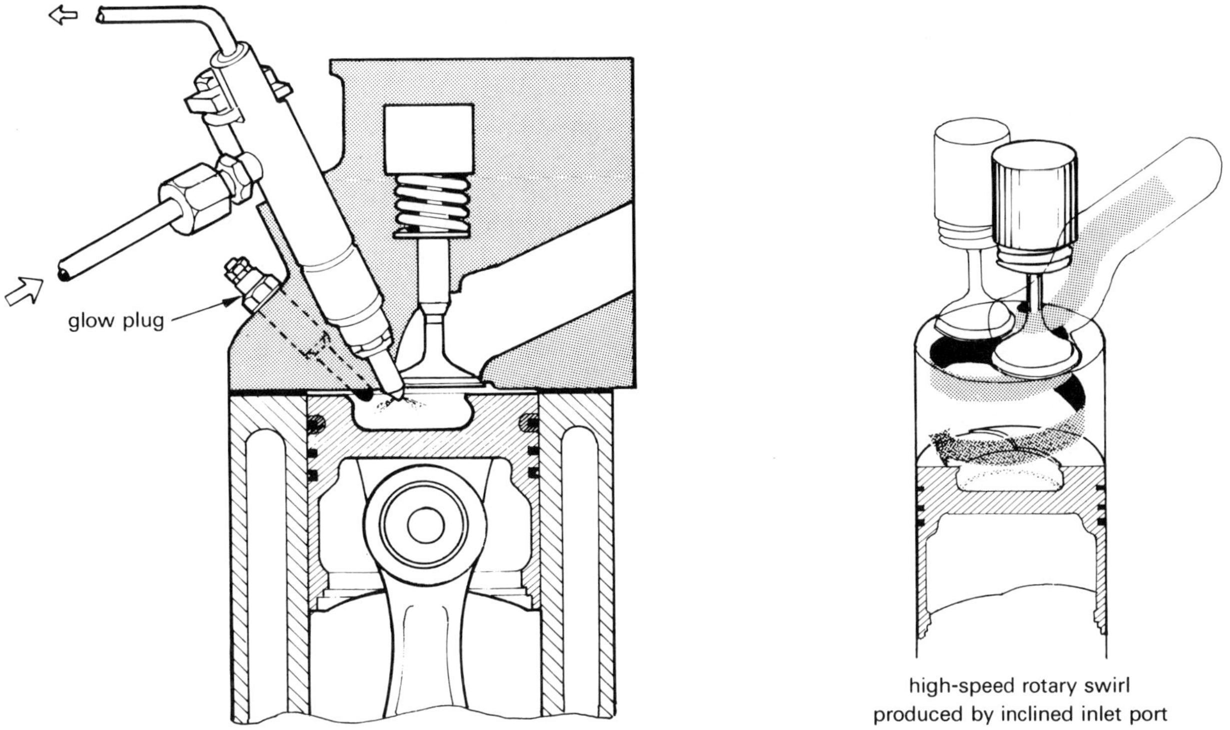

Fig. 30.4 Perkins Prima combustion chamber

these come into operation when the driver turns the ignition key. It is claimed that these aids allow the engine to be started down to temperatures well below freezing. During the cranking period the injection timing is advanced to give easier combustion, but when the engine fires the injection timing is quickly returned to normal; this reduces noise and gives smoother operation during warm-up.

The engine weighs little more than a spark-ignition engine even though it is built to withstand the stresses of turbo-charging.

Main components

The higher cylinder pressures of CI engines demand stronger components, although for production economy reasons, manufacturers of light CI units often use some engine components, such as cylinder blocks and crankshafts, that are interchangeable with the spark-ignition (SI) versions.

Pistons of aluminium alloy, often with steel inserts in the skirts to control expansion, are made longer than the SI piston so as to accommodate the combustion cavity. Three or more *piston rings* are used and the top ring groove is generally armoured by using an insert to resist wear.

Crankshafts have to be more robust to withstand the higher pressures and shock loads, and the fillets are normally rolled to improve fatigue life. Generally the *main bearings*, situated between each crank throw, are copper-lead and the *big-end bearings* are either fully copper-lead or copper-lead/tin-aluminium with the heavier-duty material fitted to the connecting rod side of the bearing.

Induction manifolds serve a different purpose because they do not have to convey a combustible charge. Air cleaning and induction silencing are still necessary, so these functions, together with other functions such as manifold heating for starting, speed governing and turbo-charging, require the use of a suitable manifold.

Valve timing The valve periods for light vehicle CI engines are similar to those used on the SI engine, because the breathing requirements are basically the same. Heavy CI engines have a much lower maximum operating speed (e.g. 2000 rev/min) because of the mass of the components, so in these the valve periods are smaller; inlet valve angles of 10° before to 40° after, and exhaust angles of 40° before to 20° after, are typical on heavy engines.

Engine lubrication

High combustion temperatures and elements such as sulphur in the fuel make the CI engine more prone to deposits of carbon, formation of gums around the piston rings, and lacquer deposits on the side of the pistons.

These conditions are minimized by using a *heavy-duty detergent oil*. Detergent additives in these special oils maintain engine cleanliness, and dispersant additives hold the carbon and soot in suspension in the oil so that the foreign particles harmlessly pass around the system until the oil is next drained.

Advantages and disadvantages of CI engines

Compared with a petrol engine of the same size, a CI engine has the following advantages:

1 Fuel economy – the high compression ratio gives a good thermal efficiency (35–40 per cent indicated) and provides the operator with approximately 30 per cent more miles to the gallon.
2 Reduced risk of fire – at room temperature, the low volatility of *Derv* (Diesel-Engined Road Vehicle) fuel makes ignition difficult.

The disadvantages are:

1 High initial cost – due to expensive fuel injection equipment and more substantial engine construction.
2 Lower maximum torque and power output.
3 Lower power/weight ratio.
4 Greater noise in operation.

31 The fuel system

Early CI engines used air pressure to deliver the fuel into the cylinder, but nowadays liquid fuel, supplied by a reciprocating plunger pump, is delivered via small-bore steel pipe lines to the injector. The modern jerk pump system, which is shown in simplified form in Fig. 31.1, is often termed *mechanical* or *solid injection*, since the ejection of the fuel is brought about by the action of the plunger on a 'solid' column of oil.

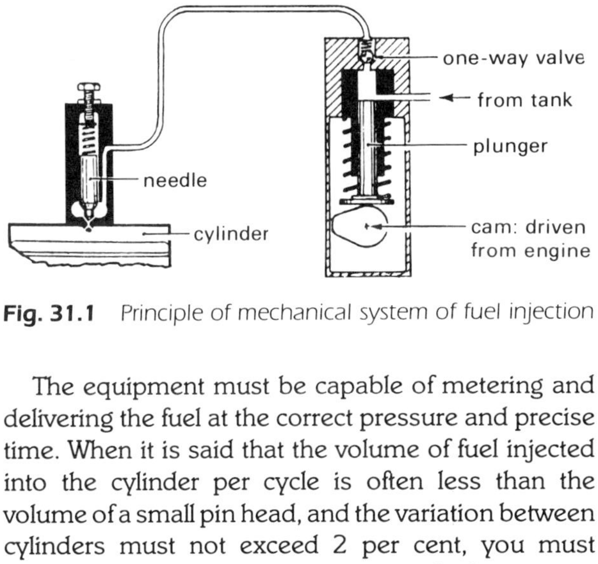

Fig. 31.1 Principle of mechanical system of fuel injection

The equipment must be capable of metering and delivering the fuel at the correct pressure and precise time. When it is said that the volume of fuel injected into the cylinder per cycle is often less than the volume of a small pin head, and the variation between cylinders must not exceed 2 per cent, you must appreciate that the injector system is a high-precision unit. Very close limits and small clearances (for example, the clearance between plunger and barrel is less than 3 μm (0.0001 in)) demand special precautions to prevent dirt from entering the system. Clean fuel, regular replacement of filters, and close attention to cleanliness during overhaul of equipment are all essential if the apparatus is to perform its exacting task.

A layout of a typical system is shown in Fig. 31.2. A low-pressure feed pump or gravity supply ensures that the injection pump receives a continuous flow of clean filtered fuel. The system incorporates many filters: a fine gauze unit(s) is fitted on the inlet side of the pump, and a felt, cloth or paper filter(s) is inserted between feed pump and injection pump. Injection, in the case of a four-stroke, occurs every other revolution and therefore the injection pump must be driven from the timing gears or camshaft at half engine speed. At the appropriate time, fuel under high pressure is conveyed to the injector by fuel lines, which should be of equal length to maintain an equal injection delay (period between the time when a high pressure builds up in the pump and the time when fuel begins to spray from the injector).

Lubrication of the main working parts of the system is usually done by the fuel oil itself: a predetermined leakage past the injector needle is directed to waste or returned to the fuel tank.

The pump will not function correctly if air is present in the system, and therefore venting screws or valves are provided at various points to allow air to be bled out. This is necessary after disturbance of supply lines or 'running-out' of fuel. A priming lever, generally fitted to the feed pump, allows the feed pump to be operated manually. With the air vent (bleed screw) open, operation of the feed pump forces a column of fuel through the line and allows the air to be driven from the system.

The injection pump

Modern CI engines are fitted with either an *in-line* pump or a *distributor* pump: the former is similar in basic construction to the pump produced by Robert Bosch over fifty years ago, whereas the distributor pump has only been in production for a few years. The comparatively low initial cost of the latter makes this type attractive to many manufacturers.

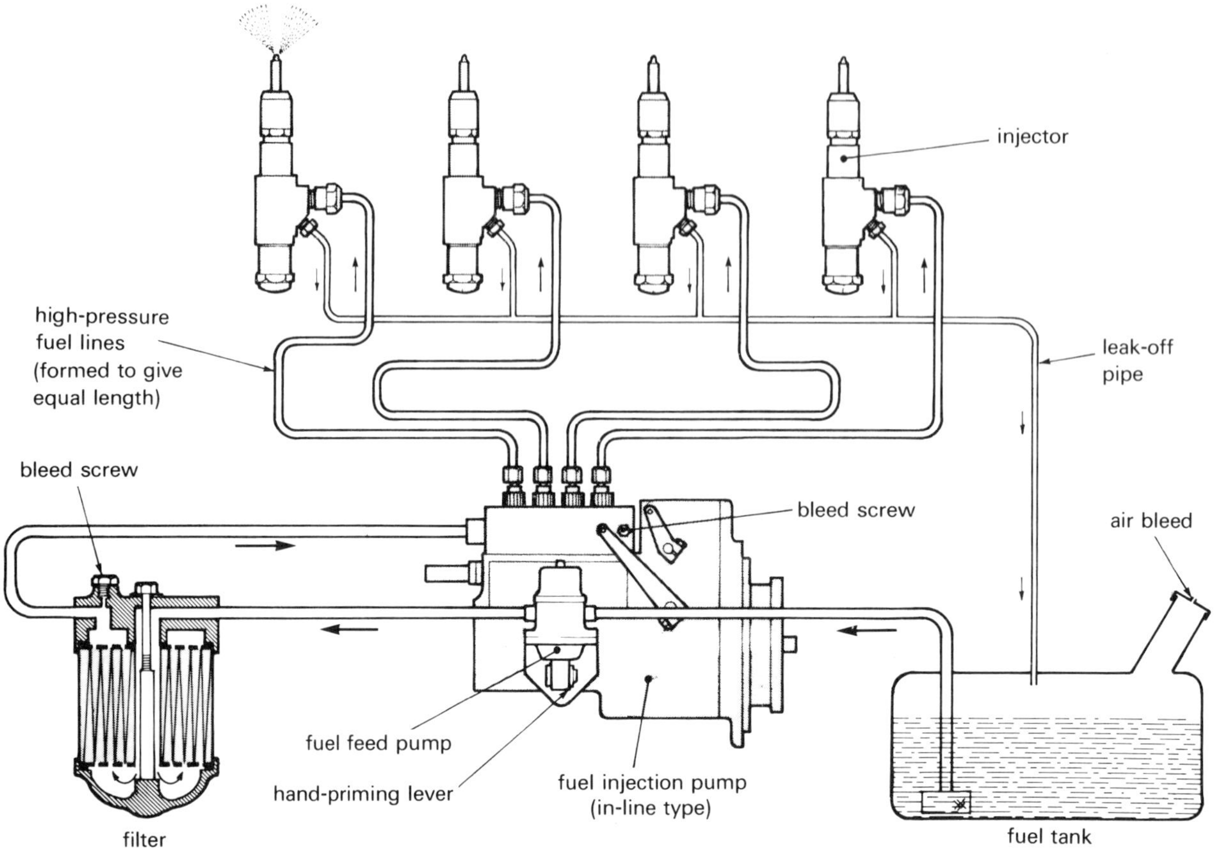

Fig. 31.2 Layout of fuel injection system

Since both pumps are being produced in large numbers, this chapter describes each type.

In-line pump

Fig. 31.2 shows an in-line, four-element enclosed camshaft type of pump, and Fig. 31.3 illustrates the construction of one pumping element. This unit comprises:

1 Pumping element

A steel plunger moving through a constant stroke reciprocates inside a close-fitting steel barrel. The plunger is partly machined away to produce a control helix and vertical groove. An inlet and spill port in the barrel communicates with a gallery fed from the fuel tank. Location of the barrel is provided by allowing a screw in the casing to register in a recess adjacent to the spill port in the barrel.

Partial rotation of the plunger varies the output from zero for stopping the engine, to maximum for starting. Between these limits a variable supply of fuel is necessary to meet the engine power and speed requirements.

The operation of the pumping element is shown in Fig. 31.3. The positions of the element are as follows:

A When the plunger is at b.d.c. the depression in the pump chamber causes fuel to enter both ports.

B This position, known as the point of port closure (or spill cut-off with this type of pump), is generally regarded as the theoretical point of injection. Both ports have been covered, and so the ascending plunger raises the pressure of the fuel to produce injection.

C Injection stops when the edge of the helix uncovers the spill port. Pressure is relieved by fuel passing down the vertical groove, around the waist of the plunger and out of the spill port.

D Rotation of the plunger causes the helix to uncover the spill port either earlier or later, to give less or more fuel respectively.

E Moving the plunger to make the vertical groove coincide with the spill port means that the port will

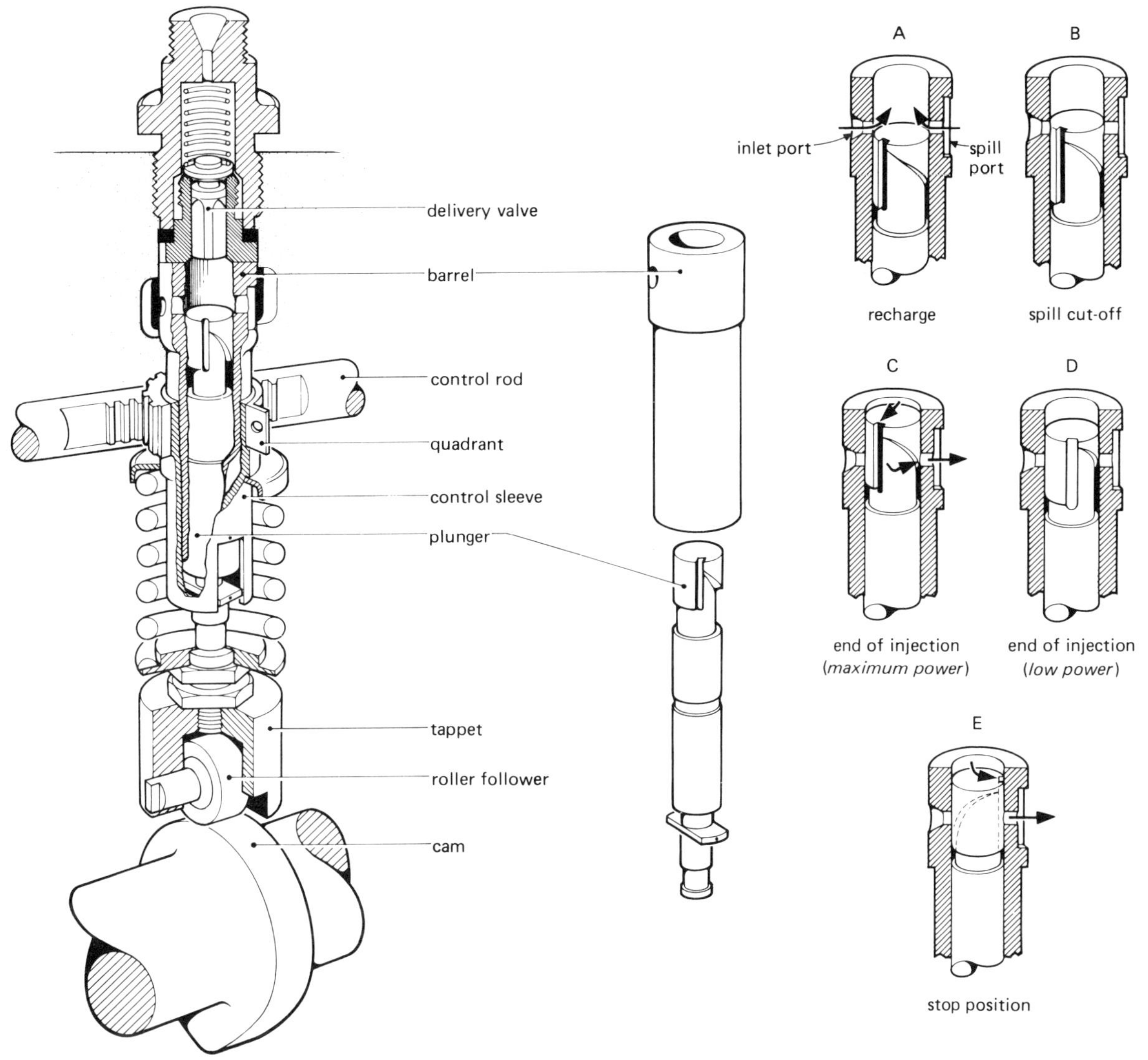

Fig. 31.3 The fuel injection pump

remain open; therefore no fuel will be delivered and the engine will come to rest.

2 Plunger control

Two lugs on the plunger fit into slots in a control sleeve on to which is clamped a toothed quadrant. This quadrant engages a rack cut in the control rod which runs the length of the pump. By moving the quadrant relative to the sleeve, the output from each element can be calibrated or equalized.

3 Drive

Symmetrical cams, set to give the appropriate firing order, act on a roller follower and tappet block. A screw or shim adjustment between tappet and plunger allows the time of the start of injection of one element to be varied in respect to the other elements. A four-cylinder engine has a *phase angle* (the interval between injections) of:

$$360°/\text{No. of cylinders} = 360°/4 = 90°$$

The operation for setting this angle is known as *phasing*.

4 Delivery valve (Fig. 31.4)

The valve performs two duties:

a The conical seat acts as a non-return valve preventing the return of fuel from the high-pressure

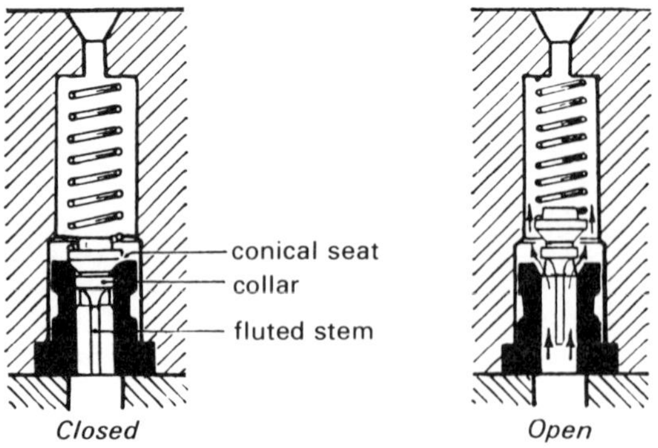

Fig. 31.4 Delivery valve

pipeline when the spill port opens. This allows the pump chamber to recharge and also enables air or gas to be purged from the pipeline via the injector.

b If air can be eliminated from the high-pressure pipeline the pump will often operate satisfactorily without a delivery valve: on the opening of the spill port the pressure at the injector will fall rapidly to pump inlet pressure giving a sharp closing of the injector. The fitting of a non-return valve, however, traps pressure in the pipeline while the injector is still open. This pressure can only be relieved by fuel continuing to pass the injector at a diminishing pressure, resulting in the last few droplets of fuel merely 'dribbling' out, causing incomplete combustion, carbon formation, smoky exhaust and high fuel consumption.

The collar below the conical seat acts as a piston and withdraws a small amount of fuel from the high-pressure pipeline as the valve closes, causing the necessary rapid pressure drop to ensure a sharp cut-off to injection.

5 Governor

Higher gas pressures in the cylinder of a CI engine usually demand stronger components than those used in a petrol engine. Strength is normally improved by increasing the component dimensions, but this also increases weight and leads to engine damage if the speed exceeds a given value which is governed by the components' strength/weight ratio.

A pump with a *rising output* characteristic, i.e. for a given control rod setting the output increases with speed, and compression pressures, which are substantially the same at *idling* as at maximum speed, makes it difficult to obtain a steady idling speed - this causes the engine either to race or to stall.

Limitation of maximum speed and control of the engine at idling speed are the two main duties performed by the governor. If the unit attends to these two duties only, it is known as an *idling and maximum speed governor*, whereas a unit which regulates throughout the speed range is termed an *all-speed governor*.

There are three main types of governor:

- *Mechanical*
- *Pneumatic*
- *Hydraulic*.

Mechanical governors The type shown diagrammatically in Fig. 31.5 is an *idling and maximum speed* governor, which is often used on heavy vehicle engines. Mounted at the end of the injector pump, it consists of two weights, which are rotated by the pump camshaft. Springs exert a force in an inward direction on the weights, and bell cranks link each weight to the bottom end of a floating lever: the top end of the lever is connected to the control rod and the centre is mounted on an eccentric, which is rotated by the accelerator pedal.

The diagram shows the governor in the *engine-stationary* position, and in this condition the weights are fully retractable to hold the control rod in the *maximum fuel* position in readiness for starting the engine. When the engine fires, rotation of the weights causes the centrifugal effect to overcome the spring and move the weight out to the position shown by Fig. 31.5b, which causes the control rod to be withdrawn to the 'idling' setting. In this position the weights are being controlled by the outer (weaker) springs only, so sensitive control is possible.

Assuming the pedal is not depressed, any increase in speed produces a slight outward movement of the weights and this moves the control rod in the direction which reduces the fuel delivery. In a similar way, stalling is prevented by the weights moving inwards, which causes the control rod to increase the fuel delivery.

Between idling and maximum speeds, the weights maintain the same position, and appear to be locked together. During this phase, downward movement of the accelerator pedal rotates the eccentric and moves the control rod in a direction which increases the fuel delivery.

As maximum speed, e.g. 1800–2000 rev/min, is reached, the high centrifugal force acting on the weights will overcome the strong outer springs and move the weights outwards (Fig. 31.5c). This motion, when transmitted to the control rod, decreases the

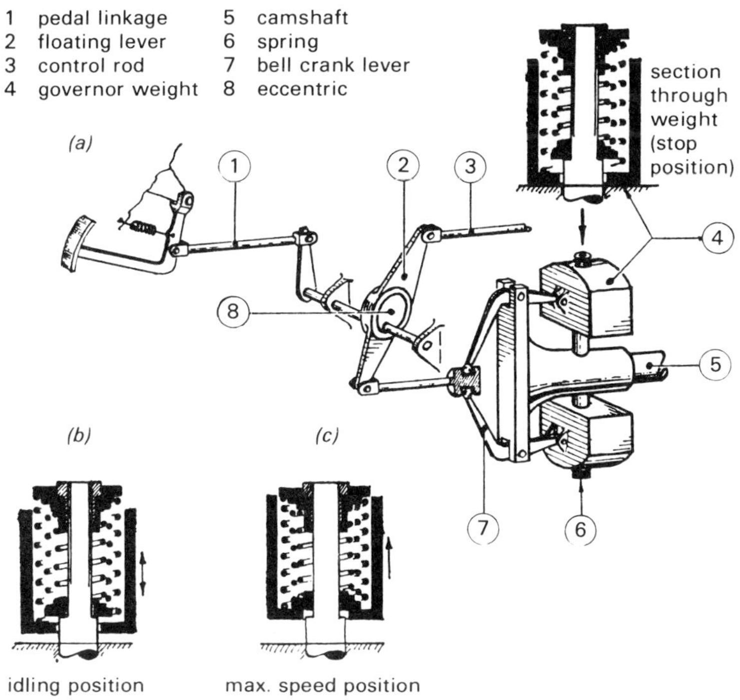

Fig. 31.5 *Idling and maximum speed mechanical governor*

quantity of fuel delivered, and reduces the engine power, irrespective of the position of the accelerator pedal.

A small quantity of engine oil, contained in the separate governor housing, lubricates the moving parts. Check the level at regular intervals, e.g. every 3000 km (2000 miles).

A number of adjusting nuts, screw and stops are employed with this type of governor. Do not disturb the settings, unless you have special tools and the necessary knowledge.

Pneumatic governors In-line pumps fitted to light CI engines often use this *all-speed* type of governor. Holding the accelerator pedal in a set position, the governor will maintain a constant speed up to a point where the load on the engine is too great.

Fig. 31.6 shows the main construction of this type. A spring-loaded diaphragm, connected to the control rod, is mounted to seal a chamber which is linked by a pipe to a venturi control unit in the inlet manifold. A butterfly valve, fitted in the waist of the venturi, is connected directly to the accelerator pedal.

When the engine is at rest, the diaphragm spring forces the control rod to the maximum fuel or excess-fuel position. (Many pumps are fitted with an excess-fuel device. This is a manually operated control fitted

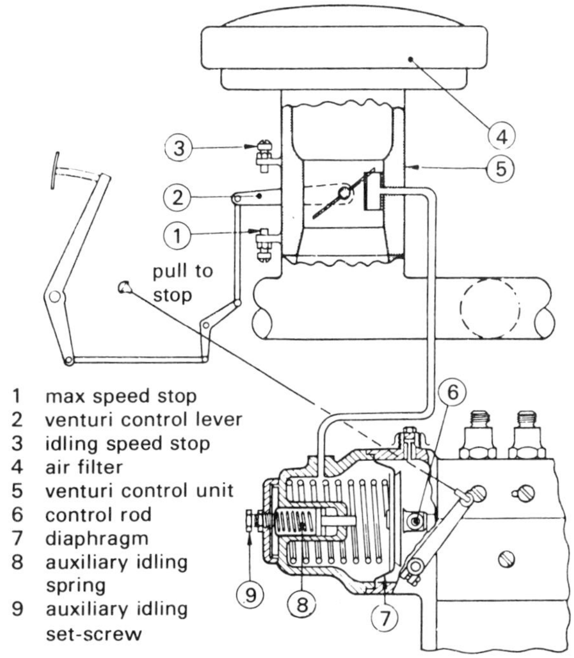

Fig. 31.6 *Pneumatic governor system*

at the end of the pump which enables extra fuel to be delivered for cold starting. It cannot be operated from the driving position.)

Closing the accelerator pedal after starting produces

a depression in the venturi and diaphragm chamber. This causes atmospheric pressure to force the diaphragm and control rod to the idling setting. With the pedal in any set position, an increase in speed intensifies the venturi depression, which reduces the control rod opening: a decrease in speed will produce the opposite condition.

As the accelerator pedal is depressed, the butterfly valve is opened and the venturi depression is decreased. This causes the spring to open the control rod and increase the engine speed until a balance is reached between the spring thrust and venturi depression.

A stop screw, acting on the lever controlling the butterfly valve, limits the maximum speed. When the valve reaches its stop, any tendency for the engine to increase speed will intensify the venturi depression and reduce the control rod opening.

A 'stop' control on the dash enables the driver to override the governor and move the control to the 'no-fuel' position.

Do not start the engine with any part of the governor system disconnected.

Hydraulic governors This type is used with in-line pumps where smooth, slow idling speeds are demanded, e.g. for coaches, but high cost tends to limit its use. A gear-type oil pump, drive from the end of the pump camshaft, supplies oil through various valves to operate a piston linked to the control rod.

Distributor pumps

Lucas

Fig. 31.7a shows a DPA-type pump suitable for a four-cylinder engine. Connected to the engine by a flange and driven by a splined shaft, this unit resembles the distributor used on a petrol engine; fuel lines to the injectors occupy the position of the h.t. leads.

A single-element, opposed plunger pumping unit supplies fuel to either a four- or six-cylinder engine; except for the cam and the number of ports, the other parts are the same for both engines. Operated by a non-rotating cam, the single element provides a correctly phased and balanced output over a very large speed range.

The pumping unit, shown diagrammatically in Fig. 31.7b, consists of two plungers mounted in a rotor which turns in a fixed hydraulic head. Ports in the head and rotor line up in certain positions to allow either inward or outward flow of oil. Fig. 31.7b shows the inlet port open. Pressure from a transfer pump fitted at the end of the main pump, directs fuel oil along the centre of the rotor to force out the plungers. Fig. 31.7c shows that the uncovered outlet port will allow the cam to force the plungers together and discharge the fuel to the injector.

Cam ring shape Dribble is eliminated by using a special cam design to stop fuel delivery sharply (Fig. 31.8). At the end of each pumping stroke the roller, which operates the pumping plunger, suddenly moves outwards and drops the pumping pressure.

This retraction arrangement overcomes the need for a delivery valve.

Fuel system Fig. 31.9 shows a typical fuel system. A feed pump supplies fuel, via a filter, to a sliding vane transfer pump which directs the oil through a driver-controlled metering valve to the rotor. Transfer pump pressure is limited by a regulating valve, and oil, which escapes from the pumping element, returns to the filter after lubricating the working parts.

The quantity of fuel delivered to the injector is governed by the position of the metering valve. The diagram shows a partly opened valve restricting the flow of oil to the rotor. This reduction in oil flow prevents the full outward travel of the plungers; hence the shortened stroke gives reduced fuel and lower engine power.

On fitting a 'new' pump, it is essential to fill the pump completely with clean fuel oil before attempting to start the engine. Various venting screws are provided to bleed the air from the system.

The absence of delivery valves in the high-pressure lines makes it necessary to bleed these lines; bleeding is carried-out by slackening the union nuts at the injectors and cranking the engine. Before operating the starting motor to purge the air, the appropriate personal safety precautions should be taken to avoid contact with the high-pressure spray.

Automatic advance mechanism An automatic advance mechanism is provided for engines that have a wide speed range. This consists of a spring-loaded piston which rotates the cam ring against the direction of rotation when the transfer pump output pressure increases (Fig. 31.10).

Governor The governor used is either a hydraulic or mechanical type.

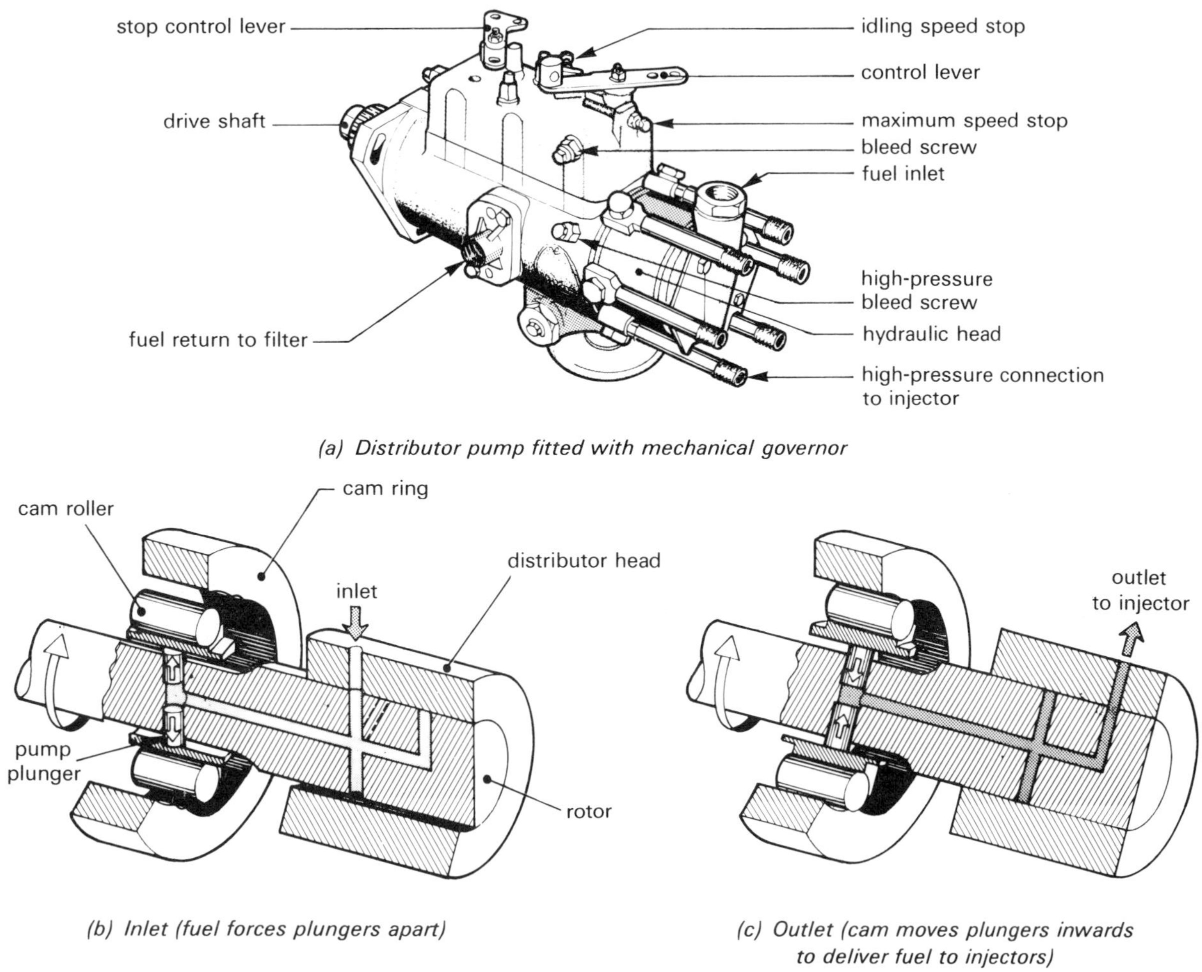

Fig. 31.7 The Lucas-CAV distributor pump

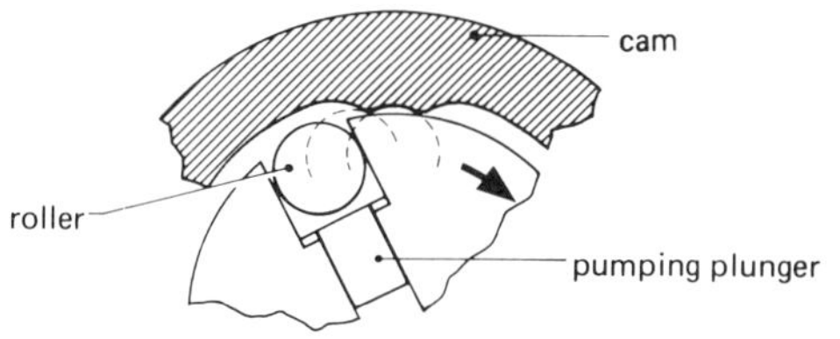

Fig. 31.8 Cam shape to retract roller

Hydraulic governor Control of engine speed is achieved by utilizing the transfer pump and the fuel metering valve (Fig. 31.11).

The driver controls the valve through a spring, so if he depresses the pedal a given amount, the valve will open wide and allow the speed to increase. The build-up in engine speed increases the transfer pump pressure; this gradually moves the metering valve towards the closed position until a point is reached at which the speed will not increase any more. These events recur throughout the range.

Mechanical governor This type is recommended when more precise control of the engine speed is required. It can be recognized by the longer pump housing which is needed to accommodate the flyweights.

Fig. 31.12 shows the constructional details of a simplified layout. Rotation of the input shaft causes the weight (1) to be thrown outwards and move a sleeve (2) against a control arm (3). The pivot on this arm causes the upper part of the arm to move in the opposite direction which pulls on the rod (4) and closes the metering valve (5). All of these movements are resisted by the control spring (6): the tension on this spring is given by the driver's control lever.

When the driver requires an increase in power,

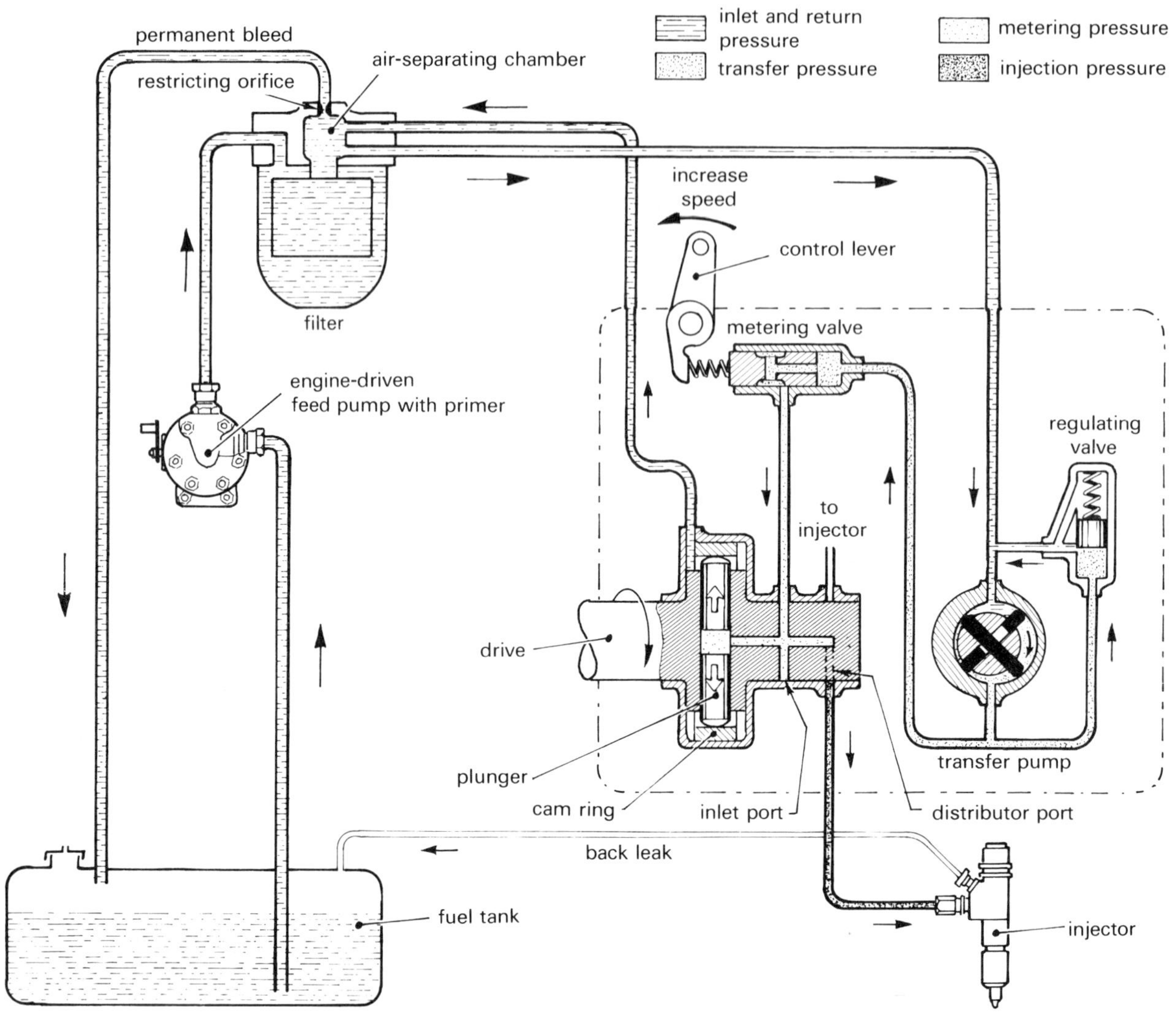

Fig. 31.9 Fuel system for Lucas-CAV distributor pump

movement of the control lever extends the spring. The larger force given by the spring resists flyweight movement until the engine speed is sufficient to enable the flyweight to move.

Under idling conditions, tension is removed from the main governor spring and a light idling spring (7) is used to resist movement of the flyweights.

At any time the engine can be stopped by a shut-off lever. This overrides the governor and moves the metering valve to the 'no fuel' position.

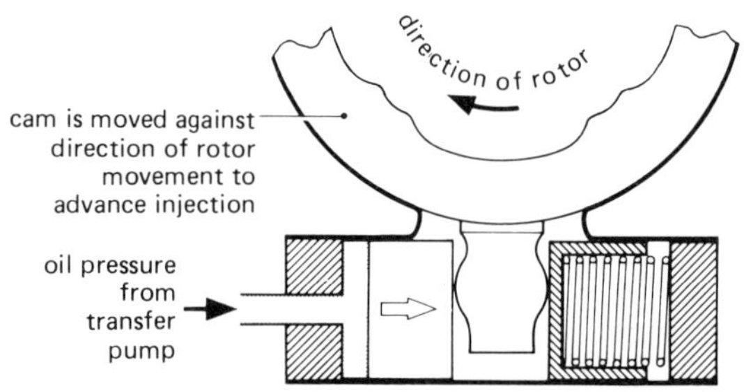

Fig. 31.10 Automatic advance mechanism

Bosch

Fig. 31.13 shows a VE-type pump fitted in a self-bleeding fuel system layout similar to that used on light vehicles.

As with other rotary pumps, this type has one pumping element and a number of high-pressure outlets, one for each engine cylinder.

In addition to the basic features associated with modern distributor-type rotary pumps, various add-

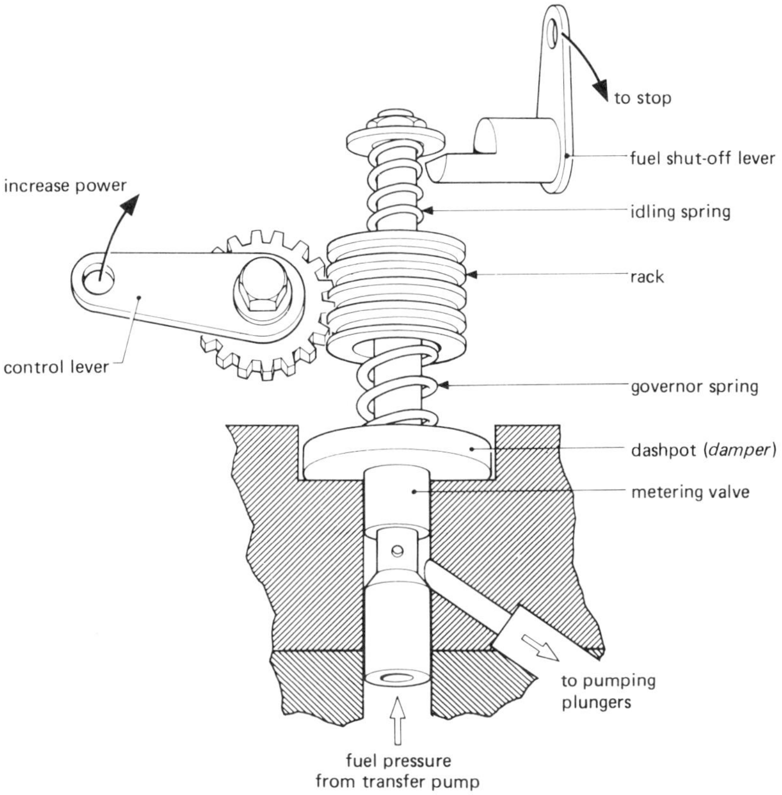

Fig. 31.11 Hydraulic governor – distributor pump

on modules can be fitted to the VE pump; these include:

- Solenoid-operated fuel cut-off to give the driver a key start/stop operation
- Automatic cold-starting module to advance the injection
- Fast-idle facility to give even running during warm-up
- Torque control for matching the fuel output with the fuel requirement.

The simplified section through the pump (Fig. 31.14) shows the layout of the basic sub-systems; these include:

- Low-pressure fuel supply
- High-pressure fuel supply and distributor
- Fuel shut-off solenoid
- Distributor plunger drive
- Automatic injection advance unit
- Pressure valve
- Mechanical governor.

Low-pressure fuel supply Driven at half crankshaft speed by a driveshaft, a transfer pump having four vanes delivers fuel to the pumping chamber at a pressure set by the regulating valve.

This fuel pressure, which rises with engine speed, is used to operate the automatic advance unit. Also it gives an overflow through the pump body which aids cooling and provides the self-bleeding feature. After passing through a small restriction at the top of the pump the surplus fuel is returned to the fuel tank.

High-pressure fuel supply Fig. 31.15 is a simplified view of the pumping chamber with part of the distributor head cut away to show the pump plunger. Besides rotating in the head to give a valve action, the plunger is reciprocated through a constant stroke to produce the high pressure; this axial movement is provided by a cam plate moving over a roller ring. The quantity of high-pressure fuel delivered to the injector via the outlet bore is controlled by the position of the control spool. This feature varies the effective pumping stroke; it increases as the spool is moved towards the distributor head.

In the position shown in Fig. 31.15a the rotation of the plunger has caused one of the metering slits to

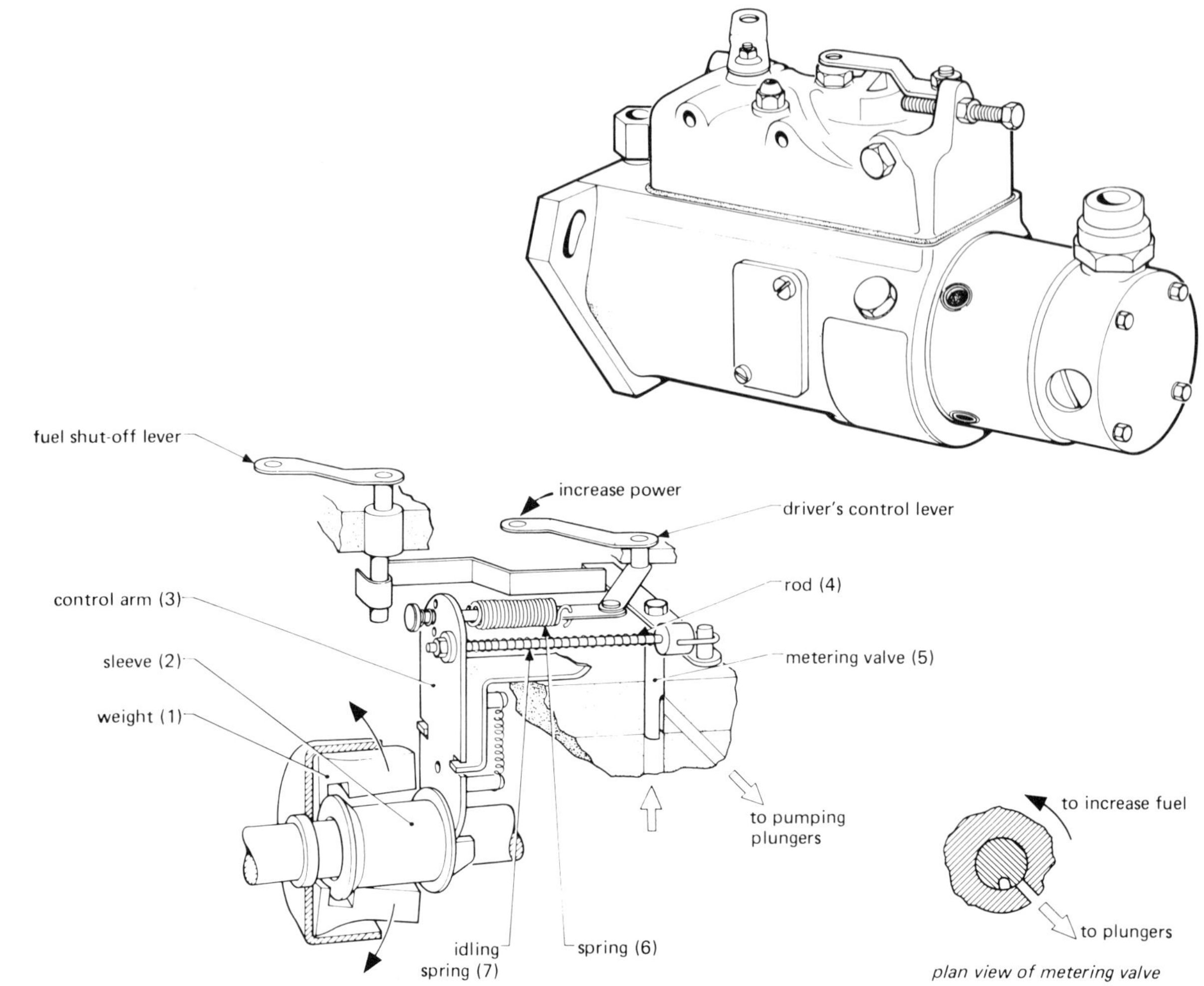

Fig. 31.12 Distributor pump with mechanical governor

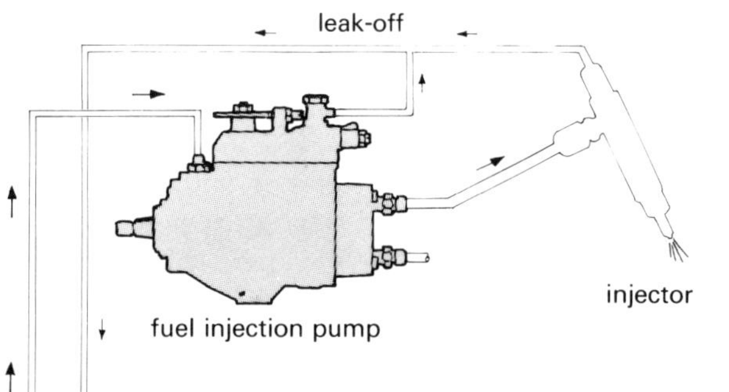

Fig. 31.13 Fuel system layout

open the inlet passage; at this point all outlet ports are closed. Prior to this, the plunger had moved down the chamber to create a condition for the fuel to enter and fill the high-pressure chamber.

Slight rotation of the plunger closes the inlet port and causes the single distributor slit in the plunger to open one of the outlet ports. Whilst in this position the plunger is moved up the chamber to pressurize the fuel and deliver it through the outlet bore to the injector.

The position of the plunger at the end of the injection period is shown in Fig. 31.15b. At this point, the control spool has already allowed a considerable movement of the plunger before the cut-off bore in the plunger has been uncovered; this port exposure instantly drops the pressure and terminates the injection. Further pumping movement of the plunger

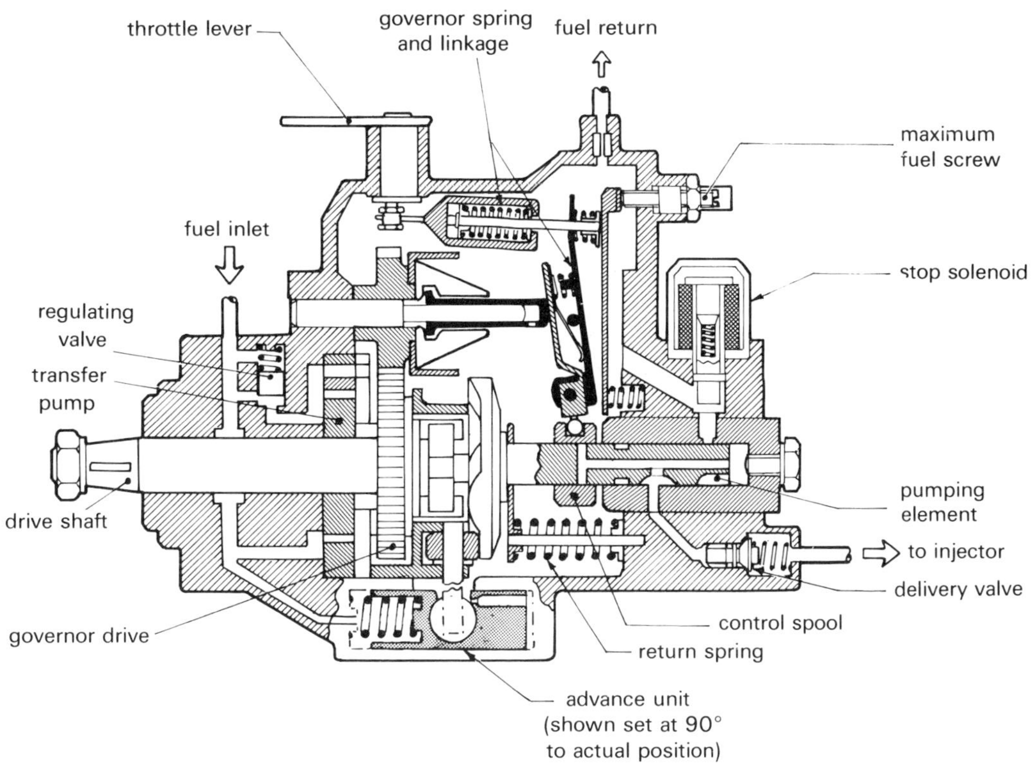

Fig. 31.14 Bosch VE-type distributor pump (simplified)

causes the fuel in the pumping chamber to be returned to the pump cavity. With the spool set in this maximum fuel position, which corresponds to the fuel requirement for starting, a movement of the control spool to an extreme position away from the distributor head reduces the output to a minimum; this is the spool setting for slow running.

Fuel shut-off The 'no fuel' or 'stop' position is provided by a solenoid-operated valve; this cuts off the fuel supply to the inlet passage when the 'ignition' key is switched-off.

Distributor plunger drive The plunger must be rotated and reciprocated and Fig. 31.16 shows how this is achieved.

Rotation of the driveshaft, at half crankshaft speed (for a four-stroke engine), is transmitted via a yoke and cam plate to provide rotary motion to the pump plunger.

Reciprocating motion is provided by the rotation of a cam plate as it moves over four roller followers fixed to a roller ring. In a pump suitable for a four-cylinder engine, four lobes are formed on the cam plate and contact between the plate and rollers is maintained by two strong plunger return springs. A yoke positioned between the driveshaft and the cam plate allows the plate to move axially whilst still maintaining a drive.

Pressure valve A delivery valve having a similar construction to that used on in-line pumps is fitted in the distributor head at the connection point to the high-pressure fuel lines. The duty of this valve is described on page 186 and its use on this type of pump overcomes the need to bleed the high-pressure fuel lines.

Automatic injection advance unit The roller ring assembly is not fixed rigidly to the casing; instead it can be partially rotated through an angle of up to 12° to allow the automatic advance mechanism shown in Fig. 31.17 to vary the injection timing.

When the pump is rotated, fuel under pressure from the transfer pump is delivered to the timing advance chamber via the pump cavity. A rise in the pump speed causes the transfer pump pressure and flow to increase; this moves the timing-advance piston against its spring, which in turn, causes the actuating pin to rotate the roller ring in a direction opposite to the direction of rotation of the driveshaft.

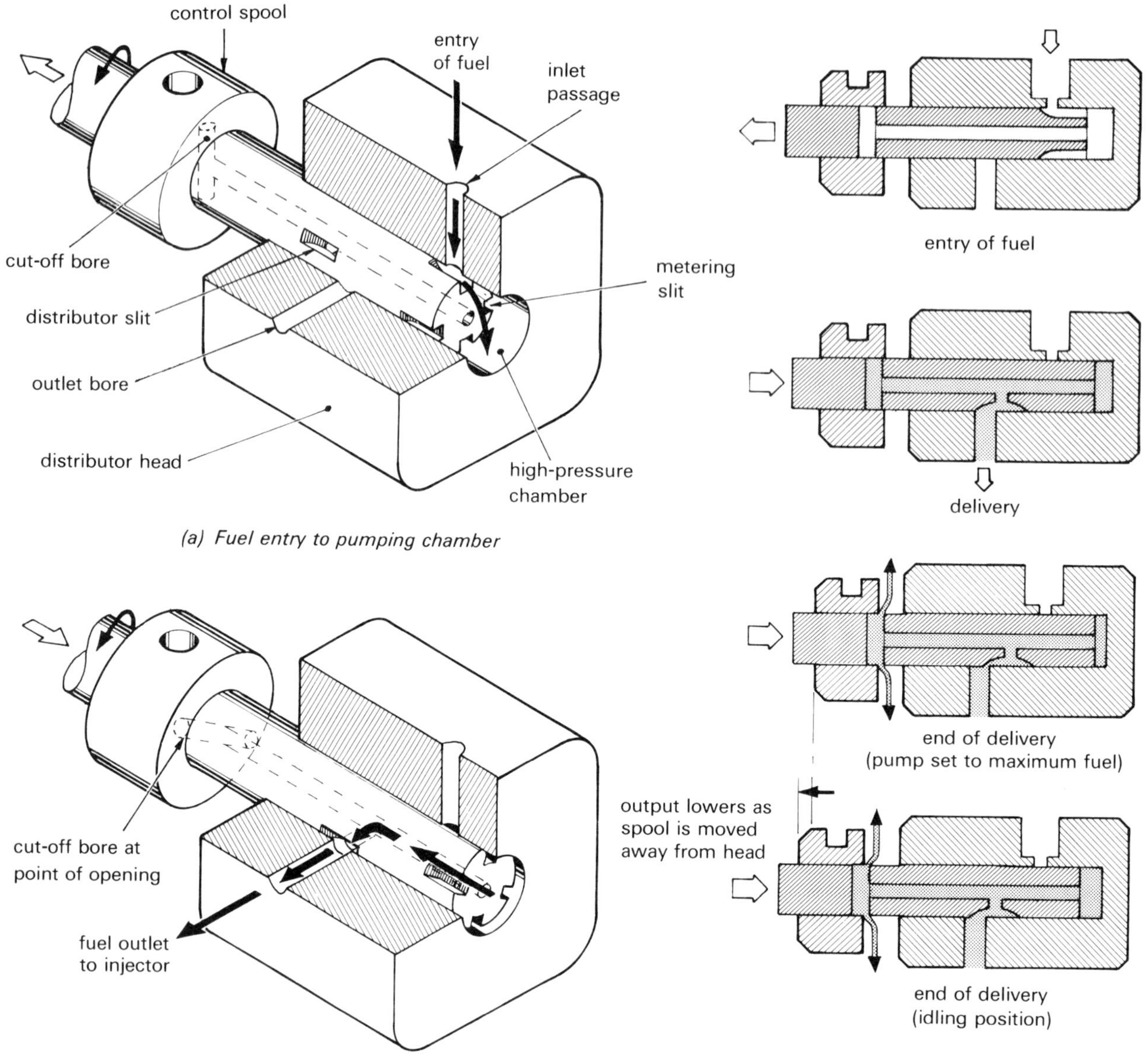

(a) Fuel entry to pumping chamber

(b) Pump at end of delivery

Fig. 31.15 Principle of Bosch VE pumping unit

Governor The VE pump is fitted with either a two-speed or an all-speed governor. The layouts of these types of governor are similar, but differ in the arrangement of the control springs.

Fig. 31.18 shows the main construction of a two-speed governor. This type controls the engine during the phases of idling and maximum-speed operation. At other times the driver has near-direct control of the quantity of fuel delivered and hence the power output of the engine.

The centrifugal governor, which consists of a series of flyweights, is driven from the driveshaft through gears having a ratio that steps up the speed. The high-speed flyweight rotation given by this ratio ensures good sensitivity of the governor, especially during the idling phase.

An increase in engine speed, and the associated centrifugal action on the flyweights, produces an outward force which pushes a sliding sleeve against a control lever system. Movement of this lever, which is connected at its lower end to the control spool on the pumping plunger, can only take place when the

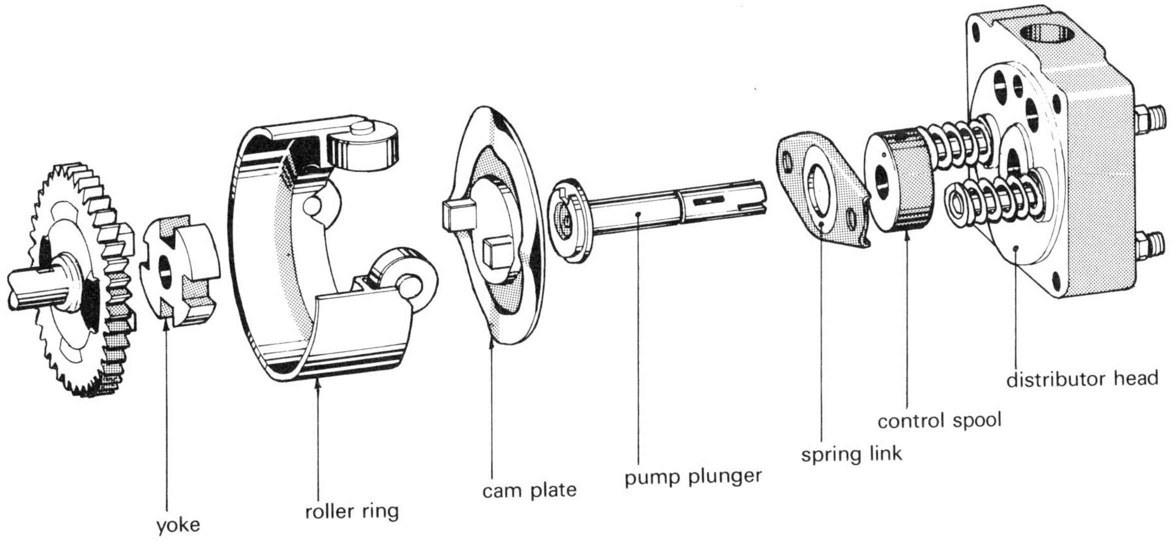

Fig. 31.16 Plunger drive

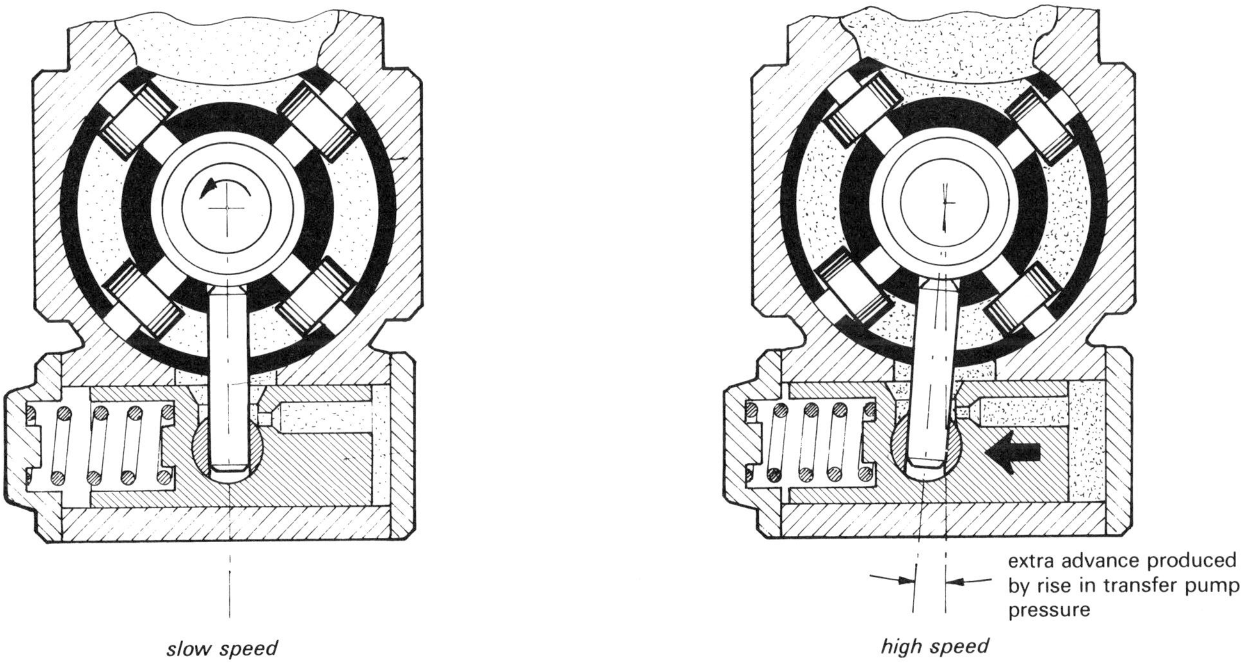

Fig. 31.17 Principle of automatic advance unit

sliding sleeve is able to overcome the reaction of the spring that is in use at that time.

Starting With the accelerator pedal half-depressed and the governor stationary, the starting spring pushes the sliding sleeve towards the flyweights and moves the control spool to the maximum fuel position.

Idling When the engine starts, the release of the accelerator combined with the outward movement of the flyweights, causes the lever to move the control spool to the minimum fuel position. When the engine is operating in this phase, smooth idling is obtained by the interaction of the flyweights and idling spring.

With the accelerator pedal lever against the adjustable idling stop, any small speed increase

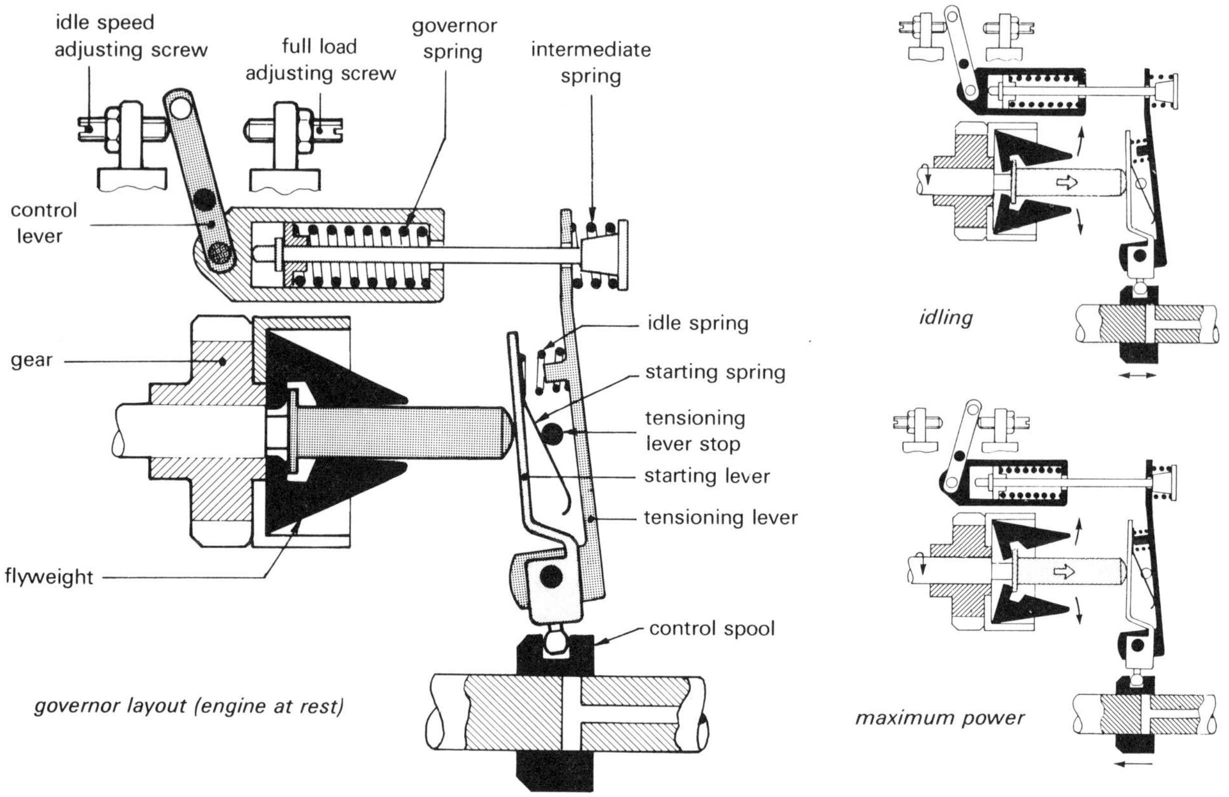

Fig. 31.18 Governor – mechanical type

causes the flyweights to exert a larger force on the sliding sleeve. This slightly compresses the idling spring and, as a result, the spool control lever moves the spool and reduces the fuel delivery.

Any slight drop in engine speed produces the opposite action, so smooth idling under governor control is obtained.

Mid-range operation Once the idling range has been exceeded the larger governor force puts the idling and starting springs out of action. At this stage the intermediate spring comes into use to extend the idle control range and so smooth the transition from idle to mid-range operation. This stronger spring provides a flexible link between the driver's pedal and the control spool lever so that when the accelerator pedal is depressed a slight delay in engine response is introduced.

Beyond this phase any movement of the accelerator produces a direct action on the control spool.

Maximum speed During mid-range operation, the preload of the main governor spring causes the spring assembly to act as a solid block. But when the engine reaches its predetermined maximum speed, the force given by the flyweights equals the spring preload. Any further speed increase allows the flyweights to move the spool control lever. This reduces the quantity of fuel being delivered and so keeps the engine speed within safe limits.

Injectors

The purpose of the injector or sprayer is to break up the fuel to the required degree and deliver it to the combustion region in the chamber. This atomization and penetration is done by using a high pressure to force the fuel through a small orifice.

Vehicles in this country use a type of injector which incorporates a valve. This *closed* system is responsive to pump pressure – raising the pressure above a predetermined point allows the valve to open, and stay open until the pressure has dropped to a lower value. The 'snap' opening and closing of the valve give advantages which make this system popular.

The complete injector, shown in Fig. 31.19a, consists of a nozzle and holder, which is clamped to form a gas-tight seal in the cylinder head. A spring,

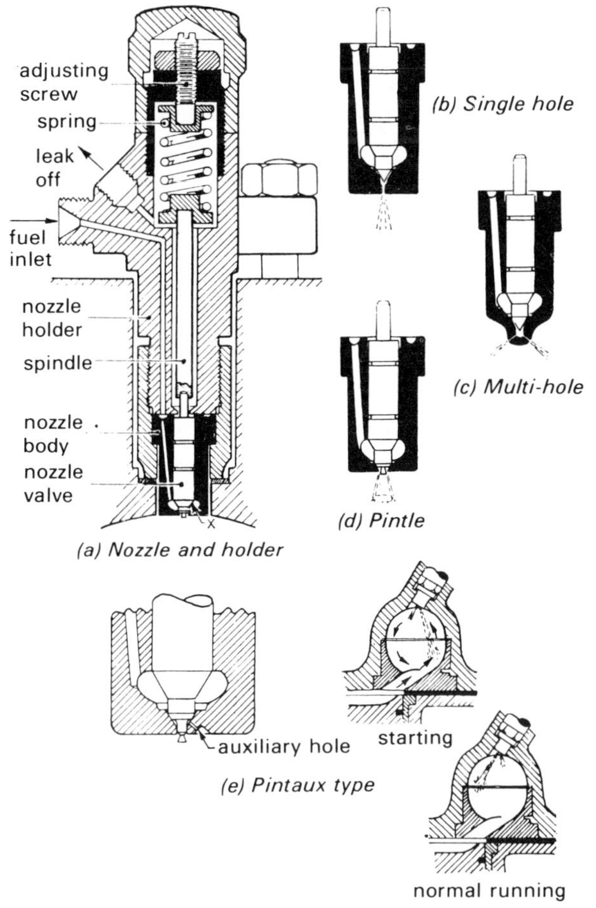

Fig. 31.19 The injector

compressed by an adjusting screw to give the correct breaking (opening) pressure, thrusts the needle on to its conical seat. Fuel flows from the inlet nipple through a drilling to an annular groove about the seat of the needle. A thrust, caused by fuel acting on the conical face X, will overcome the spring and lift the needle when the pressure exceeds the breaking pressure. The opening of the valve permits discharge of fuel until the pressure drops to the lower limit. Any fuel which flows between the needle and body acts as a lubricant for the needle before being carried away by a leak-off pipe.

Nozzle types

There are three main types of nozzle:

- *Single-hole*
- *Multi-hole*
- *Pintle*.

Single-hole (Fig. 31.19b) A single orifice, which may be as small as 0.2 mm (0.008 in), is drilled in the nozzle to give a single jet form of spray. When this nozzle is used with indirect injection systems, a comparatively low injection pressure of 80–100 bar is used.

Multi-hole (Fig. 31.19c) Two or more small orifices, drilled at various angles to suit the combustion chamber, produced a highly atomized spray form. Many engines with direct injection systems use a four-hole nozzle with a high operating pressure of 150–250 bar. A long-stem version of this type simplifies the accommodation of the injector in the head.

Pintle (Fig. 31.19d) Swirl chambers can accept a *soft* form of spray, which is the form given by a pintle type of nozzle when set to operate at a low injection pressure of 110–135 bar.

A small cone extension on the end of the needle produces a conical spray pattern and increases the velocity of the fuel as it leaves the injector. This type tends to be self-cleaning.

The elimination of heater plugs on some light indirect injection engines has been made possible by the invention of a special pintle nozzle known as the *pintaux* (pronounced 'pintawks') type, shown in Fig. 31.19e. Starting conditions produce a small needle lift, and so fuel passes through the small auxiliary hole and is directed to the hottest part of the chamber. Under normal running pressures, the full lift of the needle discharges the fuel through the main orifice.

Injector servicing

With the engine at its operating temperature, the colour of the exhaust smoke, if any, acts as a guide to injector operation and is also used to assess the condition of an engine. The general rules relating to smoke colour are:

1. *Black* This colour results when all the injected fuel is not being burnt. It means that either the engine is receiving too much fuel or is receiving insufficient air.

   The fuel will not be completely burnt if the injector fails to atomize the fuel correctly or if any engine condition, such as poor compressions or incorrect valve timing, delays the start of combustion.

2 *Blue* Smoke of this colour on any engine is a sign that lubricating oil is being burnt. Any mechanical condition that allows lubricating oil to enter the combustion chamber will result in blue smoke.

The engine operating condition at which smoke is emitted often acts as a guide to the cause, e.g. a puff of smoke emitted when the engine is started, after it has been allowed to stand idle for a few minutes, suggests that valve guides or valve stem seals are worn.

3 *White* Incorrect fuel pump timing of the engine produces this colour but it should not be confused with the natural white colour of water vapour (steam); this occurs when a cold engine is started and the steam partially condenses as it passes through the cold exhaust system.

In addition to the emission of black smoke, faulty injectors cause low power output and intense engine knocking (diesel knock).

A faulty injector can often be traced by listening to the change in engine note which occurs when one injector at a time is cut out by slackening the pipeline union. Before this is attempted, special personal safety precautions must be taken which should include:

1 Diesel fuel should not be handled if it can be avoided. Where fuel is likely to come into contact with hands, gloves or a barrier cream should be used. This prophylactic cream reduces the risk of a skin inflammation disease (*dermatitis*).
2 Fuel oil under injection pressures can penetrate the skin, so no part of the human body should be exposed to fuel spray from an injector.
3 Eye protection in the form of safety glasses should be used to prevent fuel under pressure (low or high) from entering the eyes.
4 A well-ventilated work area, with special extraction equipment, is needed to prevent atomized fuel from being inhaled into the lungs.

Injectors should be regularly cleaned, adjusted and tested. This servicing requires special cleaning tools and equipment in order to carry out the following tests:

1 *Seat tightness (front leak)* This ensures that the nozzle valve seat is sound when the pressure is below the opening pressure.
2 *Needle wear (back leak)* This indicates the amount of fuel that is passing from the needle or pressure faces to the leak-off pipe.
3 *Pressure setting* The opening pressure, sometimes called the breaking pressure, must be checked, and if incorrect, reset to the recommended pressure.
4 *Spray pattern* The spray pattern of the injector, when this is operating under its working pressure, should be observed. A poorly atomized spray of streaky appearance indicates that the nozzle orifice is partially blocked.

When refitting an injector to the engine, a new seat washer should be used and the clamping arrangement should be tightened to the correct torque.

PART EIGHT

Supercharging

32 Principles of supercharging

The power that any internal combustion engine can develop depends fundamentally upon the mass of air it can consume per minute. The normal method of filling or *charging* the cylinders of a four-stroke engine consists of allowing the pressure of the atmosphere to force air into the engine cylinders to relieve the partial vacuum formed when the pistons move downward on their induction strokes. If the air is forced into the cylinders under a pressure higher than atmospheric a greater mass of air will enter the cylinder and the engine will be *supercharged*. This can be done by using some kind of air pump to deliver air to the engine: such a device is called a *supercharger*.

A two-stroke engine, of course, does not have an induction stroke and relies upon air being forced into its cylinders by some kind of pump, but the engine is not supercharged unless the pressure of the air filling the cylinder is above atmospheric pressure. A two-stroke engine is normally *pressure-charged*, either by using a separate cylinder as a charging cylinder, as in the Clerk engine (Chapter 4), or by using the underneath of the working piston, as in the Day engine (Chapter 4), or by using some other kind of air pump such as a Roots blower (see page 177). Strictly, a two-stroke engine cannot be supercharged unless the exhaust ports are arranged to close before the inlet ports close.

Where high power output from an engine of minimum size is required, there is no doubt that supercharging is the most effective method of securing it, but fuel consumption will be high unless special arrangements are used to offset the lower compression ratio and to allow for the energy needed to drive the unit.

The position of the supercharger in the induction system depends on the fuel system used. On early spark-ignition engines the supercharger, or blower as it is often called, delivered air to the carburettor, but this system was later abandoned in favour of a position where it delivered the mixture from the carburettor to the engine. With the advent of petrol injection, the arrangement has reverted to the original: most modern superchargers only pump air.

The supercharger may be positively driven by belt, chain or gears from the engine crankshaft; also it may be driven by exhaust gases from the engine. This gas-driven type, the *turbocharger*, is favoured today because it uses the energy in the exhaust gas instead of robbing the engine of some of its power.

Some people claim that the turbocharger is the answer to emission problems as well as present day demands for a better fuel consumption, but the latter is difficult to obtain from a spark-ignition engine unless the unit is very carefully matched with the engine. This reservation is based on the fact that thermal efficiency (or fuel economy) depends on the compression ratio, and in order to avoid detonation, the ratio must be lowered when a supercharger is fitted.

Restrictions on a normal aspirated (unsupercharged) engine in respect to lead content in the petrol and NO_x emission, often necessitates a lower compression ratio than that used in the past, so this 'development' results in an inferior power output and higher fuel consumption. Framed against this background of necessary engine alteration, a turbocharger, working efficiently during the period when acceleration is required, is able to restore the engine performance to that which is considered essential to meet modern traffic conditions.

Supercharging is not limited to spark-ignition engines; compression-ignition engines may also be supercharged and may even be more suitable subjects for supercharging.

A typical example of the introduction of a supercharger into a CI engine is where there is a need to raise maximum engine power to meet a legally permitted increase in the load carried by a vehicle. Instead of fitting a larger engine, many manufacturers have uprated the power of the existing engine by using a turbocharger.

In general the problem of the compression ratio does not apply to CI engines, so the following gives a rough guide to the advantages gained by turbocharging a CI engine:

1 A four-cylinder turbocharged engine gives the power output of a six-cylinder unblown engine.
2 A six-cylinder turbocharged engine gives the fuel consumption of a four-cylinder unblown engine.

33 Types of supercharger

The Roots blower

This device was patented about 1865 by F. M. and P. H. Roots in America and was used for a number of purposes including (in very large sizes) ventilating mines. Fig. 33.1 illustrates the construction generally used for engine supercharging: depending on engine size the width of the casing would be about 150–300 mm (6–12 in).

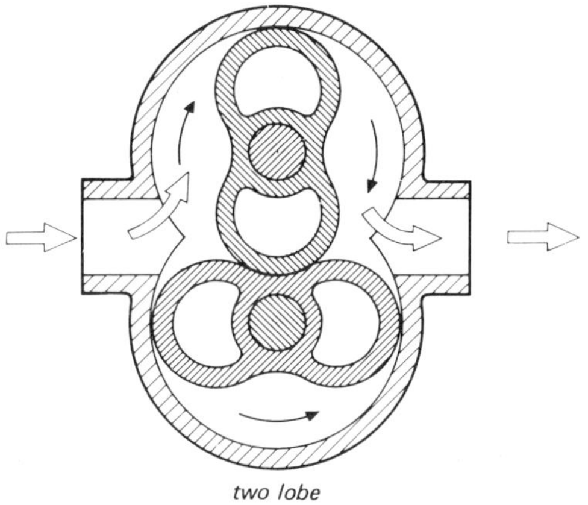

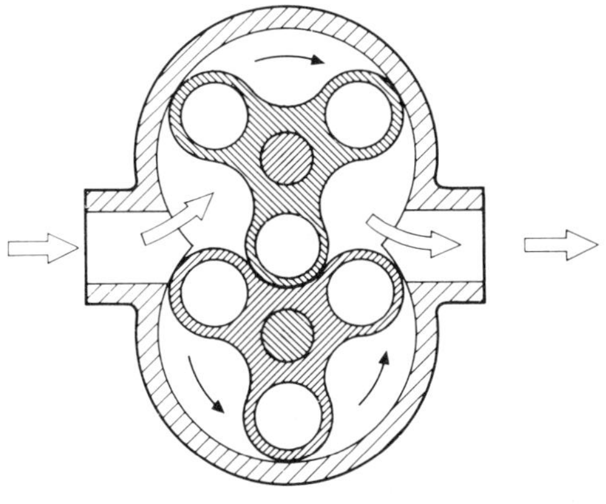

Fig. 33.1 Roots blowers

It can be regarded as a form of gear pump in which each gear (called a *rotor* in the Roots blower) has only two teeth or *lobes*. (Some Roots blowers have rotors with three or even four lobes.) The rotors have a small clearance inside the casing and are carried on shafts which are geared together outside the casing: one shaft is driven at approximately engine speed and drives the other at the same speed.

The Roots blower is an air *displacer* and not a *compressor*: the air is not compressed within the blower but simply carried round from inlet side to outlet side in the spaces between the lobes and the casing. The pressure at the output side depends upon the relative swept volumes of the blower and the engine, and the speed at which the blower is driven.

The rotors operate continuously against the full delivery pressure and thus more power is absorbed in driving this device than would be required if compression took place within the blower: at low pressures this disadvantage is small, but it increases rapidly as pressure rises.

Owing to the rotor clearance there will be some 'back leakage' of air: whilst this increases as delivery pressure rises, it decreases as speed increases. Thus this type of supercharger is mostly used for high-speed engines and relatively low supercharge pressures.

The vane compressor

In its simplest form this consists of a cylindrical casing in which is mounted eccentrically a cylindrical rotor carrying protruding vanes which divide the space between the rotor and the casing into a number of compartments which vary in volume as the rotor turns.

In Fig. 33.2a the compartment shown shaded is in the position of maximum volume and the vane (1) has just cut off communication between this

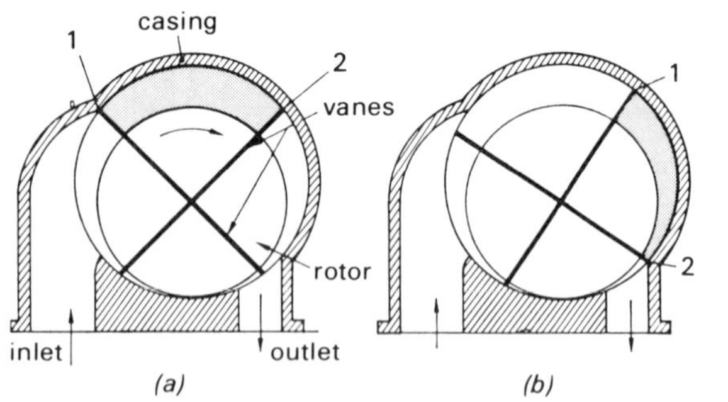

Fig. 33.2 The principle of the vane compression

compartment and the inlet port. In Fig. 33.2b the rotor has turned to the position where the vane (2) is just about to open the compartment to the outlet port, and it can be seen that the volume of the compartment is now smaller than it was in Fig. 33.2a. By altering the position of the leading edge of the outlet port the amount of compression within the compressor can be altered, and this would normally be arranged to correspond with the required delivery pressure.

Fig. 33.3 shows one practical form of this compressor which has, at various times, been marketed under the names of Centric, Arnott and Shorrocks. The vanes are mounted on a shaft which is placed centrally in the casing, each vane being carried on two ball bearings, so that the vanes are always radial to the casing. The vanes pass through slotted rods or *trunnions* carried in the rotor and there is a very small clearance between the tips of the vanes and the casing. The rotor is driven at approximately engine speed by belt, chain or gear drive.

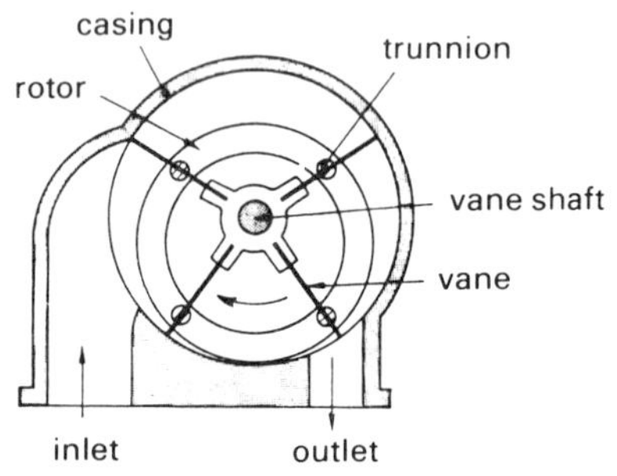

Fig. 33.3 The trunnion-type radial vane compressor

One minor disadvantage of the vane compressor is the necessity for lubrication due to the sliding of the vanes in the rotor. While this is fairly easy to arrange it is undesirable because oil mist carried into the engine may cause combustion difficulties, particularly in compression-ignition engines.

The centrifugal blower

If the Roots blower can be likened to a gear pump the centrifugal blower can be simply described as a glorified water pump (as used in an engine cooling system). The impeller is more carefully designed and made than the corresponding part of the water pump and is driven at considerably higher speed.

Fig. 33.4 illustrates the construction. The shaft is driven at some five to six times engine speed: air is drawn in through the intake, carried round between the vanes of the impeller and thrown outwards by centrifugal force into the volute casing from which it passes to the inlet manifold.

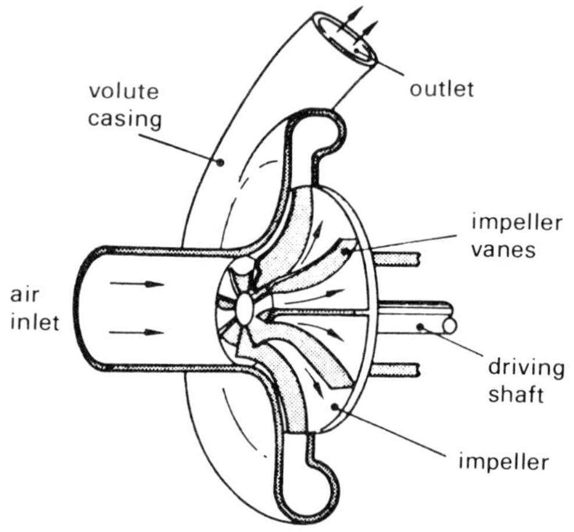

Fig. 33.4 Centrifugal blower

The chief merit of this type is that a small size of blower can deliver a considerable quantity of air. It has two main disadvantages for automobile use. The first is the very high speed at which it must be driven (which introduces problems in the drive arrangements during rapid acceleration and deceleration) and the second is that it is efficient over a very narrow range only.

The turbo-supercharger

The difficulty of driving a centrifugal blower at very high speeds, and the loss of up to 35 per cent of the engine power to provide this drive, can be neatly overcome by using the energy contained in the exhaust gas to supercharge the engine. The combination of centrifugal compressor and exhaust-driven turbine is called a turbo-supercharger or turbocharger.

Attempts to apply this principle to engines have been made from time to time since the early years of this century, but only within recent years has the method been applied to motor vehicles. The blades of a turbine have to operate at a high temperature of about 1000 °C, so only when a reasonably priced metal became available for the blades was it possible to make an economical unit.

Normally a turbine of the radial flow type is used to drive the compressor. It has a similar construction to a centrifugal blower but the gas flow through the unit is reversed (Fig. 33.5). A special material, such as nickel or nickel-chromium alloy iron, is used for the turbine; this is necessary because of the high temperatures. A cast-iron casing, formed like a snail's shell to feed the exhaust gas on to the periphery of the turbine blades, encloses the turbine. A flange on the part of the casing that forms the gas passage to the turbine connects the unit as closely as possible to the exhaust manifold of the engine. A connection at the gas outlet at the centre of the turbine casing allows the exhaust gas to be passed to a normal exhaust system.

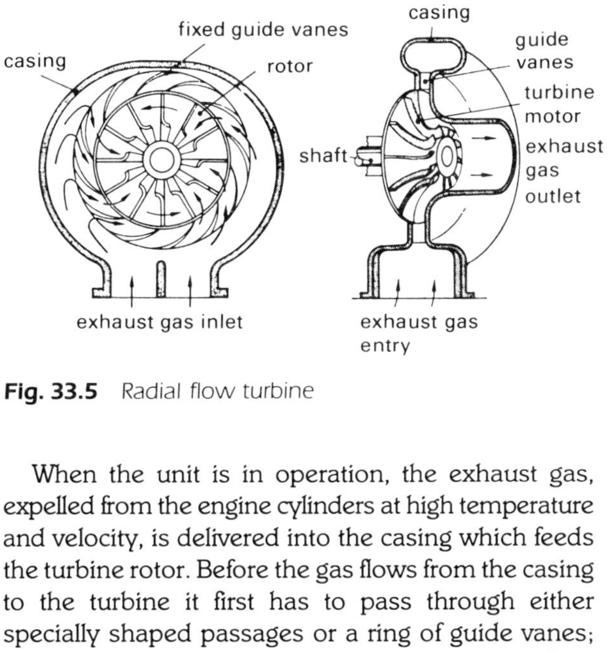

Fig. 33.5 Radial flow turbine

When the unit is in operation, the exhaust gas, expelled from the engine cylinders at high temperature and velocity, is delivered into the casing which feeds the turbine rotor. Before the gas flows from the casing to the turbine it first has to pass through either specially shaped passages or a ring of guide vanes; these ensure that the gas strikes the outer tips of the turbine blades tangentially so that the turbine can be driven in the required direction. As the gas flows towards the centre of the turbine, the blades slow it down. This action extracts energy from the gas and causes the turbine to spin at a high speed, in excess of 100 000 rev/min under engine conditions in which the temperature and/or velocity of the exhaust gas is high.

The exhaust-driven turbine is supported by a shaft which transmits the drive to the centrifugal compressor. The casings of aluminium alloy and cast iron for compressor and turbine respectively are bolted together to form a single unit; it is this assembly that is called a turbocharger or 'turbo' for short (Fig. 33.6).

In view of the high speed of operation, two plain floating bearings (often made of cast iron) are needed to support the shaft (Fig. 33.7). This type of bearing requires a pressurized supply of clean oil to resist wear and an adequate flow of oil to keep the bearings cool. The bearings are supplied with oil by a pipe connected to the engine's lubrication system. After passing through the bearings the oil is collected in a cavity, formed in the casing between the compressor and turbine, from where it is allowed to drain naturally back into the engine sump.

Gas and oil sealing is very important. Oil must not enter the compressor and exhaust gas must not pass to the bearings. The onset of either condition soon becomes apparent to the driver. Oil passing the compressor seal, especially when it is being pumped through by exhaust gas that is escaping from the turbine, produces a vast quantity of blue smoke. Any defect in the turbocharger that disturbs the action of the seals, such as shaft movement caused by worn bearings, soon shows up. Poor lubrication, caused by a lack of pressure or dirty oil, is a common cause of rapid bearing wear; this is accelerated when the shaft movement allows hot exhaust gas to escape into the bearings. When this occurs the future life of the unit is limited to a few seconds.

Fig. 33.7 also shows the bearing-and-seal arrangement. This design uses cast-iron sealing rings having a construction similar to a piston ring.

Special care of the turbocharger is needed to ensure that the bearings are not starved of oil when the engine is either started or stopped. The engine should not be accelerated either immediately after it has been started or just prior to switching off. (Bear in mind that it takes a long time for the turbine to come to rest.)

CI engines Turbocharging has been commonly used on large CI engines for many years since this type of engine is particularly suited to pressure charging. Unlike the spark-ignition type, the CI engine does not suffer from compression limitations and, in

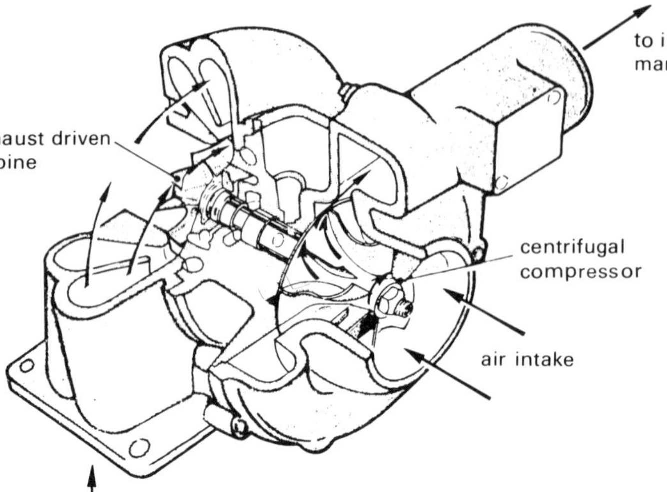

Fig. 33.6 Turbocharger

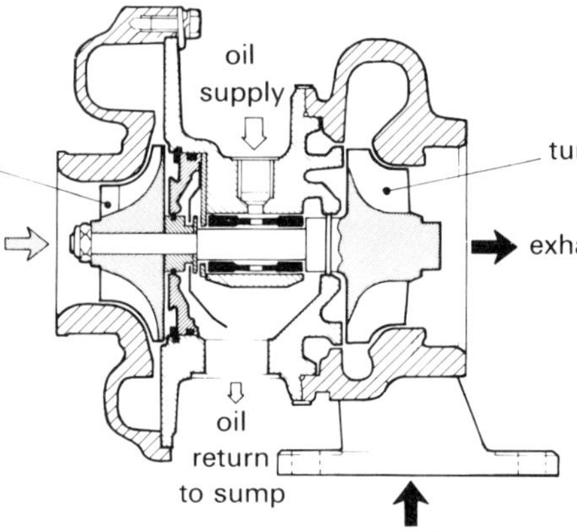

Fig. 33.7 Section through turbocharger

addition, the air-only induction system of a CI unit combined with more under-bonnet space makes the fitting of a turbocharger much easier.

Fitting a turbocharger to an engine which originally was normally aspirated gives the following advantages:

1 Greater torque and power. Since the power increase can be as high as 60 per cent, the engine is smaller and lighter than an unblown unit of equal power.
2 Fuel consumption is improved by about 10 per cent.
3 Exhaust noise is reduced owing to the smoothing out of the exhaust pulsations.

Spark-ignition (SI) engines A car fitted with a 'turbo' is associated with top-of-the-range models designed to give a performance, especially acceleration in mid-range, far above the norm. Engines that are suitable for turbocharging conversion must be structurally strong to withstand the extra load and higher speed. Also before the engine is turbocharged it must have its compression ratio lowered to avoid mechanical damage caused by combustion problems such as detonation.

Turbocharged engines normally use fuel injection systems because these arrangements overcome the pressure problems that arise when carburettors are used. Careful matching of the turbocharger to a given engine ensures that the maximum boost (pressure above atmospheric) is limited to about 1.5 bar, i.e. the maximum pressure in the inlet manifold is atmospheric plus 1.5 times atmospheric pressure.

Boost limitation The exhaust gas of an SI power unit is much hotter than that exhausted by a CI engine, so this places extra stress on the turbocharger. Furthermore, current emission regulations relating to the discharge of NO_x mean that the boost must be reduced at times when this pollutant would normally be produced. This occurs at high engine speed and/or maximum load, and so to meet this requirement, and also to minimize the risk of engine damage, a *boost limiting valve* is fitted (Fig. 33.8). Under high load conditions, this poppet-type valve, which is often called a *wastegate*, is opened mechanically so as to allow some of the exhaust gas to by-pass the turbine and reduce its pumping action. In conjunction with the turbocharger, many modern power units also use an electronic engine management system to set the ignition timing to suit the boost pressure.

An alternative method of limiting boost pressure is to use a pressure relief valve in the induction manifold; this opens when the boost reaches a predetermined maximum and allows air in the manifold to bleed to the atmosphere.

Charge-air coolers Compression of the charge and its passage through a hot turbocharger raises the temperature of the air; this results in a reduction of the density of the charge. To overcome this problem some CI and SI engines have a heat exchanger fitted between the turbocharger and engine (Fig. 33.9). This

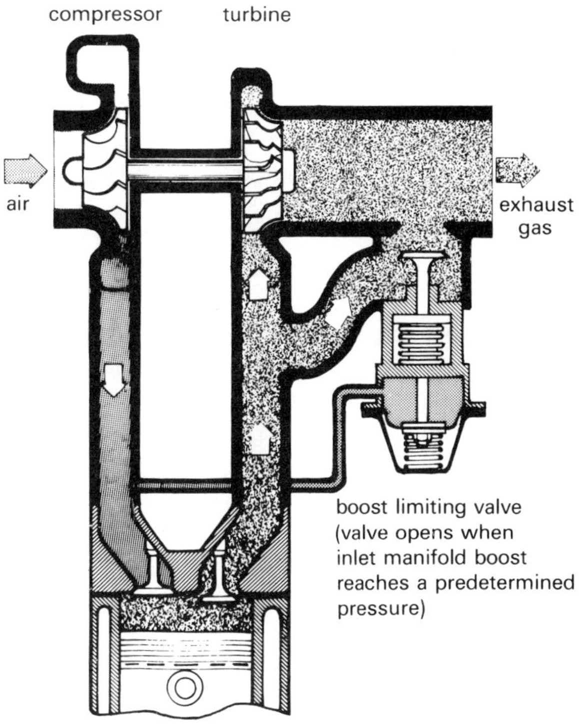

Fig. 33.8 Turbocharger with boost limiting valve

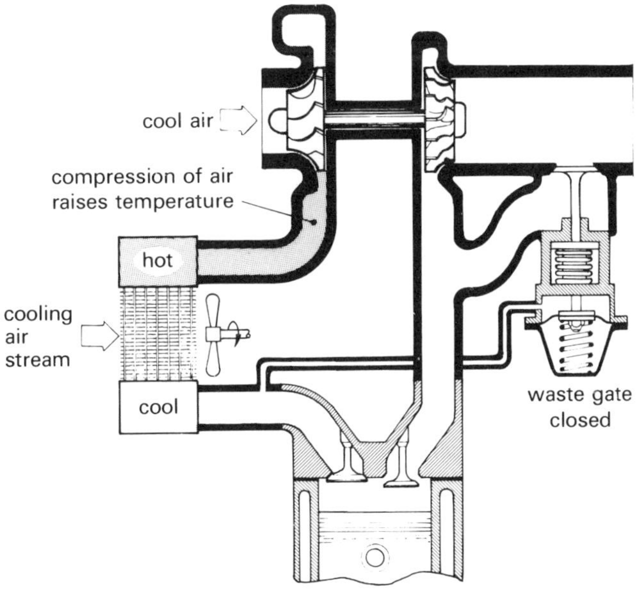

Fig. 33.9 Turbocharger with intercooler

unit, often called an *intercooler*, is generally an air-to-air exchanger and by lowering the air temperature to around 50–60 °C the volumetric efficiency is improved and the detonation tolerance is increased.

Advantages and disadvantages The advantages of turbocharging an SI engine include:

1 Higher torque for acceleration from low speed
2 Lower exhaust noise and emission
3 Better fuel economy due to a reduction in the pumping energy expended during the induction stroke. (Generally this advantage is seldom achieved in practice because many drivers alter their driving technique to take full advantage of the extra power of a turbocharged engine.)

The main disadvantages of a turbocharged SI engine, apart from the extra hazards created when incompetent drivers attempt to demonstrate the extra performance, are:

1 Higher initial cost of turbocharger and allied equipment.
2 Higher repair and servicing costs especially when other engine components are damaged by a defective unit.
3 Delayed engine response after depressing the accelerator. Since the turbine takes a time to reach its effective speed, an acceleration delay occurs with older type units.

PART NINE

Alternative types of engine

34 Propulsion units

Performance limitations of the conventional reciprocating engine, and the need to find an engine that combines economy with a pollution-free operation, urge inventors to design new forms of power unit.

Electric propulsion

For many years it was felt that the electrically propelled car had great potential, but to date, various attempts to offer this type of vehicle have not generated sufficient sales to justify widespread development of this smooth, quiet system.

Before the electric power unit becomes a serious contender to the conventional engine, a revolutionary new design of battery must be developed. It must be light in weight, low in cost and have a large energy storage capacity for its volume. Until the time when this battery is invented, the conventional lead-acid battery must be relied on to supply the propulsion energy.

Because the present day electric vehicle has to use a large number of lead-acid batteries, the space occupied by these heavy batteries is large. Operational problems include the limited range of the vehicle and its comparatively low maximum speed. Nevertheless there are many applications, such as local domestic delivery vehicles, for which an electrically propelled vehicle is ideal for the task it has to perform.

Gas turbine

The gas turbine is now firmly established in aircraft, marine and industrial applications, but is making less rapid progress in the motor vehicle field. The Rover gas turbine car was first shown in 1950, and in 1952 was demonstrated at speeds of over 240 km/h (150 mile/h). We have also seen several experimental gas-turbine-engined racing cars, but as yet no gas-turbine-engined car is on sale to the public.

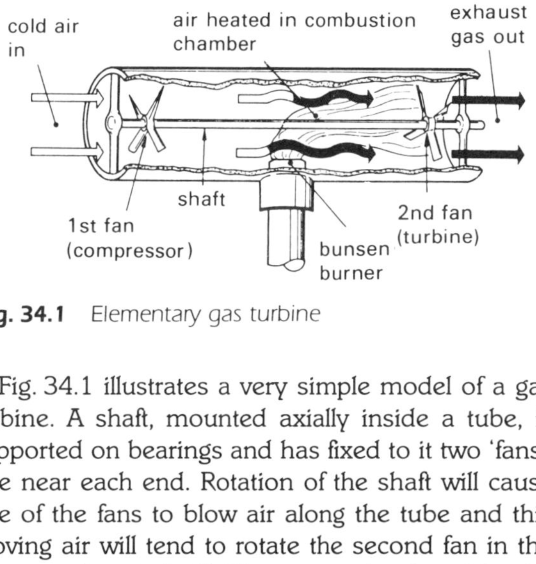

Fig. 34.1 Elementary gas turbine

Fig. 34.1 illustrates a very simple model of a gas turbine. A shaft, mounted axially inside a tube, is supported on bearings and has fixed to it two 'fans', one near each end. Rotation of the shaft will cause one of the fans to blow air along the tube and this moving air will tend to rotate the second fan in the manner of a windmill. The power developed by the second fan would only be just enough to drive the first fan if there were no friction, and clearly the device could not be used to drive any outside mechanism. However, by inserting a burner between the two fans the air passing along the tube can be heated causing it to expand. Thus it will pass the second fan travelling at a higher speed than that with which it passed the first fan, enabling the second fan to develop more power than is needed to drive the first fan.

The gas turbine can also be considered as a

development of the turbo-supercharger in which the supercharger (or compressor) delivers air to a combustion chamber in which fuel is burned continuously, the hot and expanded gases then being delivered at greatly increased speed to the turbine.

Fig. 34.2 shows, in simple diagrammatic form, the arrangement of a turbo-supercharger, and Fig. 34.3 shows the engine combustion chamber replaced by a simple combustion chamber in which the gas is merely heated by the continuous combustion of fuel. In the former case the gas does a great deal of work on the engine piston before it is released into the turbine, and consequently only a limited amount of energy is available for driving the turbine. In the gas turbine all the energy put into the gas by the combustion of fuel is available for driving the turbine, and there is usually far more than is required merely to drive the compressor. In the *jet* engine, as used in aircraft, this additional energy is used to provide a *thrust* by discharging the hot gases through a *jet pipe* after they have passed the turbine. Alternatively the shaft could be connected to some kind of machinery which is required to be driven. In this form the gas turbine would not be very satisfactory for vehicle propulsion: since the torque developed depends upon compressor speed the torque would be low at low speeds, and a

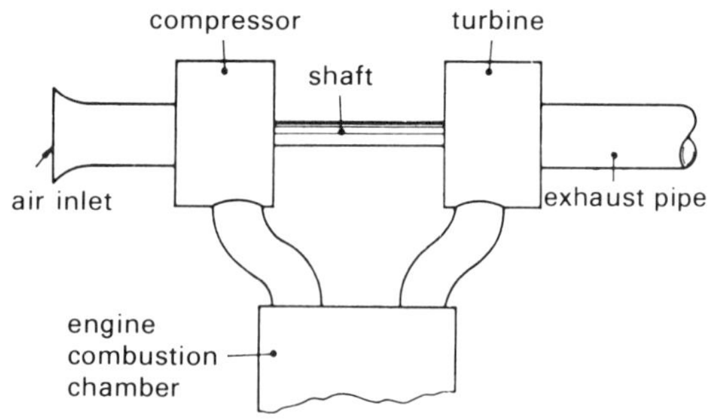

Fig. 34.2 Diagram of turbo-supercharger

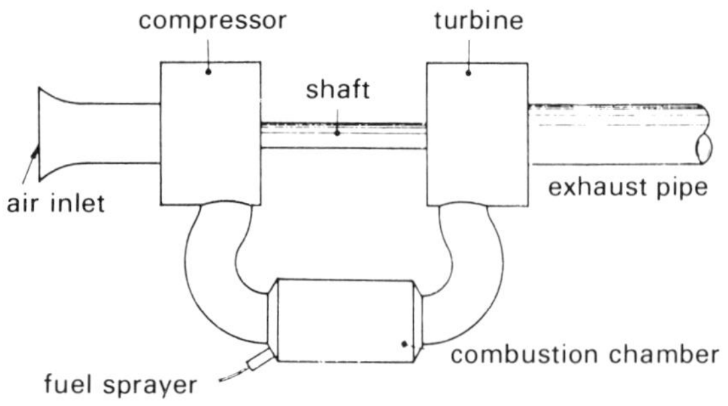

Fig. 34.3 Diagram of a simple single-shaft gas turbine

mechanism similar to the clutch and gearbox used with a piston engine would be essential.

A considerable improvement can be effected by using the arrangement illustrated in Fig. 34.4. Here one turbine is coupled to the compressor while a second turbine is mounted on a separate shaft. The *power turbine* would be coupled to the driving wheels of the vehicle, and this arrangement allows the vehicle to remain stationary with the power turbine stopped but the compressor turbine running. The torque on the output shaft when the power turbine is stationary and the compressor is running at its maximum speed is about 2–3 times greater than when the power turbine is running at full speed, giving the gas turbine something of the characteristics of an automatic variable-ratio transmission system. For example, a car which would require a four-speed gearbox when a petrol engine is used would need only a two-speed gearbox if fitted with a gas turbine.

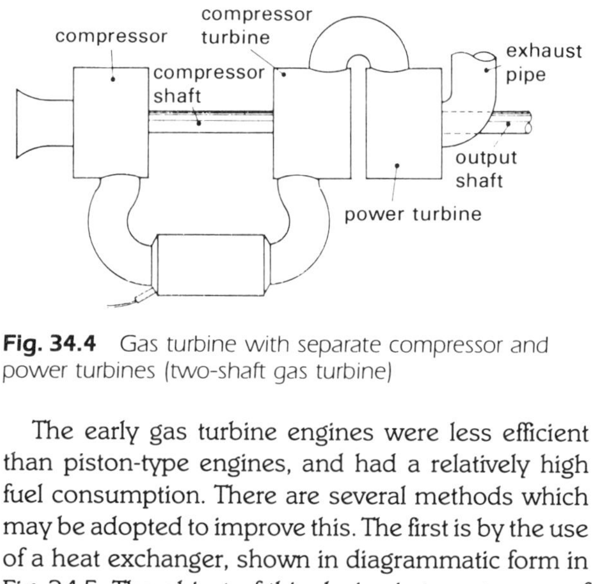

Fig. 34.4 Gas turbine with separate compressor and power turbines (two-shaft gas turbine)

The early gas turbine engines were less efficient than piston-type engines, and had a relatively high fuel consumption. There are several methods which may be adopted to improve this. The first is by the use of a heat exchanger, shown in diagrammatic form in Fig. 34.5. The object of this device is to put some of the heat normally carried away by the exhaust gas into the air leaving the compressor, thus reducing the amount of fuel which must be burned to provide a given temperature rise. By this means it has been possible to effect considerable improvements in fuel consumption, and some of the most recent gas turbines are claimed to have a fuel consumption comparable with that of a compression-ignition engine.

Several types of heat exchanger have been tried. The one most recently developed consists of two large discs, about 0.6 m (2 ft) in diameter, of a ceramic

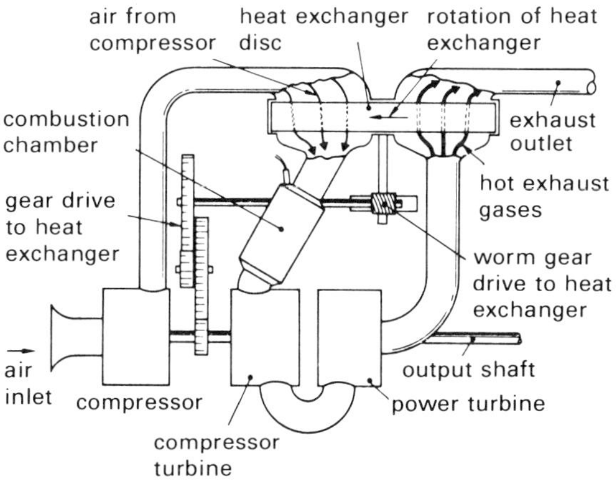

Fig. 34.5 Two-shaft gas turbine with heat exchanger

material having numerous small transverse passages through them, disposed vertically on each side of the engine and rotated at about 1/2000 of the speed of the compressor shaft. Exhaust gases from the power turbine are led through ducts to one half of each disc and in passing through the small passages in the disc give up some of their heat to the discs before being discharged through the exhaust system. As the discs rotate the heated portions are carried into ducts which direct air from the compressor through the discs before it passes into the combustion chamber, thus heating this air.

A second method by which the efficiency of the gas turbine could be improved is by raising the temperature of the gases entering the turbine. At present this is limited to about 1000–1500 °C by the inability of present materials to operate satisfactorily at higher temperatures, and even for these temperatures very special and expensive alloys have had to be developed. However, development is proceeding with ceramic materials which, if successful, should permit the use of higher temperatures.

The maximum gas temperature is limited by restricting the maximum fuel delivery to the combustion chamber to an extent which provides an air/fuel ratio considerably weaker than the chemically correct mixture.

The engine is started by driving the compressor shaft by some external means - usually an electric motor. When it is running fast enough to blow air into the combustion chamber the fuel supply is turned on and fuel is sprayed continuously through a sprayer nozzle into the combustion chamber. Combustion is initiated by a sparking plug and once started will continue as long as fuel continues to be supplied. Power output is regulated by varying the rate of fuel delivery. Compressor speed ranges from about 15 000 rev/min when idling to a maximum of about 40 000–50 000 rev/min. The power turbine speed varies from zero to about 35 000–40 000 rev/min: it is therefore necessary to have a speed-reducing gear giving a ratio of the order of about 10 to 1 between the power turbine shaft and the propellor shaft if a conventional final-drive ratio is to be used.

The chief advantages claimed for the gas turbine are:

1. High power output from a given weight of engine.
2. The torque output characteristics permit a notable simplification of the transmission system.
3. Smooth vibrationless running due to absence of reciprocating parts.
4. No rubbing parts (such as pistons) so that internal friction and wear are almost eliminated.
5. East starting.
6. Can use a wide range of fuels and does not require expensive anti-knock additives.
7. Low lubricating oil consumption.
8. No water cooling system needed.
9. Non-poisonous exhaust giving very little trouble with pollution.
10. Requires little routine maintenance.

Compared with a conventional engine the main disadvantages are:

1. High fuel consumption during part-load operation.
2. Takes a comparatively long time for the engine to accelerate from idle to normal operating speed.
3. Large space occupied by heat exchanger.
4. Characteristic whine from engine which becomes troublesome for the driver on journeys during which the speed is sustained.

Wankel engine

As long ago as 1769 James Watt took out a patent for an engine in which steam pressure drove a rotating vane in a circular housing, and there have since been many attempts to produce an internal combustion engine in which the pressure of gases after combustion acted upon rotating members enclosed in a suitable casing. For a variety of reasons none of these showed any likelihood of challenging the piston engine until, towards the end of 1959, it became known that a German engineer, Felix Wankel, had devised an engine working on the equivalent of the four-stroke cycle but employing only rotating parts. Development of this engine was taken up by the German firm of NSU and early in 1965 they produced a small car with one of these engines for sale to the public. Two years later they announced the production version of a large car having the equivalent of a 2-litre engine.

In 1965, Toyo Kogyo (Mazda) of Japan also introduced a car fitted with a rotary engine; at present it is this company that holds the world patents for the Wankel engine.

During recent years, the difficulty of meeting both the emission requirements and fuel economy regulations has caused many manufacturers to turn away from the rotary engine.

Each 'element' of this engine consists of three main parts: a housing, a rotor, and a shaft. Fig. 34.6 shows a section through one 'element'.

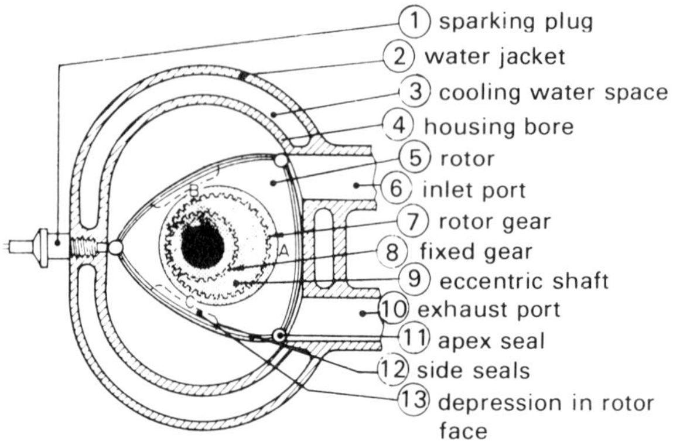

Fig. 34.6 The main working parts of a Wankel engine

The *housing* contains a shallow bore (4) of a special shape known as *epitrochoidal*. (An *epitrochoid* is the line traced out by a point between the centre and the circumference of a circle as the circle rolls round the outside of another circle twice as large.) Carried on bearings co-axial with the centre of the housing bore is a shaft on which is formed an *eccentric* (9) over which fits a *rotor* (5) shaped like an equilateral triangle but with curved sides or flanks. An internal gear (7) concentric with and fixed to the rotor engages with a fixed gear (8) secured centrally in the housing and having a hollow centre through which the shaft passes. This fixed gear has two-thirds the number of teeth in the internal gear (7) in the rotor.

The result of this arrangement is that when the shaft is rotated the rotor is not only carried round eccentrically with it but also rotates in the same direction as the shaft but at one third the speed.

In every position of the rotor its three corners or *apices* (plural of apex) touch the housing bore so that three compartments or *chambers* are formed between the housing bore, the flanks of the rotor and the end plates of the housing. These chambers are made gas-tight by special seals at each apex (11) and the sides (12) of the rotor.

In every *three* revolutions of the shaft the rotor rotates *once*, and during this time the volume of each chamber passes through *two maximum* and *two minimum* values, so that the operations of the four-stroke cycle will be carried out in each of the three chambers. The *difference* between the maximum and minimum volumes represents the *swept* volume of each chamber, and the *ratio* maximum volume divided by minimum volume is the *compression ratio*. The minimum volume is what would be called the *clearance volume* in a piston engine, and the greater part of this is contained in a depression (13) in each rotor flank.

An inlet port (6) and an exhaust port (10) enter the housing at the points indicated in Fig. 34.6 and a sparking plug (1) is fitted in the opposite side of the housing.

In Fig. 34.6 the flanks of the rotors are identified by the letters A, B and C, and in the description of the operation which follows we use these letters to refer to the chambers between the housing and the appropriate rotor flank. In the position shown chamber A is in one of its two minimum volume positions, in this case the equivalent, in a piston engine, to t.d.c. at the end of exhaust and beginning of induction. Chamber B has passed its position of maximum volume and the gas in it is now being compressed ready for ignition. Chamber C is approaching its position of maximum volume and the gas in it is expanding, the equivalent of the expansion or power stroke.

Fig. 34.7 shows the events which occur in each chamber as the shaft completes one revolution. The shaft is shown turning anticlockwise and Fig. 34.7a shows the position when the shaft has turned 90° from its position in Fig. 34.6.

Fresh mixture is being induced into chamber A which is in communication with the inlet port but cut off from the exhaust port: this is the early part of the induction stroke.

Chamber B is still decreasing in size and is approaching the end of the compression stroke.

Chamber C has just reached its position of maximum volume equivalent to b.d.c. at the beginning of the exhaust stroke, and this chamber is already open to the exhaust port.

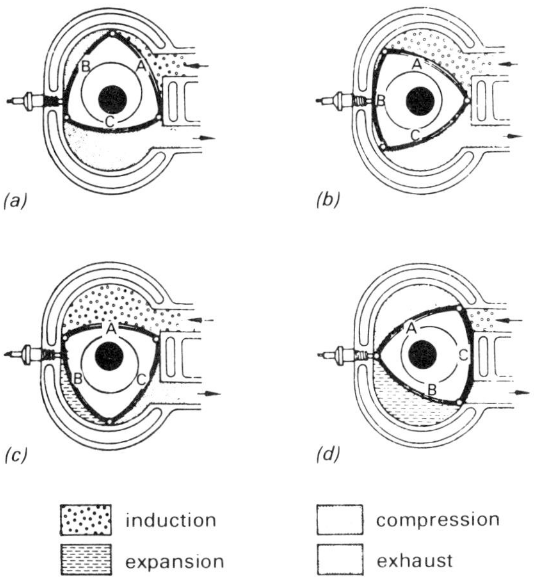

Fig. 34.7 Operation of the Wankel engine

In Fig. 34.7b the shaft has turned a further 90° and chamber A has completed about two-thirds of its induction stroke.

Chamber B is in the position of minimum volume corresponding to t.d.c. at the end of the compression stroke, the mixture has been ignited and combustion is taking place.

Chamber C has completed about a third of its exhaust stroke: its volume is decreasing and the burned gases are being forced out of the exhaust port.

Fig. 34.7c shows the position after the shaft has turned its third quarter of a revolution. Chamber A has reached its position of maximum volume, corresponding to b.d.c. at the end of the induction stroke and the inlet port is beginning to close.

Chamber B is increasing in volume at the early part of the power stroke.

Chamber C is just starting the last third of its exhaust stroke.

Finally Fig. 34.7d shows the situation at the end of one complete revolution of the shaft. This is similar to the position shown in Fig. 34.6 except that the chambers have all moved round one place. During the next revolution of the shaft chamber A will carry out the same operations as chamber B did on the revolution just described, chamber B will repeat those of chamber C, and chamber C those of chamber A.

During the following revolution of the shaft (the third) chamber A will repeat the operations just described for chamber C, chamber B those of chamber A, and chamber C those of chamber B, and the position will by then have returned to that shown in Fig. 34.6.

Thus it is seen that the engine produces one power stroke per revolution of the shaft, and it is therefore equivalent, in this respect, to a single-cylinder two-stroke engine or a two-cylinder four-stroke engine having a cylinder swept volume equal to the swept volume of one chamber.

If two of these 'elements' are used the firing frequency of a four-cylinder four-stroke engine is obtained; such an engine is called a *two-rotor* or *twin-rotor* engine.

Although air-cooled engines of this type have been built, those at present in production have the housing water-cooled. The rotor is cooled by circulating oil from the lubrication system around its hollow interior, and this necessitates the use of an oil cooler.

The main advantages of this type of engine are:

1 Simplicity. Each 'element' has only two moving parts - the rotor and the eccentric shaft - and since communication between chambers and ports is controlled by the rotor, valves and their operating gear are completely eliminated.
2 The absence of reciprocating parts should allow the engine to operate at higher speeds.
3 Since all moving parts have simple rotary motion they are easily balanced and vibration can be eliminated.
4 If peripheral ports are used they are never closed. This leads to improved volumetric efficiency. Values in excess of 100 per cent have been claimed.
5 For a given swept volume the engine is lighter and less bulky than a piston-type engine. The gain in this direction increases as the number of rotors is increased.

Stirling engine

Introduced in Scotland more than 150 years ago, this type of engine has been considered for vehicle propulsion. The clean exhaust is its main appeal, but as yet it cannot compete with conventional types of engine.

In the past the Stirling engine was called a steam engine because in the early days the pistons were powered by steam heated by an external combustion source. Modern units are far removed from the ideas suggested by the term 'steam engine'. Although combustion still takes place outside the cylinder, the fluid used to drive the pistons is hydrogen, helium or Freon.

Principle of operation

Fig. 34.8 shows a simplified two-cylinder double-acting engine which, in this case, uses a swash plate to rotate the output shaft. As both sides of the pistons are used, the piston rods are fitted with effective seals to prevent loss of the operating fluid. When a conventional crankshaft is used, the thrust from the piston rod is applied first to a crosshead and then on to a conventional connecting rod and crankshaft.

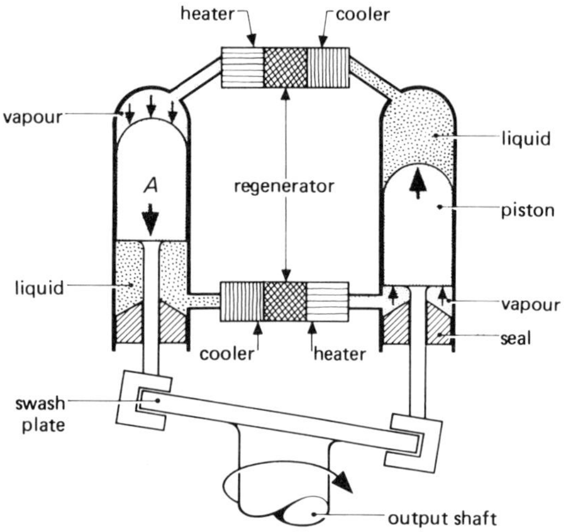

Fig. 34.8 Principle of the Stirling engine

In the diagram, the fluid above piston A has been heated and the expanding vapour is driving the piston down the cylinder. Below piston A the cooler has lowered the vapour temperature and the fluid is contracting. This action is repeated in each cylinder to provide the working power of the engine.

Unlike the normal engine, the Stirling works on a *closed cycle*: this means that the operating fluid does not leave the engine and that the energy given to the fluid conversion process is by means of a heat exchanger supplied from an external heat source.

Combustion is continuous and in many ways similar to the process used in a gas turbine. Fuel is sprayed at low pressure into a chamber and initially ignited by electrical means. The amount of heat passed to the operating fluid depends on the quantity of fuel injected, so power output can be easily regulated. At first it would appear that response would be slow, but results from prototype engines show that the time from zero torque to maximum torque is only about 0.1 s.

A prototype engine is shown in Fig. 34.9.

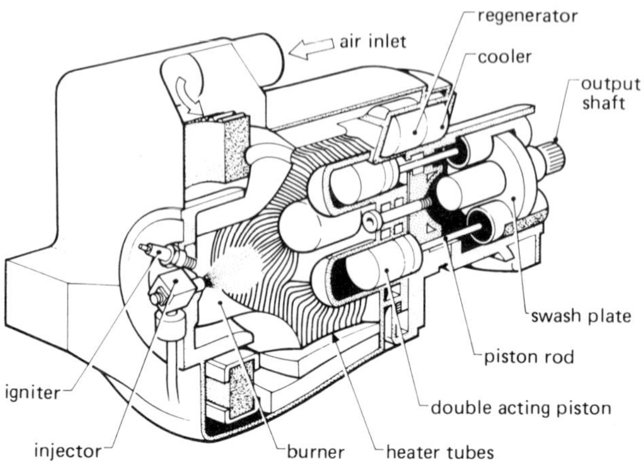

Fig. 34.9 Stirling-type engine

The advantages of the Stirling engine are based mainly on the continuous combustion process: these are:

1 *Low exhaust emission* Excess air of 20–80 per cent in the air–fuel mixture, together with adequate combustion time and no quench of the flame by 'cold' metal surfaces, gives near-complete combustion. Smoke level is also much lower.
2 *Fuel economy* Accepts a large range of liquid or gaseous fuels. There are no special fuel requirements such as octane or cetane. It is expected that a brake thermal efficiency of about 40 per cent is possible and even under part-load conditions, the efficiency is at least as good as the CI engine.
3 *Low noise level* A silencer is unnecessary because the noise level is only about 25 per cent of that emitted by a fully silenced CI engine. No valve mechanism is fitted, so mechanical noise is very low.

These advantages make the engine most attractive, but there is still much development work needed to improve the reliability of new engine components such as fluid seals. When cost and reliability approach the standard of conventional engines, then widespread use of this type of engine is expected.

PART TEN

Clutches

35 The single-plate clutch

Reason for a clutch

It has been seen that the internal-combustion engine, unlike the steam engine, does not produce high power at low speeds; therefore the engine must be rotating at a speed at which sufficient power is developed, before the drive to the wheels is established. This condition rules out the use of a dog clutch (Fig. 35.1a) since the connection of a rotating engine to a stationary transmission shaft would damage the transmission and jolt the vehicle. The clutch used must allow the drive to be taken up smoothly so that the vehicle can be gradually moved away from the stationary position.

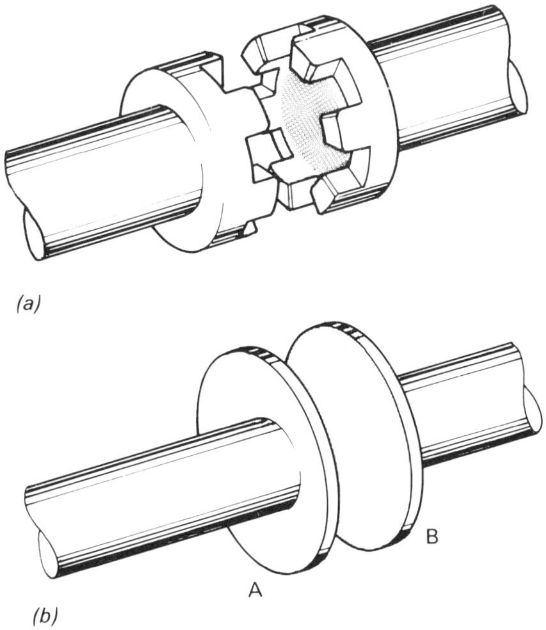

Fig. 35.1 Dog and friction clutches

Once moving, it will be necessary to change gear, and so a disengagement of the engine or transmission is required. This is also part of the clutch's function.

The clutch takes up the drive smoothly; it also disengages the drive. These two duties can be performed by various mechanisms; the friction system is considered to be one of the most effective and efficient. Hydraulic and electric systems are also employed on motor-vehicles.

The friction clutch

The operation of a friction clutch is demonstrated by the two discs shown in Fig. 35.1b. If discs A and B are connected to engine and transmission respectively, then when the discs are held apart, no drive will be made. When A is moved to contact B, the friction between the surfaces will allow a drive to be transmitted. The extent of this drive will be governed by the force which pushes the discs together; therefore if the force is gradually increased, the drive transmitted will increase proportionally. This will allow the vehicle to move smoothly from rest.

When the drive is engaged, both discs must rotate at the same speed, and so the clutch designer must ensure that there is enough pressure and friction to produce this condition. During the life of the clutch, faults may develop which restrict pressure or friction; therefore the clutch will slip and the performance of the vehicle will be affected.

The clutch shown in Fig. 35.2 is a construction developed from the previous diagram. Both sides of the friction plate are utilized, and this will mean that double the torque (turning moment) can be trans-

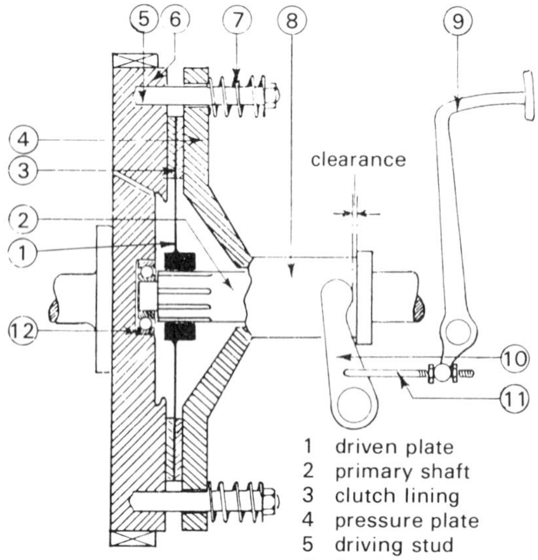

1 driven plate
2 primary shaft
3 clutch lining
4 pressure plate
5 driving stud
6 flywheel
7 thrust spring
8 withdrawal sleeve
9 pedal
10 clutch fork
11 adjustable link
12 spigot bearing

Fig. 35.2 Single-plate clutch

mitted before slip takes place. The friction plate, or *driven plate*, is sandwiched between the engine flywheel and the pressure plate, and the design is known as a *single-plate clutch*.

A *multi-plate clutch* is a unit which employs a number of driven plates. It is described in the next chapter.

Single-plate clutch

This is the most common type of clutch in use, and has the advantage of producing a quick disengagement.

Fig. 35.2 shows a simple clutch with a driven plate (1) splined to the primary shaft (2) of the gearbox. Riveted to the plate is a pair of linings or facings (3) made of an asbestos-based material, which is bonded with suitable resins to give a satisfactory coefficient of friction, and other useful properties. The pressure plate (4) is located by a number of studs (5), and is forced towards the flywheel (6) by means of springs (7). A withdrawal sleeve (8) allows the clutch pedal (9) to act on the clutch forks (10) and push the pressure plate away from the flywheel. An adjustable link (11) is set to give a small clearance between the clutch fork and the withdrawal sleeve to ensure that the full force of the springs is felt on the driven plate.

This clearance allows the pedal a small amount of free movement which, measured at the pad of the pedal, is normally about 25 mm (1 in). At the centre of the flywheel is fitted a spigot bearing (12), which locates the front end of the gearbox primary shaft and allows for the difference in speed between the two members. This bearing can take the form of a ball race or plain bush; the former is normally sealed with a metal cover to prevent the lubricant being thrown out on to the linings. The plain bearing type is generally manufactured from phosphor-bronze, and made porous so that the bush can be impregnated with graphite to make it self-lubricating.

When the driver depresses the pedal, the clutch forks push the pressure plate away from the flywheel to remove the spring pressure from the driven plate. This allows the plate to run free, and therefore disengages the engine from the transmission.

On releasing the pedal, the spring thrust forces the pressure plate towards the flywheel, and sandwiches the driven plate between the two surfaces. This movement of the pedal must be gradual, or the full force of the springs will be suddenly applied to the driven plate and cause a jolt.

It will be seen that when the pedal is fully released, the drive can take two paths – it can be transferred directly from flywheel face to driven plate, or can be taken via the studs and pressure plate to the rear face of the driven plate.

Close attention should be given to this second drive path because the existence of an excessive clearance between any two parts in this drive path leads to noise and a jerky operation of the clutch. The path, via the pressure plate, is designed to take half the transmitted torque of the clutch, so modern units use efficient arrangements to convey the drive from the flywheel to the pressure plate.

During the engagement of the clutch, the force transmitted to the driven plate must be applied gradually. Since any jerky movement of the pressure plate gives a more pronounced judder at the roadwheels, the flywheel/pressure plate connection must avoid, or limit the slip-grip action caused by friction at this point. Most probably the clutch in Fig. 35.2 will suffer from this drawback which will be evident at a time when the vehicle is moving off from rest, i.e. when the pressure plate is moving towards the flywheel and driving torque is attempting to resist this axial movement.

Borg and Beck clutches

The majority of motor manufacturers use a single-plate clutch such as the type made by Borg and Beck. This company offers a complete range of clutches to suit most vehicles. The types of spring used to provide the clamping force are:

1 *Diaphragm spring* - universally used for cars and light commercial vehicles.
2 *Coil spring* - although multi-coil spring clutches were used for all vehicles in the past, nowadays this type is only used on heavy vehicles.

Diaphragm spring (DS) clutch

This type is very compact; it has few working parts and a spring that is particularly suited to light vehicles.

Diaphragm spring action The force needed on the pressure plate is provided by a diaphragm spring. This is a circular, slightly conical, tempered steel disc with radial slits cut from its centre to give flexibility.

Whereas a coil spring gives a force that progressively increases with deflection, a diaphragm spring exerts a force which varies as its shape is altered. At first the force gradually increases like the coil spring but after the diaphragm spring has reached its 'flat' position, the force decreases and then increases once again (Fig. 35.3). In many ways the spring action is similar to that which occurs when pressure is applied to the convex end of a 'tin' can - when sufficient force is applied to move the can end past its flat position the end 'clicks' rapidly to its full inwardly dished state.

When fitted to a clutch mounted on a flywheel, the spring is compressed just beyond the point where it is flat (Point A on the graph). Set in this position, the reduction in thickness of the friction facing due to wear causes the spring to 'release' and the force to increase.

Compared with a coil spring, the diaphragm spring offers the following advantages:

1 Compact.
2 Suitable for high engine speeds. Coil springs bow outwards owing to centrifugal action and this lowers the spring force; it can also cause vibration owing to imbalance.
3 Lower pedal force. Less friction since fewer parts are needed to operate the clutch. Also the force-deflection curve suits the application.
4 Clamping force on friction facings does not decrease as the facings wear.

Pressure plate construction Fig. 35.4 shows the construction of a DS clutch. This particular design uses only five main working parts: cover, diaphragm spring, fulcrum rings, pressure plate and driving straps.

Turnover tabs on the cover hold the fulcrum rings, which act as a pivot for the diaphragm spring. As the cover bolts are tightened, the pressure plate pushes the diaphragm spring away from the flywheel and deflects the spring to a near-flat shape. In this position the reaction of the spring on the pressure plate provides the clamping force on the driven plate.

A single-plate clutch has two driving surfaces; the

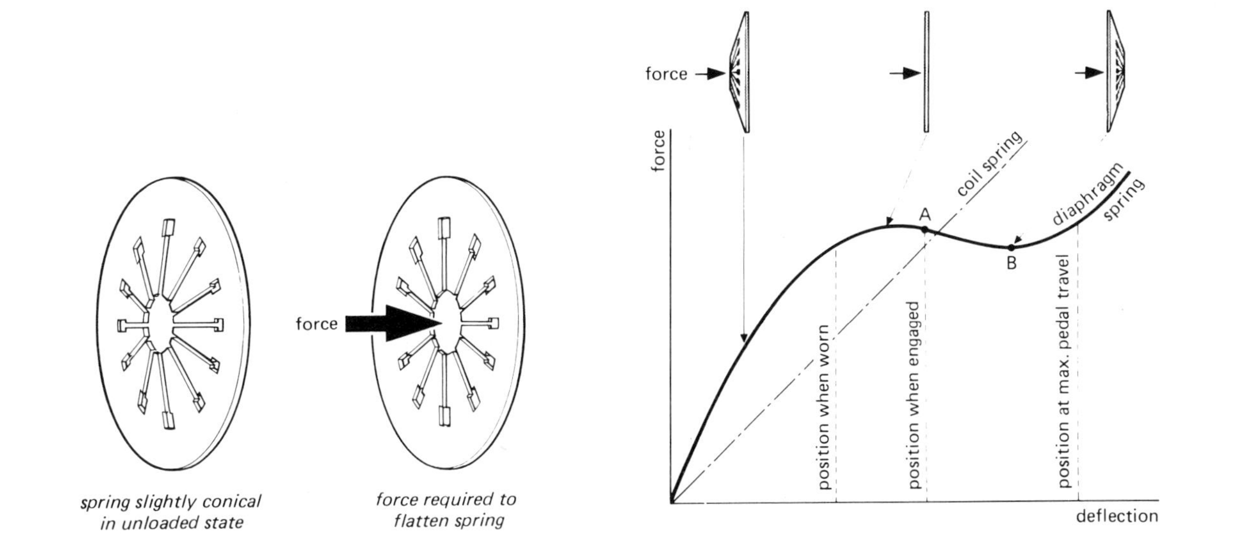

Fig. 35.3 Characteristic of diaphragm spring

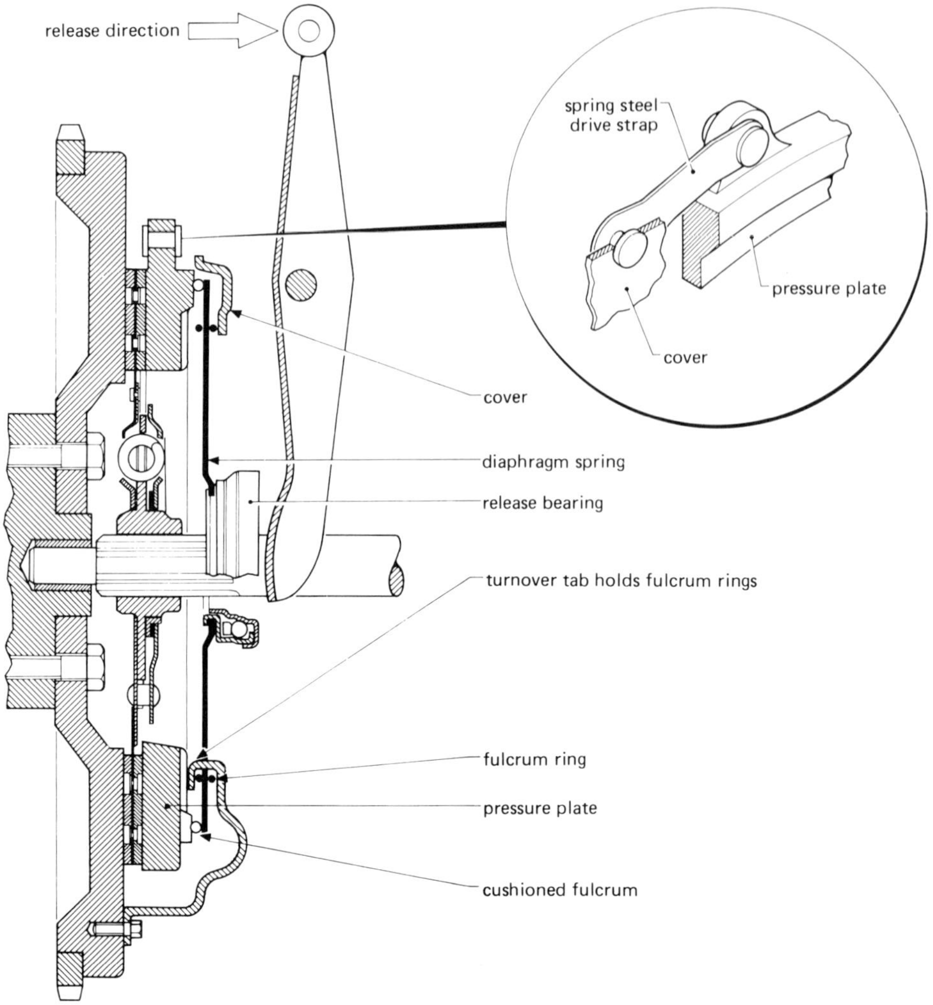

Fig. 35.4 Diaphragm spring clutch – DST type

flywheel and the pressure plate. In the clutch shown in Fig. 35.4 a *strap drive* is used to transfer the drive from the cover to the pressure plate. Three flexible steel straps working in tension allow the pressure plate to move smoothly towards and away from the flywheel, but prevent rotational movement relative to the cover.

Disengagement of the clutch compresses the spring and as the graph shows, the force required by the driver is less in the disengaged position (B) than that needed at the commencement of travel of the pedal. This is opposite to that given by a coil spring clutch.

A release bearing, attached to either the diaphragm spring fingers or to the clutch fork, moves the inner ends of the fingers of the diaphragm spring towards the flywheel to disengage the clutch. This forward movement of the inner part of the spring causes it to pivot on the fulcrum rings and move its outer part away from the pressure plate. With the spring clamping force removed, the spring straps pull the pressure plate away from the driven plate and disengage the drive to the driven plate.

Release bearing The purpose of the release bearing is to transmit the force applied by the driver from the clutch fork, which is pivoted to the bell housing, to the rotating diaphragm spring. In the past a graphite block, held to the clutch fork, contacted a flat plate secured to the diaphragm spring fingers, but today most clutches use a ball race. This race is sealed with metal end covers to retain the lubricant. In some cases the ball release bearing is connected to the clutch fork by a flexible plastic compound which allows the bearing to align itself with the centre of the

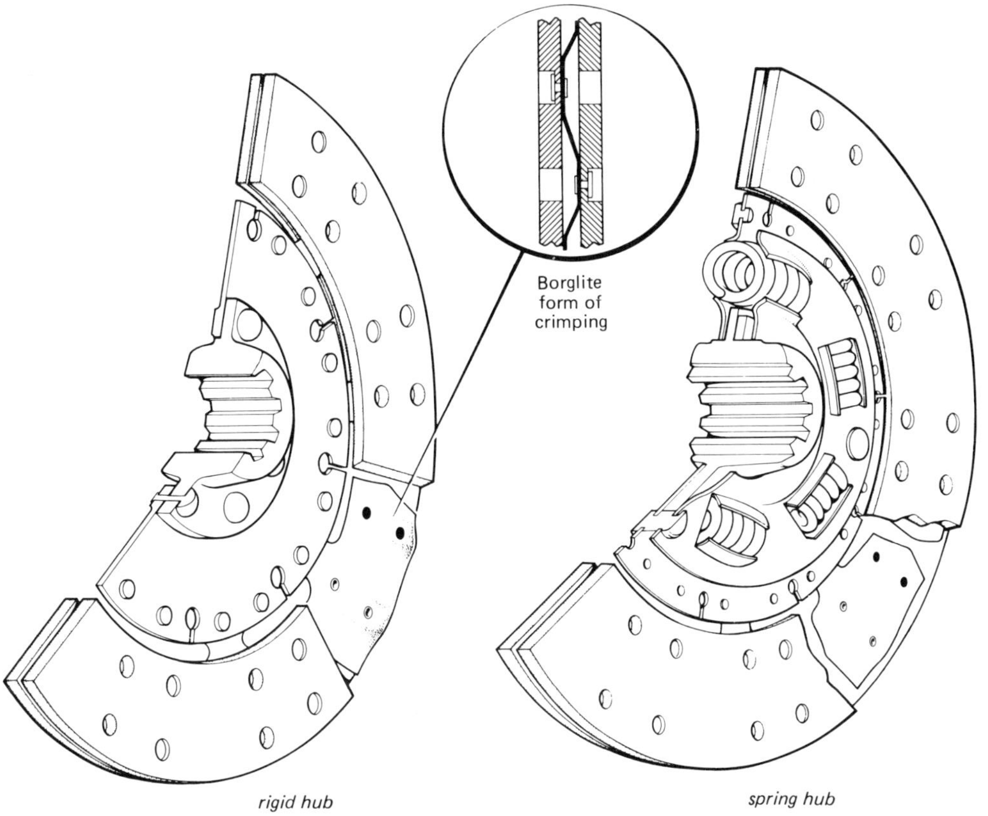

Fig. 35.5 Driven plate construction. Plate cut away to show section

clutch and give smooth clutch operation.

When the clutch is engaged, i.e. drive is taking place, a small clearance should exist between the release bearing and the diaphragm. This clearance, which is normally measured at the pedal pad and called *free-pedal movement*, is necessary for two reasons: it keeps the release bearing clear of the rotating spring and also ensures that the full spring thrust acts on the pressure plate. As the clutch facing wears, the clearance decreases, so unless periodic adjustment is made, the clearance will disappear and the clutch will start to *slip*.

The driven plate

The important features incorporated in the design of a driven plate can best be seen by considering the disadvantages of using a plain steel plate with a lining riveted to each side. These may be listed as:

1 Buckling of the plate due to heat.
2 Drag, due to the plate rubbing against the flywheel when the clutch should be disengaged.
3 Very small movement of the clutch pedal between the engaged and disengaged positions - the clutch is said to be in the *in* or *out* class - with very little control between these points.

To overcome these troubles, the plate is normally slotted, or set in such a manner as to produce a 'flexing' action. This is generally known as *crimping*, and Fig. 35.5 shows one form. Each segment is dished a small amount, so that the linings tend to spring apart when the clutch is engaged.

If the clutch is in the driving position and the pedal is depressed, the driven plate will tend to jump away from the flywheel to give a 'clean' disengagement. While in this position, the linings will be held apart, and air will be pumped between the linings to take away the heat. During engagement, axial compression of the driven plate spreads the engagement over a greater range of pedal travel and therefore makes it easier to make a smooth engagement.

Fig. 35.5 shows a driven plate which is riveted to a splined hub. This arrangement is used occasionally, but in most cases the hub is mounted independently of the main plate, and the drive between the two components made possible by either fitting a series of springs or bonding the hub to the plate by means of rubber. Rubber takes shocks and also absorbs shocks, whereas in the case of the spring hub the spring takes the shock and a special friction damper fitted between the plate and the hub absorbs the energy of the shock. Using one or other of these hub centres it is possible to absorb the torsional shocks due to clutch take-up or the more troublesome engine vibrations which cause transmission noise and rattle.

Multi-coil spring clutch

This type of clutch has been the standard clutch for a number of years, but the popularity of the DS type for cars has meant that the multi-coil spring clutch is now only used on heavy vehicles.

Pressure plate construction (Fig. 35.6) A series of coil springs is positioned between the pressed steel cover and the cast iron pressure plate. Spring steel straps are used to transmit the drive from the cover to the pressure plate. Four release levers, pivoting on fulcrum pins supported in adjustable eye bolts, connect the release lever plate to the small struts placed between the lever and the pressure plate. These struts improve the efficiency of the release mechanism and, in conjunction with the steel driving straps, allow the pressure plate to move smoothly when taking up the drive: a jerky action will cause clutch *judder*. The release bearing can be either graphite block or ball race type.

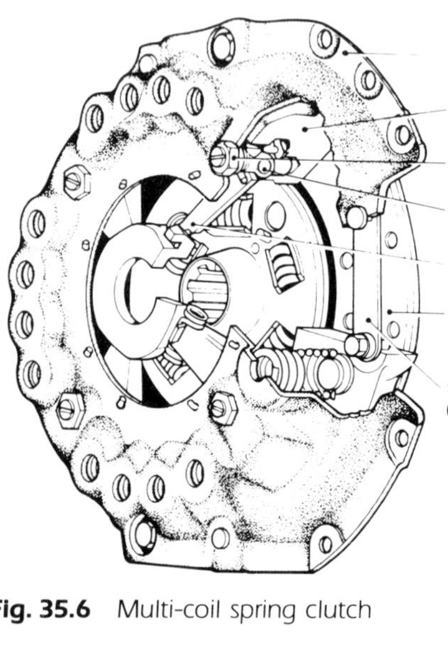

Fig. 35.6 Multi-coil spring clutch

The cover plate assembly is bolted to the flywheel; this is flat-faced for clutches having a driven plate diameter of less than 330 mm (13 in) and pot-type (recessed) for larger clutches.

Balance is most important for all clutches and since hole clearance is necessary for the cover retaining bolts, means must be provided to keep the clutch concentric with the flywheel: dowels perform this duty.

Large driven plates have a tendency to *spin*, i.e. to continue rotating after the clutch pedal is depressed. To limit this trouble, the plate should be made as light as possible. Some manufacturers use a small *disc brake* (also known as a clutch stop) to bring the driven plate to rest. This brake is mounted on the non-rotating part of the release bearing, and when the pedal is depressed approximately two thirds of its travel, the brake contacts a disc on the primary shaft.

Pedal linkage

The external linkage between the pedal and clutch may be either *mechanical* or *hydraulic*.

Fig. 35.7 shows two systems which are used. The mechanical system has an adjuster to enable the correct free pedal movement to be maintained. If the clearance is non-existent, wear on the release bearing and clutch slip will result. (This effect can also be produced if the driver rests his foot on the pedal.) When the clearance is too great, the pedal will reach the end of its travel before the clutch is disengaged, and this 'drag' will make it difficult to engage gear.

The mechanical system shown in Fig. 35.2 has a pedal mounted on the frame of the vehicle, and a clutch cross-shaft on the engine. Rubber-mounted engine bearers allow the engine to move a considerable amount in all directions. If the clutch pedal is partly depressed, any fore-and-aft movement of the engine will adversely affect the smooth operation of the clutch, and 'judder' will result. To prevent this, some manufacturers employ a tie-rod to limit the engine movement, but this arrangement defeats the object of using flexible engine mountings.

You will see that engine movement has no effect on hydraulic and cable systems.

Automatic clutch adjustment During vehicle operation the type of mechanical linkage shown in Fig. 35.7 causes the *free pedal movement* to increase when the cable stretches; this extension is quite common when a cable is used to operate a mech-

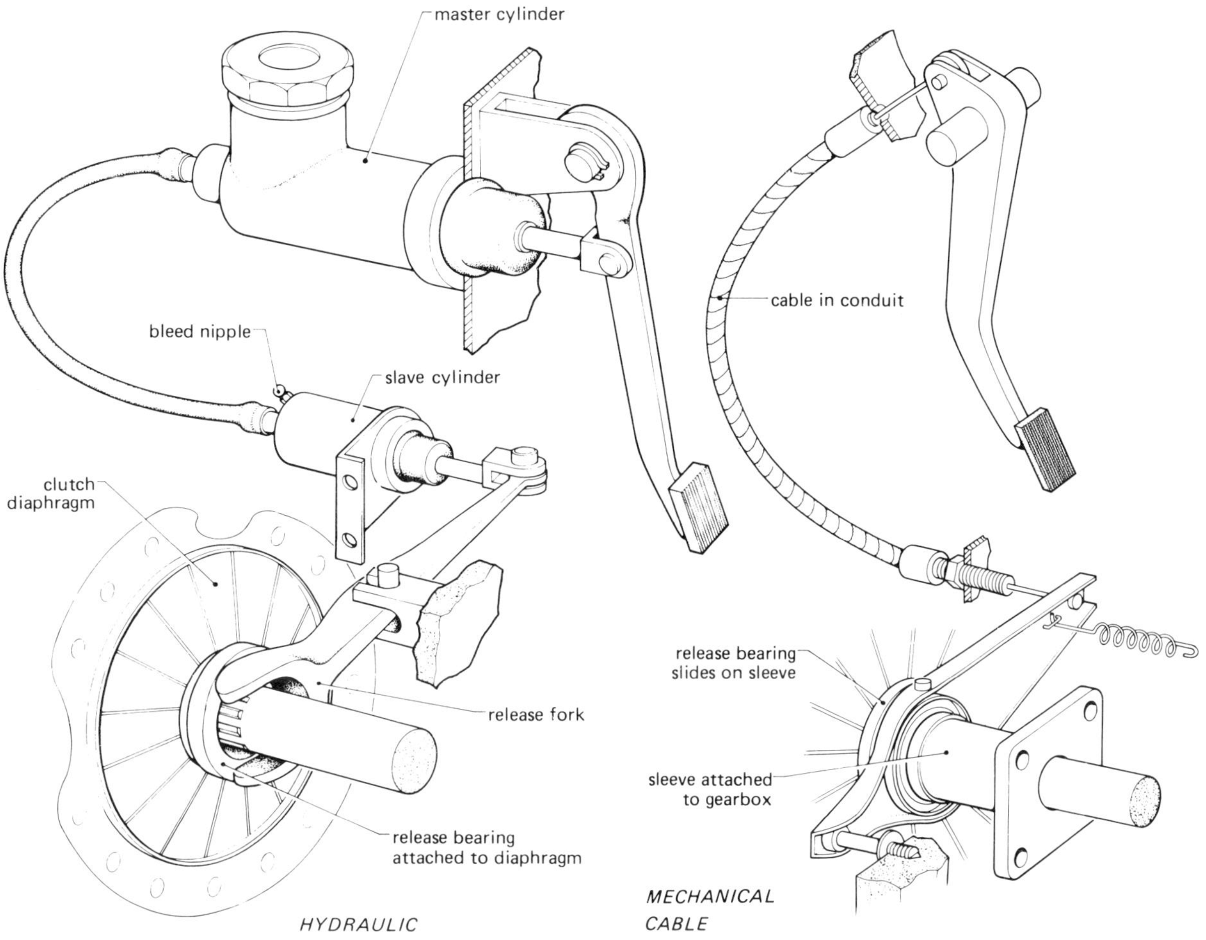

Fig. 35.7 Clutch operating systems

anism. Periodic adjustment must be made to allow for this to avoid *clutch drag*; this results when the free pedal movement is excessive. Some manufacturers consider this task is undesirable, so in these cases an automatic adjuster similar to that shown in Fig. 35.8 is provided.

The type shown uses a conventional cable arrangement from the clutch to the point where the cable normally connects to the clutch pedal. At this point the cable is attached to a ratchet; this is moved to operate the clutch by a pawl connected to the end of the pedal.

After the automatic take-up of the adjuster that follows the replacement of a cable, a downward movement of the pedal causes the pawl to lock into the ratchet and pull the cable.

When wear causes the slack movement to increase to more than the pitch of one ratchet tooth, the spring partially rotates the ratchet and causes the pawl to jump a tooth; this takes up the cable slack and resets the free pedal movement to the required amount.

Summary of clutch faults

There are four main clutch faults; these are:

- *Slip*
- *Drag* or *spin*
- *Judder*
- *Fierceness*.

Slip

Symptoms:

1. Engine accelerates without giving a corresponding increase in vehicle speed.
2. Smell of burning asbestos.
3. In very bad cases, vehicle will not drive up steep hills.

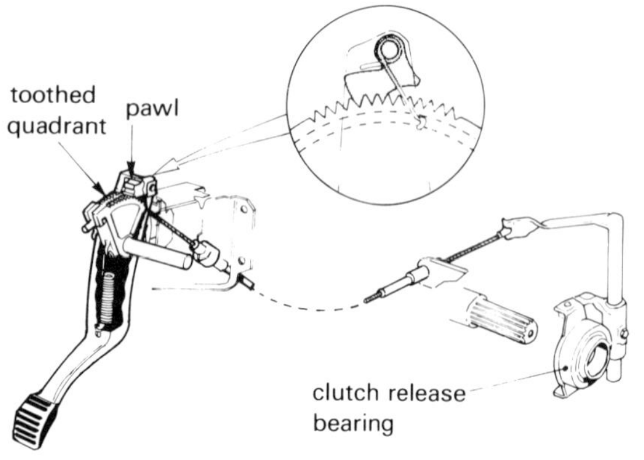

Fig. 35.8 Automatic clutch adjustment

Possible causes:

Any defect that leads to insufficient:
- pressure on clutch plate
- friction.

Such defects include:
1 No free pedal movement.
2 Partial seizure of linkage (including slave cylinder) or fouling of the pedal so as to prevent the full return of the linkage.
3 Worn clutch facings.
4 Oil-contaminated facings. (N.B. Source of oil leakage must be identified.)
5 Defective pressure plate: broken spring(s), partially seized mechanism.

Spin or drag

(In the past *spin* was a design fault associated with the weight of the driven plate and *drag* was due to a defect that could develop during the clutch life. Since the symptoms are similar, both faults are often grouped together.)

Symptoms:
1 Difficulty in engaging first gear from neutral without making a noise.
2 Jolt when changing gear with the vehicle in motion.
3 In very bad cases, vehicle will continue to move even when the pedal is fully depressed.

Possible causes:
Any defect that prevents the clean separation of the driven plate from both driving surfaces. Such defects include:
1 Excessive free pedal movement caused by the need for adjustment or lack of fluid in the hydraulic operating system.
2 Fouling or wear of the linkage such that the pedal cannot be fully depressed.
3 Oil-contamined facings causing one face to stick to a driving surface.
4 Defective driven plate, buckled plate or seizure on splines. (Sometimes a modification is recommended which involves the drilling of a hole in each rivet to avoid a 'vacuum pocket'.)
5 Misalignment of friction faces.
6 Partially seized spigot bearing.

Judder

Symptom:
A severe low-frequency vibration which normally occurs when the clutch is released and the vehicle is moved from the stationary position.
Possible causes:
Any defect that prevents the smooth engagement of the clutch. Such defects include:
1 Jerky pedal operation: sticking cable or slip/grip movement of a pivot.
2 Engine movement during the take-up of the drive. Often this is caused by a defective tie rod: this is fitted to resist engine or transmission system movement. Wear in the drive line or suspension points can also cause this fault.
3 Oil-contaminated facings such that the slip/grip take-up action is amplified.
4 Defective pressure plate. Often this is due to high friction in regions where the plate should move smoothly through the cover.
5 Misalignment of friction faces.
6 Defective driven plate: broken springs in the hub or a partially sheared plate.

Fierceness

Symptom:
A sudden move-off of the vehicle even though the pedal is gradually released.

Possible causes:
Any defect that allows the clutch take-up to be faster than the movement of the pedal. Such defects include:
1 Partial seizure of the pedal linkage or contamination of fluid in the hydraulic operating system. These defects produce an erratic movement of the pressure plate.
2 Oil-contaminated driven plate.
3 Misalignment of friction faces.

36 The multi-plate clutch

The torque transmitted by a plate-type clutch depends on four factors; these are shown in the following formula which is used to calculate the torque capacity:

$t = sp\mu r$ where t = torque transmitted (N m)
s = number of friction contacts
p = total spring thrust (N)
μ = coefficient of friction
r = mean radius (m)

There are cases where the spring thrust, friction or clutch radius must be restricted, so in these cases the number of plates has to be increased to ensure that the maximum torque can be transmitted without slip. A clutch having more than one driven plate is called a *multi-plate* clutch.

Although this type of clutch was widely used on cars up to about 1930, the numerous advantages of the single-plate clutch, especially its ability to disengage the drive completely (its resistance to drag) has meant that a multi-plate unit is now rarely used as a main transmission clutch between the engine and gearbox.

A modern use of a multi-plate type is in automatic gearboxes. This type of gearbox needs a number of clutches to hold the various gear elements, and since the clutch diameter in this application is limited, a multi-plate clutch is commonly used.

Fig. 36.1 shows the main constructional details and represents a main clutch of the type used on early motor cars. Bolted to the flywheel is a cover which engages, by means of slots, with a series of lugs on the outer plates. These plates, which may be plain steel or fitted with cork or friction material inserts, act on inner plates, which are splined to a hub. Thrust springs push the plates together to form a drive.

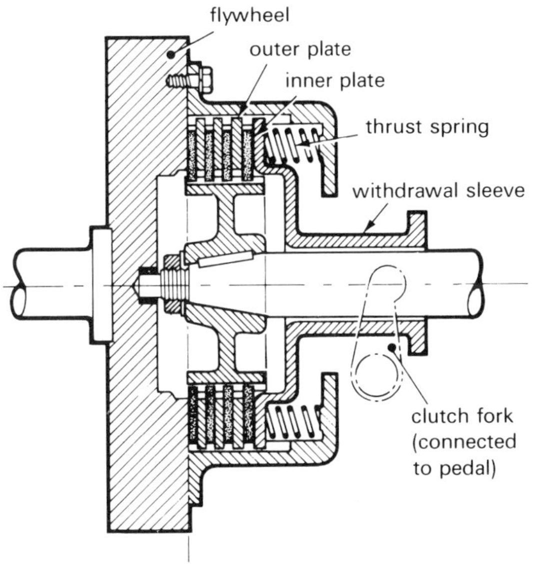

Fig. 36.1 Multi-plate clutch

To disengage the clutch, the end plate is withdrawn to compress the springs and release the other plates. It is difficult to ensure that all plates disengage, and to remedy this defect the plates are either dished or fitted with small springs to push the plates apart.

The clutch may be either *wet* or *dry*; the advantage of using the former is that it reduces the fierce engagement by allowing the engine oil to flow into the clutch housing.

The clutch employed in automatic gearboxes is generally of the wet type, and is operated by a piston controlled by hydraulic pressure. Sintered bronze plates, manufactured by partially fusing powdered bronze, or compressed paper, are used in many designs. The porous surface obtained traps the oil, to give long life and smooth operation. Other applications of multi-plate clutches are shown on pages 260 and 302.

PART ELEVEN

Gearboxes

37 The gearbox

Reason for a gearbox

The internal-combustion engine used in a modern vehicle will only operate over a limited effective speed range, e.g. 1500–5000 rev/min, and in this range will produce a comparatively low torque (turning effort). When the speed drops below the lower limit, or if the load is too great, the engine will stall and the vehicle will come to rest.

If a vehicle were not fitted with a gearbox, the following disadvantages would soon show up:

1 *Poor acceleration from rest* The clutch would have to be slipped for a considerable time to avoid stalling the engine. A road speed of about 24 km/h (15 mile/h) would have to be reached before full engagement could take place, and during this time the *driving force at the wheels* (tractive effort) would only be slightly greater than the *force opposing the motion of the vehicle* (tractive resistance). The acceleration is governed by the difference between the tractive effort and resistance, and if this difference is small, the acceleration will be poor.

2 *Poor hill-climbing ability* A gradient increases the resistance, and this will mean that as soon as a hill is approached, the engine will slow down and eventually stall. This could be overcome by employing a large engine with high torque output, but it would be uneconomic (Fig. 37.1).

3 *Vehicle cannot be driven at low speeds* As the vehicle speed is decreased the engine speed will also be decreased. Slipping of the clutch would be necessary to avoid stalling if the vehicle had to be driven at low speeds.

4 *No neutral or reverse.*

Fig. 37.1 Diagram showing how human effort can be reduced by using a lever. The gearbox acts as a mechanical lever to boost the effort of the engine

These items show the importance of fitting a gearbox.

Gear leverage

Fig. 37.2a shows a simple lever. A force applied at the end of the lever lifts a load four times as great; this shows that a small force can be amplified by using a lever system.

Fig. 37.2b shows how two discs may be used to obtain leverage. In this example a mass acting on shaft C will support a larger mass on shaft D. This arrangement may be considered as a simple gearbox, the engine connected to shaft C, and the road wheels to shaft D. In this example, the output torque is double the engine torque, and if disc B is made three times the diameter of A, the output torque will be trebled. This appears to produce something for nothing, but speed must be taken into account. It will be seen that as the torque increases, the speed decreases proportionally, and therefore the *power remains the same*, assuming the mechanism is 100 per cent

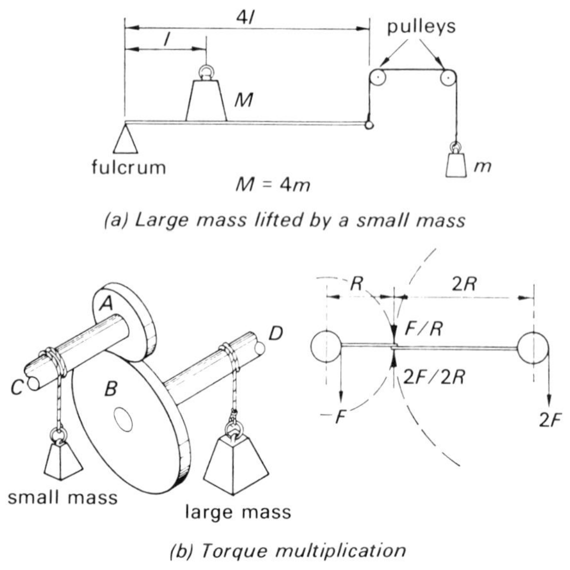

Fig. 37.2 Gear leverage

Table 37.1 Engine requirements

Operating condition	*Requirement*
Maximum vehicle speed	Maximum engine power
Maximum acceleration	Maximum engine torque
Maximum traction	Maximum engine torque
Maximum economy	Engine at mid-range speed and under light load with a small throttle opening

efficient. In Fig. 37.2b the speed ratio (movement ratio) is also called the gear ratio, and in this case is 2 : 1, which indicates that two revolutions of the input shaft are required to rotate the output shaft by one revolution.

Belts, pulleys and friction drives were used on early designs of motor car, but when Panhard introduced a sliding gearwheel arrangement, these systems gradually disappeared.

Determination of gear ratios

Achievement of a high maximum vehicle speed, combined with good acceleration and economy over the whole speed range, requires a system of gearing that allows the engine to operate at the speeds at which it develops its best performance. Maximum engine power, torque and economy all occur at different engine speeds, so this makes the task of matching the gear ratio difficult, especially when variable operating conditions and driver demands have to be taken into account.

Setting of the numerous gear ratios takes into account the engine requirement to suit a given operating condition (Table 37.1).

Today the type of engine fitted to a light vehicle generally requires a gearbox that gives four forward speeds and a reverse, i.e. a set of four different forward gear ratios in addition to reverse gear. This gives a reasonable performance to suit most driving conditions other than economy; normally this needs an extra ratio, a fifth gear, that is higher than the conventional 'top gear'. It should be noted that a *high gear* is one that has a ratio with a low numerical value, i.e. a ratio of 1 : 1 is higher than the ratio of 2 : 1. The lower the gear, the greater is the reduction between the engine and road wheels; this means that, for a given engine speed, the road wheel speed is lower.

Maximum vehicle speed This is achieved when the gear is set 'in top' and the throttle is held fully open.

To minimize friction losses in the gear that is used for most of the time, a ratio of 1:1 (direct drive) is chosen for 'top gear'. Consequently, the setting of 'top gear' is really the choosing of a final drive ratio to suit the diameter of road wheel and engine characteristic.

Fig. 37.3 shows the considerations that have to be made to ensure that a vehicle can attain a high maximum speed. It shows the balance between the *power required* and the *power available*. Data for the former are given by the brake power curve of the engine; for the latter they are obtained from calculation of the power needed to overcome the tractive resistance of the vehicle when it is moving along a level road.

The tractive resistance, sometimes called total resistance, includes:

1. *Air resistance* - due to movement of the vehicle through the air
2. *Rolling resistance* - due to friction between the tyre and road. Largely influenced by the type of road surface
3. *Gradient resistance* - increases as the steepness of the incline (hill) acts against the vehicle motion.

Fig. 37.3a indicates that the power needed to propel a given vehicle increases with the cube of the speed. This means that if the speed is doubled, the power required is eight times as great. In this example a power of 150 kW is needed to drive the vehicle at 200 km/h.

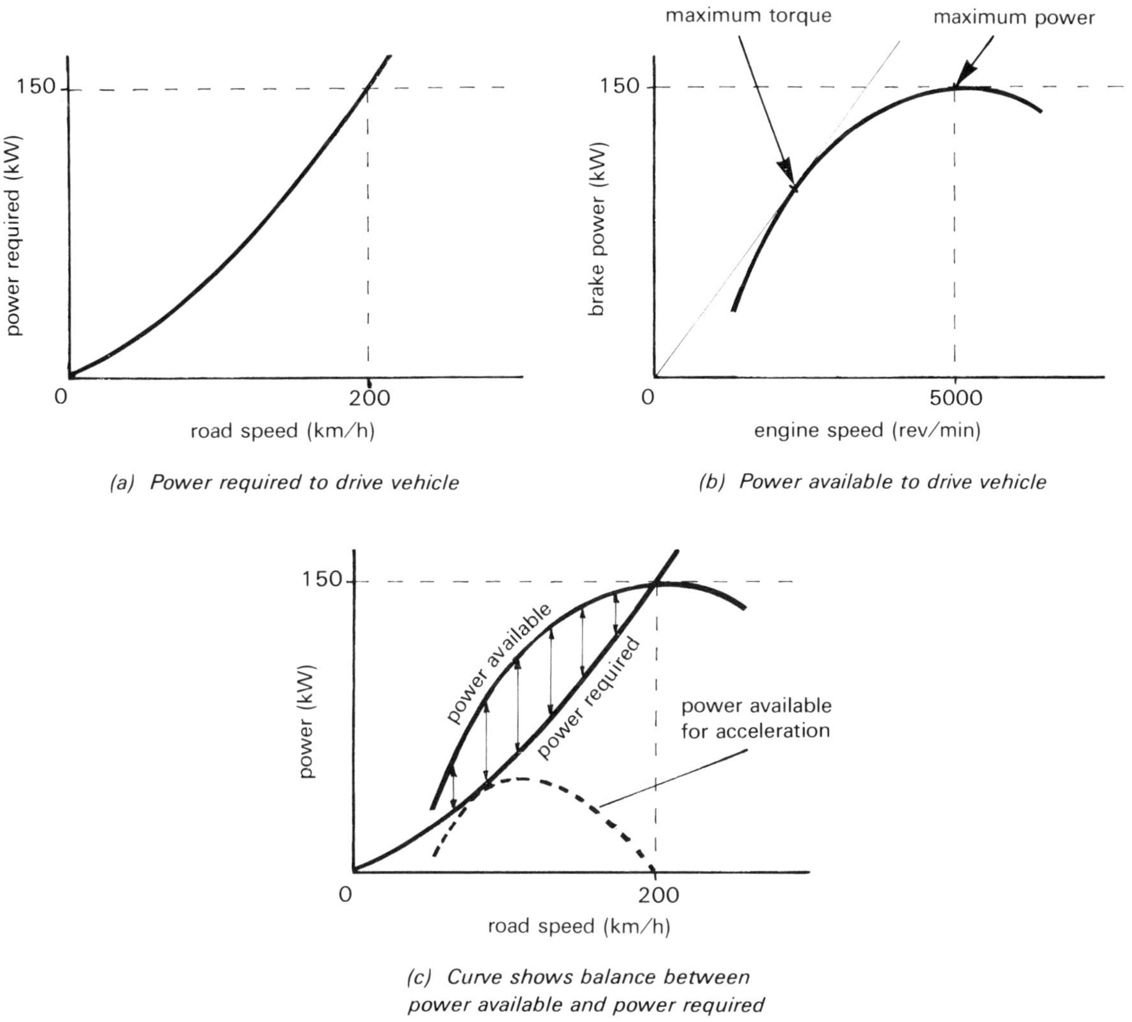

Fig. 37.3 Power balance

The power output of the engine fitted to this vehicle is shown in Fig. 37.3b. This power curve shows that the engine produces a peak brake power of 150 kW at 5000 rev/min.

If the maximum road speed is to be as high as possible, the gear ratio of this vehicle must be set so that the peak of the 'power available' curve occurs at a road speed of 200 km/h. In this case the

Overall gear ratio (Gearbox ratio × Final drive ratio)

must ensure that an engine speed of 5000 rev/min drives the vehicle at 200 km/h.

Once the relative positions of the two curves have been decided, it is then possible to examine the general performance with respect to acceleration. The vertical difference between the two curves is the surplus power available for acceleration, so this can be plotted as a separate curve to show the speed at which maximum acceleration is achieved.

Assuming friction is neglected, it must be appreciated that a gearing system neither increases nor decreases power, i.e. the power output from a transmission system is similar to the engine brake power irrespective of the gear ratio.

In view of this, a change in the gear ratio of the vehicle shown by the curves in Fig. 37.4 will cause the peak P to move horizontally from the position it occupied in Fig. 37.3c. Lowering the ratio (curve A) moves the 'power available' curve to the left and raising the ratio (curve C) moves it to the right. These two conditions are called *undergear* and *overgear* respectively.

In both conditions the maximum possible speed is

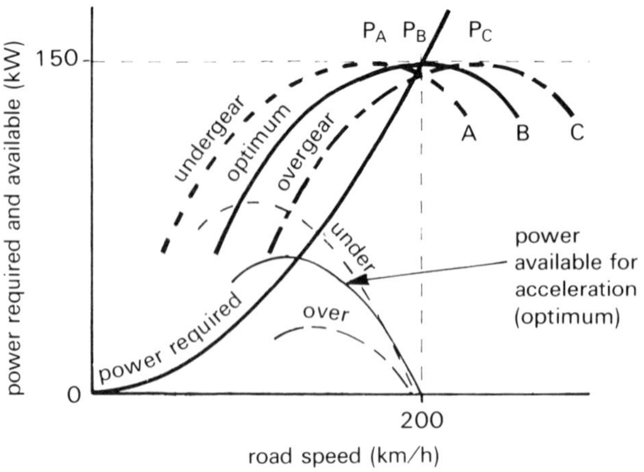

Fig. 37.4 Undergear and overgear

reduced but this is not the prime consideration. When compared with the optimum gearing needed to obtain the ideal maximum speed, the advantages of the two gearing conditions are:

Condition	*Advantages*
Undergear	1 More power for acceleration, so vehicle is more lively 2 Flexible top gear performance, so less gear changing is necessary when the vehicle encounters higher tractive resistances.
Overgear	Lower engine speed for a given road speed, so: 1 Better economy 2 Lower engine noise level 3 Less engine wear.

The advantages of undergearing can be used to obtain the disadvantages of overgearing and vice versa. A comparison of these two conditions shows that undergearing is more suitable for the average car, so undergearing to the extent of about 10–20 per cent is common. This means that the engine power peak occurs in 10–20 per cent before the maximum possible vehicle speed is reached.

Maximum traction Once the designer has set the overall top gear ratio, the bottom gear ratio (1st gear) is then decided. This gear is used when moving off and is also needed when maximum tractive effort is required so as to enable the vehicle to climb very steep hills.

Tractive effort is based on engine torque, so maximum tractive effort in a particular gear occurs when the engine is developing its maximum torque. In Fig. 37.5a, the top gear performance, which was previously plotted as a difference in power, is shown as a balance of forces. It will be seen that the driving force curve is similar in shape to the engine torque curve; the peak of the tractive effort curve occurs at a road speed controlled by the overall gear ratio and effective diameter of the road wheel. The difference between the effort and resistance curves represents the force available for acceleration.

Fig. 37.5b shows the effect on the tractive effort curve of lowering the gear ratio. In this case a bottom gearbox ratio of 4 : 1 is used to amplify the tractive effort sufficiently to meet the hill-climbing requirement.

In Fig. 37.5b the gradual engagement action of the clutch must be used to provide a tractive effort build-

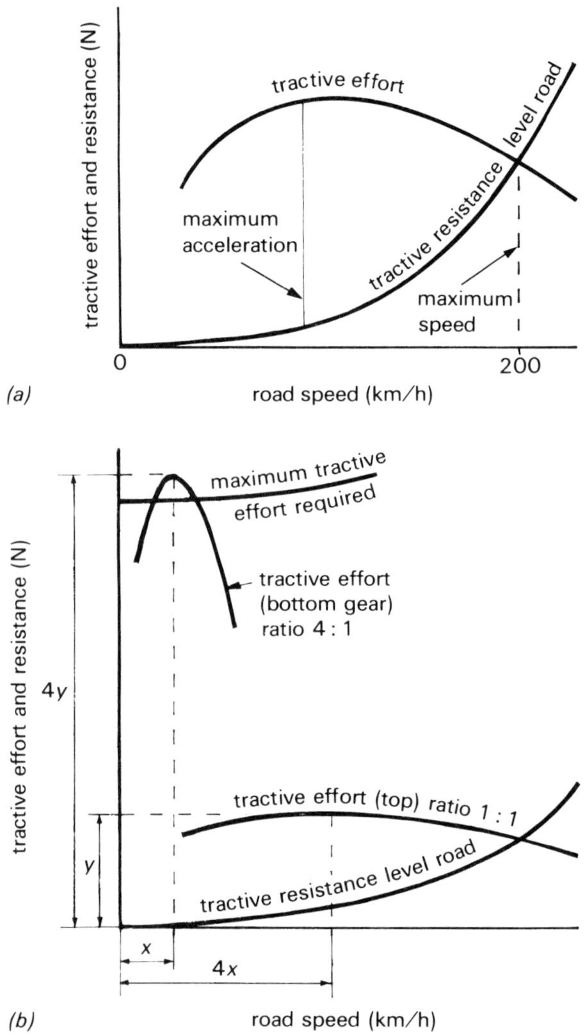

Fig. 37.5 Tractive effort curves

up sufficient to move off the vehicle on the steep hill represented by the graph. Once the clutch is fully engaged, and the engine is operating in the region of maximum torque, a small acceleration is possible – this assumes the engine speed does not drop too low.

The bottom gearbox ratio is calculated by dividing the maximum effort required by the maximum effort available in top gear.

Intermediate gear(s) Having set the top and bottom gearbox ratios, the intermediate ratios are then inserted so that they form a *geometric progression (GP)*. This means that all the individual ratios advance by a common ratio. For example, assuming the top and bottom overall ratios are 4 : 1 and 16 : 1 respectively, then the sets of ratios for the 3- and 4-speed gearbox are:

3-speed gearbox: 4, 8 and 16 (common ratio 2).
4-speed gearbox: 4, 6.35, 10 and 16 (common ratio 1.59)

To obtain optimum speed and acceleration performance, the engine should be operated in the speed range between the limits of maximum torque and maximum power. The wider this operating range, the smaller is the number of ratios required to bridge the gap between top and bottom. Most modern car engines have a narrow range, so gearboxes fitted in conjunction with these engines normally have at least four forward ratios.

Since most cars are undergeared, it is now common practice to have an extra gear, called a fifth gear, to offset some of the drawbacks associated with the undergear condition. Normally, this gear is an *overdrive*, because it has a ratio that steps up the speed; as a result it drives the output shaft faster than the engine.

Types of gearing (Fig. 37.6)

Various types of gearing are used on a motor vehicle, but gearboxes employ one or more of the following:

1 *Spur* – teeth parallel to axis, used on sliding mesh.
2 *Helical* – teeth inclined to axis to form helix.
3 *Double helical* – two sets of opposing helical teeth.
4 *Epicyclic or planetary* – spur or helical gears rotating about centres which are not stationary.

Types 2 and 3 are used on constant-mesh and synchro-mesh. Most automatic gearboxes use epicyclic gearing.

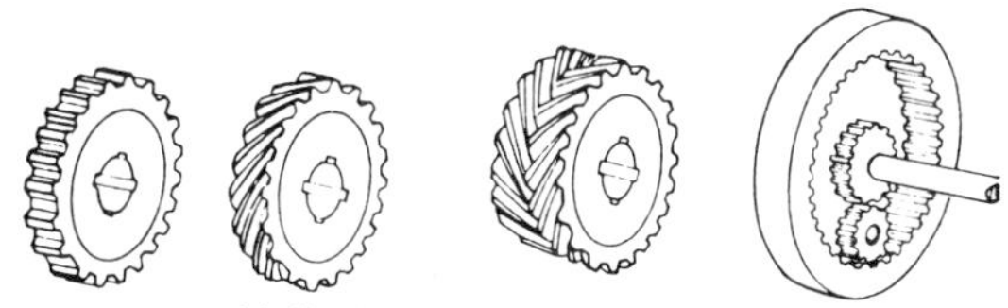

Fig. 37.6 Types of gearing

Gear material

The gear teeth have to resist severe shock loading and wear, so a case-hardened steel is used to provide a tough core and a hard surface.

Types of gearbox

Gearboxes today may be divided into two main groups: manual-change and automatic.

Manual-change

These gearboxes require the driver to have full control of the gearbox. By means of the manual control lever, the appropriate gear ratio can be selected to suit the driving conditions.

In addition to reverse, the number of 'speeds' (gear ratios) provided in a modern gearbox is 4 or 5, whereas in the past a 3-speed gearbox was commonly used for cost reasons.

The main types of manual-change gearbox are:

- *Sliding-mesh*
- *Constant-mesh*
- *Synchro-mesh*.

Today the synchro-mesh is the most common type in use. The layout of this type has been developed from the other two, and it is for this reason that the now obsolete sliding mesh type is used for initial study.

Automatic

This type covers the many systems in which the gear ratios are altered by the gearbox itself instead of relying on the driver. Once the driver has made the initial selection to determine the direction in which the vehicle is to move, and the gear range to be used, then the other decisions are made by the 'brain' within the gearbox.

In the USA a gearbox is called a *transmission*, so this explains why an automatic gearbox is sometimes called an automatic transmission unit.

Today most automatic gearboxes use an epicyclic gear system. The required gear is obtained by holding or driving a part, or parts, of the gear train by means of a friction clutch or brake. The brakes and clutches are controlled by a hydraulic system; this either incorporates its own sensing system, or uses electronics to monitor the engine and vehicle operating conditions.

In addition to the 3- or 4-speed epicyclic gearbox, most automatic systems require the fitment of a torque converter between the engine and gearbox. This is a fluid clutch that replaces the normal friction clutch and fulfils two duties; it automatically disengages the engine from the transmission when the engine speed is below about 1000 rev/min, and also provides an infinitely variable torque and speed ratio to bridge the steps between the gearbox epicyclic ratios.

In the UK the combination of a torque converter and automatic gearbox forms an *automatic transmission system*.

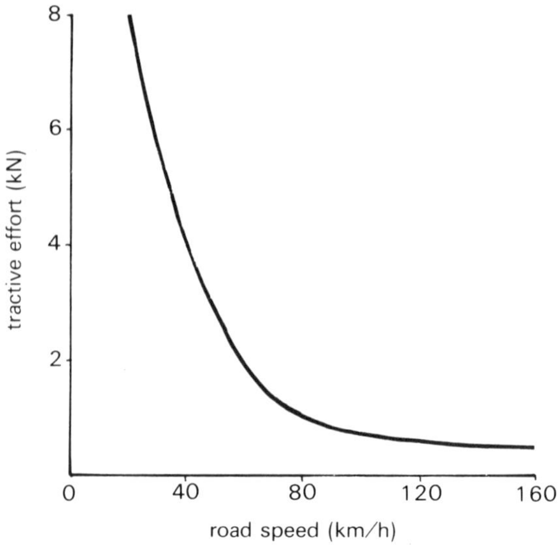

Fig. 37.7 Ideal tractive effort curve

Continuously variable transmission (CVT)

The power output of a normal engine varies with the engine speed. At slow speeds the output is very low, so if good vehicle performance is required, the engine would have to rotate much faster; it would need to operate at a speed around that at which it develops its maximum power.

This story is repeated when the torque output and fuel economy are considered. Maximum torque occurs at a different speed from that for maximum economy and neither of these speeds coincides with the point of maximum power.

This constant engine speed requirement for the achievement of any one of the three performance factors is not possible with a conventional gearbox; this is because the engine speed has to be continually changed to suit the vehicle speed. As a result, the engine only gives its best at the vehicle speed appropriate to the point of maximum engine torque, power or economy.

A gearing system having an infinitely variable ratio would give a performance as shown in Fig. 37.7; this is called an *ideal tractive effort curve*. In this case the engine is kept at the speed at which it gives its maximum power and the road speed is altered by varying the gear ratio.

An approximation to the ideal curve can be obtained from any system that operates according to the principle of the system shown in Fig. 37.8. This particular arrangement is not very practical, but the

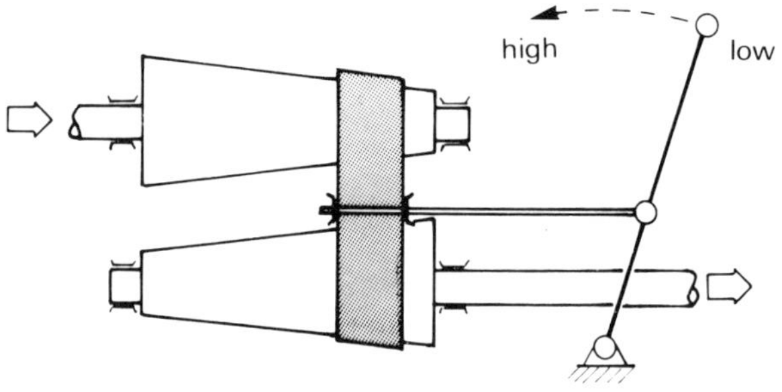

Fig. 37.8 Simple CVT system

layout shows the basic idea behind this method of gearing.

Over the years many patents for a CVT system have been registered, but very few have progressed past the initial design stages. One exception is the *belt and pulley* layout, which is called the *Variomatic CVT*. system (Fig. 37.9). This was pioneered by the Dutch manufacturer DAF and was fitted as standard to the small cars they produced in the late 1950s.

Stepped transmission

A comparison of the traditional gearbox curves with the ideal curve is shown in Fig. 37.10. This graph indicates why the term *stepped transmission* is used to describe any system that gives this kind of stepped output.

Gearboxes having only three forward gears need the engine to operate over a very wide speed range;

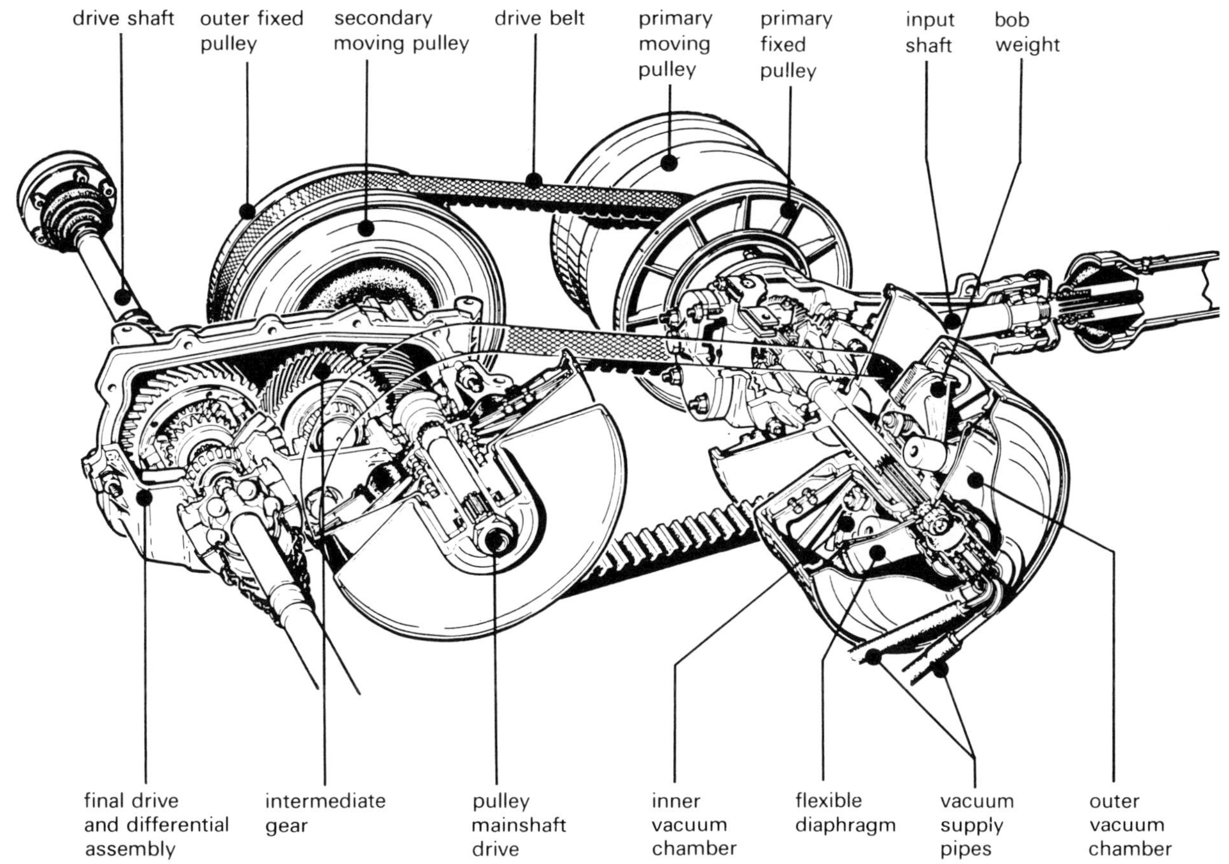

Fig. 37.9 Variomatic CVT system

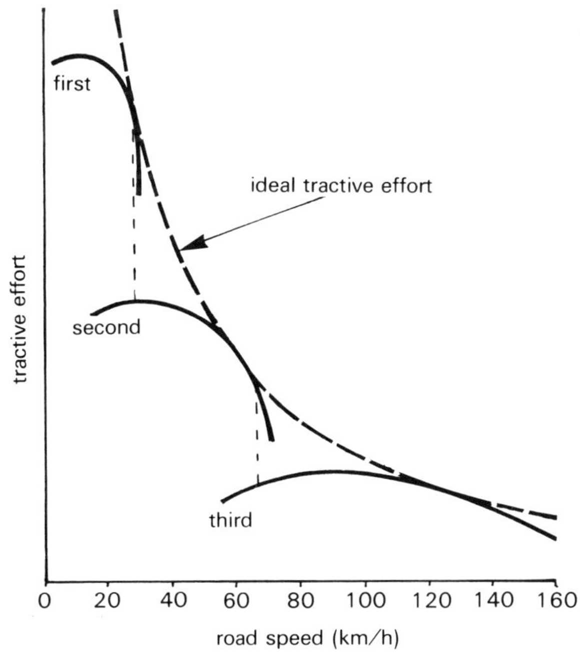

Fig. 37.10 Conventional stepped transmission

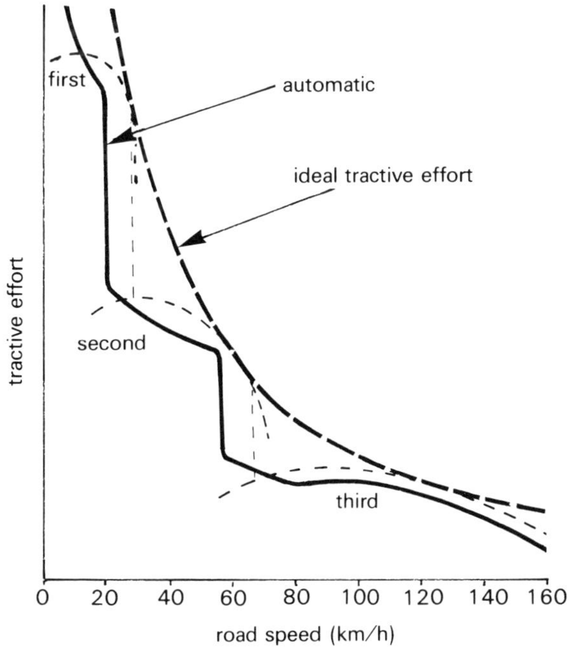

Fig. 37.11 Tractive effort with three-speed automatic gearbox

this means that the tractive effort at many road speeds is far less than the ideal. Increasing the number of ratios gives a considerable improvement, and although this is impracticable with a normal car it is used on many heavy vehicles. When the number of ratios is increased to about ten, the combination of the narrow engine range needed to 'bridge the gap' between the ratios, and the close approximation of the actual tractive curve to the ideal, allows the engine to be operated around its best speed.

Since the number of ratios in a car gearbox is restricted for reasons of weight and cost, an approach to the ideal tractive effort curve is achieved by using an automatic gearbox with a torque converter. Although the converter is very inefficient at low engine speeds, the effect of the converter's infinitely variable ratio between the limits of about 2 : 1 to 1 : 1 gives a smooth transition, and positive drive, between the stepped ratios of the main automatic gearbox (Fig. 37.11).

38 The sliding-mesh gearbox

The sliding-mesh gearbox was popular on cars up to about 1930, but today it is rarely used. Nevertheless the reader is advised to study this type because many modern gearboxes have been developed from its configuration.

The basic layout of a 4-speed and reverse gearbox is shown in Fig. 38.1. The various spur-type gears are mounted on three shafts:

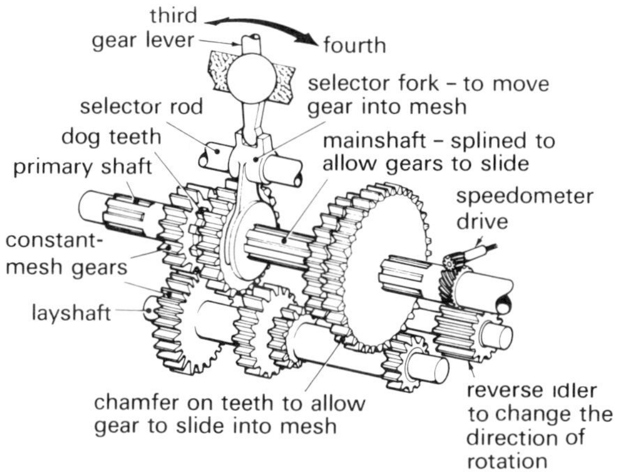

Fig. 38.1 Four-speed and reverse sliding-mesh gearbox

- *Primary shaft* (alternative names - clutch or first motion shaft)
- *Layshaft* (countershaft)
- *Mainshaft* (third motion shaft).

Primary shaft

This shaft transmits the drive from the clutch to the gearbox. At the engine end, the shaft is supported by a spigot bearing positioned close to the splines (or serrations) on to which the clutch driven plate is connected. The main load on this shaft is taken by a bearing; normally a sealed radial ball type, positioned close to an input gear called a *constant mesh pinion*. The gear is so named because it is always in mesh with a larger gear, a *constant mesh wheel*, that is part of the *layshaft gear cluster*. Note that a small driving gear is called a *pinion* and a large gear a *wheel*; the term *cog* is not used.

Layshaft

This shaft, which is normally fixed to the gearbox casing, supports the various-sized driving pinions of the layshaft gear cluster. Bearings placed between the shaft and gear cluster resist the radial load set up when the gears are forced apart. The type of bearing depends on the load it has to withstand; for a comparatively light load, a phosphur-bronze plain bush is suitable, but heavy-duty applications call for a more efficient bearing such as a caged needle roller. Gear noise occurs when a high driving torque causes the gears to move apart. To keep the shaft distance constant, a more rigid bearing is used, e.g. an uncaged needle roller bearing. In most cases the end float of the layshaft gear assembly, and in consequence the alignment of the gears, is controlled by flat thrust washers situated between the gear cluster and the casing.

Mainshaft

This splined output shaft carries spur gearwheels that slide along the shaft to engage with the appropriate layshaft gears. At the 'front' end, the mainshaft is supported by a spigot bearing situated in the centre of the constant mesh pinion. A heavy duty radial ball bearing is fitted at the other end to take the force of the gears as they attempt to move apart. The load on this bearing is at a maximum at the time when first gear is engaged and high torque is being produced.

Gear selector

The selector is a fork, mounted on a rod, that fits into a groove in the gear. The function of the fork is to slide the gearwheel along the mainshaft so that the gear can engage and disengage with the matching gear pinion mounted on the layshaft.

Striking mechanism is an alternative name for this arrangement; in a similar way, the rod is sometimes called a *rail*.

A 4-speed gearbox needs three selector forks; these slide on, or move with, three separate rods. Normally one fork controls first and second gears, and the second fork, the third and top gears. Reverse has a separate fork, longer than the other two, because it has to fit around the lower-positioned reverse gear.

A gear lever transfers the force and movement initiated by the driver to the appropriate selector fork.

Gear positions (Fig. 38.2)

Neutral

All mainshaft gearwheels are positioned so that they do not touch the layshaft gears. A drive is taken to the layshaft, but the mainshaft will not be turned in the neutral position.

First gear

The first-speed gearwheel A on the mainshaft is slid backwards to engage with pinion B on the layshaft; all other gears are positioned in neutral. In this gear, the reduction in speed that occurs as the drive passes through the constant-mesh gears, E and F, is reduced further by the first-speed gears, A and B.

The gear ratio (also called the movement ratio or velocity ratio) is given by

$$\text{Ratio} = \frac{\text{Number of teeth on driven wheel}}{\text{Number of teeth on driving wheel}}$$

$$= \frac{\text{Driven}}{\text{Driver}}$$

Since two sets of gears are used in a gearbox, then:

$$\text{Ratio} = (\text{Driven/Driver}) \times (\text{Driven/Driver})$$

N.B. THE RATIOS ARE MULTIPLIED TOGETHER.

$$\text{First gear ratio} = F/E \times A/B = 40/20 \times 40/20$$
$$= 4:1$$

The value 4 indicates that the engine will turn four times as fast as the propeller shaft. In this case the torque output from the gearbox will be four times as great as that applied at the input; this assumes that there is no energy loss to friction and oil pumping.

Second gear

The second-speed gearwheel C is slid forward to engage with the layshaft gear D; all other gears are set in the non-driving position.

$$\text{Ratio} = (\text{Driven/Driver}) \times (\text{Driven/Driver})$$

$$\begin{aligned}\text{Second gear ratio} &= \text{F/E} \times \text{C/D}\\ &= 40/20 \times 35/25\\ &= 2 \times 1.4\\ &= 2.8:1\end{aligned}$$

When the same size of gear tooth is used for all the gears in a gearbox, the total number of teeth on the various sets of gears will be the same. In this case, the total number is 60.

Third gear

In this gear position, gearwheel G is slid in to mesh with gear H. Since the size of the layshaft gear H is larger than the gear D that was used to give second, a slightly higher ratio than second is obtained.

Sometimes it is necessary when diagnosing a particular fault, to consider the path taken by the drive as it passes through the gearbox. This path is called the *power flow*. In the case of third gear the flow path is:

primary shaft; gears E, F, H, and G; and mainshaft

Top gear

In this layout, fourth gear is a direct drive; namely a gear that gives a ratio of 1:1. It is obtained by sliding gear G to engage its dog teeth with the corresponding teeth formed on the end of the constant mesh pinion E. Engagement of the dog clutch locks the primary to the main shaft and this gives a 'straight-through' drive. In this gear the power flow is not transmitted through any gear teeth so the energy loss is small, i.e. the efficiency is high. Losses would be even smaller if it were possible to reduce the churning effect of oil by the layshaft, but this oil movement is necessary to provide lubrication of the gearbox.

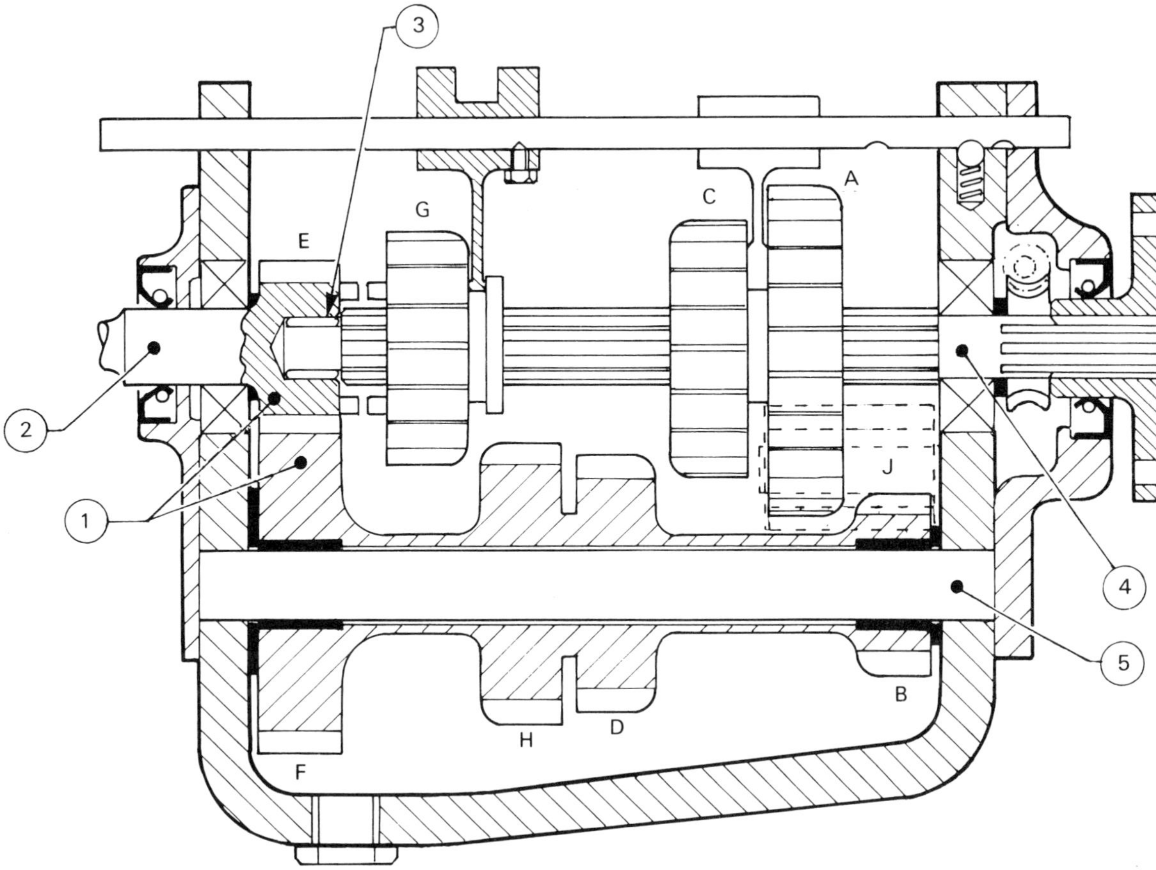

1 constant-mesh gears
2 primary shaft
3 spigot bearing
4 main shaft
5 layshaft

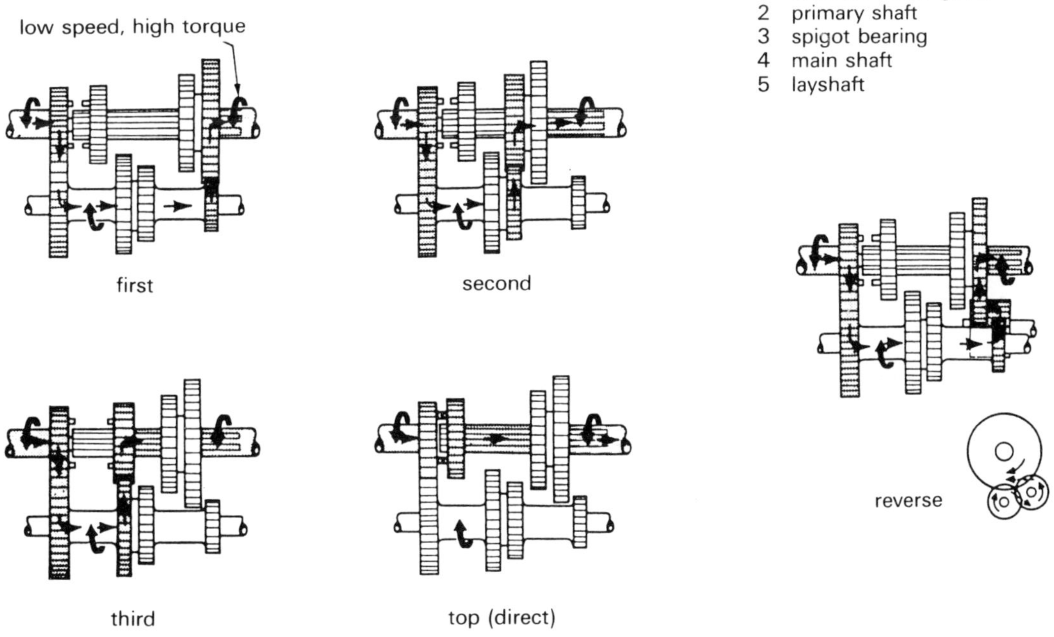

Fig. 38.2 Gear positions (sliding-mesh)

Reverse gear

Sliding a reverse gear between any two gears on the layshaft and mainshaft is the method used to change the direction of rotation of the output shaft.

The simplest arrangement uses a single reverse gear, which is mounted on a short shaft. This shaft is positioned so that the reverse can slide and mesh with the two first-speed gears as shown in Fig. 38.2. The gear ratio is:

$$(\text{Driven/Driver}) \times (\text{Driven/Driver}) \times (\text{Driven/Driver})$$

$$\text{Reverse ratio} = F/E \times J/B \times A/J$$

In this case, J cancels out, thus leaving:

$$\text{Reverse ratio} = F/E = A/B$$

This is the same ratio as for first gear, and irrespective of the size of the reverse gear J it will be seen that the ratio always remains the same. For this reason it is called an *idler* - it changes the direction, but does not alter the ratio.

With the idler arrangement, some drivers persistently slip the clutch to maintain a low reversing speed. Excessive clutch wear resulting from this practice is minimized when the reverse ratio is set lower than first gear. This is achieved by using a reverse gear arrangement as shown in Fig. 38.3. Instead of the single idler, the *compound* reverse gear has two gear pinions joined together. The reverse shaft is positioned so that the reverse pinions are able to mesh simultaneously with the appropriate layshaft and mainshaft gears.

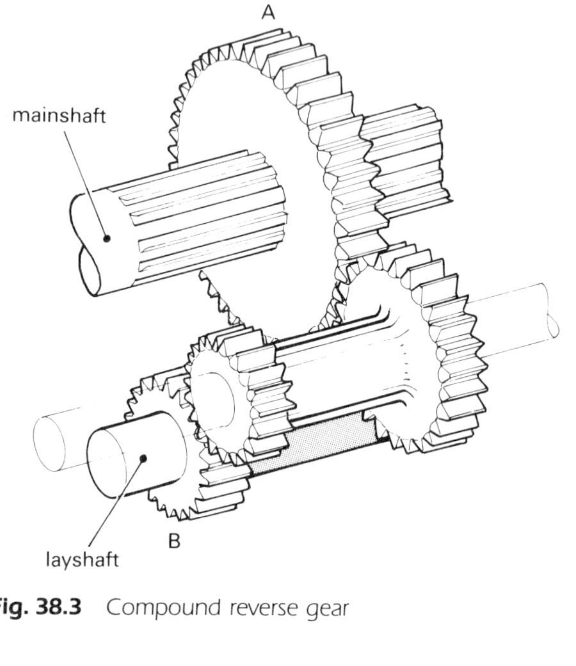

Fig. 38.3 Compound reverse gear

Gear changing

When one gear is moved to engage with another gear, noise will result if the peripheral (outside) speeds are not the same. To avoid this, the driver of a vehicle having a sliding-mesh gearbox performs an operation called *double declutching*.

On changing up, e.g. from first to second, the clutch driven plate, primary shaft and layshaft gear cluster must be slowed down by re-engaging the clutch while the lever is positioned in neutral. The engine slows down more quickly than the layshaft assembly, so a braking action on the layshaft assembly is obtained.

The opposite is required when changing down; the layshaft must be speeded up. To achieve this, the engine is reconnected to the gearbox while the lever is in neutral, and the accelerator pedal is depressed. When the layshaft assembly has speeded up by the required amount, the accelerator is released, the clutch pedal depressed, and the gear lever moved to the lower gear. Finally, as the clutch is being released, the engine must be accelerated sufficiently to match the engine speed to the road speed.

A chamfer on the entry side of each gear tooth is provided to allow for an easier entry of the sliding gear. However, if the driver makes a bad change, in addition to the loud noise created by the gear teeth rubbing against each other, burring of the teeth will occur and in some cases the teeth will chip or break off.

Selector mechanism

A fork of the type shown in Fig. 38.4 is used to slide a gearwheel along the mainshaft in order to select the appropriate gear. It is mounted on its own rod and links the driver's *gear stick* to the sliding gearwheel. Two selector forks are required for a three-speed and reverse gearbox and three forks are needed for a four-speed unit. The gear stick can be a direct-acting lever or it can be mounted remote from the gearbox.

Every gearbox must be fitted with the following:

1 *Selector detent* - holds the gears and selectors in position and so prevents gear engagement or disengagement due to vibration.
2 *Interlock mechanism* - prevents two gears engaging simultaneously; if this occurs the gearbox will lock up and shaft rotation will be impossible.

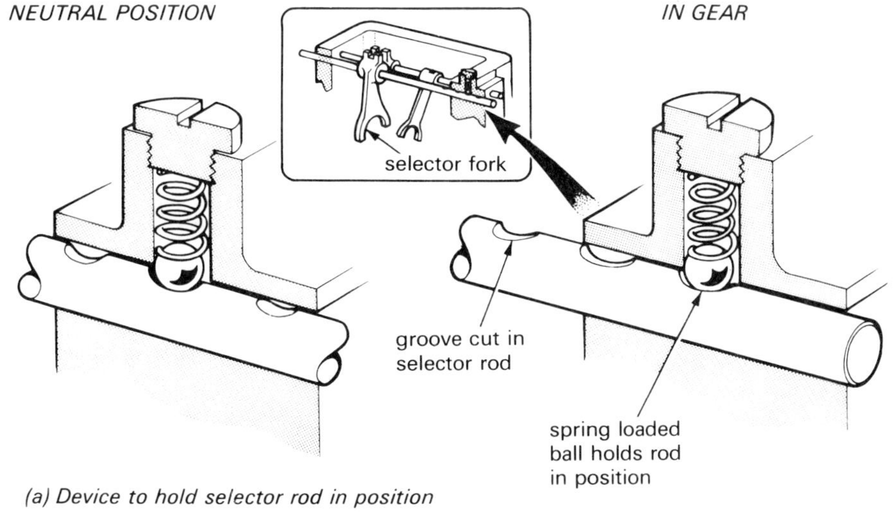

(a) Device to hold selector rod in position

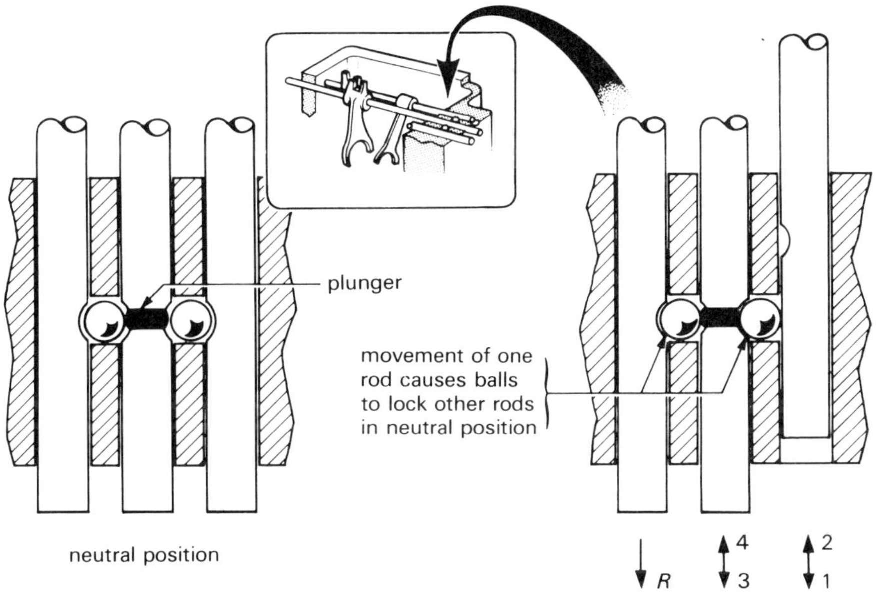

(b) Ball and plunger type of interlocking mechanism prevents two gears engaging at the same time

Fig. 38.4 Selector mechanism

(Severe damage would occur if this device were omitted during a gearbox overhaul.)

Selector detent

Fig. 38.4a shows a typical arrangement suitable for a layout having the selector fork locked to the rod. The device consists of a steel ball that is pressed into a groove in the rod by a spring. In the rod are ground three grooves; these correspond to the rod positions for neutral and for the two gears served by the selector. Retention of the rod is governed by the strength of the spring. In cases where the spring strength is adjustable, it must be set sufficiently strong to prevent the gear from jumping out of mesh, but not so strong that it makes gear changing difficult.

Each selector rod has a separate spring-loaded detent.

Interlock mechanism

Every gearbox must have some form of safety device fitted to prevent two gears, e.g. second and top, engaging at the same time. This mechanism is fitted in addition to the selector detent. Although the interlock device takes a number of different forms, the arrangement shown in Fig. 38.4b is one of the most common.

This system uses a plunger and two balls; the diameter of each ball is greater than the spaces between the rods. When one of the outside rods is moved, the full diameter of the rod presses its adjacent ball into the groove ground in the centre rod. This action retains the centre rod in the neutral position and also pushes the plunger against the second ball to lock the other rod.

From the neutral position, movement of the centre selector rod carries the plunger with it and allows the full shaft diameter of the rod to push both steel balls into the grooves machined in the outer rods.

Study of Fig. 38.4b will show the person who has to dismantle the selector mechanism that the rods cannot be driven out unless they are positioned in the neutral position.

When dismantling a gearbox for repair, it is advisable to ascertain the type of interlock used on the particular gearbox; this is because during cleaning it is easy for either a ball or the plunger to drop out. If this is not noticed when the box is reassembled, the high risk of gearbox seizure, and the effect of this on a moving vehicle, will cause many problems.

Lubrication

In the past a heavy gear oil was used as a lubricant for a sliding-mesh gearbox. If the box were in use today, then an oil similar to that used in a rear axle would be suitable; this assumes that precautions had been taken by the manufacturer to avoid corrosive attack by this oil on non-ferrous materials such as phosphor-bronze.

The gearbox oil level is set at layshaft height. Rotation of the gears on this shaft carries the lubricant around with the gears in a pump-like manner; this simple method ensures that all meshing teeth have an adequate supply of oil. Splash lubrication, set up by the churning action of the oil by the layshaft gears, is the method used to lubricate the other gearbox parts.

Some form of oil sealing arrangement is necessary at the front and back ends of the gearbox; this prevents loss of oil and possible contamination of the clutch friction facings.

Power take-off arrangement

In addition to the mechanism used for driving a vehicle along a road, a power supply is often required for operating various external items of auxiliary equipment.

A light truck having a tipping mechanism is one example, but the most varied application of power take-off units is associated with specialized off-road vehicles. Many agricultural operators regard this type of light vehicle as a transportable power source for driving saws, drills, pumps, etc.

Fig. 38.5 shows a typical power take-off arrangement that is driven from the gearbox layshaft. The extra take-off gear, which is mounted on a flange provided on the side of the gearbox, is slid in to mesh with one of the sliding-mesh gears of the layshaft by a separate selector lever.

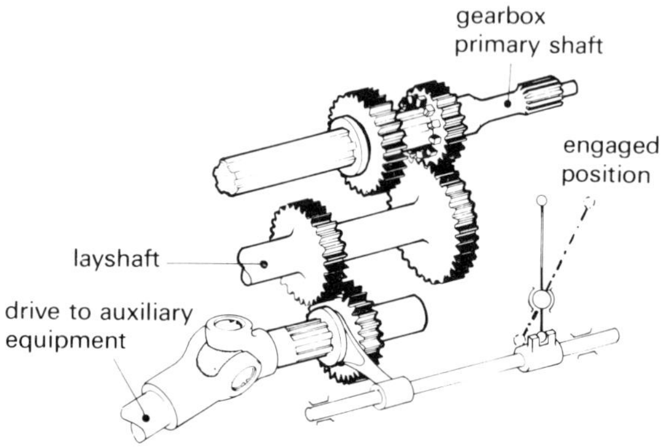

Fig. 38.5 Power take-off arrangement

Common faults and their causes

Various noises can be emitted from a gearbox. The diagnosis of all faults should take into account the driving conditions during the time that the fault is evident.

Fault	*Cause*
1 Gear whine	*a* Lack of oil
	b Geartooth wear
	c Bearing wear
	d Shaft misalignment
2 Knocking or 'ticking'	*a* Chipped gear tooth
	b Foreign matter wedged between gear teeth
	c Defective bearing
3 Jumping out of gear	*a* Defective selector detent (e.g. broken spring)
	b Shaft misalignment due to worn bearings
	c Worn gear teeth
4 Gear fails to select	*a* Incorrect adjustment between gear stick and selector
	b Defective selector

39 Constant-mesh and synchro-mesh gearbox

Disadvantages of the sliding mesh

Although the mechanical efficiency of the sliding-mesh gearbox was high, it suffered from two great disadvantages:

1 Gear noise due to the type of gear.
2 The difficulty of obtaining a smooth, quiet and quick change of gear without the application of great skill and judgement.

Gearbox designs introduced during the last 50 years have endeavoured to overcome these disadvantages. The first step in the development came when the constant-mesh gearbox was used.

Constant-mesh gearbox

This was first used on cars in the early 1930s, but gave way in a short time to other designs, although it is still used on commercial vehicles and tractors.

The main feature is the use of the stronger helical or double helical gears which lead to quieter operation. Each pair of gears is in constant mesh, and gear operation is obtained by locking the respective gear to the mainshaft by means of a dog clutch.

The layout of the box follows the sliding-mesh arrangement previously described, and Fig. 39.1 shows the main details of the third- and top-gear section. The mainshaft gearwheels are mounted on bushes or needle rollers, and are located by thrust washers. When the gear is required, a dog clutch, which is splined to the mainshaft, is slid along by the selector to engage with the dog teeth formed on the gear. This has the effect of locking the gearwheel to the shaft. There will still be noise if the dog teeth are not rotating at the same speed when the engagement is made, and so double declutching is necessary, but damage caused by a 'bad' change will be limited to the dog clutch.

Synchro-mesh mechanisms

The improvement achieved by fitting a constant-mesh gearbox was great, but a certain amount of skill was still required to produce a quick, quiet change. The difficulty was in double declutching, and the purpose of carrying out this operation was to equalize the speeds of the two sets of dog teeth before engaging the gear. It soon became apparent that some device was required to synchronize the speeds mechanically, and when the system was invented it was known as the synchro-mesh gearbox.

Constant-load synchro-mesh unit

This was the first type of synchro-mesh used. Fig. 39.2 shows the main details of the unit controlling third and top gear. Fundamentally the box is laid out in the same manner as a constant-mesh, with the exception that a cone clutch is fitted between the dog and gear members.

The female cone of this clutch is formed in a hub, which has internal and external splines. A series of spring-loaded balls is carried in radial holes in the hub, and these push outwards into a groove machined in a sleeve. The selector fork controls the position of the sleeve, which has splines of the same pitch as the dog teeth on the gear.

The initial movement of the selector and sleeve carries the hub towards the gear and allows the cones to contact. At this point, the friction between the cones adjusts the speed of the gearwheel to suit the hub and mainshaft. Extra pressure on the lever will allow the sleeve to override the spring-loaded balls, and positively engage with the dogs on the gear.

If the gear change is rushed, there will not be enough time for synchronization, and the change will be noisy. The time taken for the speed to be equalized is governed by the frictional force which exists at the

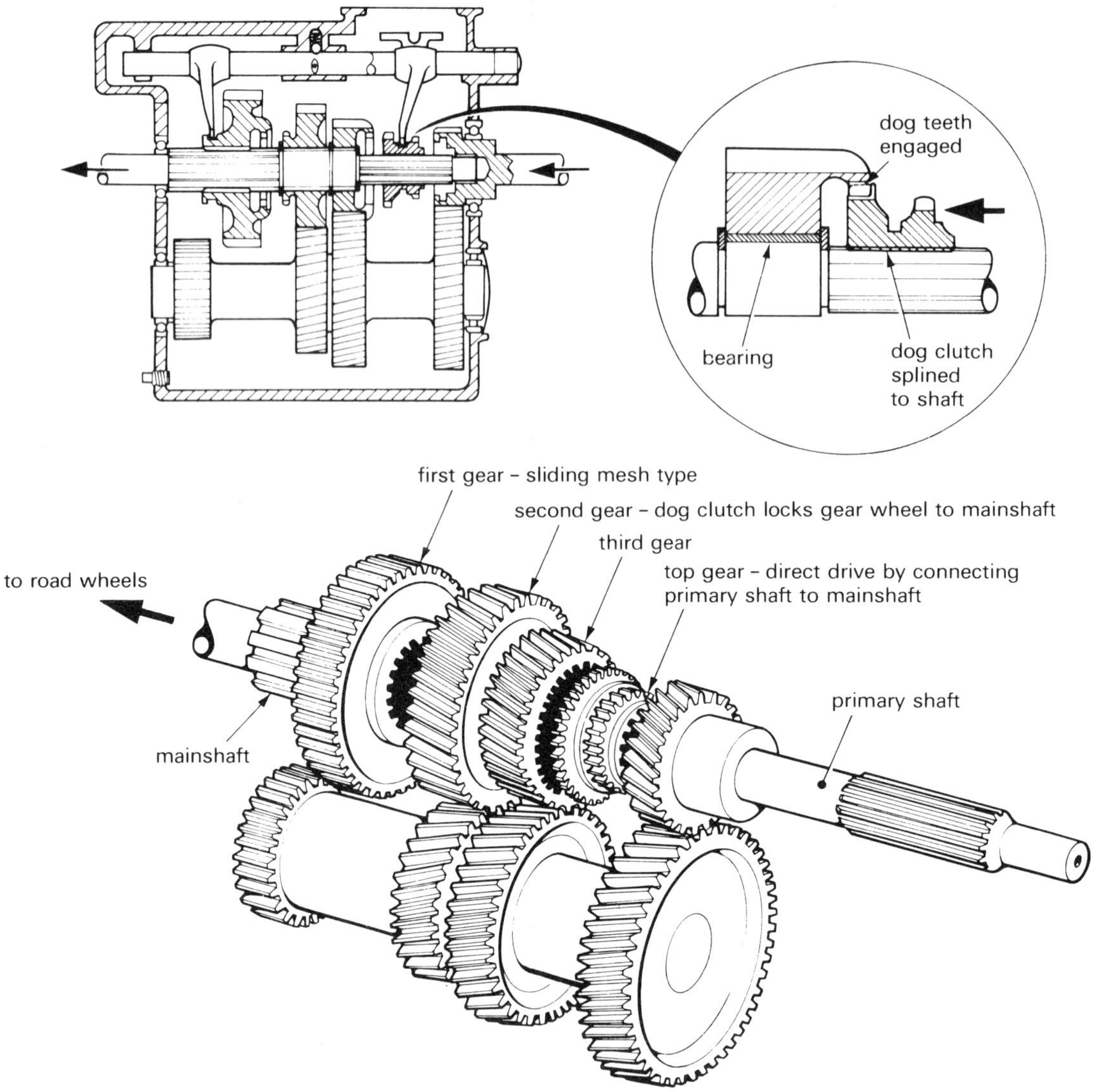

Fig. 39.1 Constant-mesh gearbox

cone faces. This force is controlled by: (a) total spring strength, (b) depth of groove in sleeve, (c) angle of cone, and (d) coefficient of friction between cones; therefore, if, because of mechanical defects, any of these factors is reduced, synchronization will take a longer time, and noise will probably be heard. This time factor has presented problems for the lubrication specialist, since the high-viscosity oil required by the gears takes a considerable time to disperse from the cones. The solution to this problem was to use a lower-viscosity oil (similar to medium engine oil - SAE 30) and provide a series of grooves on the cone face to cut through the oil film and disperse the lubricant. Up to recent times it was considered essential to drain and refill the gearbox every 8000 km (5000 miles), in order to remove the particles worn from the cones and gear teeth. With extended service schedules now in operation, this mileage has been increased considerably, and some manufacturers have recommended that after the first change no further changes are necessary.

Baulk ring synchro-mesh

The *baulk ring system*, which is sometimes called *blocker ring* or *inertia lock*, is a later development of the constant load system, and is designed to overcome the main disadvantage of the earlier

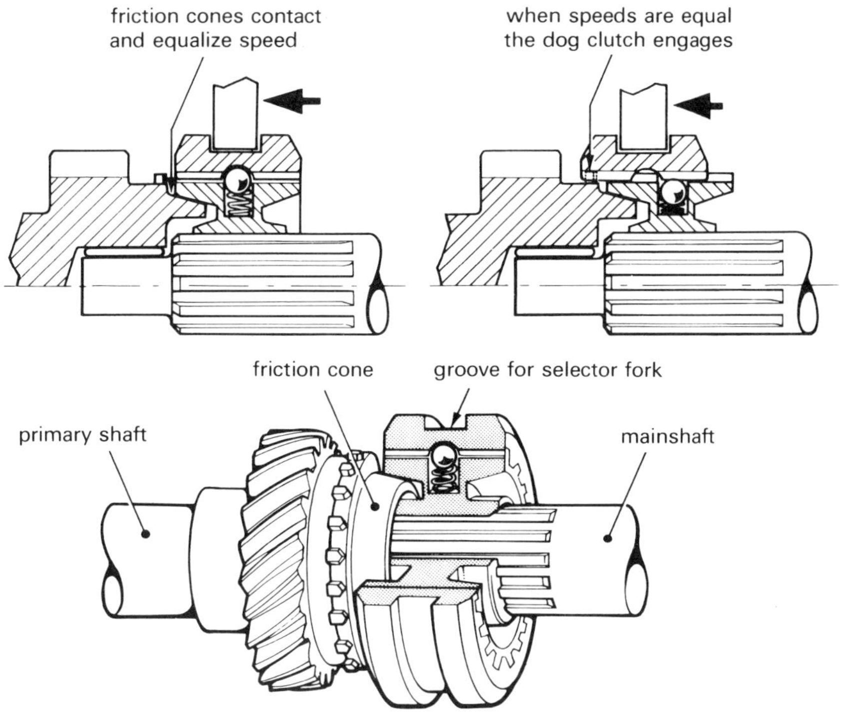

Fig. 39.2 Principle of the synchro-mesh gearbox

design - noise or crashing of the gears due to a quick change.

Two main features are incorporated in the baulk ring system:

1 The cone pressure or load is proportional to the speed of change.
2 An interception device prevents positive gear engagement until the speed of the two members is equal.

Various constructions are used to produce these features, and Fig. 39.3 shows one system in common use. Three spring-loaded shifting plates, which push out from the hub into a groove in the sleeve, fit into slots in the baulking cone. Each slot is wider than the plate; the clearance on each side is equal to half the pitch of the splines on the sleeve. The baulking cone, which is made of phosphor-bronze, has specially chamfered teeth on the outside, of a pitch similar to that of the dog teeth on the gear sleeve.

Movement of the gear lever will move the sleeve and shifting plates towards the gear selected. The plates will push the baulking cone into light frictional contact with the gear cone, and the difference in speed will allow the gear cone to carry the baulking cone around to the limit controlled by the plate (half

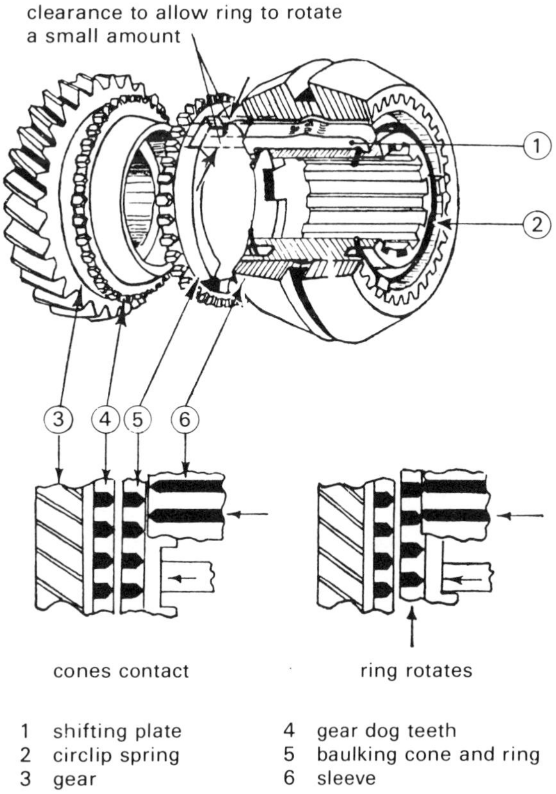

Fig. 39.3 Baulk ring synchro-mesh unit

spline movement). Extra pressure on the lever will tend to move the sleeve towards the dog teeth of the gear, but if the two members have different speeds, the dog teeth on the baulking ring will block the passage of the sleeve. In this position the splines on the sleeve are touching the teeth on the baulking ring, and therefore if a greater force is applied to the lever, a greater force will act between the cones and synchronization will be achieved in a shorter time. As the speeds become equal, the plates assume a central position in the slots of the baulking ring, and all teeth line up; therefore the sleeve can now pass the baulking cone to engage positively with the dog teeth on the gear.

Gearboxes for rear-wheel drive

Four-speed and reverse

Fig. 39.4 shows, by means of an exploded view, the gear layout of a four-speed and reverse gearbox suitable for a rear-wheel drive car. This gearbox uses a baulking-ring type synchro-mesh on all forward speeds. To simplify the construction, a sliding-mesh arrangement is used for reverse.

The type of baulking ring shown in Fig. 39.4 is different in construction to that shown in Fig. 39.3 but the basic principle is the same.

The type of unit shown in Fig. 39.4 utilizes the hub to press the synchro cones together. It uses three spring-loaded balls in the hub to lock the hub initially to the striker ring and three tangs on the baulking ring, which fit into slots in the hub, give the baulking action. Circumferential movement of the ring, relative to the hub, is arranged by having a clearance between the sides of the tangs and hub. This ensures that the ring can move each way, to the extent of half a spline, to block the passage of the striker ring as it attempts to engage with the dog teeth on the gear.

Uncaged needle rollers are used in this layout to support the layshaft gears. This bearing arrangement is now common, because in addition to having low friction, the extra rigidity of a needle roller compared with a plain bush gives a more precise gear mesh. As a consequence, the noise is reduced when the gearbox is loaded. This construction makes it more difficult to reassemble the box, but with the aid of a dummy layshaft of length equal to the gear cluster, the task is made easier.

Every gearbox must have some provision to prevent the escape of oil. For flange joining, either the fitment of a paper-based sealing joint or, on metal-

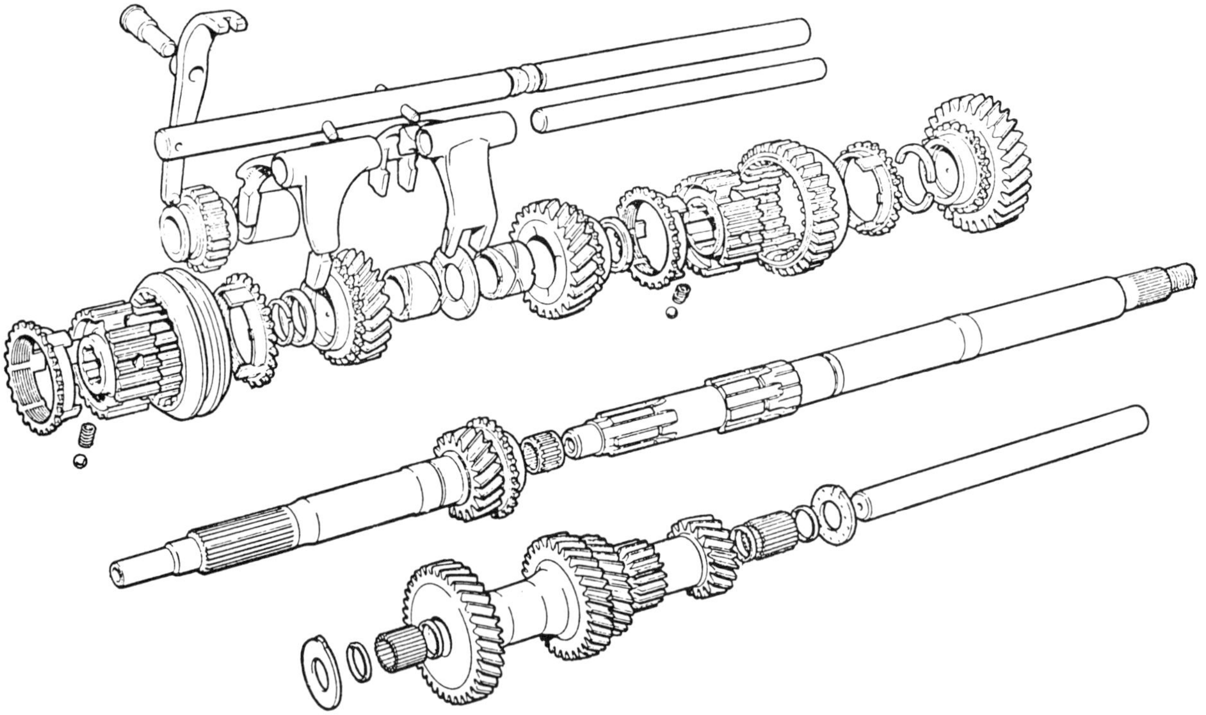

Fig. 39.4 Gear layout – four-speed and reverse

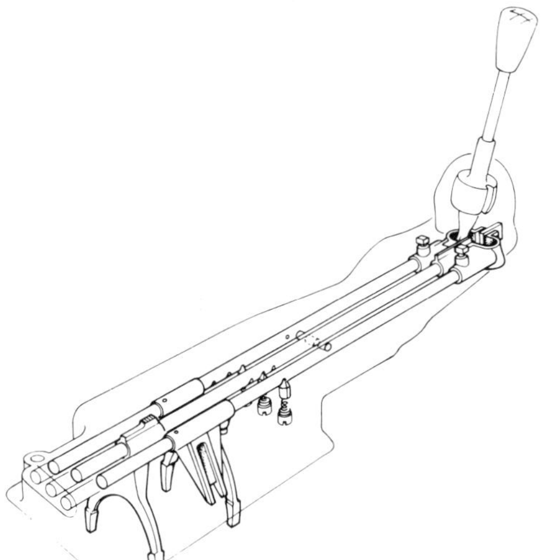

Fig. 39.5 Remote control lever

to-metal faces, the application of a sealing compound, is used.

Leakage along the shafts and through the bearings is normally prevented by fitting lip-type seals. These are positioned on the clutch side of the primary shaft bearing and at the universal joint end of the gearbox extension housing.

The *speedometer* is driven by means of a *skew gear* from the main shaft. A steel 'worm', mounted on the mainshaft and located in the gearbox extension housing, generally drives a plastic moulded pinion; this is connected to either a flexible cable or an electrical transducer.

Gear change mechanism

The seating position often means that the gearbox selectors are situated well forward of the driver's body. In the past a long 'floppy' lever was used, but nowadays the rigidity of a short lever and a *remote control* mechanism enables the driver to select the gears with greater precision.

Fig. 39.5 shows a typical arrangement.

Five-speed and reverse

During the past few years, the improved fuel consumption achieved by using a fifth-speed overdrive,

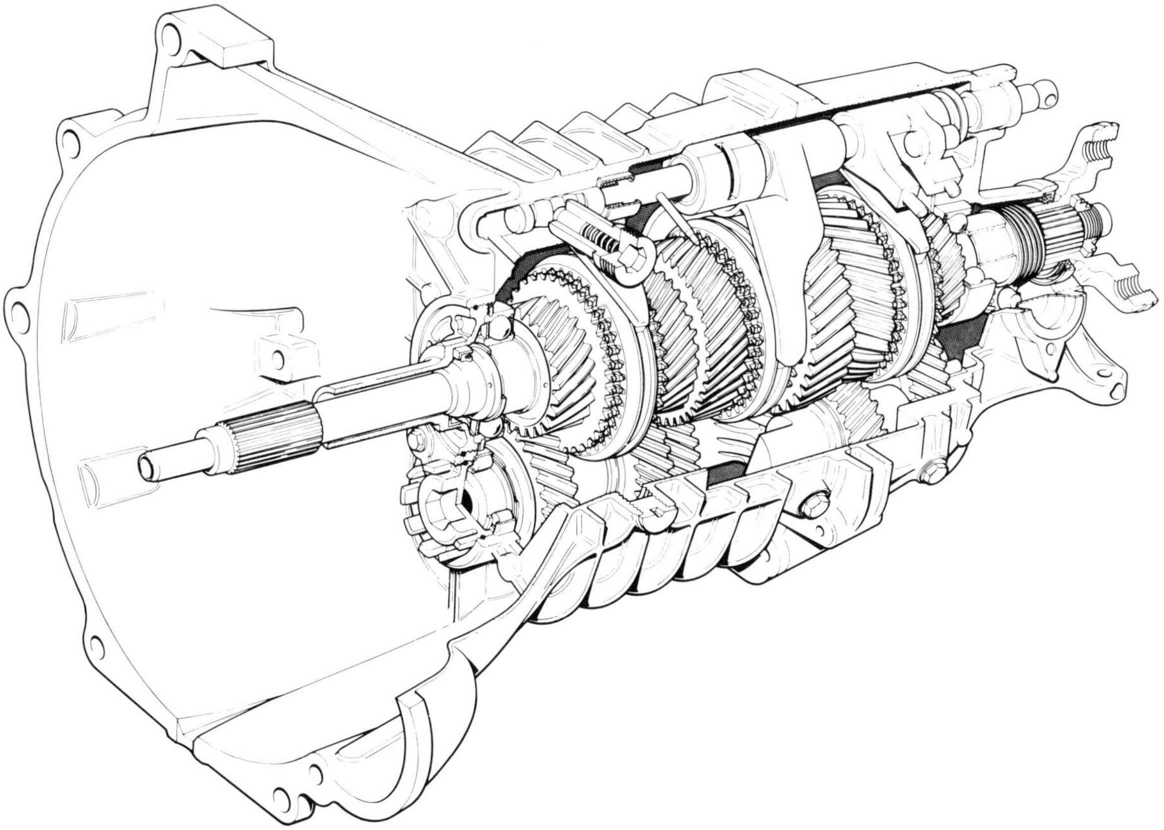

Fig. 39.6 Five-speed and reverse gearbox (Ford)

has made the five-speed gearbox a common feature of the modern car. The term *overdrive* means that the propeller shaft turns faster than the engine, so in a five-speed gearbox the fourth gear is the normal 'top' or direct-drive gear.

Fig. 39.6 shows a heavy-duty five-speed gearbox in which all gears, including reverse, are synchro-mesh. Helical gears are used throughout and each gear on the mainshaft is supported on needle rollers; this reduces noise and improves efficiency. The gearbox casing, which is ribbed to avoid distortion under load, is a lightweight aluminium alloy die casting.

In recent years, computer-controlled manufacturing processes have made it possible to produce gears with great accuracy. This allows running tolerances to be reduced, which in turn leads to a quieter operation. These developments allow the use of oils of low viscosity, so by expending less energy to rotate the layshaft the gearbox efficiency is improved. In the gearbox shown, a very thin oil similar to that used in automatic transmissions is recommended.

In common with many other five-speed gearboxes, the fifth-speed gears of the box shown are situated at the rear of the gearbox.

Reverse detent

Some 'blocker' arrangement is fitted to a gearbox to prevent the accidental engagement of reverse gear when the vehicle is moving forward. The simplest form is a spring-loaded detent; this must be overcome by the driver before the lever can be moved to the 'reverse' position. To overcome this spring the driver either has to lift the gear lever or exert extra pressure on it.

Five-speed gearboxes sometimes use a gear lever pattern of movement (gate pattern) whereby fifth and reverse are in the same plane, e.g. fifth and reverse are obtained by forward and backward movement of the lever respectively. In these cases a positive *gate lock* is often used.

In the gearbox shown in Fig. 39.6, the driver has to lift a collar on the gear lever to engage reverse.

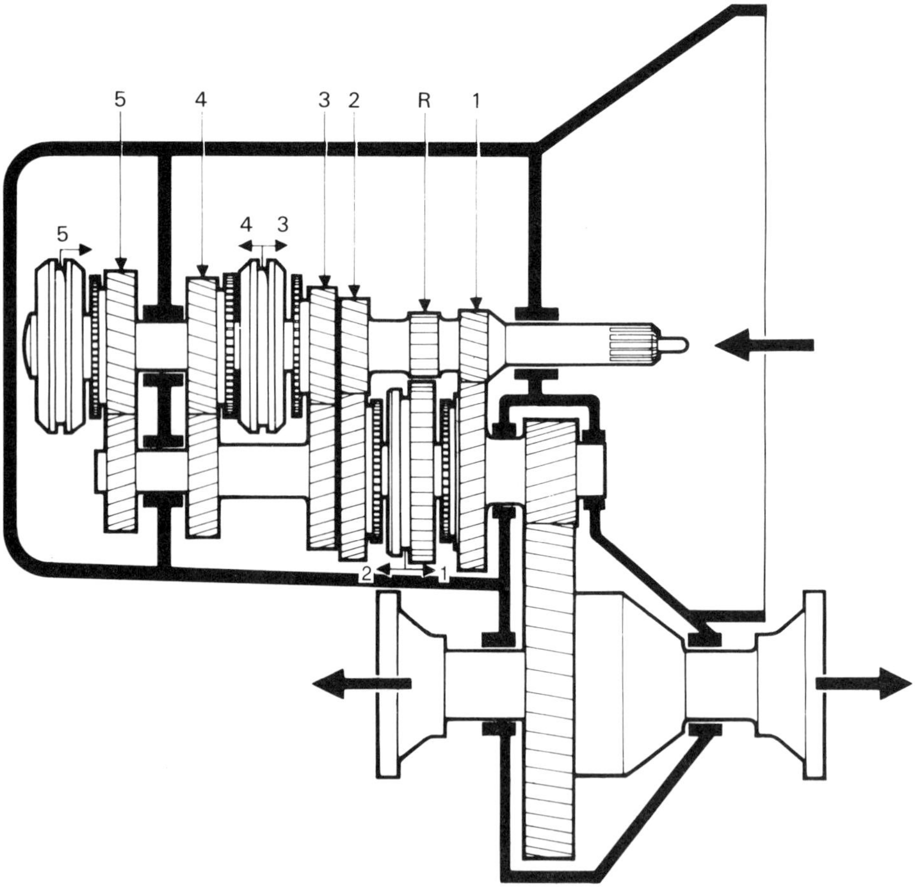

Fig. 39.7 Gear layout of a five-speed gearbox – front-wheel drive

Gearboxes for front-wheel drive

A transversely mounted engine and transmission assembly is the common arrangement for a front-wheel drive car. This compact transaxle configuration normally requires the gearbox input and output shafts to be at the same end, so a two-shaft layout is used.

Since a *simple* (single-reduction) system of gearing is used instead of a *compound* (double-reduction) system, a layshaft is not needed.

Fig. 39.7 shows the layout of a five-speed and reverse gearbox. In this design each shaft is supported by a ball race at the non-driving end; at the other end the radial load is much heavier, so a roller race is fitted. Axial thrust on each shaft is taken by a radial-type ball bearing, so this locates the shaft and maintains alignment, and takes the thrust of the helical gears. The spigot bearing needed with a layshaft-type gearbox is unnecessary with the two-shaft layout, so this gives a more rigid gear assembly, and a quieter gearbox results. A further improvement can be made when needle rollers are used to support the gears on the shafts.

Gear change mechanism

Some form of remote control mechanism is essential, because a long lever is far too flexible.

The linkage used must be capable of transmitting two distinct motions: longitudinal movement of the gear lever as it is moved from, say, first to second, and transverse movement as needed for the selection of another pair of gears.

Fig. 39.8 shows the principle of two systems in common use: a single rod linkage and a twin cable arrangement.

During operation, movement of the engine due to torque reaction is accommodated by either using a universal joint or relying on the inherent flexibility of the cable.

Speedometer drive

Front-wheel drive layouts normally have the skew gear mounted adjacent to the final drive assembly. This drive point is more accessible than the output shaft of the gearbox.

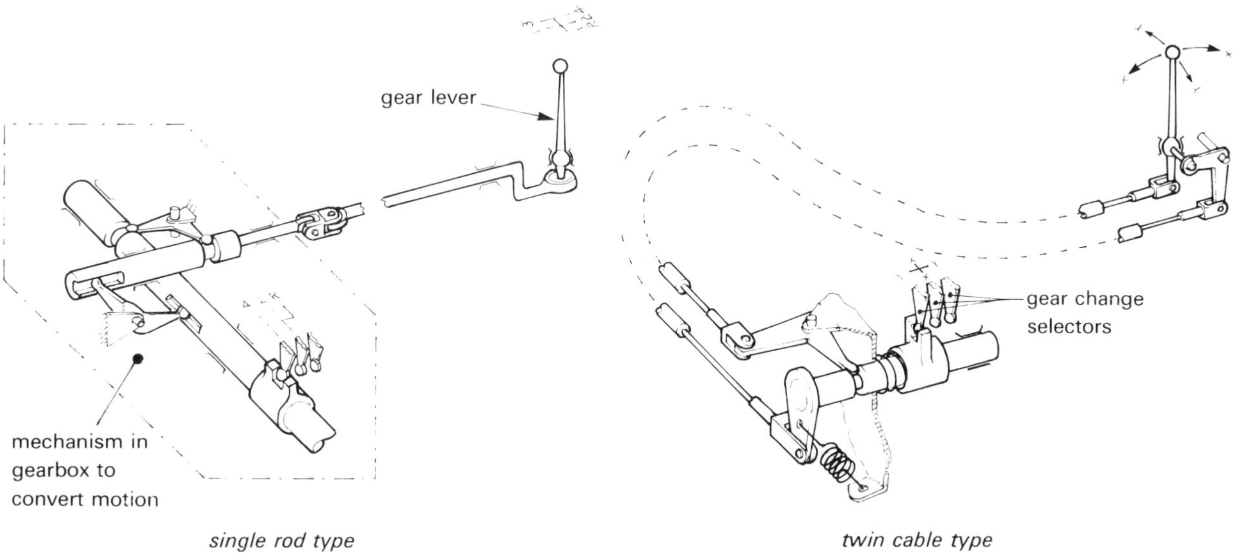

Fig. 39.8 Gear change mechanism (simplified) for transaxle

40 Planetary gearing and unidirectional clutches

Planetary gearing, also called epicyclic gearing, relates to a system in which one, or more, of the gear elements rotates around another gear in a similar manner to the revolution of the planets around the Sun. Whereas a conventional gearing system has the gears on shafts mounted in fixed positions, the planetary system utilizes the movement of some of its gears, together with their mounting shafts, to produce the required motion and gear ratio.

A planetary system can be arranged in many ways and since this type of gearing offers many advantages, it is used in a number of different automotive components, the most common being the automatic gearbox application.

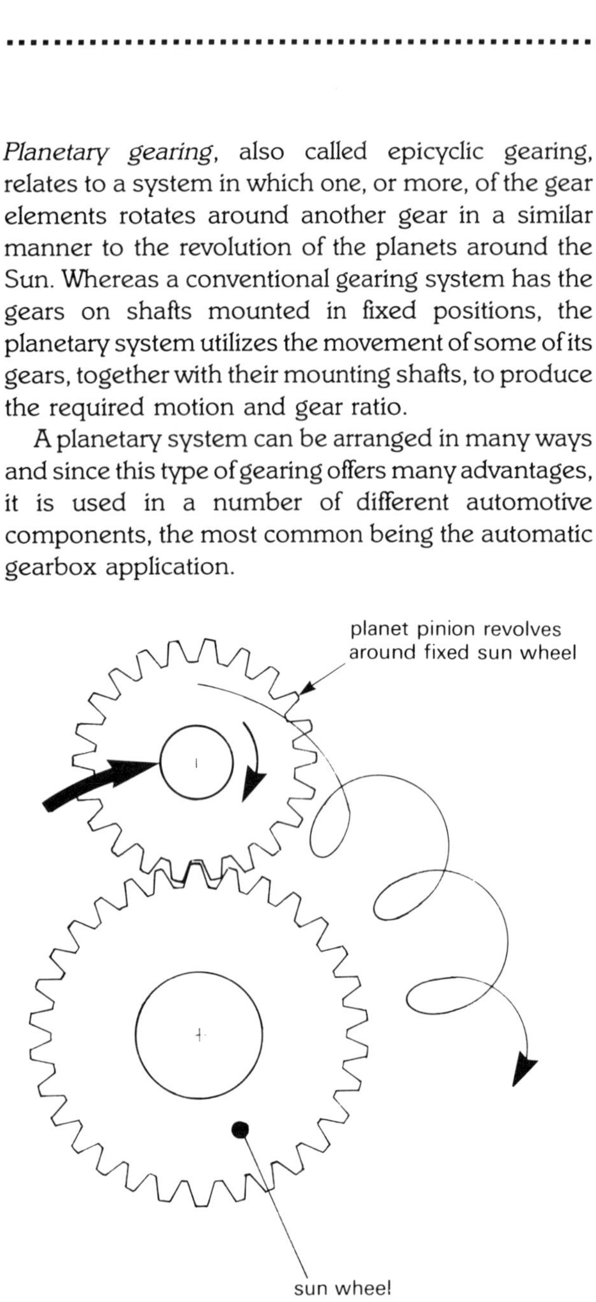

Fig. 40.1 Epicyclic gear action – planet revolving around fixed sun wheel

The dictionary states that 'an epicycle is a small circle rolling on the circumference of a greater circle', so when a gearing system uses a small pinion to roll around the outside of a larger gear the term *epicyclic* is appropriate. Fig. 40.1 shows this layout and it will be seen that rotation of the pinion drives it around the larger gear in a motion similar to that of the Earth around the Sun. This similarity is used to obtain the names *planet* and *sun* for the pinion and gearwheel respectively.

It should be noted that when a planet pinion revolves around a fixed sun, the number of revolutions it makes, relative to the observer, is always one greater than the obvious. For example, planet and sun sizes of 20 and 40 respectively causes the planet to rotate 3 times when it travels around the circumference of the sun wheel.

Another form of plentary gearing, which is shown in Fig. 40.2, meshes a planet pinion with an internally toothed *ring gear* (often called an *annulus*). When the annulus is fixed, rotation of the planet causes it to 'walk around' inside the annulus, but it takes many more turns of the planet to travel the circumference of the annulus than in the previous case.

Simple epicyclic gear train

The most common epicyclic gear arrangement, sometimes called a *Simpson gear*, uses a sun and annulus with one or more planet pinions positioned as shown in Fig. 40.3. In this arrangement the input and output shafts are connected to the sun and planet shaft respectively, and a friction brake is provided to hold the annulus stationary when a positive drive through the gear train is required. In this case the planet shaft is mounted on an *arm* (carrier plate).

When the brake is released, the natural resistance to movement of the arm will cause the planet to act as

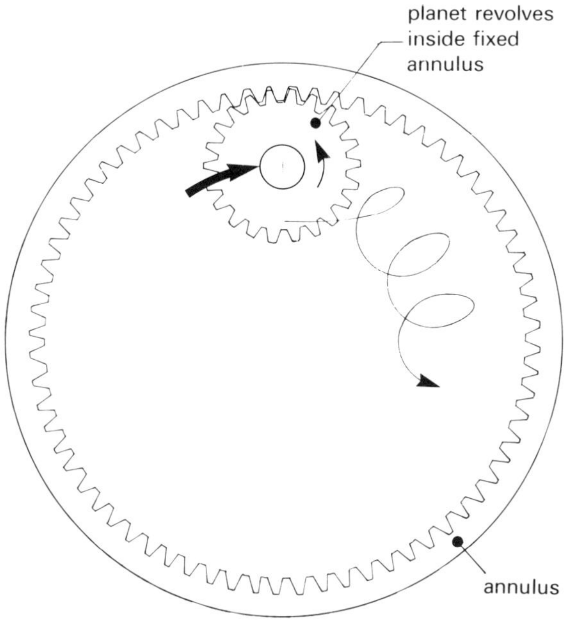

Fig. 40.2 Epicyclic gear action – planet revolving inside fixed annulus

an idler gear. As a result, the forward-moving sun wheel will cause the annulus to be driven in the opposite direction.

Brake application brings the annulus to rest and in this state the rotation of the sun wheel drives the planet pinion and causes it to 'walk around' inside the annulus; this moves the arm and output shaft in the same direction as the input shaft but at a much slower speed.

Gear leverage

All gears may be considered as levers and Fig. 40.4 shows the lever action of a planet pinion. When the annulus is held by a brake, and the sun wheel is moved through a small angle A, the planet lever pivots about a tooth on the annulus. This action causes the centre of the lever to prise forward the arm and rotate it about its axis to the extent of angle B. A comparison of the movement of the input shaft and output shaft, as shown by the angles A and B, indicates that a gear train of this size gives a reduction ratio of about 3 : 1.

With the brake released and the annulus free to rotate, the planet lever would rather turn the annulus (in the opposite direction to the sun) than move the arm forward against its resistance; as a result no drive would be transmitted.

An epicyclic unit will not transmit a drive unless one member of the unit is held stationary or is driven at a fixed speed; this action is produced by fitting either an external contracting friction brake or a multi-plate clutch.

Brake action

When an epicyclic gear train is used in any form of gearbox, the friction brake action overcomes many of the drawbacks associated with gear engagement in a conventional gearbox. The brake is both quick and quiet in its operation and during engagement the slipping action prevents a sudden jolt. This feature could be used in a transmission system to perform the function of the main clutch, but rapid wear of the brake would occur; this makes it necessary to retain

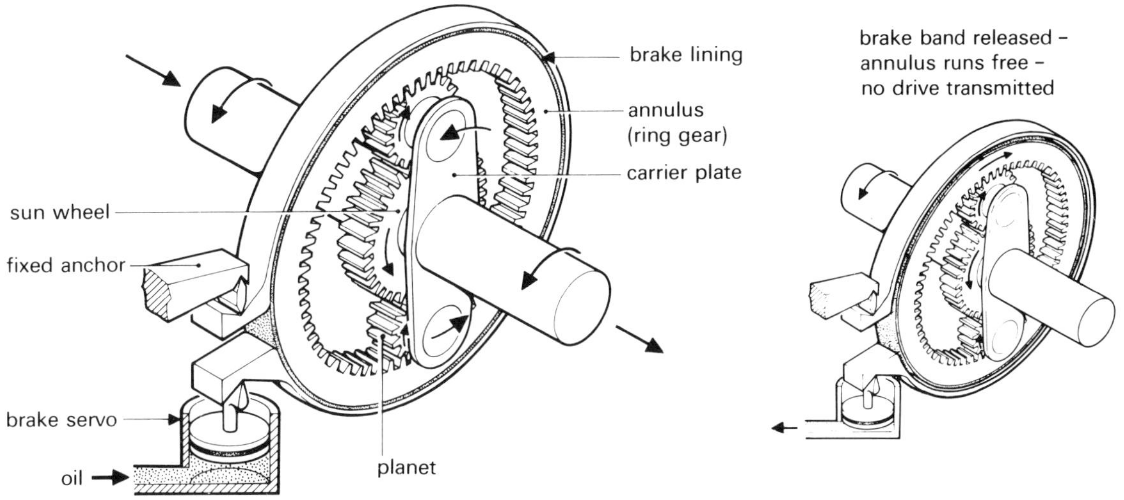

Fig. 40.3 Simple epicyclic gear train

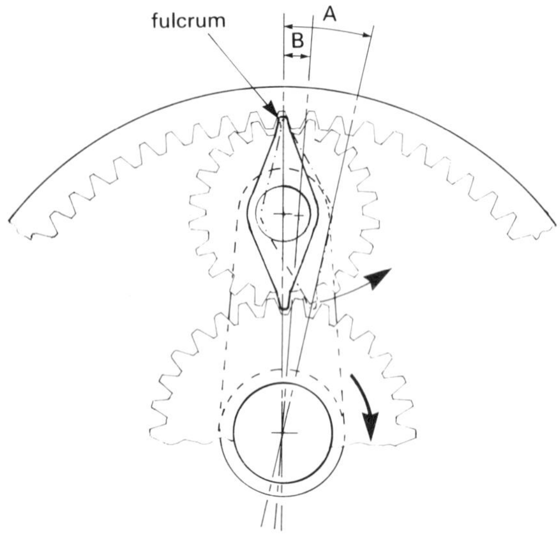

Fig. 40.4 Lever action of planetary pinion

the main clutch so that the brake can be applied at a rate that gives a smooth operation together with a good friction material life.

As gear engagement is provided by the brake mechanism, and since the need to disconnect the power is avoided, the main clutch only has to function as a device to take up the drive when the vehicle is moved off from rest. A fluid clutch satisfactorily meets this need, so this type of clutch is normally fitted in conjunction with an automatic gearbox.

Gear ratio

The number of teeth on the sun wheel and annulus of a simple epicyclic gear train governs the gear ratio; this is given by:

$$\text{Ratio} = \frac{A+S}{S}$$

where A = number of teeth on annulus
S = number of teeth on sun wheel

If $A = 100$ and $S = 20$, the ratio is 6 : 1.

Brake applied to sun wheel

There are many occasions when an epicyclic gear train is required to give a comparatively high gear ratio. This can be achieved by connecting the gear elements in the manner shown in Fig. 40.5. In this case the brake is applied to the sun wheel and the input is connected to the ring gear so as to make the annulus the driving gear.

Applying the principle of the lever makes the point of contact between the planet lever and sun wheel the fulcrum. The ratio achieved by this arrangement is:

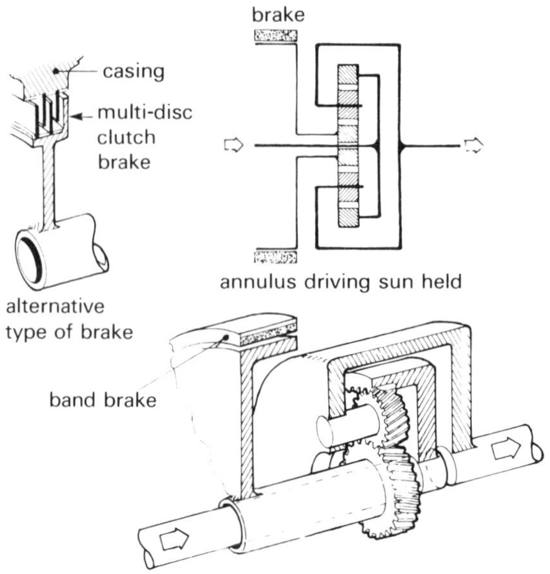

Fig. 40.5 Brake applied to sun wheel

$$\text{Ratio} = \frac{A+S}{A}$$

Using the same number of teeth as in the previous example, the ratio of this drive arrangement is 1.2 : 1. The speed ratio obtained is very close to direct drive, so this gear configuration has many uses.

Simple epicyclic drive arrangement

Using the term 'simple' in this context means that the gearing system is based on one epicyclic unit comprising a sun, planet set and annulus, i.e. it is not compounded such that one epicyclic unit is driving another.

Fig. 40.6 shows how different speeds and two directions can be obtained from a simple epicyclic unit by altering either the brake position or the input drive point.

By using a number of clutches and brakes with this simple unit, various interconnections can be made to achieve a gearbox speed range such as first, second, direct drive top, reverse and neutral.

In applications other than a transmission gearbox, the appropriate arrangement is selected from the layouts shown in Fig. 40.6. A neutral position is not needed when a fixed ratio is required, so in this case the 'braked' element is often an integral part of the housing.

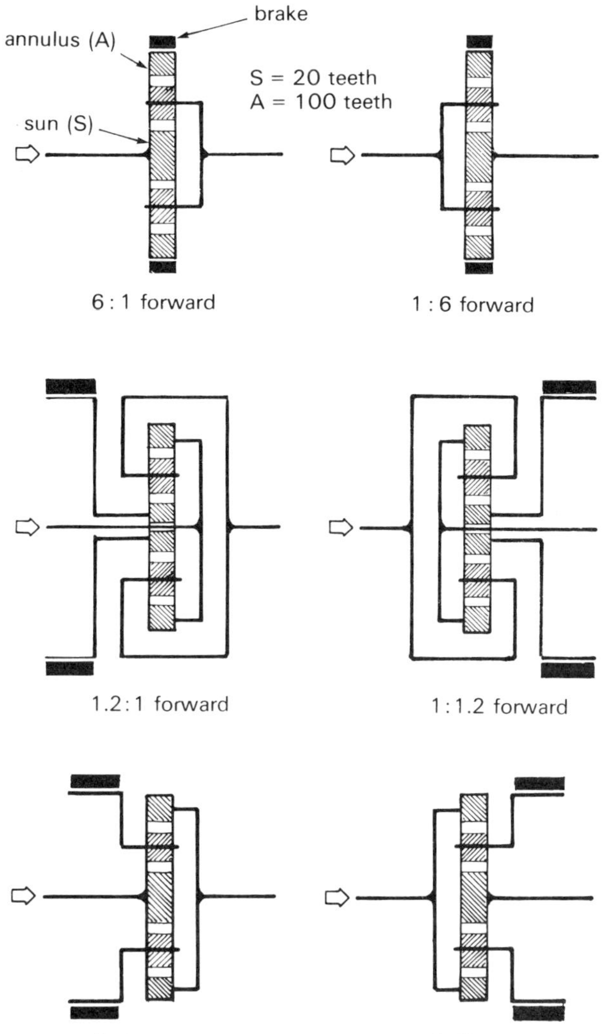

Fig. 40.6 Various methods of connecting a simple epicyclic gear train

Construction

To simplify some of the previous diagrams only one planet pinion is shown. This arrangement is impracticable because a single planet would give an unbalanced thrust on the sun and annulus members as they are forced apart under load. Fitting three equally spaced planets overcomes this problem.

Fig. 40.7 shows an exploded view of a simple epicyclic gear unit. Each planet is mounted on a ball race which is fixed to a short shaft attached to a *planet carrier*. This carrier acts as the *arm* of an epicyclic unit.

In the past spur-type teeth were used but nowadays helical gears are common. This type of gear is quieter in operation but it needs a low-friction thrust washer to resist the axial force set-up by the tooth angle.

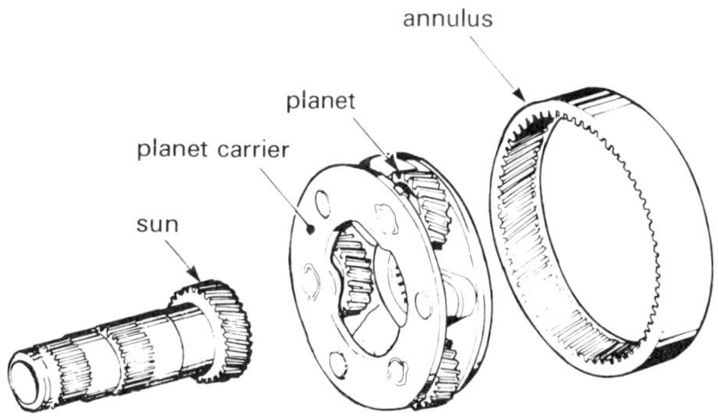

Fig. 40.7 Simple epicyclic gear train

Compound gearing

Most automatic gearboxes use two or more epicyclic units to form a compound gear train. Some of the members in these units are permanently interconnected in a pattern applicable to the make of gearbox. Engagement of a particular gear is obtained by using a combination of brakes and clutches to hold and drive the appropriate epicyclic element.

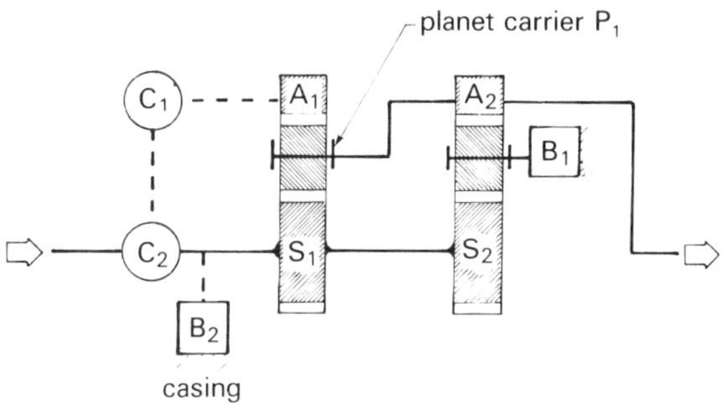

	C_1	C_2	B_1	B_2
N				
1	•		•	
2	•			•
3	•	•		
R		•	•	

Fig. 40.8 Compound epicyclic gear train (ZF)

Fig. 40.8 shows a compound gear train that uses two clutches and two brakes to give three forward gears and a reverse gear. This layout is used to illustrate the basic principle of operation of a compound epicyclic gear train (Table 40.1).

Table 40.1 Action of compound gearing system

Gear	*Clutch and brake* Engagement	*Operation*
First	C_1 and B_1	When vehicle is stationary, drive to A_1 causes the idler action of P_1 to drive S_1 in the opposite direction. This motion applied to S_2 allows the idler action of P_2 to drive A_2 forward at a slow speed. Even when the output shaft starts to move, the relative motion of the gear members remains the same
Second	C_1 and B_2	Drive to A_1 causes the planet carrier PC_1 to be driven around the fixed sun S_1 at a speed slightly faster than in first. Rear unit is inoperative.
Third	C_1 and C_2	Driving both S_1 and A_1 causes the complete unit to revolve as one; this gives a direct drive. Rear unit is inoperative.
Reverse	C_2 and B_1	Driving S_2 and holding PC_2 in the rear unit causes P_2 to act as a reverse idler. Front unit is inoperative.

Unidirectional clutches

This device is also called a *free-wheel* or *one-way* clutch. Its action is similar to that used on a bicycle - it transmits drive one way and not the other.

Free-wheel units used in the past

Up to about forty years ago a free-wheel unit was often fitted as a separate unit behind the gearbox. In this application it enabled the vehicle to be free-wheeled or *coasted* when the road conditions were favourable for its use. At the expense of extra brake wear, the use of the free-wheel unit slightly improved the petrol consumption.

The device also enabled the driver to make a gear change on a manual gearbox without operating the main clutch. Once the vehicle was in motion, release of the accelerator caused the one-way clutch to disconnect the roadwheels from the gearbox; this removed the driving load from the gears and so allowed easy movement of the gear change lever.

At times when the driver did not require the free-wheel feature, the unit was locked by a gear to give a *fixed-wheel* condition. This was necessary to obtain reverse, so provision was made to lock the unit automatically when this gear was selected.

Modern applications

Nowadays the unidirectional clutch is used as a part of a component, e.g. automatic transmission and overdrive units, to limit the movement of a particular member to one direction. In these examples the clutch provides a simple means for either driving or holding one part of an epicyclic train so that it can only move one way.

Types of unidirectional clutch

The two main types of unidirectional clutch are:

- *Roller type*
- *Sprag type*

Roller type This uses a number of parallel rollers; they are sandwiched between the inner member and the inclined cam faces of the cylindrical outer member (Fig. 40.9). The type shown uses a series of concertina-shaped strip springs to wedge the rollers between the two faces.

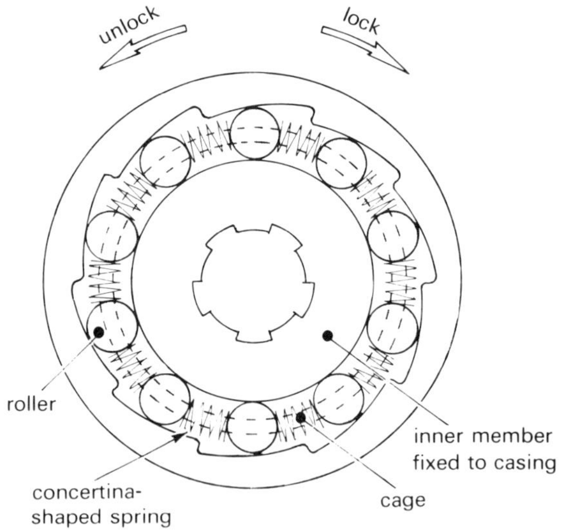

Fig. 40.9 Unidirectional clutch – roller type

The application shown in Fig. 40.9 is a torque converter; in this case the hub of the unidirectional clutch is mounted on the casing so as to control the direction of movement of the converter stator (see page 254).

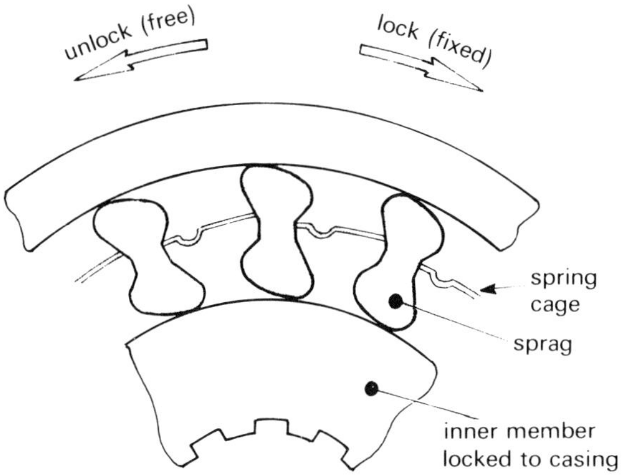

Fig. 40.10 Unidirectional clutch – sprag type

Sprag type. The main features of this design are shown in Fig. 40.10. A number of inclined wedges called *sprags* are held in a spring cage and positioned between inner and outer cylindrical tracks. The spring is designed to twist the sprags in the 'wedge direction' and keep them in contact with both the tracks.

When the inner member is rigidly secured to a component casing, an anticlockwise movement of the outer track releases the wedge action of the sprags and as a result allows the outer track to move freely.

By reversing the direction of motion of the outer track, the friction between the sprag and the track greatly increases the wedge action and prevents any slip taking place between the members. In this case the clutch will be *fixed* when the outer track is turned clockwise, and *free* when moved anticlockwise.

On a drawing the operating direction can be shown by using the symbols used on electrical diagrams to represent direction of current flow.

Unidirectional clutch faults

The two operating states of a unidirectional clutch are *free* and *fixed*, so the two main faults are 'free when it should be fixed' and vice versa, i.e. slip and seizure respectively.

A broken spring is a common cause of slip, and a fractured roller or sprag causes seizure; this is because the unit wedges in both directions.

Normally the remedy in these cases is to replace the complete free-wheel unit, that is provided it is accessible.

41 The automatic gearbox

In the past the automatic gearbox was considered by many to be a luxury, but today many drivers regard this unit as a desirable essential; this popularity has led to the introduction of many different types of box. Two-pedal control, i.e. the elimination of the clutch pedal, together with automatic gear selection reduces driver fatigue because it overcomes the need for tedious clutch and gear change operations. Consequently the simpler control of the vehicle enables the driver to concentrate on the other essentials of vehicle handling, especially the safety aspects.

Some early designs of automatic transmission were criticized because they robbed the driver of gearbox control. Modern units have overcome these drawbacks by allowing drivers to override the gearbox when they wish to take full command. More recently, the incorporation of electronic gearbox control has enabled the gearbox to be operated in a mode to match the driver's technique and performance requirement, e.g. sports mode and economy mode.

Up to about the mid 1970s, the majority of automatic gearboxes used on light vehicles were made by Borg Warner. This company offered a range of units to suit engine sizes between 1.5 and 5 litres. Based on production numbers, one of the most successful gearboxes was the Model 35 and its derivatives.

Today, most vehicle manufacturers produce a gearbox either of their own design, or made under licence to some other manufacturer.

Automatic transmission units

Other than the *continuously variable transmission (CVT)* systems, most modern automatic transmissions consist of two main units: a fluid clutch and a main gearbox.

The function of these main parts is as follows:

1 *Fluid clutch* - automatically disconnects the drive when the engine speed is low and gradually connects it as the vehicle is moved from a stationary position.

It is either a *fluid coupling* or *fluid converter*; the latter is often called a *torque converter* because, in its simple form, it can double the output torque from the engine. In addition the converter is able to reduce the torque amplification gradually as the output speed of the converter increases.

2 *Main gearbox* - automatically provides a series of *stepped ratios* to enable the vehicle to overcome road conditions needing a large driving effort; also it allows the vehicle to operate over a wide roadspeed range.

In addition to the provision of neutral and reverse, the modern gearbox also has a positive parking lock and a means for the selection, and 'hold', of a particular gear.

Fluid clutch

The arrangement was originally intended for marine use, and was invented by Dr Fottinger in 1905, but it was not until 20 years later that a British engineer (H. Sinclair) introduced the fluid drive on a motor vehicle. With this drive the transmission is connected to the engine by hydraulic means, which provides an effective damper to the torsional vibrations set up by the engine, a smooth take-up to move the vehicle from rest, and, in the case of the converter, a device to amplify the torque.

The hydraulic coupling

This unit is often known as a *fluid flywheel*, since it usually forms a component part of the main engine flywheel.

Principle of the fluid coupling flywheel You will understand the principle of operation if you consider the following steps.

Fig. 41.1a shows a 'grapefruit' member which is filled with a fluid and rotated. The fluid is thrown tangentially outwards and upwards.

If a plate is held over the member (Fig. 41.1b), the fluid will strike and tend to rotate the plate in the same direction. The force and torque acting on the plate will depend on the speed of rotation – the greater the speed, the greater will be the force, since more energy is given to the fluid.

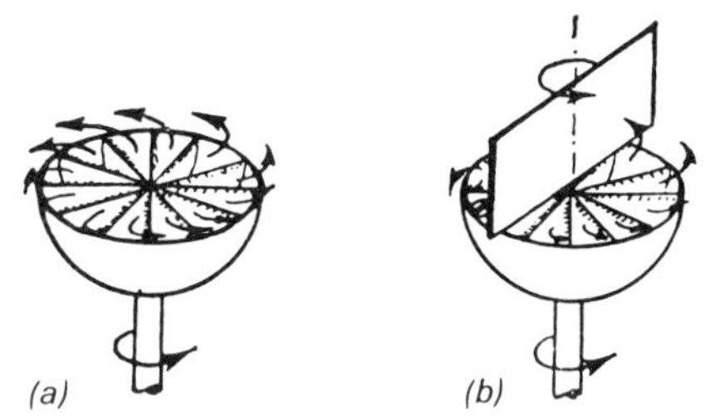

Fig. 41.1 Principle of the fluid flywheel

In Fig. 41.2a the plate is replaced by another 'grapefruit' member which receives the drive and re-diverts the fluid back to the driving member. The fluid 'circuit' can be traced by following the path taken by a particle of fluid. Rotation of A will cause the particle to move outwards from point 1 to 2. This is brought about by the resistance of the fluid to movement in a circular path. (A body will move in a straight line unless acted upon by a force. So that if you consider a plan view of A, you will see that the straight-line path will take the particle to the outside.) The farther from the centre the particle moves, the faster it will have to travel; consequently, energy will be extracted from the engine.

The shape and outward motion will force the particle upwards and cause the tangential flow to strike the vanes of member B, the force of impact being governed by the speed difference of the two members. At position 3, the energy possessed by the particle is less than at 2, since the force of impact causes energy to be given up in the form of heat. This fact means that the energy remaining in the particle is less than that given to it by the engine, and therefore the output speed will be lower than the input.

The fluid following the particle will push it, against its natural tendency, to point 4. During this movement, the linear speed of the particle is decreased and energy is given up to drive the output shaft. If the

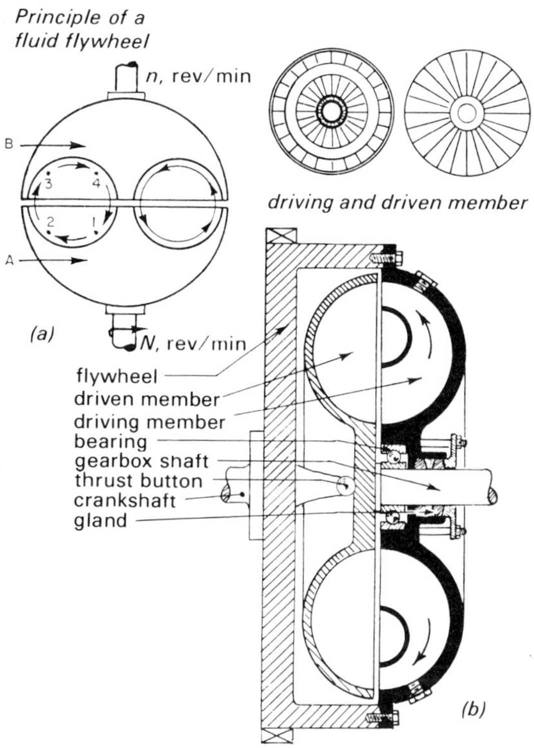

Fig. 41.2 The fluid flywheel

speed of both members is the same, the outward force at point 4 will be equal to that at point 1, and no circulation will take place. Since the operation depends on fluid passing from one member to the other, it is fortunate that the driven member B rotates more slowly.

As the particle moves from 4 to 1, the engine will have to supply extra energy to speed up the fluid. This is necessary because the fluid is rotating more slowly than the driving member and is acting as a brake on the engine. A fluid converter uses an extra member to overcome this braking effect.

Fig. 41.2b shows the main constructional details of a fluid coupling. Oil of a type similar to light engine oil (SAE 30) is introduced to the level of the filler plug, which leaves an air space for expansion. To allow for the difference of speed, a bearing is fitted between the driven and driving members, and a guide ring is sometimes incorporated to provide a smooth flow path for the fluid.

When the engine is driving, the oil circulates in the direction shown: fast when the driven member is stationary, but slowing down as the two members approach the same speed. Under 'overrun' conditions, the direction of oil flow in the axial plane is reversed, and a drive from the output shaft to the crankshaft provides 'engine braking.'

Apart from a periodical check on the oil level, the coupling needs very little attention. If a fault arises, it is normally either: (a) over-heating, due to excessive slip, which may be caused by low oil level, or (b) noise, due to bearing wear, which allows the faces of the members to touch.

The hydraulic converter

This component fulfils one of the most important tasks in a modern automatic transmission system and in its simplest form provides an infinitely variable ratio up to approximately 2 : 1. It is also responsible for duties performed by the fluid coupling. More complicated types are capable of providing greater torque ratios.

Principle of operation Before considering the operation of the converter, an appreciation of the following experiments may be of help.

Anyone who has used a hosepipe will realize that the water leaves the nozzle at a considerable speed. A moving fluid possesses energy and this means that it is capable of doing work. Directing a flow of fluid on to a plate causes the fluid direction to change and the extent of this change governs the force which acts on the plate (Fig. 41.3).

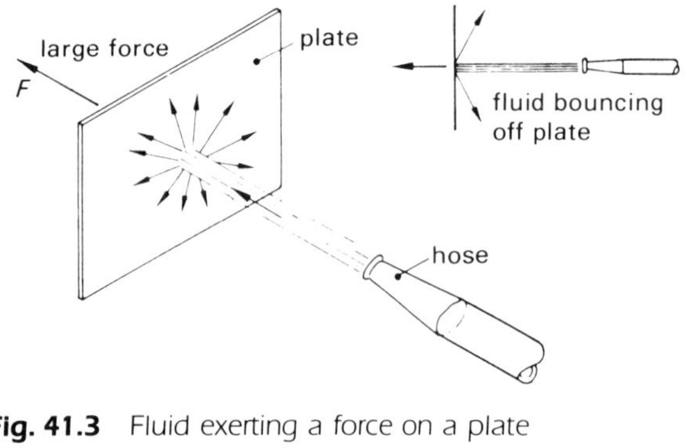

Fig. 41.3 Fluid exerting a force on a plate

Varying the speed of the fluid by adjusting the tap also alters the force of impact; a low fluid speed contains very little energy and therefore only exerts a small force on the plate.

The force of impact can be varied by moving the plate in a manner such that it is tending to travel with the fluid. During this movement the speed of the fluid relative to the plate is slower than in the previous case, so the impact force will be less. If it is possible to move the plate at the same speed as the fluid is moving, then no impact will take place, consequently no force will be produced on the plate.

Fig. 41.4 shows a curved plate. In this case the fluid strikes the plate with a velocity v and the curved surface redirects the fluid back at a velocity $-v$ towards its source. Double the force is obtained with this arrangement, since the plate not only stops the fluid but also has to send it back at the same speed.

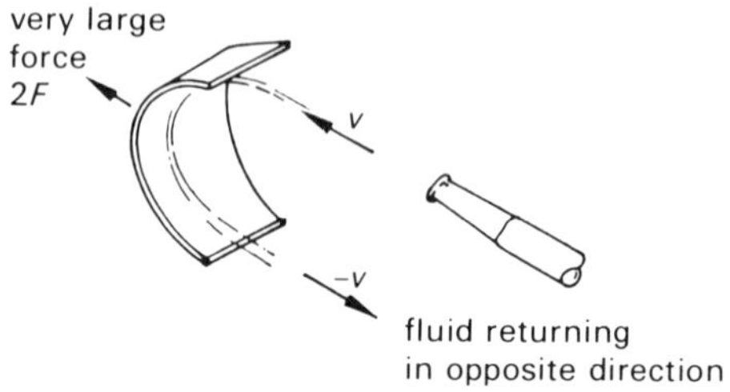

Fig. 41.4 Effect of a curved plate

Mounting a series of these plates on to a shaft (Fig. 41.5) forms a simple turbine. Each plate, which is now called a vane, will extract the energy from the moving fluid to produce a turning action on the shaft.

Consideration of the previous experiments will show:

1 Low velocity of the fluid produces a small turning movement or torque.
2 High torque occurs when the fluid velocity is high and the shaft speed is low.
3 Gradual decrease in torque occurs as the shaft speed increases.

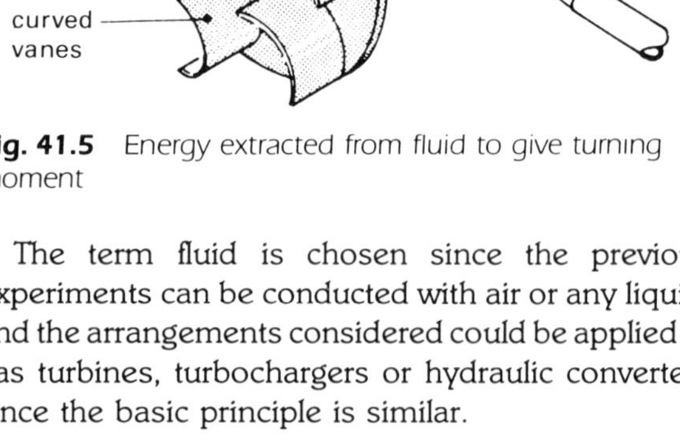

Fig. 41.5 Energy extracted from fluid to give turning moment

The term fluid is chosen since the previous experiments can be conducted with air or any liquid, and the arrangements considered could be applied to gas turbines, turbochargers or hydraulic converters since the basic principle is similar.

Having established the basic idea of the turbine, the means of fluid supply will now be examined. From the previous description it will be appreciated that the energy is contained in a particle of fluid by virtue of its velocity, so a compact device is needed to eject fluid at high velocity. In the case of the torque converter this is provided by a centrifugal pump, which in principle is similar to the type used in a cooling system or supercharger.

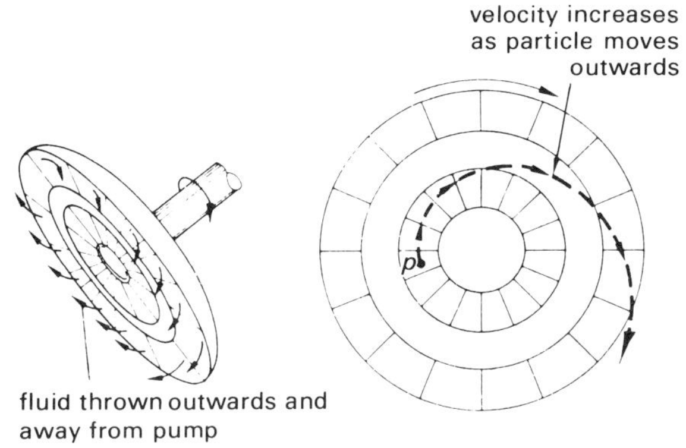

Fig. 41.6 Pump action

Fig. 41.6 shows the basic features of the pump. Rotation of the member will cause the particle of fluid *p* to move in the path shown. Initially the movement is towards the outside of the member and the greater distance it has to travel in one revolution at this point shows that its velocity has been increased. Work has to be done by the engine to achieve this increase in the energy content of the particle. Shrouding the pump with a container causes the fluid to move away from the pump as shown in Fig. 41.7. Observation of the diagram shows the fluid moving around the circumference and it is the velocity in this direction that governs the driving force that is being delivered by the pump. The resemblance between the flow given by the pump and the nozzle shown in Fig. 41.5 should now be apparent and the effect of engine speed on the pump's performance will be easy to understand - low engine speed will give a small flow (low velocity) of fluid and consequently each particle will contain only a small amount of energy.

Placing the pump adjacent to the turbine (Fig. 41.8) and making the pump act as the container shows how the drive from the engine is achieved. The pump imparts energy to the particle as it moves towards the outside. This energy is extracted by the turbine by moving the fluid inwards towards the shaft, thus causing it to slow down. To summarize this stage:

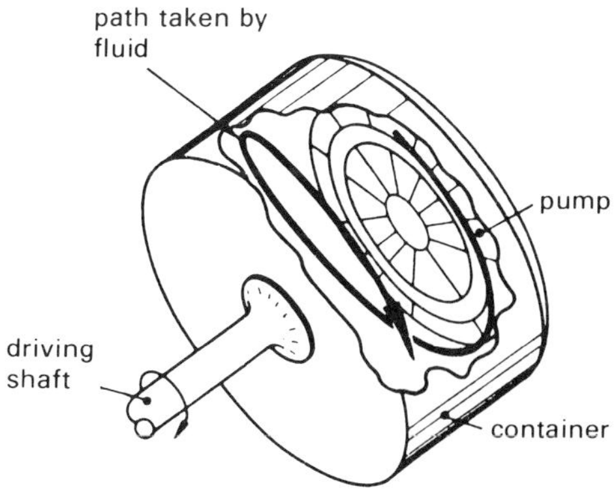

Fig. 41.7 Container fitted to pump

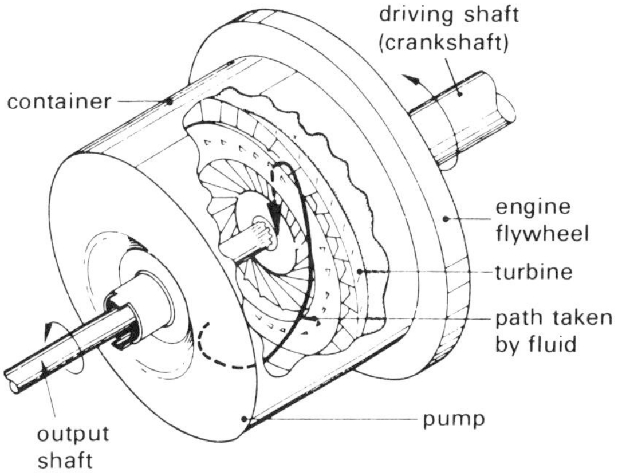

Fig. 41.8 Fluid path from pump to turbine

Pump:	Speeds up particle	Imparts energy
Turbine:	Slows down particle	Extracts energy

Fig. 41.8 is similar to a fluid coupling and in this form the torque output is always less than the torque input.

Applying the earlier analogy of the hosepipe would indicate that when the turbine is turning slower than the pump, the force of impact of the fluid striking the turbine vanes should give a torque increase. This would be true if the fluid did not have to return to the pump, but since this fluid is having to recirculate then the speed of the fluid compared to the speed of the pump vane must be considered. To give a high impact force on the output member, the pump must be turning at a higher speed than the turbine, so when the fluid is returned to the pump it is travelling slower than the pump vane. In this condition the vane will strike the slower moving fluid particle and this will give a force which is acting against the motion of the pump, i.e. the fluid will tend to act as a brake on the pump. The greater the speed difference between pump and turbine, the greater will be the impact force on the turbine but this will also result in a larger braking action on the pump. To obtain a torque multiplication, this braking action must be eliminated. This is achieved by fitting a separate member called a *stator* (stationary member) between the turbine and the pump. Mounted on an extension of the gearbox casing, the stator consists of a number of vanes which are shaped in the manner shown in Fig. 41.9. Fluid returning to the pump is redirected such that it now enters the pump at a suitable speed and direction. If the path of the oil returning to the pump is considered, the system could be shown diagrammatically in the manner shown in Fig. 41.10.

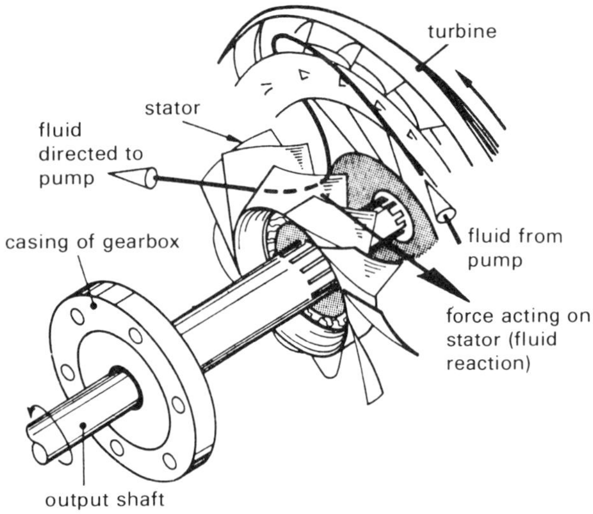

Fig. 41.9 Action of stator

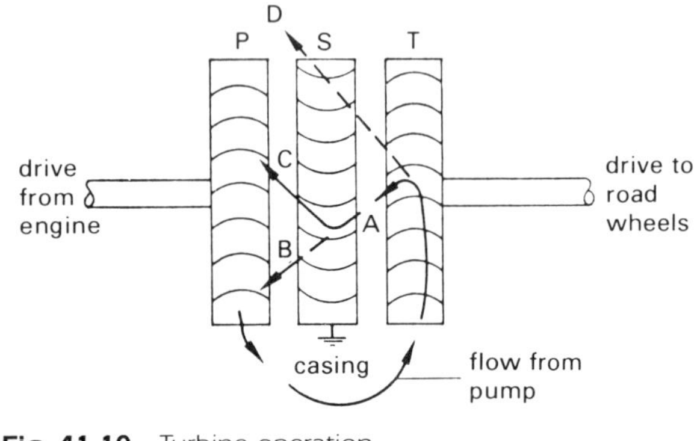

Fig. 41.10 Turbine operation

Fluid from the pump acts on the turbine T, but if the turbine is turning slower than the oil, then the oil will be deflected in the path marked A, the stator will act as a reaction member and will direct the fluid from path B to path C; a suitable direction for the pump to receive the fluid. As the turbine speed is increased the fluid path A will gradually change until it is taking path D. Fluid attempting to flow in path D will now be obstructed by the stator vanes and the disturbance caused by this obstruction would give a considerable drop in efficiency. To limit this drawback a unidirectional clutch (free wheel) is fitted between the stator and the extension of the gearbox casing. As soon as the fluid strikes the back of the stator vanes the clutch will unlock and the fluid will then pass along path D to the pump. Once this occurs the unit will act similarly to a fluid coupling.

Fig. 41.11 shows the torque output in relation to turbine speed for a given speed of the pump. It will be seen that the output torque is approximately equal to input torque when the turbine speed is about 90 per cent of the pump speed.

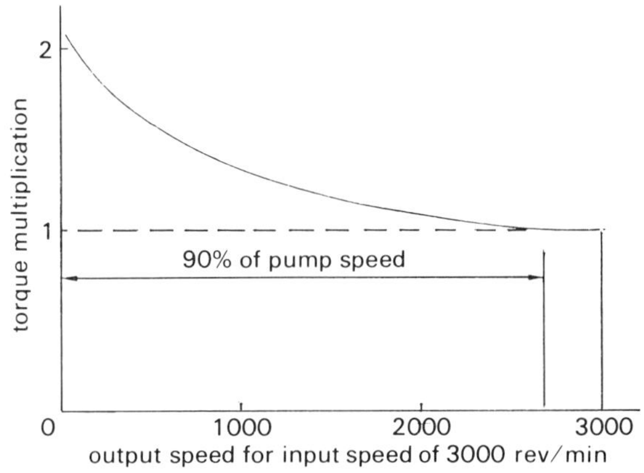

Fig. 41.11 Torque output variation

Three-element torque converter Fig. 41.12 shows the arrangement of a three-element single-stage converter. This type is quite common and is used in conjunction with many different types of automatic gearbox. Examination of Fig. 41.12 shows that the free wheel is the only mechanical feature which can produce faulty operation of the converter. A fault in the free-wheel can give:

1 *Slipping stator* – causes the fluid to enter the pump at the incorrect angle; therefore full torque multiplication cannot be obtained. This fault is detected by a stall test.
2 *Seized stator* – fluid striking the back of the stator vanes cannot make the stator free-wheel at the appropriate time so the fluid acts as a severe brake on the engine and produces overheating of the converter.

The fluid for the converter is normally supplied by

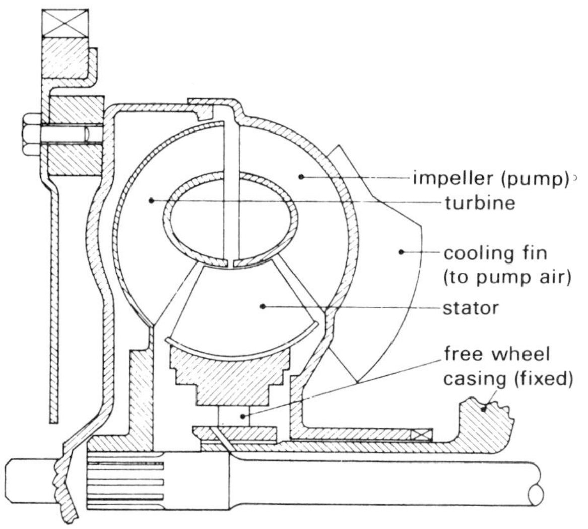

Fig. 41.12 Three-element single-stage converter

the automatic gearbox and it is generally a low viscosity mineral oil which contains additives to improve lubrication and resist frothing. Cavitation noise caused by air in the converter is minimized by pressurizing the fluid to about 138 kN/m^2 (20 lbf/in^2).

A *stall test* indicates if slip is occurring at the converter free wheel or in the gearbox. A general outline of the method is as follows:

1. Check that the engine is warm and that it is in good condition. A low engine power will give a false result.
2. Connect a tachometer to the engine.
3. Chock the wheels and apply the handbrake and *footbrake*.
4. Run the engine, select 'L' (or position recommended by the manufacturer), fully depress the throttle and note the maximum engine speed, which should be about 1800 rev/min.

Typical result	*Probable cause*
Below 1000 rev/min	Converter free-wheel slip
About 1200 rev/min	Engine not giving full power
Above 2000 rev/min	Gearbox clutch or band slip

During this test the engine power is absorbed by the converter fluid and this causes a rapid temperature rise. The duration of the test should be less than 10 seconds.

Multi-stage converter When greater torque multiplication is required a multi-stage converter is often used. This type uses a series of turbines and stators and Fig. 41.13 shows the main features.

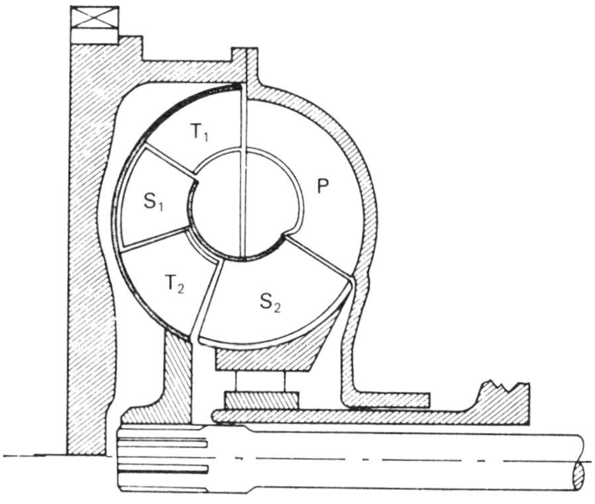

Fig. 41.13 Layout of multi-stage converter

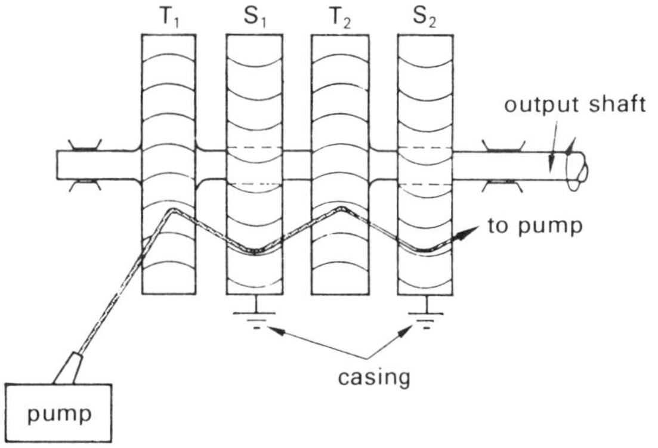

Fig. 41.14 Principle of multi-stage converter

Under conditions of low turbine/pump speeds the fluid will follow the path shown in Fig. 41.14 and by this means the energy in the fluid is gradually extracted as it passes through each stage.

Main gearbox

The majority of automatic gearboxes use planetary gearing to give three forward ratios and one reverse. However, with fuel economy in mind, some modern boxes incorporate an overdrive gear (fourth speed) and an extra clutch to lock the fluid converter when it is not needed.

Selector features

The gear selector, which is either floor-mounted between the front seats or attached to the steering column, is marked with a series of letters. In the past, a typical gear selector layout was L D N R P; these letters indicated the following:

L - Lock-up (or *low*) A low gear having no free-wheeling action which is suitable for engine braking. No automatic up-changes (up-shifts) occur whilst the selector is in this position.
D - Drive This is the normal automatic mode which covers the full range of forward gears.
N - Neutral As the name suggests, no drive is applied. This gear position is used when the vehicle is being towed; the distance should not be longer than about 5 km owing to the lack of lubrication.
R - Reverse Backward movement of the vehicle is achieved in this position.
P - Park A parking pawl, mounted to the casing, is moved into a gear connected to the gearbox output shaft. This provides a powerful lock that positively prevents movement of the transmission.

For safety reasons the selector mechanism incorporates an *inhibitor switch* to prevent operation of the starting motor in any position other than P and N.

Alternative selector layout To give the driver extra control of the gearbox, some units have selector positions: P R N D 2 1. In this case the extra position '2' gives automatic changes from 1 to 2 and 2 to 1 with engine braking in 2. The position '1' is similar to 'L' in the basic layout.

Gearbox systems

For the purpose of study the gearbox is divided into separate systems; these are:
- *Mechanical gear train*
- *Clutch and brake band actuation*
- *Hydraulic control of actuators*
- *Speed and load sensing.*

Mechanical gear train

In view of the many different gear layouts used today, a few are considered in this book to demonstrate that the basic principles are similar. Each one of the units considered is used in conjunction with a single-stage hydro-kinetic fluid converter; in some cases the converter is fitted with a lock-up clutch.

Borg-Warner (model 35)

This uses two epicyclic gear sets but the front set does not have an annulus (Fig. 41.15). Instead it uses a compound planet arrangement in which the front set of planets mesh with three gear members: the front sun, rear annulus and rear planets.

Gear	C_1	C_2	B_1	B_2	F/W	Ratio
L1	•			•		2.39
L2	•		•			1.45
D1	•				•	2.39
D2	•		•			1.45
D3	•	•				1.00
R		•		•		2.09

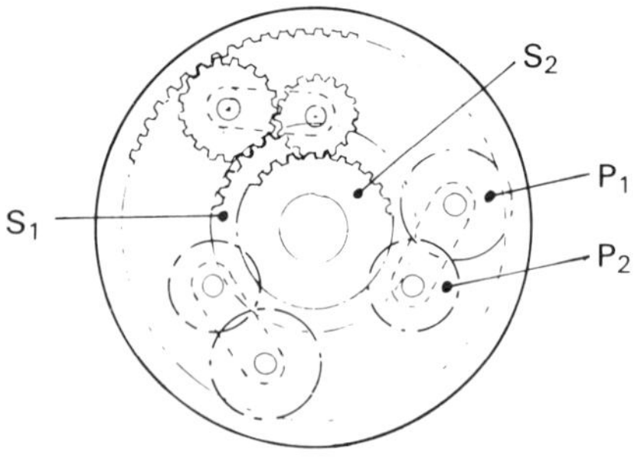

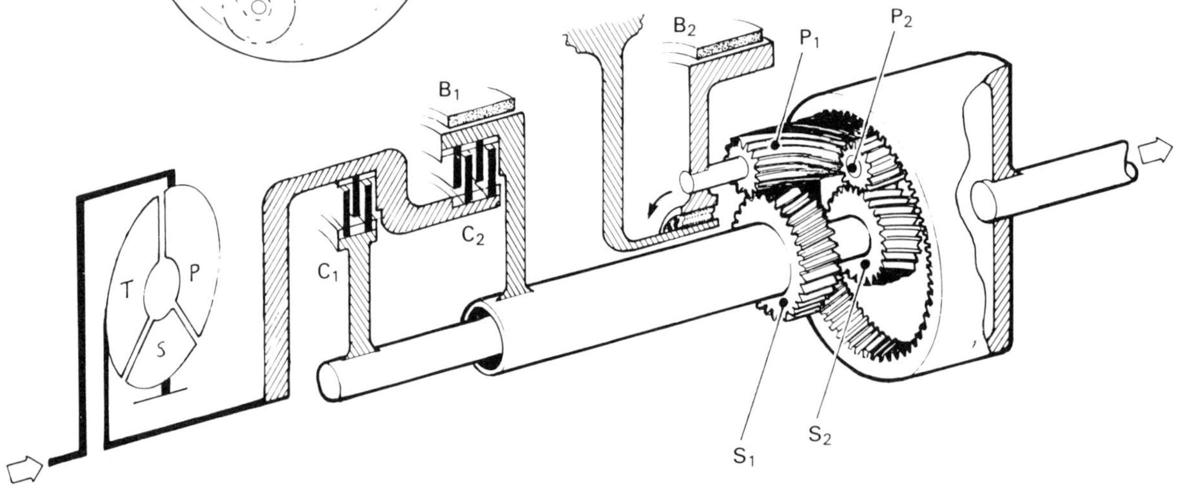

Fig. 41.15 Gear train layout (BW35)

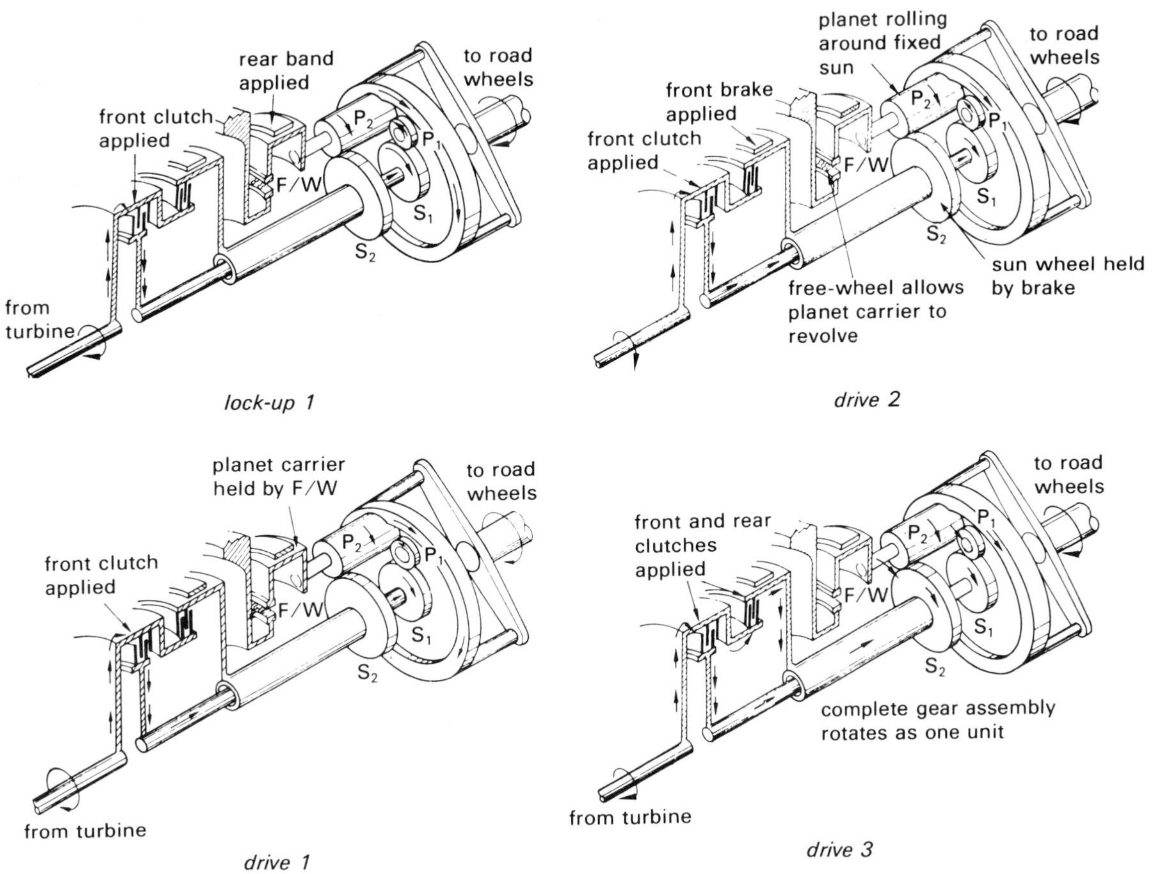

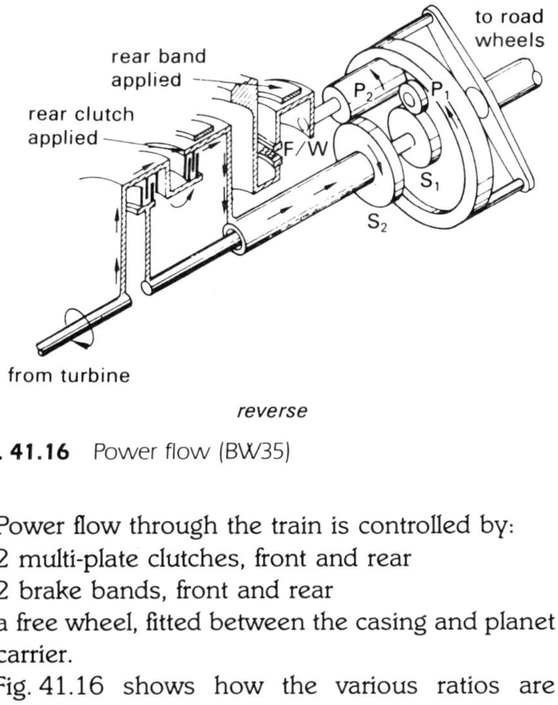

Fig. 41.16 Power flow (BW35)

Power flow through the train is controlled by:

- 2 multi-plate clutches, front and rear
- 2 brake bands, front and rear
- a free wheel, fitted between the casing and planet carrier.

Fig. 41.16 shows how the various ratios are obtained.

L – Lock-up This first gear applies the drive to sun S_2 and the rear brake holds the planet carrier.

Forward motion of sun S_2 transmits the drive through both planets, which act as idlers, to the annulus and output shaft. Engine braking is obtained because the rear brake puts the free wheel out of action.

D1 – Drive 1 This is similar to 'L' but in this case the planet carrier is held by the free wheel instead of the rear brake.

Under over-run conditions, the free wheel unlocks and prevents the road wheels from driving the engine.

D2 – Drive 2 Drive is applied to S_2, and S_1 is held by the front brake.

Rotation of P_1 by P_2 causes it to roll around the fixed sun S_1 and drive the planet carrier in the forward direction. This drives the annulus faster and gives second gear.

Since the planet carrier is moving in a forward direction, the free wheel offers no resistance.

D3 – Drive 3 Drive is applied to S_1 and S_2 by locking both clutches.

This action causes the annulus and both sets of planets to rotate as one unit and give direct drive.

R – Reverse Drive is applied to S_1 and the rear band is applied.

The locked planet carrier causes the planets to act as idlers, so the forward-moving sun S_1 moves the annulus in the reverse direction.

N – Neutral All clutches and brake bands are released, so no drive is transmitted.

Borg Warner (model 55)

This model is designed for larger engines, having a capacity of about 3 litres. Fig. 41.17 shows how it is

Gear	Fwd clutch	Direct clutch	F_1	F_2	B_1	B_2	B_3	Ratio
1	•						•	2.452 Lock up
2	•				•			1.452 Lock up
D1	•			•				2.452
D2	•		•					1.452
D3	•	•						1.000
R		•					•	2.212

possible to use a multi-plate clutch to give a braking action. Merits of this design include a smoother gear change and elimination of periodic adjustment of the brake bands.

The selector position '2' gives a *second gear hold* combined with engine braking. The unit has two drive clutches, three brake clutches and two free wheels.

The basic layout of the epicyclic gear train is similar to that employed by many manufacturers, e.g. Toyota and German-made ZF boxes used by Jaguar and Rover.

First gear Drive is applied to A_2, and the front planet carrier is held either by the free wheel F_2 (D1) or the brake B_3 (1).

Forward motion of A_2 drives both S_1 and S_2 in the opposite direction; this causes P_1 to act as an idler and drive A_1 in a forward direction.

Second gear Drive is applied to A_2 and the sun S_2 is held by either the free wheel F_1 (D2) or the brake B_1 (2).

Forward motion of A_2 'walks' the planet P_2 around the fixed sun S_2; this action causes the planet carrier and output shaft to move as given by the ratio $(A+S)/A$.

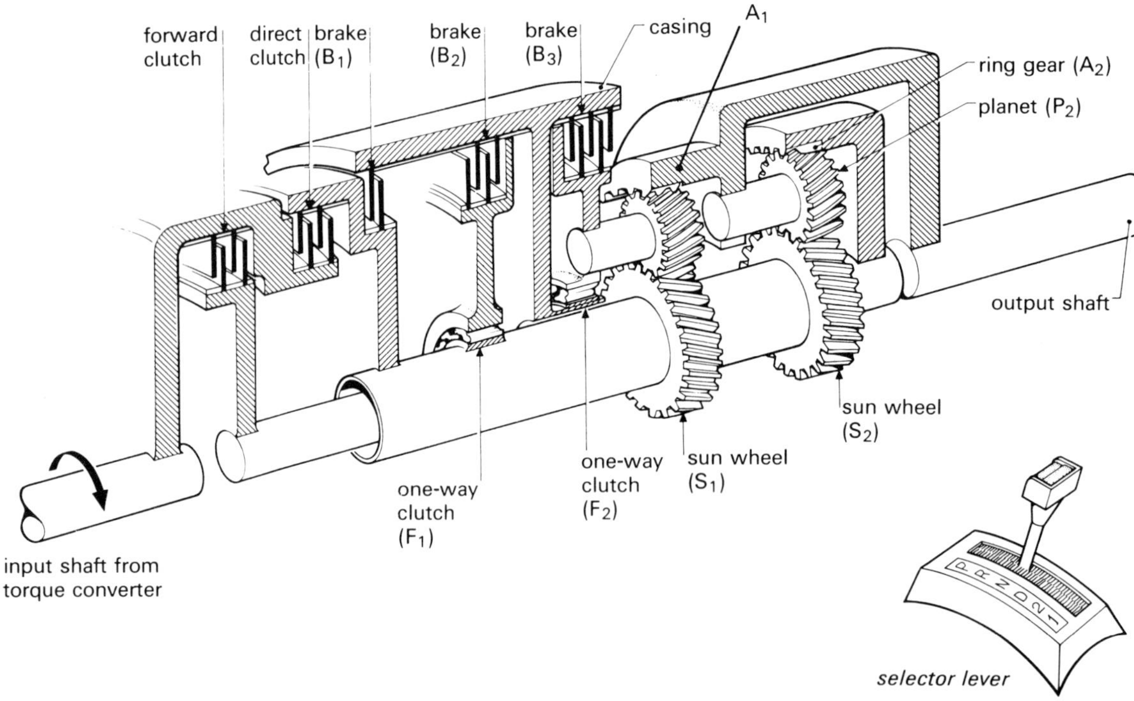

Fig. 41.17 Gear train layout

Third gear Drive is applied to A_2 and S_2. Direct drive is obtained because both gear members rotate at the same speed.

Reverse gear Drive is applied to S_1 and the front planet carrier is held by the brake B_3. The forward-moving sun will drive A_1 in the reverse direction; the planet P_1 will act as an idler.

Ford

The gear train layout used in the Ford C3 Bordeaux gearbox is similar to that used by other American manufacturers, e.g. Chrysler 'Torque Flite'.

A five-position selector, marked P R N D 2 1, controls the action of two clutches, two brake bands and a free wheel.

Fig. 41.18 shows the layout of the compound gear train.

First gear Drive is applied to A_1 and the rear planet carrier is held by either the free wheel (D1) or the rear brake B_2 (1).

The resistance to movement of the output shaft, which is connected to the front planet carrier, causes the front planet to drive the two suns S_1 and S_2 in the opposite direction. With the rear planet acting as an idler, the annulus A_2 is driven around slowly in a forward direction.

Second gear Drive is applied to A_1 and the front brake B_1 is applied to hold S_1. The speed reduction is given by the front gear set, the ratio being $(A + S)/A$.

Third gear Drive is applied to A_1 and S_1. This causes both A_1 and S_1 to rotate at the same speed; direct drive is obtained.

Reverse gear Drive is applied to S_2 and a brake holds the rear planet carrier. Reverse is obtained by the rear gear set, the planet P_2 acting as an idler.

Volkswagen

The compact VW transaxle is used by other manufacturers of front-wheel drive cars. This layout is shown in Fig. 40.8 and the gear train is described on page 248.

Overdrive gear set

A four-speed gearbox having an overdrive top gear can be achieved by adding an extra gear set at the rear of a normal three-speed epicyclic train.

In the overdrive layout shown in Fig. 41.19, fourth gear is obtained by holding sun S_3 by a brake; the drive is applied to the planet carrier PC_3.

Driving the planet carrier around the fixed sun causes A_3 to rotate at an overdrive speed as given by the ratio $A/(A + S)$.

In other gear positions the overdrive gear set is put out of action by either the lock-up clutch or the free wheel; the former is used for reverse and also for gear positions designed to give engine braking.

Clutch and brake band actuation

A number of brake bands and multi-phase clutches are fitted in an automatic gearbox. Clutches are used

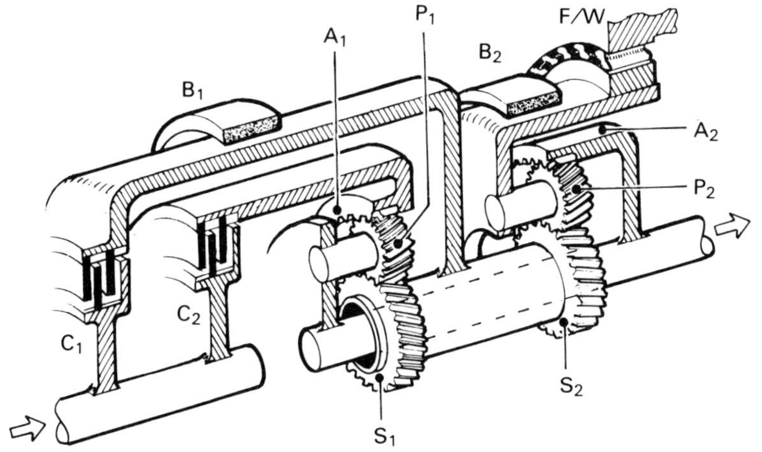

	C_1	C_2	B_1	B_2	F/W
1		•		•	
2		•	•		
D1		•			•
D2		•	•		
D3	•	•			
R	•			•	

Fig. 41.18 Gear train layout (Ford C3)

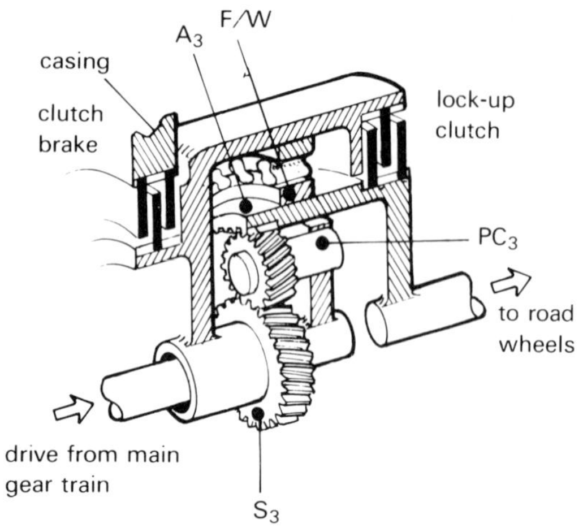

Fig. 41.19 Overdrive gear set (ZF)

to connect the gear train to the input shaft and band brakes are fitted to hold a part of the train stationary.

In a number of modern gearboxes, clutches are sometimes used instead of band brakes; this makes the gearbox more compact and lighter in weight; also it overcomes the need for adjustment that must be carried out periodically to compensate for friction lining wear.

Oil pressure, produced by a pump driven at engine speed from the fluid converter, is distributed by control valves to the appropriate clutch and brake when actuation of these parts is required.

Multi-plate clutches Numerous *wet-type* multi-disc clutches are fitted and most of these operate on the same principle.

Fig. 41.20 shows a typical construction. Two sets of steel plates, inner and outer, are connected alternately by protruding tabs to the hub and drum respectively. Bonded to the faces of the inner plates is a friction

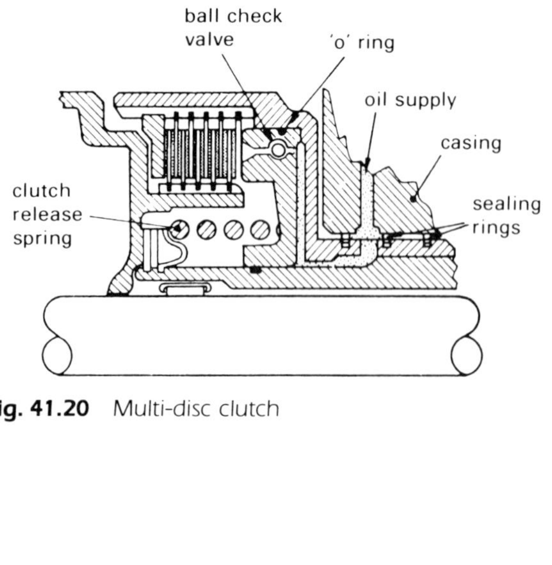

Fig. 41.20 Multi-disc clutch

material having either a hard or comparatively soft texture. A hard facing is made of a cellulose compound, or synthetic fibre, bonded together with a phenol resin to obtain a suitable friction value. A soft facing, which is based on a compound of paper, is more porous and elastic. Paper-based facings normally give a smoother and quieter take-up when operated over a wide range of temperature and pressure.

When the clutch is to be engaged, oil under pressure is supplied through a drilling, in either the casing or the shaft, to the clutch operating cylinder. Leakage of oil between moving parts, and loss of pressure needed to operate the clutch, is prevented by fitting a number of synthetic rubber 'O' rings and square-section, cast-iron seals.

Torque transmitted by a given multi-plate clutch depends on *friction value* and *operating pressure*, so one of these is the cause when *clutch slip* occurs.

Drag between the plates when the clutch is disengaged is an energy drain, so provision is made to separate the plates. A large clutch release spring returns the operating piston and in some cases the steel outer plates are slightly dished.

Immediately after the clutch has been disengaged, the centrifugal action of the residual oil in the operating chamber would apply the piston and cause slight drag. This is prevented by having a *check valve* to release the pressure.

When oil pressure operates the clutch the check valve is held closed, but removal of this pressure allows the ball to move outwards, owing to the centrifugal effect, and open the release hole.

Some automatics have a clutch which is operated by two pistons, one small and the other large in area. This construction, which is shown in Fig. 41.21, allows the thrust on the plates to be varied to suit the conditions; this achieves a smoother operation.

Brake bands Normally two or more brake bands are fitted in a gearbox to prevent rotation of the gear members. An external contracting, single-band type brake is used and Fig. 41.22 shows the basic construction. The band is anchored at one end to the casing to resist the rotational force and is actuated by a hydraulic servo at the free end. Maximum braking torque is achieved by siting the anchor in such a position that the rotational force of the drum produces a self-wrapping action.

The spring steel brake band is internally lined by bonding to it a hard friction material. When the brake is applied this material makes contact with a steel

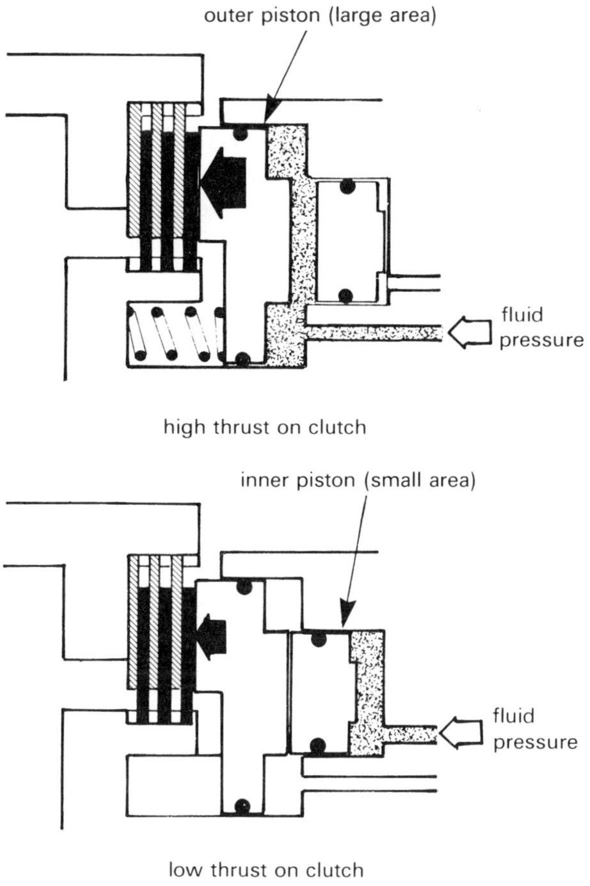

Fig. 41.21 Clutch operation with two pistons

drum connected to a part of the gear train.

An adjuster to compensate for wear is normally provided at the band anchor. The screw for this adjuster is often carried through the gearbox casing to simplify maintenance. Adjustment of the band normally involves releasing the lock nut and then tightening the screw to a given clearance or torque setting.

Various types of servo actuator are used to suit the gearbox; the common types are:

- *Single-acting*
- *Delay*
- *Double-acting*.

Single-acting servo The basic type shown in Fig. 41.22 consists of a single hydraulic piston fitted in a cylinder that is fixed to the gearbox casing. Fluid leakage past the piston is prevented by a synthetic rubber 'O' ring and a spring is fitted in the chamber to return the piston.

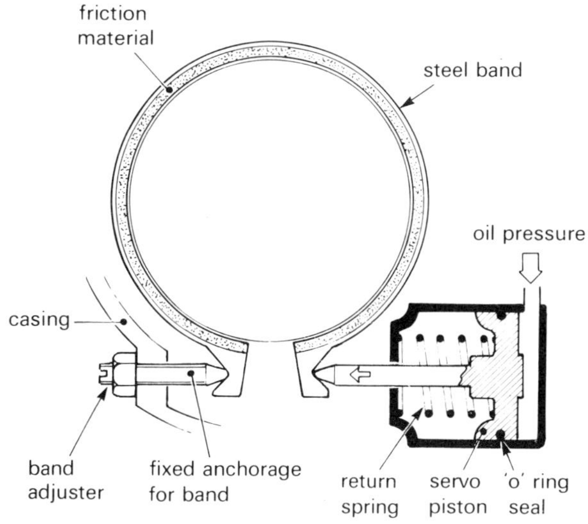

Fig. 41.22 Brake band actuation by single-acting servo

The force exerted by a servo must be sufficient to prevent slip of the drum, so bands controlling gears having a high torque output are fitted with a servo cylinder which has either a large area or a high operating pressure.

Delay servo When the vehicle is in motion, a sudden engagement of a brake causes a sudden jolt. Harsh changes and jolts should not occur with well-designed automatics because they incorporate special devices to minimize this problem; one of these is the delay servo shown in Fig. 41.23.

The delay feature is achieved by using a second

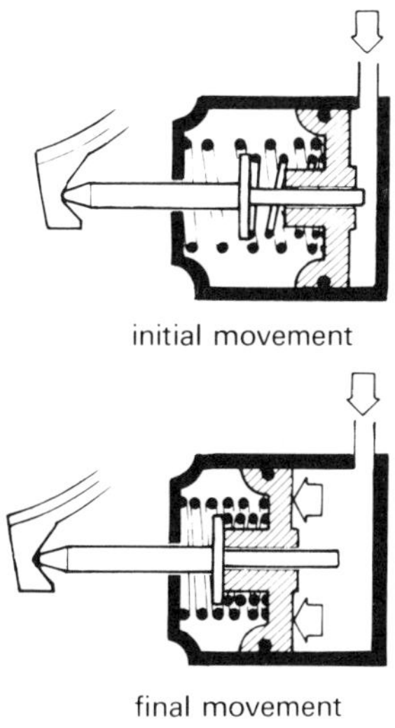

Fig. 41.23 Delay servo

(inner) spring and a piston rod that can slide through the piston, instead of being rigidly attached as in the basic type.

When fluid pressure is applied, the initial piston movement compresses the inner spring. Once this has taken place, positive contact between the piston and rod is made and the full force is applied to the brake in the normal way.

The cushioning action of the inner spring gives a gradual application of the brake; also it damps the vibration produced by the band during the initial stage of engagement.

Double-acting servo On this type fluid can be directed to one or both sides of the piston. In Fig. 41.24a fluid under pressure supplied at A will act on the small-area part of the piston and apply the servo in the normal way. When an equal fluid pressure is applied to A and B (Fig. 41.24b), the piston will move in the 'release' direction. This is due to unequal thrust produced on the piston by the fluid pressure acting on the side of larger area.

A double servo is used on a Borg-Warner 35 box to provide a smooth change between second and third; in this case application of the brake gives second gear.

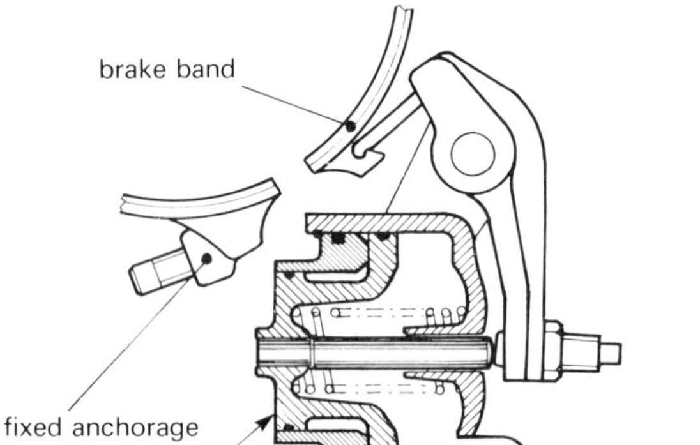

(a) Brake being applied *(b) Brake being released*

Fig. 41.24 Double-acting servo

When a certain speed is reached, a change to third is made by directing fluid from a single line to the appropriate clutch and also to the 'release' side of the servo. Besides reducing the number of hydraulic lines, this arrangement gives a quick down-change because as soon as the fluid is released from one side of the servo, the second gear immediately comes into operation.

The majority of automatic gear changes are made whilst the drive is being transmitted through the gearbox. If a brake, or clutch, were released before the next gear units were partly applied, then the engine would suddenly speed up. This condition is called *flare-up*; it is prevented by overlapping the engagement of the new gear with the disengagement of the old gear.

If the period of overlap is too great, then a harsh gear change condition, sometimes called *tie-up*, results. This occurs because the two 'gears' oppose each other, i.e. each gear attempts to drive the output shaft at a speed dictated by its ratio, with the result that the gearbox partially locks up.

Clutch and brake slip This fault produces similar symptoms to those given by slip of a main clutch. If the fault is allowed to persist for a period of time, the accelerated wear of the friction facing, together with the effects of the high temperature caused by the slip, will only be remedied by fitting a new clutch or brake unit.

Slip of the clutches and brakes can be diagnosed when a stall test is carried out. A test using each gear position will indicate which part is faulty. Manufacturer's recommendations relating to the time taken for the test, must be observed.

Hydraulic control of actuators

The clutches and brake bands of most automatic gearboxes are operated by hydraulic means. A pump, driven at engine speed from the torque converter generates the fluid pressure; this pressure should be sufficiently high to lock the clutch plates or brake drum, but not so high as to waste energy. Distribution of fluid to the clutches and brakes is by means of a control valve, called a manual valve, linked to the driver's selector lever.

Valve operation Although many valves are used in automatic gearboxes, the basic principle of operation of each valve is the same. At this stage the construction and operation of some of the valves is considered.

Regulator valves (relief valves)

The main purpose of this type of valve is to limit the hydraulic line pressure to a given maximum. The three main types, shown in Fig. 41.25, are: ball, piston and spool.

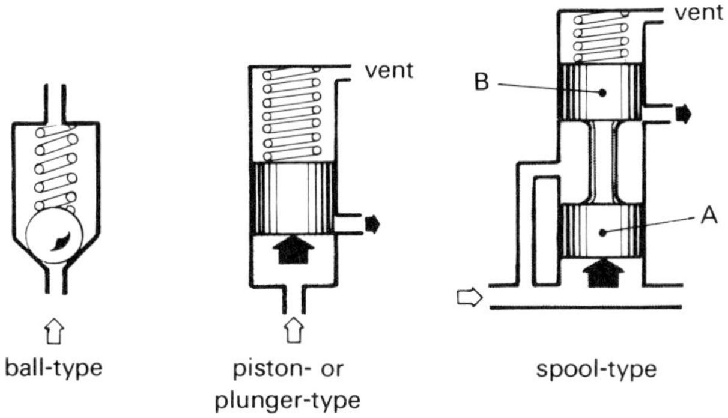

Fig. 41.25 Types of regulator valve

Ball- and *plunger-type valves* are similar to those used on engines. When the pressure acting on the valve produces a thrust greater than the force exerted by the spring, then the valve lifts and so prevents any further increase in the pressure. The operation is based on the relationship:

$$\text{Thrust} = \text{Pressure} \times \text{Area}$$

Simple spool valves are similar to the plunger type except that they are waisted (reduced in area at the centre) to control the fluid flow. In Fig. 41.25 the areas of A and B are the same, so the pressure in the waist region will exert an equal thrust upwards and downwards; as a result the pressure in this region will not affect the movement of the valve.

When pressure is sufficient to lift the valve, a port is uncovered. Fluid will then flow to the waisted part of the valve from where it is either returned to the reservoir or passed to a line that acts on another part of the system.

Differential spool valves have two spool 'plungers' of different diameter similar to the type shown in Fig. 41.26. As fluid pressure begins to build up, the spool will move to the right; this is because the effective area of B exposed to the fluid is larger than that of A. The actual movement of the spool is controlled by the spring strength. The spring is situated in a chamber that is vented to avoid a pressure build-up behind the valve. When the pressure reaches a set value dictated by the spring,

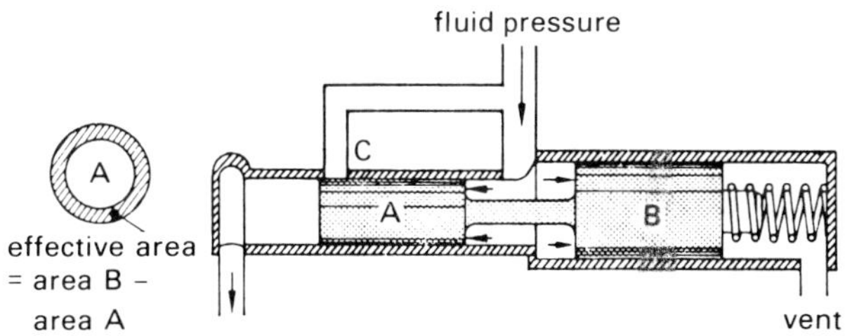

Fig. 41.26 Simple pressure regulator valve

the spool uncovers the port C; this spills out the fluid and prevents any further rise of pressure.

Control valves

The purpose of this type of valve is to direct the fluid to a hydraulic line that either activates or controls the appropriate clutch and/or brake.

Manual control valve   This valve is linked by a rod or cable to the driver's selector lever, so its position is dictated by the gear position set by the driver.

Fig. 41.27 shows a very simple manual valve used to control forward and reverse movement of a vehicle. In this case, operation of the lever moves the valve and uncovers the port that supplies either the forward or reverse gear actuators. Positive location of the valve is obtained by a selector ball detent arrangement.

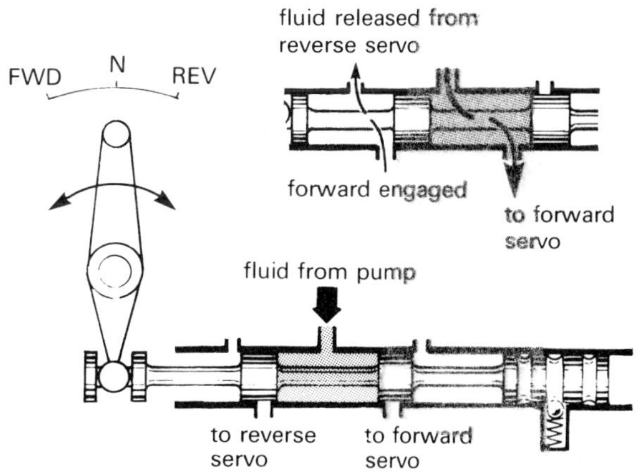

Fig. 41.27 Simple manual valve to control forward and reverse gears

Pressure sensitive control valves   The basic system for the automatic control of the gear changes must sense the speed of the road wheels. A hydraulic system of control uses a governor for this purpose. Mounted on the output shaft, the governor generates a fluid pressure that increases as the road speed rises.

Control valves, connected to the governor fluid

circuit, sense the pressure in this circuit. At the appropriate time, the valve operates and directs the fluid through the lines to engage the new gear.

Fig. 41.28a shows a spool valve with a fluid connection at each end. If the areas of the faces A and B are equal, the valve will:

1 move to the right when the fluid pressure at C is greater than at D and vice versa.
2 remain in a central position when:

$$\text{Pressure at C} = \text{Pressure at D}$$

Any pressure given by the fluid at E will have no effect on the movement of the valve because the thrust tending to move the valve to the right will be balanced by an equal and opposite thrust acting on the other face. In many ways this is similar to the action of the two springs.

Making the area of one spool larger, as shown in Fig. 41.28b, will cause the pressure acting on the larger area to give a greater thrust on the valve than that given by a similar pressure at C.

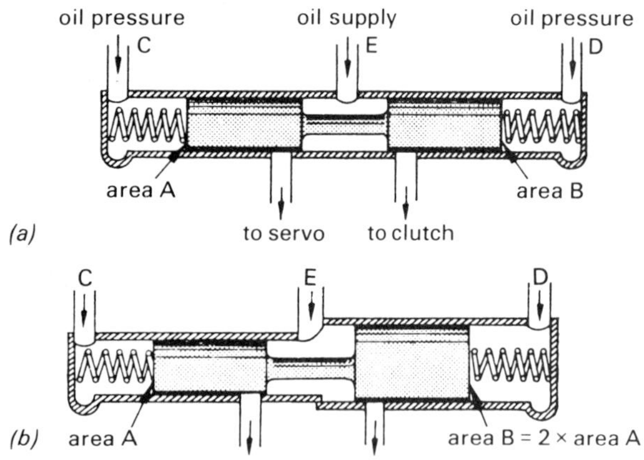

Fig. 41.28 Spool valve controlling flow to either servo or clutch

Basic hydraulic system

The hydraulic system in this chapter does not represent an actual system used in any particular make of gearbox; instead it illustrates the basic principles of automatic control by using a much simplified system. From this base the reader should be able to comprehend the construction and operation of a modern unit.

The system is considered in a series of stages.

Stage 1

Fig. 41.29 shows the layout of a simple fluid supply system used to operate a clutch and torque converter.

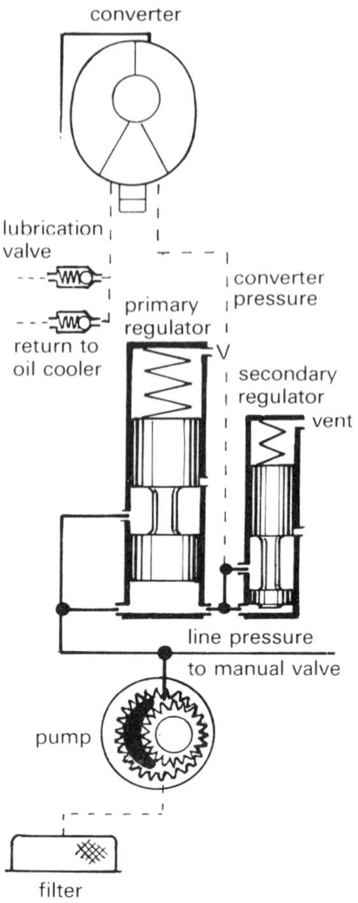

Fig. 41.29 Hydraulic circuit – stage 1

Pump Normally an internal-external gear pump (see page 161) is used; this pump is driven at engine speed from tangs formed on the torque converter casing. The fluid is drawn from the reservoir and passed to the pump through a fine screen-type filter so as to remove any small dirt particles.

Since most modern automatic gearboxes have only one pump at the input it is not possible to tow-start the engine. This is because the engine and pump will be stationary, so pressure will not be generated in the system to activate any clutch and brake – neutral will result.

Primary regulator This spool valve controls the *line pressure* that is applied to the *manual valve*. In the system shown, the line pressure always remains constant. (This is unsuitable for a modern gearbox because the high pressure needed to prevent slip under conditions of high torque will give harsh changes when the engine is lightly loaded.)

Fluid released from the valve is passed back to the inlet side of the pump instead of discharging into the reservoir. This saves energy that would be used if the fluid had to be drawn through the fine-mesh filter screen.

Secondary regulator Line pressure is too great for the converter, so the secondary regulator acts as a pressure reducer. A portion of the fluid returning from the converter passes either to an oil cooler, or to a line that provides low-pressure lubrication of the gearbox.

Manual valve Under the control of the driver this valve distributes the fluid to the various valves, clutches and brakes. In the position shown in Fig. 41.30, the fluid is delivered to the *front clutch* at *line pressure*.

Sequence of clutch and brake application The order in which the clutch and brake actuators are applied depends on the type of gear train. The layout used for this consideration has a *drive range* as shown in Table 41.1.

Table 41.1 Clutch and brake operation

Gear	*Front clutch*	*Rear clutch*	*Brake band*
D1	●		
D2	●		●
D3	●	●	●

The table shows that the three gears in the drive range all use the front clutch; the particular gear in this range is determined by the actuation, or non-actuation, of the rear clutch and brake. Use of the front clutch in this manner is common to many other gearboxes. It should be noted that wear on the forward clutch is minimal because, unlike the rear

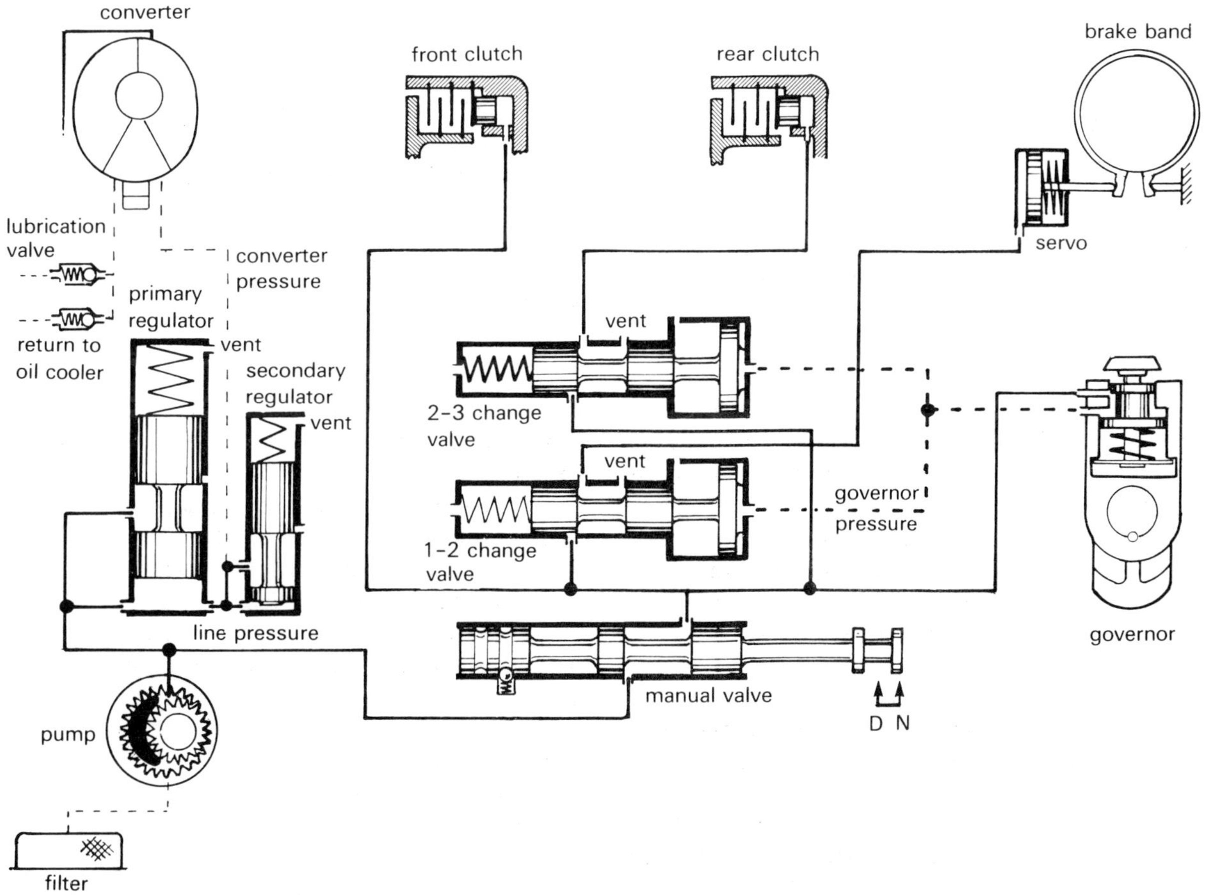

Fig. 41.30 Hydraulic circuit – stage 2

clutch, the application of the forward clutch is only carried out when the vehicle is stationary.

Speed and load sensing Gear changes must be made when a predetermined road speed is reached, so some form of sensor is necessary to 'tell' the gearbox when to change the gear.

Speed sensing Automatic gearboxes use one of the following to determine the speed of the output shaft:

- *Mechanical governor*
- *Hydraulic governor*
- *Electronic speed sensor*.

The speed sensor must work in conjunction with a *shift valve* (*change valve*); the duty of this valve is to direct the fluid to the clutches and servos so as to change the gear when a suitable road speed is reached.

Mechanical governor This centrifugal system consists of two bob-weights which throw out against the resistance of a spring as the speed of rotation increases. The governor is driven from the output shaft, and movement of the flyweights is made to operate a hydraulic spool valve that directly controls the gear changes.

Today very few gearboxes use this system of speed sensing.

Hydraulic governor This type was developed from the previous type. It uses a spool valve, sensitive to a centrifugal effect given by its rotation, to generate a fluid pressure that increases with speed.

Electronic speed sensor An indication of the road speed is needed for many electronic systems used on modern vehicles. These data are provided by some form of magnetic sensor fitted close to the output shaft of the gearbox. Signals from the sensor are passed to an ECU (electronic control unit) for processing so that the data can be used for a particular electronic system such as gearbox control.

Stage 2

Fig. 41.30 superimposes a set of *change valves* and a *hydraulic-type governor* on the previous layout.

Change valve Two valves are shown in Fig. 41.30; one valve controls the up and down changes between first and second (*1–2 shift valve*) and the other the changes between second and third (*2–3 shift valve*).

A simplified shift valve is shown in Fig. 41.31. Line pressure from the manual valve acts at A and the line supplying the appropriate clutch or servo is connected at port B. Movement of the valve against the spring is produced by pressure of the fluid from the governor at port C. When the *governor pressure* is sufficient to overcome the spring, the shift valve moves and connects port A with port B. This allows line pressure to act on the servo or clutch so as to produce the change of gear.

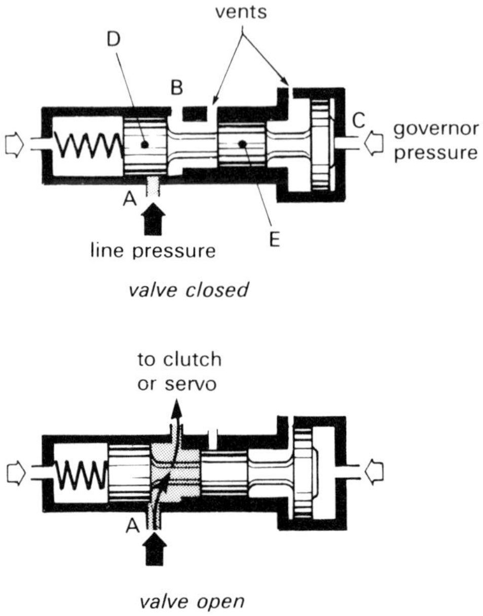

Fig. 41.31 Gear change (shift) valve

By fitting a stronger spring to the 2–3 shift valve (or making the valve smaller in area), the 2–3 valve is made to open at a much higher governor pressure than the 1–2 valve.

The diameter of the plunger D is larger than plunger E. This feature is effective after the valve has opened and its use prevents a gear *hunting action*. This action is a severe vibration of the vehicle caused by the shift valve oscillating between the open and closed positions. Once the valve starts to open, the extra thrust on piston D adds to the thrust given by governor pressure; as a result of this added thrust the valve is opened fully. This arrangement makes the up-changes occur at higher road speeds than the down-changes.

Hydraulic governor This valve is a form of regulator valve; its duty is to alter the fluid pressure in the

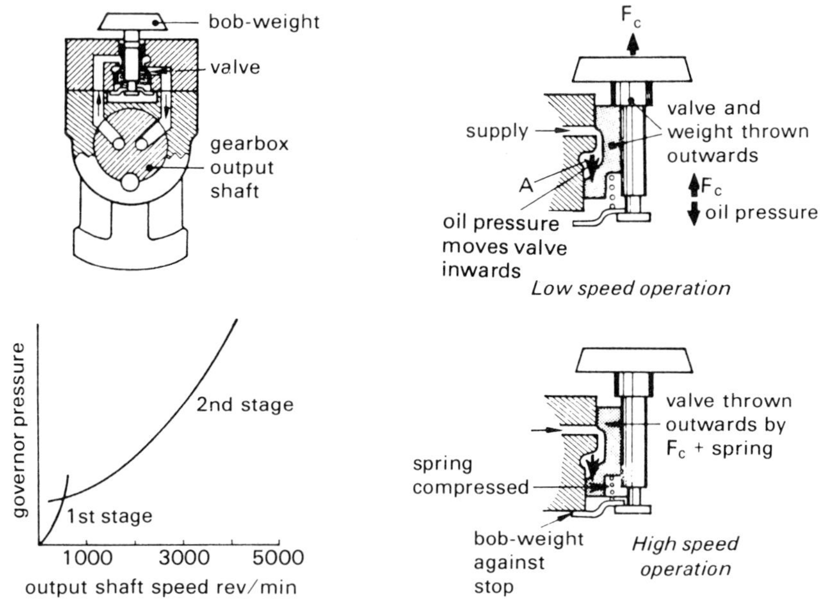

Fig. 41.32 Governor operation

governor line in keeping with the road speed: as speed increases, the pressure rises.

The principle of operation of a governor is shown in Fig. 41.32. Fitted in the governor body is a valve and in the centre of this slides a bob-weight. A spring pushes the valve outwards towards the bob-weight. This construction gives a two-stage characteristic and, as shown by the graph, the high-pressure rise, in relation to a small speed increase during the first stage of its operation, gives good sensitivity; this ensures that the 1–2 gear change always occurs at the same speed.

At speeds less than about 500 rev/min the first stage applies. During this stage the bob-weight and valve act as a solid unit, so under bob-weight control the valve opens and closes the supply port as the weight moves outwards and inwards.

The change in governor output pressure is achieved by the action of the fluid on face A. Following an increase in speed, the valve moves outwards and the supply port is opened. This causes the fluid pressure acting on face A to rise until it produces a sufficient thrust to move the valve inwards and close the supply port.

A further increase in speed causes this operation to be repeated, but in view of the greater centrifugal effect, the pressure in the valve chamber must reach a higher value before it can close the supply port.

The second stage starts when the bob-weight stop contacts the governor body. From this point onwards, the balance between the opening and closing of the valve is achieved when:

Centrifugal force on valve = Thrust of fluid on face A

Operation of Stage 2 layout The automatic selection of three gears can be achieved with the layout shown in Fig. 41.30.

When 'D' is selected, the manual valve supplies line pressure to the front clutch and with the aid of a free wheel, first gear (D1) is obtained. Line pressure is also applied to the other valves in the system, but all of these valves are inactive. As soon as the vehicle moves, rotation of the output shaft produces a governor pressure and this acts on both shift valves. At a speed controlled by the strength of the spring in the 1–2 shift valve the governor will generate sufficient pressure to open the valve. This action allows line pressure to act on the brake servo and give second gear (D2).

A further increase in road speed causes the governor pressure to rise further until it reaches a point where it overcomes the stronger spring in the 2–3 shift valve. At this speed the opening of the 2–3 valve causes line pressure to act on the rear clutch and give third gear (D3).

Load sensing

The road speed at which the up and down gear changes occur varies with the load of the engine, i.e.

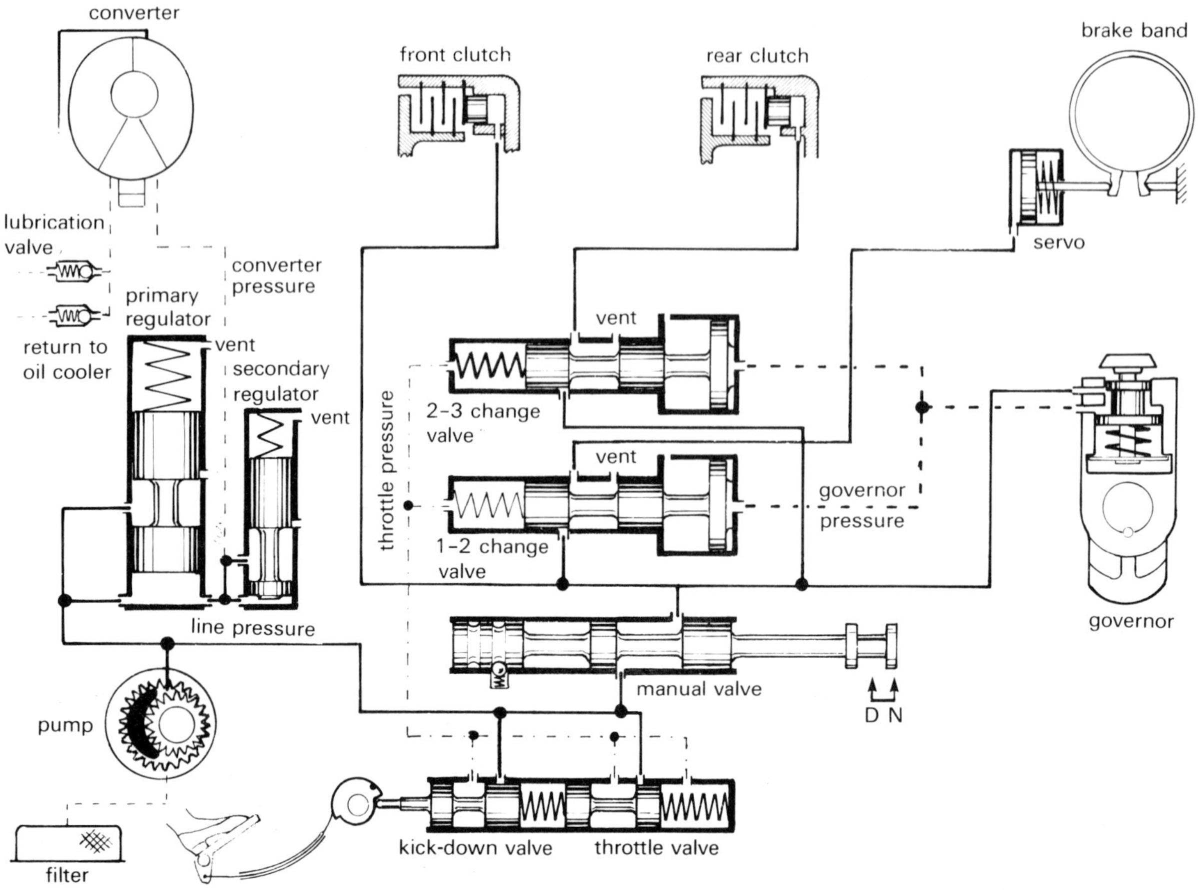

Fig. 41.33 Hydraulic circuit – stage 3

the speed at which the changes occur increases with movement of the accelerator pedal. This feature mimics the technique used by a driver: light throttle pressure, early changes; heavy throttle pressure, delayed changes.

Various methods are used to sense the 'driver's mood' and most automatics use one of the following:

- *Throttle valve, manually operated*
- *Throttle valve, vacuum-operated*
- *Electronic control.*

Throttle valve, manually operated The function of this valve is to produce, in a separate line, a fluid pressure that increases as the accelerator pedal is depressed. This *throttle pressure*, applied to the end of the shift valves, opposes the *governor pressure* that acts on the other end of the valves and, as a result, alters the point at which the shift valve operates.

The throttle valve is normally operated by a cable; this connects a lever on the side of the gearbox to some part of the carburettor throttle linkage. As the throttle is opened the cable is pulled; this action rotates a cam in the gearbox and compresses a spring that acts on the valve.

The *Stage 3* layout (Fig. 41.33) shows how a simple throttle valve is positioned in the fluid system to control flow, and pressure, to the *throttle line*.

With the throttle closed, line pressure from the manual valve cannot pass the spool of the throttle valve, but if the throttle is opened a given amount, the spring on the cam side pushes the valve to the right. This partially uncovers the port that connects with the manual valve, and allows fluid to spill past the throttle valve so as to generate a pressure in the throttle line. A connection between this line and the chamber at the end of the throttle valve causes pressure to be felt 'behind' the valve. As pressure builds up, the valve moves slightly to the left; this reduces the flow from the main line until a balance is reached, as with many other valves, between the valve opening and the output pressure.

Depressing the accelerator farther down repeats

this sequence; the only difference being that the increased spring thrust, resulting from the new position of the cam, requires a greater pressure in the throttle line to 'balance' the valve. So the more the throttle is depressed, the greater the pressure in the throttle line.

Besides acting as the main controller of 'change speeds', the throttle line pressure can also be used to vary the line pressure to suit the load on the engine. This is achieved by applying throttle pressure to the end of the main regulator valve.

Throttle valve, vacuum-operated Inlet manifold depression gives a good indication of the engine load, so some automatics have a throttle valve that is controlled from a vacuum chamber situated on the side of the gearbox.

Electronic control A sensor, controlled by either manifold pressure or throttle position, signals the engine load to an electronic control unit (ECU). After processing the signal, the ECU energizes a solenoid to move the appropriate shift valve and change the gear. Some gearboxes that use this system also have a vacuum chamber, connected to the manifold, to vary the line pressure to suit the conditions.

Kick-down

There are occasions, such as when overtaking, when a driver requires acceleration. On an automatic this is achieved by full depression of the accelerator pedal to a position called *kick-down*.

In a basic system, a movement of the pedal to this position opens a *kick-down valve* in the hydraulic system; this causes a sudden drop in pressure in the *throttle line*. Up to a certain speed in each gear position, this drop in pressure produces an immediate down-change of gear.

In Fig. 41.33 the kick-down valve is situated between the throttle valve and the cam. When the pedal is fully depressed the cam moves the kick-down valve to the right so as to cause a pressure increase in the throttle line.

Additional features

The modern automatic has a number of features in addition to the basic system previously described. These are fitted to make the gear changes smoother and the vehicle more economical, and to give the driver more control over the gearbox. The extra parts include the following:

Modulator valve This type of valve is fitted to reduce the pressure in a part of a given line to suit some special operating condition, e.g. fitted in the throttle line supply to the regulator valve. In this case it modulates (makes less violent) the rise in throttle pressure that acts on the regulator.

Accumulator The purpose of a hydraulic accumulator (sometimes called a *damper*) as applied to an automatic gearbox, is to reduce the rate at which a clutch or brake is actuated. It cushions the shock by delaying the full application of the fluid pressure.

Fig. 41.34 shows an accumulator consisting of a spring-loaded piston in a chamber that is connected in the fluid line between the shift valve and clutch (or brake servo). When the shift valve is opened, the time taken for the fluid to compress the accumulator spring ensures that the pressure rise in the line is gradual.

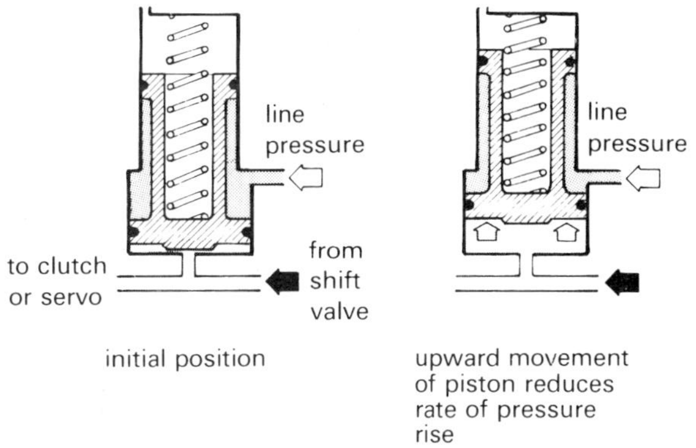

Fig. 41.34 Accumulator action

Torque converter lock-up In the interest of economy most modern automatics have a *converter clutch lock-up valve*; this provides fluid pressure to the converter clutch when the road speed in fourth gear is sufficient for the converter to be by-passed.

Automatic transmission fluid

The fluid used in an automatic gearbox must be capable of performing many functions; it must operate the torque converter, actuate the clutches and brakes, and lubricate the various gears and bearing surfaces.

Each one of these duties demands special properties, so it is essential that the fluid actually used contains a wide range of additives which will more than satisfy the recommended specification.

In traffic conditions a torque converter gets very hot, and although an oil cooler is now common, the

Fig. 41.35 Hydraulic circuit for a modern automatic (Jaguar ZF)

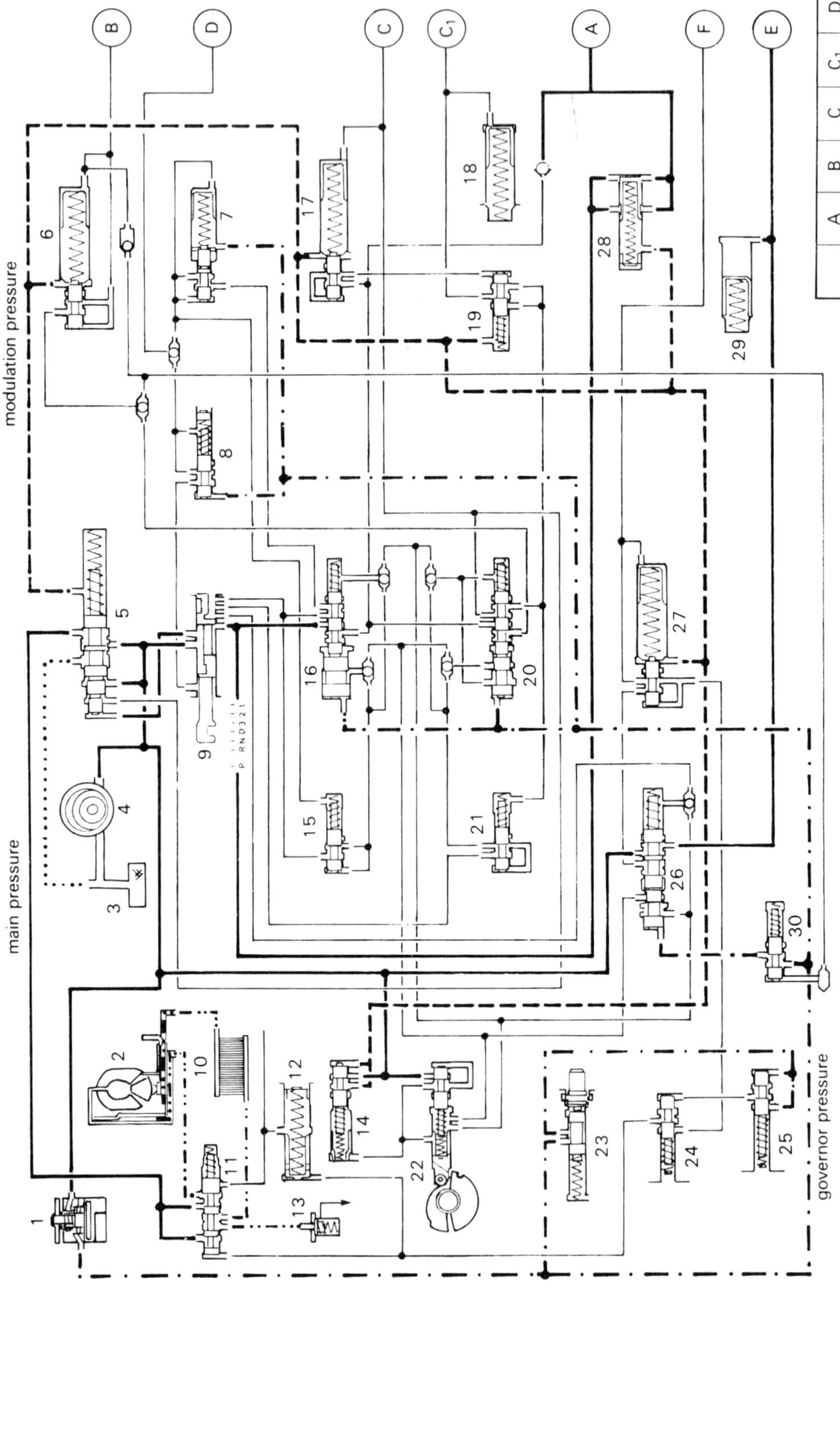

	A	B	C	C_1	D	E	F
D_1	•					•	
D_2	•		•	•		•	
D_3	•	•	•			•	
D_1	•	•	•				•
R		•			•	•	
P, N						•	

1 governor
2 torque converter
3 sump
4 pump
5 pressure regulating valve
6 B clutch valve and damper
7 D clutch valve and damper
8 reverse gear inhibit
9 manual valve
10 oil cooler
11 converter and reversing valve
12 converter clutch damper
13 one-way valve
14 modulator valve
15 1st gear inhibit
16 1–2 shift valve
17 C clutch valve
18 C_1 clutch damper
19 C_1 clutch valve
20 2–3 shift valve
21 2nd gear inhibit
22 throttle valve
23 thermo valve
24 converter clutch lock-up control valve
25 converter clutch hysteresis valve
26 3–4 shift valve
27 F clutch valve and damper
28 A clutch damper
29 E clutch damper
30 F clutch inhibit valve (reverse)

high temperature reached by the fluid must neither cause excessive oxidation nor vary the viscosity adversely to make the gear changes harsh or noisy.

Additives, to resist wear of the rubbing surfaces and foaming of the fluid, are also essential because a long period between fluid changes is demanded.

Another important consideration is the need to use a fluid that gives a stable friction value. A fluid of the incorrect type can cause the clutch and brake unit to engage either too quickly or too slowly, depending on the type of fluid used.

Most modern gearboxes use a fluid made to the GM specification DEXRON® II. Older gearboxes often used an ATF (automatic transmission fluid) to the Ford standard F. It should be noted that the fluid used must be of a specification to suit the design of the box; clutch and brake problems occur if the wrong type of ATF is used.

Fluid level is indicated by a dipstick. Usually it is marked to show the level when the fluid is hot. Since the reservoir level varies with the position of the selector, it is essential to set the lever as recommended and run the engine at the suggested speed. A number of faults arise if the level is either higher or lower than that recommended.

Typical modern hydraulic system

Fig. 41.35 shows a layout of a hydraulic system of a modern transmission unit. The system operates three forward speeds, an overdrive fourth gear and a lock-up converter clutch besides giving a neutral and reverse.

Most of the hydraulic valves are grouped together in a valve block as shown in Figs. 41.36 and 41.37.

Faults and adjustments

All manufacturers provide in their literature suitable test procedures for the diagnosis of a wide range of faults. This material should be consulted if a fault, other than the simple ones listed here, is evident.

Regular servicing should avoid problems in the areas shown in Table 41.2.

Test sequence Assuming a particular fault is not identified from the previous table, then a test programme is carried out. The sequence is:

1 *Road test* – should follow the procedure suggested by the manufacturer.
2 *Stall test* – this checks the torque converter for stator slip, and the brake bands and clutches for slip.

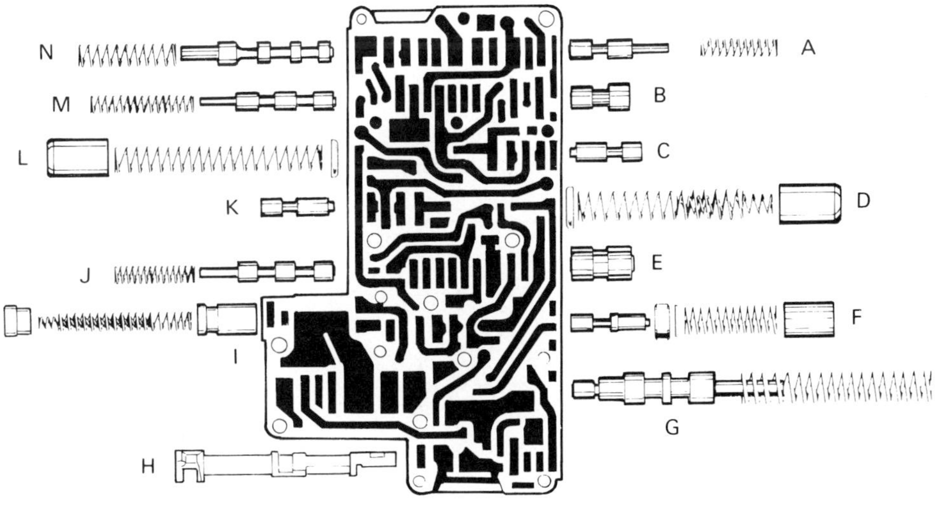

A = 'C1' clutch valve
B = 2-3 shift valve piston
C = 'C' clutch damper valve
D = 'B' clutch damper
E = 1-2 shift valve piston
F = 'D' clutch valve and damper
G = pressure regulating valve
H = manual valve
I = 'A' clutch damper
J = 1-2 shift valve
K = 'B' clutch damper valve
L = 'C' clutch damper
M = 2-3 shift valve
N = converter pressure valve

Fig. 41.36 Hydraulic valve block (Jaguar ZF)

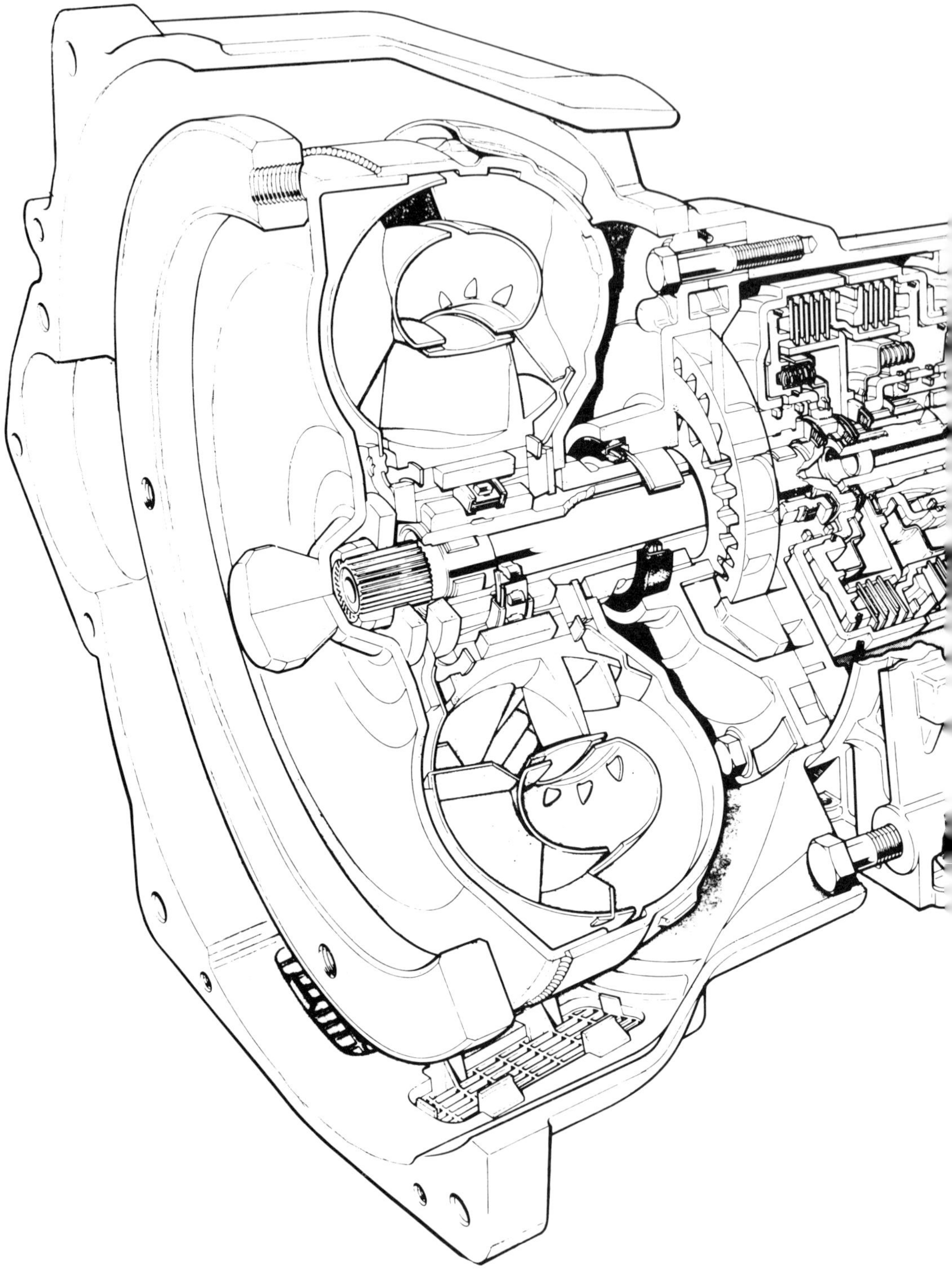

Fig. 41.37 Borg Warner Model 55 automatic gearbox

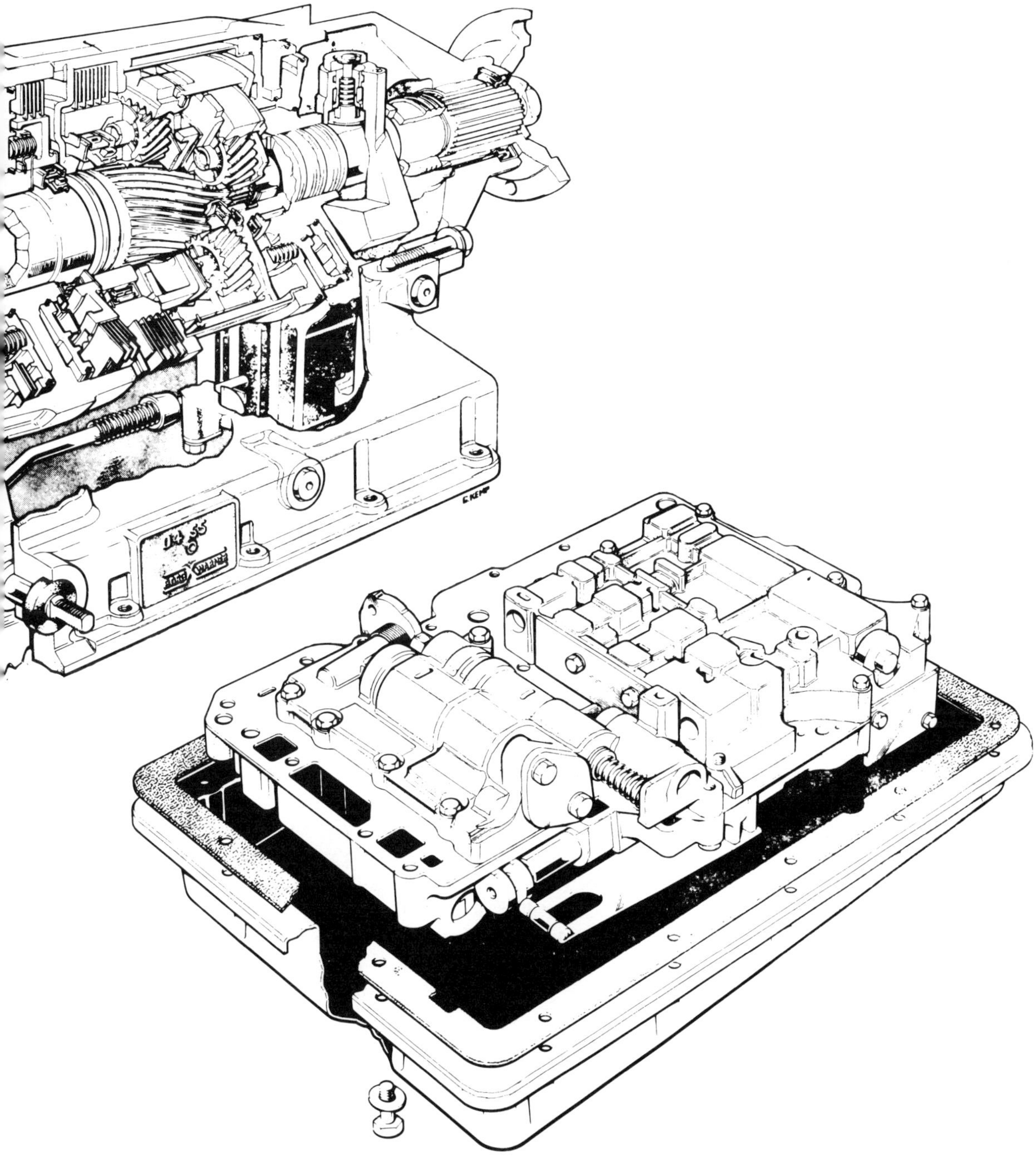

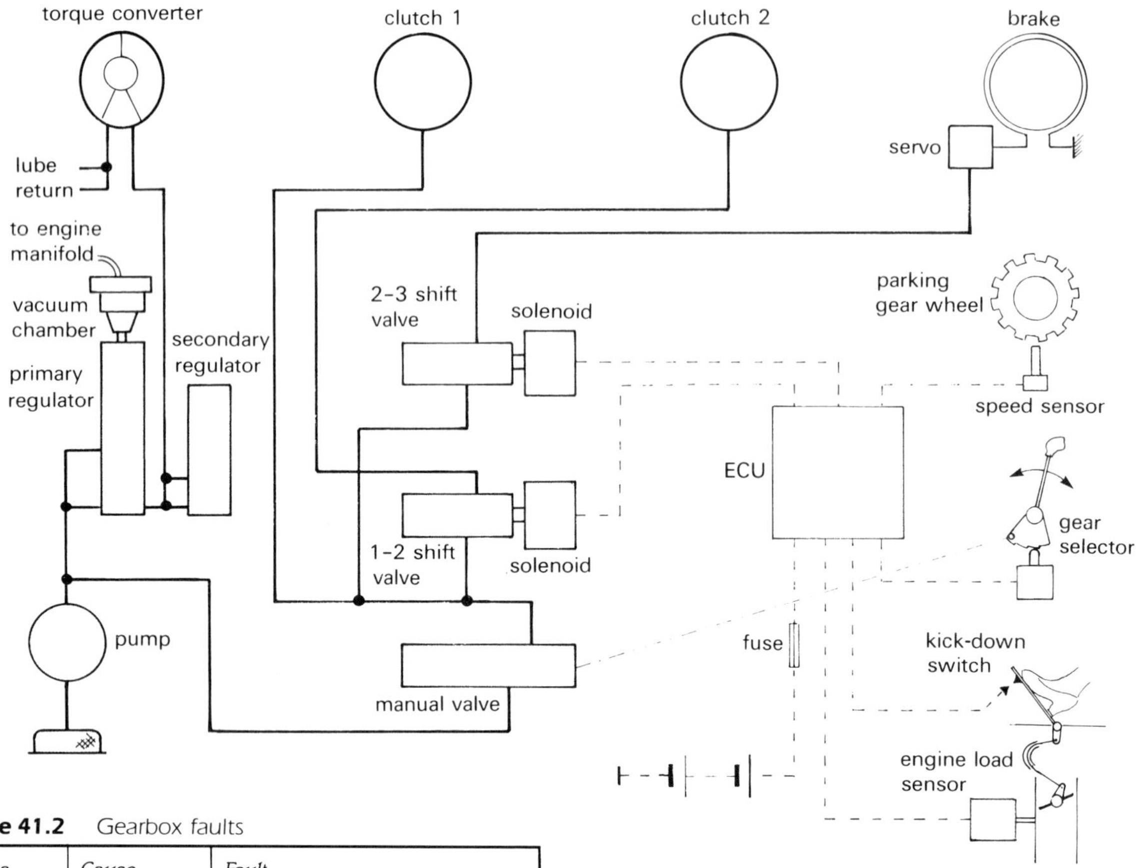

Fig. 41.38 Electronically controlled automatic gearbox

Table 41.2 Gearbox faults

Area	*Cause*	*Fault*
Fluid	Low level	Starvation of the pump prevents the hydraulic system from operating correctly; faults range from non-engagement of gears to erratic operation.
	High level	Governor strikes oil surface, so change speeds are affected.
	Poor quality	Rough gear changes. If oil is contaminated with grit particles, then a valve may stick; the resulting fault is related to the part of the system controlled by the valve.
Selector lever	Incorrect adjustment	Some gears, particular those at the end of the travel, will not select. (Lever should be adjusted when it is at mid-point of selector range.)
Throttle cable	Incorrect adjustment	Road speed at which changes take place will be either too high or too low.
Starter inhibitor	Incorrect adjustment	Starter will be active in positions other than P and N.

3 *Pressure test* – pressure take-off points are provided so that a pressure gauge can be used to check the pressure in the lines under given conditions.

Electronically controlled automatic gearbox

Electronic sensing of the operating conditions that affect the behaviour of an automatic has many merits over 'mechanical' sensing. For this reason a number of modern units now use electronic devices both to sense the conditions and to control the shift valves.

Fig. 41.38 shows a layout of an electronically controlled automatic gearbox. Clutch chambers and brake servos are hydraulically operated, as in a conventional system, so pressure regulators are needed as well as a manual valve to distribute the fluid. In this system the number of hydraulic valves is considerably reduced; this is because the various tasks performed by these valves is now undertaken by the ECU.

The interface between the electronic and hydraulic systems occurs at the shift valves. Each hydraulic shift

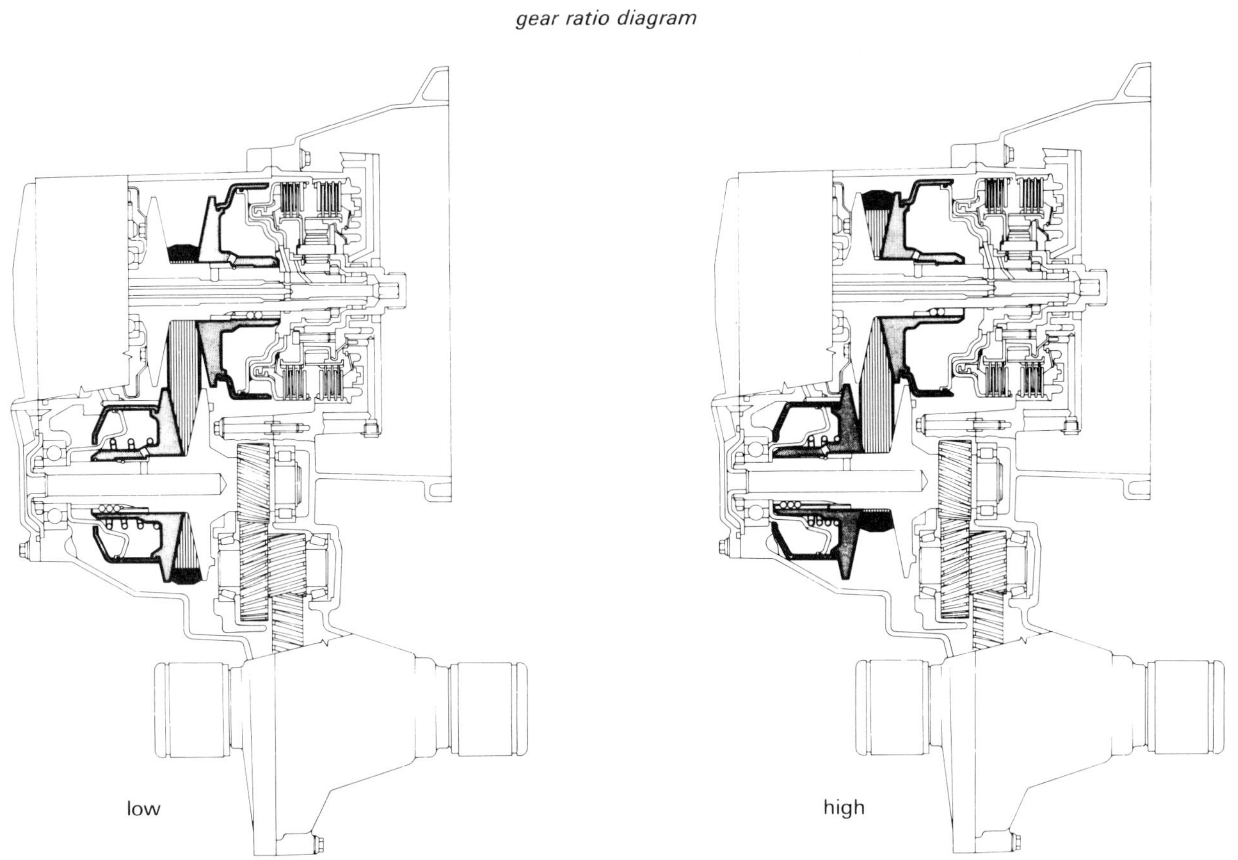

Fig. 41.39 Ford CTX transaxle

valve is controlled by a solenoid; this is energized by the ECU when operation of the valve is required.

In the event of electrical failure, which can be signalled to the driver by a diagnostic warning light, the system is arranged so that it goes into a *limp-home* mode. This means that the vehicle can still be driven in either third or reverse gear.

Continuously variable transmission (CVT)

In 1987 Ford and Fiat introduced a CVT type transmission system. This unit was developed from the Van Doorne system which was used on small DAF cars in 1955 (see page 228).

The Ford CTX (Continuously variable TransaXle) uses the same belt-drive principle as the Van Doorne unit, but instead of using two rubber drive belts in tension, the Ford CTX employs a single steel belt that works in compression. The transmission unit is suitable for front-wheel drive cars up to 1.6 litres and it is claimed that it is more economical than a conventional automatic gearbox.

Fig. 41.39 shows that the system is built around an enclosed vee belt which runs between two adjustable cone-shaped pulleys. The primary pulley is connected to the engine via a single-stage epicyclic gear train having compound planets, and the secondary pulley is connected through a pair of reduction gears to the final drive.

One half of the primary pulley, which is connected directly to a hydraulic servo, is moved when a change in gear ratio is required. By altering the width of the pulley, the effective diameter can be gradually changed whilst drive is being transmitted. Belt tension is maintained by varying the diameter of the secondary pulley; this is achieved by using a strong spring to push the two halves of the secondary pulley together.

Two multi-plate clutches control the epicyclic gear train to give forward and reverse motion; one clutch connects the sun wheel and planet carrier together to give forward, the other locks the annulus to the transmission casing to provide reverse. These clutches are immersed in oil and are designed to slip at low speed; this feature allows the car to 'creep' when manoeuvring.

The thrust belt consists of a large number of vee-shaped steel elements retained on two steel bands. When the belt is driving, each steel element transmits thrust by pushing against its neighbour.

The hydraulic system and driver controls are similar to a conventional automatic transmission. When the driving mode is selected, hydraulic pressure from an engine-driven pump acts on the appropriate clutch actuator. With the vehicle in motion, ratio changes are brought about by the action of the hydraulic valve unit; this varies the pressure that is applied to the pulley servo in accordance with the signals it receives from the engine- and road-speed sensors.

When the vehicle is accelerated or decelerated, the behaviour of the engine initially seems strange to drivers who normally use conventional systems. Since the CVT system selects the best ratio to suit the conditions, the engine does not have to be accelerated through each gear before the vehicle reaches its operating speed.

42 The overdrive

To overdrive means to drive faster. In the case of the transmission, the propeller shaft rotates faster than the engine when the overdrive system is in use.

Used mainly when the vehicle is cruising, the overdrive gear gives an improved petrol consumption, reduced engine wear and less engine noise.

Today most overdrive systems are built into the gearbox. This comparatively cheap arrangement gives an extra gear and, where a four-speed box is used, the fifth gear is an overdrive.

In the past many three- and four-speed gearboxes had a separate overdrive unit fitted, as an optional extra, behind the gearbox. The common type used was made by Laycock and this was developed from the de Normanville unit. This is the type considered here.

Mounted either in place of the rear extension housing (or in front of the final drive) the overdrive transmitted the power in all gear positions. This meant that it had to be strong to withstand the torque developed in the low-ratio gears.

Used in these locations, the overdrive gear set could also be engaged when the intermediate gears were selected in the main gearbox. This was a useful feature because its use converted a normal three-speed gearbox into a five-speed one. Road conditions for which top gear was too high and second gear was too low could be matched by using second gear and overdrive to give a ratio that fell between the main ratios. Operated by an electrical switch, the overdrive could be quickly engaged without having to use the clutch. Torque developed by the lower gear(s) was too great for the overdrive unit, so an external inhibitor switch was fitted to disengage the overdrive during reverse and low gear operation.

Laycock-type overdrive

The simple epicyclic gear set used in this type of overdrive gives either direct- or overdrive gear and is operated electro-hydraulically. The electrical system, controlled by the driver, activates a hydraulic system to move a cone-type clutch forwards and backwards to either hold or drive a sun gear.

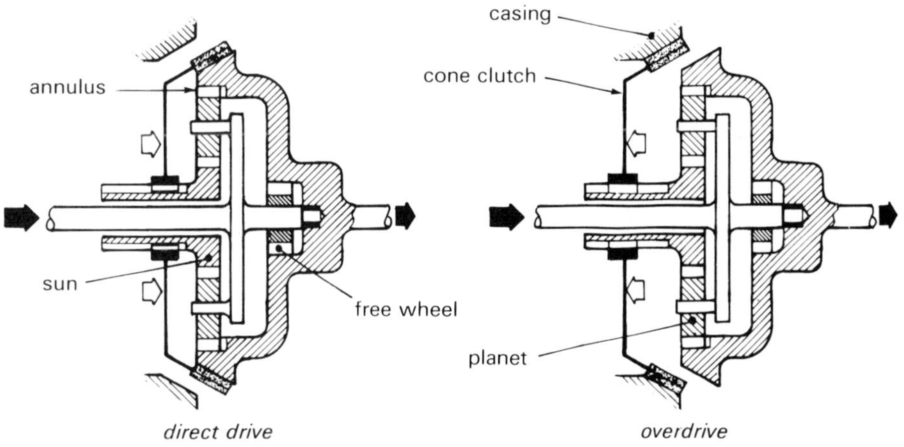

Fig. 42.1 Overdrive gear train layout

The layout of the gear set is shown in Fig. 42.1. Drive is applied to the planet carrier and the sun wheel, which is mounted on a bearing on the input shaft, can be connected by a friction cone-type clutch to either the annulus or the casing.

Direct drive

Spring pressure pushes the cone clutch to the right to connect the sun wheel to the annulus and lock up the complete gear assembly. Drive from the engine is taken through the free-wheel unit: the cone and gear unit accommodate overrun and reverse conditions, since the free wheel will not transmit a drive in the reverse direction.

Overdrive

Oil, under pressure from a pump driven by the input shaft, acts on pistons which force the cone clutch against the casing to prevent the sun wheel from rotating. The input shaft will move the planets forward and cause them to rotate around the fixed sun. This motion will drive the annulus slightly faster than the input shaft, to give overdrive.

The overdrive may be engaged without using the main clutch, and thus the operation is quick.

PART TWELVE

Propeller shafts and drive arrangements

43 Propeller shafts and universal joints

Universal joints

A universal joint, sometimes called a *carden joint*, allows the drive to be transmitted through a variable angle.

The need for a universal joint is shown in Fig. 43.10. This is a rear-wheel drive layout having the gearbox mounted on the frame and the rear axle bolted to the roadsprings.

Road shocks will deflect the springs to the position shown. This will alter the angle of the propeller shaft, relative to the gearbox and final drive, and unless a universal joint is fitted to each end of the propeller shaft, the shaft will bend and fracture.

A similar problem occurs on front-wheel drive vehicles. In this case, one end of the driveshaft is connected to a fixed transaxle assembly and the other end is attached to a wheel hub that moves up and down with the suspension spring. In addition to giving flexibility in this direction (plane), the outer universal joint must allow the roadwheels to be steered through a large angle of about 25°.

Requirements

A simple joint consisting of a flexible fabric disc, secured to the shaft by metal forks, meets the basic requirements, but this type is only suitable for driving a low-speed shaft through a very small variable angle. This type of joint is still used for some engineering applications, but it fails to meet the requirements expected of a modern universal joint; these are:

1. *Strength* - high torque must be transmitted with the minimum energy loss due to friction.
2. *Compactness* - space is limited so the joint must be small and robust.
3. *Large drive angle* - modern road springs allow large wheel deflections so the joint must be able to accommodate the large drive angle given by this movement.
4. *Shaft balance* - severe vibration occurs if the shaft runs out-of-true, so the joint must maintain good alignment.
5. *Operating speed* - the joint must operate efficiently at high speed under conditions of high torque and variable drive angle. This requirement must be combined with the need for the joint to have a long life and minimum maintenance.

Basic types of universal joint

Cross-type joint

This type of joint is commonly used today. Often it is called a *hooke-type coupling* because it was developed from the joint claimed to have been invented by Robert Hooke in the seventeenth century.

The joints in Fig. 43.1a and 43.1b, which show the basic and developed form respectively, have two yokes set at 90° to each other. These yokes are joined to each other by a cross-shaped *trunnion block*, and, in modern joints such as those made by Hardy Spicer, contact between the two parts is made by needle roller bearings held in a hardened steel cup retained in each arm of the yoke. Alignment of the trunnion is achieved by making the bottom of the cup contact the end of the block.

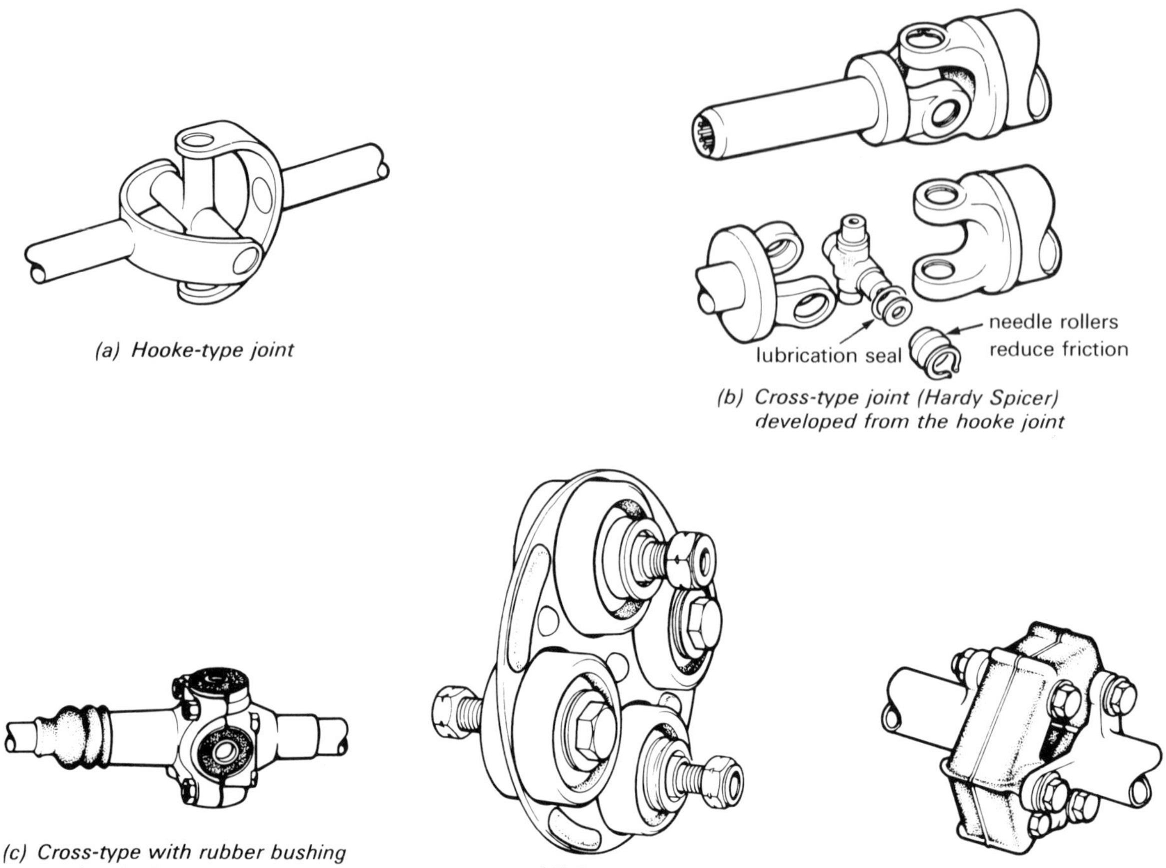

(a) Hooke-type joint

(b) Cross-type joint (Hardy Spicer) developed from the hooke joint

(c) Cross-type with rubber bushing to absorb vibrations

(d) Layrub

(e) 'Doughnut' rubber coupling

Fig. 43.1 Types of universal joint

Lubrication of the bearings is provided by using a special viscous oil, similar to that used in a final drive, which is contained in a reservoir formed by drilling out the centre of the trunnion arms. The oil is introduced either by a lubrication nipple or by pre-filling for life the small reservoir. An oil seal, retained on each arm of the block, presses against the cup and prevents the escape of the lubricant.

The cups are retained in the yoke either by circlips or staking. Peening over the edge of the yokes to stake the cups makes replacement of worn parts in the joint more difficult; in this case it is recommended that the complete shaft assembly is replaced when the joint is worn.

The merits of this type of joint are:

1 Compactness
2 High mechanical efficiency
3 Ability to drive through a large occasional 'bump' angle (maximum about 25°)
4 Accurate centring of shaft, hence the joint is suitable for high speed operation.

Lubrication failure, especially in cases where a 'grease' nipple in the trunnion block is missed when the vehicle is serviced, causes the needle rollers to indent (brinell) the bearing surfaces. This type of wear gives a slight angular movement; this allows the slackness in the joint to produce a noise, often described as a 'clonk', during the change-over from drive to over-run and vice versa. Failure to rectify this fault accelerates the rate of wear and this soon leads to misalignment and severe vibration.

Rubber joints

One disadvantage of the cross-type joint is its lack of flexibility in relation to the absorption of torsional shocks and drive-line vibrations, especially when a comparatively rigid transmission system is used. A smoother and less harsh drive can be obtained by using one or more rubber joints in the transmission driveline. Various types of rubber joint are in use; these include:

- *Moulton*

- *Layrub*
- *Doughnut.*

Moulton-type This rubber trunnion type joint (Fig. 43.1c) is based on a hooke-type coupling, but uses moulded rubber bushings to transmit the drive between the trunnion and yokes. The synthetic rubber mouldings require no lubrication and their resilience damps the torsional shocks that are generated when the drive is transmitted through an angle.

Layrub-type Originally made by the Laycock Company and constructed of a series of rubber bushings, the name 'layrub-type' is used to describe this joint, which is shown in Fig. 43.1d.

It consists of a number of moulded rubber blocks, with specially shaped cavities at the ends, which are sandwiched between two steel pressings. Each shaft is connected by means of a fork to alternate rubber blocks.

This construction allows the rubber blocks to deform and allow the drive to be transmitted through a small angle. In addition, the blocks accommodate small axial and angular movements for shaft length alteration and torsional damping.

Although the coupling is rather large in diameter, the layrub-type offers the following advantages:

1 No lubrication required.
2 Capable of driving through bump angles up to about 15°.
3 Allows for axial movement, hence no splining of the shaft is necessary.
4 Flexibility damps shocks and insulates vehicle from transmission noise.

Doughnut-type Although rather large, the great flexibility of this type provides a soft cushion to absorb the majority of torsional shocks produced by the action of other joints or by vibration from either the engine or road wheel.

Fig. 43.1e shows that the synthetic rubber coupling is near-circular in shape and is moulded around cylindrical steel inserts that are bolted alternately to the three-arm forks attached to the shafts.

The merits of this coupling are similar to the layrub-type.

Speed variation of a hooke-type coupling

When a hooke-type coupling is transmitting a drive through an angle, the output shaft does not rotate through 360° at a constant speed. Instead the speed varies every 90° of rotation, and the rate of movement for one revolution is fast, slow, fast, slow (Fig. 43.2).

This cyclic speed variation, and its associated vibration, is insignificant when the drive angle is less than about 5°, but becomes much more intense as the angle is increased.

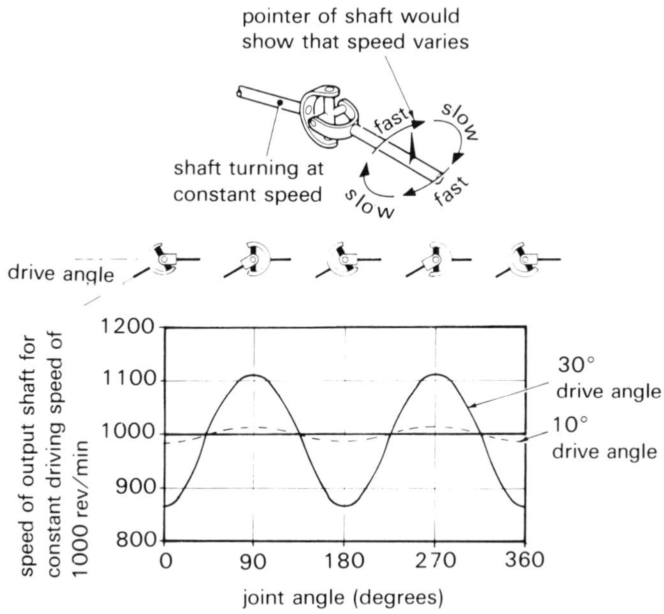

Fig. 43.2 Speed variation with hooke-type joints

One method of achieving a constant speed output from the propeller shaft is to mount two hooke-type couplings either back-to-back or positioned in a certain way at each end of the propeller shaft. In both configurations the relative positions of each coupling must be arranged so that the speed change of one coupling is counteracted by the other. The 'phasing' of hooke-type couplings, as applied to two separate driveline layouts, is shown in Fig. 43.3.

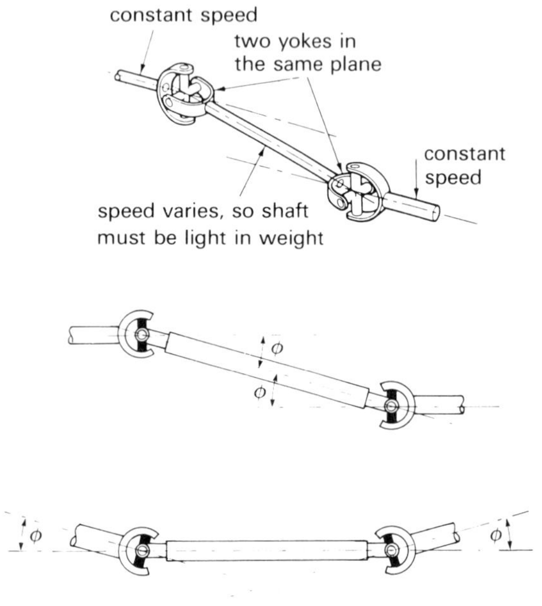

Fig. 43.3 Phasing of hooke-type couplings

This diagram shows that to restore a constant speed, two conditions must be satisfied:

1 Yokes at each end of the propeller shaft must be positioned in the same plane.
2 Drive angle of each coupling must be equal.

Constant-velocity joints

A constant-velocity (CV) universal joint is a type that provides an output shaft speed equal to that of the input in all shaft positions within the working range of the joint.

Today these joints are often used in drive lines because of their smooth operation. They were originally introduced for situations where the drive angle was large, e.g. at the wheel end of a front-wheel drive vehicle when the steering is set in the full-lock position.

Geometry of a typical CV joint can be appreciated if the action of a hooke-type coupling is considered. The speed variation of a hooke-type joint is caused by the alteration in leverage between the yokes and trunnion block. When the coupling is viewed from the side and the input shaft is rotated through 90°, the trunnion block is seen to oscillate back and forth from the vertical through an angle equal to the shaft drive angle.

A *constant angular velocity* condition occurs with this joint when the trunnion block is at the mid-point of its rocking movement. If this condition could be maintained, a constant-velocity universal joint would be obtained, i.e. constant-velocity conditions are achieved when the connecting device between the driving and driven 'yokes' is positioned in a plane that bisects the angle of drive.

The various CV joints in use have a construction which is based on either the twin hooke-type coupling arrangement or the 'angle bisect' principle.

Types of constant-velocity joints

Constant-velocity joints in use are:

- *Tracta*
- *Rzeppa*
- *Weiss*
- *Tripode*.

Tracta CV joints

The need for CV joints was discovered in 1926 when the first FWD vehicle was made in France by Fenaille and Gregoire - the Tracta (traction-avant) car. When the second car was made, the drive line incorporated CV joints, and the type used on that car is now known as a *Tracta* joint. This type is now made by Girling and Fig. 43.4 shows the main details.

Reference to the line sketch shows that the operating principle is similar to two hooke-type joints: the angles are always kept constant and the yokes are set in the same place.

The joint is capable of transmitting a drive through a maximum angle of about 40° and its strong construction makes it suitable for agricultural and military vehicles, but the friction of the sliding surfaces makes it rather inefficient.

Rzeppa-type CV joint

This joint was patented by A. H. Rzeppa (pronounced Zeppa) in America in 1935. The development of this type is in common use today; it is called a *Birfield* joint and is made by Hardy Spicer.

Fig. 43.5 shows the construction of a Birfield joint and the line sketch shows the principle of operation.

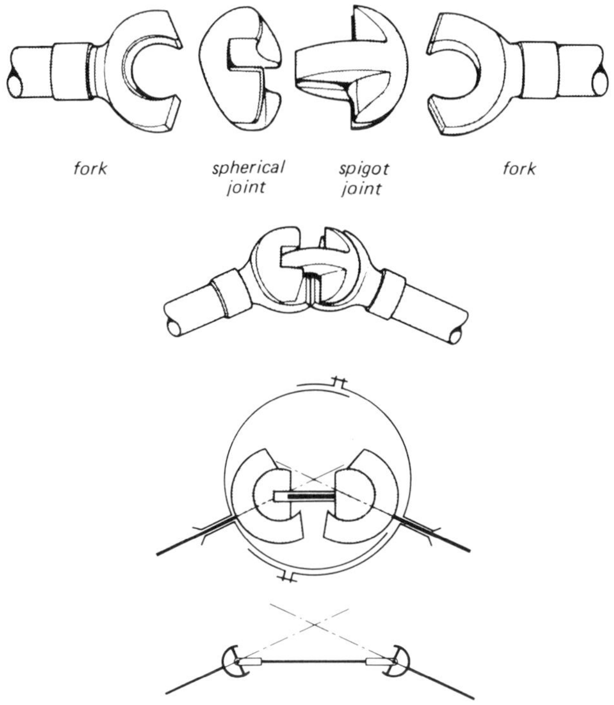

Fig. 43.4 Tracta CV joint

Constant velocity is achieved if the device (steel balls in this case) connecting the drive shaft to the driven shaft rotates in a plane that bisects the angle of drive: the Birfield joint achieves this condition.

Drive from the inner to outer race is by means of longitudinal, elliptical grooves which hold a series of steel balls (normally six), that are held in the bisecting plane by a cage. In the original Rzeppa joint the cage position was controlled by a small plunger fitted between the two shaft centres, but noise was a problem. In the Birfield design the plunger is omitted. Instead the balls are made to take up their correct positions by offsetting the centres of the radii for inner and outer grooves.

A Birfield joint has a maximum angle of about 45°, but this angle is far too large for continuous operation because of the heat generated. Lubrication is by grease – the appropriate quantity is packed in the joint 'for life' and a synthetic rubber boot seals the unit.

Weiss-type CV joint

This type was patented in America in 1923 by Weiss and later developed by Bendix. It is now known as a

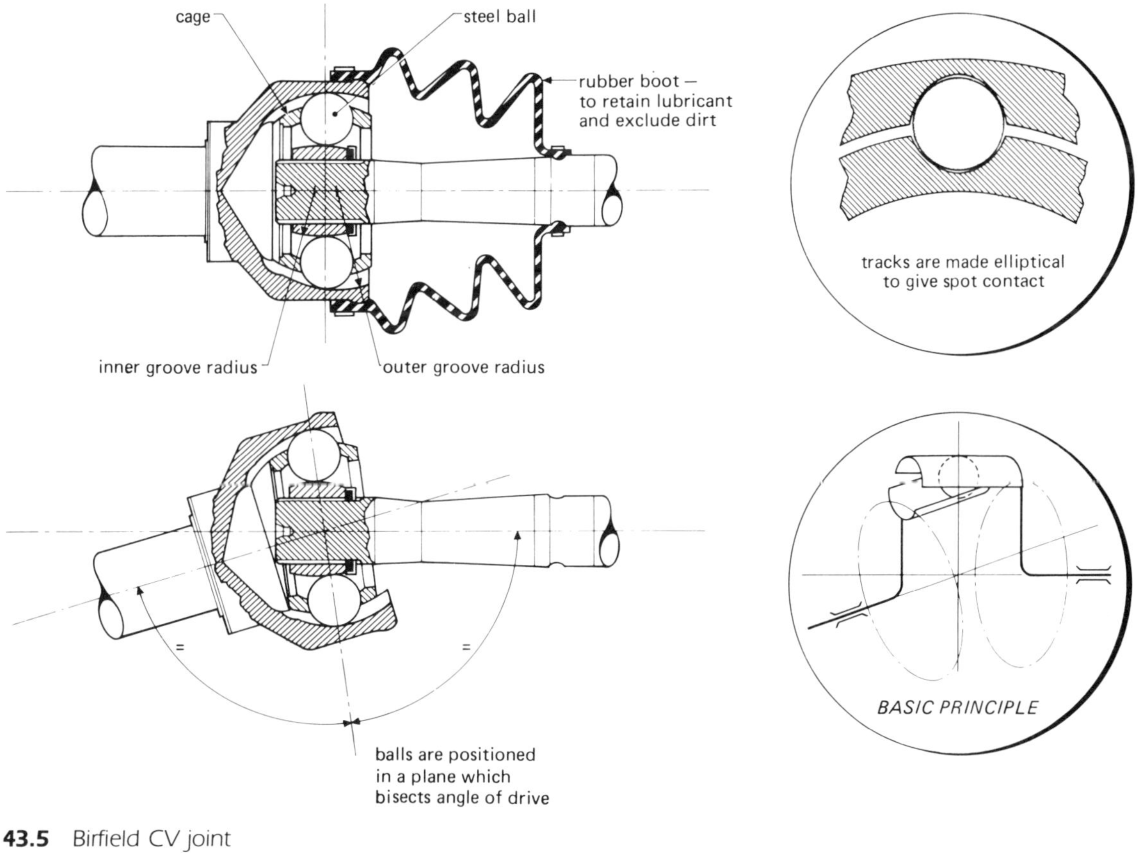

Fig. 43.5 Birfield CV joint

Bendix-Weiss and produced in the UK by Dunlop. Fig. 43.6 shows a simplified drawing of this type. The two forks have grooves cut in their sides to form tracks for the steel balls. There are four tracks, so four balls are used to transmit rotary motion and a fifth ball placed at the centre of the forks locates the two forks and resists the inward force. The driving balls work in compression, so two balls take the forward drive and the other two operate when reverse drive is applied. The complete joint is contained in a housing filled with grease. Maximum angularity is about 35°.

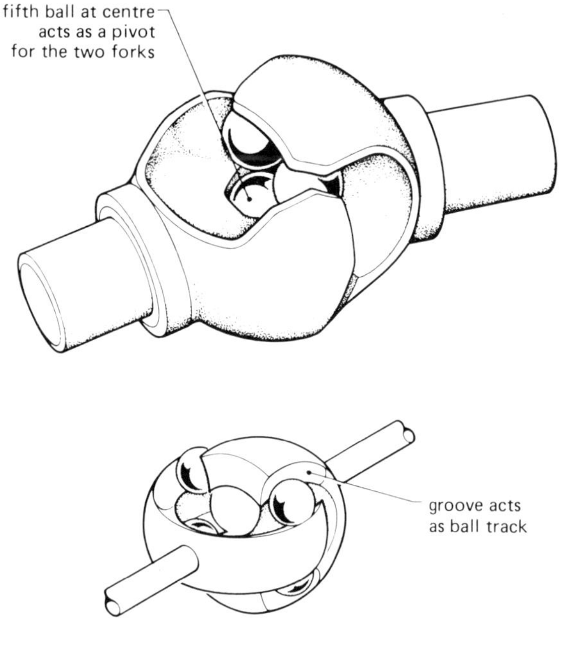

Fig. 43.6 Weiss CV joint

Constant velocity is achieved in a manner similar to the Rzeppa - the balls always take up a position in a plane that bisects the angle of drive.

Tripode-type CV joint

This compact joint has been developed over the past 25 years and, in view of its ability to operate efficiently at high speed, its use is becoming more common.

Besides providing good resistance to high-speed centrifugal effects, the construction of this joint, combined with the reduced working clearances achieved by modern production techniques, produces a transmission drive line with good noise-vibration-harshness (NVH) performance.

In the construction shown in Fig. 43.7 the three-armed support (tripod) carrying the spherically

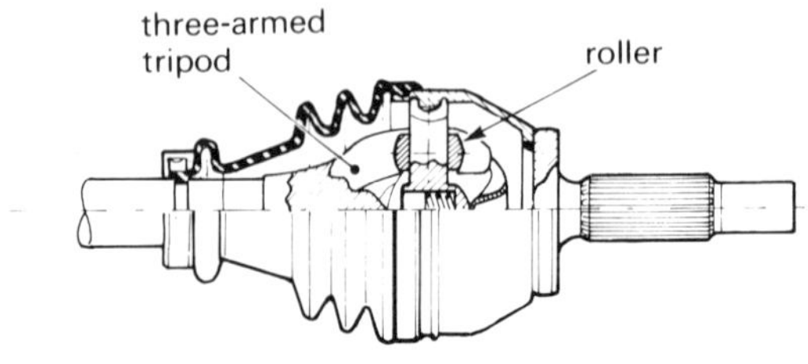

Fig. 43.7 Tripode CV joint

shaped rollers is fixed to the outer housing. On both sides of each driving fork, which also has three arms, grooves are cut to form a bearing track for the rollers.

Drive through the joint is produced by the force exerted by the side of the driving fork on the rollers. This force is transmitted to the tripod and joint housing.

Alteration of the drive angle causes the roller to move backwards and forwards along the grooved track as the joint rotates through one revolution. A small clearance is given between the roller and track to permit this movement.

Homokinetic motion (constant velocity) is achieved with a Tripode joint because of the path taken by the rollers with respect to the contact point on the track.

This type of fixed joint accepts an occasional drive angle up to about 45°.

Plunge joints

Suspension movement causes an alteration in length to many drive line shafts, so a special CV joint, called a *plunge joint* must be used to allow for this. The alternative is to use splines to allow the joint to slide back and forth along the shaft but lubrication often causes a problem.

Two plunge joints in common use have been developed from the fixed Birfield and Tripode types.

Birfield plunge joint Development work on the original Rzeppa joint by Löbro in Germany in the late 1950s produced the type known in this country as the Birfield CV joint. In 1960 the Löbro company redesigned this fixed-type joint and produced it in a plunge form to satisfy the needs of front-wheel drive vehicles that were gaining favour at that time.

The constructional features of this joint are shown in Fig. 43.8. In this type the grooves that act as the tracks for the balls are straight instead of curved; this allows the shaft length to vary up to about 50 mm.

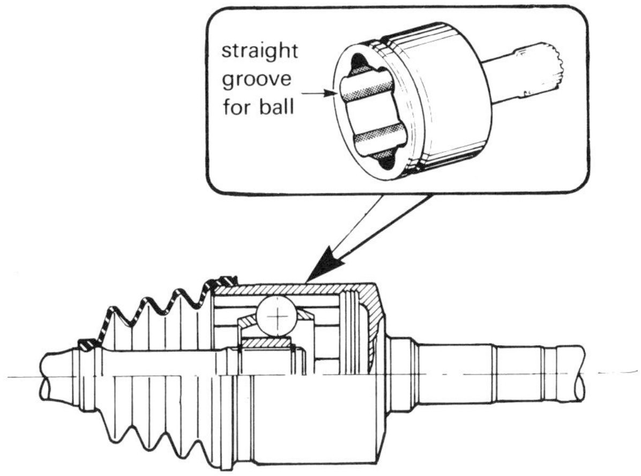

Fig. 43.8 Birfield plunge joint

When the shaft angle is increased, the rolling action of the balls in the inner and outer grooves positions the cage in the bisecting plane to give a CV drive.

The maximum drive angle of 22° is considerably less than that in the fixed type, so, in front-wheel drive vehicles, the plunge joint is positioned at the engine end of the drive shaft.

Tripode plunge joint   Comparing the plunge joint in Fig. 43.9 with the fixed type shows that the basic construction is similar. The main difference is that the plunge joint has roller tracks that are straight instead of curved.

The type shown permits a maximum plunge of 55 mm and a drive angle up to 25°, but, as with all types of CV joint, the working angle should be less than 10°.

Later types of tripode-type plunge joints use either ball or needle roller bearings to reduce the friction between the contact surfaces. These improve the NVH performance and allow the joint to operate efficiently at high speed.

Propeller shafts and drive shafts

Power from the gearbox to the rear axle is transmitted by a *propeller shaft*; sometimes this is called a *carden shaft*.

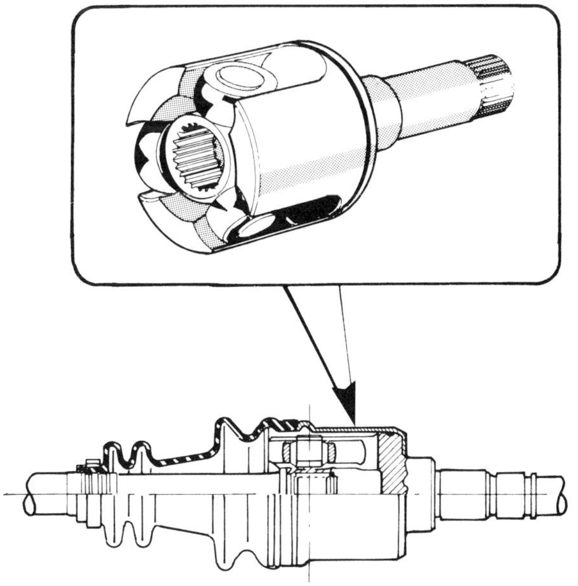

Fig. 43.9 Tripode plunge joint

The propeller shaft is normally tubular in section and is made in a one- or two-piece construction; the latter arrangement is supported at the mid-point by a rubber-mounted bearing.

Short *drive shafts* are used in both front- and rear-wheel drive arrangements for the transmission of power from the final drive assembly to the road wheels.

Propeller shafts

The shaft must be strong to resist the twisting action of the driving torque and it should be resilient to absorb the torsional shocks. Also it must resist the natural tendency to sag under its own weight because vibration occurs when the centre of gravity does not coincide with the axis of the shaft.

A tubular-section propeller shaft is normally used because it has the following merits:

1 low weight
2 large resistance to misalignment, especially sag
3 good torsional strength (a tubular shaft is only slightly weaker than a solid shaft of similar diameter).
4 low resistance (low inertia) to changes in angular speed which arise when a hooke-type coupling is used to drive the shaft.

A propeller shaft often rotates at high speed, especially when the overdrive gear is in use, so the shaft must be made, and repaired, to meet good balance limits as well as design features. During

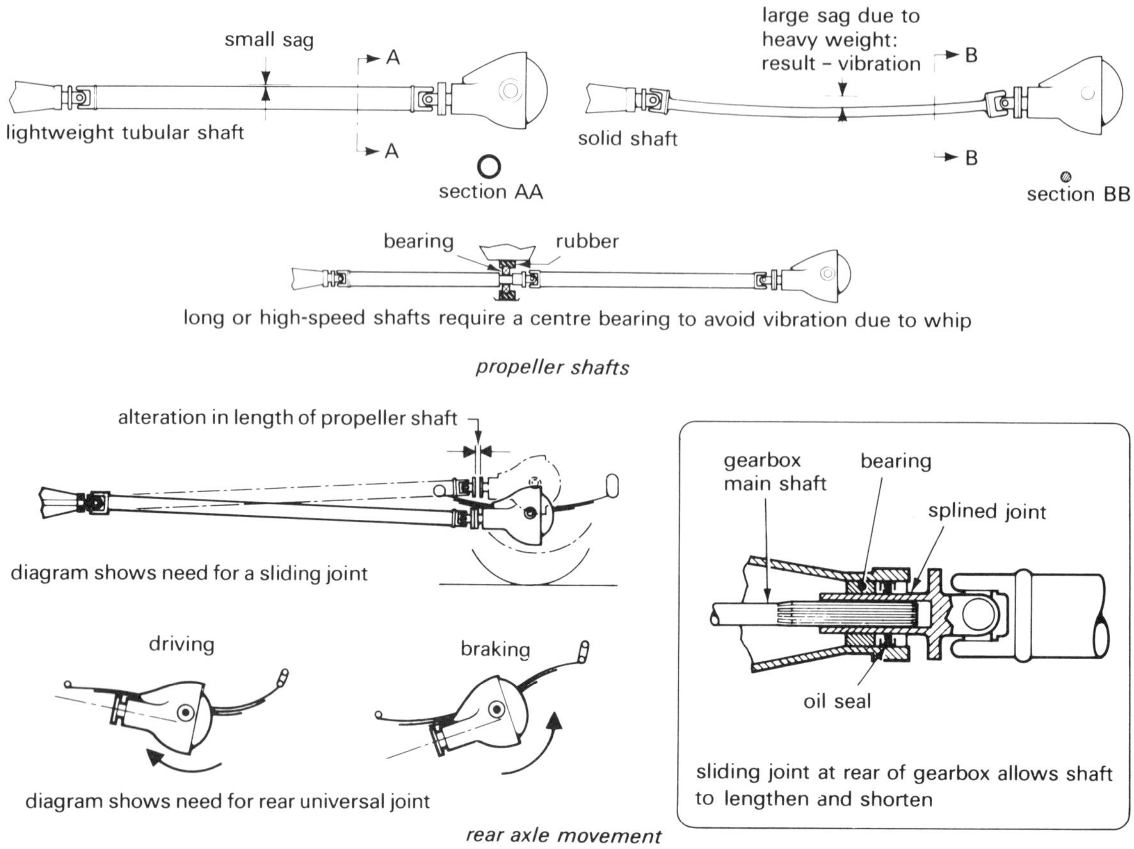

Fig. 43.10 Propeller shaft features

manufacture, steel shafts are balanced and during this operation small patches, are spot-welded on to the 'light parts' of the tube to correct the imbalance.

Even when the static alignment of a shaft is perfect, the weight of the shaft makes it bow at the centre. When this is excessive, rotation of the shaft causes the bow to increase owing to the centrifugal effect. This deformation, or *whip* of the shaft, sets up a vibration that becomes severe as it approaches the *whirling speed*. The critical speeds at which this condition occurs depends on two vital dimensions: the mean diameter of the tube and the length of the shaft. Whirling speed is raised when either the diameter is increased or the length is decreased.

Many rear- and four-wheel drive vehicles need a longitudinal propeller shaft that has to span a great distance between the gearbox and final drive. In these cases the drive line is normally divided and a bearing is fitted to support the shafts in the region where they are split (Fig. 43.10). This bearing is mounted in rubber to absorb any vibration that would otherwise be transmitted to the body.

Although axle movement is restricted to the rear shaft and universal joints are fitted to accommodate this movement, extra joints are needed on the front shaft to allow for the slight flexing of the vehicle body. Since it is nearly impossible to maintain the correct

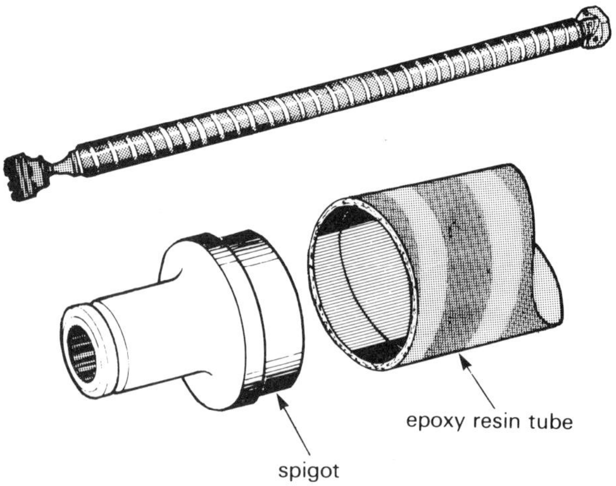

Fig. 43.11 Composite propeller shaft

drive angles of hooke-type couplings fitted to a two-piece shaft layout, many modern arrangements use one or more CV joints.

The *composite* propeller shaft shown in Fig. 43.11 is a modern alternative to the divided arrangement. The tubular shaft is made of epoxy resin, which is strengthened with glass and carbon fibres, and bonded to a steel spigot for connection to the universal joints.

Compared with a conventional two-piece steel-shaft arrangement, the advantages of the composite shaft are:

1 Weight reduction by about 50 per cent
2 High internal shock absorption
3 Good noise, vibration, harshness (NVH) performance
4 Exceptional corrosion resistance.

Drive shafts

Since these shafts are comparatively short in length, they are often made solid to give clearance to allow for movement of the suspension. Where space is not restricted the lightweight tubular section is often used.

The short distance between the road wheel and the final drive housing, combined with a large road wheel movement due to suspension deflection, means that both the maximum drive angle of the universal joints and the length variation of the shaft are great. A CV joint at each end of the drive shaft meets the angle requirement and a plunge CV joint accommodates the length change. Rear-wheel drive vehicles having independent rear suspension need a drive shaft to link the road wheel to the fixed final drive assembly. On many of these vehicles a plunge-type CV joint is fitted at each end of the driveshaft.

44 Drive arrangements

Rear-wheel drive

Torque reaction

'To every action there is an equal and opposite reaction.' This statement means that every component that produces or changes a torque will also exert an equal and opposite torque tending to turn the casing. For example, when the engine crankshaft exerts a torque in a clockwise direction, the cylinder block will tend to rotate in an anticlockwise direction. (Some early designs of engine used a fixed crankshaft with rotating cylinders.)

A further example of torque reaction is shown in Fig. 44.1a, which represents a tractor with its rear driving wheels locked in a ditch. In this situation the driver must be careful, because torque reaction is likely to lift the front of the tractor rather than turn the rear wheels.

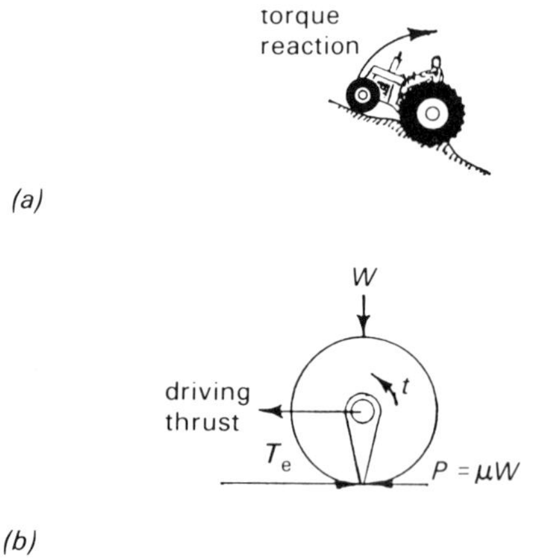

Fig. 44.1 Lever action of a wheel

When the law stated above is applied to rear axles, you will see that some arrangement must be provided to prevent the axle casing turning in the opposite direction to the driving wheels.

Driving thrust

A torque t applied to the wheel, which may be considered as a lever (Fig. 44.1b), produces a *tractive effort* T_e at the road surface, and an equal and opposite forward force at the axle shaft. This driving thrust must be transferred from the axle casing to the frame in order to propel the vehicle. The maximum tractive effort is limited by the adhesive force P of the tyre on the road; this force depends on the coefficient of friction μ and the load W on the wheel.

Types of drive

A drive system is an arrangement for transmitting the driving thrust from the road wheels to the vehicle body. The system used must also incorporate a means for resisting the movement of the main components due to torque reaction.

In the past, leaf springs were normally used for the rear suspension of a vehicle, so these springs were often utilized to provide the thrust and torque reaction functions of the drive system.

On pre-1950 motor cars there were three notable drive systems in use and since that time these basic systems have been modified to suit modern requirements. The drive systems were:

- *Hotchkiss open-type*
- *Four-link (semi-Hotchkiss)*
- *Torque tube*
- *de-Dion.*

Although these systems are rarely used today in their original form, the object of including them here is to enable the reader to appreciate the arrangement used to resist the various forces associated with the propulsion of a vehicle. This knowledge should help in the diagnosis of many of the 'knock, clonk and vibration' faults associated with drive systems.

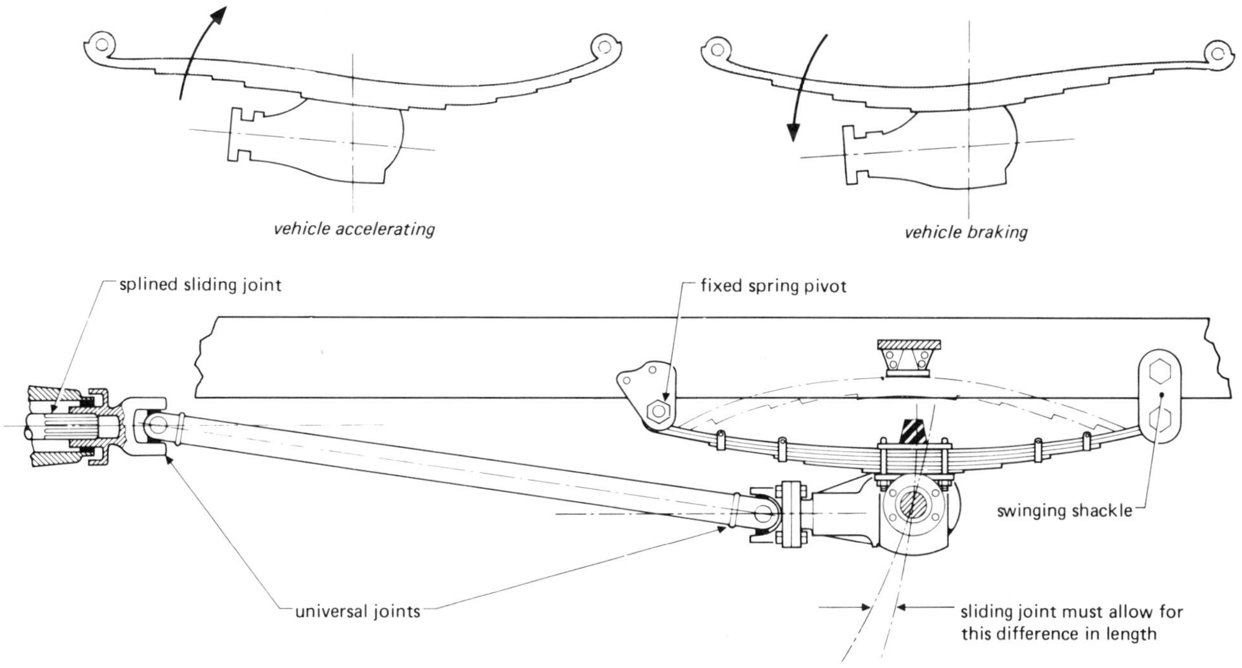

Fig. 44.2 Hotchkiss open-type drive (light commercial vehicle)

Hotchkiss open-type drive

This once popular arrangement is shown in Fig. 44.2.

Two rear leaf springs, longitudinally mounted, are connected to the frame by a 'fixed' pivot at the front, and swinging shackles at the rear. At each end of the exposed or 'open' type propeller shaft is fitted a universal joint, with provision made for alteration in shaft length, which occurs when the springs are deflected.

Torque reaction is resisted by clamping the axle to the springs by means of 'U' bolts. Under heavy driving conditions the springs will deflect up at the front and down at the rear, and vice versa during braking. This movement will help to damp driving shocks and improve transmission flexibility. Since the axle continually moves up and down, the need for a rear universal joint is clear.

Driving thrust is transferred from the casing to the spring by the friction between the two surfaces, and then transmitted through the front section of the springs to the vehicle frame. If the 'U' bolts become loose, the spring centre bolt (axle location bolt) will have to take the full driving thrust, and the high shearing force will quickly fracture it.

Four-link (semi-Hotchkiss) drive system

Some car manufacturers use helical springs in conjunction with a live rear axle. Helical springs cannot take driving and braking thrust, torque reaction or give lateral support to the rear axle, so extra parts must be fitted to fulfil these functions.

In the layout shown in Fig. 44.3 the rear axle is positioned by upper and lower trailing suspension arms; these arms transmit driving thrust and prevent rotation of the axle casing. Sideways movement of the axle is controlled by a transverse stabilizer called a *Panhard rod*; this rod connects the rear axle to the vehicle body. Rubber mountings at each connection point reduce noise transmission, eliminate the need for lubrication and also provide slight flexibility to allow for drive movement and geometric variations during spring deflection.

At first it would appear that the helical spring gives a reduction in the unsprung weight, but when the weight of the additional locating arms and rods needed for this arrangement is added, the unsprung weight difference is very small. However, the accurate positioning of the axle in this system is an advantage.

Torque-tube drive

Whereas the Hotchkiss drive uses stiff springs to resist torque reaction and driving thrust, the torque tube or enclosed drive relieves the springs of all duties other than their intended purpose. This means that a 'softer' ride can be achieved by using either 'softer' springs or another form of spring, e.g. helical.

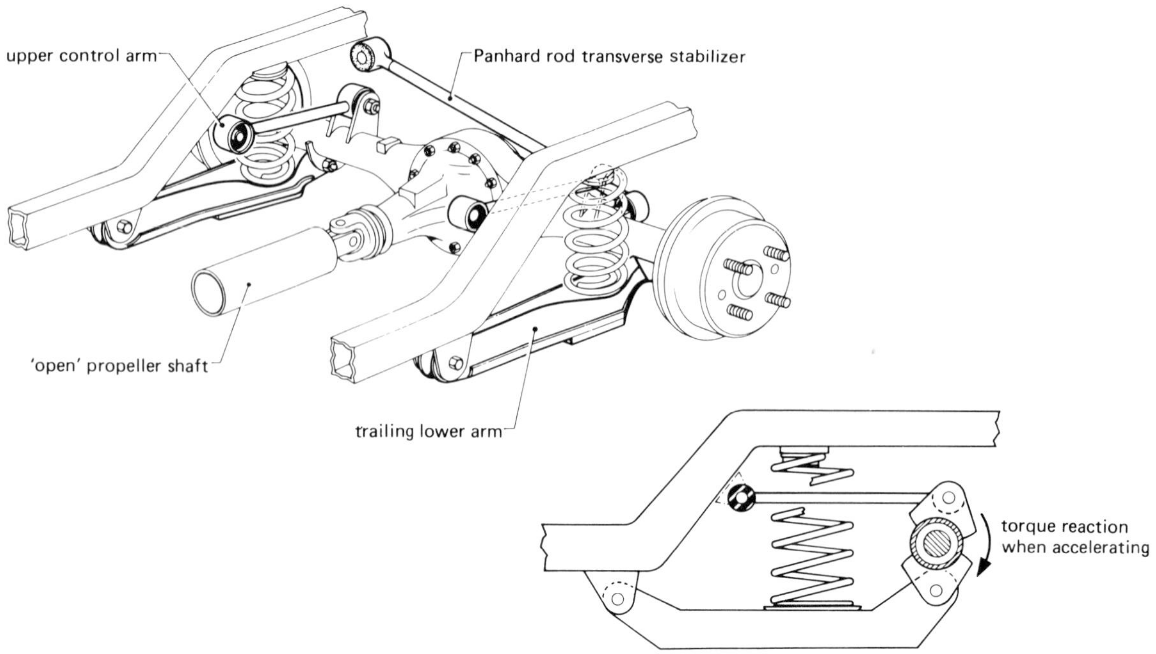

Fig. 44.3 Helical spring and four-link suspension

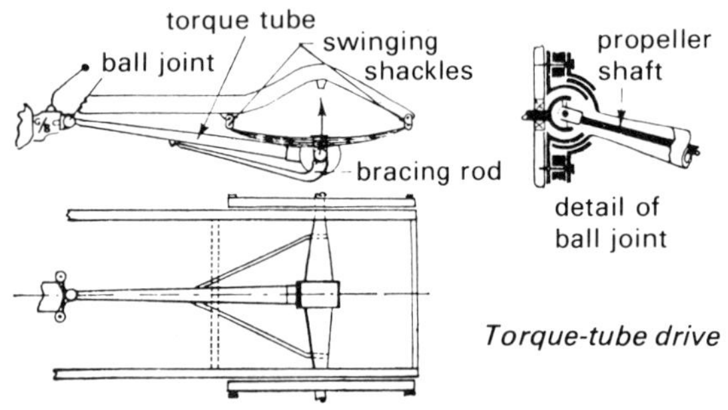

Fig. 44.4 Torque tube drive

Fig. 44.4 shows a layout using laminated springs, which are connected to the frame by a swinging shackle at each end. Bolted rigidly to the axle casing is a tubular member, which is located at the front by a ball and socket joint, positioned at the rear of the gearbox or cross-member of the frame. Bracing rods, connected between the axle casing and the torque tube, strengthen the construction. A small-diameter propeller shaft is fitted inside the torque tube and splined to the final-drive pinion. Mounted in the centre of the ball joint is a universal joint to allow for angular deflections of the drive.

In this arrangement the torque reaction and driving thrust are taken by the torque tube. When the forward thrust from the ball is taken on the rear housing of the gearbox, means must be provided to transfer this force through the gearbox mountings to the frame.

Helical or torsion bar springs may be used as alternatives to laminated springs, and in these cases, side movement of the axle must be controlled by fitting some form of transverse stabilizer, e.g. a Panhard rod, between the frame and the axle.

de-Dion drive

The de-Dion axle is often regarded as the halfway stage between the normal axle and independent suspension. Many of the advantages of the latter are achieved in this layout, but since the rear wheels are still linked by an axle tube, the system cannot be classed as independent.

Fig. 44.5 shows a basic arrangement: laminated springs are connected to the frame by a 'fixed' pivot at the front and a swinging shackle at the rear. Each spring carries a hub mounting, which is rigidly connected to a tubular axle beam, to support the wheel on a stub axle shaft. Bolted to a cross-member of the frame is the final-drive unit, and from this the drive is taken through two universally jointed shafts to the wheels. The main propeller shaft is fitted with a universal joint at each end to allow for flexing of the frame.

Torque reaction of the final-drive casing is taken by the frame, and the driving thrust is resisted by the springs.

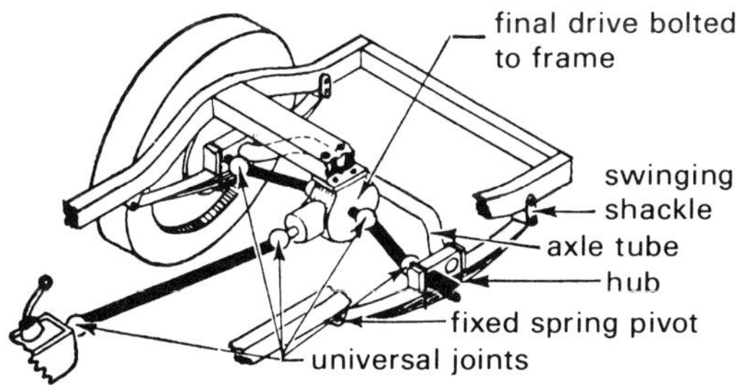

Fig. 44.5 de-Dion drive

The major advantage of this layout is the reduction in unsprung weight. This ensures that wheel spin is reduced by allowing the light driving wheels to follow the contour of the road surface closely. Wheel spin is caused, too, by the tendency of the normal axle to rotate around the pinion when a high propeller shaft torque is exerted. This lifting effort of the wheel is eliminated with the de-Dion system, and weight is equally distributed.

Other rear-wheel drive arrangements are discussed in Chapter 61.

Front-wheel drive

Advantages

Concentrating the engine and transmission system in one unit and placing the assembly at the front of the vehicle to give front-wheel drive (FWD), produces the following advantages:

1 Compact construction.
2 Flat floor of passenger compartment - no propeller shaft tunnel or gearbox bulge.
3 Good traction because majority of weight is taken by the driving wheels.
4 Engine can be mounted transversely, which either reduces bonnet length or increases the size of the passenger compartment.
5 Good steering stability - driving thrust of the wheels is aimed in the direction that the vehicle is

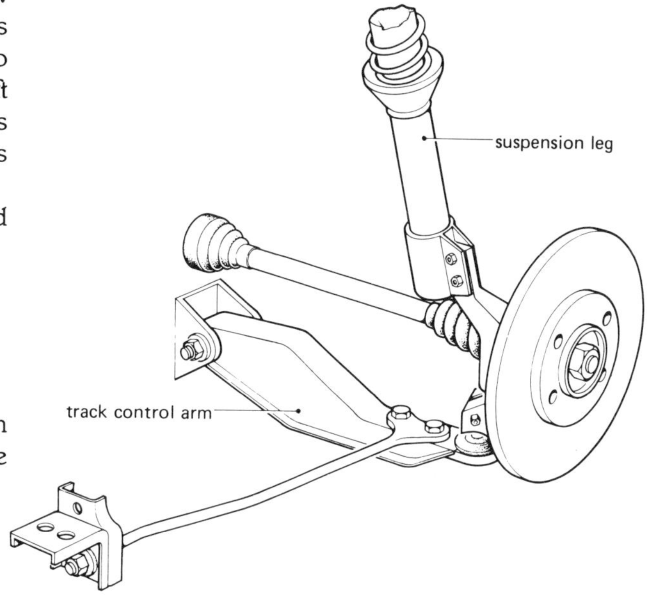

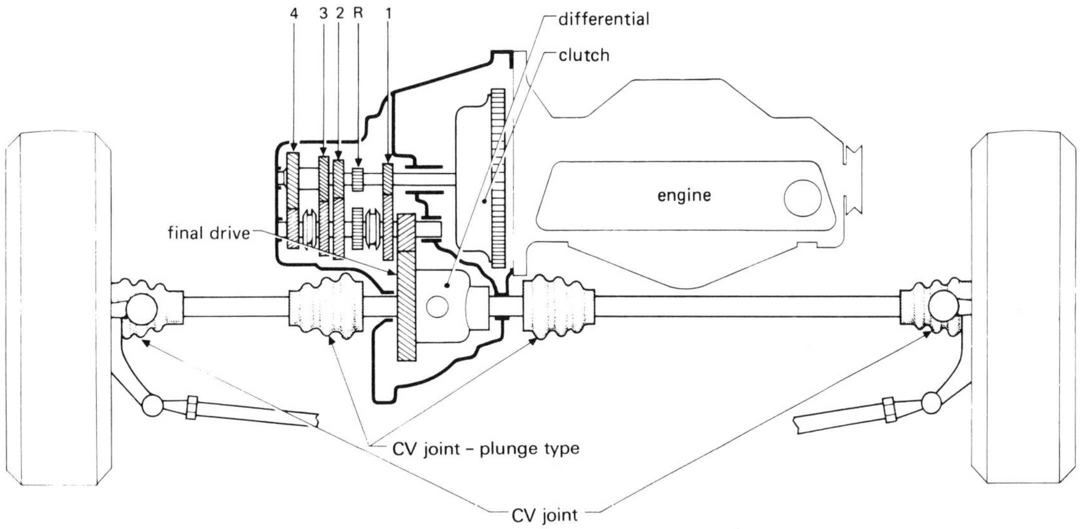

Fig. 44.6 Front-wheel drive layout

intended to follow. Also the layout rarely suffers from the oversteer characteristics described on page 316.

Although FWD vehicles need more complicated drive shafts, the many advantages make the layout very suitable for a small car.

Transverse engine layout

Fig. 44.6 shows a typical FWD layout. The transversely mounted engine is bolted to a unit called a *transaxle*. This unit gets its name from the two words *transmission* (a term used in America for gearbox) and *axle*. In the past the general term 'rear axle' often applied to the gearing as well as the axle casing, but nowadays the various drive arrangements make it necessary to restrict the use of the name 'rear axle' to the member which performs the axle duty. The two main components inside the axle of a rear-wheel drive vehicle are the final drive, e.g. crown wheel and pinion, and the differential.

The four-speed gearbox in Fig. 44.6 has two shafts: an input shaft splined to the clutch and a mainshaft connected to the final drive pinion. A bevel gear final drive is necessary when the engine is mounted longitudinally but in the layout shown, the engine position allows the use of normal helical gears. Drive from the final drive is transmitted through the differential to the two drive shafts. Speed variation due to drive shaft angularity is prevented by using constant velocity universal joints at each end; the inboard joint at each side being of the plunge type.

During acceleration, or at times when the engine is used as a brake, torque reaction tends to make the engine rotate about the crankshaft, so suitably spaced engine mountings are fitted to resist this movement.

Driving thrust and braking torque are taken by the suspension members in particular the tie bar fitted from the track control arm to the body.

PART THIRTEEN

Rear axles

45 Final-drive gears

Purpose

The purpose of a final drive, as applied to a rear axle, are:

1 To transmit the drive through an angle of 90°.
2 To gear down the engine revolutions so that a 'direct top' gearbox ratio may be employed. In the case of cars this requires a final drive ratio of approximately 4 : 1.

These functions can be performed by bevel or worm gears.

Bevel gears

The geometry of a bevel gear layout may be considered by referring to Fig. 45.1. This represents two friction cones: A forming the crown wheel and B the pinion. To avoid slipping and wear, the apex of the pinion must coincide with the centre line of the crown wheel. If the pinion is incorrectly positioned, it will be seen that the peripheral speeds of the crown wheel and pinion will not be equal. Mounting the gear in the correct position will show that the angle of the bevel is governed by the gear ratio.

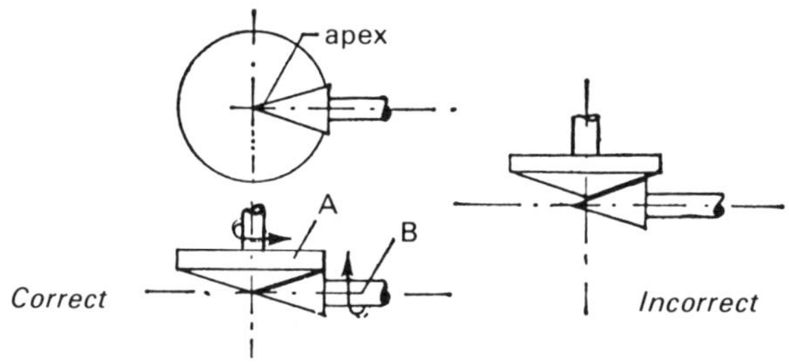

Fig. 45.1 Friction cones representing bevel-gear drive

Types of bevel gear

Straight bevel

Fig. 45.2 illustrates the main features of the bevel type of gear. Tapered teeth, generated from the centre, are machined on the case-hardened steel gears, and then ground together to form a 'mated pair'. The direction of rotation of the axle shaft will be determined by the position of the crown wheel relative to the pinion. (It is possible on some vehicles to fit the crown wheel on the wrong side, and this gives one forward and several reverse ratios.)

Adjusters in the form of distance pieces, shims or screwed rings enable the correct mesh and backlash to be set. When the clearance between the teeth (backlash) is too small, expansion due to heat and wear caused by lack of lubrication will result, whereas excessive clearance produces slackness and noise. Each manufacturer recommends a suitable backlash; this is normally in the region of 0.15 mm (0.006 in) for cars and 0.25 mm (0.01 in) for heavy vehicles.

Spiral bevel

Although the straight bevel was cheap to produce and mechanically efficient, the meshing of the gears caused an objectionable noise, which was reduced when a *helical* form of tooth was employed. Naturally, it is impossible to generate a helix on a tapered pinion, so the gear is known as a spiral bevel.

The construction of the gear is shown in Fig. 45.3. A number of teeth, generated from the centre of the crown wheel, form, in the case of the pinion, a left-handed spiral. This direction causes a large outward thrust on the drive and a smaller inward thrust on the

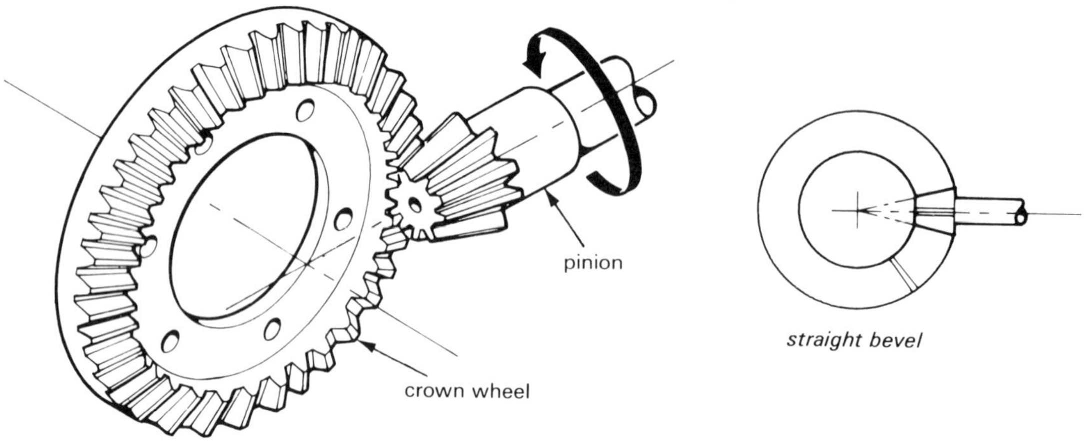

Fig. 45.2 Straight bevel

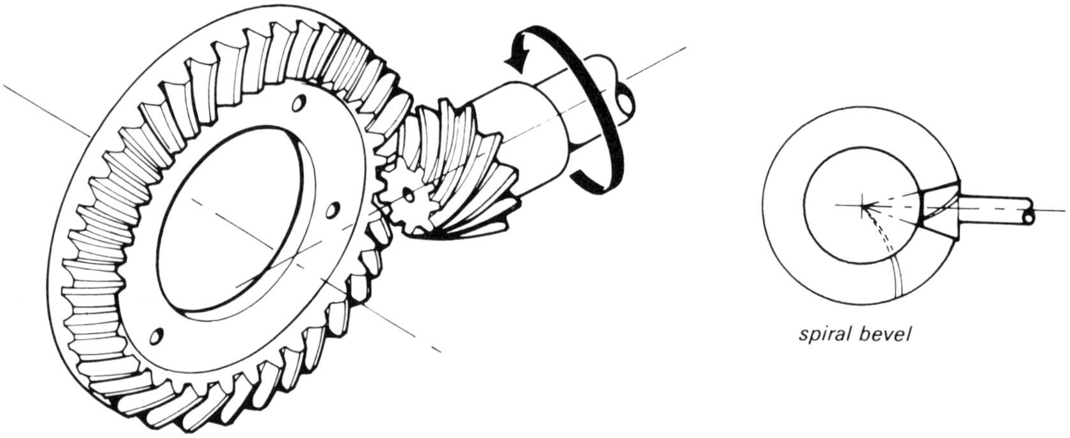

Fig. 45.3 Spiral bevel

over-run; therefore pinion bearing wear will increase the backlash rather than cause seizure of the gear.

You will see that, since the crown wheel teeth are inclined to the pinion, the tooth pressures are much higher. The *straight*, i.e. no additives, high-viscosity gear oil, which was satisfactory for the straight bevel type, gave poor results when used in spiral bevel units. The oil film broke down under the high loads, and allowed rapid wear and scoring to take place. Special oils to lubricate the surface after rupture of the oil film were developed and these are known as EP (extreme pressure) lubricants. They contain various additives such as sulphur, chlorine and phosphorus compounds, which react chemically, at high temperatures, with the metal surface to form a compound of low frictional resistance.

Hypoid

This type of gear, shown in Fig. 45.4, is the popular form in use today.

The pinion axis is offset to the centre line of the crown wheel. It can be placed above or below the centre, but in the case of cars is always placed below to give a lower propeller shaft and a reduction in the tunnel height. Pinion offset varies with the application but an offset equal to one-fifth of the wheel diameter is often used.

It will be seen that, by lowering the axis, the tooth pitch of the pinion increases, and so, for a given ratio, the pinion diameter can be larger (30 per cent for normal offset). This gives a stronger gear and is the main reason for its adoption on commercial vehicles.

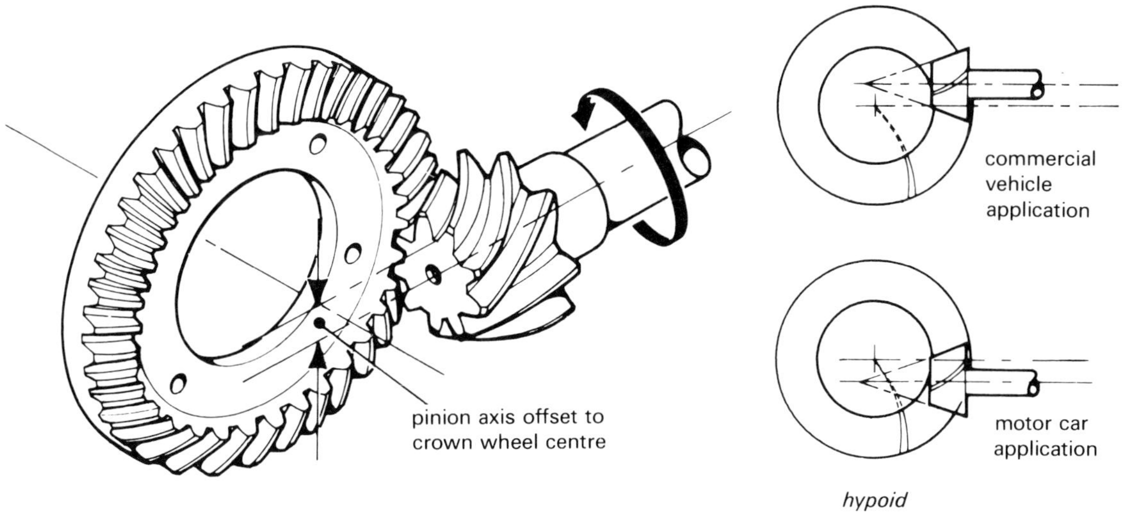

Fig. 45.4 Hypoid bevel

It is often said that a hypoid is halfway between a normal bevel and a worm drive. In the former case a rolling action takes place, whereas in the latter the motion is all sliding. By increasing the sliding motion in the hypoid gear, meshing noise is reduced, but the high temperature and pressure of the oil film is a strain on the lubricant. To deal with this, a special extreme pressure oil is employed which contains more active agents than those used with the normal spiral bevel. These oils contain expendable EP agents to resist scuffing and wear at high temperature, and a fatty acid to improve boundary lubrication at low temperature.

Worm and wheel

Today this expensive form of drive is rarely used as a final drive on light vehicles, but it is still used on heavy vehicles.

It is included in this chapter because this important type of gear has a number of other applications on motor vehicles.

Various arrangements, as shown in Fig. 45.5, can be used to give a very quiet and long-lasting gear, but efficiency is not so good as with the bevel (94 per cent against 98 per cent).

The gear ratio of a worm and wheel is given by:

$$\text{Ratio} = \frac{\text{Number of teeth on wheel}}{\text{Number of starts on worm}}$$

You can see that this type of gear provides a large reduction in a small space.

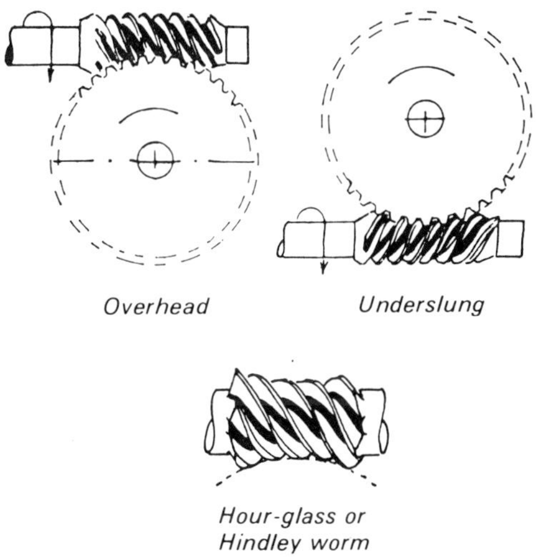

Fig. 45.5 Worm drives

The worm may be mounted below (underslung) or above (overhead) the wheel.

An hour-glass or Hindley worm embraces more teeth than the straight worm but adjustment is more critical.

Friction caused by the sliding action of the worm is reduced by using a worm wheel of phosphor-bronze and a worm of case-hardened steel, but even with this material the unit gets rather hot. A large, well-cooled sump is used to reduce oxidation of the oil. This occurs at high temperature and causes the oil to thicken. To improve the boundary lubrication, a vegetable-based oil is sometimes used as an alternative to straight gear oil.

Bevel-drive adjustment

Current types of bevel drive operate under severe conditions, but will give satisfactory service provided the gear is adjusted correctly. Every manufacturer gives detailed information of special tools, clearances, etc., and you should refer to this before attempting such an important overhaul.

The following general points should be a guide:

Noise from the final drive is caused by bearing defects or incorrect meshing (the latter may be caused by bearing wear). Generally noise or 'whine' on the drive occurs when the gear is too deep in mesh, and noise on the over-run is caused by insufficient depth of mesh. Whenever a final-drive noise develops, determine the cause as soon as possible. Misaligned gears will 'mate' to the new position, and make it impossible to obtain a quiet operation even when the adjustment is corrected. If repair is delayed, you will probably require a new crown wheel and pinion as well as a bearing.

Preloading of bearings Adjustable taper roller bearings are commonly used in a final drive assembly to support both the crown wheel and pinion.

If each bearing was set by adjustment to give a small clearance between the rollers and tracks, then end-float and misalignment of the gear would result. This would lead to gear noise and premature bearing failure, due to incorrect meshing and uneven loading respectively. Even when the clearance is eliminated, the elasticity of the bearing material will still give similar effects unless this is taken into account.

To overcome these problems, all adjustable bearings used for final drives are preloaded. This means that the bearings are forced together, by means provided for the adjustment, to the position occupied by the bearings when the unit is under full load.

In the case of pinion bearings the extent of this pre-compression is indicated by the torque required to rotate the pinion in its housing; for example, a torque of 1.36 N m (12 lbf in) is required to rotate the pinion. The recommended preload normally states the conditions under which it should be measured, e.g. with the bearings dry and the oil seal removed.

Fig. 45.6 shows two ways of checking the preload of a pinion.

Pinion adjustment The method of adjustment of the pinion bearing preload depends on the type of spacer used to hold the bearings apart. Fig. 45.7 shows the two main arrangements: a *rigid spacer* and a *collapsible spacer*.

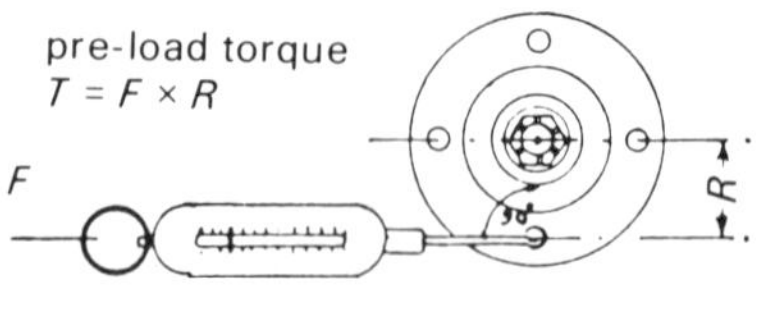

Checking preload with spring balance

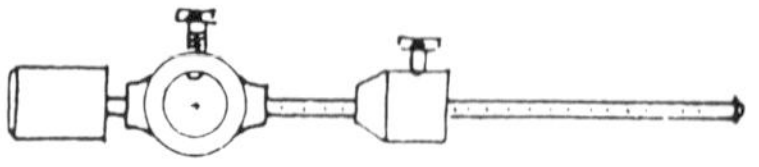

Preloading gauge

Fig. 45.6 Measurement of bearing preload

A bearing arrangement controlled by a *rigid spacer* has a series of shims to control the static load on the bearings (Fig. 45.7a). On this type the pinion nut has to be fully tightened before the preload is measured; if it is found to be incorrect then shims have to be either added or removed.

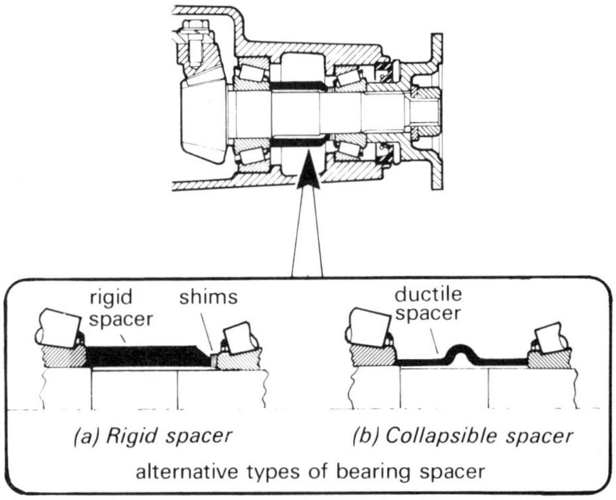

Fig. 45.7 Pinion bearing adjustment

This type of spacer simplifies the task of renewing the pinion oil seal, because the bearing preload is not altered by seal replacement provided the pinion nut is fully tightened.

Time is saved on the preload operation when a *collapsible spacer*, as shown in Fig. 45.7b, is used. In addition to holding the bearings in alignment, this ductile steel spacer collapses under pressure from the pinion nut and so allows the preload to be set in one operation. With the pinion, bearings and seal in position, the nut is tightened only sufficiently to give the recommended preload. When the torque is correct, the nut is locked in position. A ductile material has very little elasticity, so if the bearing preload is too high a new spacer is required.

Greater care must be taken with units having a collapsible spacer when a new pinion seal has to be refitted to cure an oil leak. Prior to unlocking the nut, the precise position of the nut relative to the pinion must be noted. Unless this action is taken, the crown wheel will have to be removed so that the pinion bearings can be preloaded.

The pinion position, relative to the crown wheel, must be set accurately in order to position the apex point of the pinion correctly. Noise, in the form of a 'whine', and possible breakage of the gear teeth is likely to occur if this is not done.

In the past, the initial position of the pinion was set by aligning it with the edge of the crown wheel teeth. Nowadays a special 'dummy pinion' jig is used for this operation.

Crown wheel adjustment After setting the pinion, the crown wheel is fitted to the assembly. When doing this it is important to fit the correct bearing cap, because these are not interchangeable.

Backlash is adjusted at this stage by moving the crown wheel towards or away from the pinion. Before taking the final measurement of backlash with a clock gauge, it is wise to check that the crown wheel run-out is within the limits recommended.

Meshing of the gear is carried out by applying a smear of marking compound to the driving side of a few crown wheel teeth and then turning the pinion in the d.o.r. whilst applying a resistance to the crown wheel.

The marking obtained indicates the mesh of the gears with respect to pinion position and backlash. Fig. 45.8 shows the change, in relation to the correct marking, when the pinion position is changed. It should be noted that the 'correct' marking shows only a limited contact between the teeth under light-load test conditions. When full load is applied to the gear, the contact area covers the complete tooth.

Crown wheel preload Housing rigidity and accurate gear alignment are essential with a hypoid gear if the low noise level feature of this type is to be achieved. This means that slackness in the crown wheel bearings must be eliminated by preloading.

Various methods are used to measure the preload; the most common is to measure the spread (outward deflection) of the bearing caps when the adjusters are screwed together. For example, a preload cap-spread of 0.1 mm (0.004 in) means that after preloading the side bearings, the distance between the bearing caps has increased by 0.1 mm.

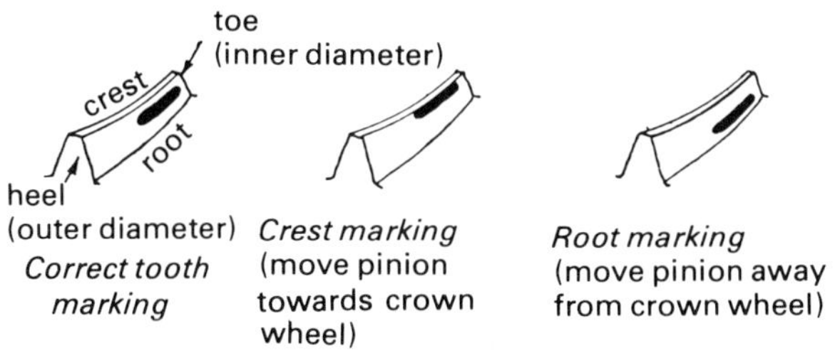

Fig. 45.8 Crown wheel tooth marking

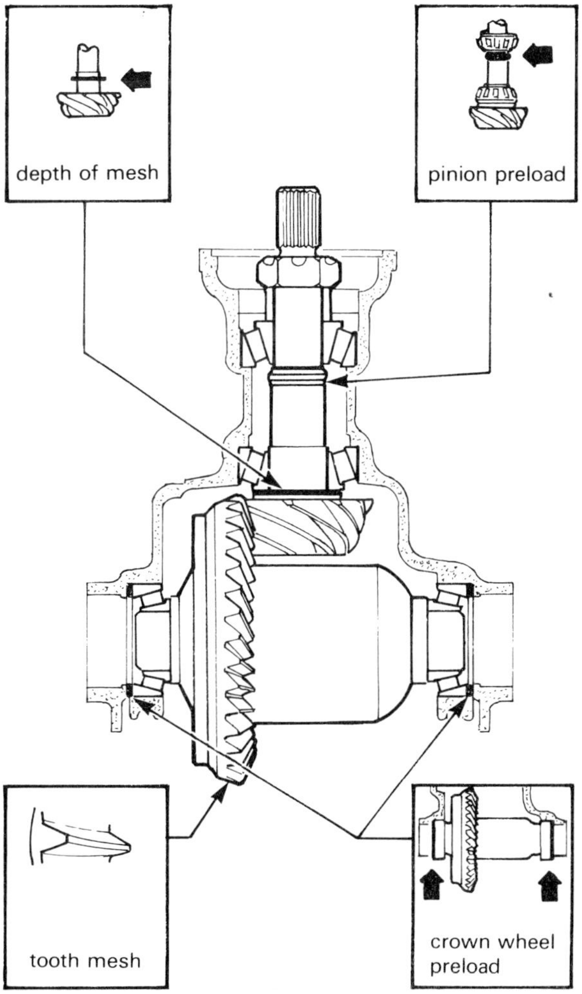

Fig. 45.9 Final drive adjustments

When shims are used to control the backlash, the preload is achieved by increasing the shim thickness on each side of the crown wheel; the shims are added after the backlash has been set. Axles having this construction need a special tool to 'spread the caps' before the crown wheel assembly is removed or refitted.

A visual summary of the main adjustments is shown in Fig. 45.9.

46 The differential

Purpose

If both rear wheels were connected to a common driving shaft, two effects would soon show up: (a) rapid rear tyre wear, and (b) difficulty in steering from the straight-ahead position.

Fig. 46.1 shows that the outer wheel must travel a greater distance when the vehicle is cornering: therefore if the wheels are interconnected, the tyres will have to 'scrub' over the road surface and tend to keep the vehicle moving straight ahead. These effects can be minimized by driving one wheel and allowing the other to run free, but the unbalanced driving thrust and unequal cornering speeds make the arrangement unpopular. The solution to the problem came in 1827 when Péqueur of France invented the differential. This mechanism allows the wheels to rotate at different speeds, but still maintains a drive to both wheels.

Principle

Consider the two discs, shown in Fig. 46.2a, to be linked by shafts to the wheels and interconnected with a lever. When a force (F) is applied to C at the centre of the lever, each disc will receive an equal share – i.e. half the force applied. The movement of the discs will depend on the resistances (R) opposing the motion of the shafts. If a larger resistance acts on disc B, the lever will tilt, and push disc A forward a greater amount. This condition is shown in plan view in Fig. 46.2b and you will see that:

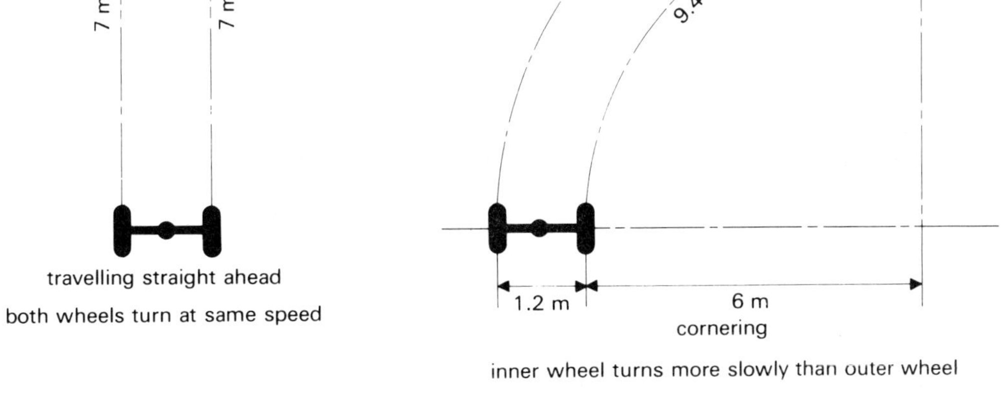

Fig. 46.1 Need for differential

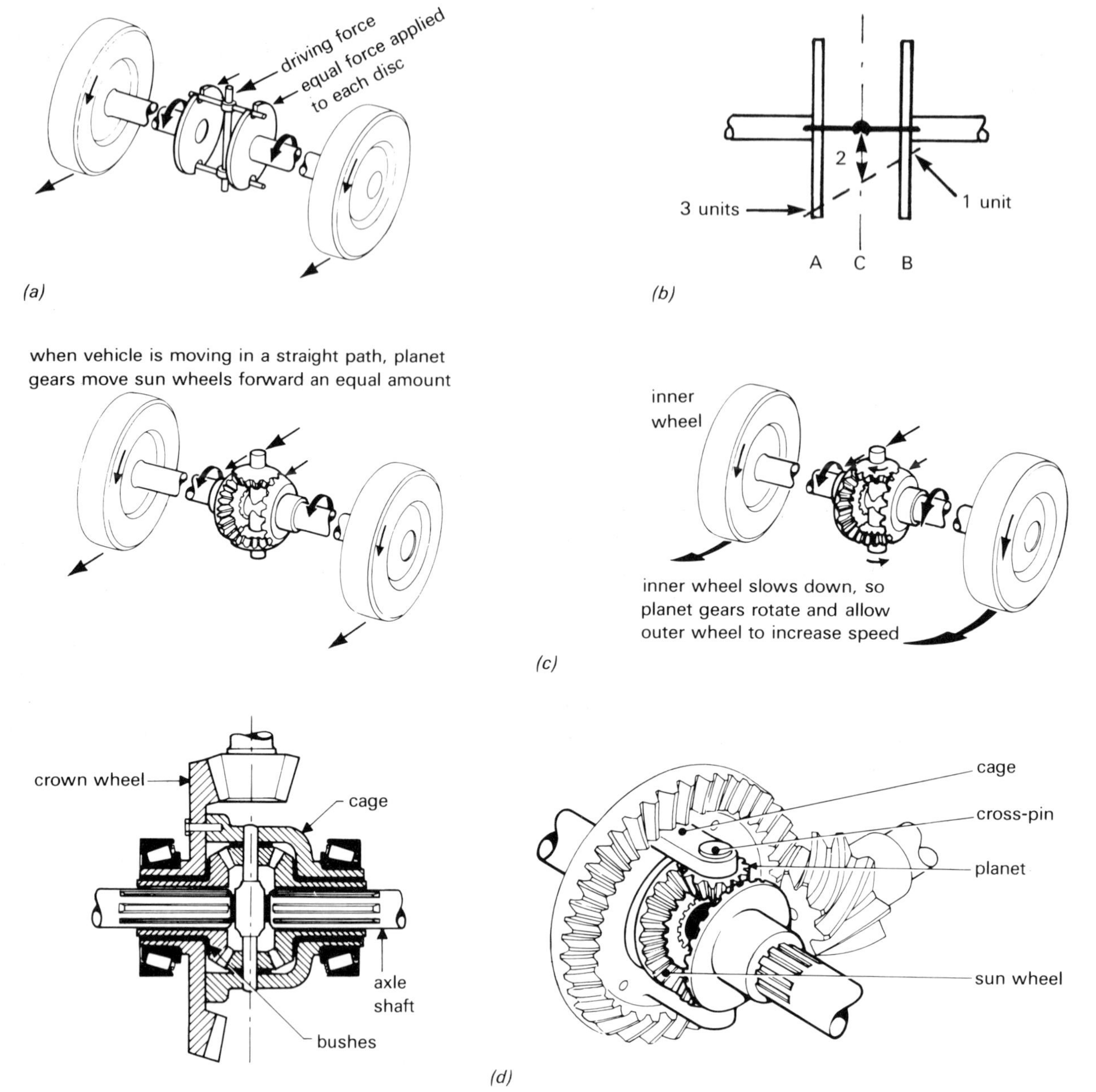

Fig. 46.2 Action of differential

Increase in distance moved by A
 = Decrease in distance moved by B

Increase in speed of A
 = Decrease in speed of B

Therefore

$$A + B = 2C$$

In Fig. 46.2c the disc system is replaced by bevel gears, which are called sun wheels (discs) and planets (levers). The drive is applied to the cross pin and will push the planet gears forward to exert an equal torque on each sun wheel irrespective of the speed. When the vehicle turns a corner, the inner wheel will slow down and cause the planets to rotate on their own axis to speed up the outer wheel. Straight-ahead motion of the vehicle will allow the whole unit to rotate at the same speed.

The complete differential is shown in Fig. 46.2d. This arrangement shows a crown wheel bolted on to a differential cage. This cage supports the sun wheels on plain bearings and transmits the drive to the cross pin.

Light cars only need two planet gears, but four gears are necessary on heavier vehicles to reduce tooth pressures.

Lubrication is provided by the final-drive oil, which can splash through holes in the differential cage.

Differential lock

If one driving wheel of a two-wheel drive vehicle loses adhesion, the propelling force is considerably reduced and this results in the vehicle being immobilized. On these occasions the differential action is undesirable, so, on vehicles designed to operate over poor surfaces, a differential locking arrangement is often fitted.

Differential action can be prevented by locking together any two individual units of a differential and Fig. 46.3 shows one arrangement. Splined to a differential sun wheel is a sliding dog clutch member, which engages with dog teeth formed on the cage of the differential. The clutch is engaged by means of a fork and movement of this can be made by a lever fitted on the outside of the axle.

In the engaged position, the sun wheel, and consequently the rear wheel connected to this sun gear, is made to turn at the same speed as the cage. Locking one sun gear to the cage in this way ensures that the other sun gear also turns at the same speed.

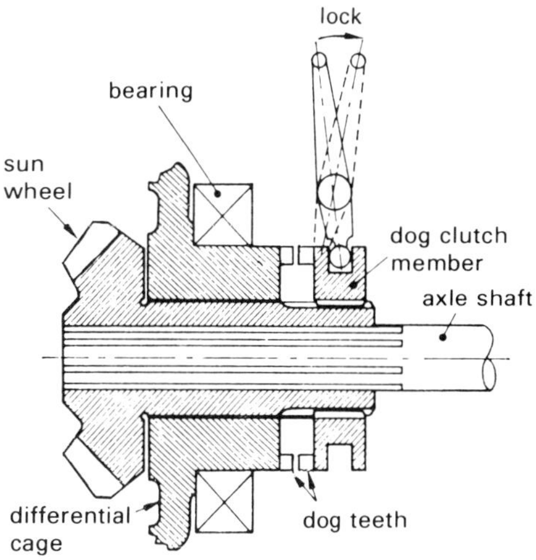

Fig. 46.3 Differential lock

Limited-slip differential

Although a high mechanical efficiency is desirable for the majority of mechanical components, it is a disadvantage in the case of a differential. In addition to the reduced traction over slippery surfaces, a 'low friction' differential fitted to a high-powered vehicle limits acceleration and causes excessive tyre wear. Observation of such a vehicle during acceleration shows that torque reaction of the engine tends to lift the left-hand driving wheel off the ground and when this is accompanied by an uneven road surface, the presence of excessive wheel spin is apparent. In order to restrict these drawbacks, the differential action is opposed by artificially increasing the friction between the sun wheel and the differential cage. Arrangements having this feature are called *limited slip differentials*.

There are two basic types of differential which give this, or a similar action; these are:

- *Mechanical limited-slip differential*
- *Visco-differential.*

Mechanical limited-slip differential Fig. 46.4 shows the layout of one type of differential, which is bolted to the crown wheel.

A multi-disc clutch pack fitted behind each sun wheel has the inner and outer plates splined to the sun and cage respectively. Since bevel gears are used, an axial thrust will be developed which will be proportional to the torque applied by the crown wheel to the differential. Under low torque conditions, the differential will function in the normal way, but if the torque is increased, the clutch pack will be loaded and this will resist rotation of the sun gear at a different speed from that of the cage (Fig. 46.5).

To increase the load on the clutch pack further many designs incorporate:

1 A Belleville disc-spring washer between the cage and the clutch discs of each pack to provide an initial load on the discs.
2 Angled cam faces between the cage and the cross pins. Driving thrust exerted by the cage on the pin causes the pin to force the planets against the side gear ring. When four planets are used, two separate pins, flexibly linked at the centre and having opposite cam faces, cause two planets to act on one clutch pack and the other two to exert force in the opposite direction.

The action of this type of differential should be remembered by the mechanic. With only one driving wheel lifted off the ground, a drive transmitted to this wheel from the engine can cause the vehicle to be driven forward.

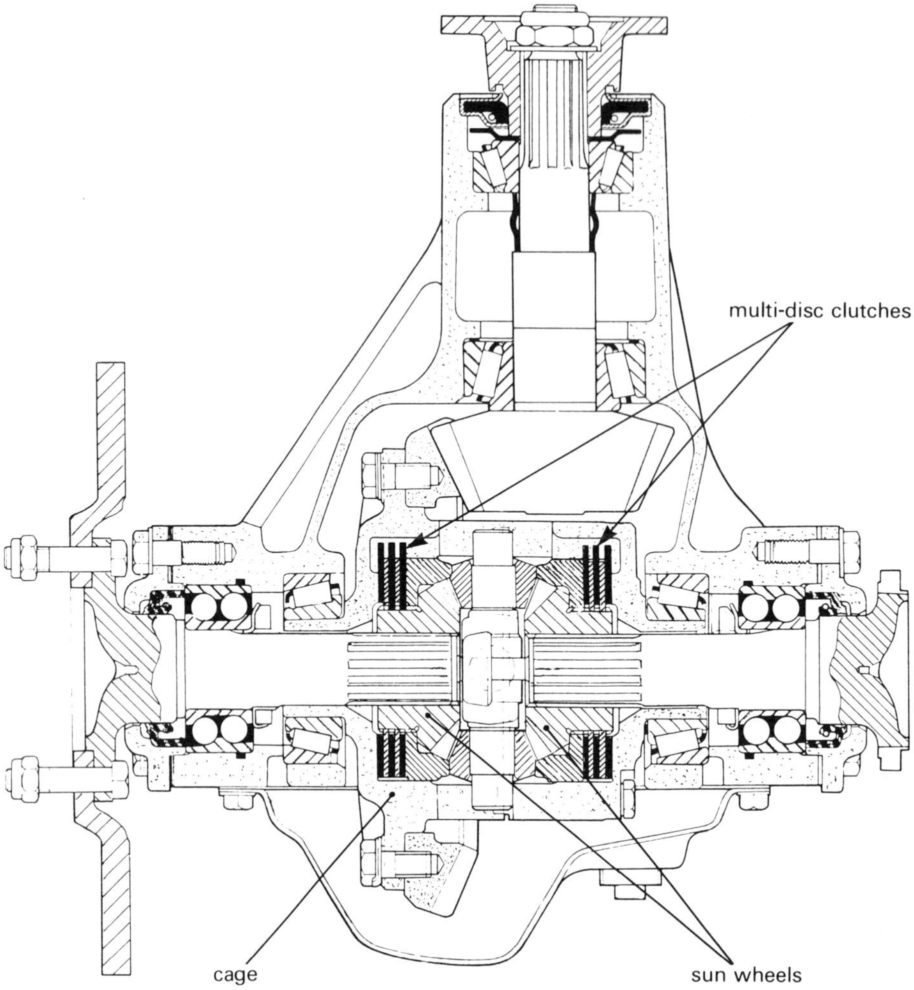

Fig. 46.4 Final drive assembly with limited-slip differential

Visco-differential

This type, which has been used more and more each year since 1969, combines a standard differential with a *viscous coupling*. This transmission application uses the coupling as a *viscous control* device to regulate the speed difference between the two driving wheels.

Whereas the locking action on a mechanical limited-slip differential depends on input torque, the viscous type depends on the speed difference of the driving wheels. Very little resistance is offered when the speed difference is small but the resistance progressively builds up as the difference increases. Compared with the mechanical limited-slip differential, it is claimed that this type gives lower tyre wear, easier steering and lower stresses in driveline components.

The basic construction of a visco unit is similar to a multi-plate clutch. It consists of a housing and hub, between which is sandwiched a series of perforated metal plates (Fig. 46.6). Alternate plates are attached to the hub and housing; this also acts as a container for the viscous silicone fluid. A heat resistant seal, made of fluorinated rubber, isolates the silicone fluid from the lubricating oil in the final drive assembly and ensures that the fluid will suffice for the life of the coupling.

The operating principle of the coupling is based on the fact that the *shear force* (drive) given by the fluid on a plate depends on the *shear rate* (speed difference). By using a high-viscosity fluid, the coupling can be designed to output a comparatively high torque, which progessively builds up as the shear rate is increased.

The relationship between torque and shear rate is

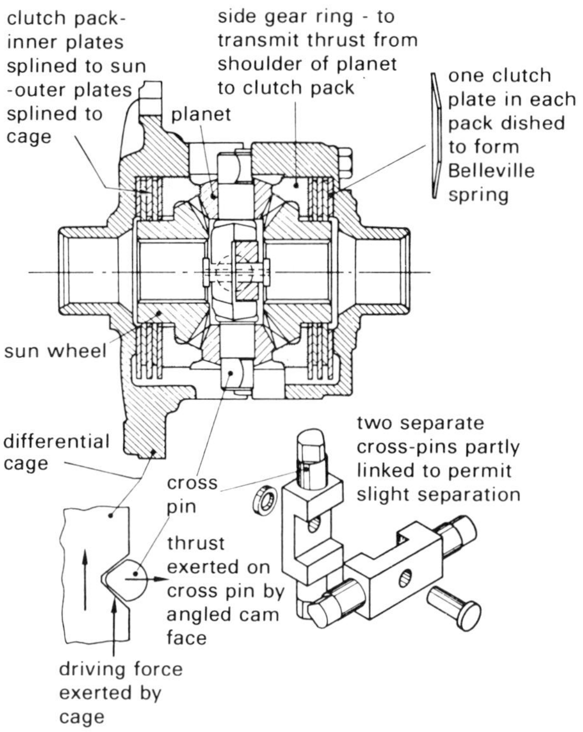

Fig. 46.5 Limited-slip differential

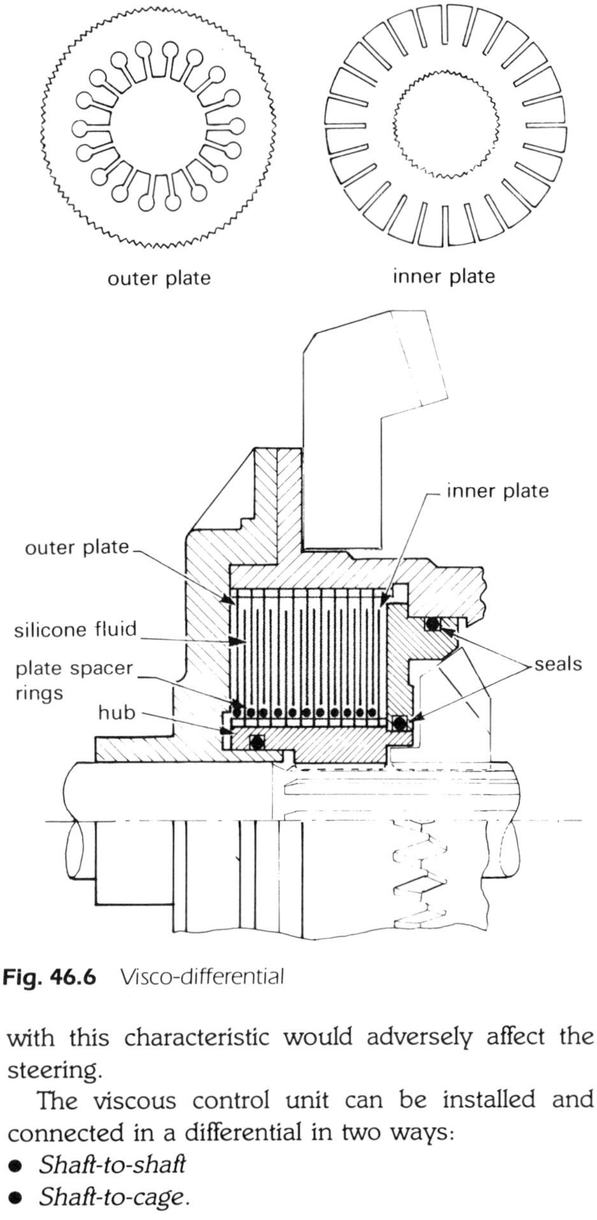

Fig. 46.6 Visco-differential

affected by a number of factors such as fluid viscosity, the gap between plates and plate perforation. By varying these design factors, the coupling can be 'tuned' to suit the application.

Prolonged slipping of the coupling generates heat and this causes the fluid to expand and take up some of the space occupied by the air. Where spacers are not used to hold the plates apart, the combination of the increase in air pressure due to fluid expansion, and the reduction in fluid pressure in the gaps, causes the plates to be pushed together. This gives metal-to-metal contact and at this point (*hump*) the coupling temporarily departs from its *viscous mode* operation. As it passes through this phase the torque output rises considerably; in some designs it can be six times as great.

Viscous couplings can be used to control front, centre and rear differentials. Each application has different requirements in view of its position, speed of operation and type of vehicle to which it is fitted. For example, a viscous control unit at the rear of a powerful car requires some form of overload protection, so the 'hump' is used to fill this need. On the other hand, a viscous coupling used for a front-wheel drive car should have no 'hump', because a unit with this characteristic would adversely affect the steering.

The viscous control unit can be installed and connected in a differential in two ways:

- *Shaft-to-shaft*
- *Shaft-to-cage*.

Shaft-to-shaft viscous coupling Fig. 46.7 shows a bevel-type differential having a viscous control unit connected between the two axle shafts, the hub holding the inner plates joined to one shaft, and the housing with its outer plates connected to the other shaft. Space is made available for the unit at the centre of the crown wheel by moving the differential to one side.

When the car is moving with no wheel slip in a straight line, the control unit is out of action. But at times when the shaft speed is different, such as when one driving wheel loses adhesion or when wheel spin

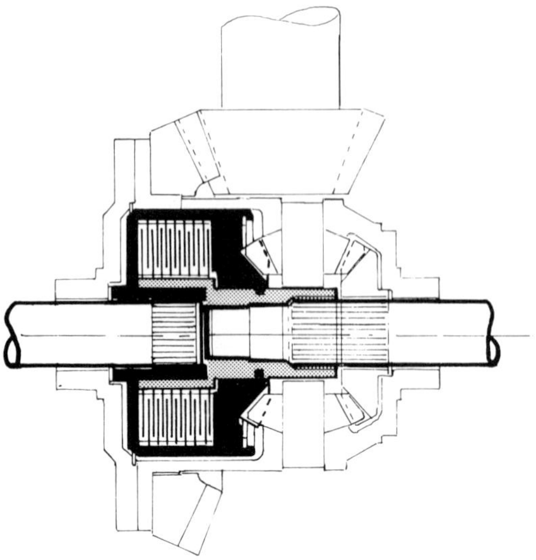

Fig. 46.7 Shaft-to-shaft viscous coupling

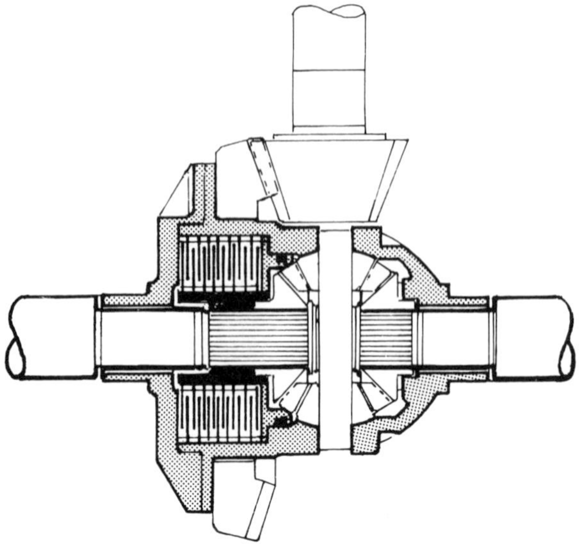

Fig. 46.8 Shaft-to-cage viscous coupling

occurs during acceleration, the resistance offered by the unit ensures that the driving torque to both wheels is maintained.

Shaft-to-cage viscous coupling The cheaper arrangement shown in Fig. 46.8 has the housing of the viscous control unit integral with the differential cage and the hub connected to one axle shaft.

The speed difference between the cage and axle shaft is half the difference of the road wheels, so compared with the shaft-to-shaft type, the shaft-to-cage unit must have a torque characteristic about three times greater to achieve the same locking effect.

47 Rear axle construction

In cases where the rear suspension is non-independent, the type of axle used is either a *dead axle* or a *live axle*. The former only has to support the weight of the vehicle, whereas the latter has to fulfil this task and, in addition, contain a gear and shaft mechanism to drive the road wheels.

Axle casing construction

The type of axle casing dictates the method that must be used to remove the final drive assembly. Today the casing used will be either a *banjo*- or *carrier*-type. In the past a type known as a *split* (*trumpet*) casing was occasionally used. These three types are shown in Fig. 47.1.

Banjo axle

The tubular axle section is built up of steel pressings welded together and suitably strengthened to resist the bending load. The centre of the casing, combined with the axle tube on one side, resembles a banjo, hence its name.

The final drive assembly, which is mounted in a detachable malleable iron housing, is secured by a ring of bolts to the axle casing. When this assembly is in position, the axle shafts are slid in from the road wheel end of the casing.

On some banjo axles a domed plate is bolted to the rear face of the casing. Removal of this plate allows the final drive gears to be inspected and, in cases where the axle shaft is secured to the differential, the access point enables the axle shaft to be unlocked from the sun gear (side gear).

A lubricant level plug, set at a height about one-third up the crown wheel, is screwed into the domed cover or the final drive housing. The level is normally just below the axle tubes, so lubrication of the hub bearings is by splash caused by rotation of the crown wheel. Overfilling should be avoided because this swamps the oil seals and caused oil to enter the brakes.

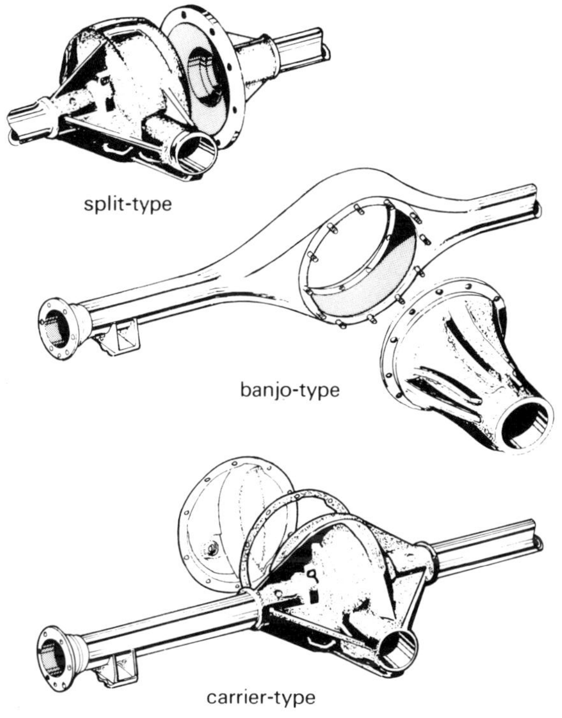

Fig. 47.1 Types of rear axle casing

When in operation the final drive gets hot, so some form of air vent is provided to release the pressure in the axle casing; this prevents oil being forced past the seals.

Carrier axle

This type is more rigid than a banjo casing and is often used to support a hypoid gear. The final drive assembly is mounted directly in a rigid malleable cast iron carrier, into which the axle tubes are pressed and welded. Extra rigidity is obtained by using reinforcing ribs that extend from the pinion nose to the main carrier casting. Access to the final drive gear is by means of a domed plate at the rear of the casing.

Axle shafts

The axle shaft transmits the drive from the differential sun wheel to the rear hub.

The various types may be compared by considering the stresses the shaft has to resist. Fig. 47.2a shows a line sketch of a simple shaft which is subjected to:

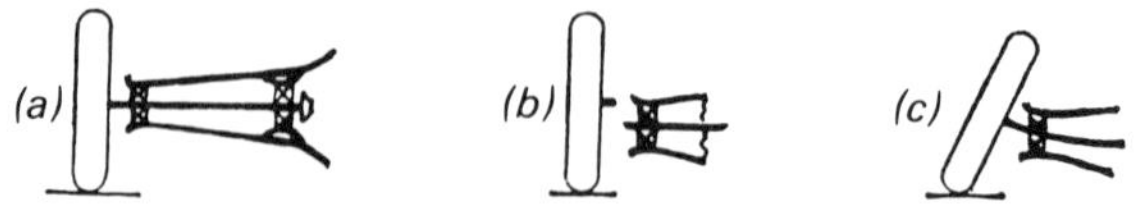

Fig. 47.2 Effect of load on simple axle shaft

1 Torsional stress due to driving and braking torque.
2 Shear stress due to the weight of the vehicle (Fig. 47.2b).
3 Bending stress due to the weight of the vehicle (Fig. 47.2c).
4 Tensile and compressive stress due to cornering forces.

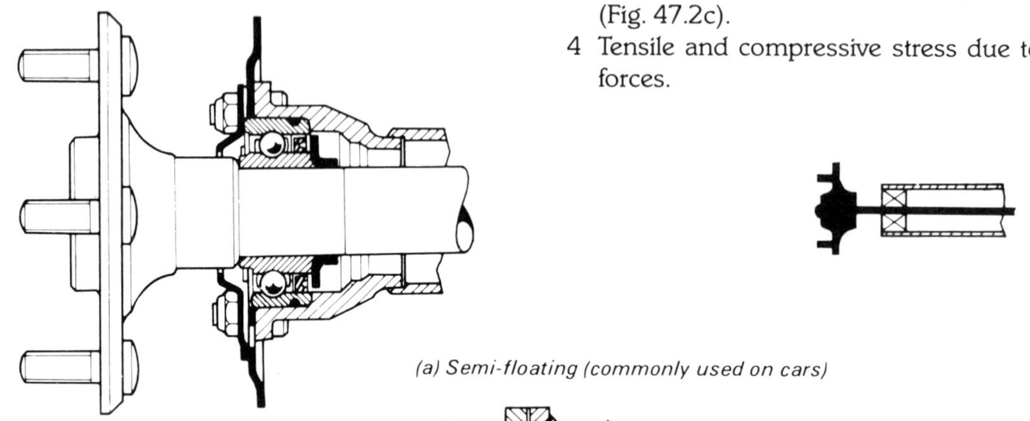

(a) Semi-floating (commonly used on cars)

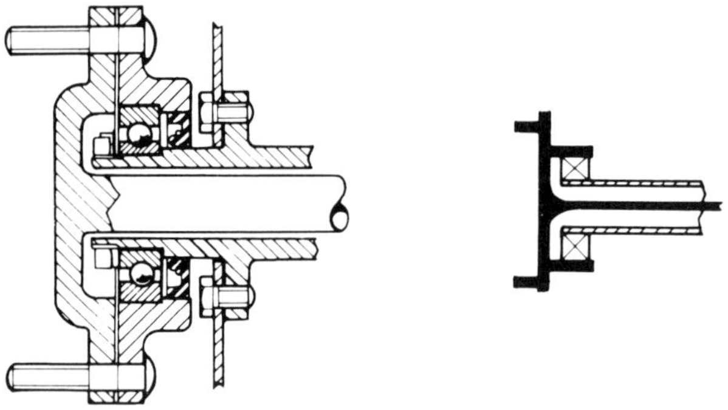

(b) Three-quarter floating (rarely used today)

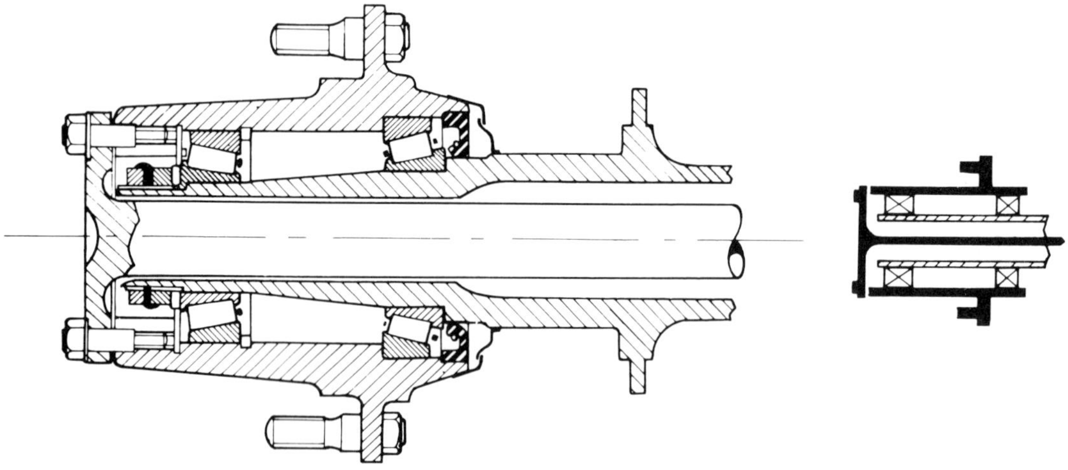

(c) Fully floating (commonly used on heavy vehicles)

Fig. 47.3 Rear hub construction

Types

Axle shafts are divided into three main groups, depending on the stresses to which the shaft is subjected:

- *Semi-floating*
- *Three-quarter floating*
- *Fully floating*.

Semi-floating

Fig. 47.3a shows a typical mounting of an axle shaft suitable for light cars. A single bearing at the hub end is fitted between the shaft and the casing, so the shaft will have to resist all the stresses previously mentioned. To reduce the risk of fracture at the hub end (this would allow the wheel to fall off), the shaft diameter is increased. Any increase must be gradual, since a sudden change in cross-sectional area would produce a stress-raiser and increase the risk of failure due to fatigue. (Fatigue may be defined as breakage due to continual alteration of the stress in the material.)

Although the final-drive oil level is considerably lower than the axle shaft, the large amount of 'splash' would cause the lubricant to work along the shaft and enter the brake drum. Sealing arrangements normally consist of an oil retainer fitted at the hub end (the lip of the seal is positioned towards the final drive).

Fully floating

This is generally fitted on commercial vehicles where torque and axle loads are greater.

The construction shown in Fig. 47.3c consists of an independently mounted hub which rotates on two bearings widely spaced on the axle casing. This arrangement relieves the shaft of all stresses except torsional, so the construction is very strong. Studs connecting the shaft to the hub transmit the drive and when the nuts on these studs are removed, the shaft may be withdrawn without jacking up the vehicle.

Three-quarter floating

Having defined the semi- and the fully floating shaft, any alternative between the two may be regarded as a three-quarter floating shaft. Fig. 47.3b shows a construction which has a single bearing mounted between the hub and the casing. The main shear stress on the shaft is relieved but all other stresses still have to be resisted.

Axle shaft material

A tough, hard material must be used to withstand the various stresses, resist spline wear and provide good resistance to fatigue. A medium carbon alloy steel containing such elements as nickel, chromium and molybdenum is the usual choice.

48 Four-wheel drive

A four-wheel drive, sometimes called an FWD or 4 × 4 (four by four), vehicle transmits a drive to all of its four wheels instead of two wheels as is common with the majority of light vehicles.

There are two main traction problems with a two-wheel drive (4 × 2) vehicle; these are:

1 Loss of traction during cross-country operation.
2 Loss of adhesion during acceleration.

The intended use of the vehicle governs the type of 4 × 4 drive system that is offered by the manufacturer.

Vehicles in the Land-Rover category are designed to work efficiently both on and off the highway. This type of vehicle uses two-wheel drive on 'hard' surfaces and restricts the use of four-wheel drive to cross-country work where the vehicle is likely to encounter muddy surfaces that could lead to a complete loss of traction from the two driving wheels.

High-performance cars that are built for highway operation use four-wheel drive at all times; this gives improved handling and extra safety. In this case, the full weight of the car is spread over all of the driving wheels. This full utilization of the vehicle weight considerably increases the tractive effort that can be applied to the car; as a result a high rate of acceleration can be achieved.

Cross-country operation

The limitation of a conventional two-wheel drive vehicle is soon appreciated when a 4 × 2 vehicle encounters the rough terrain and muddy conditions normally associated with cross-country operations. In these situations the rear wheels soon lose their adhesion; consequently the traction available is insufficient to provide a drive. The normal differential adds to these difficulties since if the adhesion of one driving wheel is lost, the drive to the other wheel is generally too low to propel the vehicle. Although a differential locking device reduces the chance of this occurrence, there are still many occasions when both driving wheels are rendered ineffective. It is to overcome this situation that four-wheel drive is provided – when either front or rear axles lose traction the other axle maintains a drive to move the vehicle out of the difficulty.

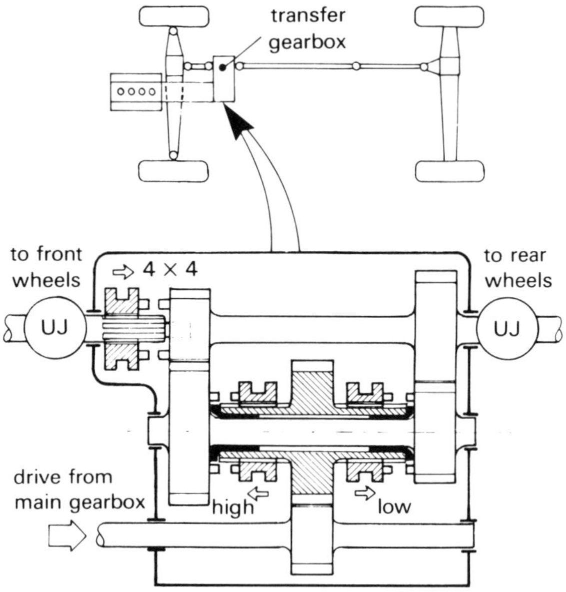

Fig. 48.1 Four-wheel drive layout for off-highway use

Fig. 48.1 shows a typical layout of a four-wheel drive vehicle. Mounted behind the main gearbox is a transfer gearbox and this unit divides the drive between the axles and normally incorporates extra gears to provide a very low ratio. The transfer box is generally controlled by two gear levers; one for selecting four-wheel drive and the other to give 'high' or 'low' gear. Drive to the front wheels is transmitted via the final drive, differential and drive shafts. Steering movement of the front wheels is accommodated by fitting extra universal joints at the wheel end of the drive shafts and in order to overcome vibration, the constant-velocity type of joint is generally used.

The four-wheel arrangement shown in Fig. 48.1 should not be operated on a hard-surfaced road because of the risk of transmission 'wind-up'. When a vehicle using this arrangement turns a corner the mean speed of the front wheels is higher than that of the rear wheels, so the speed difference causes the propeller shafts to deflect. Drivers disregarding this instruction often cause serious injury to mechanics who are performing a wheel change operation. To overcome the wind-up problem and make the vehicle suitable for four-wheel operation on hard surfaces, a third differential 'between' the driving axles would be needed. This differential is rarely fitted to a vehicle designed for cross-country work, since the differential would defeat the main purpose of four-wheel drive - if one wheel lost adhesion the vehicle would be immobilized.

Hard surface operation

Tractive and braking effort are limited by the adhesive force at the road wheels; this force is calculated from:

$$\text{Adhesive force} = \text{Coefficient of friction} \times \text{Load on wheel}$$

By using this expression, it will be seen that the total weight of the vehicle must be fully utilized if the maximum adhesive force is to be obtained.

When only two driving wheels are provided, the loss of 'weight' on the non-driving wheels proportionally reduces the adhesive force; as a result the maximum tractive effort that can be safely applied to the vehicle is limited.

Beyond the point where the tractive effort equals the adhesive force, the driving wheel spins; this spinning not only limits the rate of acceleration, but can also lead to loss of control of the vehicle with dangerous results. This effect also relates to braking; skidding occurs beyond the point where the braking force equals the adhesive force.

By spreading the driving and braking forces over all of the wheels, greater safety is achieved. Backing up this force distribution with a device for sensing the approach of wheel spin or slip and then adjusting the effort accordingly, leads to even greater safety.

Cars of this type are not designed for serious off-road use, but they do give the all-weather driver an added sense of security for tackling icy or wet conditions.

One of the first persons to appreciate the importance of 'all-wheel drive' was Harry Ferguson, the inventor of the lightweight tractor. It was in 1954 when he took out a patent, the 'Ferguson formula' (FF), and this was used on the Jensen car in the early 1960s.

Ferguson Formula

Fig. 48.2 shows a typical layout as applied to a Jensen car. A transfer box, containing the sensing mechanism, is adjacent to a centre (third) differential, and propeller shafts transmit the drive to front and rear axles. A Maxaret unit prevents the possibility of all wheels locking during braking. Driven by the transmission it senses the mean deceleration of front and rear road wheels and just before the wheel locks the hydraulic pressure to the brake is relieved.

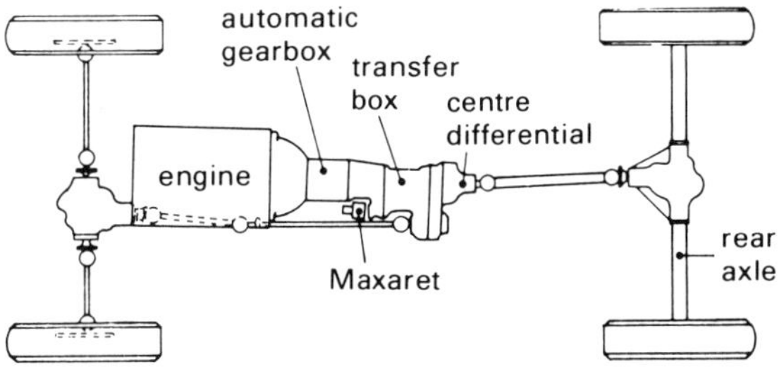

Fig. 48.2 Transmission layout (Jensen)

The basis of the Ferguson control system consists of a master differential and two one-way multi-disc clutches; the unit divides the torque in the ratio of 37 per cent front and 63 per cent rear, besides allowing for a speed variation between front and rear wheels. The differential is a simple epicyclic gear train which has the input shaft driving the planet carrier, the annulus and sun being connected to rear and front wheels, respectively. Application of a force to the centre of the planet gives equal force at the annulus and sun. Since the force at the annulus is acting at a larger radius, the torque is greater, so this member is connected to the rear wheels. As with a normal differential, speed variation can take place. In this arrangement the difference in speed between the output shaft and the planet carrier causes the planet to rotate around the slower-moving gear, a motion which speeds up the other member. The extent of this speed variation is limited by the control unit and in this particular design the front wheels are only permitted to over-run the rear ones by 16.5 per cent, whereas the rear wheels can only over-run to the extent of 5.5 per cent. Once this limit is reached a multi-disc clutch locks the planet carrier (input shaft) to the sun wheel (front output shaft), i.e. the centre differential is put out of action. This action can occur during acceleration or when the vehicle is braking.

Four-wheel drive using viscous coupling

Unless some form of control is used to limit the action of a central differential, the loss of adhesion at any one wheel will cause the vehicle to become immobile. On the other hand, if a central differential is not fitted, then driveline components will be stressed because of transmission wind-up. A compromise is achieved by using a viscous coupling. This can be arranged in one of two ways: it can be used either as a control unit to limit the speed difference between front and rear wheels, or as a viscous transmission to connect the front drive system to the rear when the conditions demand it.

Four-wheel system derived from the standard drive The parallel axis of the engine makes it necessary to use a transfer gearbox; this is positioned behind the main gearbox (Fig. 48.3).

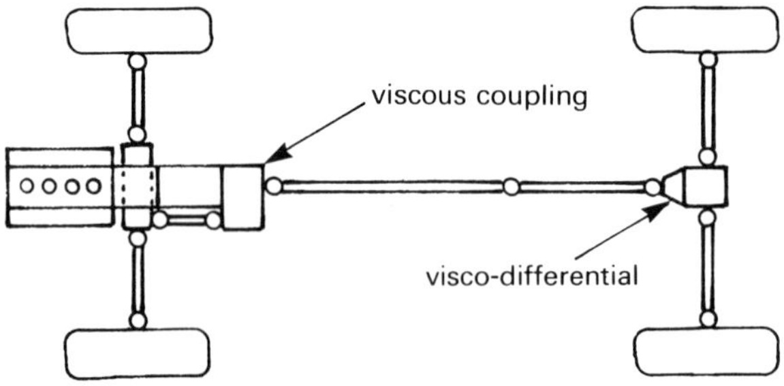

Fig. 48.3 4 × 4 layout – longitudinal engine (viewed from underside)

The Ford 4 × 4 uses this construction; it has an epicyclic gear set as a central differential and a viscous coupling to control the speed difference between the front and rear wheels (Fig. 48.4). The input shaft of the differential is connected to the planet carrier and the front- and rear-wheel output shafts are linked to the sun wheel and annulus respectively. This gives a torque split between the front and rear axles of 34 : 66 – a typical proportion for this type of car.

At the front a standard differential is fitted on the right-hand side of the engine and a drive shaft, fitted with CV joints, transmits the power to the right-hand front wheel. The drive train to the left-hand front wheel necessitates an intermediate shaft which, owing to the location of the engine, has to pass through the engine sump before connecting with the left-hand drive shaft.

The rear axle is driven in the normal manner; this includes a two-piece propeller shaft and a visco-differential.

Compared with a rear-wheel drive 4 × 2 vehicle, the front axle steering geometry must be modified to take into account the different driving conditions.

On this car the power take-off to the offset output shaft, which connects with the front axle shaft, is by means of a silent chain. The viscous coupling is a shaft-to-shaft design (see page 303). Connected 'between the axles' in this manner the coupling senses the speed differences between the axles and gives a torque difference twice the value of the torque transmitted by the coupling.

Control of drive by the coupling to the front and rear wheels varies with the operating conditions. At times when the mean speed of front and rear wheels is nearly equal, the coupling is ineffective, but as the speed difference increases, the coupling resistance and associated locking rise progressively.

Complete loss of adhesion at any wheel causes a large slip in the coupling; as a result the coupling exerts a high locking torque on the gears so as to maintain a drive to the wheels that are attempting to spin. Extended use of the coupling under maximum slip conditions, would soon cause it to overheat, so to give it protection, the coupling is designed to go into the 'hump mode'.

Four-wheel system derived from the front-wheel drive The basic arrangement is comparatively easy to adapt to a 4 × 4 system because the complicated part of the assembly, namely the front drive layout and steering geometry, is already in place.

Most front-wheel drive cars use a transaxle, so the drive is taken from this unit through a central differential to the rear wheels (Fig. 48.5).

The torque distribution proportion varies with the make of car. It depends on the weight distribution, in particular the location of the engine, and the characteristics that are desired. The front/rear proportion varies from about 56 : 44 to 35 : 65.

Most central differentials use an epicyclic gear set. This is controlled by a viscous coupling mounted in one of three ways:

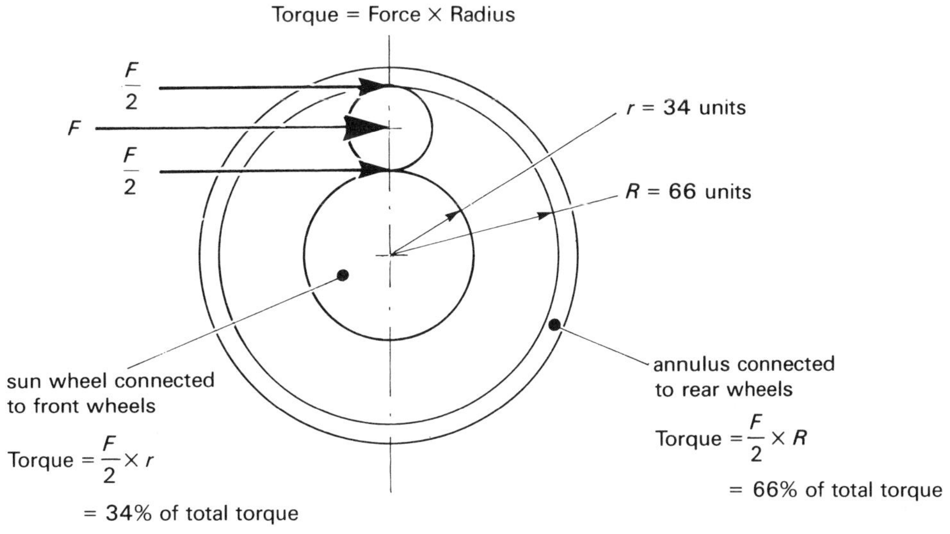

epicyclic differential to divide front/rear torque in the proportion 34/66

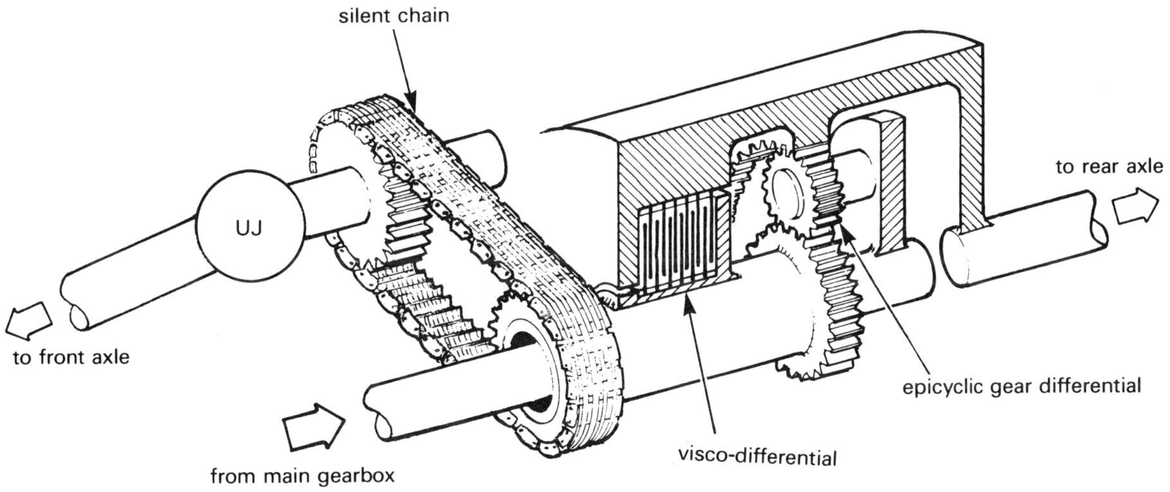

Fig. 48.4 4 × 4 transfer box (Ford)

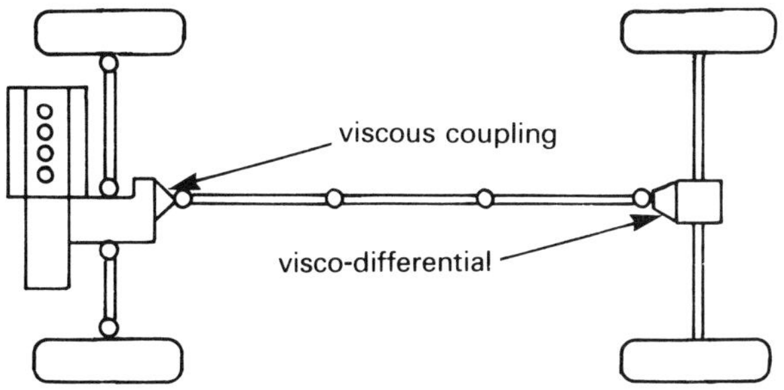

Fig. 48.5 4 × 4 layout – transverse engine

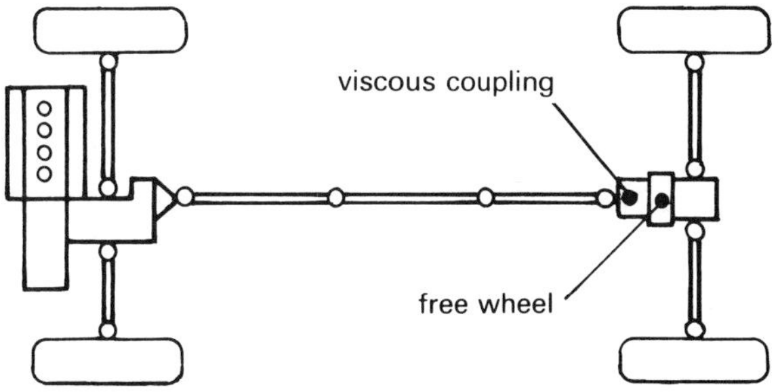

Fig. 48.6 4 × 4 layout – viscous transmission (VW)

- Shaft-to-shaft between take-offs to front and rear axles
- Between input drive and take-off to either the front axle or rear axle.

Variable four-wheel drive This system achieves the benefits of four-wheel drive, but without the draw-backs of a permanently connected arrangement.

The *viscous transmission* layout shown in Fig. 48.6 uses a propeller shaft and viscous coupling to interconnect the two axles. A central differential is unnecessary, but a free wheel is fitted in the rear axle final drive gear.

During normal operation, when the axle speeds are nearly equal, the coupling ensures that the rear wheels are only driven to a very limited extent, i.e. the vehicle behaves like a normal front-wheel drive system. But when the front wheels lose their adhesion, the difference in speed causes the coupling to make use of the rear wheels for propulsion.

The free wheel ensures that no drive is transmitted from the rear axle to the engine during overrun, so this feature makes the arrangement suitable for use with anti-lock brake systems. A free wheel locking device, operated by an electro-pneumatic system when reverse gear is selected, makes four-wheel drive available for reversing.

PART FOURTEEN

Steering systems

49 Directional control

The fact that vehicles today often travel at high speeds illustrates the need for good steering, but perhaps even more important are efficient maintenance and fault rectification. Steering fault diagnosis can often place great demands on the mechanic's basic knowledge, and so it is essential in the interests of road safety that he has a thorough understanding of this subject.

The steering mechanism must enable the driver to maintain easily the straight-ahead motion of the vehicle, even when bumps are encountered at high speeds, and enable the path of the vehicle to be changed with the minimum amount of driver's effort.

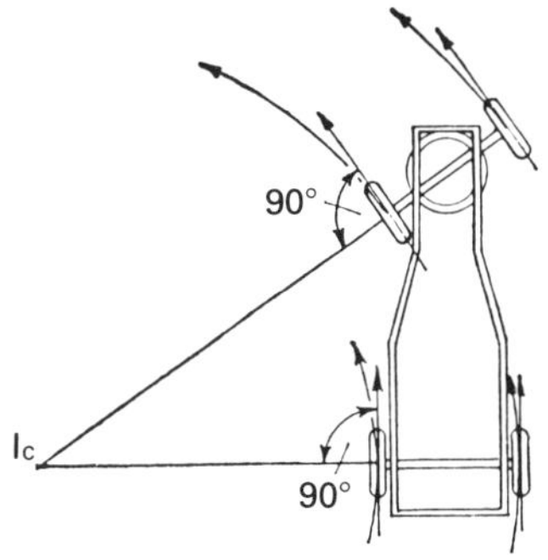

Fig. 49.1 Swinging beam system

Swinging beam system

The geometry of steering may be understood by considering the layout of a vehicle, as shown in Fig. 49.1. A swinging axle beam, mounted on a turntable on the frame, turns the wheels and allows the vehicle to move around an imaginary centre I_c. You will see, in the position shown, that all wheels are at right angles to radial lines drawn from I_c, and each wheel forms a tangent to the curved path that the wheel is actually taking. The natural tendency of a wheel is to travel in a straight path, and it is obvious that a curved path will cause greater tyre wear. This wear can be kept to a minimum if misalignment is limited.

Ackermann layout

Many of the disadvantages of the swinging-beam system were overcome in 1817 when a Munich carriage builder named Lankensperger first introduced the fixed-beam, double-pivot system. In 1818, his agent in England, Rudolph Ackermann, took out a patent on the system which is known as the Ackermann layout and is widely used.

The steering layout should position the wheels at angles shown in Fig. 49.2. In 1878, Jeantaud demonstrated that this condition must be fulfilled if tyre wear is to be minimized. The diagram shows that the difference in the front-wheel angles depends on the ratio T/W. The Ackermann layout does not fully achieve the conditions shown by Jeantaud in all wheel positions; normally it is accurate only when the wheels are straight-ahead and in one position in each left-turn and right-turn wheel setting.

The main details of the layout are shown in Fig. 49.3. Each wheel hub is mounted on a stub axle which swivels on either ball joints or a king pin to give the steering action. Linkage connecting the two stub axles together comprises two track arms and a track rod. The Ackermann layout is obtained by positioning the king pin (or swivel joints) and the track rod joint on an imaginary line which is inclined to the centre

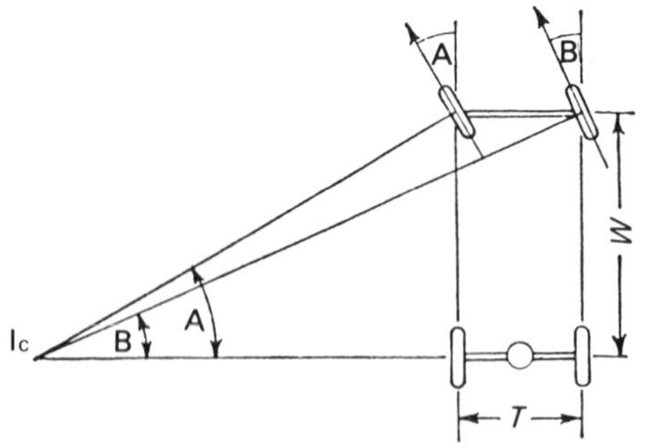

Fig. 49.2 Diagram showing why the inner wheel must be moved through a larger angle

line of the vehicle. This means that when a track rod is fitted to the rear of the swivel centres, it is shorter than the distance between swivel centres. Although this is satisfactory on heavy vehicles, it is not used on cars because of the engine position: instead the track rod is set forward of the wheel centres.

Fig. 49.4 shows the steering angles when the track rod length is equal to the swivel centre distance; both wheels will move through the same angle. Insetting the track arms and shortening the track rod to give an Ackermann layout will cause the angles to vary when the wheels are turned. The action will be seen if the track rod is moved a given distance to the left: the left-hand track arm approaches its effective crank angle (rod and arm at 90°) and this gives the arm a small angular movement, whereas on the right-hand side the arm moves away from the effective crank angle and increases the arm and wheel angle. The difference in angle between the two wheels increases as the wheel is turned.

It should be noted that the essential feature of the system is the position of the track arm connection in relation to the steering swivel axis and *not* the shape of the track arms. However, any distortion of the arms due to vehicle impact will cause misalignment and incorrect wheel angles. The setting of the track arms can be checked either by placing the wheels on turntables and measuring the angles, or by checking the amount that the wheels move out-of-parallel when one wheel is steered through a given angle – this latter test is called '*toe-out on turns*'.

Oversteer and understeer

Slip angle

Previous steering geometry has been based on a vehicle using 'hard' tyres. Low-pressure tyres, used on modern vehicles, take a different path when

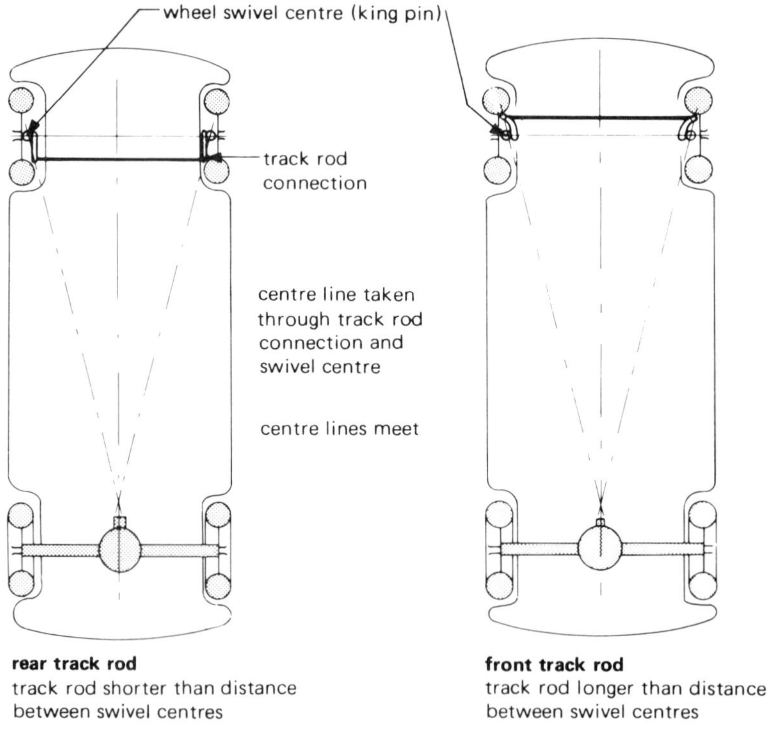

Fig. 49.3 Ackermann steering system

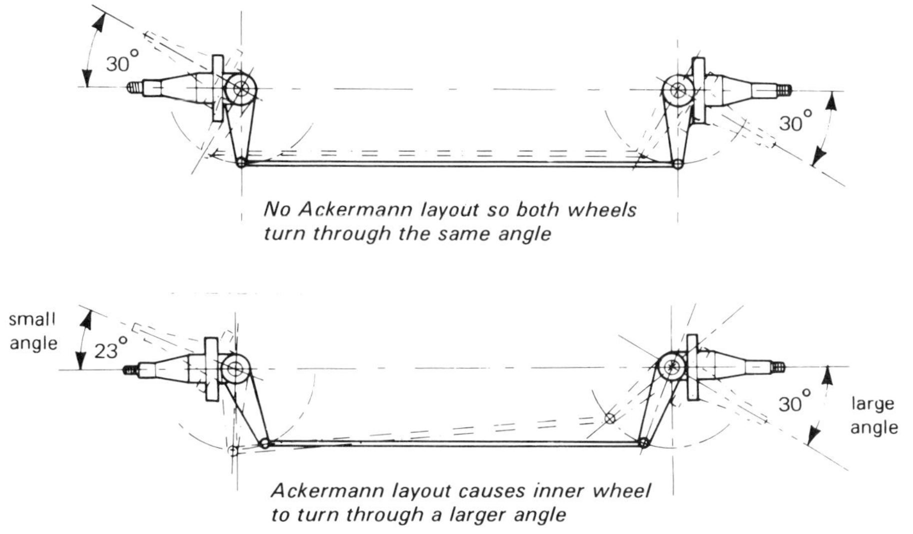

Fig. 49.4 Effect of Ackermann layout

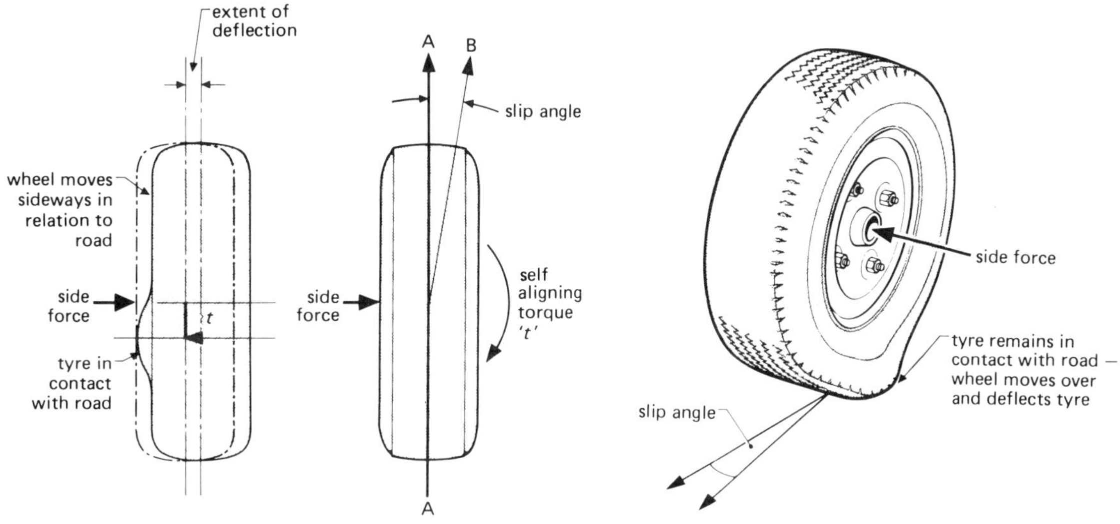

Fig. 49.5 Slip angle of a tyre

subjected to a side force. Fig. 49.5 shows a plan view of a wheel travelling in the direction A.

If a side force acts on the wheel, tyre deflection will cause the wheel to take the path B, although the wheel is still pointing in the original direction. The angle between the path that the wheel is actually taking, and the plane of the wheel (AA), is termed the *slip* angle. (This term is misleading since no slip is actually taking place: 'creep angle' might be a better term, but 'slip angle' is in common use.)

The slip angle caused by the deflection of the side wall of a given tyre is proportional to the side force acting on the tyre. This statement is true up to the point where adhesion is lost and the tyre starts to slide sideways. During the non-slip phase the following applies:

$$\text{Cornering power (N/degree)} = \frac{\text{Side force (newtons)}}{\text{Slip angle (degrees)}}$$

The cornering power (c.p.) of a tyre is governed by:

1 Inflation pressure – an increase raises the c.p.
2 Tyre construction – a radial-ply tyre has a higher c.p. than a diagonal-ply tyre.
3 Tyre size – a low-profile tyre has a smaller wall so a higher c.p. is achieved.
4 Camber (tilt) of wheel – tilting the wheel away from the side force increases the c.p.

5 Load on the wheel - if the load is varied from the normal, then c.p. will decrease.

(See Chapter 63 for further details about tyre construction.)

Self-aligning torque

When a side force acts on a tyre the wall deflects to give a slip angle. Observing a wheel that is rotating in this condition shows that the side deflection of the tyre is greater at the point where the tyre leaves the road as it revolves (Fig. 49.5). This effect produces a *self-aligning torque* (*t*) which attempts to turn the wheel to align itself with the actual direction that the wheel is taking.

The effect of this torque is felt by a driver of a vehicle which is travelling along a highly cambered road - the steering will 'pull' to one side.

Oversteer and understeer

Tyre slip angles affect the steering characteristics of a vehicle by causing either oversteer or understeer. A side force, caused by wind, road camber or cornering forces, produces a slip angle at each tyre.

When the rear slip angles are greater than the front (Fig. 49.6), the vehicle will turn more sharply than normal - a condition called *oversteer*. To correct for this, the driver has to straighten up the steering wheel - a difficult feat under cornering conditions and when this is linked with the fact that the sharper turn causes a further increase in the cornering force, then it will be apparent why oversteer is regarded as dangerous. A mixed tyre arrangement which uses radial ply on the front and diagonal ply on the rear, increases the oversteer effect, so for safety reasons it is made illegal to use this tyre arrangement.

Oversteer can also be caused by:

1 Moving the centre of gravity towards the rear by increasing the load on the rear of the vehicle.
2 Lower tyre inflation pressure at the rear than that recommended.
3 Large load transference from inner rear wheel to outer rear wheel when cornering.

Understeer is produced when the front slip angles are greater than the rear. This tends to cause the vehicle to move away from the side force and makes the vehicle take a path of larger radius than normal. A reasonable degree of understeer is desirable, but if it becomes excessive, 'hard' steering results.

Steering mechanism

The mechanism used to link the driver's steering wheel to the road wheels depends on the type of suspension system. Modern layouts have been developed from the basic system shown in Fig. 49.7. Today the use of this system is confined to light trucks.

Steering for a light truck

The steering column, set at an angle to suit the driving position, contains a shaft (mast), which is connected to the spoked steering wheel, either by a taper and key or by splines. At the lower end of the column, a steering gearbox is fitted and this moves a *drop arm*. A ball joint on the bottom end of the arm joins it to a *drag link*, which pushes and pulls a *steering arm* mounted on a *stub axle*. Construction of stub axles

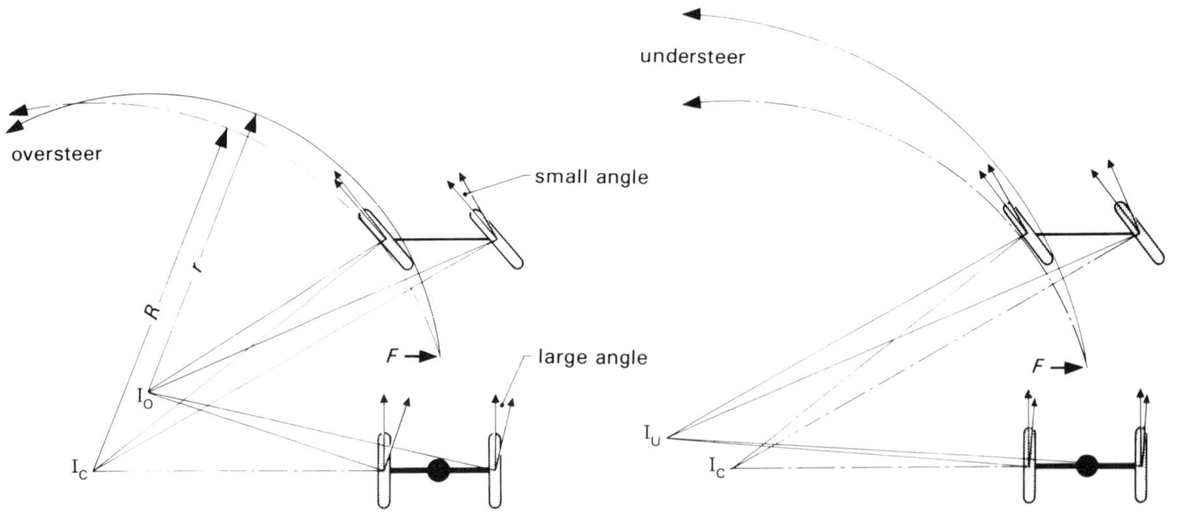

Fig. 49.6 Effect of slip angles

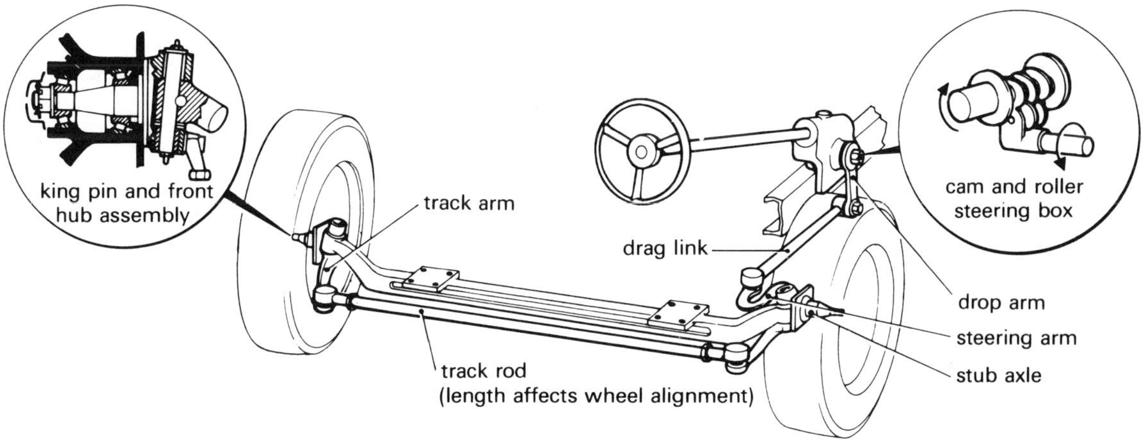

Fig. 49.7 Steering layout – light truck

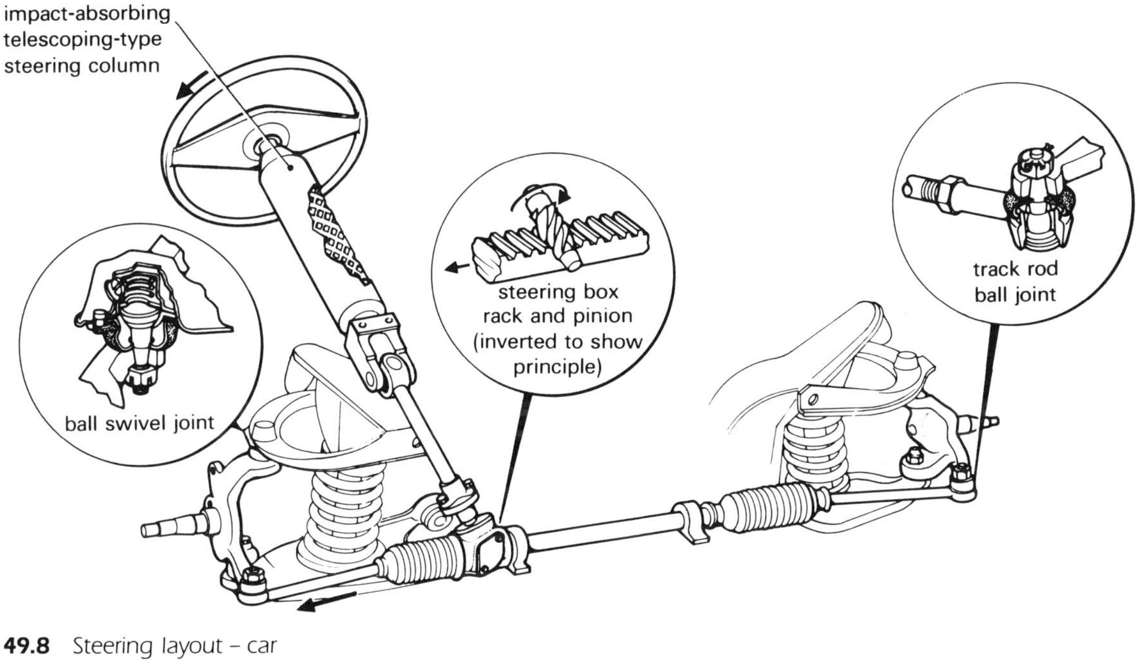

Fig. 49.8 Steering layout – car

takes various forms; the arrangement shown is called a *reversed-elliot* type. Phosphor-bronze bushes in the stub axle provide for steering movement of the road wheels around a *king pin*, which is retained in the *axle beam*.

In the UK the steering box controls the right-hand (off-side) wheel and movement of the other wheel is by means of the track rod linkage.

Steering for a car

Fig. 49.8 shows a typical layout for a motor car. Independent suspension is normal for this type of vehicle, so a split track rod, made up of two or three separate pieces, is necessary to avoid a change in the steering geometry when one or both wheels strike a bump in the road. A rack-and-pinion type steering box naturally takes the place of the centre link of a three-piece track rod and also eliminates the need for a drag link. These features, together with its comparatively direct action, make this type of box attractive for a car.

Steering movement of the wheels is normally provided by widely spaced *ball swivel joints* situated at the outer connection points of the suspension arms.

50 Camber, castor and swivel-axis inclination

Centre-point steering

Fig. 50.1 shows a vertical wheel and king pin arrangement; this has the following disadvantages:

1 Large splaying-out effect of the wheel - the wheels are pushed along by the force *F* which is opposed by the resistance *R*. These two forces produce a couple ($F \times x$) which becomes very large when the front brakes are applied.
2 Heavy steering owing to the distance between king pin and wheel centre. The wheel has to be moved in an arc around the pin, the radius being *x*.
3 Large bending stress on the stub axle and king pin.

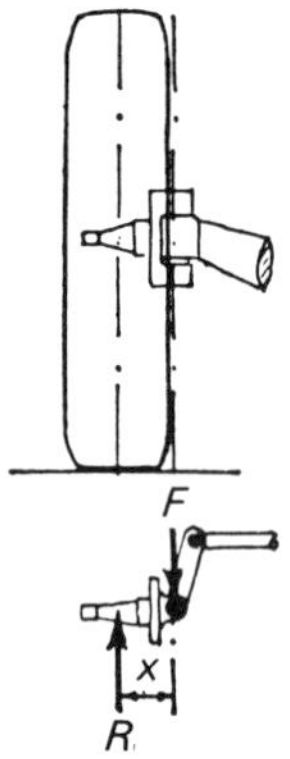

Fig. 50.1 Layout showing need for centre-point steering

To overcome these problems, the wheel and king pin are so arranged that the 'offset' distance *x* is reduced. When the offset is eliminated, i.e. when the centre line of the wheel meets the centre line of the king pin at the road surface, the condition is termed *centre point steering*. This can be obtained by (a) camber, (b) swivel-axis inclination or (c) dished wheels.

Centre point steering appears to be ideal, but the 'spread' effect of the pneumatic tyre causes the wheel to 'scrub' and give hard steering and tyre wear.

Camber

The downset of the stub axle tilts the wheel outwards at the top; the angle formed between the vertical and the wheel is termed the *camber angle*.

The bending stress on the stub axle and the 'splaying out' couple are reduced, but the different rolling radii of the tyre produce a cone effect, which causes tyre wear and another splaying out action. Providing both camber angles are equal, this action will be balanced, but if this is not so the vehicle will pull to the side of greater camber. (Unequal tyre pressures can also produce this effect.)

Some independent suspension systems vary the camber angle as the spring deflects: this changes from the positive form shown in Fig. 50.2 to the negative form, which is the inner tilt of the wheel.

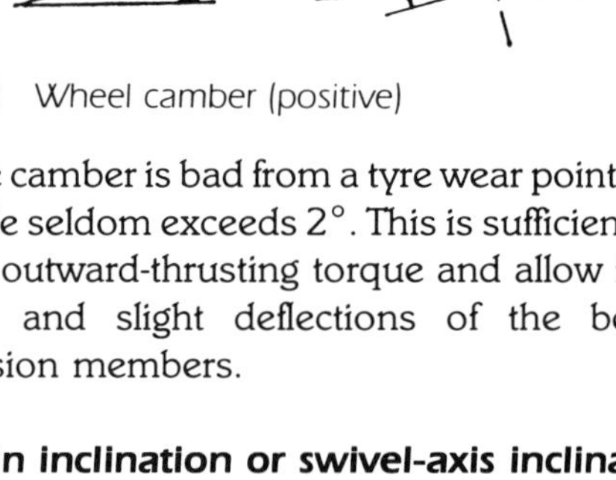

Fig. 50.2 Wheel camber (positive)

Since camber is bad from a tyre wear point of view, the angle seldom exceeds 2°. This is sufficient to give a slight outward-thrusting torque and allow for road camber and slight deflections of the beam or suspension members.

King pin inclination or swivel-axis inclination

Tilting the king pin outwards at the bottom produces an angle between the king pin centre line and the vertical, which is known as king pin inclination (k.p.i.)

or swivel axis inclination (Fig. 50.3). most layouts require k.p.i. of between 5° and 10°, in order to obtain the required offset. The larger angles are used when the designer moves the wheel away from the king pin to accommodate brakes, bearings, etc.

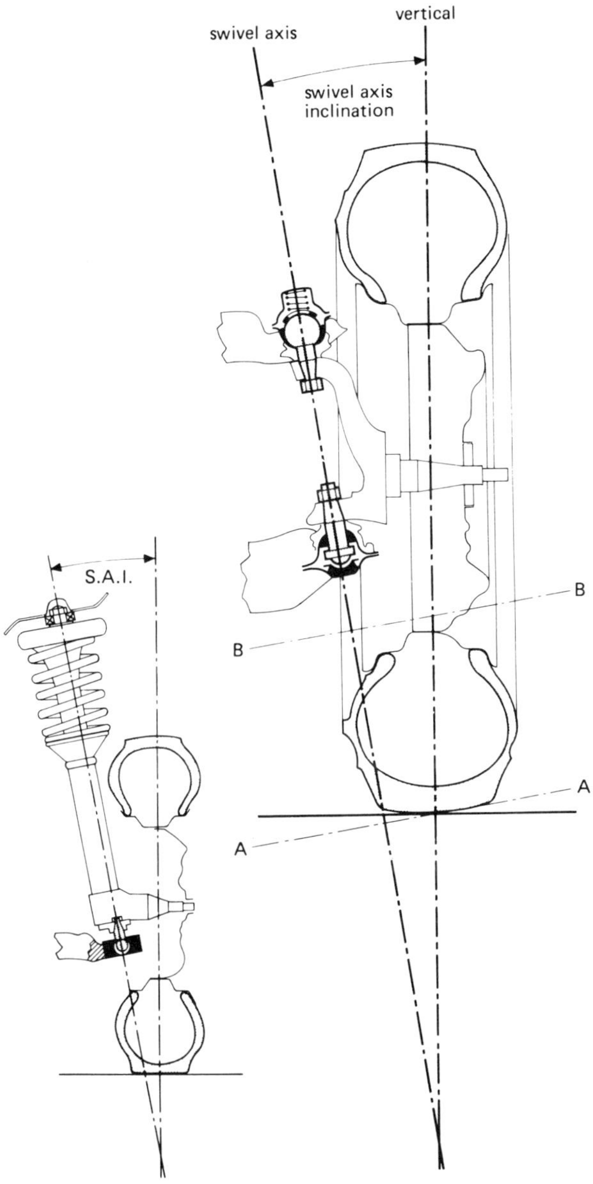

Fig. 50.3 Swivel axis inclination

As the wheel is turned it will move in the plane AA, and will lift the front of the vehicle: this produces a self-centring action. When vertical king pins are used, a simple yoke-and-pin type of steering joint can be used at each end of the track rod. An inclined king pin causes the joint to move in the plane BB, upwards at one end of the track rod and downwards at the other. A ball joint is necessary at each end of the rod to allow for this motion.

Dished wheels

By slightly dishing the wheel (Fig. 50.4), the amount of camber and k.p.i. may be reduced. The light pressed-steel wheel must not be excessively dished, or the strength will be diminished.

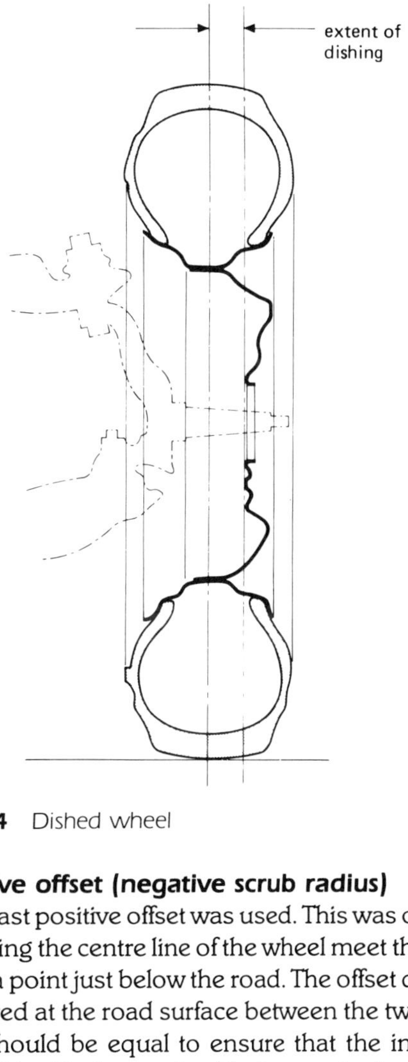

Fig. 50.4 Dished wheel

Negative offset (negative scrub radius)

In the past positive offset was used. This was obtained by making the centre line of the wheel meet the swivel axis at a point just below the road. The offset distance, measured at the road surface between the two centre lines, should be equal to ensure that the inward or outward 'pull' of one wheel balances the 'pull' of the other wheel.

When one front tyre deflates, the positive offset on that side will increase. This will cause the vehicle to pull violently to that side and will make it difficult for the driver to maintain control especially if the brake is applied (Fig. 50.5).

Road safety is improved when *negative offset* is

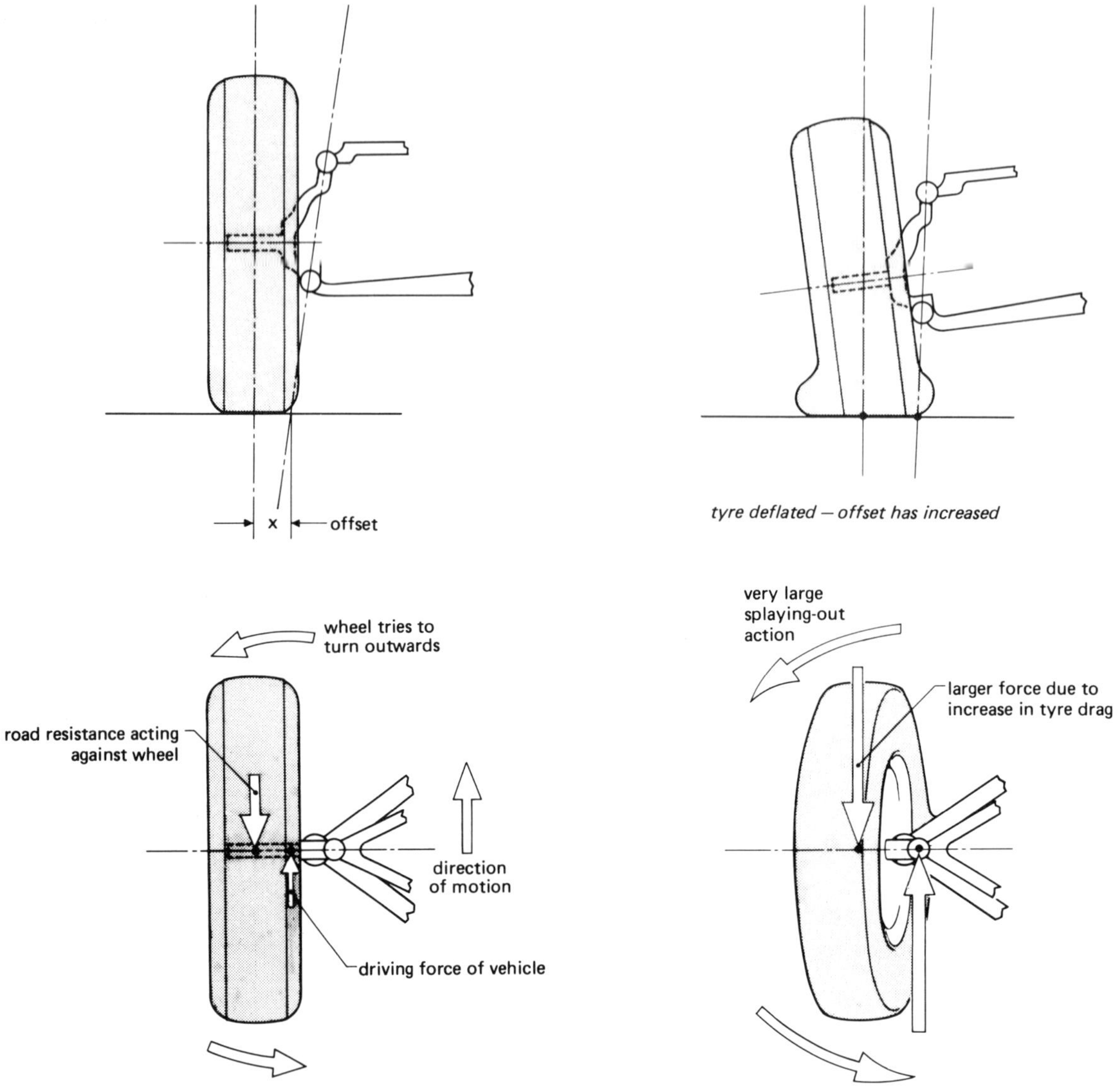

Fig. 50.5 Positive offset

provided by the manufacturer. It is obtained by inclining the swivel axis more than in the past in order to give an intersection point well above the road surface.

With this geometry the effect of tyre deflation is to shorten the offset. Although the rolling resistance is increased, this shorter offset robs the deflated tyre of leverage to change the direction of the vehicle.

A further safety feature of negative offset is achieved when the front brakes are unbalanced owing to poor adhesion of one wheel or failure of one front brake. Even under these adverse conditions the vehicle can still be brought to rest in a straight line. For example, if the right-hand front brake failed (Fig. 50.6), the geometry would cause the braking action of the left-hand wheel to steer the wheels to the right to counteract the loss of brake drag on the right-hand side of the vehicle. This inbuilt action relieves the driver of a difficult control problem. Also it makes the negative offset layout particularly suited to a diagonally connected split-line braking system.

Castor

Wheel castor is given to enable the driver to 'feel' the straight-ahead position so that he may steer a straight path. When cornering, a torque must be exerted on the steering wheel to overcome the self-centring or

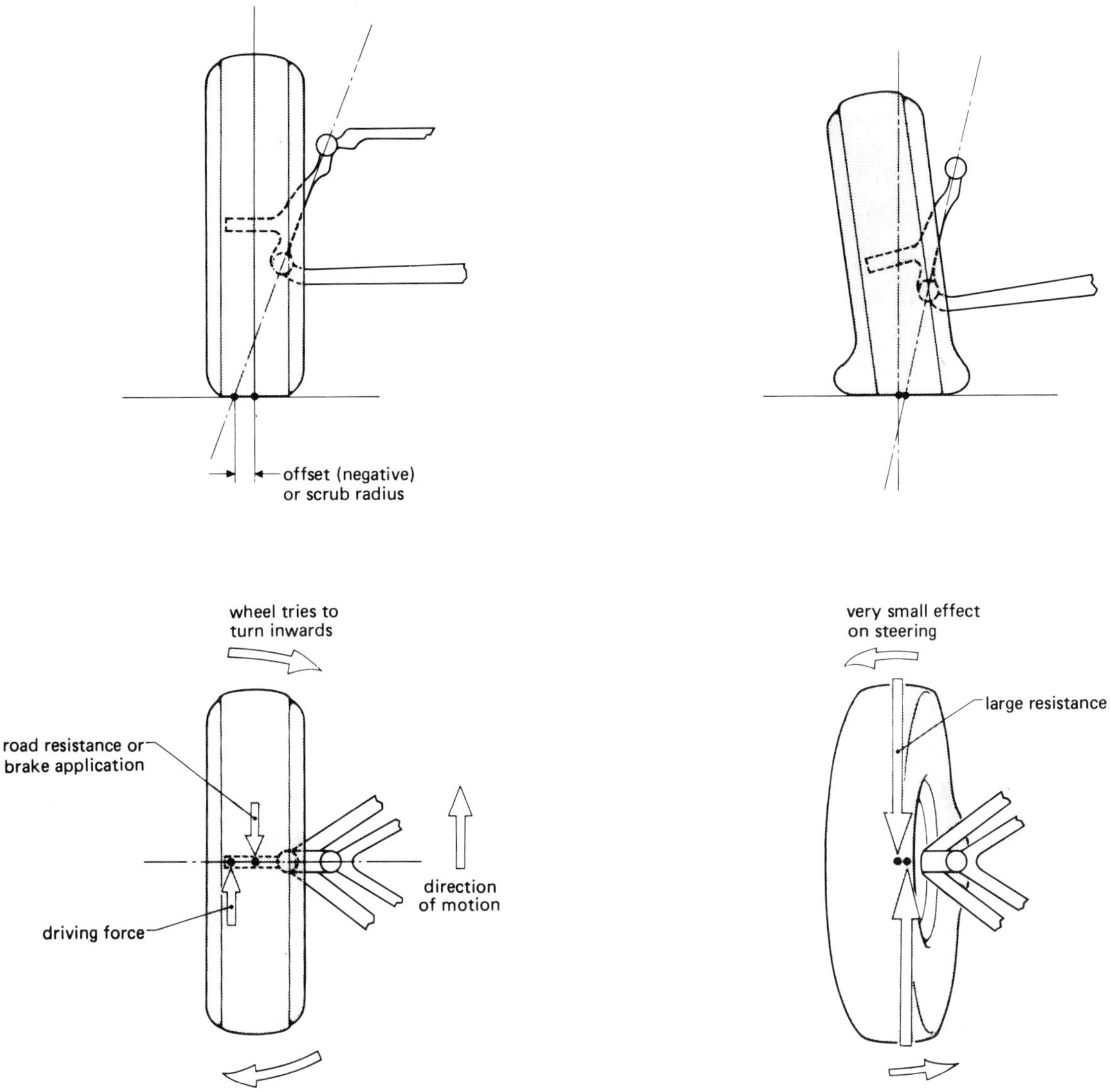

Fig. 50.6 Negative offset

castoring action, which tends to keep the wheels pointing straight ahead. From this introduction it will be seen that too much castor produces hard steering, whereas too little causes 'wander'.

The action of this steering feature may be understood by considering the operation of a simple furniture castor (Fig. 50.7a) fitted to a tea trolley. When a force is exerted on the trolley, it moves in the direction of the force. The effect of this force on the castor is shown in the diagram: the force *F* acting at the pivot and the resistance acting at the wheel produce a couple which rotates the castor to a position where the wheel is following the line of thrust.

On a motor vehicle the pivot centre line is normally made to strike the road in front of the centre of contact of the wheel. In this case the wheel will follow the path taken by the pivot centre line, which will always be in front of the vehicle.

Castor can be obtained by mounting a vertical king pin in front of the wheel (Fig. 50.7b) or by inclining the king pin forward at the bottom (Fig. 50.7c) to give castor angle. The latter is simpler, and most manufacturers use this arrangement. The angle is generally between 2° and 5°; but once again reference must be made to the manufacturer's recommendation for castor angle and tolerance (normally $\pm\frac{1}{2}^\circ$ for all steering angles).

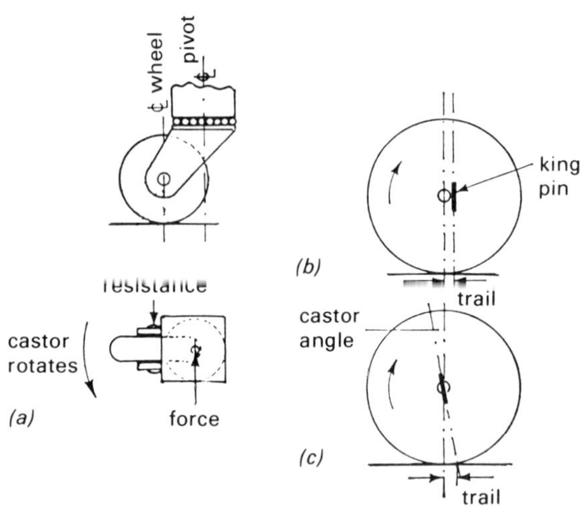

Fig. 50.7 Wheel castor

The inclining of the swivel axis or the tilting of the king pin can be arranged by using one of the following methods:

1 Upper independent suspension members mounted slightly to the rear of the lower members.
2 Tilting the axle beam of a light truck by
   a fitting wedges between axle and spring,
   b mounting the axle towards the front of a laminated spring,
   c inclining the laminated spring

Fig. 50.7 shows the pivot centre line making contact with the road in front of the centre line of the wheel; this is called *positive castor*. This geometry is common for a rear-wheel drive arrangement, but is unsuitable for front- and four-wheel drive vehicles. This is because the conditions shown in Fig. 50.7 are then reversed. Most vehicles that use the front wheels as driving wheels have *negative castor*, but since this is only a general rule, reference to the manufacturer's data should be made when checking the steering geometry.

Wheel alignment

Both wheels should be parallel when the vehicle is travelling forward in a straight path.

To achieve this basic condition, an allowance must be made for the change in alignment between the stationary and moving positions. The forces on a moving wheel take up the clearance in each ball joint and produces a slight deflection of the linkage. This action causes the road wheels to splay out or in, depending on the geometry of the steering and whether the drive is by the front or rear wheels.

Most rear-wheel drive vehicles have the wheels set so that the distance, measured at hub height, between the front of the wheel rims is 1.5–3.0 mm shorter than that measured between the rear of the same wheels. When the wheels are set in this way they are said to *toe-in*. Fig. 50.8a shows a toe-in setting of 3 mm.

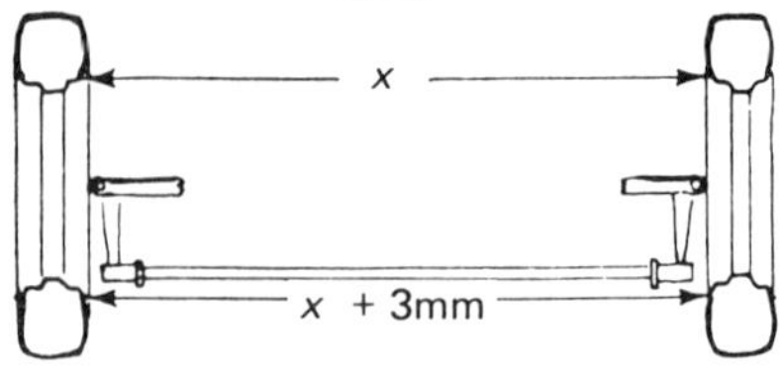

(a) Wheel alignment

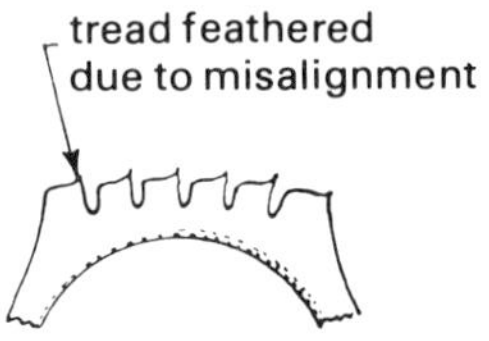

(b) Section through tyre tread

Fig. 50.8 Wheel alignment

In a similar way, if the distance at the front of the wheels is greater than that at the rear of these wheels, then the expression *toe-out* is used.

The toe setting is determined by the manufacturer by carrying out tests to find the value that gives minimum tyre wear.

An alternative method that is independent of the wheel size expresses the toe setting as an angle, e.g. with the steering set straight ahead, the wheels should toe-out so that each wheel makes an angle of 7′ 30″ with the longitudinal axis of the car.

In the past a trammel gauge was employed to measure the wheel alignment, but nowadays either an optical or an electronic gauge is used. Prior to *tracking* the wheels, certain preliminaries should be undertaken; these include:

- checking the tyre pressures
- setting the correct load on the vehicle
- positioning the wheels in the straight-ahead position
- moving the vehicle forward to 'settle the steering'
- checking the run-out (buckle) of the wheel; the maximum run-out should be positioned so that it does not affect the alignment measurement.

Left- and right-handed screw threads on the track

rod are provided to alter the length of the rod for correction of the alignment. Equal adjustment of the two outer rods is necessary if the layout has a rack-and-pinion steering box. Failure to do this will have two effects: firstly it will alter the position of the steering wheel with the result that the horizontal spokes will be misaligned, and secondly the steering locks will be unequal.

Incorrect wheel alignment can be recognized by the wafer edge of rubber left on the side of the tread pattern after it has been scraped over the road surface (Fig. 50.8b). An examination of the position of the feather gives an indication of the fault; a wafer of rubber on the inside of the tread suggests that the wheels are toed-in and vice versa. This *feathering* of the tread is often more pronounced on one tyre than the other.

The combination of road camber and the action of the tyre's self-righting torque causes both front wheels to turn towards the near-side. This movement affects the steering geometry and shows why a manufacturer may recommend a toe setting that appears to contradict basic theory. Interaction between the various geometry aspects of a modern vehicle makes it difficult to pinpoint the cause of a fault from a given symptom, so after carrying out the preliminaries, a *front-end check* covering all aspects of geometry is advised.

An adjuster for rear wheel alignment is provided on some vehicles fitted with independent rear suspension.

51 Steering components

Steering gearbox

The steering gearbox provides the driver with a lever system to enable him to exert a large force at the road wheel with the minimum effort, and to control the direction of vehicle motion accurately.

The overall ratio between the steering wheel and the road wheel varies from about 18 : 1 to 35 : 1, depending on the load on the road wheels and the type of steering.

As the ratio is lowered, a large number of turns is required to move the wheel from lock to lock: this makes it difficult to make a rapid change in vehicle direction.

By varying the efficiency, the degree of *reversibility* (a reversible gear transmits motion from steering wheel to drop arm and vice versa) can be controlled, to enable the driver to 'feel' the wheels, yet not subject him to major road shocks.

Types of steering gear

Over the years a number of different types of steering box have been used; these include:

- *Worm and sector*
- *Screw and nut*
- *Recirculating ball*
- *Cam and peg*
- *Worm and roller*
- *Rack and pinion*

Worm and sector

This type has been developed from a worm and wheel which was one of the earliest designs of box.

A case-hardened steel worm and sector are located by bearings in a malleable iron or light alloy casing. Fig. 51.1 shows the worm connected to the inner column and the sector forming a part of the rocker shaft.

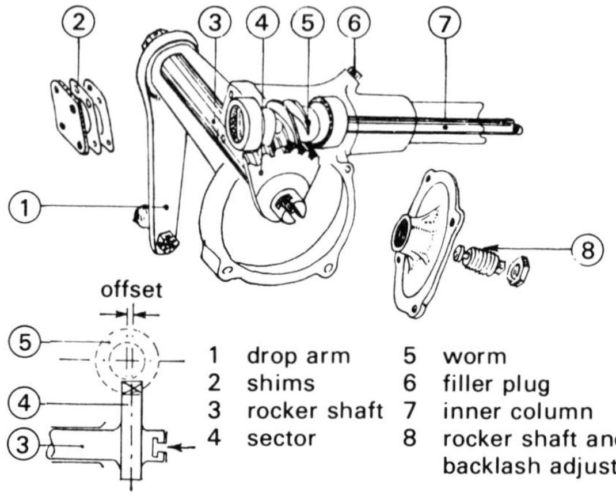

1	drop arm	5	worm
2	shims	6	filler plug
3	rocker shaft	7	inner column
4	sector	8	rocker shaft and backlash adjustment

Fig. 51.1 *Worm and sector steering box*

Most steering boxes are provided with the following adjustments:

1 End-float of inner column - generally shim adjustment.
2 End-float of rocker shaft - shim or screw adjustment.
3 Backlash between gears - gears can be moved closer together.

The greatest wear takes place in the straight-ahead position of the box, so the gear is normally made with a larger backlash in the lock positions. This reduces the risk of seizure at full lock when the box is adjusted to compensate for wear. It is essential to reduce end float and backlash to a minimum, but tight spots must be avoided.

Steering box lubrication is provided by filling the box to the level of the plug with normal gear oil.

Screw and nut (Fig. 51.2)

A phosphor-bronze or steel nut is screwed on to a multi-start Acme thread formed on the inner column. Rotation of the nut is prevented by a ball fitted in the rocker arm. Axial thrust of the column is taken by a

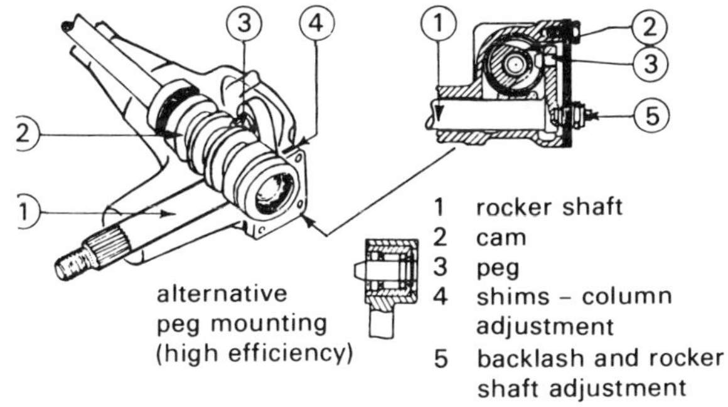

Fig. 51.2 Screw and nut

single ball race fitted at the top end, and the nut sliding in the housing supports the lower end. The end float of the inner column is adjusted by the nut at the top end.

Recirculating ball

A higher efficiency (90 per cent as opposed to 50 per cent) is achieved by using a nut with steel balls acting as 'threads'. The type shown in Fig. 51.3 uses a half nut with a transfer tube which feeds the balls back to the nut. A peg on the nut is located in the rocker arm.

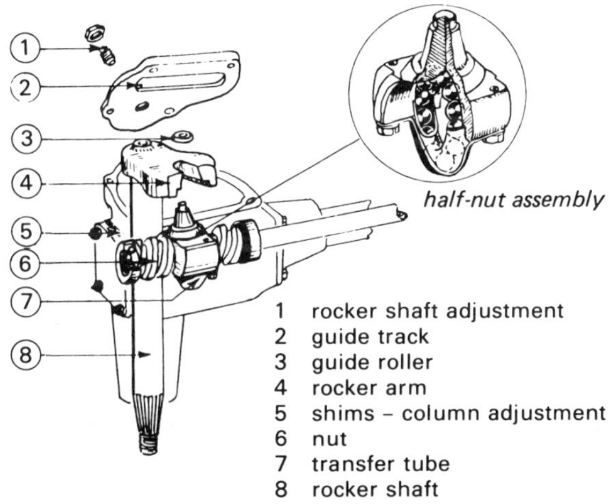

Fig. 51.3 Recirculating ball (half-nut)

Cam and peg (Fig. 51.4)

A tapered peg in the rocker arm engages with a special cam formed on the inner column. The end float of the column is controlled by shims, and an adjusting screw on the side cover governs the backlash and end float of the rocker shaft.

A modified form known as the *high efficiency cam and peg gear* uses a peg which is allowed to rotate in bearings in the rocker arm.

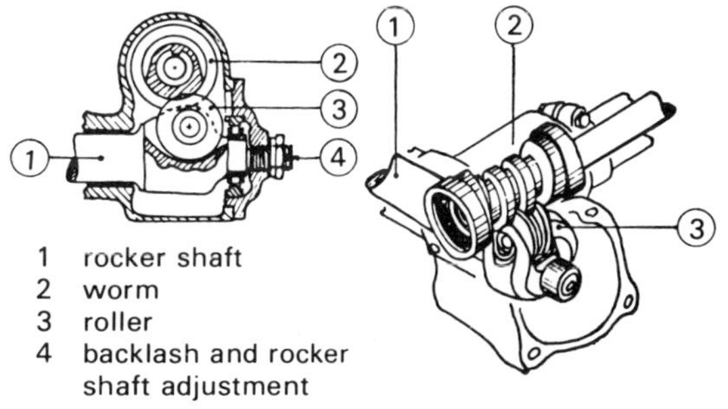

Fig. 51.4 Cam and peg

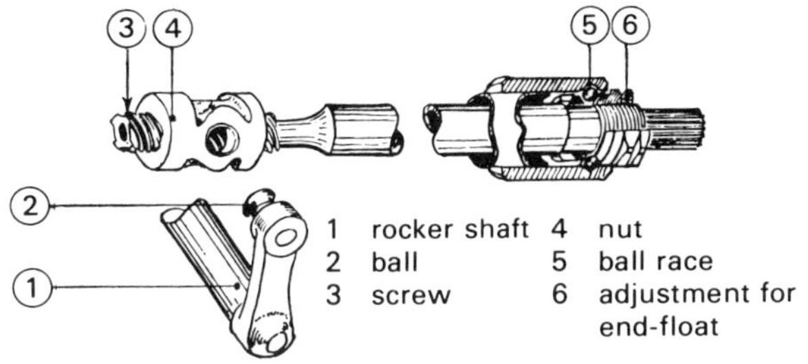

Fig. 51.5 Worm and roller

Worm and roller (Fig. 51.5)

A roller follower fitted to the rocker shaft engages with an hourglass worm. The small offset of the roller to the worm enables an adjusting screw to control backlash and end float of the rocker shaft.

Rack and pinion (Fig. 51.6)

This type is often used with i.f.s. where the rack acts as the centre section of a three-piece track rod. The pinion is normally connected to the column by a universal joint to enable the box to be centrally mounted. Each end of the rack has a ball connection to the track rod. An adjustable spring pad rubs on the underside of the rack; this reduces backlash to a minimum and also acts as a damper to absorb road shocks.

Steering joints

The various steering levers are connected by ball joints which allow universal movement. Many new models use self-lubricating types similar to the one in Fig. 51.7. Fitted on either side of the plated steel ball is a split moulded bearing which is compounded with a specially developed metallic lubricant.

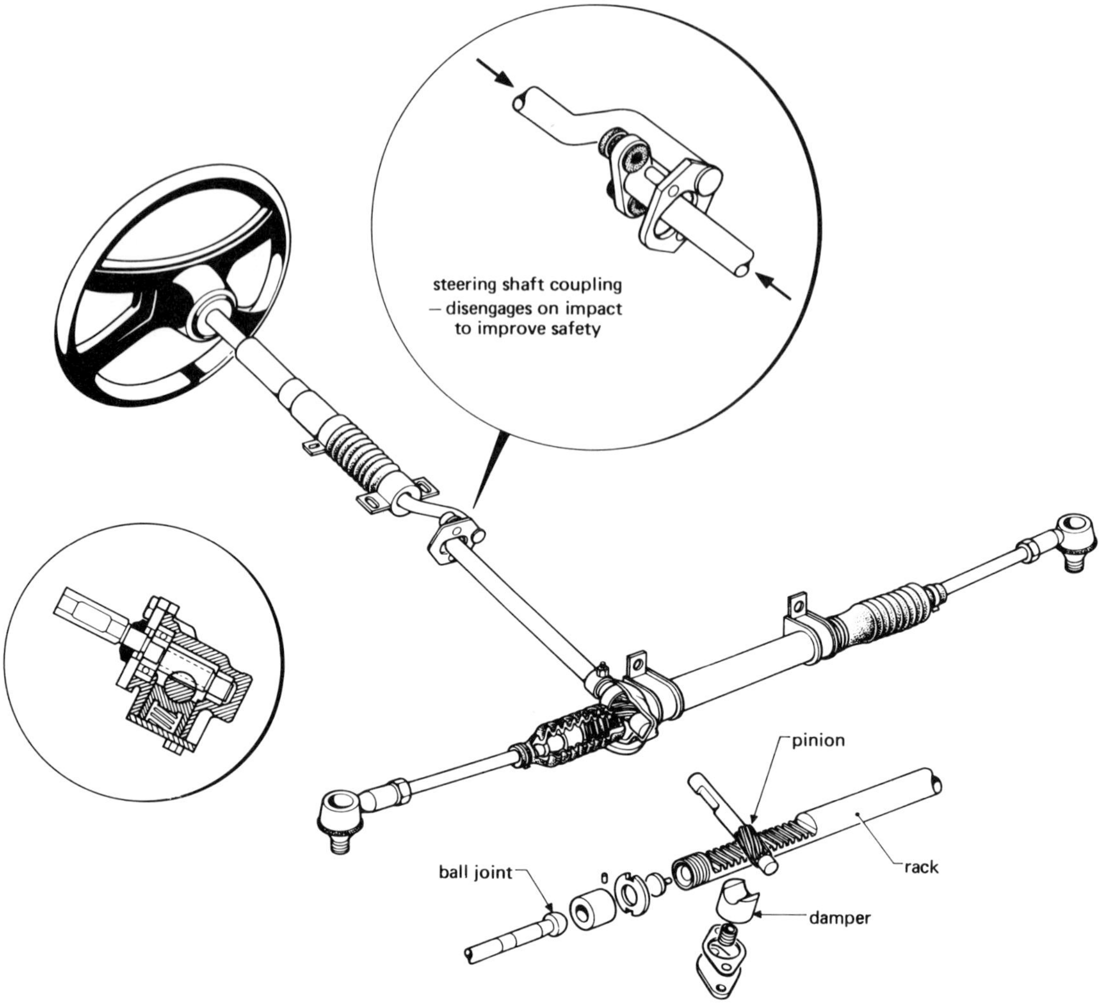

Fig. 51.6 Rack and pinion

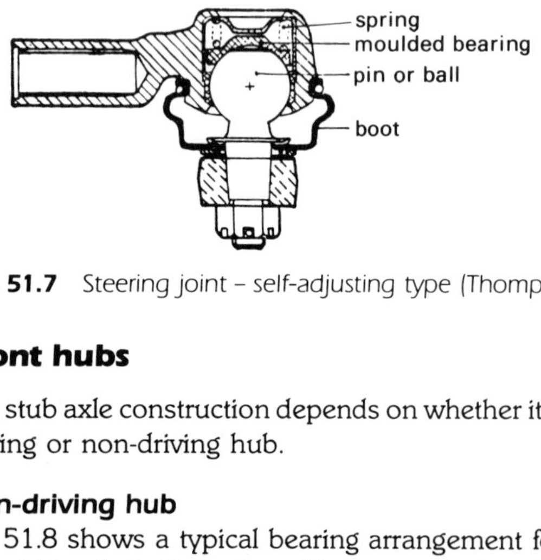

Fig. 51.7 Steering joint – self-adjusting type (Thompson)

Front hubs

The stub axle construction depends on whether it is a driving or non-driving hub.

Non-driving hub

Fig. 51.8 shows a typical bearing arrangement for a non-driving hub. Two adjustable taper roller bearings, spaced as widely apart as possible, are fitted between the stub axle and the malleable iron or steel cast hub.

The hubs are lubricated by packing each bearing and half the cavity between the bearings with a grease of a type to suit the brake system. Whereas a lime-soap grease of medium consistency is suitable for a shoe brake, a grease base having a high melting point, such as lithium or bentone, is essential for hubs used with disc brakes. A lip-type synthetic rubber seal prevents the escape of grease.

This type of hub is adjusted by tightening the nut until all clearance is eliminated and then slackening it off one 'flat'. To achieve a more accurate setting, some manufacturers recommend that the adjusting nut is tightened to a specified torque before releasing the nut a given amount. This method should not be confused with preloading. A hub bearing has to

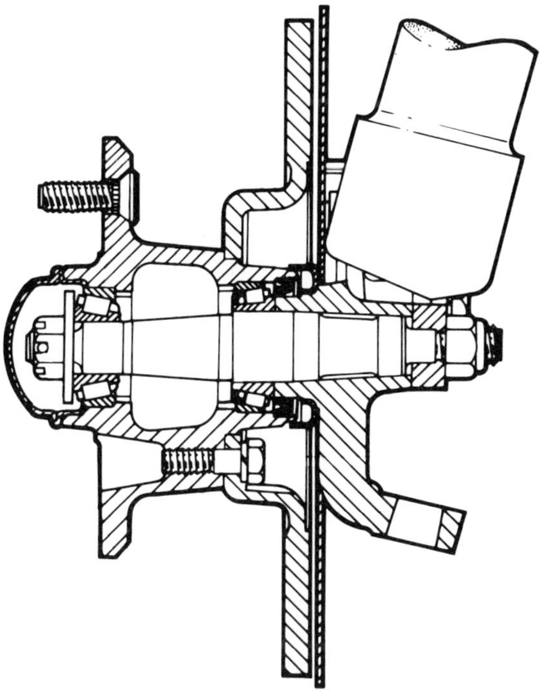

Fig. 51.8 Front hub - taper roller type

withstand heat from the brakes, so it needs a running clearance; this grade of fit is quite different to that obtained after a bearing has been preloaded.

An alternative bearing arrangement is shown in Fig. 51.9. This uses two angular contact type ball races which are held apart by a rigid spacer. The nut on this non-adjustable hub must be tightened fully to the correct torque loading.

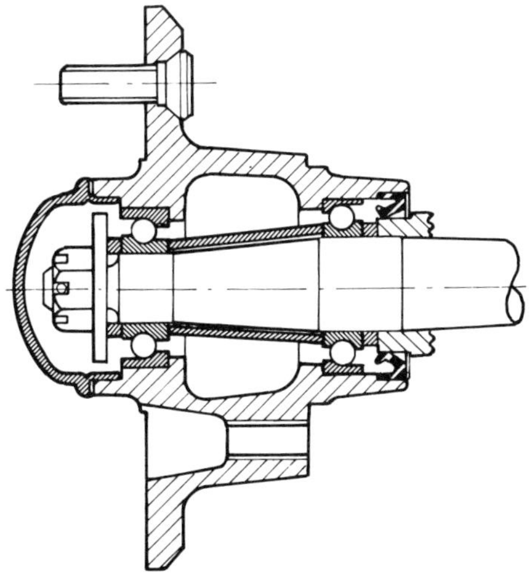

Fig. 51.9 Front hub - ball-bearing type

Driving hub

A typical hub arrangement used for a front-wheel drive car is shown in Fig. 51.10. The stub axle housing contains two bearings and these support both the wheel hub and driving shaft. The type of bearing used depends on the load carried by the road wheel.

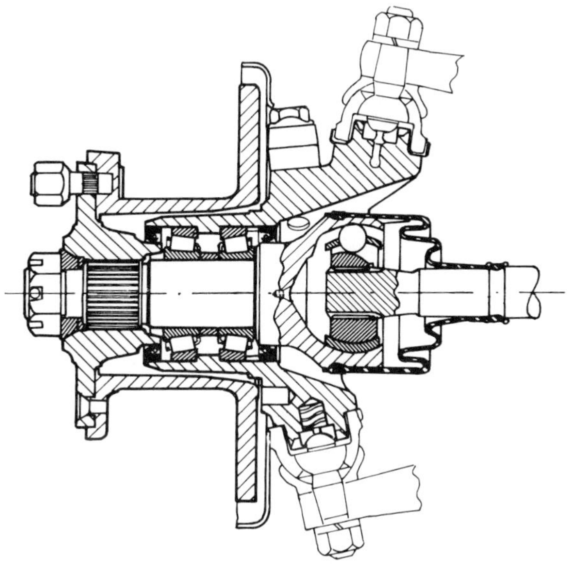

Fig. 51.10 Front hub - front-wheel drive

Power-assisted steering

With increased loads on steered wheels and wider section tyres the large effort required at the steering wheel makes the driver's job very tiring and difficult. Improvements such as an increase in the mechanical efficiency of the steering system or lower steering box ratios help to reduce driver fatigue, but if the latter is not limited the number of turns made by the steering wheel to move it from lock to lock becomes troublesome. When steering effort exceeds a safe maximum some method must be found and power assistance is the answer.

The power assistance arrangement should obey certain requirements:

1 It must be 'fail-safe' - if the power system fails the driver must still be able to retain effective control.
2 The degree of assistance should be proportional to the effort applied by the driver and the driver must be able to retain the 'feel' of the wheels.

On light vehicles, hydraulic power is used to help the driver steer the vehicle.

Hydraulic system

Hydraulically operated power-assisted systems are based on either a constant pressure or constant flow layout - the former employs a hydraulic accumulator to store the pressure, whereas the latter has fluid flowing around the system continuously until assistance is needed.

Fig. 51.11 shows the essential components required to operate a constant flow system. In addition to the normal steering components the system requires a pump, control valve and ram cylinder.

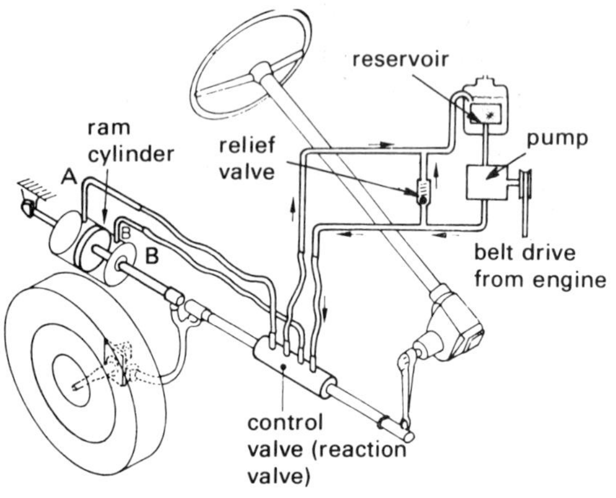

Fig. 51.11 Power-assisted steering layout

Pump This is generally an eccentric rotor type (Fig. 26.4) driven by a vee belt from the engine crankshaft. Attached to the pump is a fluid reservoir and this normally contains a low-viscosity mineral oil similar to the type used in an automatic transmission. A pressure relief valve limits the maximum pressure, 7 MN/m^2 (1000 lbf/in^2), and when this is reached the oil passes from the pump outlet back to the reservoir.

Control valve (or reaction valve) In Fig. 51.11 the valve is positioned between the two halves of the drag link, but in modern applications it can be included together with the ram cylinder in the steering box.

A spool-type valve, attached to one half of the drag link, is held centrally in the valve body by means of two reaction springs. Flexible pipes connect the valve to the ram cylinder, pump and reservoir.

Ram cylinder A double-acting ram piston connects with the steering arm to supply the appropriate force to assist the drive. The force must act in either direction and this is achieved by venting one side and pressurizing the other side of the cylinder. Assistance given by the ram will depend on the fluid pressure supplied to it by the control valve.

Operation of the system will depend on the torque that the driver applies to the steering wheel. Under low torque conditions no assistance is required and this condition is represented by Fig. 51.12. Low torque on the wheel will be insufficient to overcome the reaction springs in the control valve, so fluid will be able to flow back to the reservoir. During this phase the control valve remains in the 'neutral' position and offers no resistance to the oil flow to or from the ram cylinder.

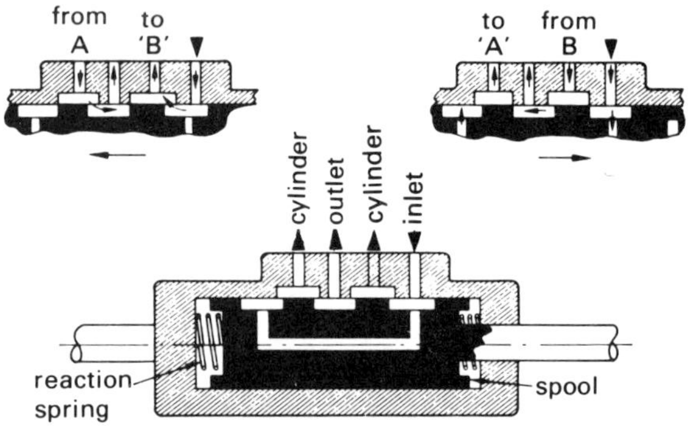

Fig. 51.12 Principle of the control valve

When the torque applied by the driver reaches a predetermined value, which is governed by the strength of the reaction springs, the greater force

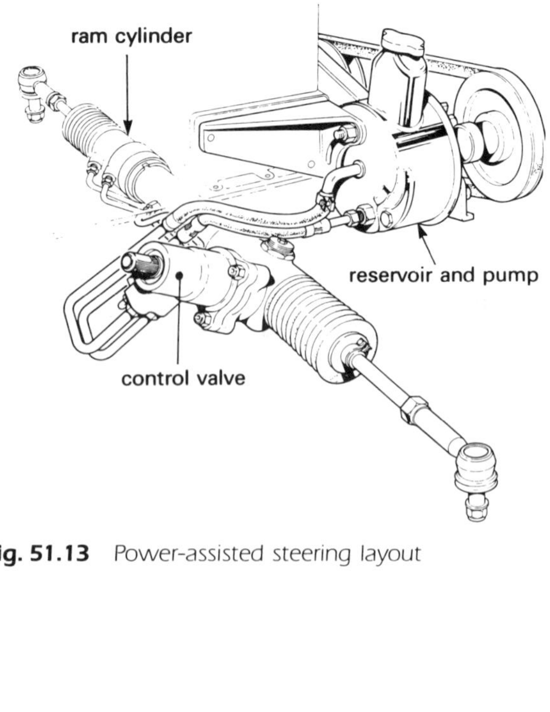

Fig. 51.13 Power-assisted steering layout

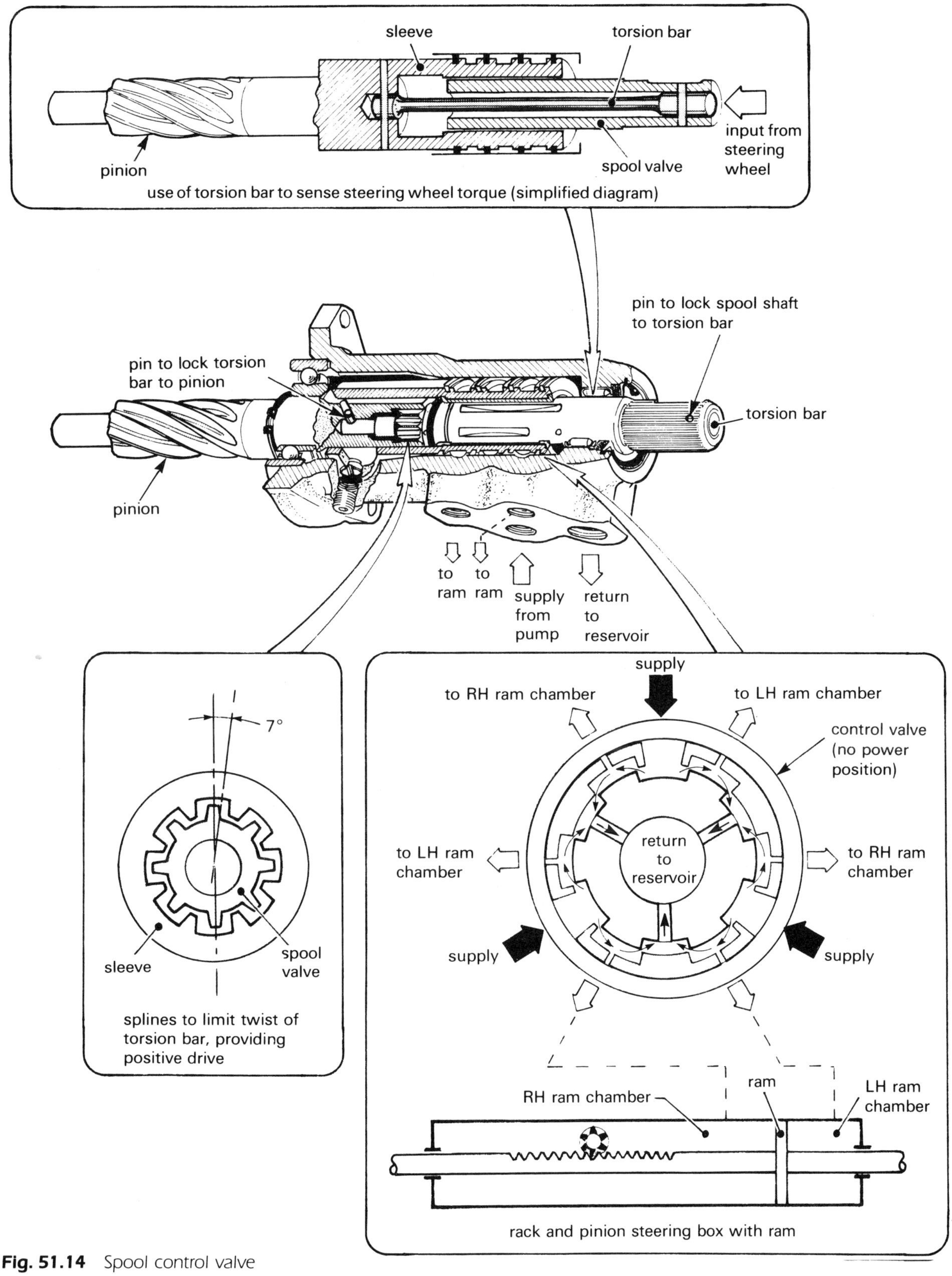

Fig. 51.14 Spool control valve

acting in the drag link is capable of moving the valve in relation to its body. This action of the valve connects one side of the ram to the pump and the other side vents to the reservoir. Interruption of the fluid circuit in this way causes the pump pressure to build up quickly. As this occurs the thrust on the ram will also build up until a point is reached where the movement of the ram overcomes the resistance of the road wheel. At this stage the force in the drag link will reduce, the valve will return to the 'neutral' position and the pressure in the complete system will drop to the original value. Consideration of this sequence shows that steering wheel torque opens the valve and the ram exerts a proportional thrust to close the valve.

Turning the steering wheel the opposite way produces a similar action with the exception that the control valve movement now directs fluid to the other side of the ram piston.

The design of reaction valve and spring takes many different forms. One manufacturer uses a small torsion bar to transmit the drive between the inner column and the gear in the box. When the torque exceeds a given amount the bar twists, moves a valve and activates the ram.

Power-assisted rack and pinion steering

The power steering system, shown in Fig. 51.13, is applied to a rack-and-pinion steering box.

The engine-driven hydraulic pump, which is integral with the reservoir, supplies oil to a control valve situated in the housing that supports the pinion shaft. Movement, imparted to the control valve from the shaft in the steering column, is via a torsion bar. This sensing bar moves the control valve, which in turn directs oil to one side or the other of the ram piston attached to the steering rack.

Control valve Fig. 51.14 shows a simplified view of a rotary-type spool valve controlled by a torsion bar that is interposed between the steering shaft and the pinion of the steering box. The spool valve is a shaft with six flutes and this is encased by a sleeve, which has six internal axial grooves. Radial ports in the sleeve and shaft pass the oil from the supply to the lines connected to the ram chambers.

A series of splines between the shaft and sleeve limit the twist of the torsion bar to about 7° in each direction; below this angle the torque applied by the driver to the steering box pinion is transmitted by the torsion bar. This fail-safe feature provides a mechanical drive from the steering shaft to the pinion in the event of a failure of the power system.

The amount of twist of the torsion bar, and the movement of the spool valve, are proportional to the effort applied by the driver. Initial power assistance is obtained at about 0.5° deflection of the bar and this power rises progressively as the bar moves to about 4°; the point of maximum assistance.

With the valve positioned in the 'no-power' position as shown in Fig. 51.14, all the ports are open, so oil is allowed to pass through the valve and return to the reservoir.

When the steering wheel is turned against a resistance at the road wheel, the torsion bar is deflected; this allows the spool to rotate relative to the sleeve, cutting off the oil flow both to the reservoir and one side of the ram. At the same time the other side of the ram is subjected to oil pressure which builds up sufficiently to move the road wheel and return the torsion bar to the no-torque position. During this stage the oil displaced from the unpressurized side of the ram is returned to the reservoir.

On occasions when the resistance to road wheel movement is excessive, the oil pressure builds up to its maximum. At this pressure a relief valve, fitted adjacent to the pump, opens and allows oil to return to the pump inlet.

52 Four-wheel steering

It has been known for many years that the front-to-rear wheel alignment plays a significant part with respect to directional stability of a vehicle. This is clearly shown up when a car has worn suspension bushes; often the lack of stability makes it dangerous to drive.

Up to the mid-1950s the majority of cars suffered from oversteer and a contributory cause of this dangerous condition was the uncontrolled movement of the rear axle due to spring deflection initiated by body roll. The obvious dangers associated with an oversteer characteristic are now well known, so in order to improve stability, most modern cars are designed to understeer when they are driven at high speed.

One step towards an understeer characteristic can be achieved by using body roll movement to 'steer' the rear wheels. This action is produced either by misaligning the rear axle or, in the case of a car fitted with independent rear suspension, by utilizing the change in geometry of the suspension system.

Although improved safety with respect to stability is associated with understeer, this condition increases the steering effort that must be applied by the driver. Also since the manner in which a vehicle handles changes with road speed, the response to a steering movement, both in time and direction, is difficult to predict.

These difficulties can be overcome by using four-wheel steering (4WS). In 1988 two car manufacturers, Honda and Mazda, offered 4WS on some of their models.

Advantages of 4WS

Compared with a conventional two-wheel steer system, it is claimed that 4WS offers the following advantages:

1 Superior cornering stability.
2 Improved steering responsiveness and precision.
3 High-speed straightline stability.
4 Notable improvement in rapid lane-changing manoeuvres.
5 Smaller turning radius and tight-space manoeuvrability at low speed.

Relative wheel angles

The direction that the rear wheels steer in, relative to the front wheels, depends on the operating conditions.

During low-speed manoeuvring, wheel movement is pronounced, so the rear wheels are made to turn in the opposite direction; this simplifies the positioning of the car in situations such as when parking in a confined space (Fig. 52.1). Since the rear wheels are made to follow the path on the road taken by the front wheels, the rear of a 4WS car does not cut in in the normal way, so the risk of hitting an obstacle is greatly reduced.

At high speed, when steering adjustments are subtle, the front wheels and rear wheels turn in the

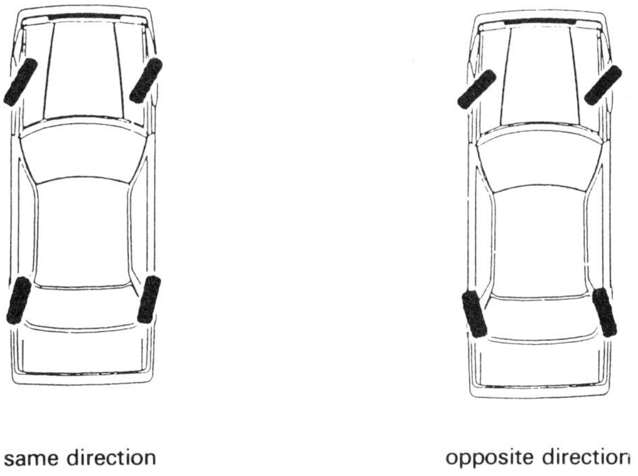

Fig. 52.1 4WS wheel angles (angles shown are exaggerated)

same direction. This action causes the car to move in a crab-like manner rather than move in a curved path; this arrangement is beneficial at times when the car has to change lanes on a high-speed road. The elimination of the centrifugal effect and, in consequence the reduction of body roll and cornering force on the tyre, improves the stability of the car and makes it easier, and safer, to control.

Wheel angle control

The two basic systems in use vary in the way that the wheel angles are controlled. Honda uses mechanical rear steering whereas the Mazda arrangement has a rear 'gearbox' that is both power-assisted and electronically controlled.

Both systems use a power-assisted rack-and-pinion steering box for the front steering gearbox with a take-off connection to control the rear steering gearbox via a centre, or transfer, shaft. Two or more universal joints on the shaft are provided to allow for small drive angle variations.

Honda 4WS system

This system is *steer angle* dependent, i.e. the movement of the rear wheels is controlled by the angular movement of the front wheels.

For steering wheel angles up to about 130°, the rear wheels are so arranged that they turn through a small angle in the same direction as the front wheels. As the front wheels are steered beyond that angle, the rear wheels gradually straighten up and then turn through a comparatively large angle in the opposite direction (Fig. 52.2).

Control of the rear wheel angles is obtained by an epicyclic gear mechanism in the rear steering gearbox (Fig. 52.3). Meshing with a fixed annulus is a large planet gear that is driven by an eccentric on the input shaft. A short shaft, integral with the planet and offset from the centre of the planet, transmits a drive by a slider and guide to a stroke rod; this is connected to the rear wheel track rods (Fig. 52.4a).

Slight rotation of the input shaft causes the planet to rotate and this slightly moves the offset output shaft in the same direction as the input (Fig. 52.4b). As the input shaft carries the offset shaft towards the t.d.c. position (Fig. 52.4c), the stroke rod moves back to the central position – at this point the rear wheels are set in a straight-ahead position.

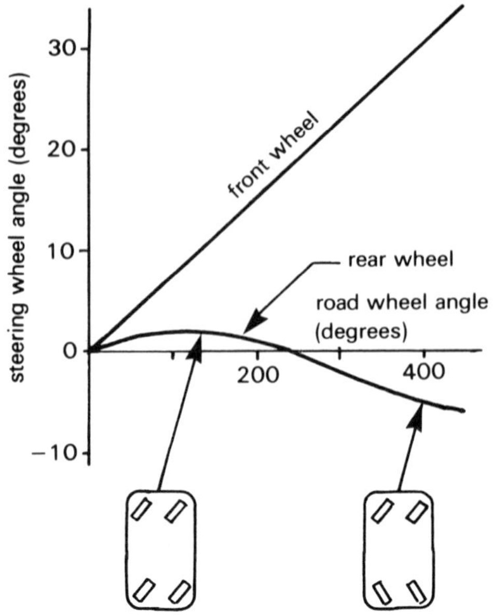

Fig. 52.2 Movement of steered wheels (Honda)

Further rotation of the input shaft and planet to the full-lock position (Fig. 52.4d), gives a maximum displacement of the stroke rod and a corresponding movement of the rear wheels.

The rear gearbox is maintenance-free; it is packed for life with grease. The centre shaft couplings are splined to both steering gearboxes and a master spline at each connection point ensures that the units cannot be incorrectly phased on assembly.

Mazda 4WS system

In this system the rear wheels are steered by a hydraulically operated power unit, which is electronically controlled according to steering wheel angle and vehicle speed (Fig. 52.5).

The Mazda arrangement is more complicated than the previous system, so has more things to go wrong, but as is normal with other advanced control systems, special fail-safe devices are incorporated. In this system it takes the form of a centring lock spring and special safety solenoid. In the event of hydraulic or electronic failure, these devices set the rear wheels to the straight-ahead position.

This system uses two electronic sensors, fitted to the transmission output and speedometer drive; these measure the vehicle speed. The signals are passed to an electronic control unit (ECU), which has a built-in memory that is used to 'tell' the hydraulic

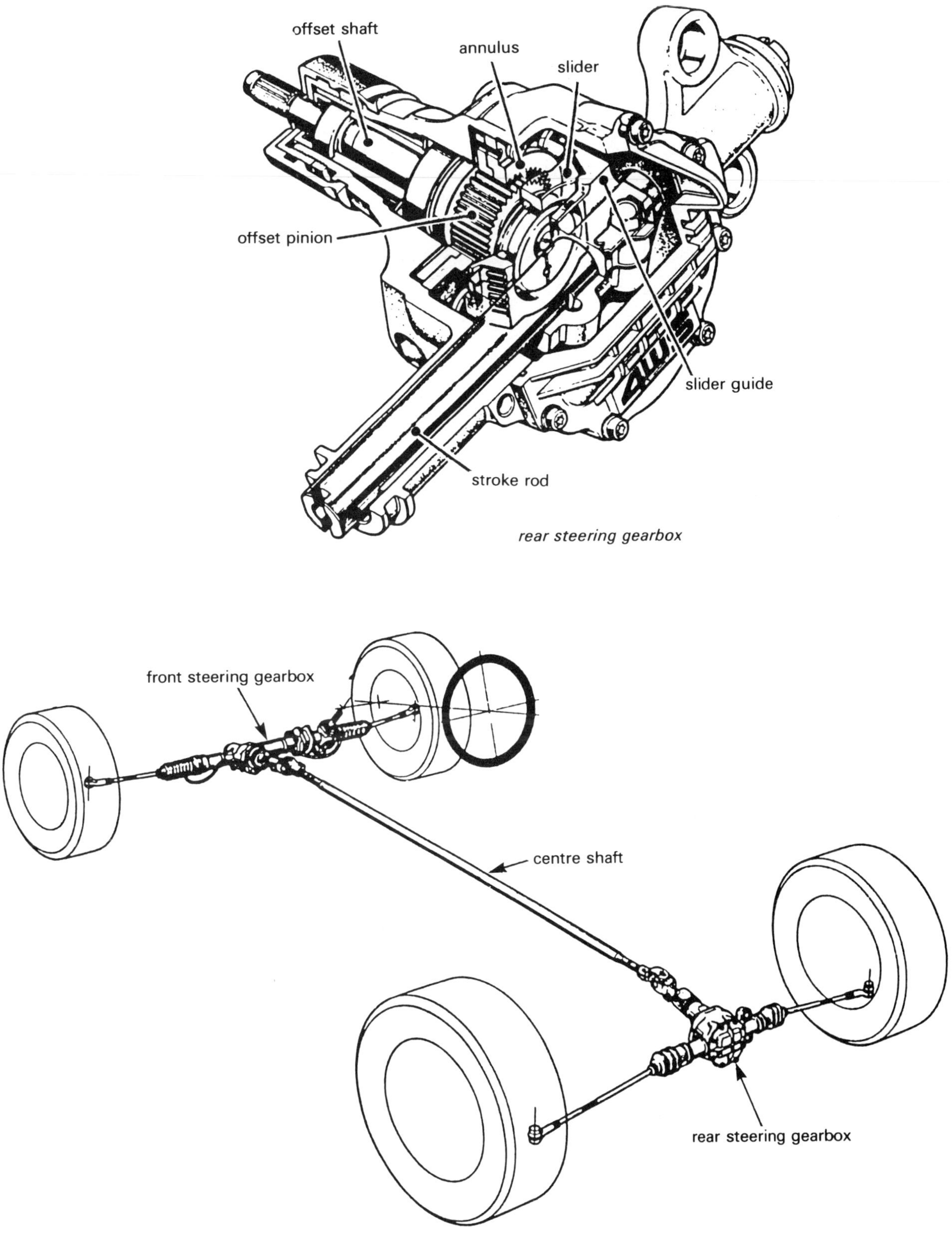

Fig. 52.3 Layout of Honda 4WS

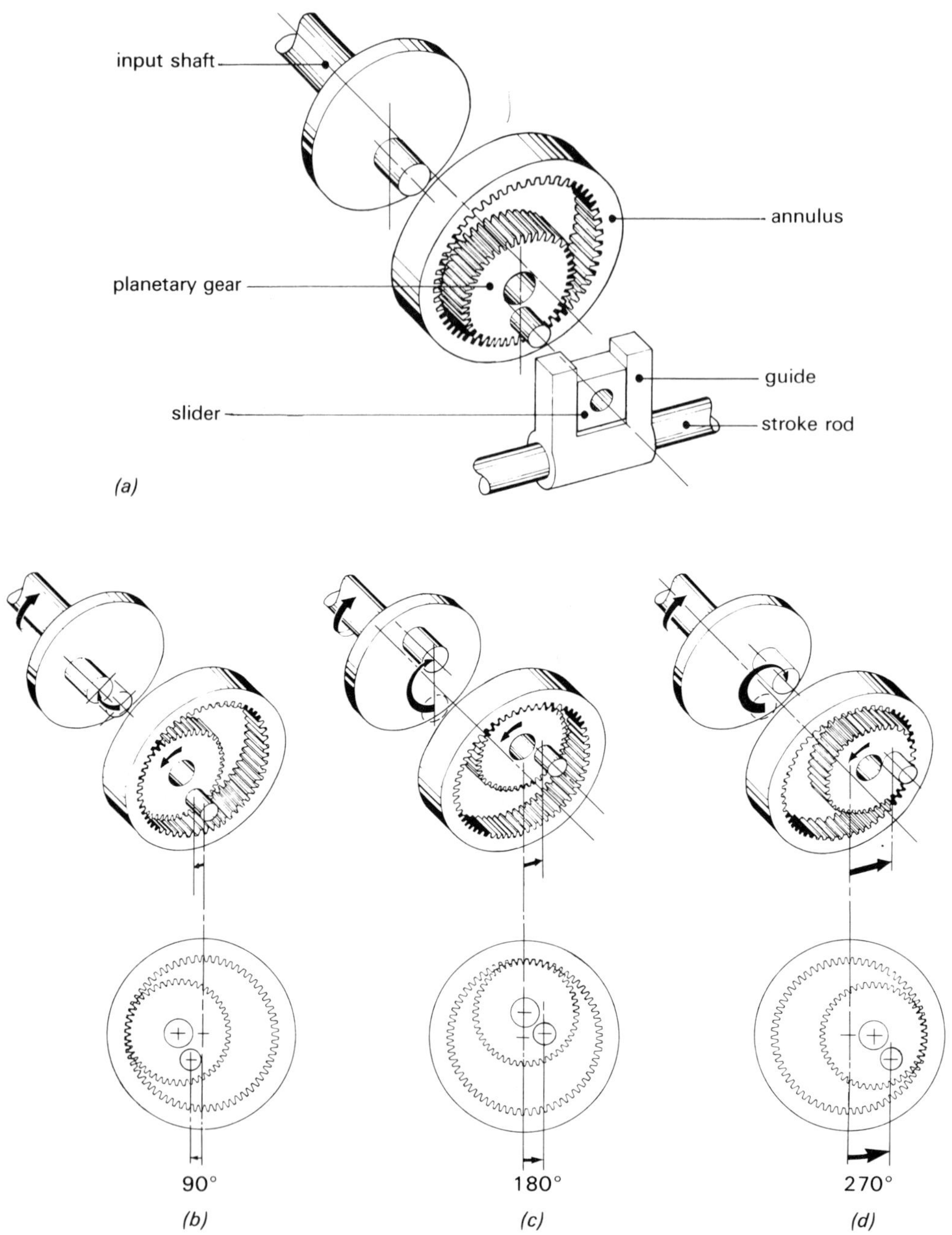

Fig. 52.4 Epicyclic gear action (Honda)

system the direction and angle to set for the rear wheels.

At speeds considerably less than 35 km/h (22 m.p.h.), the rear wheels are steered in the opposite direction to that of the front wheels. As 35 km/h is approached, the rear wheels are moved to the straight-ahead position. Above this speed the rear wheels are steered in the same direction as the front wheels, but the angle is limited to 5°.

Main components

The block diagram (Fig. 52.6) shows the main components used in this system to steer the rear wheels; these include:

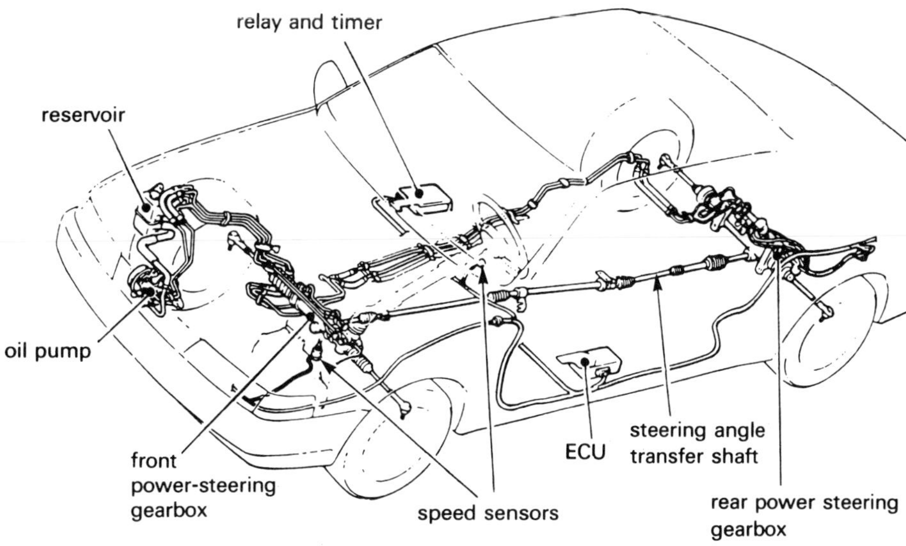

Fig. 52.5 Mazda 4WS layout

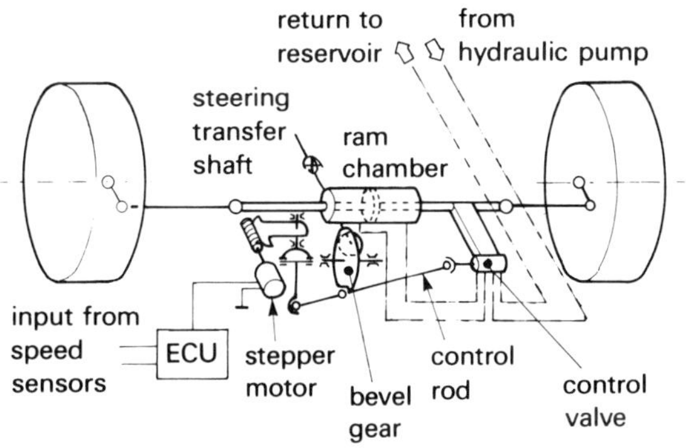

Fig. 52.6 Schematic diagram showing layout of rear-wheel steer unit (Mazda)

1. *Vehicle speed sensors.*
2. *Steering phase control unit* - conveys to the hydraulic ram control valve the required stroke and direction of movement.
3. *Electric stepper motor* - alters the yoke angle and bevel gear phasing according to the signals it receives from the ECU.
4. *Rear steering shaft* - transmits the position of the front wheels to the bevel gear in the steering phase control unit.

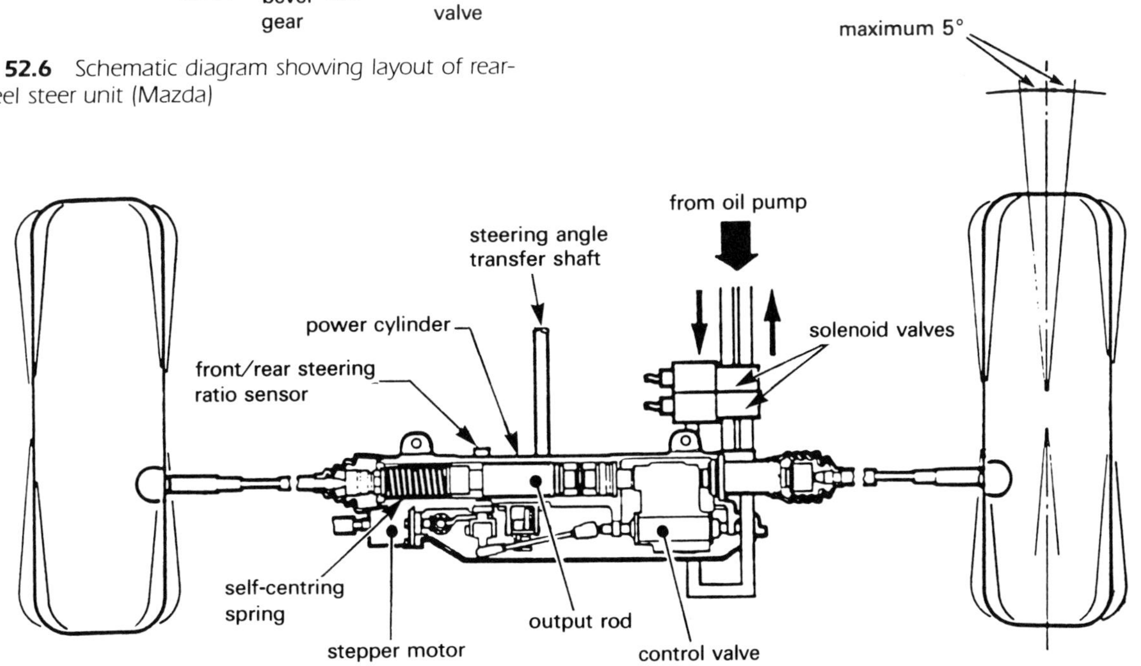

Fig. 52.7 Steering phase control unit

5 *Control valve* - controls the hydraulic pressure supplied to the ram cylinder.
6 *Hydraulic ram cylinder* - steers the rear wheels according to the requirements.

Steering phase control unit This unit, which is shown in Fig. 52.7, alters the direction and angle of the rear wheels. From the electrical pulses passed from the ECU to the stepper motor, and the movement from the steering shaft to the bevel gear, the position of the hydraulic control valve is altered to suit the conditions.

PART FIFTEEN

Vehicle structure

53 General frame construction

The chassis frame

The chassis frame is the structural member to which the main components, such as the engine, transmission and body, are attached, i.e. it is the skeleton of the vehicle.

Although a separate frame is used on trucks, the majority of modern cars combine the frame with the body; this *integral construction* produces a stronger and lighter vehicle which is cheaper to produce when a large number of similar units is made.

Whether it is a car or a truck, the vehicle structure has to withstand various static and dynamic loads. The deflections of a simple rectangular frame, consisting of two side members connected together with two cross members, are shown in Fig. 53.1. These deflections are caused by the following forces:

1 Weight of the components and passengers gives a bending action; this causes the frame to sag (Fig. 53.1b).
2 Road shocks, caused by vehicle motion, or impact with an obstacle, produces two deflections of the frame:

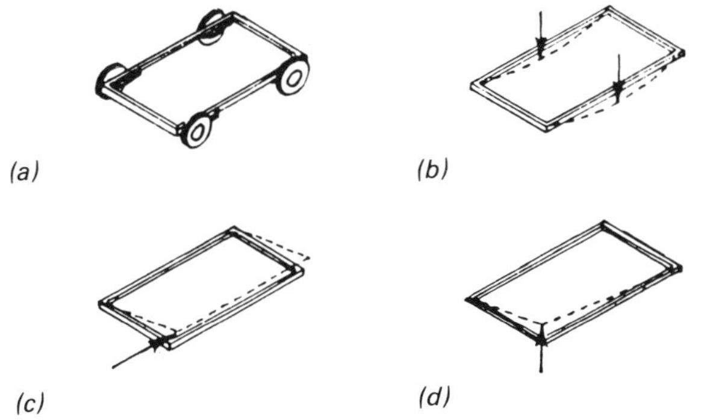

Fig. 53.1 Distortion of a simple frame

a *Lozenging*, the term used when the rectangular frame is pushed into a parallelogram shape (Fig. 53.1c).
b *Twisting* due to a deflection at one corner of the frame (Fig. 53.1d).

The body and frame must be as light in weight as possible, yet must offer maximum resistance to these deflections.

Frame sections

The ideal frame section has a good resistance to bending and torsional effects. Three main forms (see Fig. 53.2) are in general use:

1 *Channel*
2 *Tubular*
3 *Box*

When a beam bends, the material is subjected to a tensile and a compressive stress. Fig. 53.3a shows the top layers of the material being compressed, whereas the bottom layers are being extended. The mid layer is not subjected to tension or compression, and is known as the neutral axis. Some idea of the amount of stress can be given by referring to the cross-section of the beam shown in Fig. 53.3b; the length of each horizontal line represents the stress taken by that layer.

From the above it will be seen that a slightly deeper channel or box section can give the same resistance to bending as a much heavier solid rectangular section. In a similar manner, the distance between the top compression member and the lower tension member is the main factor governing the strength of the tubular frame shown in Fig. 53.3c.

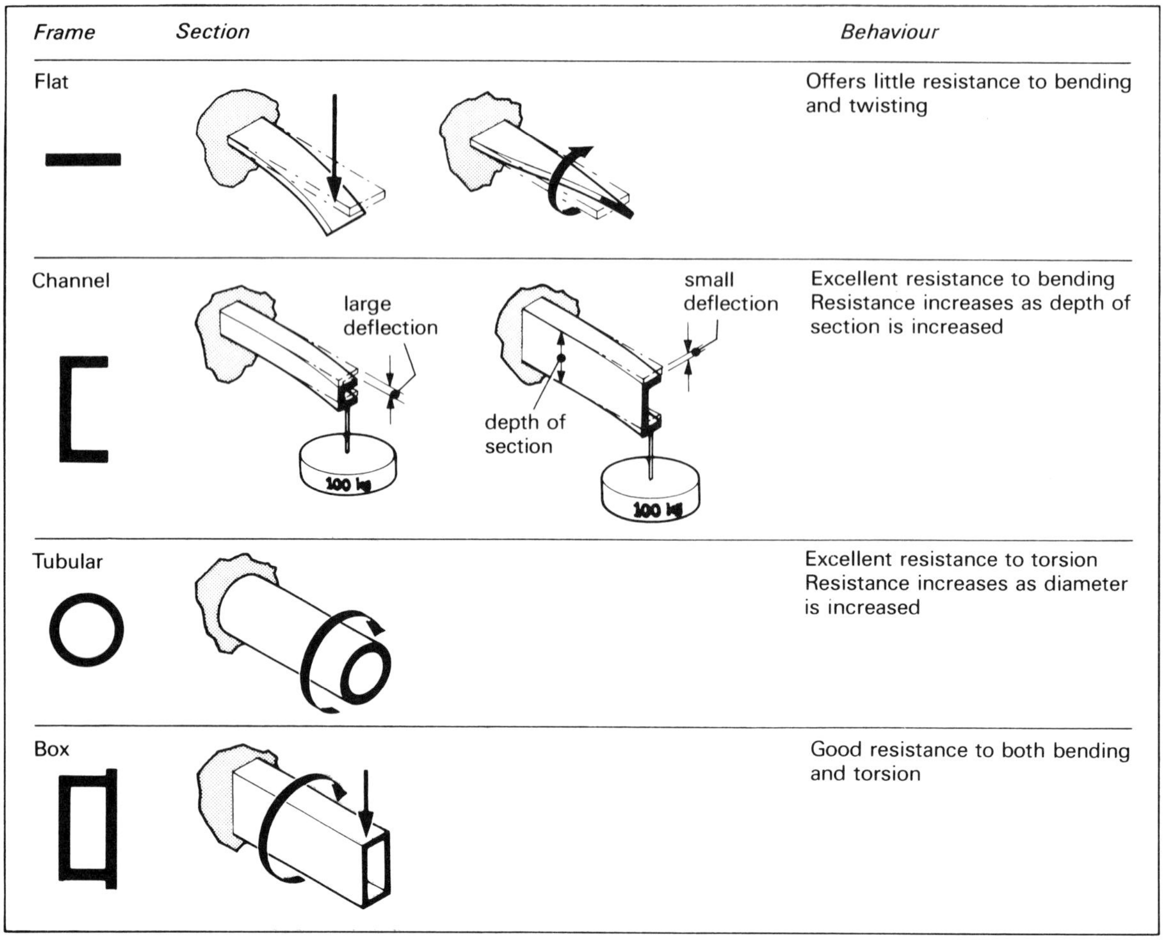

Fig. 53.2 Frame sections

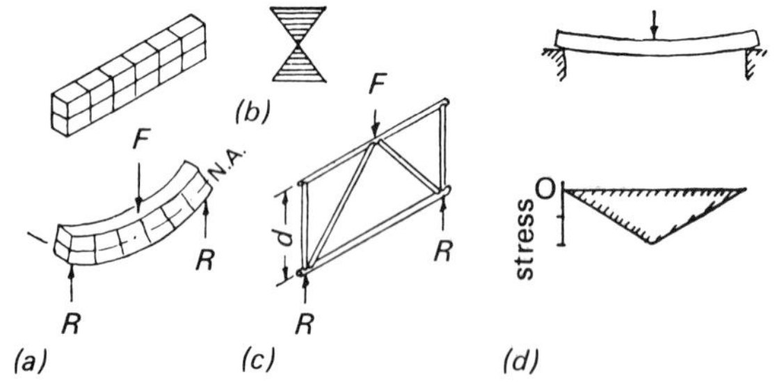

Fig. 53.3 Effect of load on frame members

Whenever a hole has to be drilled in a chassis member, either to reduce weight or to attach something, it should be positioned in a low-stress region, e.g. the neutral axis.

When a force is applied to the centre of a beam (Fig. 53.3d), there is a tendency for the material to break at the centre: this is because the stress is greatest at this point, as shown in the diagram. In order to resist this stress, either the beam must have the same deep section throughout its length, or a varying section of a depth proportional to the stress: the latter design would be much lighter.

Frame design

Fig. 53.4 shows a pictorial view of a frame suitable for a light truck or minibus that uses a non-independent suspension system. It consists of two channel-shaped side-members, which are held apart by a series of cross-members. These are positioned at points of high stress and are cold-riveted to the side-members. The depth of the channel section must be sufficient to minimize deflection. Most frames used on light vehicles are made of low-carbon steel having a carbon content of 0.15–0.25 per cent.

Since the load at each point of the frame varies, a reduction in weight can be achieved by either

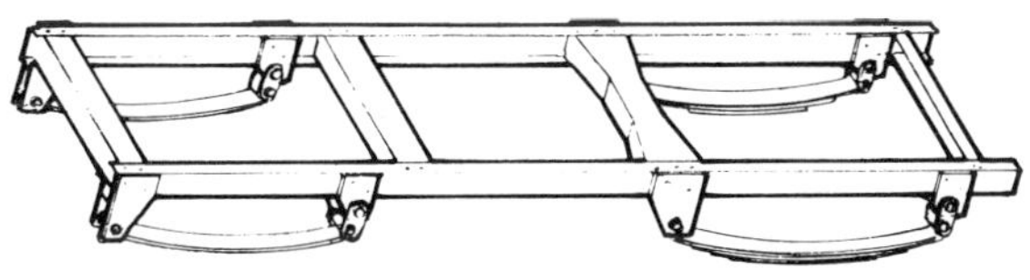

Fig. 53.4 Frame for light truck

reducing the depth of channel, or having a series of holes positioned along the neutral axis in the regions where the load is not so great.

Lozenging of the frame is prevented by fitting gusset plates to reinforce the joins between the side- and cross-members, or by adding an 'X' type (cruciform) bracing between two or more of the cross members.

The frame shown in Fig. 53.4 has a very poor resistance to torsion, so the body has to fulfil this role. Often the body is not designed to resist these stresses, so problems occur; these include movement between doors and pillars, broken windscreens and cracking of the body panels.

Body jigs for pressing the integral bodies are very expensive, so it is common to use a separate chassis frame when the production of a given model is limited in number. Since most of these cars have independent suspension, the frame needs to be extremely rigid at the points where the body and main components are attached. This is achieved by using box-section members welded together and suitably reinforced in the regions of high stress (Fig. 53.5).

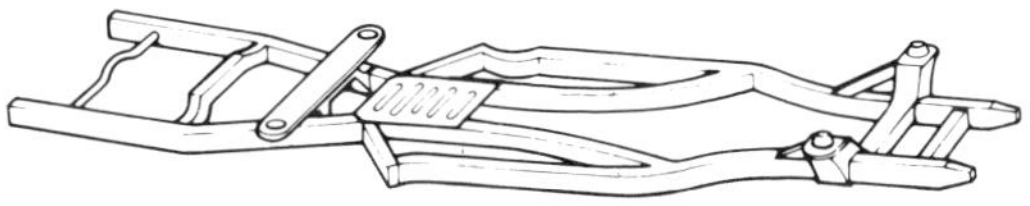

Fig. 53.5 Box section frame

A backbone frame, shown in Fig. 53.6, is an alternative construction to the conventional rectangular frame. This consists of two longitudinal box-section members welded together at the centre and separated at the front and rear so as to accommodate the main components. A series of out-rigger frame members are welded to the spine to support the floor of the body.

Energy-absorbing frame

In the past it was considered that safety for the occupants of a car involved in a collision could be improved by making the chassis frame a very stiff

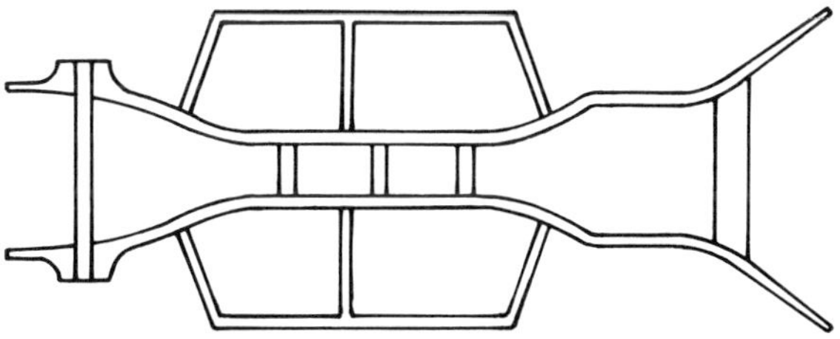

Fig. 53.6 Backbone-type frame

structure. This is a false view because on impact a 'tank-like' structure subjects the occupants to an extremely high deceleration; as a result the force acting on the human body as it smashes into a hard surface is likely to cause death or serious injury.

Most modern frames overcome this problem by constructing the front- and rear-end of the frame in a manner that will allow it to crumple in a concertina manner and absorb the main shock of the impact. It is true that body panels in the vicinity of these *crumple zones* are generally damaged beyond repair, but this is a small price to pay if injury to the occupants is minimized.

Fig. 53.7 shows the principle of a frame designed to absorb the energy of front- and rear-end impacts.

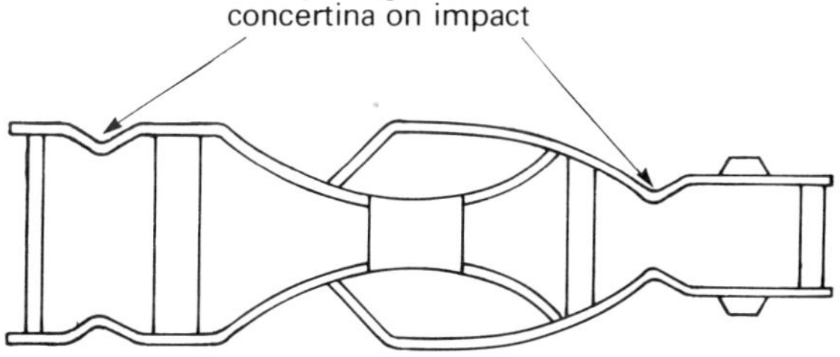

Fig. 53.7 Energy-absorbing frame

Frame repairs

Whenever the vehicle has been involved in a major collision, it is necessary to check the frame alignment. A visual check generally reveals major misalignment, but if this is not obvious a frame check will be required. This is conducted as follows (see Fig. 53.8):

1 *Wheel base check* Set the front wheels in the straight-ahead position and check the wheel base on each side.

2 *Alignment* After checking to verify that the rear wheels are parallel with each other, hold a cord or straight edge against the rear wheel: then turn the

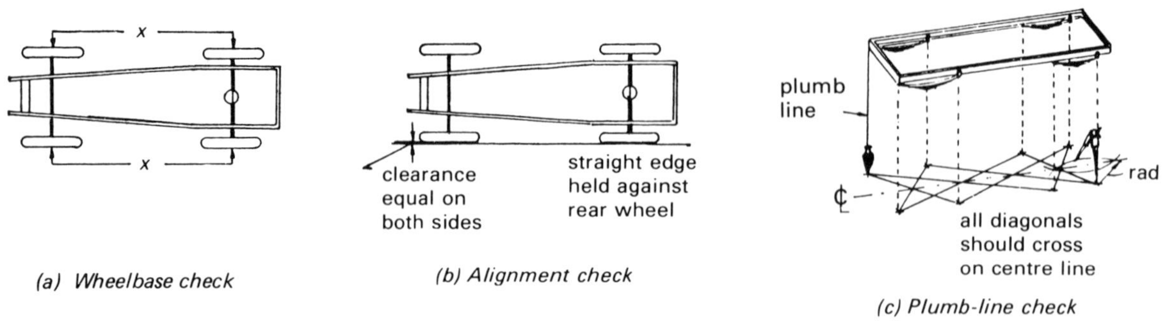

Fig. 53.8 Chassis checking

front wheel until it is parallel with the cord, and note the clearance (if any) between the wheel and cord. This should be the same on both sides.

3 *Plumb-line check* Drop a plumb line from the outside of each fixed shackle of the spring to give eight chalk marks on the floor. Connect up the points as shown; all diagonals should cross the centre line if the frame is correctly aligned.

The tolerance for each check depends on the size of frame, but 6 mm is often laid down as the maximum.

Frame straightening is carried out by using jacks and chains, and is a specialized repair. Unless the frame has been heat-treated it is possible to heat the damaged member to ease the straightening operation.

Cracks can be repaired by welding, and when a reinforcement plate is fitted, it is advisable to taper the end to avoid any sudden change in cross-sectional area.

54 Integral construction of frames

In the earlier work on frames it was said that the body shell resisted torsional movement of a simple frame, but defects in the construction soon showed up because the shell was not designed to withstand these stresses. About 1934, the development of the all-steel body made possible the elimination of a separate frame, the body shell being capable of withstanding the various frame stresses when suitably designed. This *frameless* or *integral* arrangement gives a stiff, light construction, which is particularly suitable for mass-produced vehicles, and since 1945 most light cars have been built with this construction.

Fig. 54.1 shows some of the forces that act on a car body and the manner in which the various body panels are arranged to form a *unitary* structure of sufficient strength to resist these forces.

The diagram shows that the weight of the occupants causes a 'sagging effect' which is resisted by the floor and roof panels. Since these two members are widely spaced, thin sheet metal can be used to form a box-like structure that is both strong and light in weight.

Torsional stiffness of the body is achieved by strengthening the scuttle at the front and by using cross ties, or fitting a ribbed metal panel, behind the rear seat squab.

The thickness of the material depends on the stress taken by the panel: structural members such as sills, rails and pillars are often about 1.1 mm (0.045 in) thick, whereas panels such as the roof are 0.9 mm (0.035 in) thick. Component attachment points require reinforcement with thicker material, and in some cases a separate sub-frame is used to mount such items as engine and i.f.s. members. This frame is sometimes connected to the body by rubber insulation mountings.

Since extremely good ductility is essential for the pressing of the panels, a very low (0.1 per cent) carbon steel is required. The low strength [278 MN/m^2 (18 tonf/in^2)] of this material means that structural members must be stiffened by forming the thin steel sheet into intricate sections, and spot-welding into position. Some idea of the number of separate pressings can be gained by the fact that approximately 4000 spot welds are used on a modern car body.

A modified construction is needed when the roof cannot be fully utilized as a compression member. This occurs on drop-head coupé models, and in situations where either a sunshine roof, or very thin door pillars are used. In these cases the required strength is achieved by using a strong *underbody frame*. In addition, extra stiffness is given to the bodyshell parts that are subject to torsion.

A bodyshell is normally constructed in one of two ways: it is made either by spot-welding the panels, pillars and pressings together so as to form a strong box, or by building a *space frame* (Fig. 54.2). The latter structure gives a skeleton of high structural strength on to which is attached the steel, aluminium or glass-reinforced plastic (GRP) body panels, doors, roof, etc. Of the three materials used, steel is the most common for vehicles made in high volume; this is because production costs are lower once the initial outlay on expensive body jigs and robots has been recovered.

To avoid vibration of the panels, which gives an objectionable noise called *drumming*, a sound-damping material is stuck on the inside of the panels.

Safety in the event of an accident

Safety of a modern car is improved by enclosing the driver and passengers in a rigid cell. At the front and rear of this safety compartment are attached sub-frames; these are designed to concertina on impact as shown in Fig. 54.3. The crumple zones of the body are intended to absorb the shock of a collision and, in

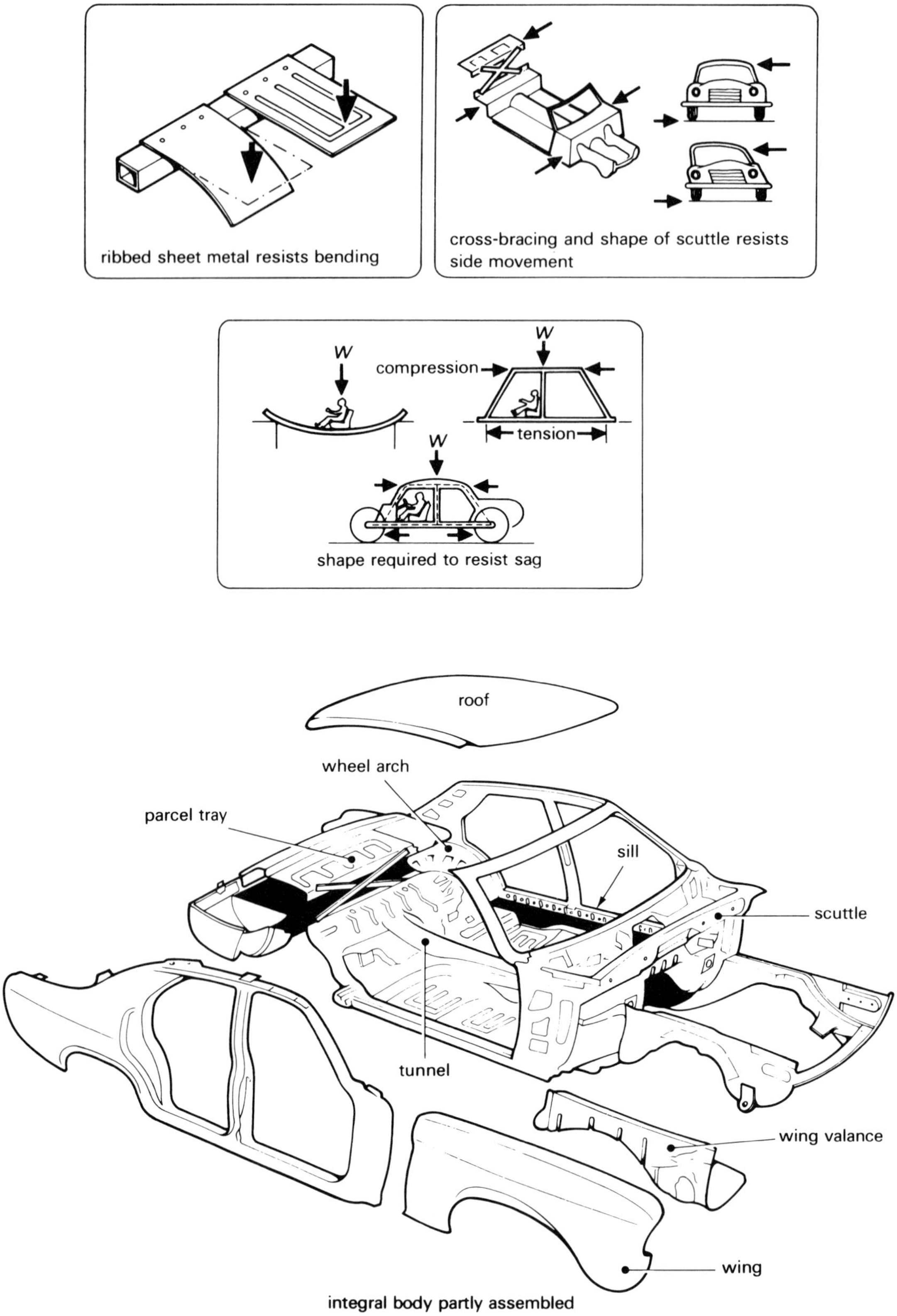

Fig. 54.1 Integral body construction

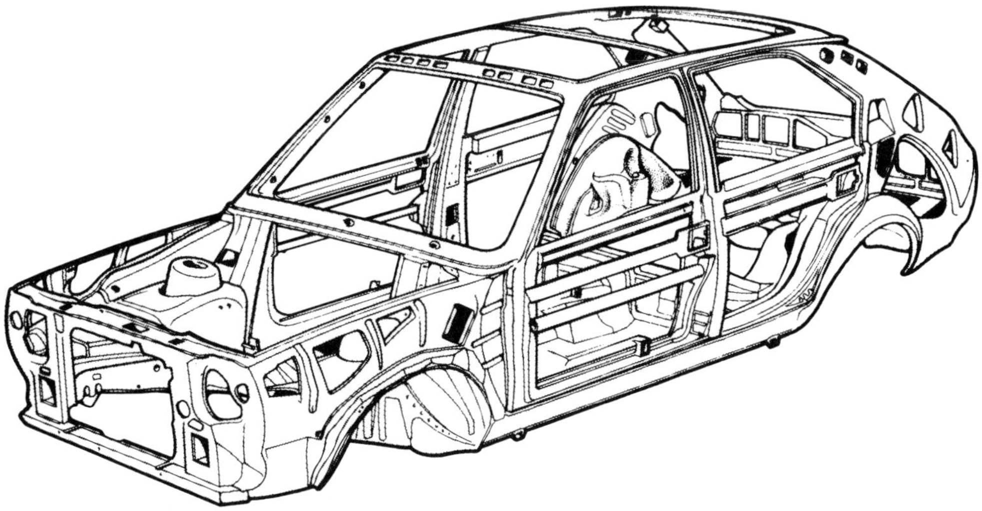

Fig. 54.2 Space frame

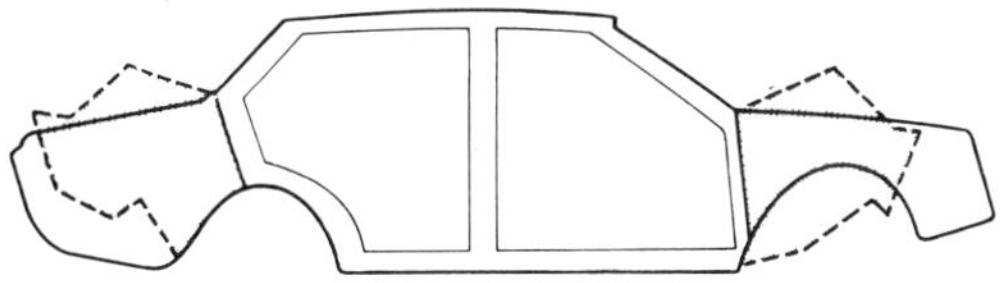

Fig. 54.3 Crumple zones

consequence, reduce the rate of deceleration that is experienced by the occupants.

Nowadays, before vehicles can be sold, at least one vehicle in a model range must be submitted to an approved centre for an impact test. To pass this severe destructive test, the level of safety of the occupants must reach a given standard. In addition the doors must remain closed during impact and must be capable of being opened after the test. The inclusion of this test feature shows why special 'anti-burst' locks are now in common use.

On modern vehicles safety belts, or some other approved body restraint system, must be provided for the driver and all passengers. These belts, which must be securely anchored to suitable strengthened parts of the body, reduce the risk of front and rear seat occupants being thrown through the windscreen, or thrust against the body fittings, when the car slows down faster than the occupants.

Internal body trim, fittings and controls must all conform to safety standards and changes during recent years in the design of parts such as steering wheels and control knobs has materially reduced the risk of human injury.

Extra protection for the occupants during roll-over of a drop-head model car is given by incorporating in the body structure a strong tubular bar to take the place of the metal roof panel. This tube, set across the car, can be either fixed rigidly or made to move to its protection position automatically when roll-over of the car is sensed.

Body shape

Body shape is dictated by a number of factors: it must appeal to the buyer as regards its comfort and shape, and should have a good performance in relation to the ease in which it passes through the air.

The aerodynamic shape of cars is expressed as a *drag coefficient*. The lower the value of this coefficient, the easier the car slips through the air; as a result the fuel economy is improved. Today fuel cost and conservation are important, so greater attention is now paid to the air resistance of a car.

Air resistance

Under normal conditions the power required to drive a vehicle through the air consumes most of the engine energy, so ways for reducing this power drain need to be sought.

The resistance is given by the expression:

$$\text{Air resistance} = \text{Drag coefficient} \times \text{Area} \times \text{Velocity}^2$$

This expression shows that the air resistance increases with the square of the velocity of the vehicle

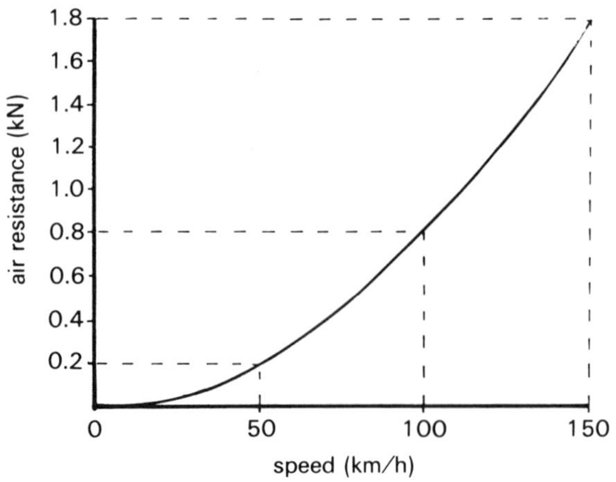

Fig. 54.4 Force required to overcome air resistance

relative to the air, so the resistance becomes very great when the vehicle speed is high (Fig. 54.4). Compared with the resistance at 50 km/h, the resistance at speeds of 100 km/h and 150 km/h is 4 and 9 times as great respectively.

Wind tunnel tests enable the air resistance of a vehicle to be measured. Since the cross-sectional area of the vehicle and its relative velocity to the air are known, it is possible to calculate the aerodynamic drag constant (C_d).

A low C_d is obtained when the body is *streamlined* to enable it to pass through the air with the minimum disturbance. Since most of the resistance is caused by the low-pressure region at the rear of the vehicle, the aim is to return the air to this region with the minimum of turbulence after it has flowed over the body.

Resistance is directly proportional to the cross-sectional area, so a low and sleek sports-type car would have good performance in this respect.

Various refinements are made to the body to reduce air drag; these include the recessing of protruding items such as door handles and the shaping of the body below the front bumper to form an air dam (Fig. 54.5a).

Air flow control devices are sometimes fitted by the car manufacturer, or car owner, to the rear of the vehicle. According to their shape and position, these devices either smooth out the air flow to reduce the disturbance, or act as a *spoiler* to deflect the air upwards so as to increase the adhesive force acting on the rear wheels (Fig. 54.5b). Although these arrangements are beneficial on racing cars, their purpose on many production cars may be regarded as 'image creation' embellishments.

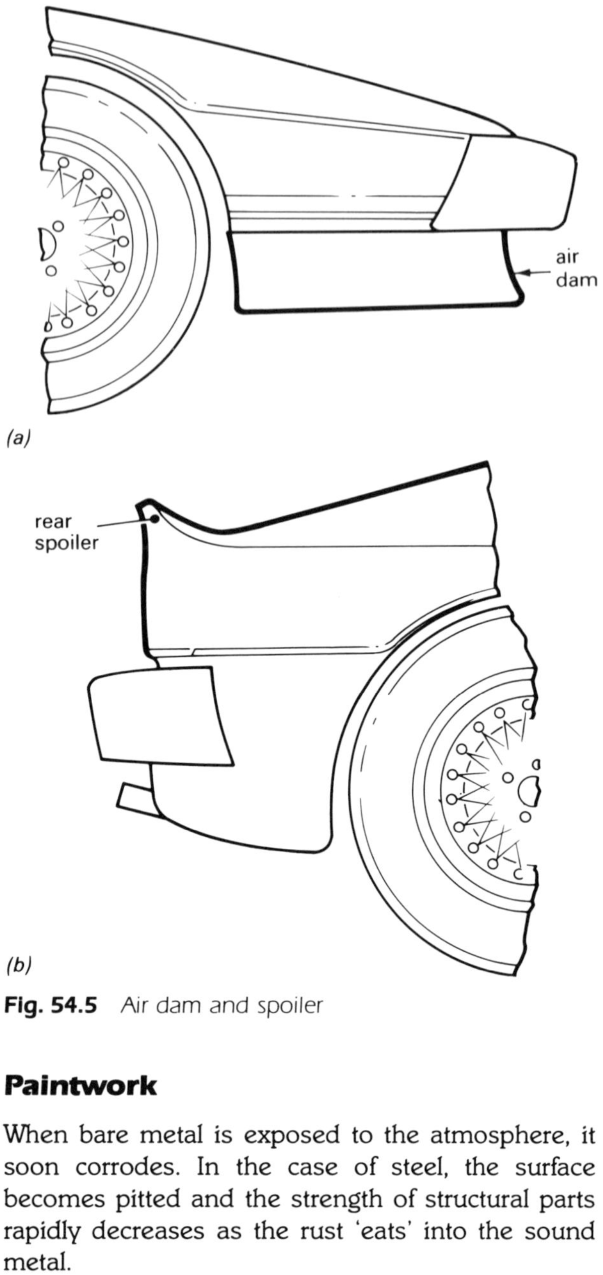

Fig. 54.5 Air dam and spoiler

Paintwork

When bare metal is exposed to the atmosphere, it soon corrodes. In the case of steel, the surface becomes pitted and the strength of structural parts rapidly decreases as the rust 'eats' into the sound metal.

One way of reducing corrosion is to coat the surfaces with paint; this acts as a barrier between the metal and oxygen in the atmosphere.

Corrosion also occurs when two dissimilar pieces of bare metal are joined by water. In this case the action is similar to that of a battery; the passage of electricity from one 'plate' to the other 'plate' causes an *electrolytic* action; this corrodes the metal. Salt, mixed with the moisture that connects the two metals, increases the electrical flow; this is the reason why corrosion is accelerated in winter.

The car owner regards the paintwork as an embellishment of the bare metal, so the manufacturer selects a colour, finish and texture to anticipate the personal preferences of the potential owner of the car.

Special underbody corrosion treatment is given to vehicles that are likely to be exposed to severe operating conditions. This includes electrolytic coating of the underbody parts, wax injection of box members and dipping of the bodyshell in a protective solution to seal all exposed joints.

55 Heating, ventilating and air conditioning

The time is long gone when the occupants of a vehicle had to suffer the extremes of temperature and exposure to the elements. Over the years, heating and ventilating systems have evolved from the days when the open window was the only source of cool air to the sophisticated air conditioning systems in use today. Beside the cooling aspect, the occupants of a modern vehicle demand a heating system that maintains a comfortable temperature whilst still providing a steady supply of fresh air to prevent the driver falling asleep.

Systems in use today can be divided into two main classes; these are:

- *Basic heating and ventilating system*
- *Full air conditioning.*

Basic heating and ventilating system

In this country this system is the common arrangement used on the majority of vehicles, other than the luxury class which nowadays fits an air conditioning system as standard. When the basic system does not satisfy the potential owner, most manufacturers offer an air conditioning system as an optional extra.

The basic arrangement is normally a fresh air system with provision for heating and ducting the air to discharge at foot or face level. Controls allow the occupants to both set the temperature of the air and select the discharge points to suit the conditions. These controls are connected to two or more flaps, which direct the air through the appropriate ducts. An electrically driven fan is fitted in the system to speed up the air movement at times when the natural circulation is poor.

System layout

A fresh air circulation system is not effective unless the air can pass through the body compartment and flow out to make more room for fresh air to enter. Fig. 55.1 shows the paths taken by the air as it flows towards the various discharge vents in the body. Ideally these should be situated at points where the

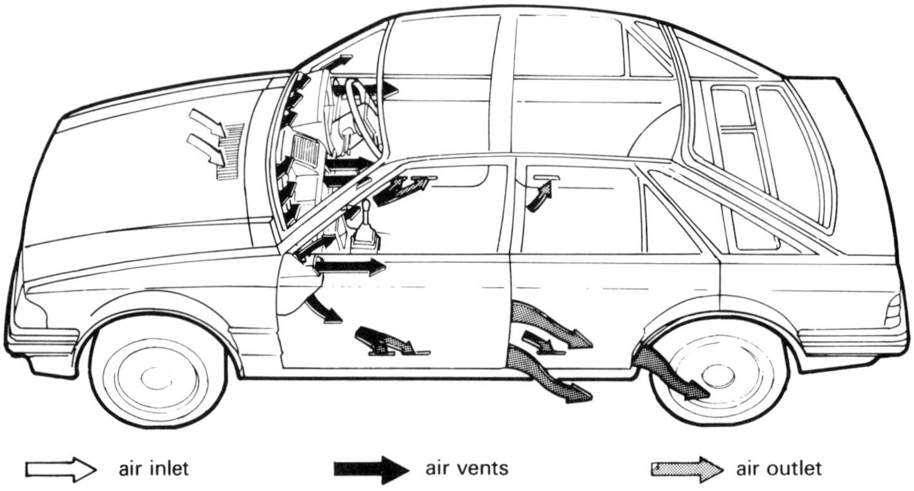

Fig. 55.1 Air flow through vehicle

external air pressure is low to allow for a natural circulation. In a similar way, the inlet point should be in a region of high pressure; the pressure increase at these points is normally caused by the combination of vehicle movement through the air and change in the air flow direction.

The cut-away drawing (Fig. 55.2) shows a typical ducting arrangement for a car. A manual control system allows the driver to direct the air so that it can be discharged from one or more outlets. These are:

- along the length of the front windscreen (for head level ventilation and defrost or demist) and two small side outlets (for side window demist)
- at independently controlled central and side vents
- in the foot-well region.

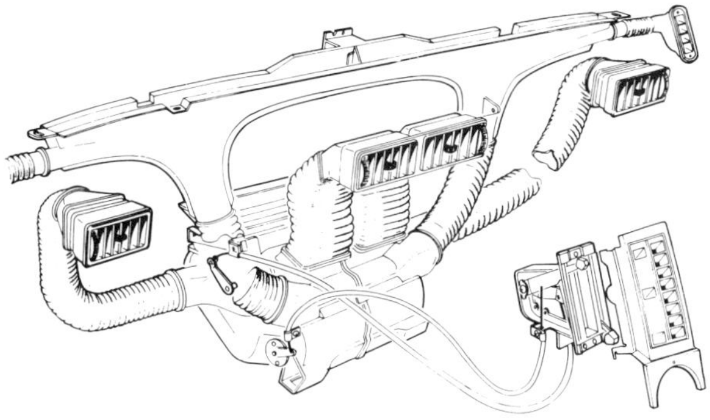

Fig. 55.2 Air duct layout

The second control lever operates a flap valve which directs all, or a part of, the intake air through a heat exchanger when warm air is required. Although exhaust gas from the engine has occasionally been used at a heating source, most manufacturers prefer to utilize the engine coolant heat because it is considered to be a safer system.

Fig. 55.3 shows how the valves control the path taken by the air; in the position shown part of the air flow is directed through the heater matrix. This provides warm air at the windscreen and from any vent that is opened by the front-seat occupants.

Naturally the heater will not function until the coolant reaches a sufficient temperature. Assuming the thermostat in the cooling system is serviceable and of the correct type, the heater should soon become effective after a cold-engine start. To meet this requirement the coolant flow to the heater should be taken from the hottest part of the system - on the engine side of the thermostat.

The multi-point switch that controls the fan enables

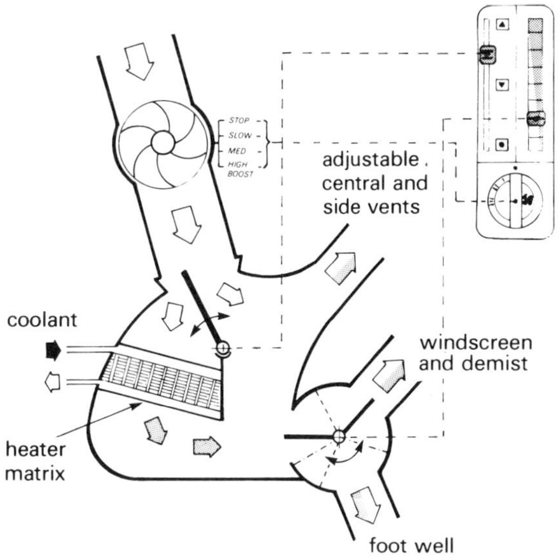

Fig. 55.3 Heating and ventilating system

the speed to be controlled by varying the resistance of the circuit; this alters the voltage applied to the permanent-magnetic-type motor.

On some systems an extra control lever is fitted to allow the outside air to be cut off and the inside air to be used instead. This recirculated-air setting is useful when the outside air is heavily polluted.

Air conditioning

Although the climatic conditions in countries like the USA force the majority of manufacturers to offer cars with air conditioning (AC) as standard, the temperate conditions enjoyed by the UK generally persuade potential owners to regard an AC system as a luxury. Nevertheless, more and more cars are now being sold with AC because modern *environmental control systems* (or climate control systems) offer the following features:

- Cooling in summer to a set temperature
- Dehumidification of (removal of moisture from) air in summer
- Dehumidification of air in winter, so preventing misting of windows
- Heating in winter to a set temperature
- Control of the flow of air.

These features can be achieved by fitting, in the basic system previously described, an extra heat exchanger. This additional matrix contains a substance that is very cold compared with the hot coolant in the normal heater matrix, so the air passing through it will be cooled. When this cooled air is discharged to the interior of the car, the effect will be similar to the action of the heater except that the air temperature will go down instead of up. Whereas most basic heating and ventilating (non-AC) systems rely on the driver to control the temperature, the sophisticated AC system used today controls the temperature automatically to suit the setting selected by the driver.

An AC system is the apparatus needed to maintain a very low temperature in the cooling matrix (evaporator) placed in the ventilating air stream.

Heating and cooling principles

Before considering the layout of an AC system it is advisable to consider some fundamentals in order to understand the purpose of the various parts of the system.

Compression and expansion

When a gas is compressed, an increase in the pressure and temperature occurs (Fig. 55.4a). The heat generated by this action causes the fins on the compressor to get hot; this heat is radiated to the atmosphere.

Likewise expansion of a gas decreases the pressure and lowers the temperature (Fig. 55.4b). In this case the fins of the cylinder would feel cold to the touch; this is because the cylinder absorbs the heat from the human hand. A similar expansion effect is produced when the air is released from a tyre valve: the high pressure air rushing out from the tyre causes a sudden drop in pressure at the end of the valve; on some days the accompanying drop in temperature sometimes causes frost to form on the metal valve.

In both the compression and expansion actions heat passes from the HOT to the COLD substance.

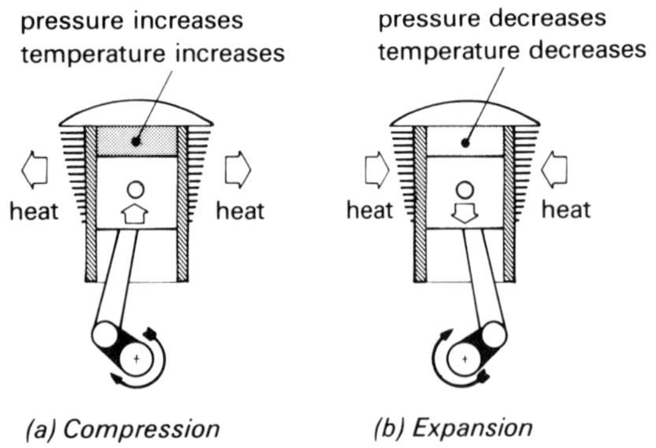

Fig. 55.4 Heat flow during compression and expansion

Evaporation

A large quantity of heat, called latent heat, is required to cause a liquid to change its state to a gas. In the case of water, the change to steam requires a large quantity of heat. If this natural evaporation is accompanied by an expansion process, then a large amount of heat from an external source will be needed.

This expansion takes place in the AC evaporator matrix situated in the duct that supplies air to the interior of the car (Fig. 55.5). As the liquid is pumped through the expansion valve, the drop in pressure, together with the heat required to change its state, makes the matrix very cold. This causes heat to be extracted from the air as it is pumped to the car interior.

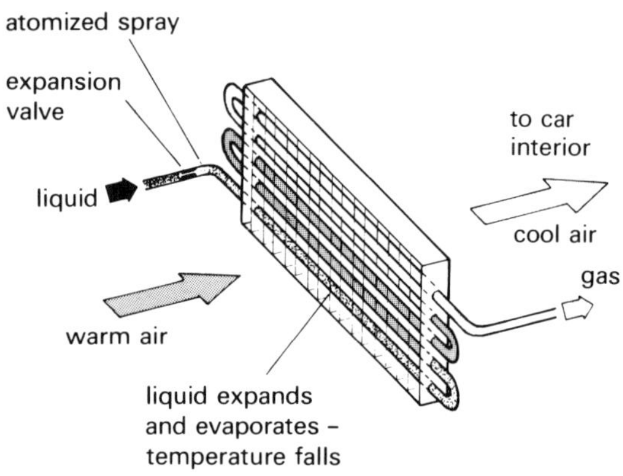

Fig. 55.5 Evaporator action

Refrigerant

A refrigeration system relies on the properties of the working substance to have both a low boiling point and the ability to change its state readily from a liquid to a vapour and vice versa.

In the past this requirement was met by a special chlorofluorocarbon (CFC) substance called Freon-12 or *Refrigerant-12* (generally referred to as R-12). Under standard conditions this substance boils at −30 °C, i.e. at atmospheric pressure and a temperature of −30 °C, it evaporates from a liquid to a vapour. Refrigerant-12 requires only 1/14th of the heat needed by water to change its state, and since it maintained good stability during repeated changes, this comparatively cheap refrigerant was commonly used for the majority of automotive AC systems.

All refrigerants must be handled with great care in view of the very low temperature at which they are stored.

Manufacturer's recommendations should be sought and be understood before handling refrigerant or working on any part of an AC system.

Modern refrigerants Chlorofluorocarbons (CFCs) have been identified as the primary cause of damage to the Earth's ozone layer. In view of legislation and various international agreements introduced to combat this environmental damage, CFC-type refrigerants and flushing agents are no longer acceptable for AC systems.

From about 1993 new *ozone-friendly* chemicals, such as HFC 134a, are to be used in all new automotive AC systems. The refrigerant HFC 134a cannot be used in a system designed to operate with R-12; this is because the properties of HFC 134a are different. As a result, the cost of the new refrigerant and system will be higher than at present.

In the next few years the supply of R-12 will cease. During this change-over period existing AC systems will be charged, after repairs have been carried out, with a new refrigerant that is specially blended to be more ozone-friendly. This *drop-in* replacement for R-12 has a number of operational limitations, one of these being that gas leakage through the flexible hoses is increased.

Effect of pressure on refrigerant As with other liquids the temperature at the change of state (boiling point) depends on the pressure. The pressure-temperature dependence is useful in the operating cycle because by varying the pressure of the refrigerant as it flows around the system, the temperature can be controlled to suit the need of a particular part. To summarize this pressure and temperature relationship:

Pressure	*Temperature*
low	low
high	high

Vapour-compression AC system

The vapour-compression refrigeration system, shown in Fig. 55.6, consists of:

1 *Compressor* - driven by the engine, it compresses the refrigerant vapour to raise its pressure and temperature.

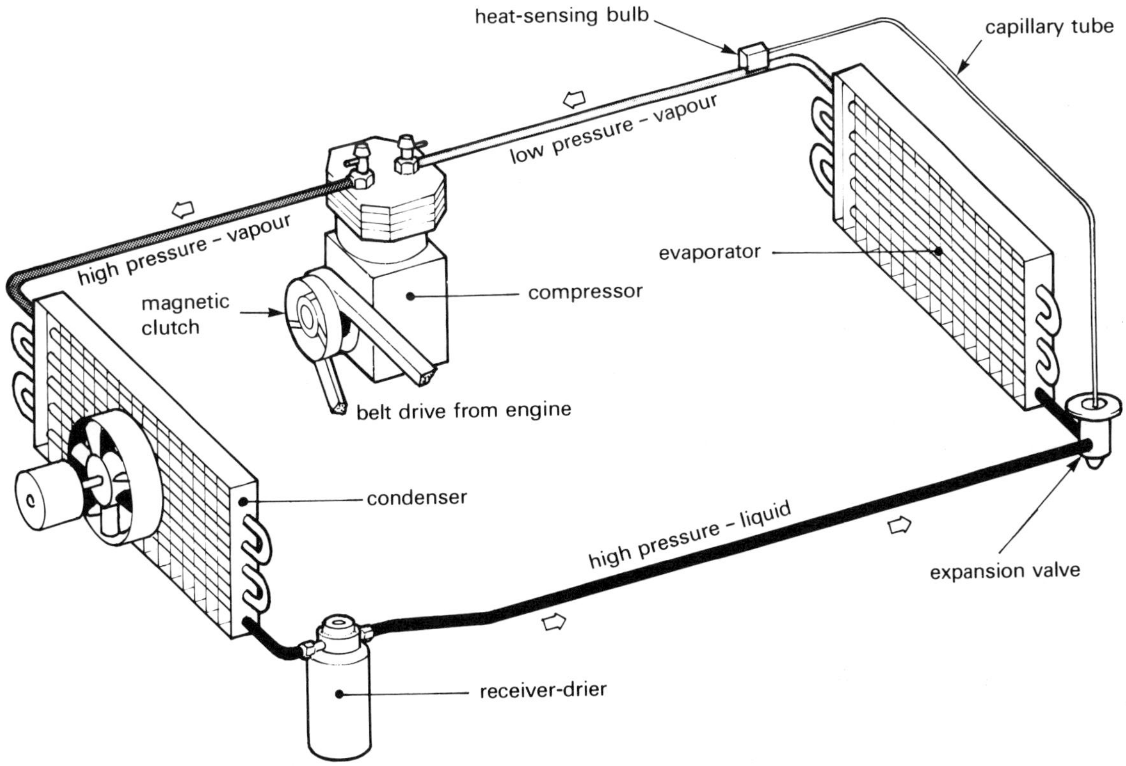

Fig. 55.6 Air conditioning system

2 *Condenser* - a heat exchanger that uses external air flow to cool the vapour to change its state to a liquid.
3 *Receiver-drier* - acts as a liquid storage tank and also contains chemicals that remove water from the system.
4 *Expansion valve/orifice* - meters the refrigerant and lowers its pressure to allow it to expand.
5 *Evaporator* - a heat exchanger to extract heat from the ventilating air flow as it passes to the car interior.

Basic operation

The refrigeration cycle is a continuous process and is based on the temperature changes as the refrigerant flows around the fluid circuit. These temperature changes are brought about by pressure variations and changes in state of the refrigerant.

When the driver selects the AC mode, an electromagnetic clutch, contained in the compressor drive pulley, is engaged. Rotation of a compressor (using R-12 refrigerant) draws in vapour at a low gauge pressure of 1-2 bar (15-30 lbf/in^2) and compresses it to a pressure of 8-18 bar (118-265 lbf/in^2); the actual pressure depending on the ambient temperature and compressor speed. This pumping action increases the temperature of the vapour to 70-110 °C and causes it to flow to the condenser; this is normally situated in front of the engine cooling radiator.

As the hot vapour flows through the condenser tubes, its heat is radiated to the air that passes over the condenser fins and tubes. This action cools the refrigerant vapour and causes it to condense, i.e. the refrigerant changes its state from a vapour to a liquid. By making the condenser tube of sufficient length, the refrigerant leaving the condenser is in a totally liquefied state.

Any water in the system chemically reacts with refrigerant to form corrosive acids, so this moisture, and other contaminants, are removed at this point in the fluid circuit to avoid damage. This duty is performed by passing the liquid refrigerant through silica gel desiccant crystals in the *receiver-drier* (sometimes called a *dehydrator*).

In addition to the drying and filtering duty, the receiver-drier also acts as a high-pressure container to store liquid refrigerant in reserve for its supply to the evaporator.

On leaving the receiver-drier the clean dry liquid passes to the expansion valve; this meters and atomizes the flow of refrigerant, at low pressure, into the evaporator. The quantity delivered by the valve depends on the temperature at the evaporator outlet. In the system shown, a temperature-sensing bulb and capillary tube signal to the expansion valve when the flow should be increased or decreased. A low output temperature indicates that the valve must reduce the flow and vice versa.

After the flow has been regulated by the expansion valve, the atomized refrigerant passes through the evaporator. Here the heat is extracted from the air that is flowing over the tubes and fins. This cools the air and, as a result of heat transfer from air to refrigerant, the liquid will first change into a vapour and then be heated to a superheated state well above the boiling point of the refrigerant as it passes through the evaporator.

Having completed the cooling function the low-pressure vapour is returned to the inlet side of the compressor to allow the cycle to be repeated.

Air conditioning components

Although the basic operation remains the same, the construction of the five basic components needed to perform the refrigeration cycle varies slightly between the different manufacturers of AC systems.

Compressor

The compressor used in a particular system can be one of two types:

- *Reciprocating piston*
- *Rotary vane.*

Both types are driven from the engine crankshaft via a single- or multi-vee belt. The belt drives a free pulley that is engaged to the compressor by an electromagnetic clutch. When air conditioning is required, the driver operates a switch that activates a relay. This energizes the clutch windings and connects the vee pulley to the compressor drive shaft until the driver, or an automatic control system, switches off the AC unit.

On some systems such as the CCOT described on page 354, the compressor is *cycled* (automatically switched on and off) by a *thermostatic* switch to maintain a nearly constant temperature at the outlet of the evaporator.

For safety purposes, a compressor is fitted with a valve or switch to relieve the pressure if the outlet temperature reaches a given maximum, e.g. 93 °C; this represents a pressure of 28 bar (400 lbf/in^2).

Fig. 55.7 shows the construction of a magnetic clutch having a stationary winding around an iron

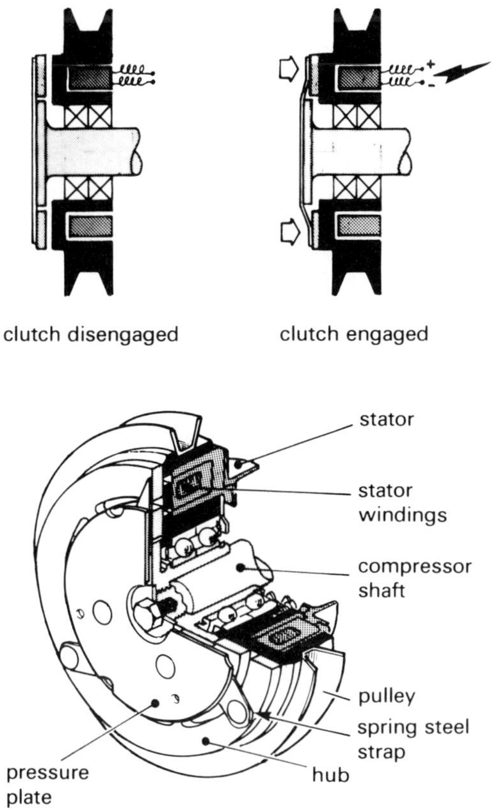

Fig. 55.7 Magnetic clutch

core; this becomes a strong magnet when current is passed through the winding. This action attracts a pressure plate attached to the compressor shaft and causes it to make frictional contact with the driving pulley. Three spring steel straps are fitted; these transmit the drive between the pressure plate and compressor hub, and provide a spring action to release the clutch when the compressor is switched off.

A small quantity (e.g. about 175 ml) of special moisture-free refrigerant oil is used in an AC system to lubricate the seals, gaskets, valves and compressor. Most of the oil circulates around the system and only a small amount (about 20 ml) is retained in the compressor. On most compressors the refrigerant must be discharged before the oil level can be checked, so the oil check is only carried out when the system fails to operate satisfactorily.

When the refrigerant is exhausted prior to carrying out a repair of a component, the quantity of oil discharged with the gas should be measured. This indicates the amount of oil that must be introduced when the system is recharged.

Reciprocating piston compressor There are two main types:

- *Crankshaft*
- *Swash-plate.*

Crankshaft-type This is similar to a small engine (Fig. 55.8a) and is made as a single, twin or vee-twin two-stroke unit.

Reed valves are used to control the intake and exhaust flow of the gas refrigerant and lubrication is provided either by a separate built-in oil pump or by utilizing the pressure difference between input and output lines.

A plug on the side of the compressor is provided for the insertion of an oil-level dipstick. Removal of this plug will allow refrigerant to escape.

Often Schrader-type valves are fitted adjacent to the inlet and outlet pipe connectors. These valves are provided for the attachment of special equipment which allows the technician to check the low- and high-pressure sections of the refrigerant circuit; also the valves allow the system to be topped up, or drained.

In the past these valves were opened to the atmosphere to exhaust the system, but nowadays environmental considerations make it necessary to drain the refrigerant into a sealed container to allow it to be reclaimed.

Swash-plate-type This compact type (Fig. 55.8b) uses either three or five pairs of pistons, positioned radially around the driving shaft. These pistons are moved back and forth in an axial direction by a *swash-plate* (Fig. 55.8c); this is driven by a shaft connected to the electromagnetic clutch.

Reed valves at the end of each cylinder control the vapour flow to and from the low- and high-pressure connectors situated at the rear of the compressor.

A small internal oil pump lubricates the compressor.

Rotary pump The rotary vane type compressor shown in Fig. 55.8d is becoming popular because of its low-friction, high pumping output, and low noise level.

Construction and operation of the pump is similar to other vane pumps used on vehicles. In the layout shown two through-vanes, driven by a rotor, form four separate pumping chambers.

Condenser

The heat exchanger shown in Fig. 55.6 is a long finned tube that is designed to use the cool air flowing

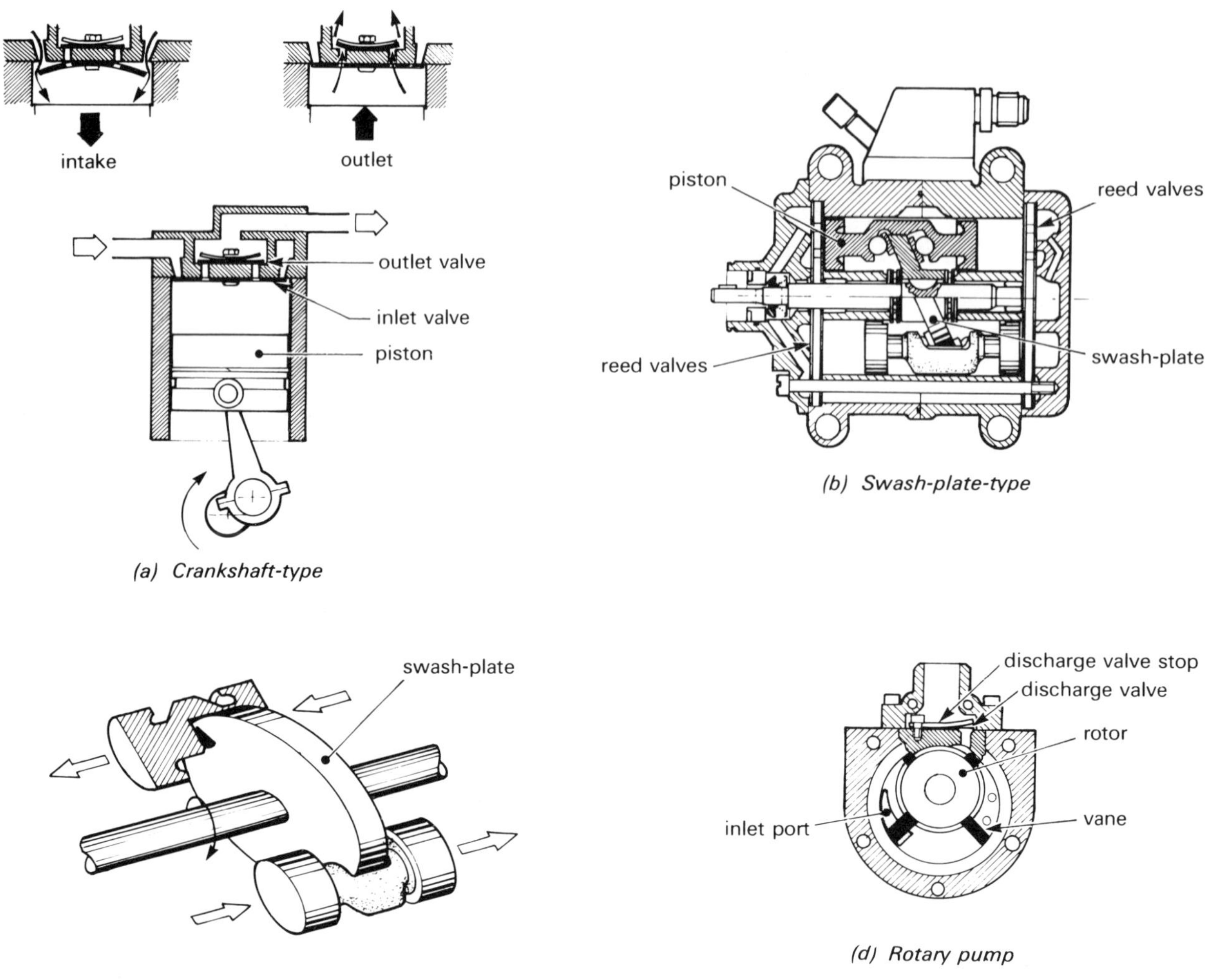

(a) *Crankshaft-type*

(b) *Swash-plate-type*

(c) *Action of swash-plate*

(d) *Rotary pump*

Fig. 55.8 Types of compressor

around the tube to change the refrigerant from a superheated vapour to a liquid.

To ensure a good supply of cooling air, the condenser is normally placed in front of the engine cooling radiator. At times when the air speed from the forward motion of the car is insufficient, an extra fan, placed in front of the condenser, is brought into operation. Since this fan is electrically driven, it can be cut out by a thermal switch when extra cooling of the condenser is not desired.

As the refrigerant flows downwards from the compressor connection at the top, the vapour gradually condenses. Generally two-thirds of the condenser contains vapour; the bottom of the condenser holds warm liquid refrigerant.

Receiver-drier

Although this unit acts as a temporary storage container for the refrigerant, its main duty is to remove moisture and dirt from the liquid. Besides causing corrosion, water in the system allows the formation of ice; this blocks the orifice of the expansion valve and obstructs the flow of refrigerant.

Water is removed by passing the refrigerant through either a sieve desiccant or silica gel crystals. Most modern AC systems have large silica gel driers, because new ozone-friendly refrigerants, and associated lubricants, are extremely *hygroscopic* (able to absorb moisture).

Fig. 55.9 shows the main details of a typical receiver-drier. Liquid refrigerant from the condenser

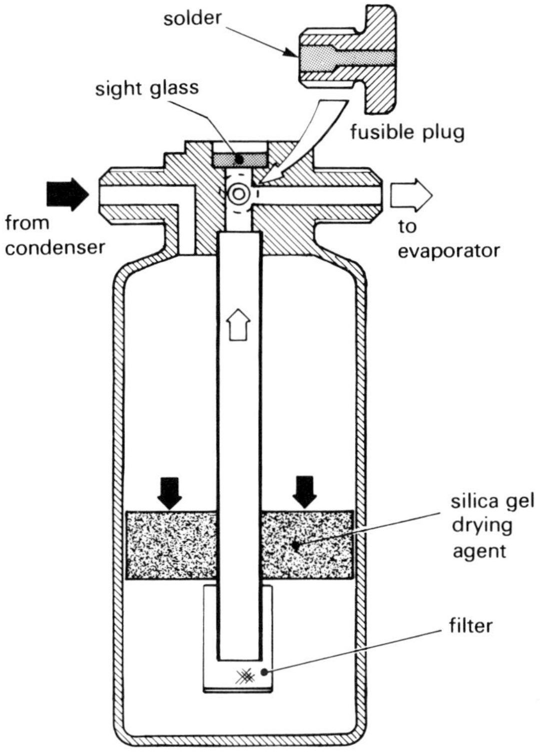

Fig. 55.9 Receiver-drier

collects in the aluminium alloy container and then passes through the desiccant and filter on its way to the evaporator.

Often the receiver-drier incorporates a *sight-glass* and *fusible plug*.

Sight-glass This feature allows the condition of the refrigerant to be observed. The main indications are:

Condition	*Indication*
Foamy or containing bubbles	Low in refrigerant and air in system
Streaky	Lack of refrigerant and excess oil is circulating
Cloudy	Circulation of desiccant

Fusible plug Damage to a system by an abnormally high pressure, caused by poor condenser ventilation or a high cooling load, is prevented by this *melt-plug*.

A special solder filling of a hole in the plug melts when the temperature reaches about 100 °C. At this temperature the fluid pressure is about 29 bar.

Thermostatic expansion valve (TXV)

The valve shown in Fig. 55.10 is a thermal expansion type. This uses a diaphragm-controlled ball-and-needle valve to regulate the flow of refrigerant into the evaporator. As the liquid refrigerant is forced through the orifice, it becomes atomized. This causes a sudden drop in pressure and the accompanying expansion produces the required low-temperature cooling vapour.

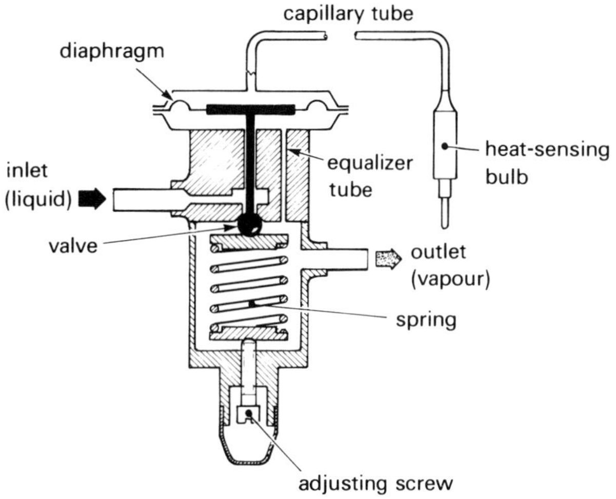

Fig. 55.10 Expansion valve (TXV)

Diaphragm movement adjusts the flow of refrigerant to equal the rate at which the evaporator can vaporize it. Too much refrigerant causes a severe build-up of ice around the evaporator coil and too little means that the compressor is supplied with vapour that is not in a superheated state.

The control system is a sensing bulb, attached to the evaporator outlet pipe, and a capillary tube. When the cooling load on the evaporator is increased, the gas in the bulb senses the rise in temperature. As a result, the gas expands and moves the diaphragm downwards; this increases the valve opening and allows more refrigerant to flow.

Valves-in-receiver (VIR) type On some AC systems the thermostatic expansion valve is fitted in the receiver-drier unit. By mounting the evaporator adjacent to the evaporator, this layout overcomes the need for a separate capillary line to the TXV unit.

This type of receiver-drier also incorporates a *pilot-operated absolute pressure suction throttling valve (POA valve)*. By means of a capsule, this valve controls the flow of refrigerant from the evaporator to maintain a constant pressure difference across the evaporator of 2 bar (30 lbf/in^2).

Evaporator

Whereas a condenser gives up the heat of the refrigerant vapour to the air, the duty of an evaporator is just the opposite - it takes the heat of the air that is passing through it and uses the heat to vaporize the refrigerant.

Besides acting as a heat exchanger to cool the air as it enters the car interior, the evaporator also dehumidifies the air. This feature is advantageous at times when excess moisture in the air causes windows to mist up and cause discomfort for the occupants in the car. Heating the air after removing the moisture provides a ventilating air flow of excellent quality.

Frosting and icing of the evaporator fins reduce the air flow through the unit; as a result, the effectiveness of the cooling system decreases. This is minimized by the incorporation of either an *evaporator thermostat switch* or *evaporator pressure regulator*. Fitted to the evaporator outlet, this switch or regulator cuts out the compressor when it senses that the temperature is less than about 4°C.

Using the AC in humid ambient conditions causes water to drip from the evaporator. A tray fitted below the evaporator collects this fall-out and passes it to a drain pipe. An indication of the amount of water passing to waste from this pipe is often seen on the ground after the vehicle has been parked for a while.

An evaporator is generally made of aluminium alloy and uses either the tube or film construction (see page 145). Heat transfer depends on the surface area of the exchanger so a large number of fins are used to improve the efficiency.

Fig. 55.11 shows an evaporator and the fittings that are used to house it in the ventilating air duct.

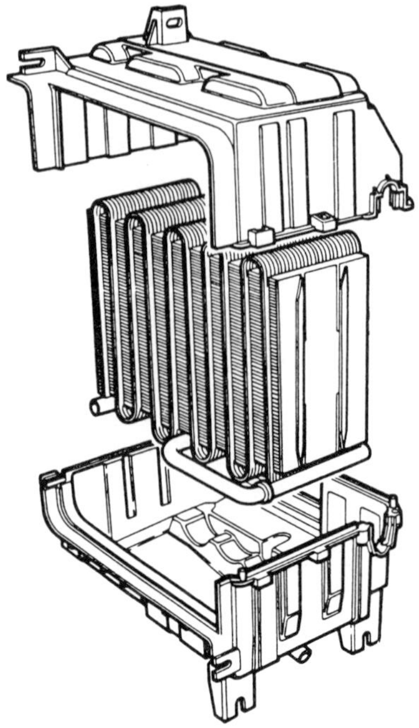

Fig. 55.11 Evaporator

Electrical control

Apart from the ventilation control of air flow, the basic AC system is fitted with a number of electrical switches, sensors and relays to both improve its working efficiency and safeguard the system in the event of overload.

Pressure switches   Most systems have switches in the low- and high-pressure lines which break the circuit to the magnetic clutch at times when a given pressure is exceeded.

Idle-up control   The AC system places an extra load on the engine when the compressor is in operation. This would show up when the engine is slow-running and would result in a change in speed, or possibly stalling, of the engine. To overcome this problem, an *idle-up* device is fitted to open the throttle slightly when the magnetic clutch engages the compressor.

Thermostatic switches   An extra switch is often fitted when more than one fan is used to cool the AC condenser and the engine and oil cooling radiators. For economy reasons only one fan is in use during normal operation; the other fan is brought into use when the cooling load is high.

Cycling clutch orifice tube system (CCOT)

This system, which is shown in Fig. 55.12, differs from the conventional AC system previously described. It has an *orifice tube* instead of a TVX unit and an *accumulator*, containing the desiccant and positioned in the low-pressure line, instead of a receiver-drier.

In this design, the flow of refrigerant through the expansion orifice into the evaporator is controlled by a cycling action of the compressor. Magnetic clutch operation for this action is governed by the signals received from two sensors: a pressure switch in the high-pressure line and a thermostatic switch at the outlet of the evaporator.

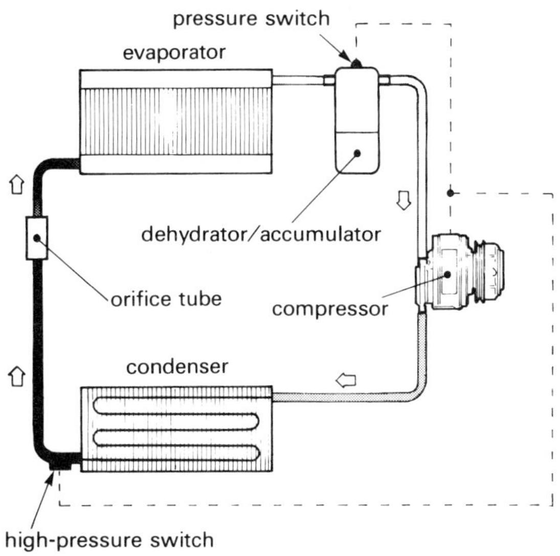

Fig. 55.12 Layout of CCOT system

Climate control system

Recent developments in electronic control have made it possible to use these devices to control the air temperature and quality in accordance with the wishes of the car occupants. As a result many cars in the luxury category are now fitted with a sophisticated *climate control system*.

Once the driver has selected a given temperature, a computer automatically maintains that temperature by controlling the path taken by the ventilating air as it flows to the interior of the car. Under hot or humid ambient conditions, the air is cooled or dehumidified respectively by ducting it through an evaporator, whereas at times when the ambient temperature is low the air is heated.

One annoying driving problem is the misting up of the windows that occurs on 'damp' days. The cold glass condenses the moisture in the air in the car and a dangerous loss of visibility often results. A modern climate control system, which is sometimes called an *environmental control system*, overcomes the problem by passing the ventilating air through the evaporator to remove the excess moisture before heating it to the required temperature.

Filtering the air to remove traffic pollutants before it enters the interior is another feature of some modern climate control systems.

Fig. 55.13 shows the layout of a *full air-mix* type of system suitable for automatic temperature control. In this type the temperature and air flow pattern are controlled by three flap valves. Movement of the

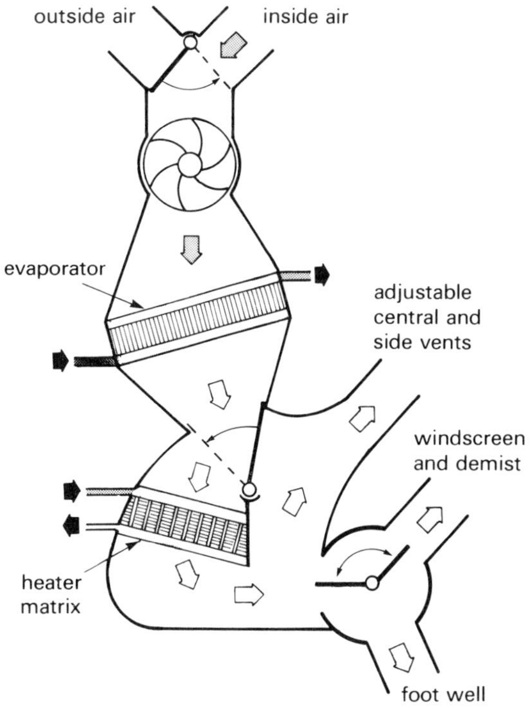

Fig. 55.13 Climate control system

valves is achieved by using either electric motors or vacuum cylinders; the latter arrangement is controlled by small valves opened by electric solenoids.

Control systems

For fascia layout and ease of operation, electrical switches are provided to enable the driver to select the

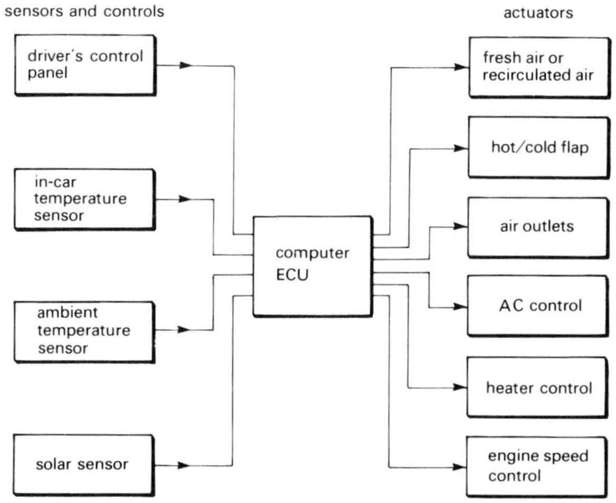

Fig. 55.14 Climate control system

required temperature and distribution of the air that enters the car interior. After the target temperature, which is often displayed in digital form, has been set by the driver, the computer controls the heating and cooling sections of the system so that the air inside the car is kept at a temperature very close to the set value. The in-car temperature is constantly monitored by an electrical sensor, and by signalling this information to the computer it is possible for the computer to make the necessary temperature adjustments.

Natural heating of the car is either by the direct rays of the sun or by the heat contained in the incoming air. Since the amount of heat entering the car governs the amount of cooling or heating required, sensors are fitted to measure this input.

Fig. 55.14 shows the main items of an automatic electrical control system.

PART SIXTEEN

Suspension systems

56 Types and characteristics of metal springs

Purpose of a spring

Mounting the axle directly to the frame of a vehicle would subject the occupants and general components to severe shocks. This can be seen from Fig. 56.1a, which shows the upward movement of the frame when the wheel strikes a bump. In this case the vertical acceleration would cause considerable discomfort: most probably the reluctance of the vehicle frame to move upwards quickly (inertia) would buckle the wheel. (You will have observed this effect when riding a bicycle over a bump.)

A spring fitted between the wheel and the frame allows the wheel to move up and down without causing similar movement of the frame (Fig. 56.1b). This spring absorbs road shocks, and allows the wheel to follow the irreglar contour of the road surface: for this purpose the wheel assembly should be as light as possible.

Road shocks can be further reduced by fitting a 'spring' to the wheel - a pneumatic tyre.

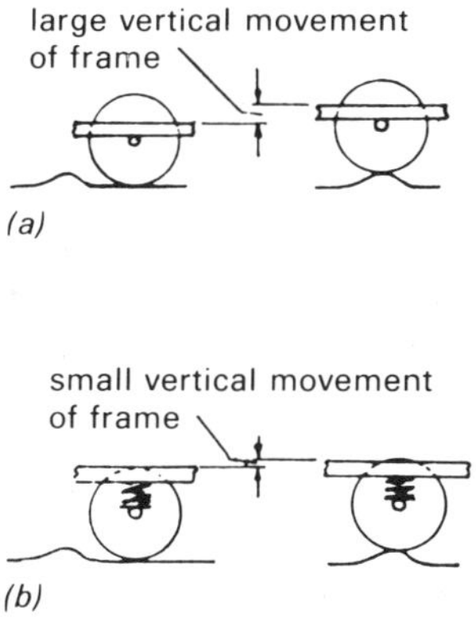

Fig. 56.1 Effect of fitting a spring

The mechanism introduced between the wheel and the frame is termed the suspension system, the main items being the spring and damper.

Types of spring

Various forms of springing can be used. These are:
1 Steel - laminated, helical and torsion bar.
2 Rubber.
3 Pneumatic.

Laminated or leaf springs

Low cost and simple connection to axle make this type very popular for rear suspension.

The main details of a semi-elliptic spring are shown in Fig. 56.2. A main leaf, rolled at each end to form an eye, has a number of leaves clamped to it. To ensure a constant stress throughout the spring the leaves are graduated in length. Rebound clips transmit the load to some of the lower leaves during the return motion of the spring, and eliminate the need for fitting a large number of leaves above the main plate. Rubber bushes, fitted in each eye, allow for movement of the spring, and act as noise insulators. Alteration in spring length is accommodated by a swinging shackle.

The stiffness or rate [force required to deflect the spring 1 m (1 in)] of a leaf spring is governed by:
1 Length of spring - shorter spring, higher rate.
2 Width of leaf - wider spring, higher rate.
3 Thickness of leaf - thicker leaf, higher rate.
4 Number of leaves - greater number, higher rate.

To obtain a 'soft' ride, a low-rate spring is required, and this will deflect a larger amount under a given load. Normal springs have a constant rate, and give a deflection which is proportional to the load (Hooke's

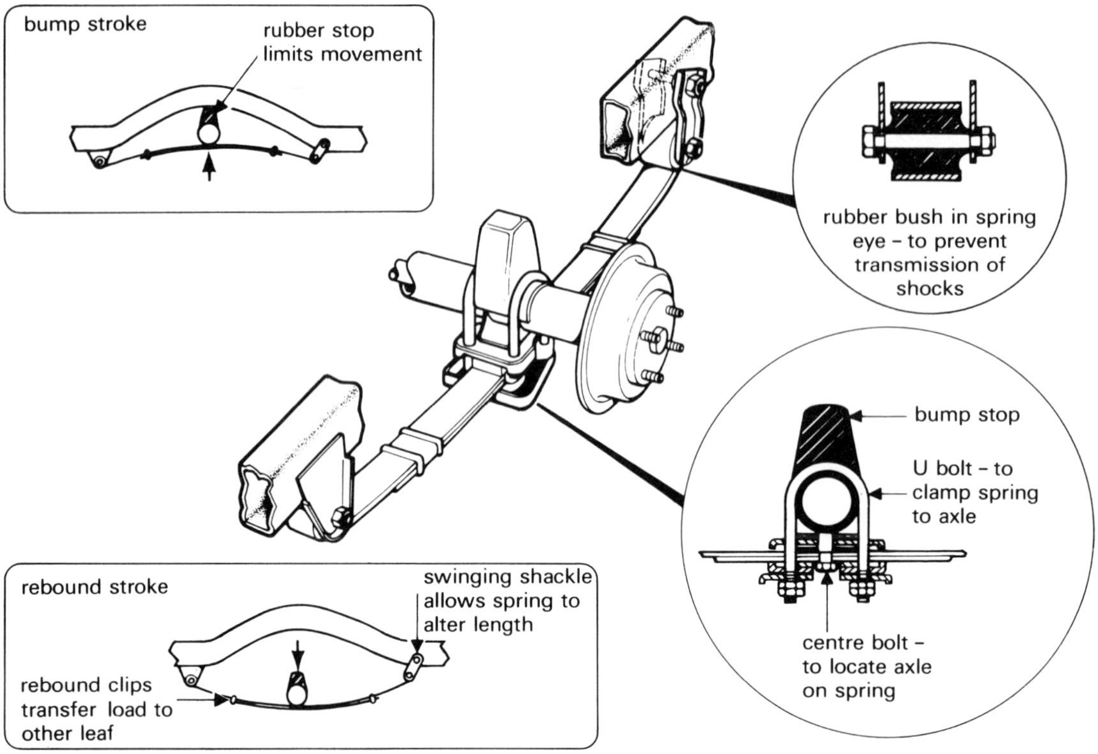

Fig. 56.2 Laminated or leaf spring

Law). However, if the lower leaves are set to a reverse camber, a stiffening-up of the spring will occur as deflection increases: this is called a *progressive* or *variable-rate* spring.

As the laminated spring deflects, the plates or leaves slide over each other and cause inter-plate friction. Although this has a beneficial damping effect, the 'hard' ride, noise and wear make it necessary to reduce this friction as much as possible. Old designs of spring had to be sprayed with penetrating oil, but nowadays special features are incorporated to eliminate the need for periodic attention. These are:

1 Synthetic rubber buttons fitted at the ends of the leaves.
2 Reducing the number of leaves (as the number is reduced, the width must be increased).
3 Inter-leaf plates of low friction material.

To gain the full advantage of item (2) many springs in use today have only one leaf. Overstressing at the centre of the spring is avoided by using a tapered leaf; thin at the ends and thick in the centre.

Helical spring

This spring is often used in conjunction with independent suspension, although the absence of inter-plate friction has prompted some manufacturers to use it at the rear (Fig. 56.3). Coil and torsion bar springs are superior to the leaf springs as regards energy storage (energy stored in a given weight of spring), but whereas the leaf spring fulfils many duties, the other types require extra members and these add to the basic weight.

The rate of the spring is governed by the length and diameter of the wire. The wire is wound in the form of a coil, so the length will be controlled by the diameter of the coil and the number of active coils.

Helical springs are commonly used to support the dead axle fitted at the rear of a front-wheel drive car. Fig. 56.4 shows a typical layout in which the spring acts directly on to the axle. Axle location at the top and bottom is provided by the top mounting of the damper (shock absorber) and the trailing arm. Besides providing fore-and-aft stiffness, the lower arm in conjunction with the damper tube absorbs the braking torque.

Lateral movement of the axle is resisted by a steel Panhard rod. The rod has a rubber bush at each end where it is bolted to the axle and body.

On high-performance cars, a rear stabilizer bar is often fitted; this reduces roll when the car is cornering.

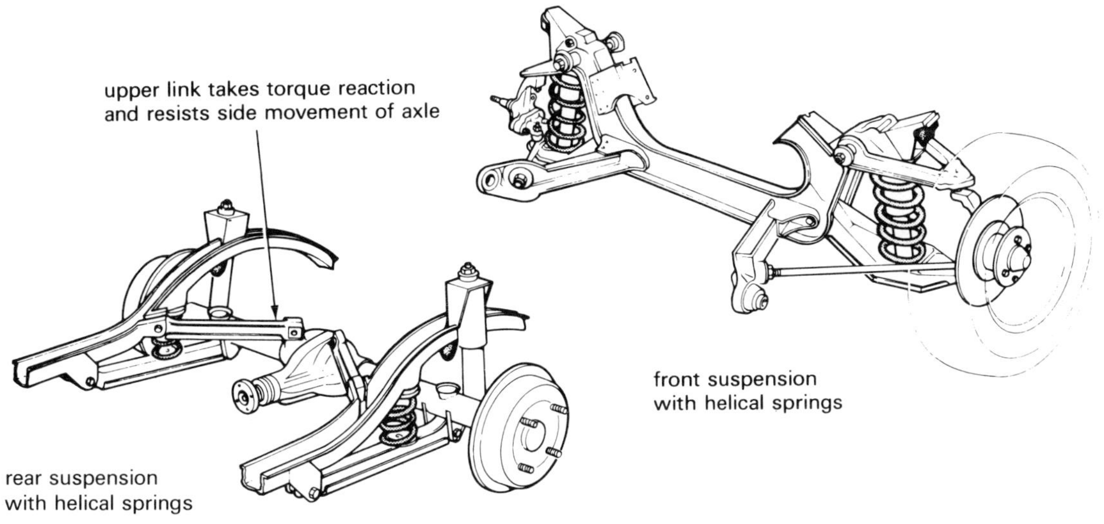

Fig. 56.3 Helical spring suspension

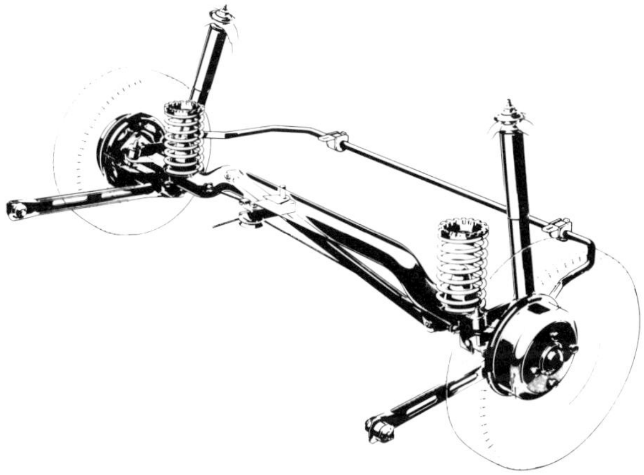

Fig. 56.4 Rear suspension for a dead axle

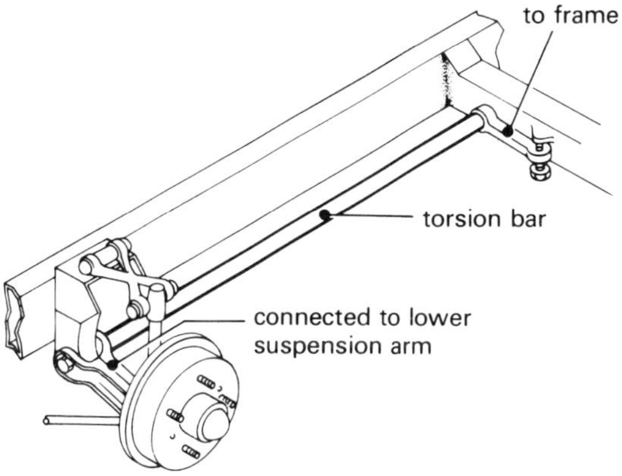

Fig. 56.5 Torsion bar springing

Torsion bar

This is a straight bar of circular or square section anchored to the frame at one end, and at the other end connected by a lever or wishbone-shaped member to the wheel.

Fig. 56.5 shows a torsion bar suspension system used on a car. At each end of the bar the diameter is increased and serrations connect the bar with the levers. Adjustment is provided at the frame end to 'level' the suspension.

Since the coil spring is a form of torsion bar, the rate of both springs is governed by the same factors – length and diameter. If the length is increased or the diameter is decreased, the rate will decrease, i.e. the spring will be softer.

Spring material

The material used for springs must be capable of withstanding high stresses and resisting fatigue. Early designs used high-carbon steel, but nowadays low-percentage alloy steels such as silico-manganese are used.

57 Rubber suspension

Much weight can be saved with this form of suspension, because rubber can store more energy per unit mass than any other type of spring material.

Rubber springs, loaded in compression or shear, may be used to act as the main suspension spring, or fitted in conjunction with metal springs to modify the suspension characteristics. Many suspension arrangements employ a large rubber 'bump' stop to stiffen the suspension spring at maximum deflection.

Fig. 57.1 is a simplified drawing of a rubber suspension system which is similar to the type used on a popular small car. The spring is positioned between the frame and the top link of the suspension system. By connecting the spring to a point near the link pivot, deflection of the spring can be reduced to a minimum, without reducing the total wheel movement.

This design of spring gives a *rising-rate* characteristic, i.e. it is 'soft' for small wheel movements but becomes harder as the spring deflects.

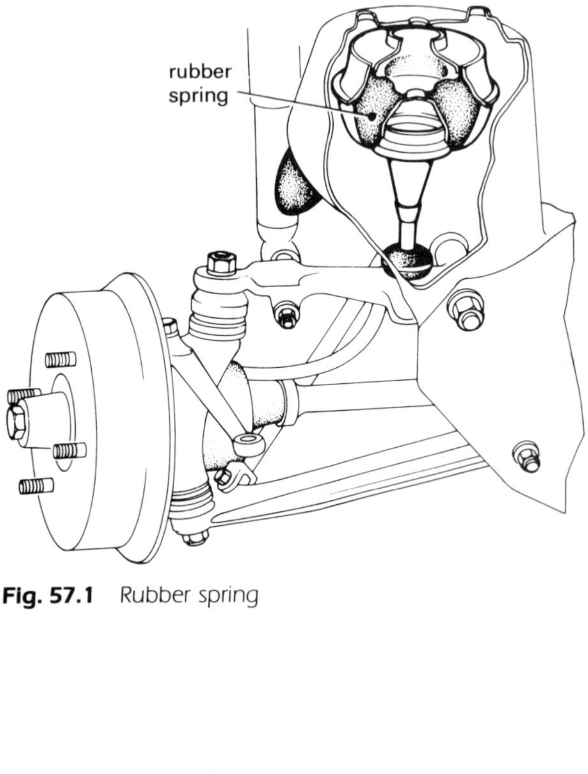

Fig. 57.1 Rubber spring

The energy released from the spring after deflection is found to be considerably less than that imparted to it. This internal loss of energy is termed *hysteresis*; it is an advantage, since lower-duty dampers may be used.

Some rubber suspension systems have a tendency to 'settle down' or 'creep' during the initial stages of service, and allowance must be made for this.

Hydrolastic suspension

This system is a development of the previous suspension arrangement and is intended to improve the vehicle's resistance to pitch – the tendency of the body to oscillate in a fore-and-aft direction. This movement is produced if the front springs compress and the rear springs extend simultaneously. The continuous forward and backward pitching motion gives a most uncomfortable ride and this would be serious if the frequency of vibration of front and rear springs were the same.

The Hydrolastic layout on a vehicle consists of rubber displacer units (Fig. 57.2) which are interconnected and mounted between the frame and the independent suspension linkage controlling the wheel. The interconnection is made with two pipes; one to link the left-hand side units together and the other pipe to do a similar job on the right-hand side. The system is pressurized with a liquid (water + alcohol + anti-corrosive agent) after the air has been extracted.

Each displacer unit comprises a rubber spring, metal separating member holding two rubber damper valves, rubber diaphragm attached to the suspension linkage holding the wheel, and a metal body which is secured to the frame of the vehicle.

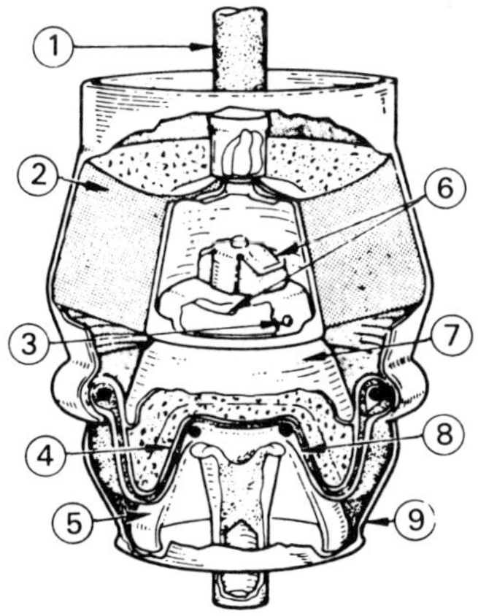

Fig. 57.2 Hydrolastic displacer unit

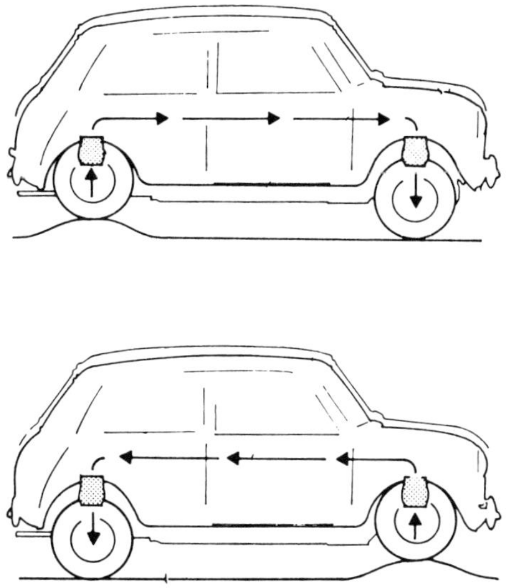

Fig. 57.3 Action of Hydrolastic units

Road irregularities normally tend to cause the vehicle to pitch, roll and bounce, so the operation of the system under these conditions will be considered.

Pitch A sudden upward movement of the front wheel causes the diaphragm to displace the fluid through the damper. This action will in turn force fluid along the pipe to the rear unit where it will move the diaphragm and raise the rear of the car to the level of the front (Fig. 57.3). When the front wheel descends, the fluid is returned and the vehicle settles back in its normal riding position. During this sequence the fluid has to pass the damper valve in each unit, so restriction to fluid flow at the valves and in the pipeline damps out the pitch oscillation tendency.

Roll When a vehicle is cornering, centrifugal action causes the body of the vehicle to tilt or roll outwards and this action is apparent when 'soft' conventional springs are used. The hydrolastic system is 'soft' when a single wheel is moved, but if, as when cornering, the two outside suspension units are loaded, a stiffening of the system occurs. Under this type of loading fluid is not displaced from one unit to the other. Instead the increased fluid pressure deflects the rubber springs and these provide a marked resistance to the roll of the body.

Bounce This condition causes the four wheels to deflect at the same time, so all the hydrolastic units will perform similarly to the outer units when they were reacting to roll.

58 Air suspension

Hydropneumatic suspension

This system differs from the normal metal/spring suspension in a number of ways. One main difference is that the suspension unit is supported by a mass of gas which remains constant irrespective of the load carried by the wheel. Gas pressure increases progressively as the volume is reduced and this desirable feature means that the suspension stiffens as the load on the wheel increases. Fig. 58.1 shows the basic principle of a hydropneumatic system. Liquid is used to transmit the force from the suspension piston to nitrogen gas in a closed container.

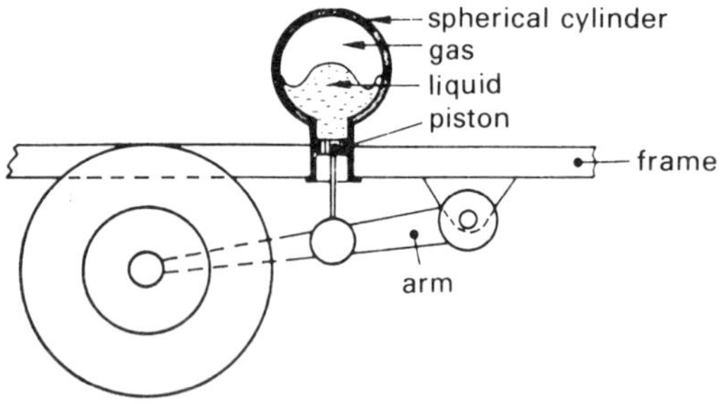

Fig. 58.1 Hydropneumatic suspension system

Hydragas suspension

This system has been developed from the Hydrolastic type of suspension, the main difference being that the rubber spring is replaced by a pneumatic spring.

Fig. 58.2 shows the layout of one of the four Hydragas units fitted to a vehicle. The units are interconnected in pairs by a fluid pipeline which links the front unit with the same side rear unit. This line equalizes the liquid fluid pressure in the two units to reduce the fore and aft pitching motion which is most noticeable on short-wheelbase cars.

The unit consists of three main parts; a nitrogen gas spring, a fluid displacer and a damper valve block.

Bump movement of the wheel suspension arm deflects the diaphragm and compresses the fluid. If the pressure in the other displacer unit is similar, then the fluid will flow through the damper, the gas will be compressed and the resistance to the suspension arm will increase progressively. On rebound the fluid will flow in the opposite direction. Energy which causes the spring to oscillate is absorbed as the fluid is passed through the damper.

The action of the interconnecting fluid lines in relation to pitch, roll and bounce is similar to that in the Hydrolastic system.

Citroën system

This pneumatic spring system has a hydraulic control which not only allows the driver to adjust the ground clearance of the vehicle but also maintains this set clearance irrespective of the load carried.

Fig. 58.3 shows the layout of a system similar to that used by Citroën. Each wheel is mounted on a suspension arm which is supported by a pneumatic spring. Connected transversely, between the suspension arms at the front and rear, are anti-roll bars, and these are linked to height correctors by means of control rods. An engine-driven pump supplies oil under pressure to a hydraulic accumulator and this is connected to the height control or levelling valves.

As the vehicle is loaded the downward movement of the vehicle structure causes rotation of the anti-roll bar. This moves the slide valve and uncovers the supply port to allow oil to flow from the accumulator to the suspension cylinders. When the vehicle reaches a predetermined height, which can be varied by a selector inside the car, the anti-roll bar and control rod move the slide valve back to the 'neutral' position. A decrease in load gives a similar sequence except that

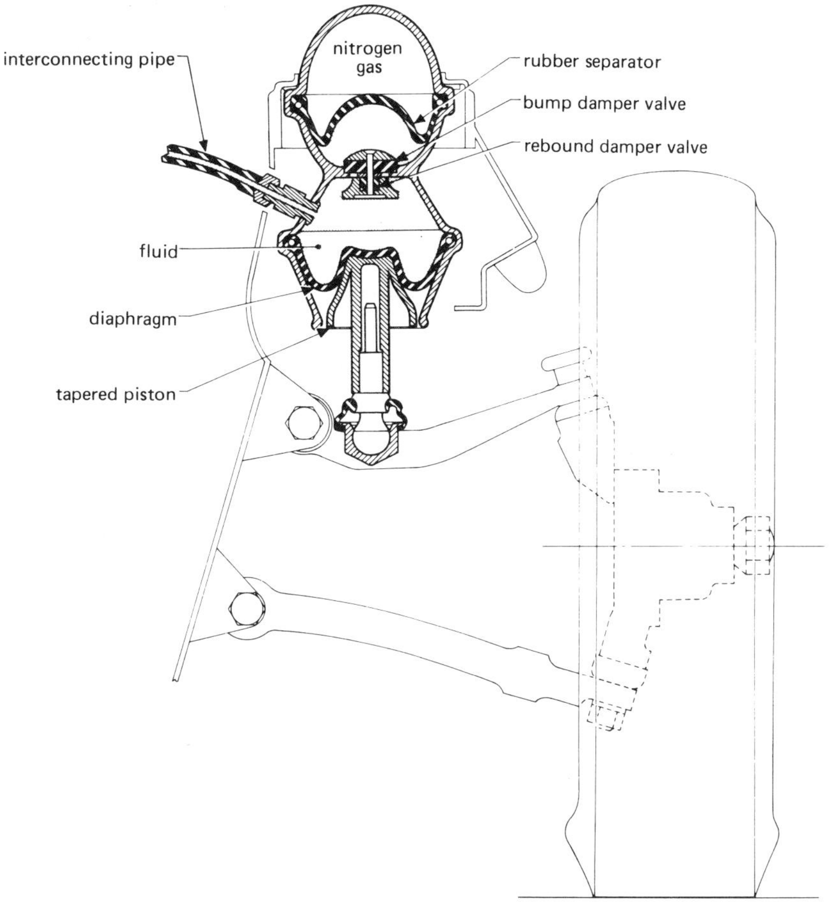

Fig. 58.2 Hydragas suspension

valve movement causes the oil in the suspension cylinder to be discharged back to the reservoir.

A delay device or dashpot is incorporated in the height corrector valve to prevent rapid oil flow past the valve when the wheel contacts a bump or dip in the road. Without this dashpot, sudden movements of the wheel would mean that the valve is continually working and this would give an unsatisfactory operation.

Damping is provided by partially separating the oil in the spherical chamber from the cylinder in which the piston slides. Small holes, closed by disc-type valves, allow the oil to flow to and from cylinder and sphere in a manner similar to other hydraulic dampers.

In the Citroën application the hydraulic pressure is also used for power assistance in steering, braking and gear changing. One model employs a braking

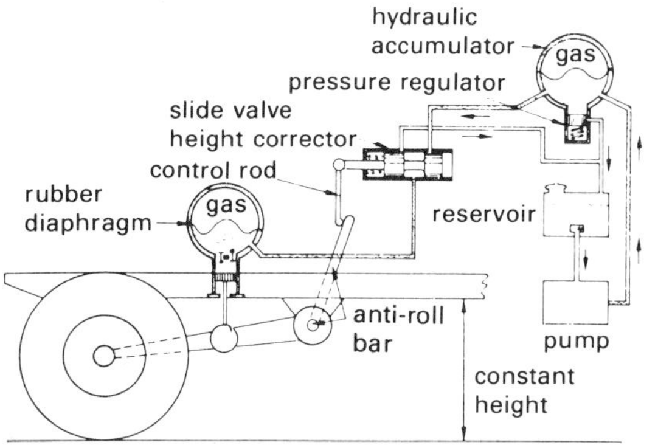

Fig. 58.3 Layout of hydropneumatic suspension system

system with independent front and rear brake circuits, interconnected with the hydraulic suspension system. This arrangement proportions the effort applied at each brake to the load carried by the wheel.

59 The damper

Purpose of a damper

When the wheel strikes a bump, energy is given to the spring, which is deflected. When the bump is passed, rebound or release of the stored energy will take place, and will carry the spring past the normal position to set up an oscillating motion. This action is similar to the movement of a pendulum - a freely suspended pendulum will oscillate for a considerable time after being struck (Fig. 59.1). In order to give a comfortable ride, some device must be fitted to absorb the energy stored in the spring and so reduce the number of oscillations occurring between the initial bump and the return of the spring to the rest position. This is the duty performed by the damper (often misleadingly called a shock absorber).

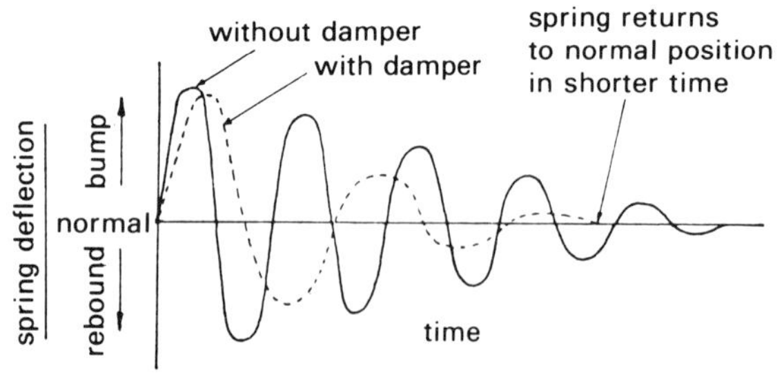

Fig. 59.1 Spring oscillation curves

Early designs of damper utilized the friction between two sets of plates, one set attached to the frame and the other set connected to the axle. This type converts the 'spring' energy to heat.

Hydraulic dampers

Hydraulic dampers are the main type in use today. They dissipate the energy by pumping oil through small orifices. Resistance of the hydraulic damper increases as the speed of spring deflection increases, whereas the friction damper gives a constant resistance. The resistance to spring movement can apply to the rebound stroke only (single-acting), or to both the bump and rebound strokes (double-acting), or it can offer a differential action by resisting both strokes but exerting a greater action on the rebound.

The two main types of hydraulic damper are: (1) lever and (2) direct acting (telescopic).

Lever type

Fig. 59.2 shows a diagrammatic view of a lever-type damper which is mounted to the frame and connected by a lever and link to the axle. The horizontal cylinder contains two pistons which are fitted with recuperator and pressure valves. A thin, mineral-base, damper oil is introduced to the level of the bottom of the filler plug.

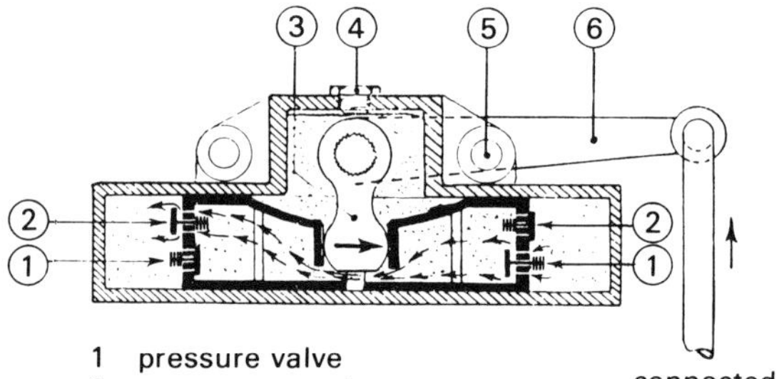

1 pressure valve
2 recuperator valve
3 rocker
4 filler plug
5 frame attachment point
6 lever arm

Fig. 59.2 Lever-type damper

Bump movement of the axle operates the damper pistons and displaces oil from one chamber to the other. Oil exerts pressure to open the pressure valve, flows through an orifice to provide resistance, and passes through the open recuperator valve to break down the depression created in the other chamber.

Rebound of the spring produces a similar action in the opposite direction.

The actual damper differs from Fig. 59.2 in various ways - the valve construction and position are different, and in some designs the cylinders are mounted vertically.

Telescopic direct-acting type

As the name suggests, this type of damper is directly connected between the body and the suspension member that moves with the road wheel.

It is made either as a separate bolt-on unit or as an integral part of a suspension system. Direct-acting dampers are made in two basic forms: single-tube and twin-tube.

Single-tube (Fig. 59.3) The operating cylinder is a single tube, closed at the bottom end by a cap, which is attached to an eye or stem for connection to the moving part of the suspension. In this tube slides a piston fitted with two-way reed valves which cover a series of holes through which oil can pass. The piston is attached to a rod that passes through a guide retained in the top of the tube. The rod is attached to the car body and rubber bushes at the attachment points isolate road shocks and allow for slight angular movement of the damper body. A rubber seal, held in place by fluid pressure, is positioned adjacent to the guide to prevent the escape of fluid. An abutment plate fitted to the piston to limit the stroke provides a flat surface for contact with the top seal when the damper is fully extended. A tubular shield, attached to the rod member, is fitted to protect the rod from dirt and stone damage.

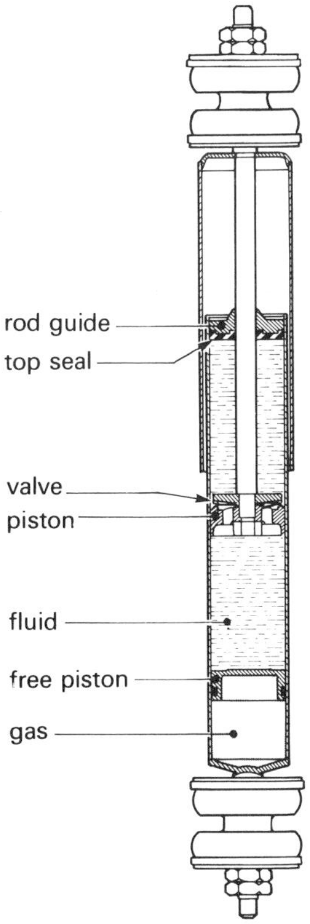

Fig. 59.3 Single-tube telescopic damper

A chamber at the base of the damper, sealed by a free piston, contains a quantity of inert gas. When the damper is completely filled with thin-type mineral oil, this gas is in compressed state.

On the bump stroke the inner movement of the piston displaces oil from the bottom to the top chamber. The energy needed to pump the oil through the drillings and deflect the piston valves provides the damping action. By varying the size of the 'bump' and 'rebound' orifices, the resistance of each stroke can be altered to suit the requirement.

For a given downward movement of the piston, the volume available for the oil in the upper chamber is less than that required by the oil, so when this action occurs the free piston moves slightly downwards.

The single-tube construction has a number of merits: it can displace a large quantity of fluid without fluid aeration or noise and be consistent in service even when installed at a large angle to the main suspension movement. Denting of the single working tube by stones must be avoided, so this is a good reason for placing the damper at the centre of the helical suspension spring.

The potential energy stored in a deflected suspension spring is converted into heat by a damper, so with this design good heat dissipation from the damper to the air flow is achieved.

On MacPherson suspension systems (see page 369), a single-tube, gas-pressurized damper is used as a main suspension 'leg'. In this application a more robust rod guide is fitted to enable it to act as a top bearing to accommodate steering movement.

Twin-tube (Fig. 59.4) An extra tube is used on this type to form both a reservoir for the fluid and an overflow region to accommodate fluid due to rod displacement and expansion.

The base valve at the bottom of the working chamber has three duties to perform: it controls the outward flow of fluid into the reservoir during the

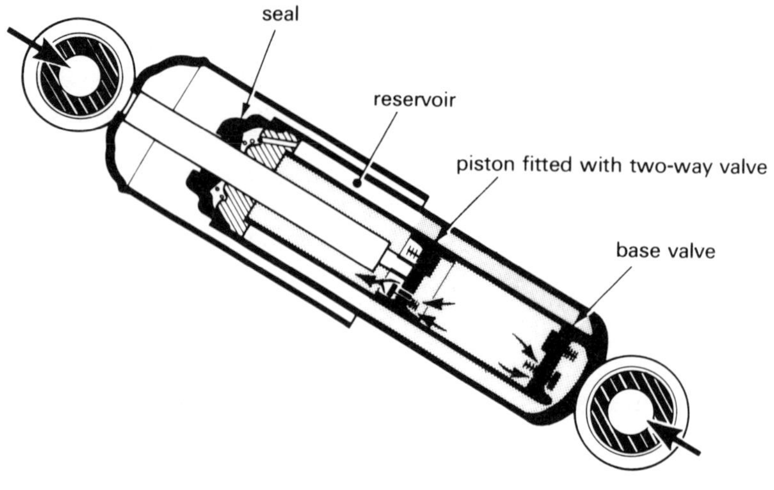

Fig. 59.4 Twin-tube telescopic damper

'bump' movement, the fluid return on the 'rebound' stroke, and the bleed leakage for reducing the damper's resistance during slow-speed suspension movements.

The life of a direct-acting damper depends on its fluid-sealing ability, so great attention is paid to the design of bearing and seal. The hard chrome-surfaced piston rod works in a sintered iron guide bearing, which is lubricated by a small amount of fluid forced through the bearing on each stroke. A multi-lipped rubber seal placed on the outside of the bearing cuts the fluid off the rod and returns it to the reservoir.

Baffles in the reservoir prevent violent movement of the fluid which would otherwise cause aeration. Any fluid in an aerated state that is allowed to enter the working chamber would render the damper ineffective, so for extra protection a fluid containing special anti-foam additives is used.

The long stroke of a direct-acting damper allows the working pressure to be much lower than the lever-type damper. In view of this, the direct-acting type is more reliable, and is also cheaper to manufacture.

Although the best performance is achieved when the dampers are mounted vertically, many suspension layouts in use today, especially those used to support a live rear axle, have the dampers set 'diagonally' to improve the stability of the suspension.

Dampers do not offer any resistance to slow-speed body roll, because the inherent bleed feature in the damper allows this movement to take place. High-speed roll is resisted, but manufacturers generally fit a transverse stabilizer, often called an anti-roll bar, to control the roll motion.

Although sophisticated equipment for oscillating the dampers to test their operation is available, most manufacturers recommend the simple test carried out by pressing down with your hand on the corner of the car. By noting the number of oscillations made by the spring for the body to come to rest, the effectiveness of the dampers can be ascertained.

60 Independent front suspension (IFS)

Disadvantages of beam axles

In order to appreciate the advantages of independent suspension, you should consider the disadvantages of the beam axle. These are:

1. Small maximum spring deflection, therefore 'hard' springing – vertical axle movement is limited by the clearance between the axle and the engine.
2. Steering geometry is not accurately controlled.
   a Fig. 60.1a shows the alteration to camber angle which occurs when one wheel strikes a bump. Sudden changes in camber angle cause the wheels, which are acting as a form of gyroscope, to 'flap' around the king pin. (This fault is known as *wheel shimmy*.)
   b Fig. 60.1b shows the difference in castor angle when the spring is deflected.
3. High unsprung weight – maximum wheel adhesion is not obtained.
4. The engine normally has to be situated behind the axle to give clearance. If the engine can be moved forward, it may be possible to accommodate all the passengers within the wheelbase – this gives greater comfort.
5. Poor 'roll stiffness' at the front tends to produce *oversteer* – the front springs have to be mounted close together.

To overcome these disadvantages, independent front suspension (IFS) is employed. This term is used to describe any system for connecting the wheels to the frame in which the movement of one wheel has no effect on the other wheel.

Types

Many types of IFS have been used in the past, but nowadays, in the UK, most designs fall under the heading of transverse link system.

Transverse link system

Wishbone type

The main details of this system are shown in Fig. 60.2. Two links, often parallel in the normal ride position, are made in a wishbone shape to provide fore-and-aft stiffness and resist braking torque. Each wishbone has three bearings, two inner bearings connecting with the frame and an outer one attaching to the stub

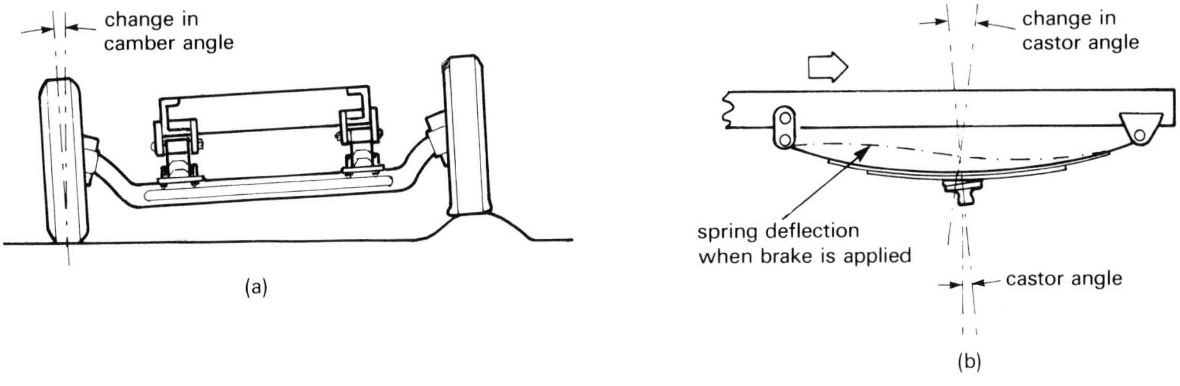

Fig. 60.1 Steering geometry alterations

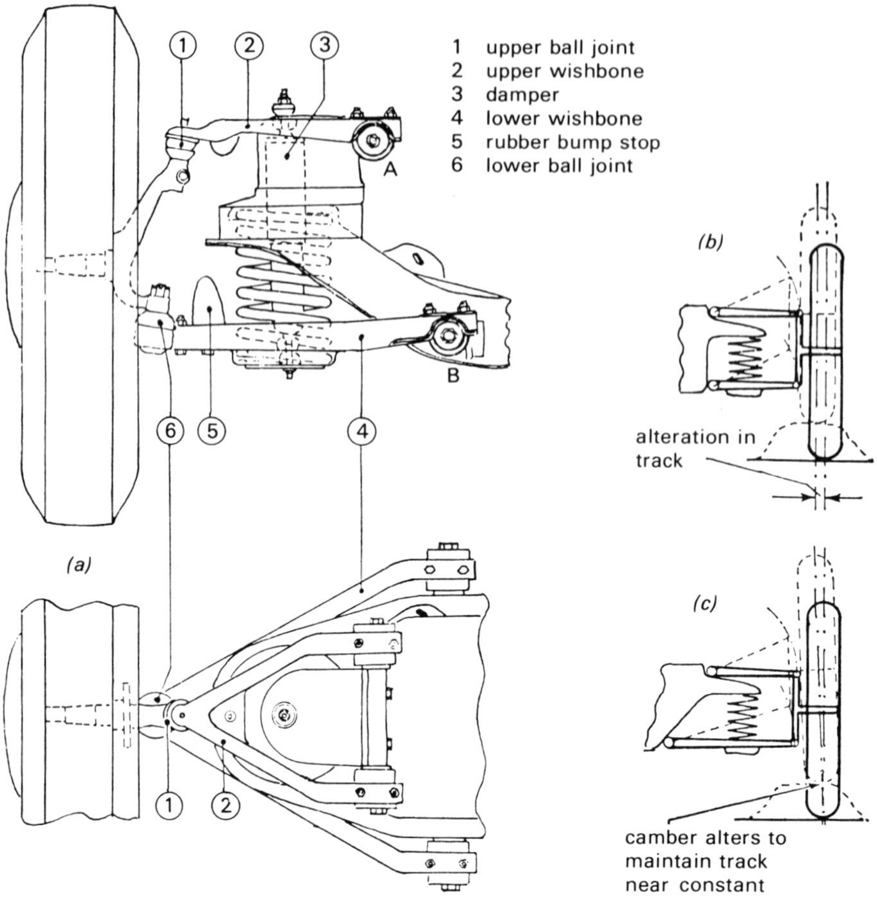

Fig. 60.2 Transverse link system – wishbone type

wishbone, and the upper end to a point on the frame just above the upper wishbone.

Since the lower wishbone pivot points on the link-type system are normally set parallel to the road, the front of the car 'dives' towards the ground when the brakes are applied. This problem can be minimized by using *anti-dive geometry*; this is achieved by placing the rear pivot point of the lower wishbone higher than the front pivot.

When the front brakes are applied, the action of the braking torque on the inclined wishbone generates a vertical force that counterbalances the extra load transferred from the rear to front wheels.

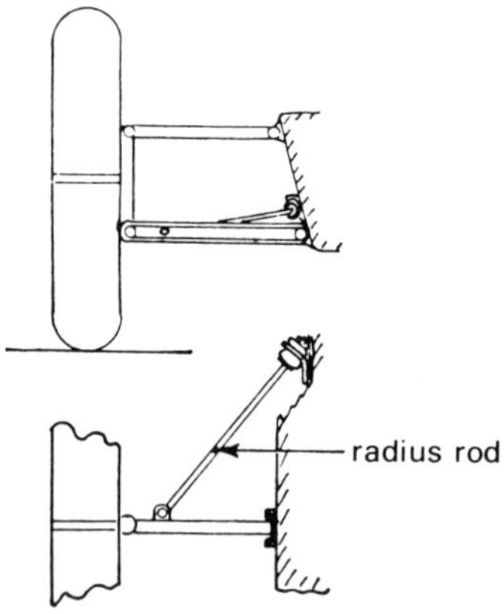

Fig. 60.3 Double link and radius rod

Double link and radius rod (Fig. 60.3)

Two links, mounted in the normal manner, connect the stub axle carrier to the frame. A semi-trailing radius rod, fitted between the lower link and the frame, resists longitudinal dynamic loads and braking torque. The spring can be positioned above the top axle carrier. Rubber or plastic (PTFE) bushes are fitted at the inner ends of the wishbone, and in many cases a ball joint at the outer end enables the stub axle to swivel. Springing can be provided by using coil springs in the position shown or above the upper wishbone, or a torsion bar at points A or B.

Early designs used wishbones of equal length, but the track variation (Fig. 60.2b) caused considerable tyre wear. This can be reduced by using wishbones of unequal length, the longer one at the bottom (Fig. 60.2c), but changes in camber angle will now have to be tolerated.

A constant castor angle is achieved with this design by mounting the top wishbone slightly behind the lower one.

In order to obtain the maximum wishbone length without restricting engine space, the wishbone axis is sometimes inclined towards the rear.

Either a piston-type damper is mounted to incorporate the inner bearings for the top wishbone, or a telescopic damper is fitted in the centre of the coil spring. In cases where torsion bar springing is used, a large movement of the telescopic damper can be achieved by mounting the damper diagonally: the lower end is connected to the outer end of the lower link, or a torsion bar can be connected to the inner ends of the lower link.

MacPherson type

Fig. 60.4 shows the main details of this type of suspension. A long telescopic tube, incorporating the damper, is pivoted at the top end and rigidly connected to the stub axle at the lower end. Track control is maintained by a single transverse link, attached to the frame by rubber bushes and connected to the stub axle by a ball joint. The coil spring is located between the fixed and floating suspension members. Both front suspension lower links are interconnected by a stabilizer bar, which also provides the required fore-and-aft stiffness.

As in many other suspension systems, castor, camber and swivel-axis inclination are set in production and cannot be altered.

The swivel-axis inclination is the angle formed between the vertical and the line taken from the centre of the strut thrust bearing to the centre of the ball joint which connects the strut to the track control arm. For tyre clearance purposes the strut is set to a smaller angle than the swivel axis inclination.

Setting the top of the strut towards the vehicle centre allows the manufacturer to obtain the advantages of a negative offset (negative scrub radius) for the steering.

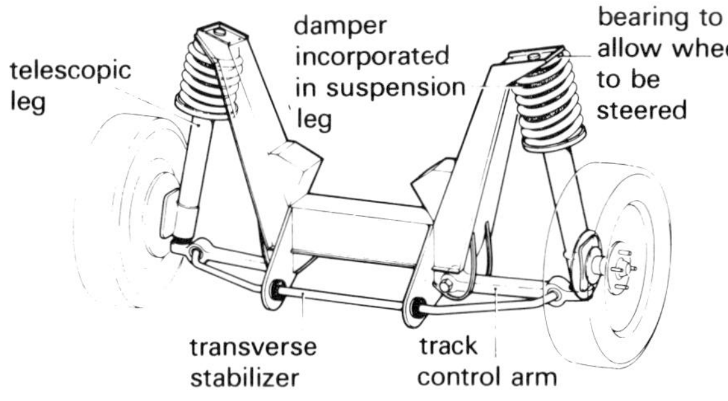

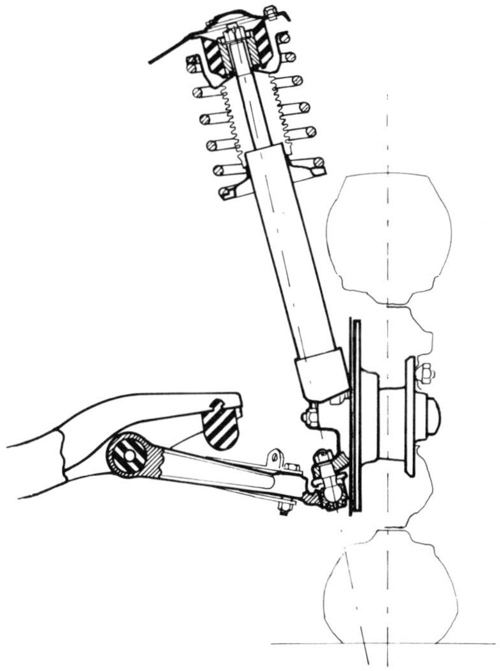

Fig. 60.4 MacPherson type

Swinging arm

Two leading or trailing arms of equal length, i.e. longitudinally mounted links, are connected between the frame and the stub axle carrier. The spring can be mounted above the top arm, or a torsion bar may be connected to the arm at the frame location point (Fig. 60.5).

Track-rod linkages

The one-piece track rod used with beam axle layouts is unsuitable for the majority of IFS systems, because wheel alignment would be altered if one wheel was deflected up or down. The stub axle carrier in many IFS systems moves in an arc, so in order to eliminate misalignment, the track rod connection must move in a similar arc. This condition can be met by using a

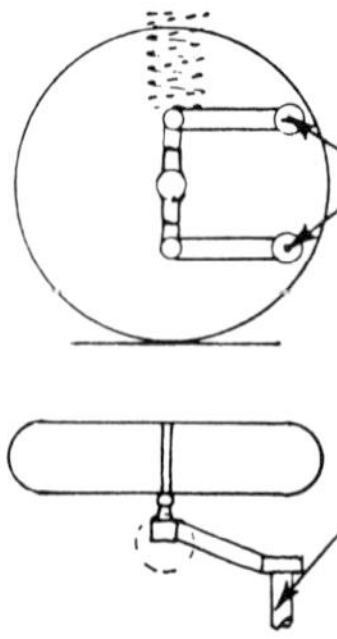

Fig. 60.5 Swinging arm

three-piece track rod similar to the types shown in Fig. 60.6.

Adjustment of the rack and pinion system is provided on both outer track rods. When resetting the wheel alignment it is necessary to adjust each rod an equal amount.

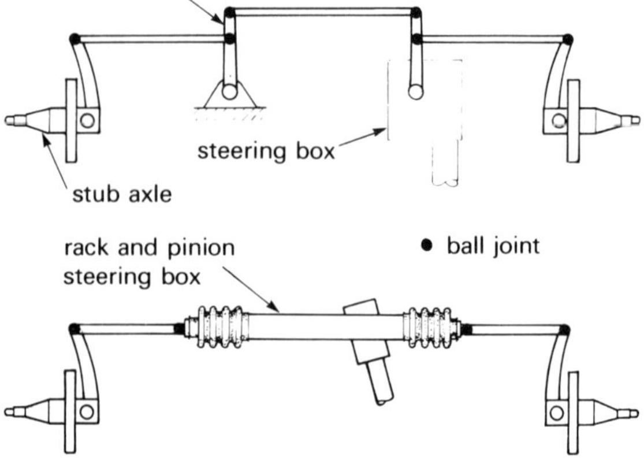

Fig. 60.6 Track-rod linkage for IFS

Unequal adjustment of the track rods results in:

1 Incorrect spoke position of the steering wheel.
2 One steering lock being greater than the other.

61 Independent rear suspension (IRS)

Many of the advantages of IFS apply to IRS, but the most important item is the reduction of unsprung weight. The final-drive unit and the brakes are the major weight items, so if 'inboard' brakes are fitted to a frame-mounted final drive, as much as 50 per cent reduction in unsprung weight can be achieved.

The similarity between IRS and the de-Dion drive often makes it difficult to draw a dividing line, but if the definition of independent suspension is remembered, a division can be made.

The following systems are examples of IRS.

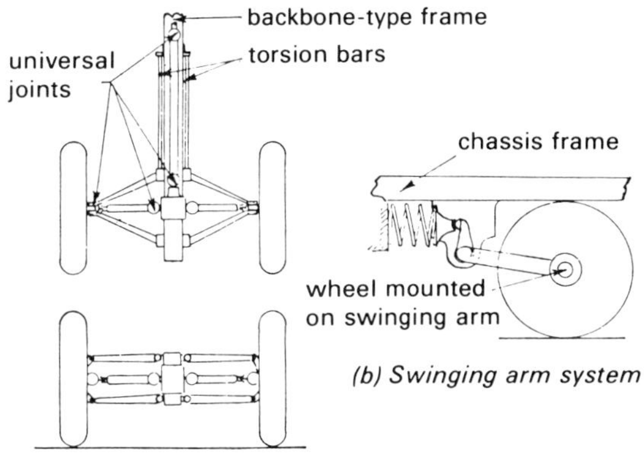

Fig. 61.1 Independent rear suspension

Parallel link system

Two wishbone-shaped links, mounted transversely, connect the wheels to a backbone-type frame (Fig. 61.1a). Springing is provided by longitudinally mounted torsion bars which connect with the lower wishbone. The drive is transmitted from the final drive through 'wide-angle' universal joints.

Swinging arm

An alternative method of mounting the wheels is provided by the trailing arm system shown in Fig. 61.1b. A spring, mounted as shown, or a torsion bar, acting at the pivot, may be used. One popular light car uses a rubber spring instead of the normal metal type shown.

Swinging half-axles

This system consists of two axle tubes which are jointed to the final-drive housing to allow the wheel to rise or fall (Fig. 61.2). To allow for the change in drive angle, universal joints are fitted at the centre of each axle joint.

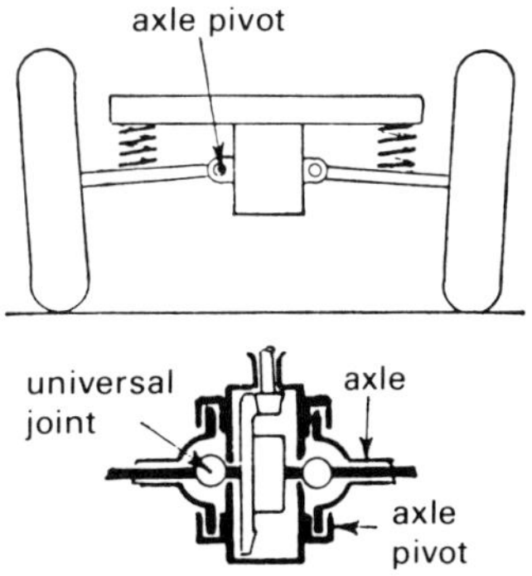

Fig. 61.2 Swinging half-axles

Transverse link and coil springs

An arrangement similar to that shown in Fig. 61.3 is used on Jaguar cars. The road wheel is located in the transverse plane by a tubular suspension link at the bottom and a drive shaft at the top. Non-plunge universal joints are fitted at each end of the drive shaft to allow for shaft movement.

Longitudinal stiffness and driving thrust duties are performed by a longitudinally mounted radius arm

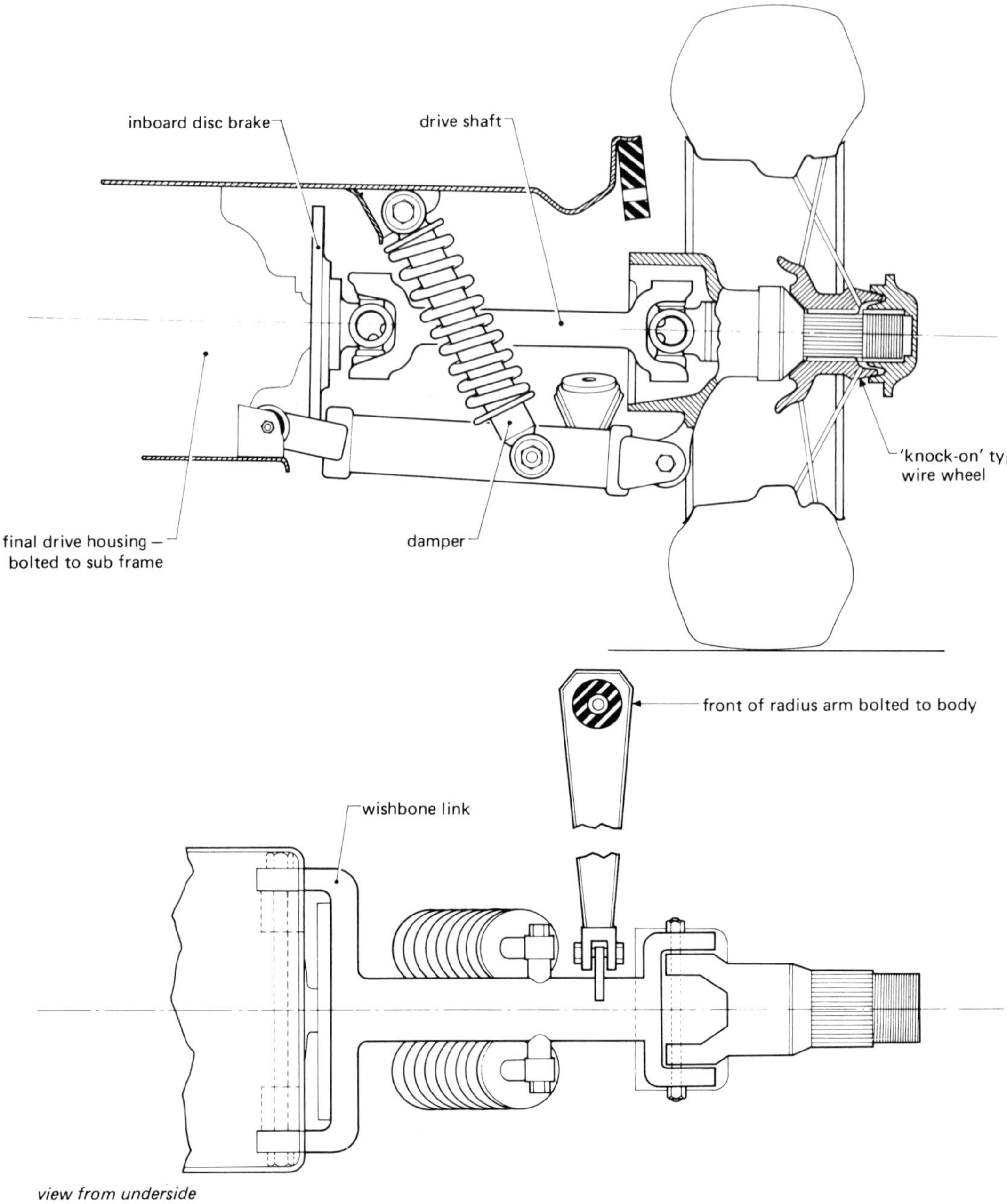

Fig. 61.3 Transverse link and coil springs

which connects the wheel end of the link to the vehicle body.

Torque reaction caused by driving the wheels and torque from the inboard disc brakes are both taken by the final drive housing. This housing is bolted to a sub-frame on to which the complete rear suspension is built.

As the two 'links' are of different length, a variation occurs in the rear wheel camber angle when the springs are deflected. Camber angle is adjusted by shims fitted between the brake disc and the universal joint flange.

Semi-trailing arm and coil spring

This system supports the rear hub by a wishbone-shaped suspension arm mounted diagonally on to a sub-frame which also carries the final drive housing.

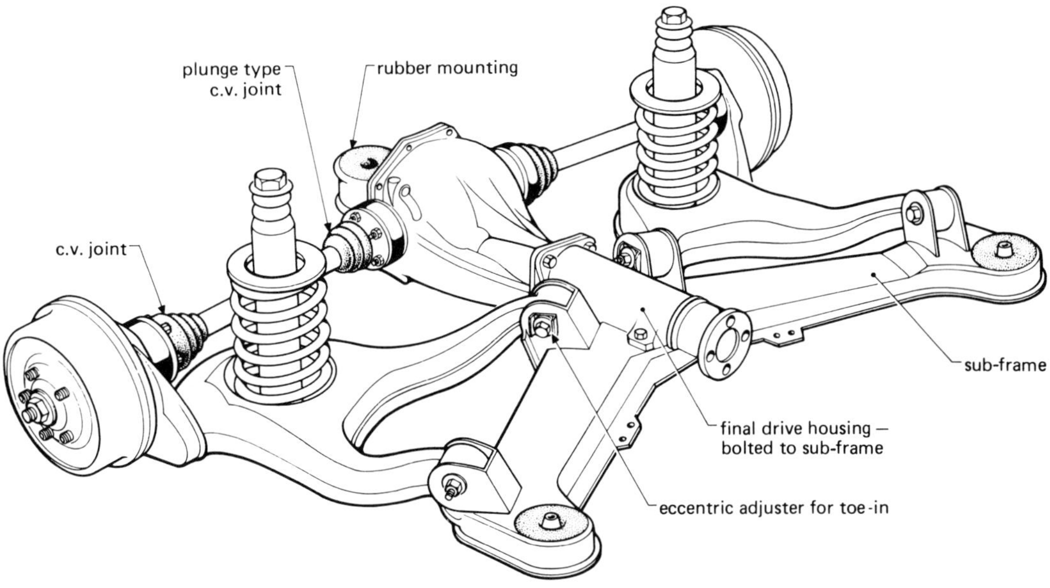

Fig. 61.4 Semi-trailing arm and coil spring

By arranging the suspension arm in this way, the wheel is supported both laterally and longitudinally (Fig. 61.4).

Driving torque reaction is transferred from the final drive housing to the sub-frame and, as in similar arrangements, the construction prevents the tendency of the right-hand rear wheel to lift during hard acceleration. Braking torque and driving thrust are taken by the suspension arms.

Rubber is extensively used at the suspension arm pivots and for mounting the springs and sub-frame to the body: this gives flexibility, reduces the transmission of vibration and noise, and overcomes the need for lubrication. To allow for alteration of wheel alignment (toe-in), one mounting of each suspension arm is fitted with either eccentric adjusters or shims.

Transverse link and strut

The layouts shown in Fig. 61.5 for independently mounting each rear wheel are similar in basic construction to the MacPherson IFS system.

The stub axle is rigidly attached to a long vertical strut that incorporates the suspension damper, and a transverse link, in the form of a wishbone or arm, controls the wheel track. Resistance to rearward movement of the road wheel is provided by a longitudinal tie bar; this links the lower end of the strut with the car body so as to give a stable three-point mounting for the suspension.

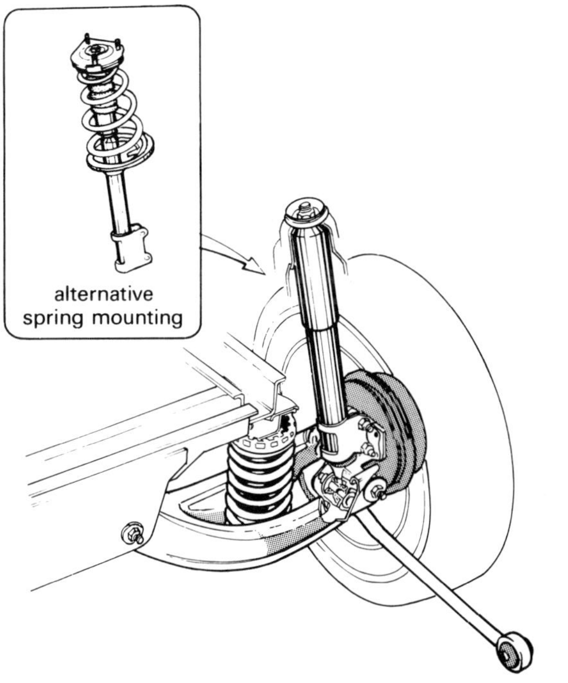

Fig. 61.5 Strut-type IRS

The helical spring can be fitted in one of two positions: it can be mounted either on the transverse wishbone or around the strut in true MacPherson style.

On some systems a spring having coils of differing radii is fitted to improve comfort by providing a variable rate. As this progressive-type spring is being compressed, the larger diameter end-coils close up. This action shortens the effective working length of the spring and stiffens the suspension.

As with many other IRS systems, the camber angle changes as the wheel is deflected. This feature is utilized to alter the tyre's cornering power, so by careful design excellent handling characteristics of the car can be achieved.

Since rubber is used at all mounting points, slight deflection of the suspension layout causes a change in wheel alignment. To allow for this, it is common to set the rear wheels so that they toe-in a small amount. On non-adjustable layouts it is important to replace all washers and spacers in the place from which they were removed. Other layouts often incorporate an eccentric bolt on one of the track control arms.

PART SEVENTEEN

Wheels and tyres

62 Wheels

A wheel must be light, to enable the tyre to follow the contour of the road, and strong, to resist the many forces acting on it. It must be cheap to produce, easy to clean, and simple to remove. These conditions are met by the detachable-disc type of pressed steel wheel: either this is bolted to the hub, or the disc is permanently connected to the hub with a detachable rim: the latter arrangement is used on some Continental cars. In both cases spherically seated nuts ensure a rigid mounting. To reduce the risk of the wheel accidentally coming off, some manufacturers use 'left-hand' threads on the nearside and 'right-hand' threads on the offside; the direction of thread is generally indicated by an L or R on the nut.

The wire or spoke-type wheel, which was common up to about 1935, is still used on many sports and racing cars. Modern designs use tangential spokes to transmit driving and braking forces; inner and outer sets of spokes are connected to the hub shell as wide apart as possible to provide lateral stiffness of the wheel. A *centre-lock* or *knock-on* mounting is used whereby the drive is taken by serrations and the shell is located between two cones, one formed on the hub and the other on a single large nut screwed on to the hub to retain the wheel. Left-hand threads on the offside hubs and right-hand threads on the nearside hubs are generally used to prevent the nut working lose. This quick-change type of wheel is strong and light, and ensures good circulation of air to the braking system, but is difficult to clean.

During recent years, the lightness of the wire wheel and the ease of cleaning of the pressed steel wheel have been combined in a wheel made of light alloy. Although much more costly, the appeal of attractive styles of aluminium and magnesium alloy wheels has made this type popular for use on the high-performance type car. A light alloy wheel is cast to shape and then machined to give the final finish.

The three main types of car wheel are shown in Fig. 62.1.

Types of rim

Well-base

Fig. 62.2 shows the main details of this construction, which is generally used for car tyres. The drop centre or well enables the tyre to be pressed into this recess, so that the opposite side (adjacent to the valve) may be levered over the rim flange. The air pressure in the tyre causes the bead to ride up the slight taper (5°) and 'lock' the tyre to the rim. Rim diameter is stated on the tyre. For example, a tyre 145 × 12 is fitted to a 12-inch diameter rim; the 145 refers to the nominal sectional width of the tyre in millimetres.

Flat-base three-piece rim

The stiff, heavy-bead tyres used on heavy vehicles require a detachable-flange type of rim. A split lock ring, like a large circlip, holds the flange in position. When the flange is pushed towards the tyre, the lock ring may be removed.

Semi-drop centre

This two-piece rim is a compromise between the well-base and the flat-base rim, and is suitable for light trucks. A split, detachable flange simplifies removal and the slight taper enables the tyre to 'lock' to the rim. The small well must be used when the tyre is being removed.

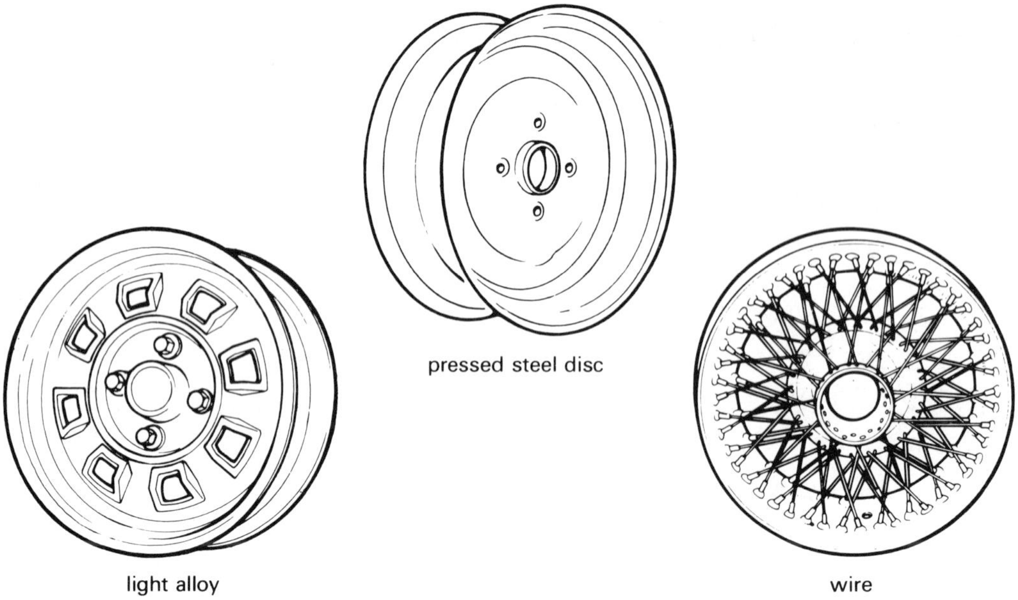

Fig. 62.1 Types of car wheel

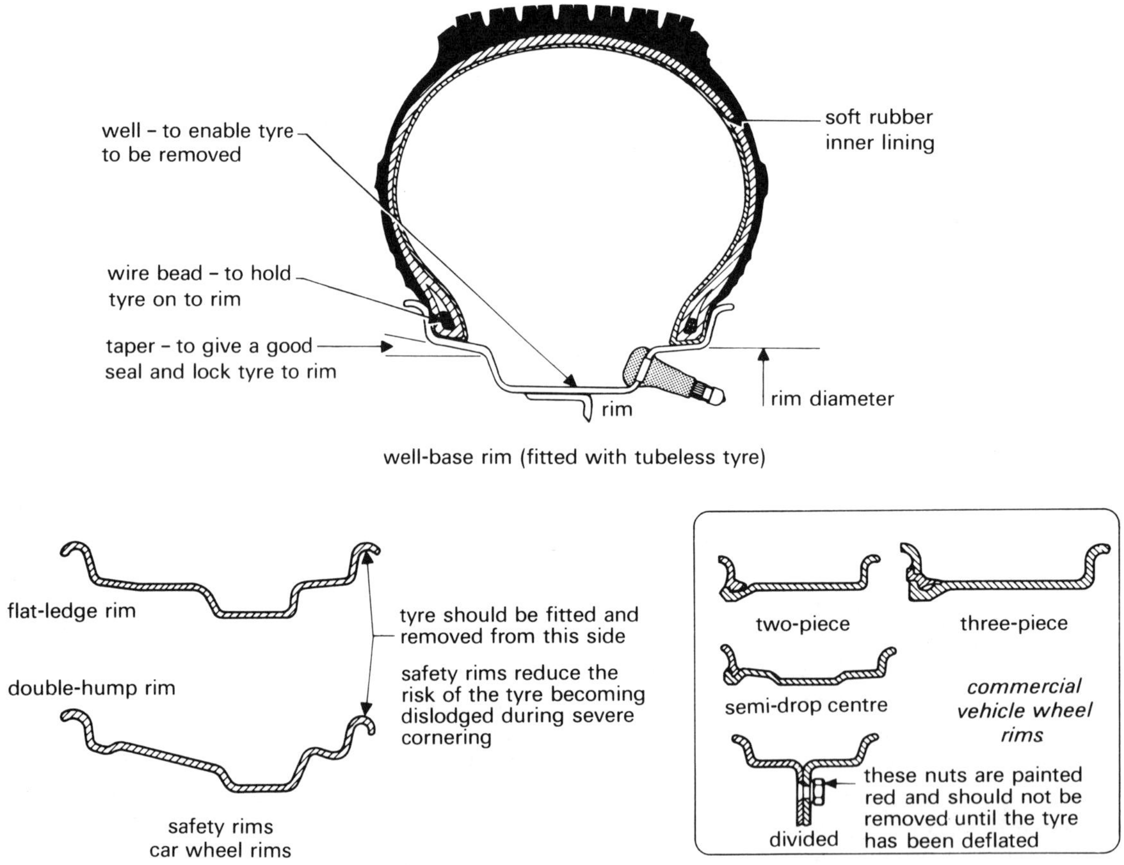

Fig. 62.2 Types of wheel rim

Flat-based divided type

Mainly used on military vehicles, this type of rim is made in two sections and bolted together by a ring of nuts, adjacent to the rim. On no account must you remove these nuts when changing a wheel.

Space-saver wheels

A spare wheel has to be carried on a vehicle in case one of the tyres develops a puncture. Various locations are used to carry this wheel, the most common place being the luggage space at the rear of the car.

To reduce the space taken up by the spare wheel in order to improve the luggage capacity, some manufacturers fit a special small-diameter, thin spare wheel.

This *space-saver* type wheel should only be used in an emergency to allow the car to be driven to a repair location. Sometimes this type is called a 40–40 wheel; the car should not be driven faster than 40 m.p.h. (65 km/h) over a distance greater than 40 miles (65 km).

63 Tyres

The dictionary definition of a tyre is a band of iron, steel, rubber, etc., placed round the rim of a wheel to strengthen it or reduce vibration.

The pneumatic tyre

Early tyres certainly strengthened the wheel, but did little to improve the comfort. It was not until R. W. Thomson in 1845 invented the pneumatic tyre that high-frequency vibration could be reduced. This idea was developed by J. B. Dunlop in 1888 for cycle use and was quickly applied to motor vehicles. An air bag or inner tube was contained in a cover so that the vehicle was 'floated' on a cushion of air.

The modern tyre, besides improving comfort, increases the adhesion between road and wheel to give satisfactory grip for braking and steering in various driving conditions.

Tyre construction

Fig. 63.1 shows the constructional details of a tyre.

The *casing* must resist the expansion of the tube,

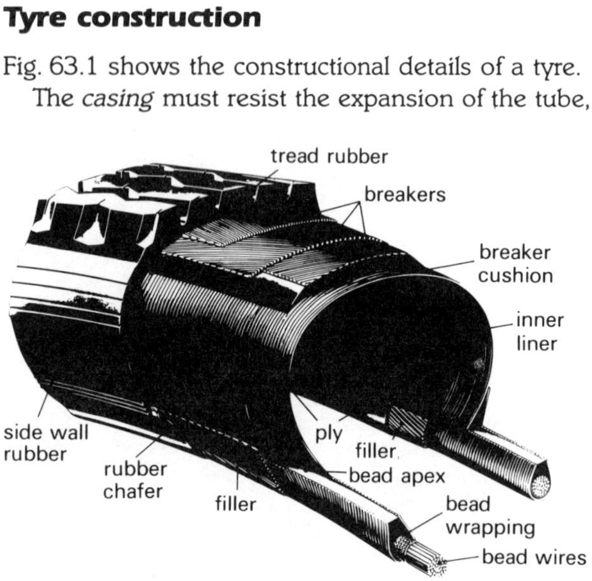

Fig. 63.1 Tyre construction

especially when the tube is subjected to road shocks. Fracture of the casing causes the tube to blow out, i.e. burst the tyre; thus the strength of the tyre is governed by the construction of the casing. It is usually made up of four or six layers of fabric; each layer consists of a series of rubberized cords laid side by side to form a sheet. A strong casing is produced by placing a number of sheets or plies together. The number of plies and the relative angle between the cord and the tyre bead governs the tyre characteristics. In the past cotton was the main cord material, but modern tyres use man-made fibres such as rayon and nylon. These materials are much stronger and offer greater resistance to the heat built up by the flexing of the tyre. Heavy vehicles and some cars require an even stronger material, so steel cords are needed in these cases.

The *bead* consists of a number of hoops of steel wire which are responsible for retaining the tyre on the rim. The casing is wrapped around the bead wire and moulded into shape.

The *tread* is bonded to the soft rubber which encloses the casing. The material used is natural or synthetic rubber, which is compounded with chemicals such as carbon-black to produce a hard, abrasion-resisting substance. Various tread patterns are used to wipe water and grease off the road, so that the teeth, formed by the zig-zag circumferential grooves, can 'bite' into the surface. An excellent grip, especially on soft surfaces, can be obtained by transversely slotting the tyre to form bold tread bars, but when used on hard roads this type is rather noisy and generally causes the 'heel and toe' form of tread wear.

The tread pattern should be selected to suit the operating conditions:

1 *Normal touring* requires a tread which gives quiet operation and provides effective adhesion under wet, dry and greasy conditions.

2 *High-speed operation* places emphasis on the heat generated, the noise level and the resistance of the tyre to *aquaplane*. This is a condition where a water layer completely separates the tyre from the road and totally destroys the adhesion.
3 *Cross-country* tyres have to bite into mud, snow and loose surfaces, so heavy tread bars are needed. Extensive use of this type of tyre on hard highways causes noise and component wear.

Tubeless tyres

Recent years have seen the introduction of this type of tyre, which is now basic equipment for most new cars. The inside of the casing and outer surface of the bead is lined with a soft rubber, which forms an air-tight seal with the rim and eliminates the need for a separate tube. Two main advantages are claimed:
1 Better air sealing qualities are obtained providing the cover is correctly sealed to the rim.
2 The soft inner liner of the cover provides a puncture-sealing arrangement. On fitting or removing this type, be very careful to avoid damaging the bead.

Resistance to a side force

The extent of the side deflection produced when a side force acts on the tyre governs the actual path taken by the wheel (see Chapter 49). This side deflection (Fig. 63.2) is increased when the inflation pressure is low but is controlled to a very large extent by the construction of the tyre.

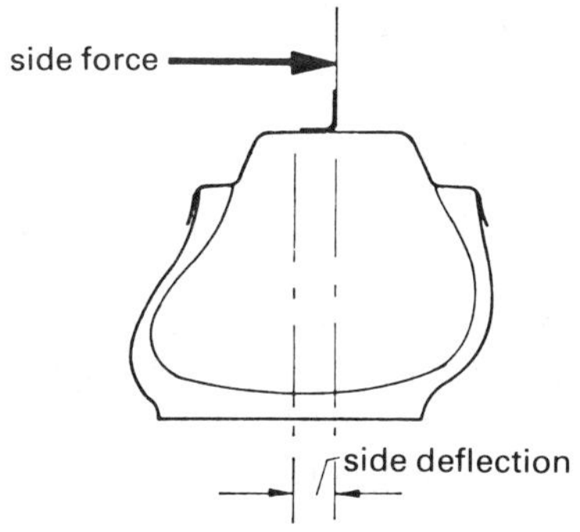

Fig. 63.2 Effect of a side force on a tyre

Diagonal-ply Fig. 63.3 shows a tyre with the carcass composed of several layers of casing plies. The cords forming one ply run at an angle of about 100° to the cords of the adjoining ply, and each cord of the ply forms an angle of approximately 40° to the bead. This

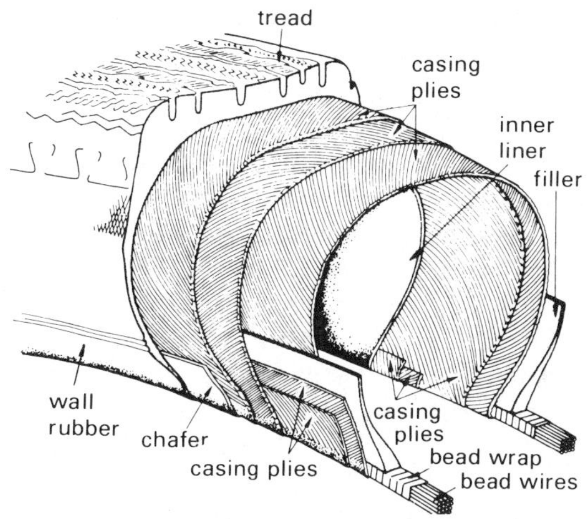

Fig. 63.3 Tyre construction (diagonal-ply)

type is also called *cross-ply* and has been in use for a long period.

Radial-ply Arranging the cords in a manner such that they form an angle of 90° to the bead (i.e. the cords are radially disposed to the wheel) gives a construction which is called *radial-ply* (Fig. 63.4). This type has the great advantage that it offers a large resistance to side deflection and its effect on the vehicle handling is very noticeable. A vehicle fitted with cross-ply tyres often suffers from side drift when it corners and gives the impression to the driver that the tyres are sliding on the road, whereas radial-ply considerably reduce this effect. In a similar manner if the vehicle oversteers badly then this could be reduced by fitting radial-ply to the rear axle. These 'stiffer' tyres will reduce the slip angle of the rear

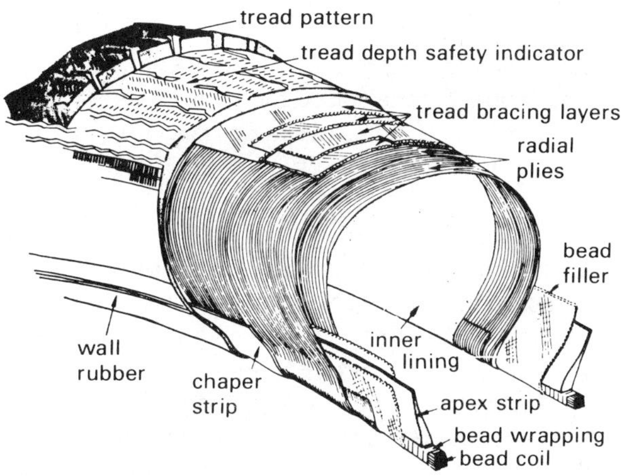

Fig. 63.4 Tyre construction (radial-ply)

wheels and will produce a safer and more stable condition. This mixing of radial-ply and cross-ply tyres on a vehicle must be used with great care and under no circumstances must radial-ply be fitted to the front axle when cross-ply are used on the rear wheels. Mixing in this configuration is illegal, because of the oversteering problem, so care is needed to identify the tyres correctly. Today, radial-ply tyres can be identified by the word 'radial' which is moulded on the wall of the tyre. Prior to the general introduction of this method the means of identification was to note the type letters and refer to the manufacturer's catalogue.

Diagonal- and radial-ply advantages

Diagonal-ply:

- Lighter steering at low speed
- Not so critical with respect to steering geometry
- Smoother ride at low speed
- Cheaper.

Radial-ply:

- About 80 per cent longer life
- Lower rolling resistance, so fuel consumption is improved
- Side deflection is reduced so the vehicle corners without drift
- Full width of tread is held on the road when the car is cornering, so grip is improved, especially on wet roads.

Belted and braced tyres

Fig. 63.4 shows that in addition to the radial cords a number of layers of breakers are used. These bracing layers act as a belt around the circumference to stabilize the tread and resist the enlargement of the tyre on inflation, since the radial cords do not offer the same resistance to tyre enlargement as the cross-ply cords. Textile materials used for these breakers give a tyre which is sometimes called a *belted* type whereas steel cord breakers give a *braced* tread type.

Tyre profile

The 'section height' to 'section width' of early tyres was approximately equal and this aspect ratio was considered as 100 per cent. As increased demands were made for improved vehicle stability, the cornering power of tyres was raised by reducing the aspect ratio. Today low-profile tyres having ratios of about 80 per cent and ultra-low-profile types of 70 per cent are in common use. These 'wide oval' tyres, and other similarly shaped tyres, can be identified by relating the dimensional markings on the tyre to the manufacturer's handbook.

Valves

Fig. 63.5 shows the main details of the Schrader one-way valve used in conjunction with a tubeless tyre. In this case the valve is inserted in the rim, whereas in separate-tube designs it is vulcanized, or bolted, to the tube.

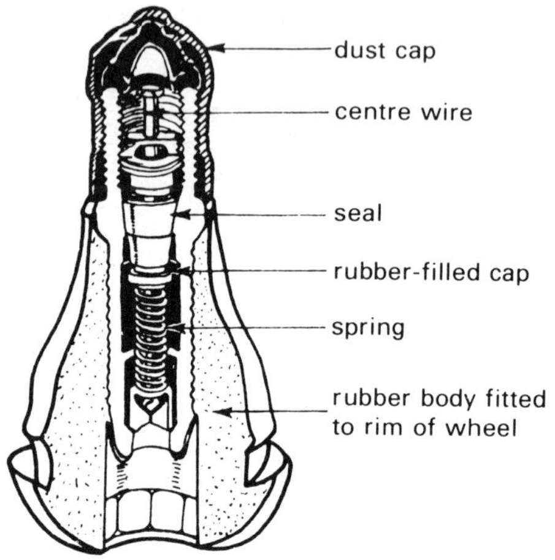

Fig. 63.5 Schrader valve

The core is threaded into the stem until the tapered hard rubber section forms a seal with the stem. Spring pressure forces the centre wire and rubber-filled cup on to its seat to prevent an outward flow of air. A dust cap acts as an extra air seal and prevents entry of dirt.

Inflation pressure

This is governed by the load carried by a given tyre, although manufacturers vary the pressure to modify the steering characteristics. Large-section tyres use a lower pressure for a given load, and so greater comfort is achieved, but the resistance to rolling is increased. The pressure recommended by the manufacturer is applicable to a *cold* tyre, i.e. it is the pressure *before* the tyre is used, and takes into account the average pressure rise [28 kN/m^2 (4 lbf/in^2)] which is caused by the temperature increase during use. Recommended pressure is normally suitable for sustained speeds up to 130 km/h (80 mile/h), but a higher pressure is advised if speeds or loads are increased.

Underinflation or overloading leads to rapid wear on each side of the tread and internal damage to the casing, whereas overinflation wears the centre of tread.

Tyre marking

Local statutory regulations based on EEC recommendations specify the classification markings that must be shown on a sidewall of a modern tyre. If the tyre is to be sold in the USA, additional data must be included.

Fig. 63.6 shows an example of the markings required on a tyre intended for sale in Europe and the USA. The symbols on this tyre are:

Fig. 63.6 Side wall markings

1	Type of tyre: tubeless or tubed type.
2	Size designation: 175 - width of tyre in mm S - speed rating 14 - rim diameter in inches.
3	TWI stands for *tread wear indicators*.
4, 5	EEC standards: 88 - maximum load per wheel
5	S - speed rating.
6	Casing material and number of plies.
7	Tread material and number of plies.
8	Maximum inflation pressure when cold, stated in p.s.i. (lbf/in^2)
9	US standard for giving maximum load per wheel.
10	Manufacturer's name.
11, 12	EEC approval marks.
13	US approval marks.
14	Manufacturer's coding.

Speed rating

Tyres are marked to show the maximum recommended speed. The speed marking, indicated by the letters S, H and V to represent 'speed', 'high speed' and 'very high speed' respectively, is incorporated in the size marking, e.g. 165 SR 13. This is a radial tyre of size 165 × 13 and reference to the manufacturer's manual shows that the speed marking 'S' represents a maximum speed of 180 km/h (113 m.p.h.). The alternatives for this size of radial tyre are:

Size	*Speed limit*
165 HR 13	210 km/h (130 m.p.h.)
165 VR 13	over 210 km/h

The tyre manufacturer should be consulted if a tyre is required for a vehicle which is to be subjected to either a high maximum speed or a high consistent speed.

Ply rating

This is an indication of the strength of a tyre and nowadays does not represent the number of layers that form the casing of the tyre.

Ply rating combined with tyre size are two factors which determine the legal maximum load (plated load) that can be carried by each axle of a commercial vehicle.

Tread wear indication

The tread of a modern tyre has a pattern which changes when the tread has worn down to a given limit. This safety feature allows an observer to determine the danger limit quickly without having to use a tyre tread depth gauge.

Interchanging tyres

At present there are various recommendations regarding the interchanging of tyres, so tyre manufacturers' guidance should be followed.

Generally cross-ply tyres should be interchanged at frequent intervals if uniform tyre wear is to be achieved. Longitudinal changes (L/H front to L/H rear, etc.) and diagonal changes (R/H front to L/H rear and spare to L/H front, etc.) are normally advised.

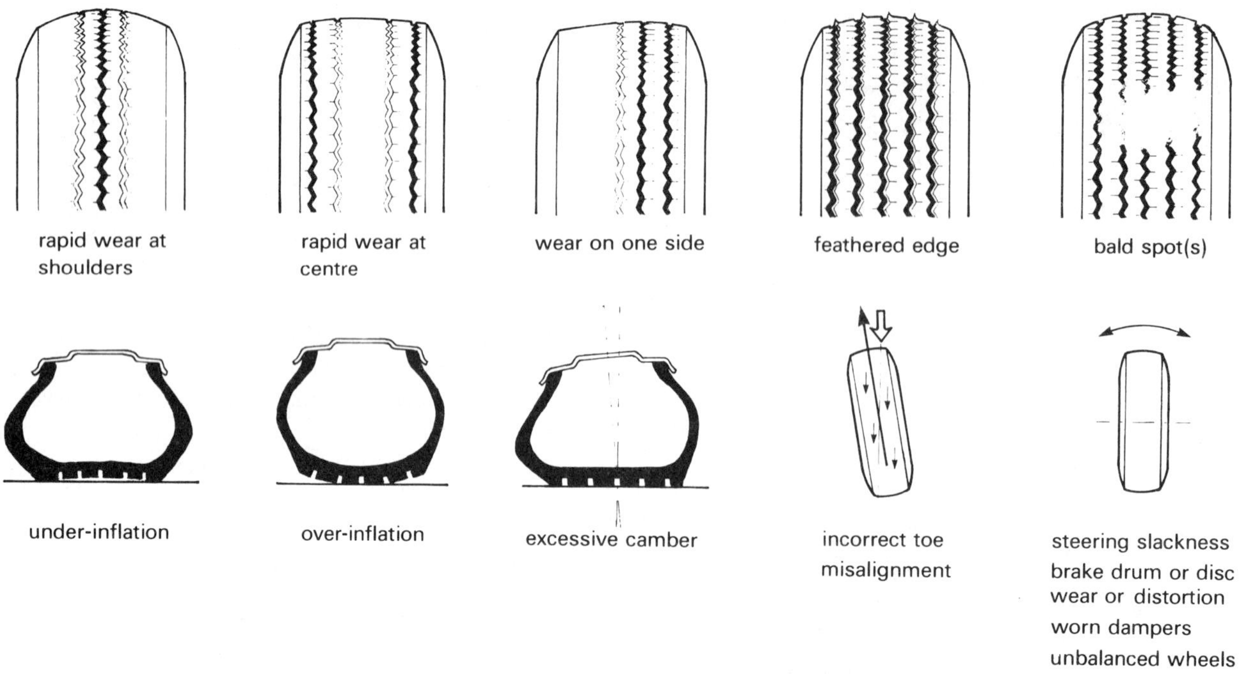

Fig. 63.7 *Abnormal tyre wear*

Some manufacturers of radial-ply tyres recommend that the front tyres are not moved to the rear because of the different wear pattern of front and rear tyres.

Legal requirements

Legislation exists to prevent the use of a vehicle which is fitted with defective tyres. The current regulations specify the various tyre faults and these include:

- The use of unsuitable tyres
- Underinflation
- Break in the fabric in excess of 2.5 cm (1 in) or 10 per cent of the section width
- Lumps or bulges
- Exposure of cords
- Tread depth less than minimum. The law requires a tread depth of at least 1.6 mm across the central $\frac{3}{4}$ of the breadth of tread around the entire circumference of the tyre.

The preceding list of faults is intended to outline the requirements and it is recommended that the current regulations be read in order to acquire an accurate understanding of the legal requirements.

Tread wear patterns

The type of irregular wear pattern of a tyre tread is often a good indication to a specific mechanical fault in the steering, braking or suspension system.

The life of the tread on a tyre depends on the service it receives and on the manner in which the car is driven. The life is shortened if the tyre is subjected to rapid acceleration and braking, high speed cornering and excessive scrubbing by abuse of the power steering.

Fig. 63.7 shows the main types of abnormal wear and the cause(s) of each.

64 Wheel balancing

Independent suspension systems and modern, small-diameter, large-section wheels demand extra care and attention to wheel balancing.

Although tyre manufacturers balance the tyre and tube, it is often necessary to balance the complete wheel assembly to ensure that the wheel rotates free from vibration.

Wheels out of balance may be the result of: (a) static unbalance, or (b) dynamic unbalance.

Static unbalance

This is apparent when the wheel is mounted on a horizontal shaft having low-friction bearings (Fig. 64.1a). The wheel always comes to rest in one position, which indicates the presence of a 'heavy spot' directly below the wheel axis. (A bicycle wheel shows this action clearly.) A wheel used in this condition will tend to cause the unbalanced centrifugal force to lift the wheel off the road periodically (Fig. 64.1b), and produce conditions known as 'wheel hop', 'shimmy' and uneven tyre wear.

Dynamic unbalance

Fig. 64.1c shows a wheel in good static balance, but when this wheel is rotated, the centrifugal forces acting on each 'heavy spot' will not balance, since the forces are not acting through the same line. In this diagram the wheel axis tends to rotate around the king pin in a clockwise direction, whereas the

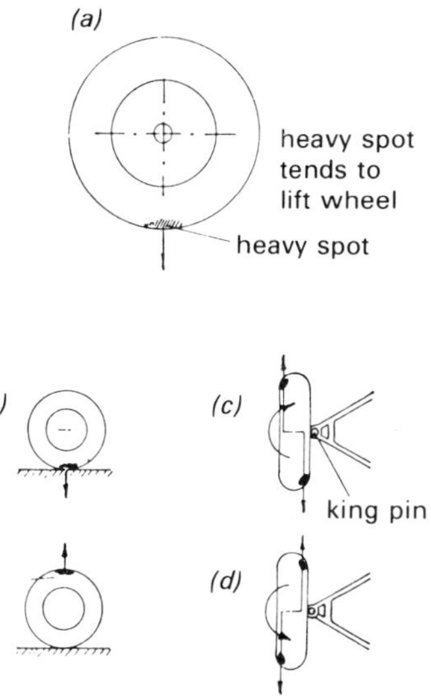

Fig. 64.1 Unbalanced wheels

opposite applies to Fig. 64.1d. Serious dynamic unbalance causes the wheel to 'flap' around the king pin, a condition known as wheel shimmy.

Wheel-balancing machines

These enable the operator to determine the position and extent of the correcting mass which must be clipped on to the rim. Simple equipment can be used to detect static unbalance, but since the wheel must be rotated to determine dynamic unbalance, more costly apparatus is necessary.

PART EIGHTEEN

Brakes

65 Braking principles

Purpose of a brake

When a vehicle is accelerated, energy supplied by the engine causes the vehicle speed to increase. Some of this energy is instantly used up in overcoming frictional and tractive resistances, but a large amount remains stored in the vehicle. This energy of motion is called *kinetic energy*.

The existence of kinetic energy is seen when a vehicle is moving and neutral is selected. The vehicle does not immediately come to rest; instead it travels for a considerable distance before it becomes stationary. In this case the stored energy is slowly being converted and used to drive the vehicle against the resistances that oppose the vehicle motion. Reliance on these resistances to slow down a vehicle would cause many problems, so an additional resistance called a *brake* is needed to convert the energy at a faster rate. The purpose of a brake is to:

convert kinetic energy to heat energy.

The speed of the energy conversion controls the rate of retardation of a vehicle, i.e. its rate of deceleration.

Heat generation at the brake is obtained by rubbing a fixed pad or shoe against a rotating object driven by the motion of the vehicle. In the early days this was achieved by pressing a wooden block against the steel tyre of a road wheel (Fig. 65.1).

Apart from its crude operation, the retardation given by this single brake was limited both by the friction between the tyre and road and by the fact that the braking action was only applied to one wheel.

Applying a brake to the other wheels increased the stopping power of vehicles, so when a braking system was introduced that utilized the full 'weight' of the vehicle by fitting a brake to all four wheels, a far more efficient system was obtained.

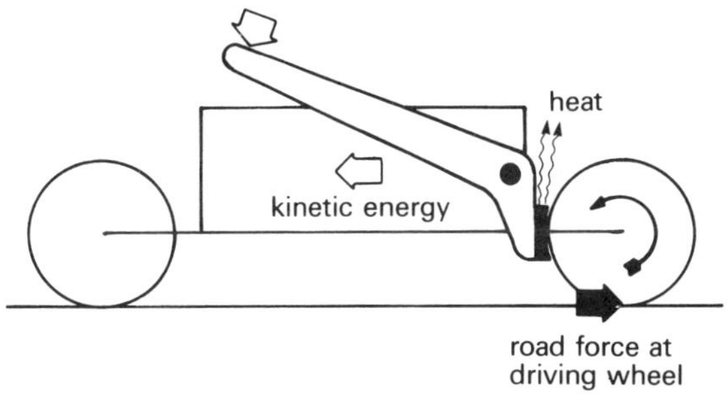

Fig. 65.1 Principle of a brake

A large step forward was made when the drum brake was introduced. This used a rotating drum on to which a brake band or shoe was pressed. Originally an *external contracting* band was used but it got very hot owing to the difficulty in radiating the heat to the surrounding air. The *internal expanding* type was a natural development and an early type is shown in Fig. 65.2.

Stopping distance and tyre adhesion

The strength of the force pressing a shoe against a drum governs the resistance to rotation of a road wheel. During this operation the road surface has to drive the wheel around. The limit of this driving force is reached when the resistance of the brake equals the maximum frictional force that is produced between the tyre and road. The latter is called the *adhesive force* and can be calculated from the expression:

$$\text{Adhesive force} = \text{Load on wheel} \times \text{Coefficient of friction}$$

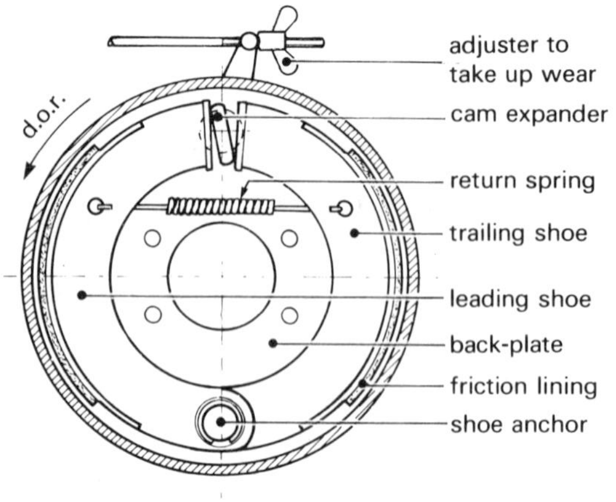

Fig. 65.2 Cam-operated brake

When the limit is reached, the wheel starts to skid, so extra force on the brake shoe will not produce any increase in the rate at which the vehicle slows down, no matter how good the braking system. This situation is apparent when a vehicle is braked on a slippery surface: slight pressure on the brake soon locks up the wheel and very poor braking results. This shows that the adhesion between the tyre and road is the main factor that controls the minimum stopping distance.

Road adhesion is affected by:

- Type of road surface
- Condition of surface, e.g. wet, dry, icy, greasy, etc.
- Design of tyre tread, composition of tread material and depth of tread.

Many people think that the shortest stopping distance is achieved when the wheel is locked to produce a skid. This idea is false because experiments show that the force needed to 'unstick' a tyre is greater than the force required to skid it over the surface. A wheel held on the verge of skidding not only gives the shortest distance; it also allows the driver to maintain directional control of the vehicle.

Types of brake

Fig. 65.3 shows the two main types of friction brake; they are:

- *Drum*
- *Disc.*

Both types use a fixed (non-rotating) shoe or pad which rubs against a moving drum or disc. The friction between the rubbing surfaces is increased by attaching a special friction material to the fixed part. In the past this friction material had a high asbestos

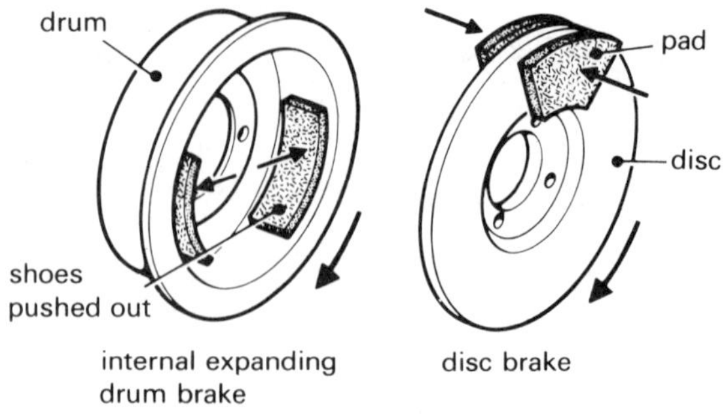

Fig. 65.3 Types of brake

content but the health dangers associated with the dust from this substance has forced manufacturers to replace it with a safe *asbestos-free* material.

There are many vehicles on the road that have asbestos-lined brakes, so great care must be exercised to ensure that dust from this material is not inhaled. Manufacturers' instructions should be understood before attempting to dismantle a brake assembly.

Drum brakes

This *internal expanding* type of brake uses two shoes that are attached to a back-plate, which is fixed to a stub axle or axle tube. Each shoe is of 'T' section and a friction lining is riveted or bonded to the outer face of the shoe. At one end of the shoe is fitted a device for expanding the shoe when the brake pedal is depressed. In a simple brake, such as that shown in Fig. 65.2, a cam is used as a *shoe expander*, but modern layouts fitted to cars use hydraulically operated pistons for this purpose.

All shoe-type brakes must have some arrangement to prevent the shoes rotating with the drum. The *shoe anchor*, which must be rigidly attached to the back-plate, takes the form of a large pin that passes through the shoes, or a housing, against which the shoes butt.

Springs pull the brake shoes on to the back-plate and also return the shoes to the 'off' position after the brake has been applied. In some cases separate springs are used to perform the retention and return functions.

The inner cylindrical surface of the cast iron drum is ground to give a smooth surface on to which the brake linings can rub. The drum is generally attached to the hub flange by small counter-sunk screws and retained securely by the wheel nuts. If possible the drums should be exposed so that a good flow of air over the drum is achieved to dissipate the heat and prevent loss of brake efficiency that occurs when the unit gets very hot.

Some form of adjuster is provided at each brake to take up excessive clearance due to wear of the friction facing. Since a large leverage is needed between the brake pedal and shoe, a large movement of the shoe would mean that the brake pedal would strike the floor before the brake is fully applied. This dangerous condition is avoided by either manually adjusting the brakes periodically, or having an automatic adjuster that continually sets the shoes so that they are always positioned very close to the drum.

Disc brakes

One problem with drum brakes is *fade*. This is a condition where the brake performance falls drastically when a brake, especially the friction linings, reaches a given temperature.

When this occurs, the driver has to apply a much larger effort and in extreme cases it becomes impossible to bring the vehicle to rest. The problem arises when a drum brake is continually operated for long periods, as when a vehicle descends a long mountain road.

The friction between the lining and drum, and the effect of this friction on the application of the shoes, govern the maximum braking torque that a drum brake can produce. Although friction materials having good anti-fade qualities are produced, there is still the chance of fade with a drum brake.

Since high temperature is the cause of brake fade, some means must be provided to dissipate the heat. This is difficult in the case of a drum brake because the heated regions of the brake are shrouded by metal.

The disc brake was developed to minimize the fade problem. It consists of an exposed disc which is attached to the hub flange; the two friction pads are pressed on to this disc to give a braking action (Fig. 65.4). The pads are moved by hydraulic pistons working in cylinders formed in a caliper that is secured to a fixed part of the axle.

When hydraulic pressure is applied to the two cylinders held in the fixed caliper, the pistons move; this action forces the friction pads into contact with the rotating cast iron disc. The sandwiching action of the pads on the disc give a retarding action and heat generated from the energy of motion is conducted to the disc.

As a large part of the disc is exposed to the air, heat is easily radiated, with the result that the brake can be used continuously for long periods before serious fade occurs. Since the friction pads move at a right angle to the disc, any drop in the friction value does not affect the force applied to the pad. As a result this type of brake is not so sensitive to heat.

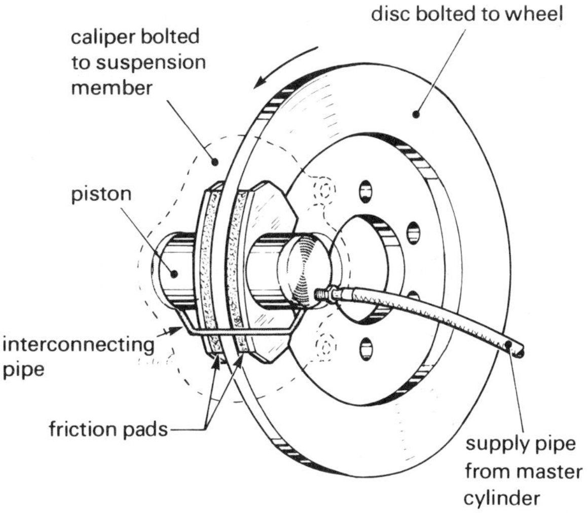

Fig. 65.4 Disc brake

No assistance is obtained from the rotating disc to aid the driver in the application of a disc brake, so to achieve a given retardation, a disc brake requires a greater pedal pressure than that needed for a drum brake.

Adjustment to take up pad wear is automatic on a disc brake, so minimum attention is required. A merit of this type is the ease with which the pads can be inspected. Assuming corrosion is not present, the pads can be easily renewed.

Brake operating systems

Today brake shoes and pads are generally operated hydraulically. Up to the mid-1930s most systems were mechanically operated and this type is considered for the purpose of appreciating the features offered by modern systems.

Mechanically operated system

Fig. 65.5a shows the layout of a simple mechanical system. Four adjustable rods or cables link the brake shoe operating levers to a transversely mounted *cross-shaft*. The footbrake and handbrake controls are connected to the cross-shaft by links having elongated holes to allow independent operation of each control.

In this system each brake will receive its share of the brake pedal force only when the mechanism is balanced; i.e. set up so that each shoe contacts the drum simultaneously. If one brake has a much

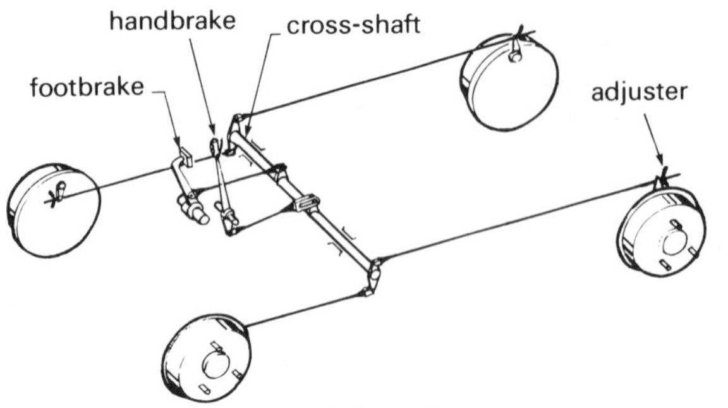

(a) Brake layout

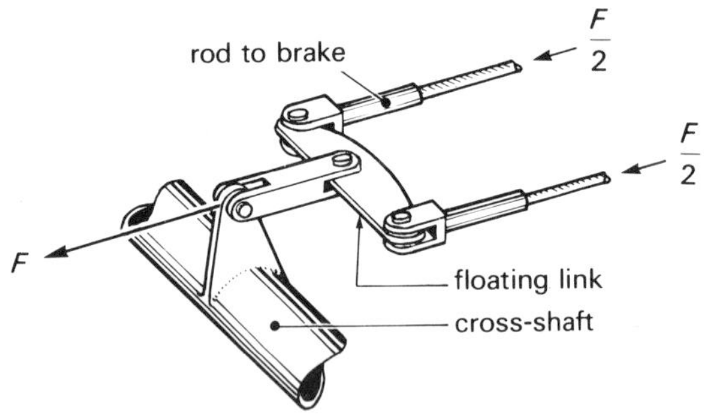

(b) Brake compensator to balance two brakes

Fig. 65.5 Mechanical brake layout

smaller shoe-drum clearance than the others, all the driver's force will be directed to that brake; as a result the unbalanced braking action will cause the vehicle to 'pull' violently to the side on which this brake is situated.

Compensation devices were fitted in the layout to overcome this problem and Fig. 65.5b shows a simple arrangement for balancing two brakes.

A *fully compensated* brake system ensures that each brake is applied at the same time and that each one receives its share of the braking effort. To achieve this condition, the mechanical system shown would require three compensators: front (to balance the front brakes), rear (for the rear brakes) and centre (to equalize front and rear).

The large force on the brake shoe is achieved by using a compound lever system. The larger the leverage, the smaller is the force needed on the brake pedal, but if a very large leverage is provided, the brake will require frequent adjustment to counteract wear of the brake linings.

By law a *handbrake* (*parking brake*) must be provided to hold the car stationary when it is left unattended. A ratchet and pawl is fitted to this hand-lever to hold the brake in the 'on' position.

Hydraulically operated system

This system has a high efficiency, is suited for vehicles that have independent suspension and is *fully compensated*.

Fig. 65.6 shows the main features of this system. The basic layout consists of a *master cylinder*, which is connected by small-bore tubing (*Bundy tubing*) to *wheel cylinders* situated at each brake. A special brake fluid completely fills all pipelines and working cylinders, and a reserve of fluid is held in a reservoir placed adjacent to the master cylinder.

When the footbrake is operated, a piston in the master cylinder pumps fluid through the lines into the wheel cylinders. As fluid enters the wheel cylinders the pistons move outwards to bring the shoe or pad into contact with the drum or disc. After all movement has been taken up, the force on the master cylinder piston pressurizes the fluid in the system and applies a force to each brake. At this stage the greater the force that is applied to the pedal, the higher is the pressure produced in the system.

The pressure of the fluid in each wheel cylinder produces a force on the piston which is applied to the brake. Since fluid may be regarded as being incompressible, and the pressure is the same throughout the system, the thrust on any part of the system is proportional to the area of that part. In view of this fundamental law, the force applied to the pistons can be varied to suit the application.

Any air that is present in the system causes the brake pedal to lose its 'solid' feel, i.e. the pedal becomes *springy*. This is dangerous because the required fluid pressure cannot always be built up before the master cylinder piston reaches the end of its travel. Although the compressibility of the air does not prevent the build up in fluid pressure, its elastic nature makes it necessary to remove the air after any part of the system has been disconnected. This operation is called *bleeding* and valves are fitted at each wheel cylinder to allow the air to be removed.

Brake fluid should have a low freezing point, high boiling point and low viscosity. Also the type of fluid used must be compatible with the type of rubber seals and hoses fitted in the system. Today many different fluids have been developed to suit particular systems, so always use the type, if not the make, recommended by the manufacturer. In the past most brake systems used a vegetable-based fluid and

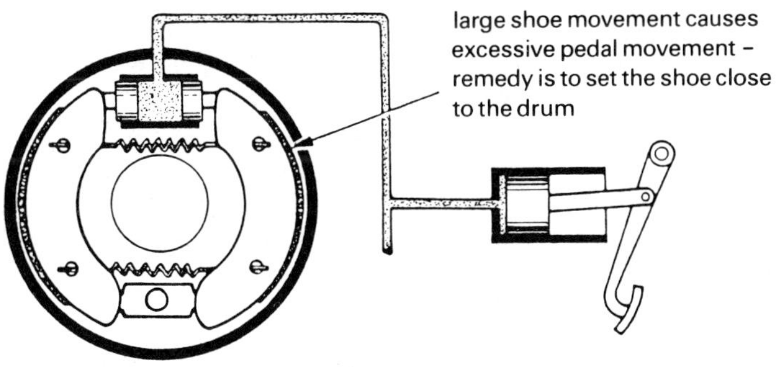

brake requiring adjustment

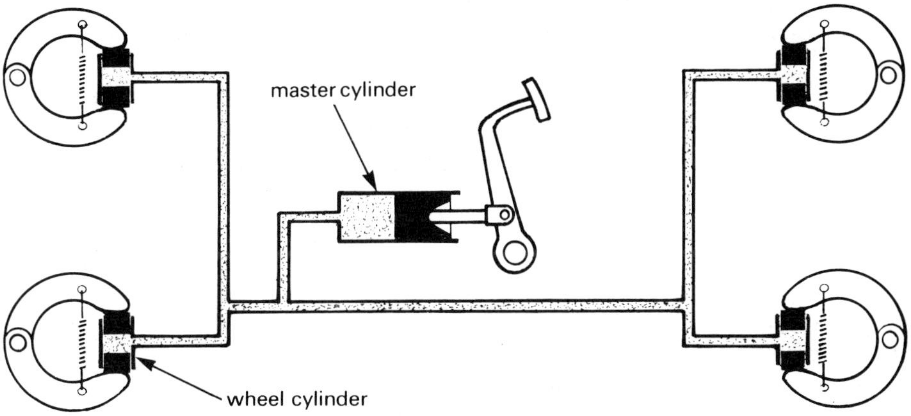

hydraulic brake layout

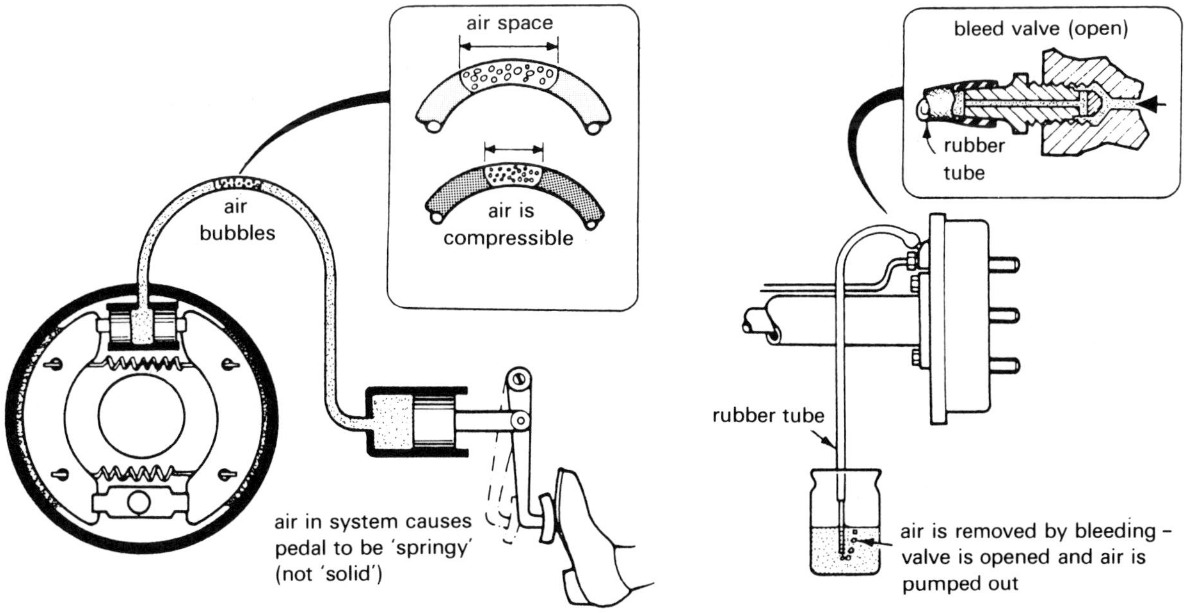

effect of air

Fig. 65.6 Hydraulic brake operation

some systems used mineral-based oils. If, by accident, a mineral oil is used to top up a common system, then damage to the rubber seals, including the wheel cylinder seals, occurs in a very short time.

Handbrake Legal requirements insist that hydraulically operated systems must be fitted with a mechanical handbrake that acts on at least two wheels. Normally a rod or cable is used to link the handbrake with a mechanical lever-type shoe-expander fitted to the rear brakes.

The single-line system shown has one major disadvantage: if a fluid leak occurs in any part of the system, then brake failure results. On these occasions it is hoped that the driver realizes that the handbrake is the emergency brake; although it acts on two wheels only, it enables the vehicle to be safely brought to rest.

For safety reasons modern vehicles use a dual-line system. This layout ensures that at least two brakes can be operated by the foot brake in the event of a leak.

Pneumatic operating systems

Today both low- and high-pressure air brake systems are used, but most light category vehicles only utilize air to boost the effort applied by the driver. A full compressed-air braking system, of the type normally associated with heavy vehicles, is too expensive, bulky and heavy for light vehicles.

Some light trucks in the 3 tonne region use a part-compressed-air system called air/hydraulic (air over hydraulic); this air arrangement, combined with a hydraulically operated system, reduces the maximum pedal pressure to a comfortable limit.

The widespread use of disc-type brakes has meant that many cars now fit a *vacuum servo* as standard equipment. This device uses manifold depression, or 'vacuum' generated by an engine-driven pump, to assist the driver in applying the brake.

Front/rear brake apportionment

When a vehicle decelerates, an inertia force acting at the centre of gravity of the vehicle causes the load on the front wheels to increase. The brake designer must take this *load transfer* into account if premature skidding of the rear wheels is to be avoided.

Some idea of the magnitude of the change in wheel loading can be appreciated by considering a typical rear-wheel drive car having a 50/50 static load distribution. Tests show that under maximum braking conditions the front wheels of the car carry about 75 per cent of the total weight. To achieve good braking with this car, the front brakes are designed to provide about 60 per cent of the retarding force. This 60/40 apportionment is chosen for safety reasons because it ensures that under maximum braking the front wheels do not lock up before the rear.

The 60/40 ratio can be achieved by using:

- discs at front and low capacity drum brakes at the rear
- larger area wheel cylinders at the front
- larger diameter drums at the front
- more effective brakes at the front.

66 Brake arrangements

Brake arrangements can be divided into two main classes: *drum* and *disc*.

Drum brakes

Although external contracting band brakes are favoured for automatic gearboxes, the *internal expanding* type is commonly used with a drum for the main braking system of a vehicle.

Today drum brakes fitted to light vehicles are hydraulically operated and are commonly used to brake the rear wheels to complement a disc system at the front. This disc/drum arrangement allows more braking to be undertaken by the front wheels; also the simpler application of a mechanical handbrake to a drum brake makes this type an obvious choice for a rear brake.

Various shoe arrangements are used; these include:

- *Leading and trailing shoe* (L & T)
- *Two leading shoe* (2LS)
- *Duo-servo*.

Leading and trailing shoe

Fig. 66.1 shows the layout of a leading and trailing shoe brake. The shoes are hydraulically expanded by two separate pistons fitted in a wheel cylinder fixed rigidly to the back-plate. Rotation of the shoes with the drum is prevented by an anchor pin that is securely mounted on the back-plate so as to resist the full braking torque. A strong spring is fitted to return the fluid and shoes when the brake is released. By mounting this spring on the back-plate side of the shoes, the shoes are pulled against high spots of the back-plate formed to provide a bearing surface for the brake shoe. A snail cam or eccentric, mounted behind each shoe, is provided for manual adjustment of the shoes. These cams can be partially rotated with the aid of a wrench without removing the drum; they are positioned so that each shoe is set as close to the drum as possible without causing excessive *binding* (rubbing).

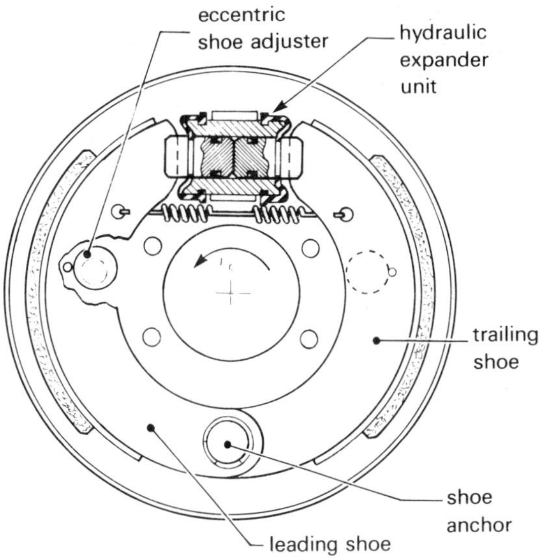

Fig. 66.1 Leading and trailing shoe brake

When the vehicle is stationary, hydraulic pressure forces both shoes outwards and after the clearance has been taken up, an equal force is applied by each shoe to the drum.

This equal force does not apply when the vehicle is moving. Instead the drag of a moving drum on the friction linings causes one shoe to be applied hard and the other to be pushed towards the 'off' position. As a result, the braking torque produced by the shoes is in the ratio 4 : 1. The shoe that does the most work is called the *leading shoe*, and this can be identified by the fact that *the leading shoe is the first shoe in the direction of rotation after the hydraulic expander*.

Conversely, the other brake shoe is called the *trailing* shoe. Since the work done by the leading shoe is much greater, the rate of lining wear of this shoe is higher than the trailing shoe. This means that unless a thicker lining is used on the leading shoe, this shoe reaches its wear limit well before the trailing shoe.

On the layout shown, the unequal wear pattern can be taken up by the floating pistons in the wheel cylinder. If this were not so, some provision would be needed to allow the brake shoes to be adjusted so that they can be centralized with the drum.

Self-servo action The effect of drum rotation on a brake shoe can be demonstrated by the arrangement shown in Fig. 66.2. Applying an equal force to the free end of the two levers, and rotating the drum in the direction shown, causes the frictional force on lever A to increase the loading on the shoe. The opposite effect is produced on lever B; here the frictional force acts against the force *F*, so this lever, which represents the trailing shoe, is not pressed so hard against the drum.

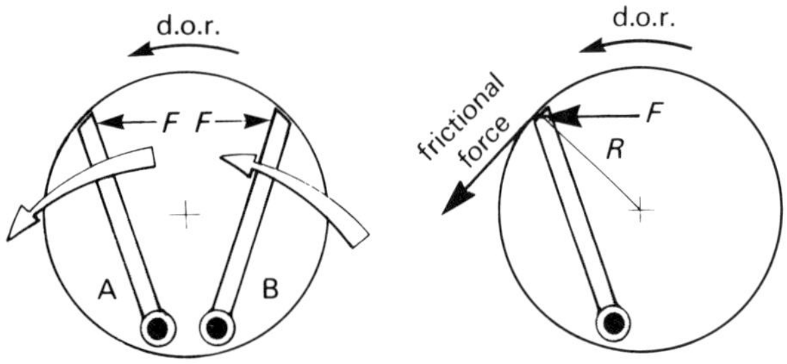

Fig. 66.2 Self-servo action

Reversing the direction of the drum changes the action and this converts the leading shoe into a trailing shoe.

The term *self-servo* is used to describe this self-apply condition and Fig. 66.3 shows the three factors that vary the amount of self-servo that acts on a leading shoe. They are:

1 *Position of shoe tip* - self-servo is increased as the leading edge of the friction lining is moved towards the hydraulic expander.
2 *Coefficient of friction* (μ) *of the lining* - an increase in the friction value gives a larger self-servo action.
3 *Position of the shoe anchor* - moving the anchor towards the centre of the drum increases the self-servo action.

Effects of self-servo To obtain an L & T shoe brake that has a high braking power (large drum drag for a given application force) a large self-servo action is required. The idea is to use the energy from drum rotation to minimize the effort applied by the shoe expander.

The negative servo action on the trailing shoe must be minimized in order to obtain a powerful brake. Often this is achieved by placing the friction lining of the trailing shoe nearer to the expander than is used in the case of the leading shoe. Attention to this feature must be given when reassembling a brake, because if the shoes are reversed a fierce braking action, and possible *grab* (lock-up) will result.

A number of problems arise when a large self-servo action is relied on to give a powerful L & T brake. Since self-servo action relies on friction, then any fall-off in the friction value seriously affects brake performance. Normally this occurs when the brake reaches a critical temperature; as a result serious *fade* occurs.

Another drawback of a brake that uses a high self-servo is the variation in braking power that occurs when the force applied to the shoe is varied. Ideally the ratio, 'drum drag/applied force' should be constant because this gives a *progressive* braking action. This was not the case with many L & T shoe brakes used in the past; instead the brake power increased with pedal pressure and often the brake locked up without warning.

Floating shoes Fig. 66.4 shows an L & T shoe brake fitted with a wedge-type adjuster. In this design the brake shoes butt against an adjuster unit that acts as a fixed anchor for the shoes. Both shoes are free to float up and down on the anchor; this arrangement allows for shoe centralization and also reduces the tendency of the brake to grab.

When this type of brake is applied, the two shoes are pushed outwards to contact the drum. Rotation of the drum forces the leading shoe directly against the anchor; the reaction of the trailing shoe is taken through the expander to the leading shoe.

Note the position of the brake linings on the shoes on this design. This asymmetrical layout means that care is needed to ensure that the linings are refitted in their correct positions after overhaul. Also it is important to check that the assembled back-plates have not been interchanged with their shoes from left to right and vice versa.

Hand-brake operation is by a lever and strut mechanism.

Two leading shoe brake

Before the universal adoption of the disc for front brakes, the 2LS was commonly used. Each shoe has

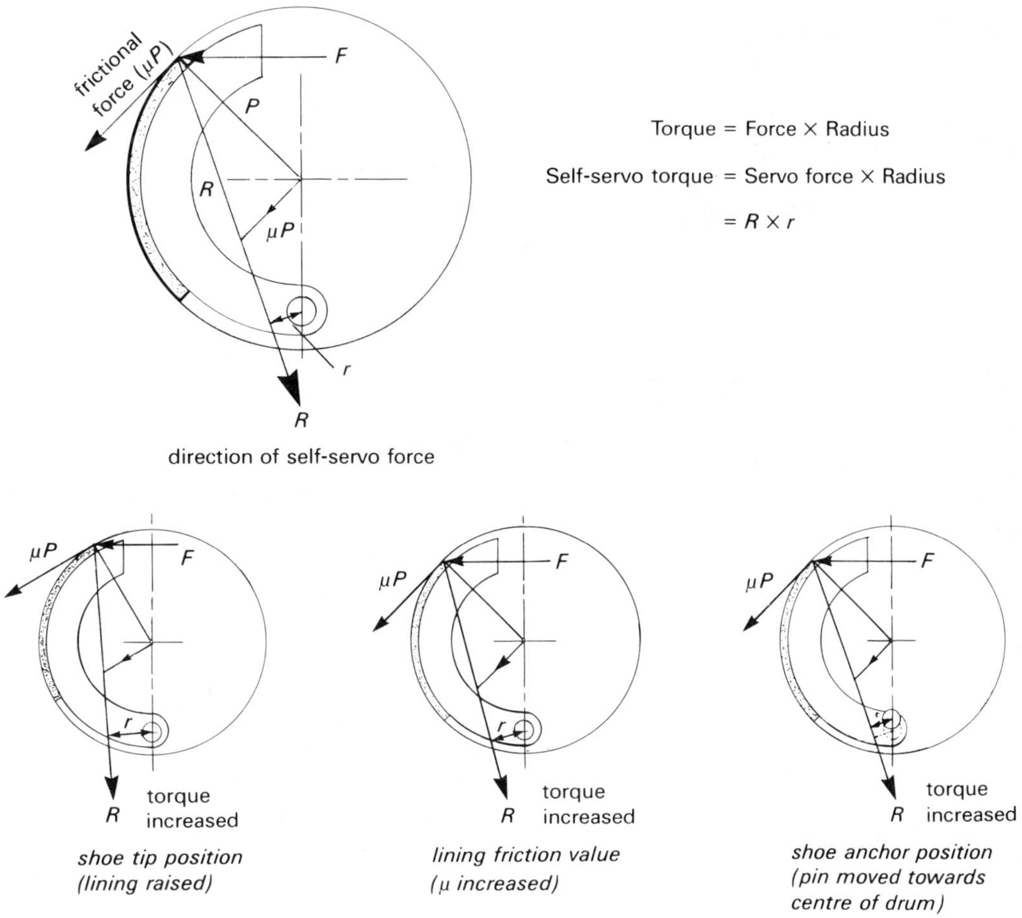

Torque = Force × Radius

Self-servo torque = Servo force × Radius

$= R \times r$

Fig. 66.3 Factors affecting self-servo torque

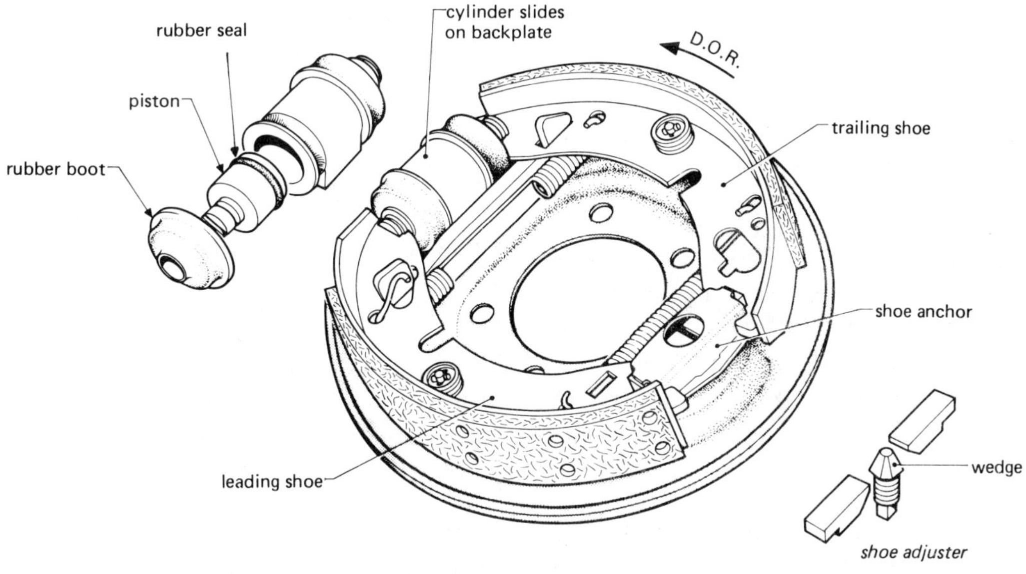

Fig. 66.4 Leading and trailing shoe brake

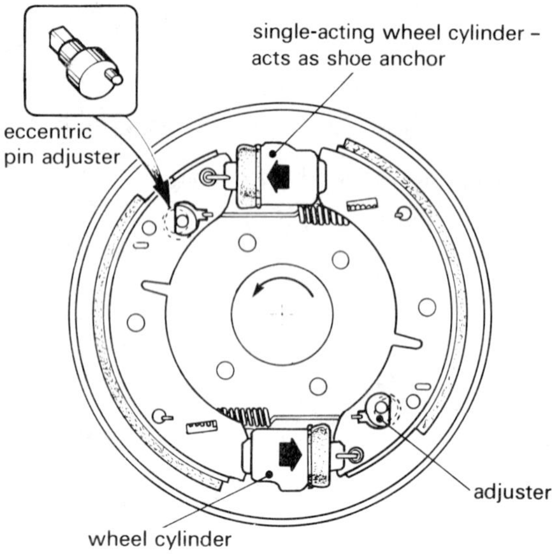

Fig. 66.5 Two leading shoe brake

its own expander, which was positioned so that both shoes were subject to a self-servo action (Fig. 66.5).

An interlinking pipe behind the backplate provides an equal hydraulic pressure to each single-acting cylinder. The cylinder housings acts as a shoe anchor for the floating shoes, so the cylinders must be rigidly attached to the back-plate.

Compared with a L & T shoe brake, the 2LS type has the following advantages:

1 *Even lining wear* - because both shoes do an equal amount of work, a longer life is achieved, the brake runs cooler and the need for adjustment is less frequent.
2 *Equal self-servo* - there are two effective shoes so a more powerful, stable brake is obtained.
3 *Greater resistance to fade* - less reliance is placed on one shoe to do the major share of the braking, so the self-servo action on this shoe can be reduced. This results in a more progressive brake and, as a result, it is less sensitive to heat.

One disadvantage of the 2LS type shows up during reversing. Unless special double-acting linkage is provided, both shoes become trailing shoes when the car travels backwards. To allow for this problem, and the handbrake provision, most manufacturers retained the L & T type for the rear brakes.

Duo-servo

This system is often called a *self-energizing* brake. It is a very powerful brake, but its effectiveness falls tremendously if the friction value decreases.

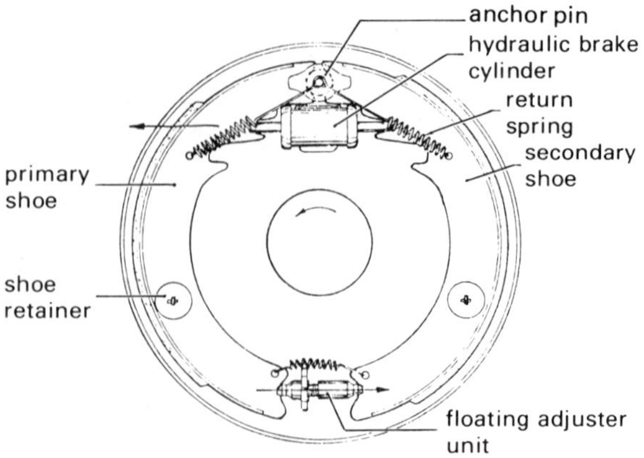

Fig. 66.6 Duo-servo brake

Fig. 66.6 shows a modern, hydraulically operated duo-servo brake. The operating principle is based on the use of drum energy to boost considerably the force applied by the driver.

When the leading shoe is pushed out into contact with the forward-moving drum, the frictional force causes it to rotate partially with the drum. The shoe movement produced by this self-wrapping action is transmitted through a floating adjuster to the trailing shoe which brings the shoe into contact with the drum. With the trailing shoe against the anchor pin and the shoe-to-drum clearance having been taken up, the force applied by the expander is supplemented by the self-energizing action of both shoes.

To minimize the delay before the self-energization assistance comes into action, the trailing shoe is held on the anchor pin by a stronger return spring; this means that the expander only moves the leading shoe. In this case only, the leading shoe is called the *primary shoe*; it gets this name because the shoe is made to contact the drum before the secondary shoe.

Automatic adjusters

Various methods are used to provide some form of automatic adjuster to set the shoes the correct distance from the drum.

Adjustment of front disc brakes is automatic, so to reduce servicing costs most modern rear drum brakes are fitted with separate automatic adjusters.

Two main types of adjuster are used: one type takes-up the wear when the footbrake is operated; the other type is controlled by handbrake operation.

Footbrake-operated Fig. 66.7 shows one of the many arrangements that uses the footbrake to take up wear. This type uses a cross-strut, retained to one

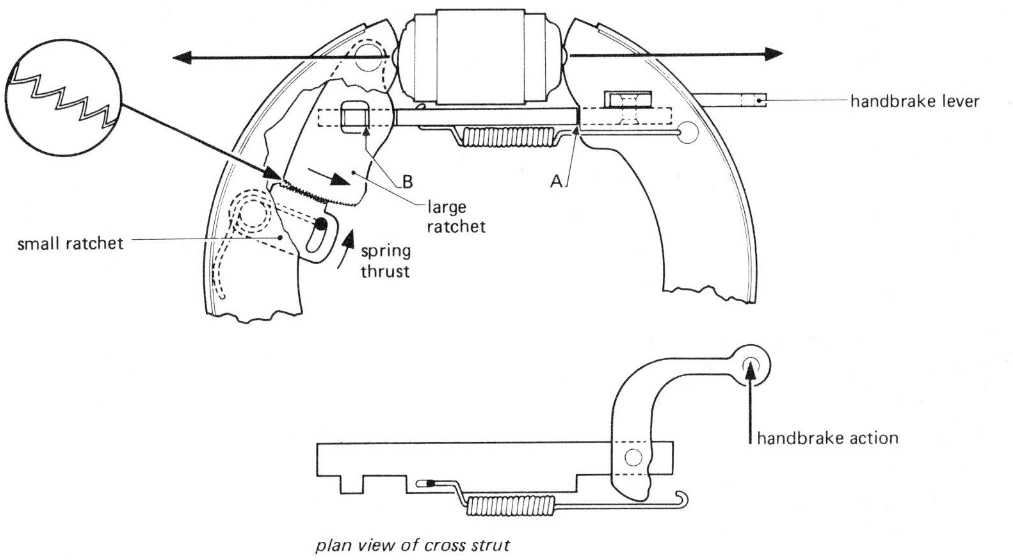

Fig. 66.7 Automatic brake-adjusting mechanism

shoe by a spring, and two ratchets mounted on the other shoe. A spring acting on the small ratchet pushes the fine saw-like teeth towards the large ratchet, which has a slot to receive the cross-strut. When the brake is set up correctly, a small clearance exists at B.

When the shoe-to-drum clearance is small, brake operation allows the ratchets to move with the shoe, but when the outward movement of the shoes exceeds the clearance at B, the strut holds back the large ratchet and causes it to jump a tooth. Consequently, when the brake is released the return movement of the shoe is less than before.

The handbrake operates in the normal way - a cranked lever acts directly on one shoe and the reaction on the pivot pushes the strut against the locked ratchets to expand the other shoe.

Handbrake-operated This system utilizes the movement of the handbrake to take up lining wear; Fig. 66.8 shows one arrangement. In this case the shoes are held apart to give the correct shoe-to-drum clearance by a threaded strut, which is expanded by a ratchet toothed wheel. This is moved by a pawl connected to the handbrake lever.

As lining wear takes place, the movement of the handbrake lever increases. When this movement exceeds a given limit, the pawl jumps a tooth on the ratchet. The handbrake action following this relative tooth-to-pawl movement lengthens the threaded strut and sets the shoes closer to the drum.

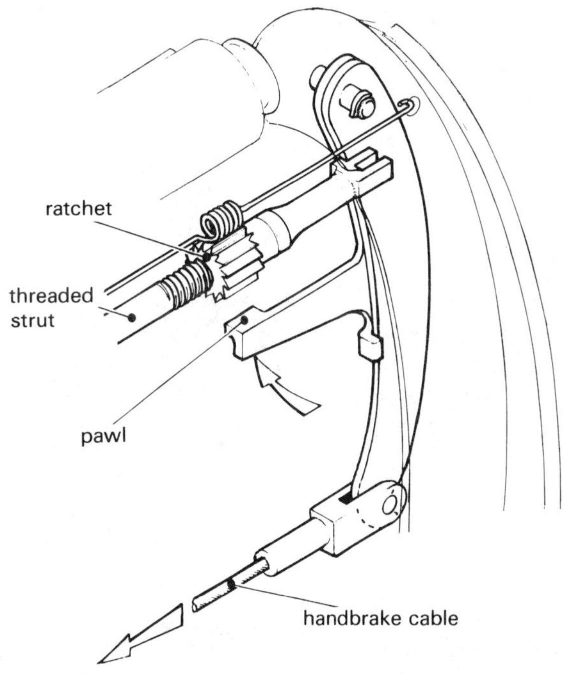

Fig. 66.8 Automatic adjuster – handbrake-operated

Duo-servo type brakes often utilize the self-wrapping movement to control the length of the floating adjuster. Some pawl and ratchet arrangements use the forward motion of the car to give the adjustment, whereas others only operate in reverse.

Shoe-release Sometimes drum removal is difficult if the drums have become grooved. To allow for this, a hole is often provided in the back-plate in the region of the automatic adjuster.

Disc brakes

Disc brakes are now favoured by many manufacturers of high-speed cars, since this type offers a number of advantages:

1 The surface on which heat is generated is directly exposed to the air, whereas in the case of a drum brake this is not so. Easier dissipation of the heat gives a greater resistance to fade. High temperature reduces the coefficient of friction of the lining, a fact clearly demonstrated when a vehicle fitted with drum brakes descends a long, steep hill; the pedal pressure at the end of the descent is much greater than at the start.
2 Independence of self-servo effect. The non-assisted brake may require more effort but its action is progressive, i.e. the brake gives a braking torque proportional to the applied force.
3 The brake is not so sensitive to friction changes.
4 Self-adjusting linings or pads are used which are easily replaced.
5 Pedal travel does not increase as the disc heats up - heating a drum causes expansion which increases pedal travel.

Two-cylinder caliper

The construction of a light-duty disc brake is shown in Fig. 66.9. A split caliper assembly, rigidly attached to the stub axle carrier, houses the cylinders and two opposed pistons, which act directly on friction pads mounted one on each side of the disc. Since the pads are visible, the degree of wear may be easily determined, and they are replaced by removing the split pins and pad-retaining springs.

A rubber seal, mounted in a groove in the cylinder, prevents fluid leakage and also retracts the piston and pad after application of the brake (Fig. 66.9b). This feature not only compensates for lining wear but also keeps each pad close to the disc.

Internal drillings link the two fluid chambers, and a rubber hose supplies the caliper with fluid from the master cylinder. A bleeder screw is fitted to each caliper.

Normally disc brakes are used for front wheels and

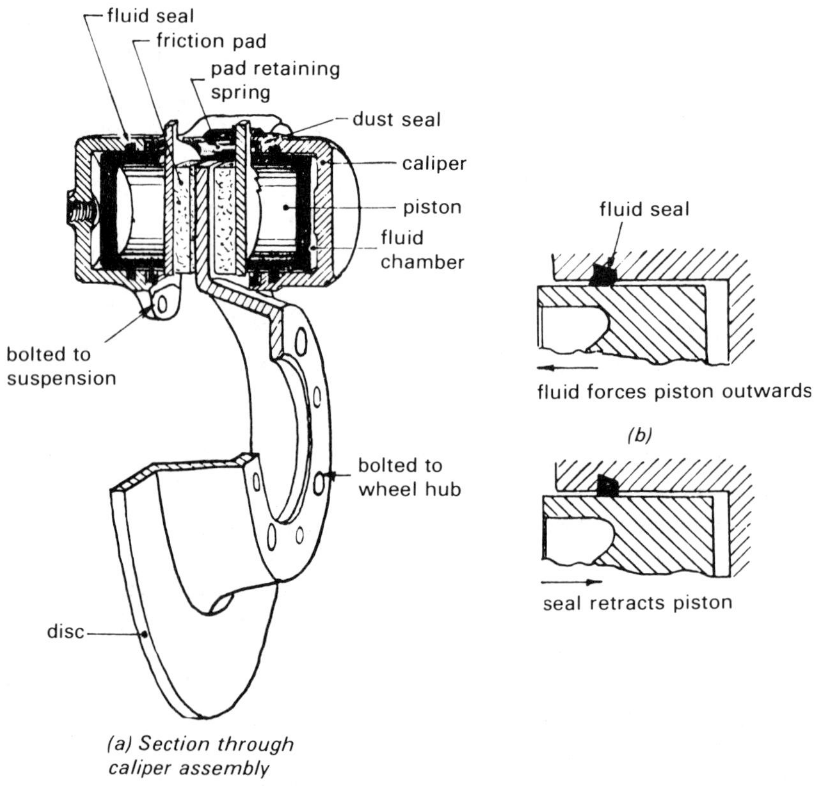

Fig. 66.9 Disc brake

conventional L&T shoe drum brakes are fitted to the rear.

Four-cylinder caliper

When greater safety is required, a four-cylinder caliper unit may be fitted. This is used in conjunction with a tandem master cylinder; one opposing pair of cylinders in the caliper is connected to the primary master cylinder piston and the other pair to the secondary. In the event of a line failure, one pair of cylinders will still remain operative.

Some disc pads are fitted with an electrical contact which rubs on the disc and activates a warning lamp when the pad needs replacing.

Single-cylinder caliper

The limited road wheel-to-disc clearance on vehicles with a steering geometry based on the negative offset (negative scrub radius) principle is often insufficient to accommodate a caliper having two opposed pistons. In these cases a single-cylinder caliper similar to Fig. 66.10 is used.

The piston housing is keyed to the pad housing, which is bolted to the wheel suspension member. Hydraulic pressure moves the piston in one direction and the piston housing in the opposite direction.

Asbestos-free friction material

In the past asbestos fibres were commonly used as a friction material. These were either woven or moulded and then bonded with resins to give a friction facing of the required form. Asbestos combines strength with good thermal/friction qualities and is relatively cheap to produce.

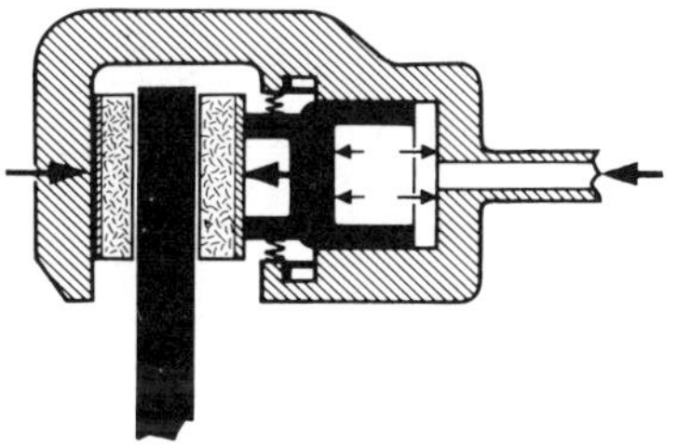

disc is sandwiched by the action of the piston and the reaction of the cylinder

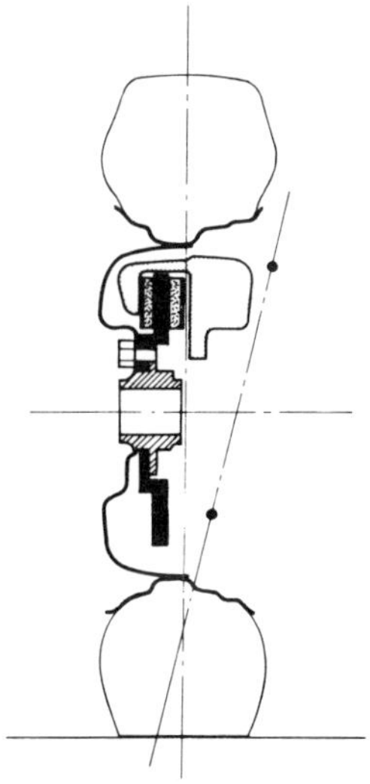

diagram shows that negative offset layout needs a compact caliper

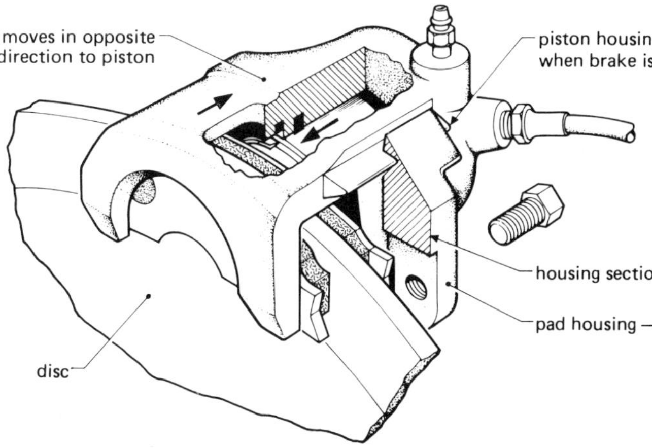

Fig. 66.10 Single-cylinder caliper

During recent years, it was found that breathing in asbestos dust increases the risk of lung cancer, so asbestos-free materials have been developed to overcome this health problem. New vehicles have asbestos-free materials for brake and clutch systems and although these materials are more expensive than asbestos, they have now been developed to a stage where they give a comparable performance to asbestos-based products.

The material used varies with the manufacturer; one company uses the man-made fibre *Kevlar*®, whereas others use a combination of steel and mineral fibres.

One problem with these new materials is the increased heat flow from the rubbing surface to the brake fluid. Vehicle tests are made in a mountainous terrain so that materials can be selected that limit the maximum fluid temperature to 175 °C. Since this is close to the 210 °C boiling temperature of normal brake fluid, various means, such as underlays, are used to act as a heat barrier.

Asbestos-free brake pads are comparatively rigid, so slots are formed to improve flexibility for bedding. The friction coefficient of these new materials is higher than that obtained from asbestos – an average of 0.43 compared to the 0.3 used in the past. Sometimes this high friction results in brake squeal, which is caused by vibration between the pad and piston. To overcome this problem the back of the pad is coated with either rubber or a copper compound.

Bonded friction facings

Up to a few years ago friction facings were secured by means of brass, copper or aluminium rivets. The modern arrangement is to use a specially cured adhesive to 'stick' the facing on to the pad or shoe. Besides giving greater strength and a more accurate profile, this method of attachment also improves heat flow and eliminates scoring of the rubbing surface by the grit-embedded rivets.

Although bonding allows more wear to take place before the shoe or pad is changed, it is recommended that as a general rule the facing should be changed when it has worn down to one-third of its original thickness.

67 Hydraulic operating systems

These popular systems offer the following advantages:

1 Fully compensated - each brake receives its full share of the pedal effort.
2 High efficiency - the efficiency of the hydraulic system is greater than that of the mechanical layout.
3 Suitable for vehicles having independent suspension.
4 Easy to alter thrust on shoe - the force exerted on a piston is governed by the piston area: the larger the area, the greater the thrust. For example, if the manufacturer requires a greater thrust on the trailing shoe, a larger piston can be used.

Layout of system

Fig. 67.1a shows a simple, single-line hydraulic layout used to operate a drum and disc brake system. Each wheel cylinder unit is fed from a master cylinder which is controlled from the footbrake. Regulations demand that a separate mechanical parking brake must be provided on at least two wheels; this provision also enables the driver to stop the vehicle in the event of failure of the hydraulic system. In the system shown, the parking brake, or handbrake, acts on the two rear wheels.

Most light vehicle braking systems used in the UK are made by either Lockheed or Girling.

Master cylinder

The brake manufacturers offer the car manufacturers two main types of master cylinder: single and dual (tandem) chambers for single- and dual-line layouts respectively.

Lockheed type

Single cylinder Fig. 67.1b shows the layout of a single-type master cylinder.

Linked to the pedal by an adjustable push rod is a piston (6), which is sealed by a main rubber cup (8), held in position by a spring, and seated on a thin washer to prevent the rubber entering the feed holes (7). An inlet port (3) allows fluid around the waist of the piston; leakage of this fluid is prevented by a secondary rubber cup (4). Outward travel of the piston is limited by a circlip, and a rubber boot (5) excludes the dirt. A 0.7 mm by-pass port (2), counterbored half-way with a drill of 3 mm diameter, is positioned just in front of the main cup. A check valve (1), fitted at the outlet end of the cylinder, ensures the non-return of aerated fluid during the bleeding process.

If you depress the pedal, the piston forces the fluid into the line until all shoe movement has been taken up. Thrust on the pedal will then pressurize the system and force each shoe against the drum. If shoe movement is too great, i.e. if the shoes require adjustment, the piston will reach the end of its travel before the brake applies. To overcome this problem, pump the pedal: a quick return of the piston creates a depression in the main chamber and causes fluid to flow through the feed holes and over the main cup to recharge the cylinder.

On release of the brake, the shoe return springs pump the fluid back to the master cylinder: any excess fluid in the line, caused by 'pumping' or expansion due to heat, is returned to the reservoir via the by-pass port. Blockage of this small port prevents the return of the fluid and leads to 'binding' brakes, so to ensure that the main cup does not obstruct the port when the brake is off, a small free pedal movement is given.

Tandem master cylinder

Pipe fracture in a layout operated by a normal type of master cylinder normally results in complete failure of the foot-brake system. When this occurs it is hoped

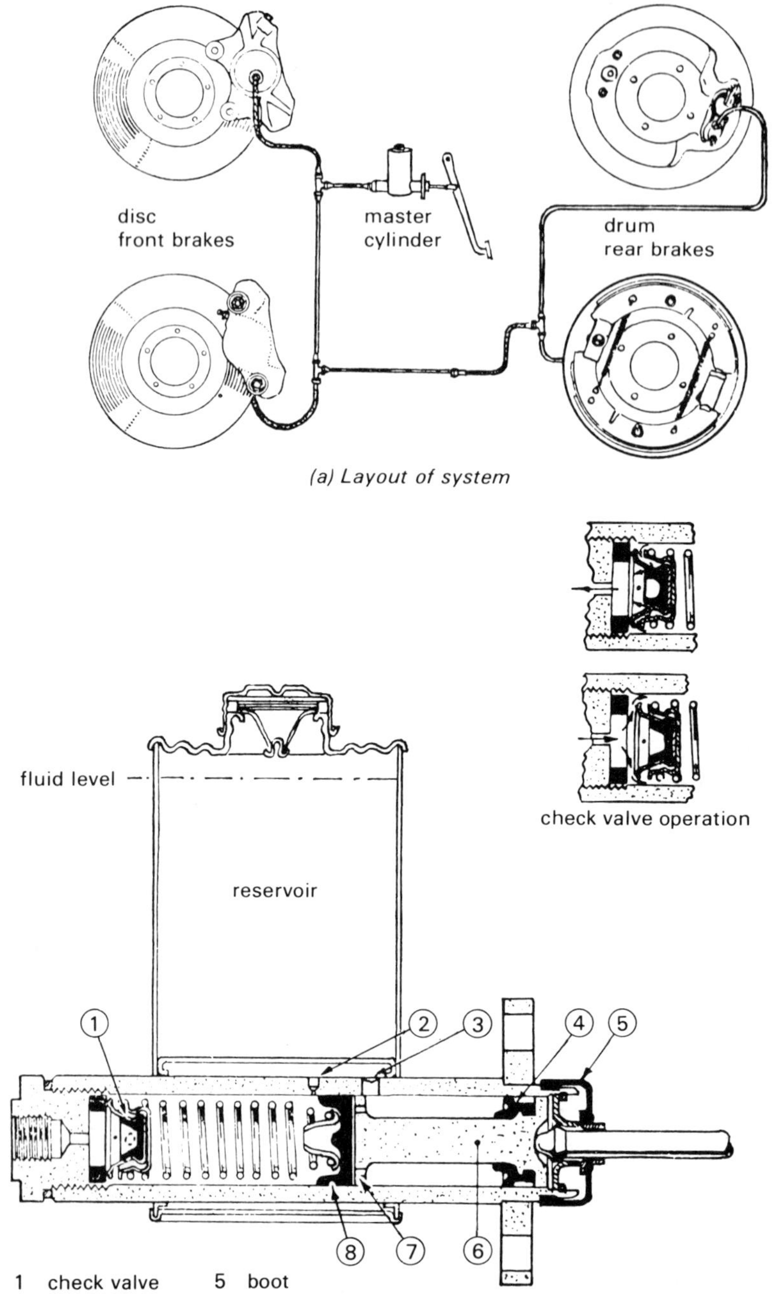

1 check valve
2 by-pass port
3 inlet port
4 secondary cup
5 boot
6 piston
7 feed holes
8 main seal

Fig. 67.1 Hydraulic brake system (Lockheed)

that the driver remembers that the handbrake is independent, because with this he is able to bring the vehicle to rest in a reasonable time. In order to avoid complete failure of a hydraulic system, a tandem-type master cylinder is fitted. With this type a failure in any one line results in a loss of only two brakes, so effective foot-brake control is retained and safety is improved.

In general the tandem cylinder may be considered as two single cylinders mounted end to end and Fig. 67.2 shows the close similarity between this conception and the actual layout.

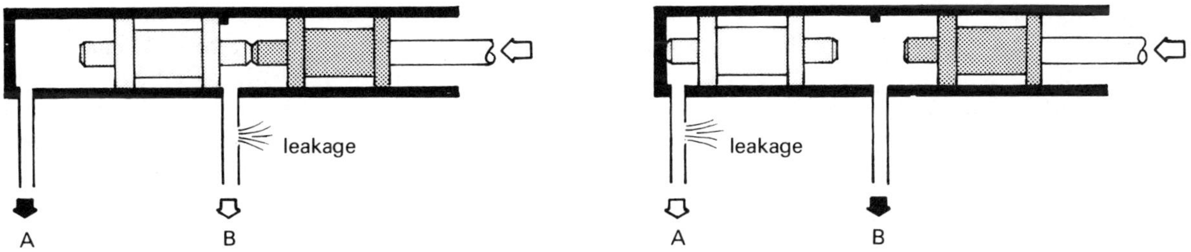

failure in line B *failure in line A*

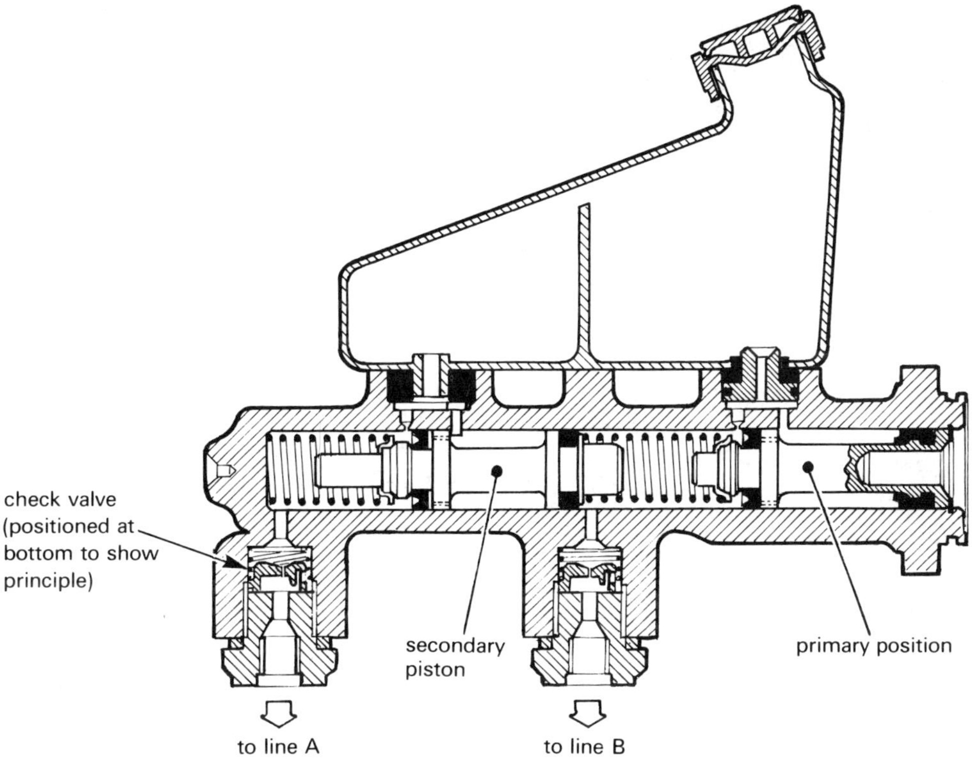

Fig. 67.2 Tandem master cylinder

The cylinder contains two pistons, one directly connected to the pedal and the other operated by fluid pressure. Each piston, which is fitted with seals to prevent fluid leakage, controls a separate line to either the front or rear brakes. At each of the two outlet points a check valve is fitted. One return spring is positioned between the pistons and a stronger return spring, acting on the independent piston, ensures that the pistons are forced back to their stops. Fluid is supplied through ports similar to the solo cylinder and the reservoir is divided into two parts to prevent a total fluid loss when one line fails.

Under normal conditions the movement of piston 1 causes an increase in the fluid pressure in the chamber controlled by piston 1. Pressure from this chamber is transmitted to the front brake line and to piston 2 which, being free to move, will pressurize the rear brake line to the same extent as the front line.

Assuming a failure occurs in the front line the movement of piston 1 will discharge fluid at the fracture and will allow the two pistons to contact. Although this stage has taken up some of the pedal travel, the remainder of the movement is available to operate the rear brakes. Repeated applications of the brake will eventually discharge all of the fluid from the portion of the reservoir which supplies the faulty section.

Failure in the rear line causes the initial pressure to move piston 2 to the limit of its travel. Once this point is reached the front brake can be operated successfully. An additional rubber seal fitted to piston 2 prevents fluid from leaking from the serviceable section to the fractured line.

Although this description of the tandem cylinder has referred to the two lines as being front and rear, it must be pointed out that other configurations exist. One common arrangement is a *split line diagonally linked* system: this connects the brakes diagonally into two separate circuits - left-hand front to right-hand rear and vice versa (Fig. 67.3). This layout, together with negative offset (negative scrub radius) steering geometry, improves the safety by overcoming the problems of front-rear systems: limited braking of the lightly loaded rear wheels and spinning of the vehicle if the rear wheels lock.

A more expensive method uses four-cylinder calipers on each front disc, each pair supplied by independent lines from the tandem cylinder. Failure of one line ensures that the other line keeps both front brakes and one rear brake in operation.

Girling type

Fig. 67.4 shows one type of Girling master cylinder. The flow of fluid from the reservoir to the main chamber is controlled by a compensating valve, which is set to open when the piston is in the fully returned position.

Wheel cylinders

A shoe-type brake assembly uses two single-acting type wheel cylinders to expand the shoes in a two leading shoe brake and a double-acting type for a leading and trailing shoe brake. These types of cylinder are shown in Fig. 67.5.

Wheel cylinders are fitted with a valve at the highest part to allow air to be bled from the system. Normally only one valve for the two cylinders is fitted on a 2LS brake assembly; this valve is provided on the upper cylinder.

Pressure regulating valve

This unit (Fig. 67.6) is fitted in the rear brake line, and is designed to reduce the risk of rear wheel skidding by limiting the pressure acting on the rear brakes.

The valve consists of a spring-loaded plunger contained within a body. Low fluid pressure will not overcome the spring, so the full pressure will act on all brakes. When the brake is applied, fluid will initially pass to the rear brakes, but when a predetermined pressure is reached the valve will close and allow any further increase of pressure to be felt only on the front brakes.

Brake pressure control valve (inertia valve)

This valve is a development of the pressure regulating valve and is particularly suited to overcome the problem of the large load variation between front and rear wheels of front-wheel drive vehicles.

The valve, which is fitted in the rear brake line(s), is an inertia-sensitive pressure-reducing valve that

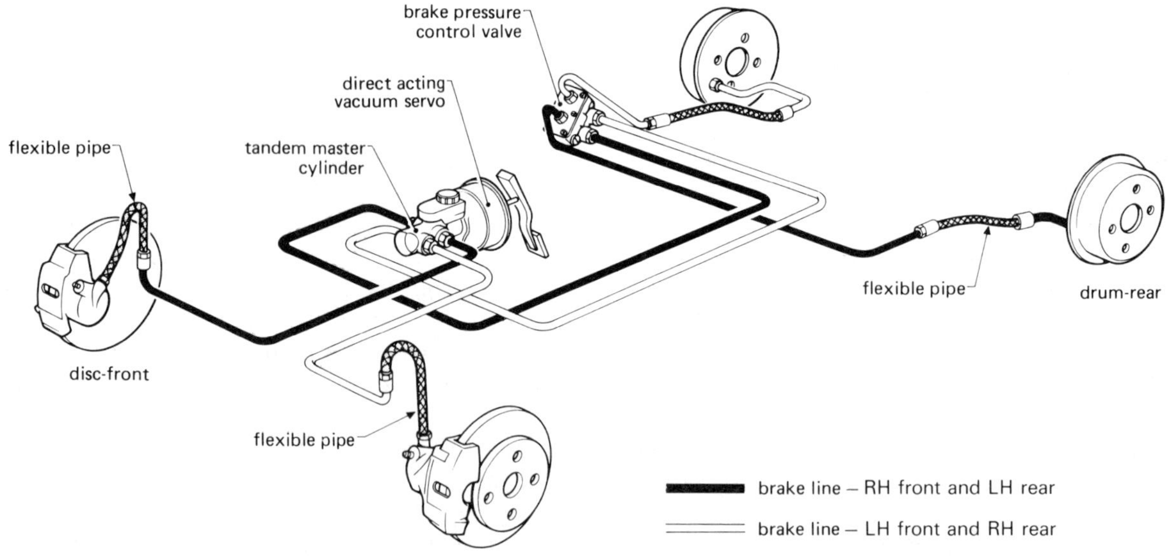

Fig. 67.3 Split-line diagonally linked system

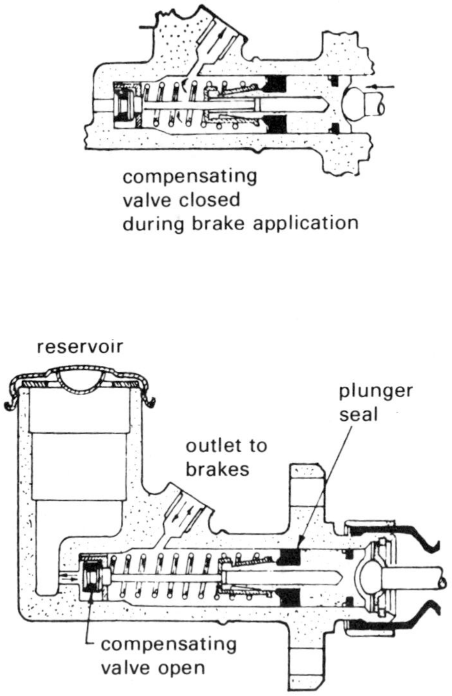

Fig. 67.4 Master cylinder (Girling)

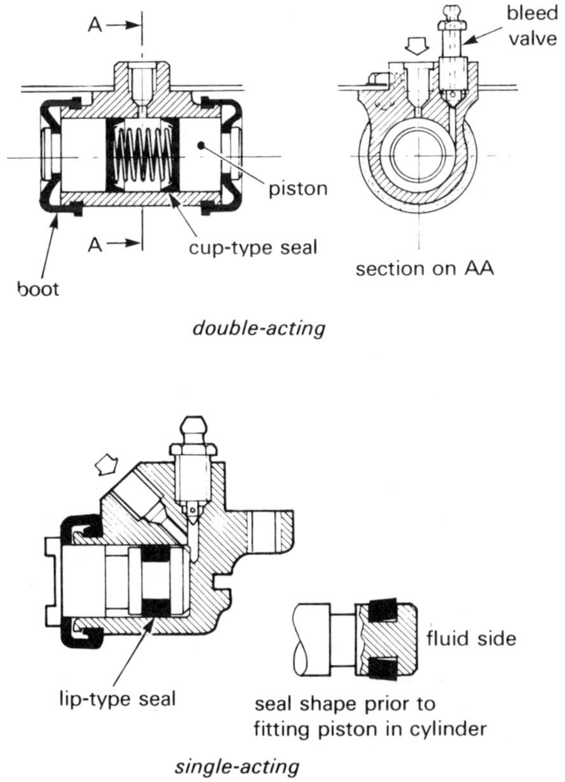

Fig. 67.5 Types of hydraulic wheel cylinder

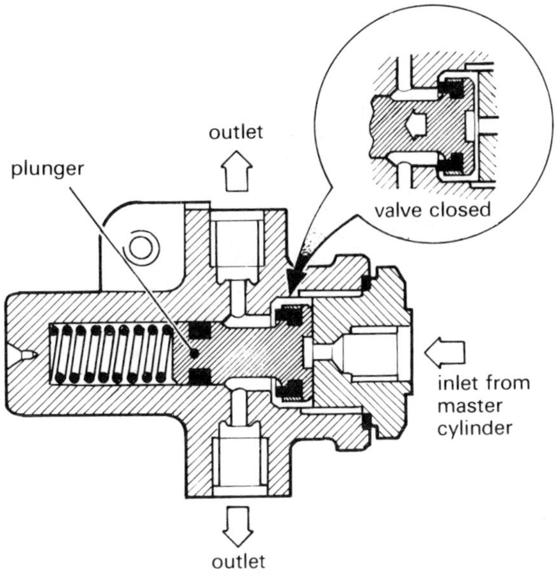

Fig. 67.6 Pressure-regulating valve

operates when the vehicle decelerates at a pre-determined rate. When this occurs, the valve temporarily shuts off the rear line and allows the extra force on the pedal to increase the front brake pressure further. After a given pressure is reached, the valve re-starts the pressure supply to the rear brakes but at a reduced rate of increase (Fig. 67.7). The unit takes into account vehicle weight transfer and attitude during braking, and is sensitive to vehicle loading and road conditions.

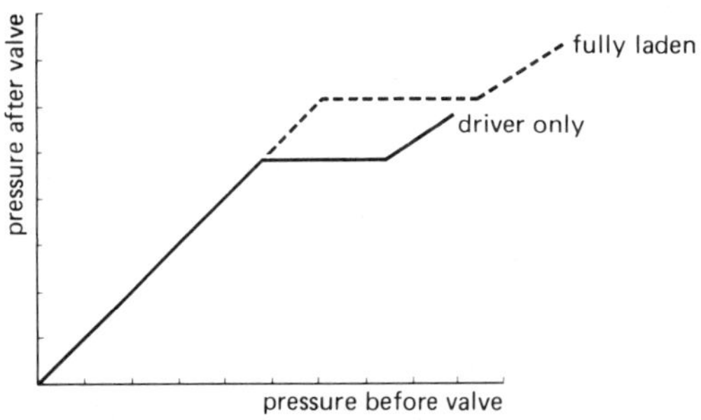

Fig. 67.7 Control valve performance

Fig. 67.8 shows the construction of a valve suitable for a common rear brake circuit; a system having independent lines has two valves mounted side by side. The unit consists of a cylinder, mounted to the car body at a given angle, which contains a stepped piston and a steel ball.

At low vehicle deceleration rates, fluid enters the inlet port, passes around the ball and through the

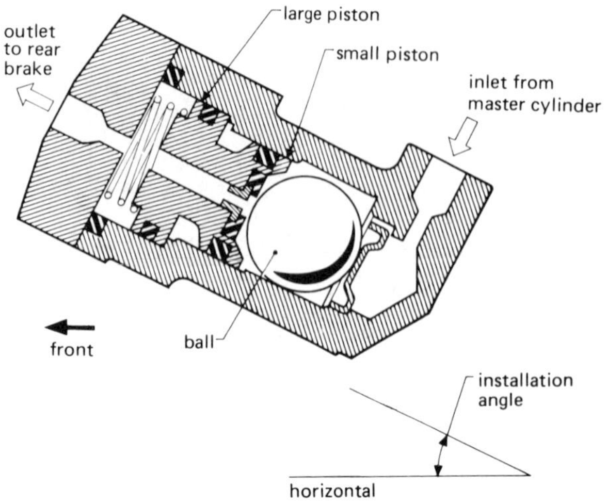

Fig. 67.8 Brake pressure control valve (inertia valve)

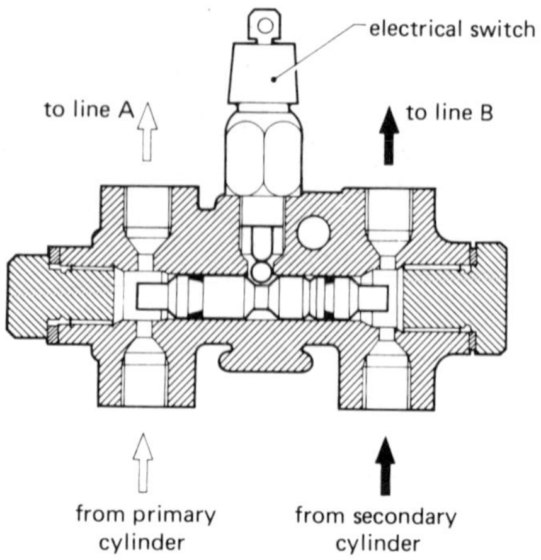

Fig. 67.9 Pressure differential warning actuator

piston drilling to the rear brakes - pressure is equal in front and rear brake lines.

When the rate at which the vehicle slows down produces an inertia force sufficient to roll the ball up the sloping cylinder, then the ball shuts off the fluid supply to the rear brakes. When this occurs, the difference in piston area causes the outlet pressure to remain constant even though the inlet pressure is being increased. At a given point, governed by the piston areas, an increase in inlet pressure starts to move the piston so as to give a proportional pressure to the rear brakes. Pressure in the two lines at this stage is given by:

INLET PRESSURE × Small area
= Outlet pressure × LARGE AREA

Pressure differential warning actuator

This is a warning device to illuminate a 'brake failure warning lamp' if the pressure difference in the two brake lines differs by more than a given amount.

Failure of one brake line causes the pistons in Fig. 67.9 to move and operate the electrical switch. The switch will remain closed until the pistons are reset.

Load-apportioning valve

The purpose of this valve is to supply, within given limits, hydraulic pressure to the rear brakes in proportion to the load carried by the rear wheels.

Besides reducing the risk of a rear wheel skid at times when the rear of the vehicle is lightly loaded, this arrangement also provides good braking when the rear wheels are heavily loaded.

One valve only is needed for a single hydraulic line layout, but where a diagonal line system is used a separate valve must be fitted in each line.

The valve housing is bolted to a fixed part of the vehicle body and either a compression or tension spring is used to sense the load on the rear wheels. This spring connects the valve-operating lever to a part of the suspension system that moves in proportion to the vehicle load.

Fig. 67.10 shows the construction of the valve. The operating lever acts directly on a piston which incorporates a ball valve. When the brake is 'off' the piston is at the bottom of the bore and the ball valve is held open by a push-rod fixed to the valve body. In this position the fluid can pass freely between the inlet and outlet ports.

Upward piston movement takes place when hydraulic pressure is applied to the valve. This is achieved by making the area exposed to the fluid on the upper part of the piston larger than the area on the bottom part.

Operation – light load

The force exerted by the external spring on the piston governs the hydraulic pressure that must be applied before the piston is raised and the ball valve is closed.

When the load is light, only a small force is exerted by the spring on the piston. As a result only a comparatively low pressure is needed to move the piston upwards so as to close the valve. When the pressure at this closure point is exceeded, full pressure cannot be applied to the rear brake; instead any further increase in pedal force will cause the

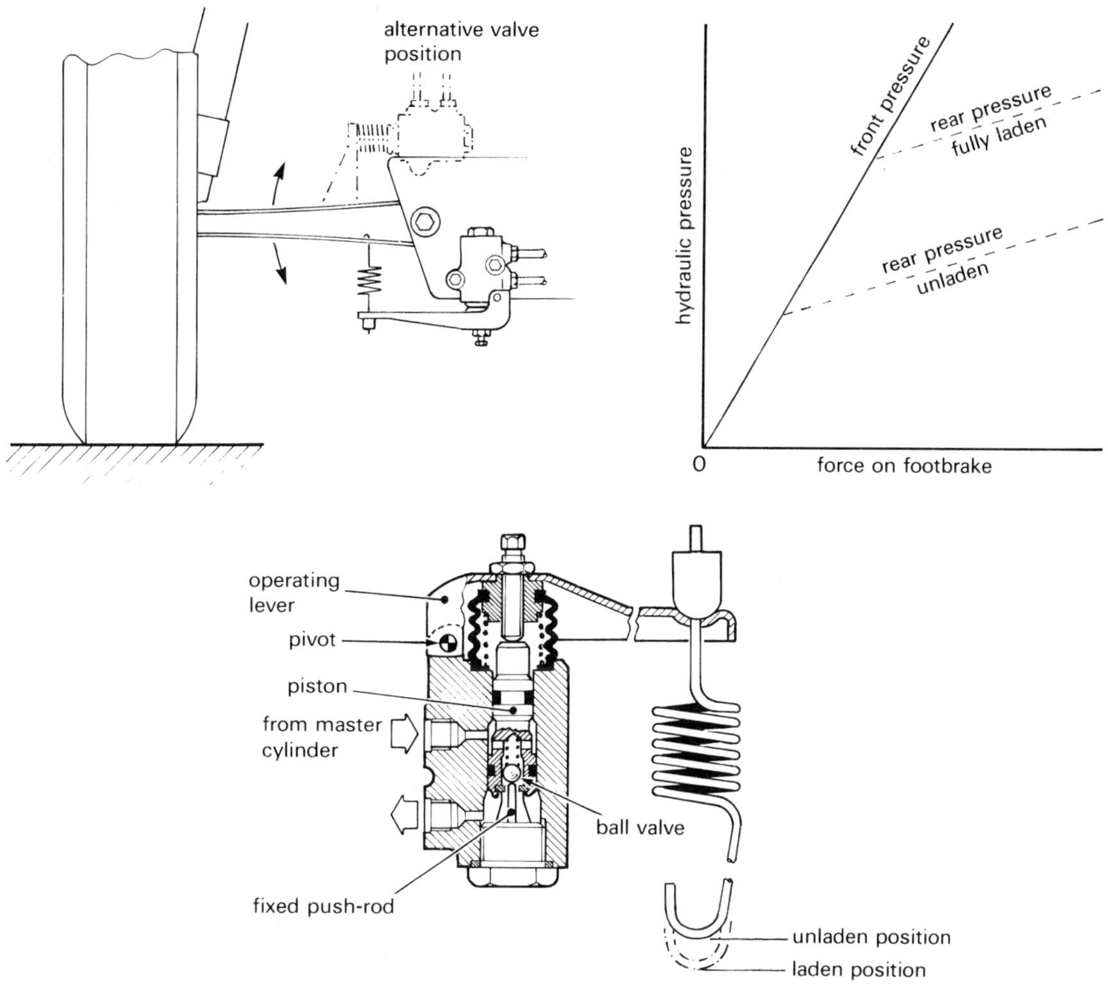

Fig. 67.10 Load-apportioning valve (Bendix)

piston to control the valve so as to supply a pressure that is lower and proportional to the pressure applied to the front brake.

Operation – heavy load

As the load on the rear wheels is increased, a deflection of the suspension causes the external spring force to increase. To overcome the extra force exerted by this spring on the piston, a higher fluid pressure must be built up before the piston is able to rise. Consequently, full pressure on the rear brakes is maintained until a much higher pedal force is applied.

Faults

Besides possible leakage of the seals, fracture of the external spring is likely. This breakage will cause the valve to reduce the pressure supplied to the rear brakes considerably. When the valve is defective, the complete unit is usually replaced.

The type shown in Fig. 67.10 has an adjusting screw between the lever and piston. This controls the point at which the valve comes into operation, i.e. the front/rear braking ratio for a given rear wheel load.

Brake fluid

A brake fluid must conform to the international standards introduced in the United States by the Society of Automotive Engineers (SAE) and Department of Transportation Federal Motor Vehicle Safety Standard (FMVSS).

The main requirements of a fluid include:

1 *Low viscosity* – the fluid must flow easily over a wide temperature range and be able to operate in very cold conditions.

2 *Compatibility with rubber components* - besides resisting corrosion of metal parts, the fluid must not chemically react with the rubber seals etc. (i.e. it must be non-injurious to the system).
3 *Lubricating properties* - to reduce friction of moving parts, especially rubber seals.
4 *Resistance to chemical ageing* - the fluid should have a long life when stored and must be stable when in use.
5 *Compatibility with other fluids* - the fluid must mix with other fluids in its class.
6 *High boiling point.*

Most brake systems use a glycerine–alcohol (glycol) fluid with additives incorporated to allow it to meet the specifications. Since a number of different fluids are available, some vegetable- and some mineral-based, the manufacturer's recommendation should be consulted BEFORE a system is serviced. This is essential because if a minute quantity of mineral oil is introduced into a system designed to operate on vegetable-based fluid, the rubber seals will be severely damaged.

Boiling point of brake fluid

Glycol-based brake fluids are *hygroscopic*; this means that they absorb water from the atmosphere over a period of time. As a result, the boiling point lowers and in extreme circumstances causes complete failure of the brakes because of *vapour locking*. This condition is produced when the fluid in a part of the system is heated above its boiling point sufficiently to vaporize the water in the fluid. Once this occurs, the elastic nature of the steam allows the pedal movement to reach its limit of travel before it can build up sufficient pressure to apply the brakes effectively.

In view of the hygroscopic nature of most brake fluids, the SAE and FMVSS specifications require the fluid to have a *wet boiling point* and *dry boiling point* in excess of the stated value. The 'wet' boiling point is the temperature at which a fluid containing 3–3.5 per cent of water boils and produces steam bubbles.

In the case of a typical fluid such as SAE J1703C, FMVSS 116 DOT 3, the 'wet' boiling point must be above 140 °C. Although this is higher than the 120 °C of fluids used in the 1950s, for safety reasons it is still necessary to change the fluid in a brake system every year. Bearing in mind that a brake fluid absorbs about 5 per cent water in this time, and that this contamination lowers the boiling point to about half its original value, then the need to change the fluid is evident.

To take account of the rising demands placed on brake fluids, especially in the field of disc brakes, the US Safety Regulations are continually being updated. Fluids that meet the new DOT 4 and DOT 5 standards are now available and some new fluids that have wet–dry boiling points of 180–260 °C permit the fluid renewal interval to be extended to 2 years.

Recently some special silicone-based fluids have been developed to overcome the hygroscopic problem, but these are not commonly used because of their cost.

Brake fluids in common use should be stored in sealed containers and should not come into contact with the paintwork of the car. If any fluid accidently drips on to the paintwork, then it should be washed off with water immediately.

Bleeding the brakes

This is necessary whenever air has entered the system. The main steps in the operation are:
1 Ensure that the reservoir is full.
2 Attach one end of rubber tube to bleeder valve and immerse other end in fluid contained in jar.
3 Open bleeder valve and slowly pump pedal until air bubbles cease to appear. Close bleeder valve as pedal is being depressed.
4 Repeat at all wheel cylinders.
5 Top up reservoir.

Faults

The main faults and their causes are as follows:

Fault	*Cause*
Pedal requires pumping	Shoes require adjustment
Springy pedal	Air in system
Spongy pedal (pedal creeps downwards)	Leakage in system, e.g. fluid passing main rubber cup

68 Servo operation

When the force that a driver can comfortably apply to the footbrake is insufficient to slow the vehicle down at the required rate, some form of assistance must be given. The boosting force used to reinforce the driver's effort is called *servo assistance*.

In the past the assistance given by brake drum rotation (*self-servo*) kept the pedal force low, but when powerful disc brakes were introduced, some other form of assistance was needed. Today servo assistance for light vehicle brakes is provided by either pneumatic or hydraulic means: vacuum assistance for medium cars, hydraulic assistance for heavy cars and vehicles fitted with anti-lock braking systems, and compressed-air assistance for some light trucks and minibuses.

Vacuum assistance

This is the most popular form of servo system. The term 'vacuum' is misleading because most systems use the depression formed in the induction manifold of the spark-ignition type of engine as a source of servo energy. In the case of a diesel engine, this energy is not available at the manifold, so an engine-driven 'vacuum' pump (exhauster) is used to provide the required assistance.

All servo systems must be *fail-safe*, so if for example, a fault develops in the vacuum unit, the main braking system must still be able to operate, albeit with a pedal force considerably greater than normal.

Progressive operation of the servo is a further requirement; this means that for light pedal pressure, the assistance given by the servo is proportional to the pedal effort. Fig. 68.1 shows the relationship between the hydraulic pressure that acts on the brake cylinders (both with and without servo assistance) and the pedal effort. The servo valves provide a gradual increase in the assistance up to the *knee point*; at this point maximum vacuum assistance is given. Any rise in output pressure beyond the knee point is a direct result of the increased pedal effort.

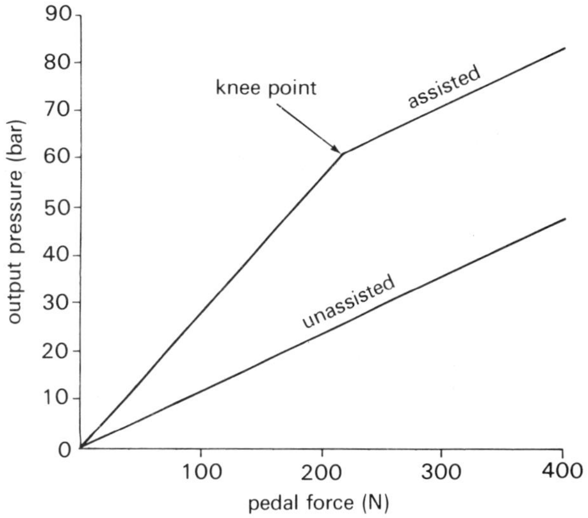

Fig. 68.1 Servo assistance

Vacuum servos in use today are called *suspended-vacuum* types, because when the vehicle is in use and the brakes are off, 'vacuum conditions' exist on both sides of the servo piston. When the brake is applied, air from the atmosphere is bled to the chamber on one side of the piston to give the pressure difference. This arrangement allows the servo to respond quickly and is a considerable improvement on the older type that was *atmospheric suspended*; in this air was present on both sides of the piston and this had to be 'drawn out' before assistance could be provided.

There are two main types of suspended-vacuum servo: indirect and direct.

Indirect type

This type is sometimes called a *remote* type because it is mounted remote from the pedal in the hydraulic line between the single master cylinder and the wheel cylinders.

Fig. 68.2 shows a servo suitable for a medium-sized car; it consists of three main items:

1 Vacuum cylinder containing a spring-loaded diaphragm.
2 Slave hydraulic cylinder.
3 Control valve actuated by hydraulic pressure.

With the engine running and the brake off, the unit will be as shown in Fig. 68.2a. The vacuum valve will be open and equal 'vacuum' pressure will be felt on both sides of the diaphragm.

Application of the pedal produces a hydraulic pressure on the brakes and also raises the valve piston in the servo. This movement closes the vacuum control valve and opens the air valve to allow a breakdown of the 'vacuum' in the outer chamber of the vacuum cylinder. Air pressure difference causes the booster diaphragm to apply a thrust on the slave-cylinder piston, which boosts the thrust given by the driver (Fig. 68.2b). 'Proportional' braking is obtained by allowing the difference in air pressure, felt by the booster diaphragm, to act on the control valve diaphragm: as the difference increases, the diaphragm will overcome the hydraulic pressure acting on the valve piston and close the air valve (Fig. 68.2c).

Release of the pedal drops the hydraulic pressure to allow the valve piston to return and open the vacuum control valve. Air is quickly evacuated from the outer chamber of the vacuum cylinder and a spring returns the diaphragm.

A hole in the centre of the slave-cylinder piston ensures that the brake can still be operated if the servo fails.

Apart from general inspection of hoses etc., the only attention the unit requires is the periodic cleaning of the air filter.

The popularity of a tandem-operated hydraulic system has made the indirect type obsolete.

Direct-acting type

The direct-acting, suspended-vacuum servo is designed as original equipment for cars and has the advantage that it can be used with either single or tandem master cylinders. It gets its name from the fact that the pedal linkage is directly connected to the servo.

Fig. 68.3 shows the position of the unit when the engine is running, the brakes are 'off' and the servo piston is '*suspended in vacuum*'. Air has been removed from both servo compartments and passed to the engine manifold through the pipe and non-return valve in the front of the vacuum chamber. Evacuation of air from behind the rubber-diaphragm-

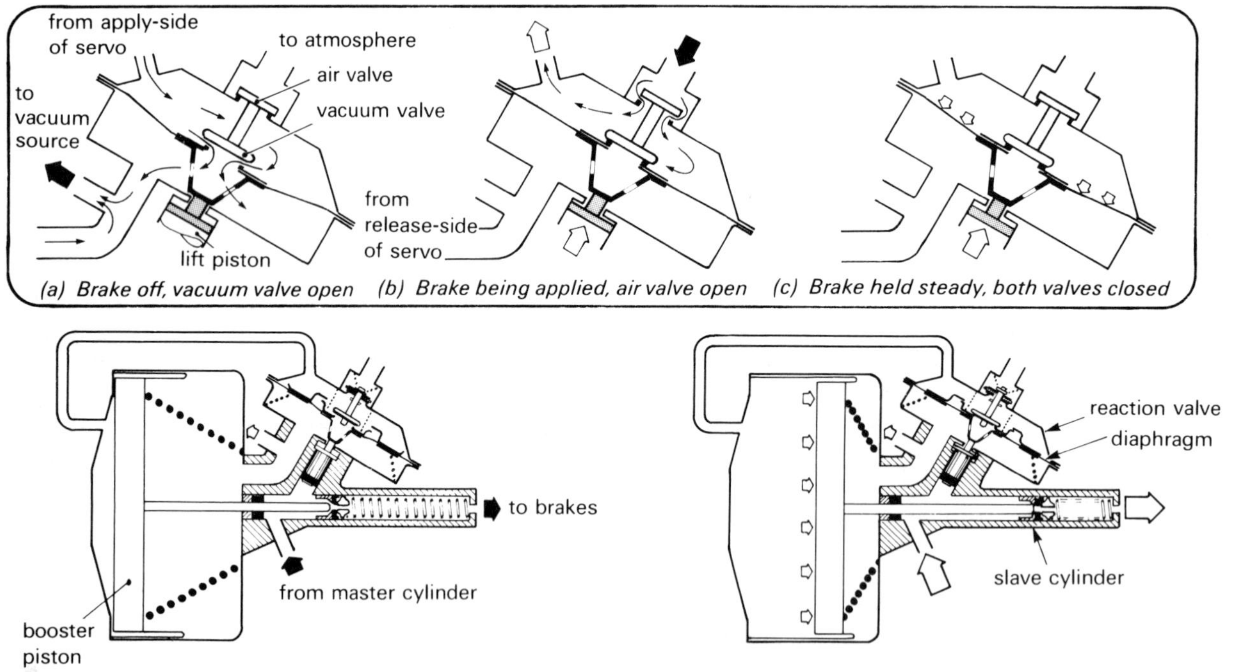

Fig. 68.2 Suspended-vacuum servo – indirect type

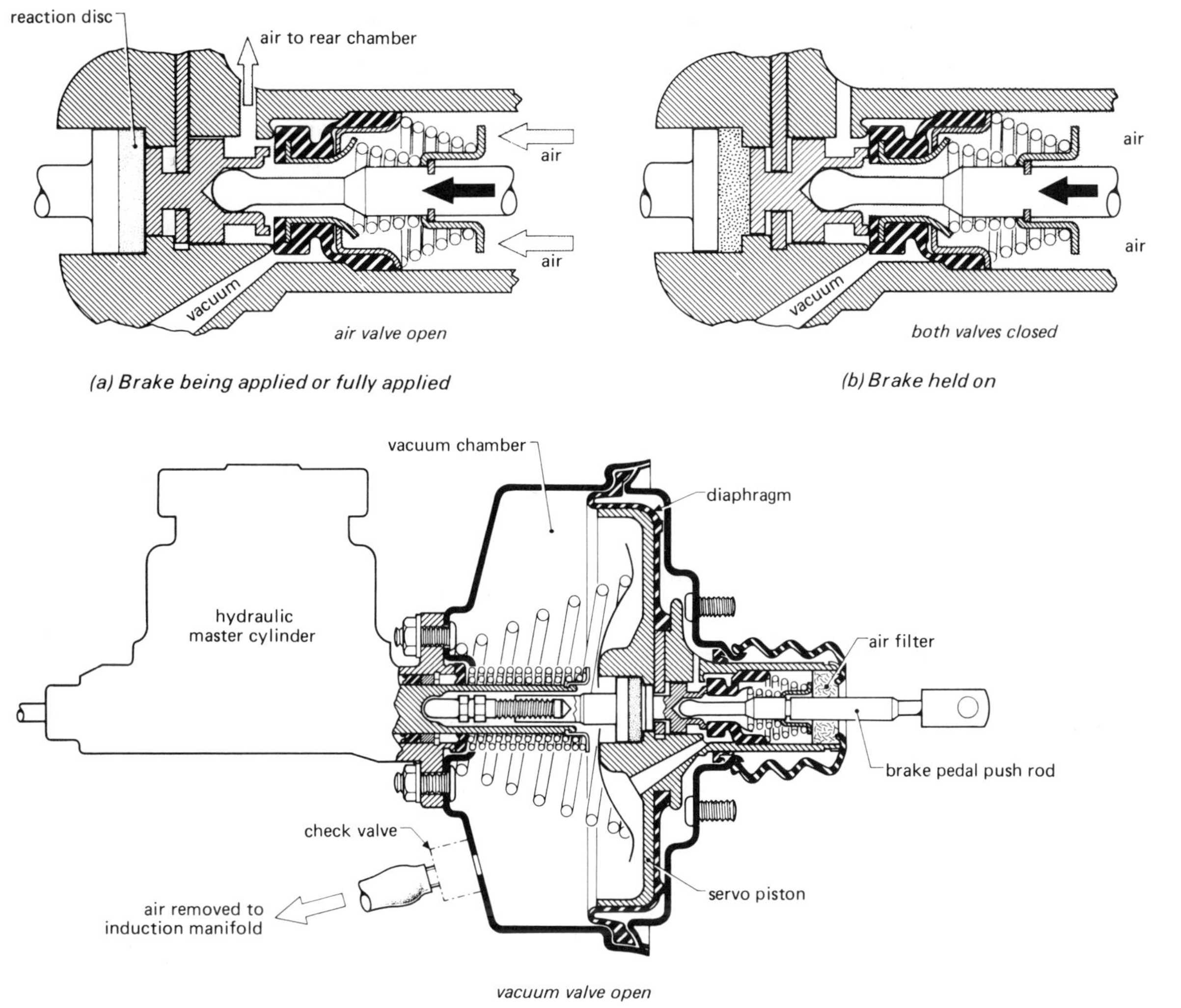

Fig. 68.3 Direct-acting servo

sealed piston is through a control valve linked to the footbrake pedal.

Movement of the brake pedal first closes the vacuum valve and then opens the air valve (Fig. 68.3a). This allows air to flow through the filter and valve to the rear chamber. Difference in air pressure between the two chambers produces a force on the piston which boosts the effort applied by the driver.

As the servo piston starts to move, it will exert a force on the master cylinder push-rod through a rubber reaction disc. Pressure on this spongy disc will cause it to squeeze back and close the air valve (Fig. 68.3b). In this position the driver's efforts will be supplemented by a servo boost which is proportional to the effort applied: this feature enables the driver to 'feel' the load applied to the brake and also gives a progressive operation.

Further movement of the pedal will reopen the air valve and the previous events will be repeated up to the 'knee point'. Beyond this stage the driver will be able to keep the air valve open; this indicates the limit of servo assistance.

Release of the brake pedal causes the air valve to close and the vacuum valve to open. This will restore the vacuum suspension of the piston and allow the spring to return the piston to the 'off' position.

In the event of vacuum failure the brakes can still be applied without servo assistance.

The assistance given by a vacuum chamber is proportional to the area of the piston, i.e. the larger the area the greater the assistance. Accommodating a very large cylinder adjacent to the footbrake on a modern car is difficult and various systems are used to overcome the problem. One arrangement is to fit a servo cylinder with two pistons mounted in tandem.

Vacuum supply

Although many servo units operate directly from the induction manifold, an improvement in performance, and extra safety, is achieved by using a vacuum reservoir to 'store vacuum'. Fig. 68.4 shows a typical system.

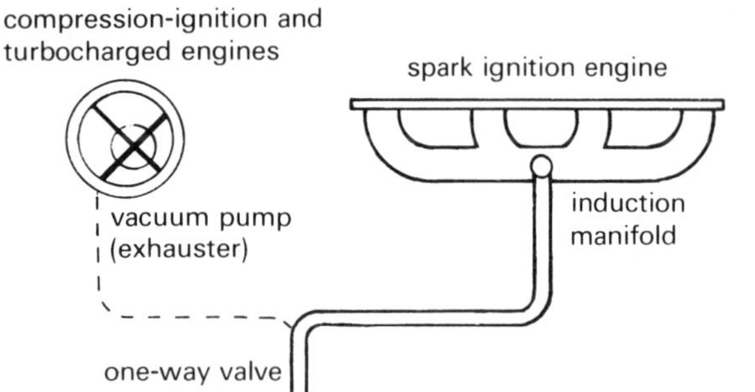

Fig. 68.4 Vacuum servo layout

The manifold depression, if any, of a diesel engine cannot be used to operate a brake servo, so a separate engine-driven pump is fitted. This pump, or exhauster, produces a 'vacuum' of about 500–600 mm (20–25 in) of mercury.

Hydraulic assistance

Unless an effective servo is fitted, a considerable effort is required to stop a heavy motor car or light truck. Since the pressure difference of a vacuum arrangement is limited, a system is needed that provides a much greater source of energy; this can be achieved with a hydraulic servo because this type of system operates on a pressure of 55–82 bar (800–1200 lbf/in^2).

The hydraulic power produced by the engine-driven pump of this system can also be used to provide for other servo needs, e.g. power-assisted steering, lifts, etc.

Fig. 68.5a shows a diagrammatic sketch of a *continuous-flow* hydraulic servo system.

Mounted behind the conventional master cylinder, the servo valve is supplied with fluid from a multi-cylinder radial pump which is driven from the engine or transmission. The diagram shows the brakes in the 'off' position, and in this state the fluid can easily pass between the master-cylinder piston and servo valve to a drilling which leads back to the reservoir.

Depression of the pedal initially closes the conical servo valve and causes the pump to build up a pressure in the region A, which will tend to force piston and valve apart. The piston has a larger area than the valve, and so the thrust exerted on the piston will be greater than that acting on the valve and brake pedal. This *area ratio* governs the assistance given to the driver. As soon as a given pressure, which will depend on the force applied to the pedal, has built up, the servo valve will partially open to maintain the pressure and give assistance. If the pedal force is exceptionally high, a pressure relief valve will open and allow fluid to escape to the reservoir.

Release of the pedal returns and opens the servo valve, releases the brakes, and restores uninterrupted flow of fluid from pump to reservoir.

Continuous-flow system with accumulator

The assistance given by the continuous-flow system depends on the pump speed, so a 'hard' pedal is felt when the pump is stationary or rotating slowly. To overcome this disadvantage, a hydraulic accumulator is normally incorporated into the system.

Fig. 68.5b shows the layout of the system, which uses the same pump and master cylinder as the previous system.

The accumulator or pressure storage cylinder

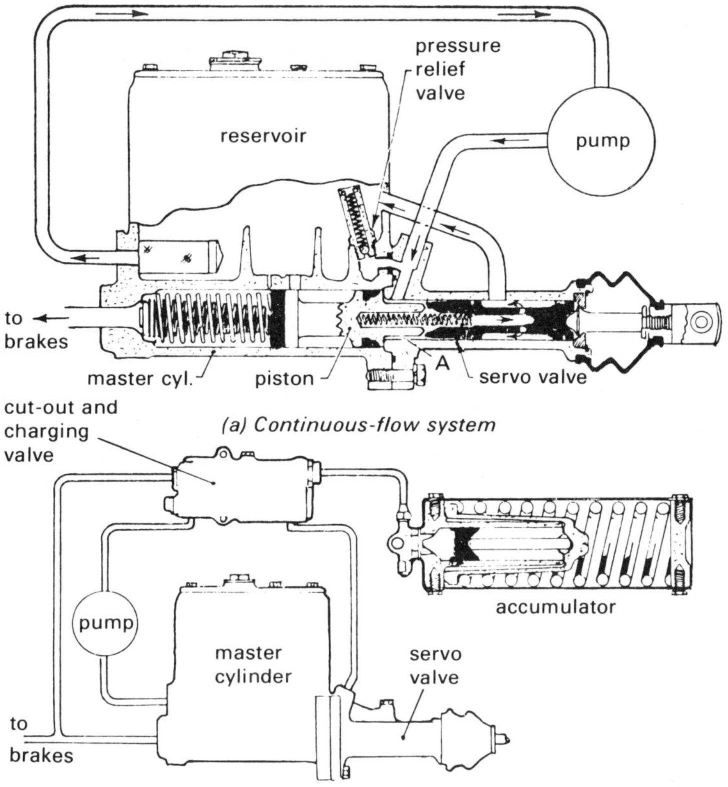

Fig. 68.5 Hydraulic servo systems

contains a spring-loaded piston or air bag, which is acted upon by the fluid: the higher the fluid pressure, the more the bag or spring is compressed. A cut-out valve maintains the accumulator pressure in the range 55–82 bar (800–1200 lbf/in^2), and a charging valve, activated by fluid pressure from the output or brake line side of the master cylinder, releases fluid from the accumulator to act on the servo valve, should the pump be incapable of supplying the necessary fluid.

Whenever you have to disconnect any part of this system, it is essential that you discharge and drain the accumulator.

Compressed-air assistance

On light trucks and minibuses, an alternative to a hydraulic servo is a *compressed-air* servo; this is generally called an air/hydraulic (air over hydraulic) or Airpac system because compressed air is used to boost the force applied by the driver to a hydraulically operated brake.

The system is based on the layout shown in Fig. 68.6. Air pressure, generated by an engine-driven compressor, is stored in a reservoir adjacent to the servo chamber; this houses a piston that operates another piston that controls the main hydraulic brake line. When the brake is applied, a valve directs compressed air into the servo cylinder to boost the pedal effort.

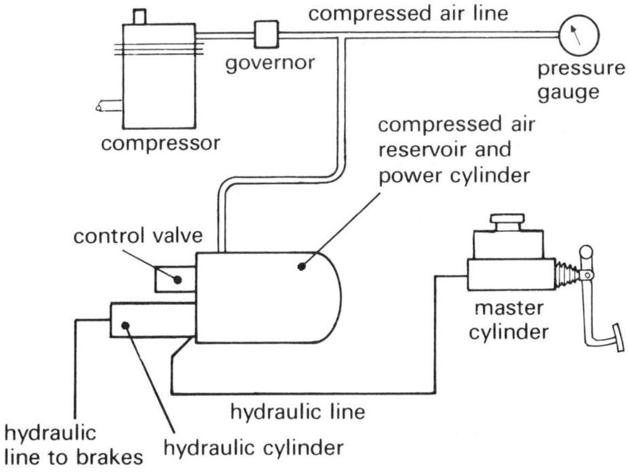

Fig. 68.6 Basic layout of air/hydraulic system

69 Anti-lock braking systems

As more and more vehicles use the road, there is an increasing need for systems that provide for greater safety. A braking system that incorporates an arrangement that prevents the road wheels from locking up and skidding should contribute to this extra safety, providing it is used correctly.

Need for an anti-lock system

A driver finds it difficult to assess the force that should be applied to the brake pedal to make an emergency stop and bring the vehicle to rest in the shortest possible distance. This is because it is an impossible task to take into account the many varying factors such as type and roughness of the road and tyre, and the condition of the surfaces, e.g. wet, dry, greasy or icy. This generally means that the driver applies either too much or too little pressure on the pedal. The effects of these actions are:

1 *Pedal pressure too high* One or more of the wheels skid over the surface with the result that:
   a Stopping distance is increased because the adhesion between a skidding wheel and the road is less than that given by a wheel that is held on the verge of skidding.
   b Directional control is jeopardized. In the case of a rear-wheel skid, the vehicle turns from front to rear.

2 *Pedal pressure too low* Stopping distance is increased, which can result in impact with an obstacle.

The adhesion between the road wheel and road surface is governed by the coefficient of friction. This varies considerably when the condition of the road surface is changed, e.g. the friction value of a dry asphalt road falls from about 0.8 to 0.15 when it is covered with black ice.

On a dry surface the adhesion varies as the percentage of wheel slip changes from 0 per cent (no slip) to 100 per cent (locked wheel sliding over the surface). Fig. 69.1 shows how the longitudinal and transverse adhesion of a tyre on a typical surface changes as the amount of slip is increased. The curve representing longitudinal adhesion for a given condition shows that a braking force set to hold a wheel at about 15 per cent slip provides the maximum retarding force.

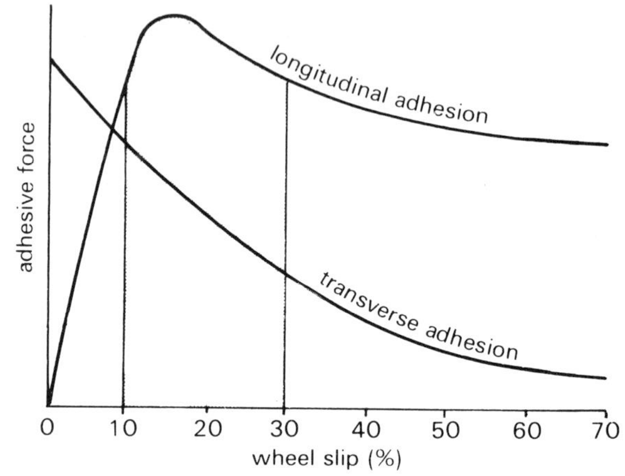

Fig. 69.1 Adhesive force and wheel slip

Vehicle stability with respect to its tendency to slide sideways is indicated on the graph by the transverse adhesion curve. The adhesion in this direction falls as soon as the wheel starts to skid, so if any serious skidding occurs, this can quickly lead to loss of control, especially when the driver is making a steering correction, or when the vehicle is cornering.

In the mid-1980s various skid-limiting systems were introduced. In the UK, these arrangements are generally called *anti-lock brake systems* (ABS).

Types of ABS

Retention of steerability in maximum braking situations and avoidance of skid danger under all road conditions are the basic requirements of a good anti-locking system. The system should be fail-safe and low in weight, and the initial cost should be reasonable.

The systems in use differ in the method that is used to sense wheel slip; the systems are:

- *Electronic*
- *Mechanical.*

Electronically sensed ABS

Whereas early designs of electronic system were intended as 'add-on' units, the latest generation of ABS units are incorporated into the hydraulic brake circuit so as to form an integrated brake system. The master cylinder and vacuum servo used in original designs are replaced by a hydraulic pump-assisted braking unit; this provides both normal braking and anti-lock functions.

Skid sensing of each wheel is provided by *electromagnetic sensors*; these measure and communicate to an *electronic control unit* (ECU) any rapid change in speed of rotation of the wheel, i.e. the sensors indicate the percentage slip that is taking place at each wheel.

Layout of system

Fig. 69.2 shows the layout of the main components of a four-wheel regulated system similar to that produced by ATE. The system may be divided into two sub-systems: hydraulic and electronic.

Hydraulic components For normal and anti-lock functions these are built as an integral unit in modular form around a driver-controlled master cylinder (Fig. 69.3). At the end of this cylinder is fitted a hydraulic booster that is supplied with fluid under pressure (between 140 and 180 bar) from an electrically driven pump. An accumulator stores the pressure and smoothes the output pulses from the pump.

Control of the pressure in the two front hydraulic

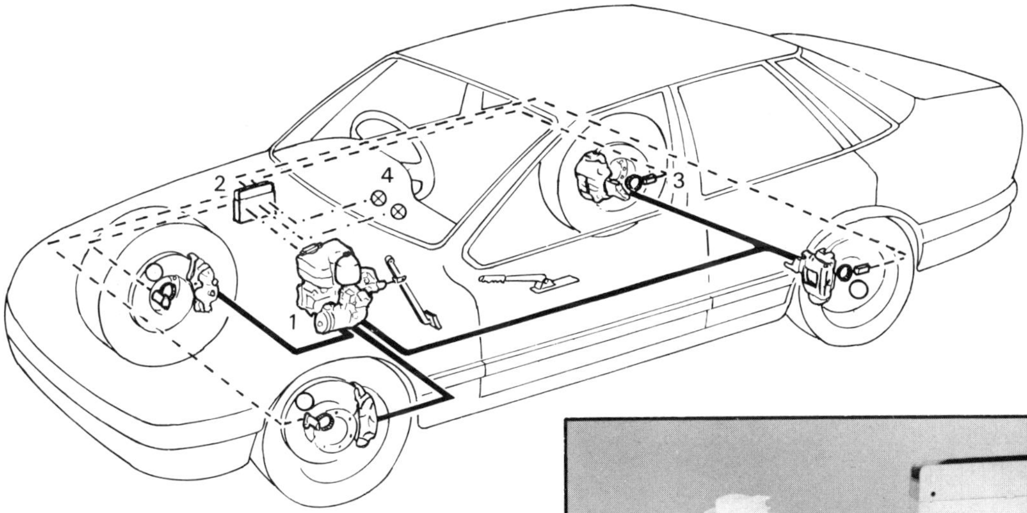

1 ABS module
2 electronic control unit (ECU)
3 road wheel speed sensor
4 warning lamps

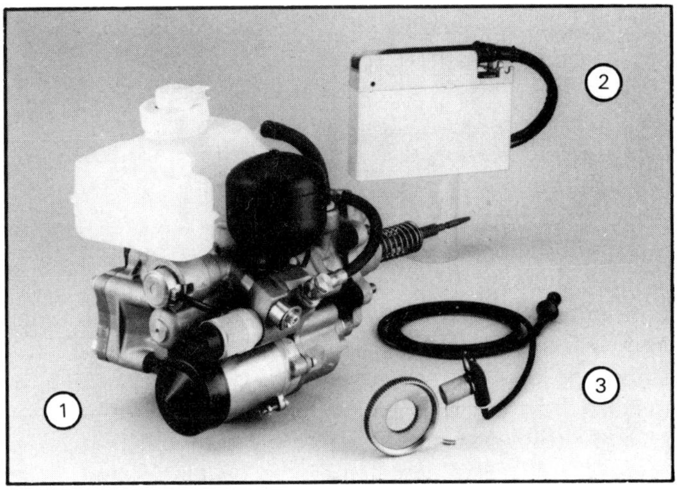

Fig. 69.2 ABS layout (ATE)

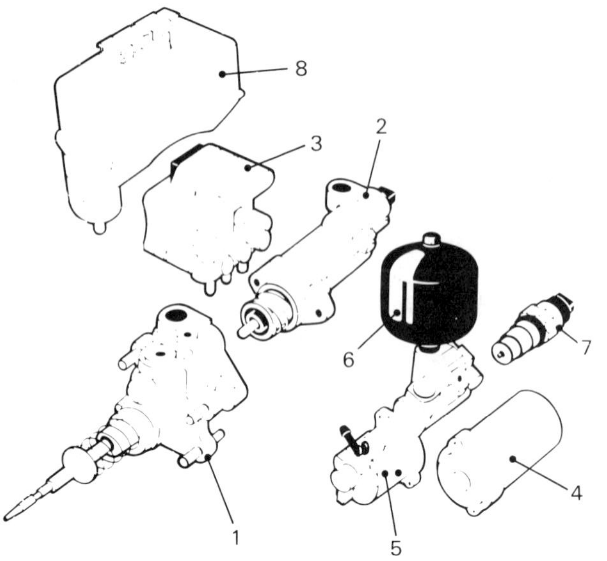

1 hydraulic booster
2 TEVES ABS master cylinder
3 TEVES ABS valve block
4 electric motor
5 high-pressure pump
6 high-pressure accumulator
7 pressure warning switch
8 reservoir

Fig. 69.3 ABS integrated hydraulic unit (ATE)

lines and the line supplying the two rear brakes is by means of valves contained in a modulator unit.

Electronic components Fig. 69.2 shows the electrical layout. Fitted to each wheel is a sensor that detects the speed of movement of a toothed gear-wheel; this gear is driven by either the drive shaft or the road wheels.

Although the system is a four-wheel sensed arrangement, the rear wheels are regulated as a pair. If a sensor on a rear wheel detects the onset of a skid, then fluid pressure to both wheels is reduced; this is called a *select-low* operating principle, because the pressure applied to both wheels is dictated by the wheel having the lowest adhesion.

Signals from the sensors are transmitted to an ECU which contains two microprocessors; these compare the input signals received from each wheel with those obtained previously. If they find that a wheel is starting to skid, an output signal is transmitted to actuate a return valve which slightly releases the pressure in that particular hydraulic line. By monitoring about 8000 sensor signals per second and taking action within in a few milliseconds, the ECU is capable of providing a lock-free state of all wheels so as to ensure safe deceleration and good vehicle stability.

Warning lamps are fitted to signal to the driver any fault in the pressure and electrical sections of the ABS. Self-checking electrical pulses continually pass around the circuit and valves to monitor the system and set the anti-lock mechanism in the fail-safe mode if a defect is detected; a lamp signals this information to the driver.

Operating principle

During an anti-lock brake operation on a triple fluid circuit system having one pair of solenoid valves per circuit, three phases of operation take place; these are:

A pressure reduction phase
B pressure maintain phase
C pressure increase phase.

Actuation of the input and return valves in each fluid circuit control these phases so as to give the maximum braking force without exceeding the skid limit for the road conditions.

Fig. 69.4 is a schematic diagram of an ABS layout using a *dynamic flow* principle. The fluid in this type of system is being pumped continuously around the circuit. From the pump it passes through the hydraulic booster and master cylinder and is then returned to the pump input for recirculation.

A. Pressure reduction When a sensor indicates that a wheel is beginning to skid, the ECU closes the input valve in the valve block and opens the return valve. This decreases fluid pressure in the appropriate line and as a result reduces the risk of a wheel locking. The surplus fluid displaced from the return valve is returned to the reservoir.

When ABS regulation commences, the ECU opens the main valve; this allows fluid under pressure in the booster chamber to flow over the cup seal of the master cylinder into the pressure chamber in the forward part. This compensates for fluid returned to the reservoir by the return valve and provides both front brake circuits with a pressure proportional to the pedal force.

During normal brake operation the rear brakes are supplied with fluid from the booster circuit at a pressure proportional to the pedal force. When ABS regulation of the rear wheels takes place, the single return valve in the line reduces the pressure equally for both rear wheels (select low principle).

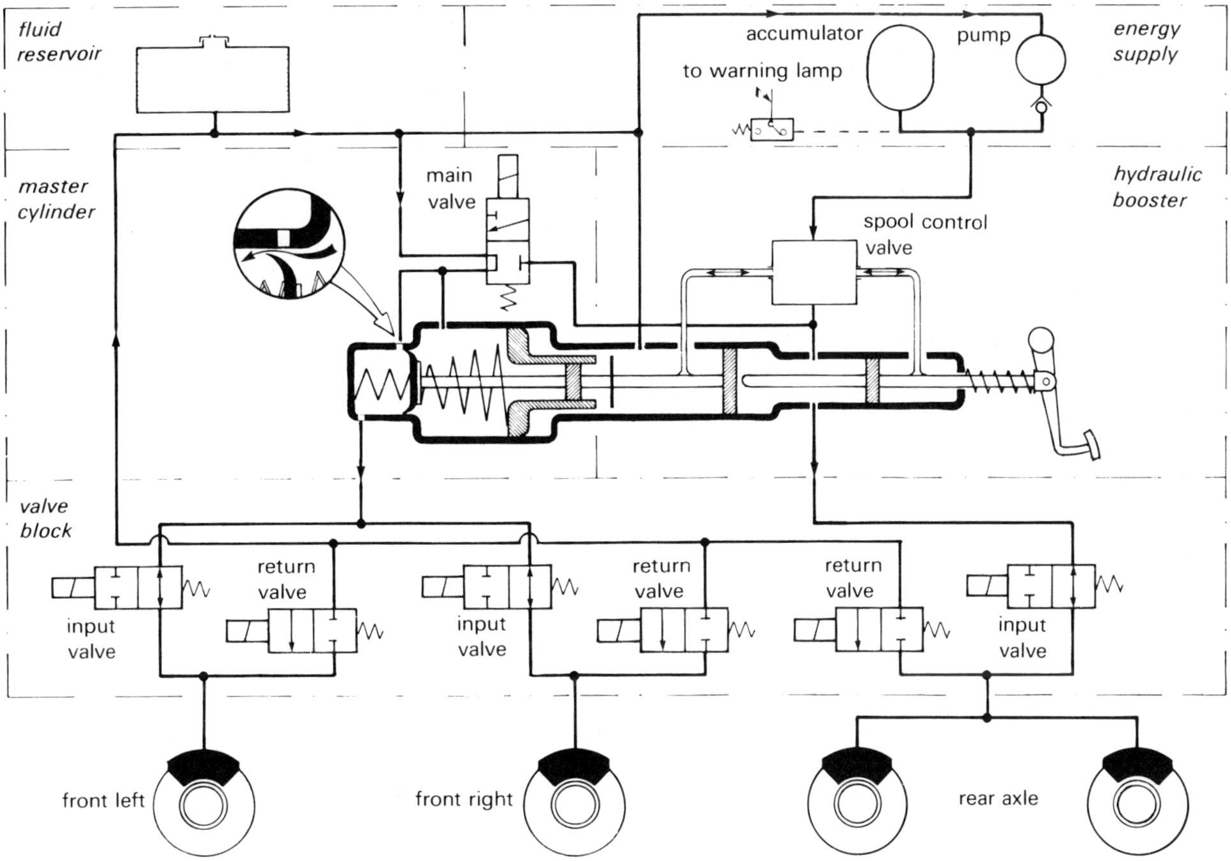

Fig. 69.4 Schematic diagram of an ABS layout

B. Pressure maintain When the ECU detects that the drop in fluid pressure is sufficient to avert wheel locking, it closes the return valve and allows the pressure to be maintained.

C. Pressure increase If the sensor signals to the ECU indicate that the wheel is running free, then an input valve in the regulated fluid line will be opened and the pressure will be increased up to a point where the percentage slip is greater than required.

This regulated cycle is repeated about 12 times per second; this frequency is designed to avoid vibration of the wheel and suspension assembly.

Regulation cycle The graph and diagram in Fig. 69.5 shows the sequence performed by an ABS arrangement as a vehicle is slowed down at its maximum rate. The simplified curves represent the braking action on one wheel of the vehicle.

Controlling the valves in the manner shown allows the braking force to be varied to keep the wheel within the slip limits dictated by the conditions. The variation in the speed of the wheel is the result of the ABS action and the points A, B, and C marked on the graph indicate the three phases of the operation. At point A, brake pressure is reduced by a specified amount, at B held constant so that the wheel can accelerate, and at

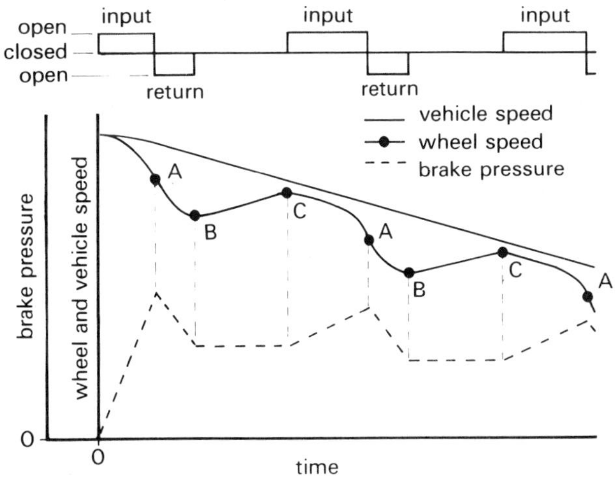

Fig. 69.5 Regulation cycle

point C brake pressure is again increased until the designed slip limit is reached.

Mechanically sensed ABS

Introduced by Lucas-Girling, and offered as an 'add-on' for the Ford Escort, this system is called a mechanical system because it senses the adhesion of the front wheels by using two modulator units belt-driven from the front drive shafts.

The modulator units adjust the braking effort on diagonally disposed wheels so that if it is more slippery on one side than the other, one rear wheel still keeps its grip. This maintains the car on a straight path even if the diagonally opposite front wheel has

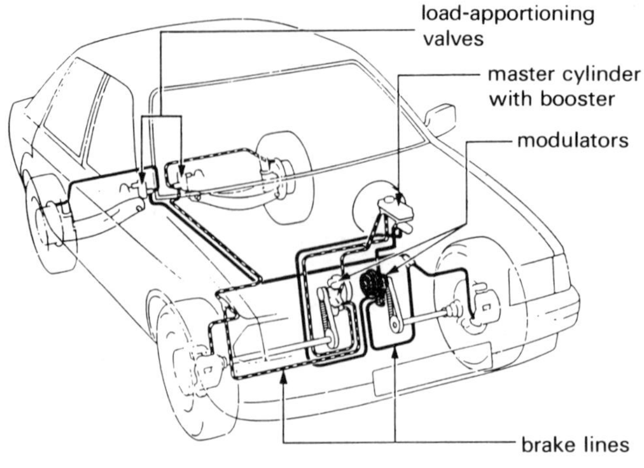

Fig. 69.6 Lucas–Girling stop control system (SCS)

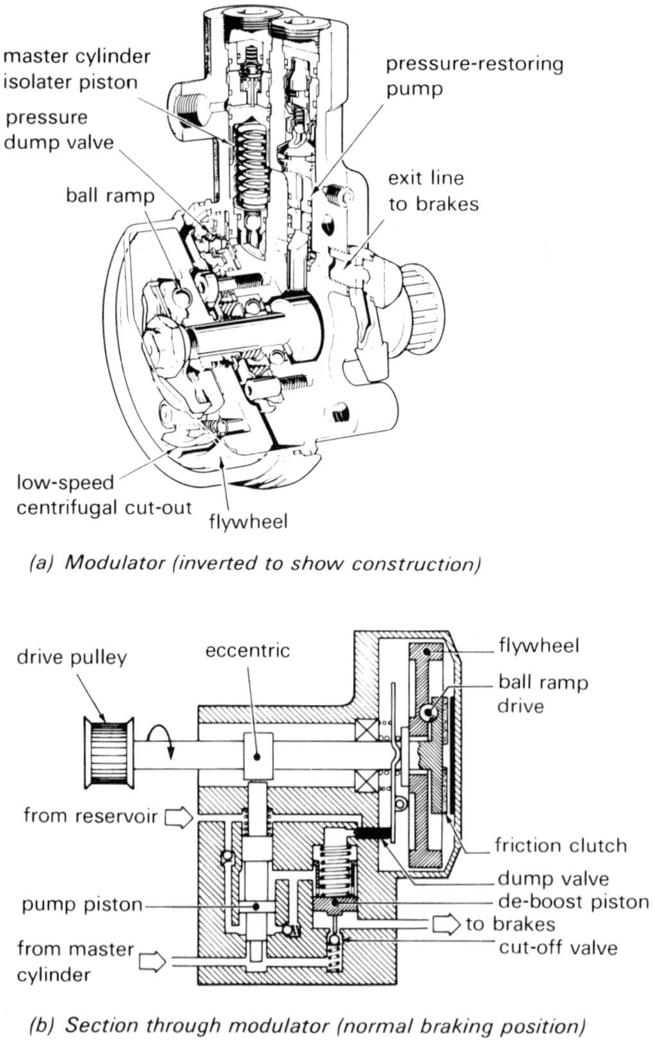

Fig. 69.7 Mechanically sensed modulator (Lucas–Girling)

not brought its anti-skid modulator into action. Once slip has been detected by a modulator, it goes through the normal ABS phases of release, maintain and apply.

Fig. 69.6 shows the layout of this system. The ABS modulator unit is placed in the diagonally split fluid circuit between the master cylinder and the wheels.

Modulator

This is a mechanically driven pump and dump valve that is triggered by a slip-sensitive flywheel and clutch arrangement (Fig. 69.7a).

Driven by a toothed belt at a proportionally faster speed than the adjacent front wheel, the 'free-running' flywheel detects any rapid changes in wheel speed such as that which occurs when a wheel starts to skid. By taking the drive through a ball ramp, the speed difference between the flywheel and driving shaft causes the flywheel to move sideways; this is utilized to operate a fluid dump valve.

Fig. 69.7b shows a diagram of the modulator set in the normal braking position.

Traction control system (TCS)

The electronic sensing arrangement that is used in ABS for the detection of slip can be further utilized to control and limit the maximum traction force used to propel the vehicle. This TCS feature, which was first introduced by Harry Ferguson in 1954 (see page 309), can be obtained by using the intelligence of a computer to block the compensating function of the differential in certain driving situations.

This partial differential lock action can be achieved by using ABS components to give a braking action on a wheel that is slipping. Should both driving wheels display excessive slip, the ECU can, in addition to acting upon the brakes, also be made to alter the engine throttle valve setting in order to reduce the driving torque.

Some of the new generation ABS arrangements are designed so that they can easily be expanded by adding a TCS module to provide the TCS function.

PART NINETEEN

Electrical equipment

70 General layout and basic principles

Over recent years the amount of electrical and electronic equipment has increased considerably and this trend is likely to continue into the next decade.

In view of the increased complexity of electrical equipment now being used on motor vehicles, the treatment of the subject in this book is targeted on the basic knowledge that a non-electrical specialist would be expected to possess. It should be noted that serious problems are likely to arise on modern vehicles if the person undertaking a repair does not understand the basic construction and operation of both the mechanical and electrical/electronic systems. This is particularly important in the field of diagnosis, because ignorance in either one of these areas of knowledge can lead to component damage and costly repairs. Whereas in the past most repair activities on a motor vehicle could be divided into either mechanical or electrical tasks, nowadays the systems are fully integrated.

In cases where further study of electrical/electronic topics is needed, the reader is referred to the companion book written by the present author entitled *Fundamentals of Automotive Electronics*.

Electrical terms

In many ways electricity behaves similarly to water – cables are like pipes and many electrical terms, expressions and components can be related to hydraulic systems. One obvious difference is that electricity cannot normally be seen, whereas opening a water tap readily shows the flow.

Part of the mystery of electricity can be dispelled if its basic behaviour is understood: this study demands a clear understanding of electrical terms. Furthermore, this basic knowledge is essential before any electrical test meter is used on a vehicle.

A comparison between water and electricity is used here to introduce and describe the basic terms and principles.

Fig. 70.1 compares the flow of water and electricity in their respective circuits and is used to introduce the main terms.

Pressure

Electrical 'pressure' is expressed in *volts* (V) and is measured by a *voltmeter*. This meter has two leads to enable it to show the difference of pressure between the points in the circuit at which the leads are connected.

A voltmeter connected across a battery indicates the *potential difference* (p.d.) between the two battery terminals – a typical reading is about 12 V.

Water pressure is measured by a pressure gauge. If a sealed water container is pressurized, the water inside will try to escape. This is similar to a battery; the 'pressure' of the electrical energy exerts a force called *electromotive force* (e.m.f.) which attempts to drive the electrical current from one battery terminal to the other.

Flow

The amount of electricity flowing through a cable is called the *current* and is expressed in *amperes* (A). It is measured by an *ammeter* which is connected so that the current passes through the ammeter as it flows around the circuit.

Electrical current flow is similar to a flow of water which is given in litres/second.

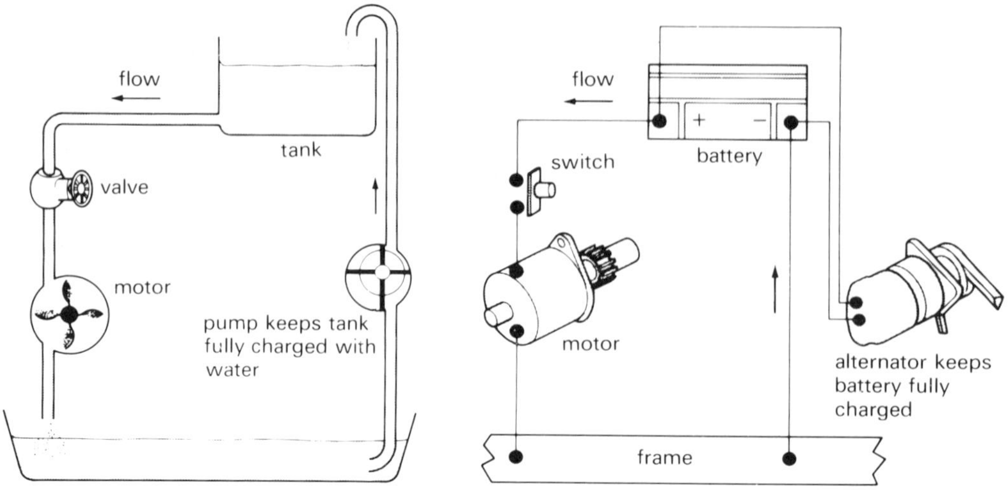

Fig. 70.1 Flow of water and electricity

Resistance

This term indicates the opposition to current flow. The unit of resistance is the *ohm* – the symbol for 'ohm' is the Greek letter Ω (omega). Cable size affects the resistance. A thin cable has a higher resistance than a thick cable, just as a small pipe has a greater resistance to water flow than a large pipe.

Resistance is measured with an *ohmmeter*.

Power

The power is the work done in a given time and is expressed in *watts* (W). The power can be calculated from the expression:

$$\text{Power} = \text{Voltage} \times \text{Current}$$

or

$$\text{Watts} = \text{Volts} \times \text{Amperes}$$

Ohm's law

The relationship between the three electrical terms is given by Ohm's Law; this may be written as:

One volt is required to force a current of one ampere through a resistance of one ohm.

The law can also be expressed as:

$$\text{Voltage} = \text{Current} \times \text{Resistance}$$

or

$V = IR$ where V = voltage
I = current in amperes
R = resistance in ohms.

The various electrical units are summarized in Table 70.1.

Table 70.1 Summary of electrical units

Quantity	*Unit*	*Symbol*
'Pressure' or potential difference	Volt	V
Current (flow)	Ampere	A
Resistance	Ohm	Ω
Power	Watt	W

Conductor

A material which freely allows a flow of electrical current is called a *conductor*. The low resistance of copper makes this metal suitable to use as a cable.

Insulator

An insulator is a material, such as rubber or PVC plastics, that resists or opposes the flow of electricity. A cable is covered with an insulation material to prevent electrical energy taking a shorter path than that intended.

Electrical symbols

Diagrams of electrical systems are shown either in pictorial form or theoretical form; in the latter graphical symbols are used to indicate the various items that make up the circuit.

Many separate parts are used in an electrical system of a motor vehicle, so a convention is needed to enable people to understand the various graphical symbols. In the UK the British Standards Institution

Table 70.2 Electrical symbols (BS 3939-1985)

Description	*Symbol*
Direct current Alternating current	—— ∼
Positive polarity Negative polarity	+ −
Current approaching Current receding	⊙ ⊕
Battery 12V (Long line is positive)	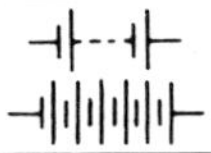
Earth, chassis frame Earth, general	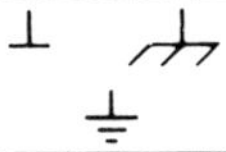

Conductor (permanent) Thickness denotes importance Conductor (temporary)	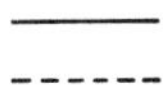
Conductors crossing without connecting	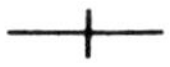
Conductors joining	
Junction, separable Junction, inseparable Plug and socket	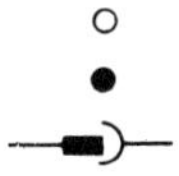

Variability: applied to other symbols	
Resistor (fixed value)	
Resistor (variable)	
General winding (inductor, coil)	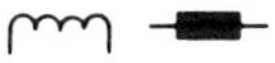
Winding with core	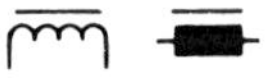
Transformer	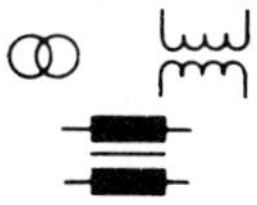

Diode, rectifying junction	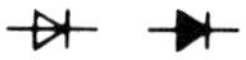
Light emitting diode	
Diode, breakdown: Zener and avalanche	
Reverse blocking triode thyristor	
Transistor pnp Transistor npn	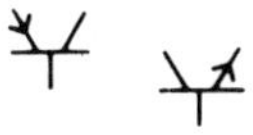

Description	*Symbol*
Lamp	⊗
Fuse	
Switch ('make' contact, normally open)	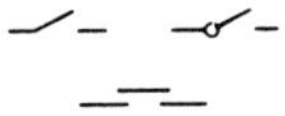
Switch ('break' contact, normally closed)	
Switch (manually operated)	
Switch (two-way)	
Relay (single winding)	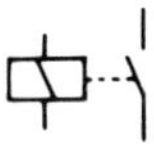
Relay (thermal)	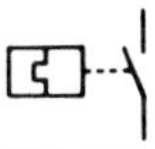
Spark gap	
Generator ac and dc	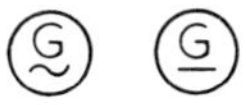
Motor dc	Ⓜ
Meters: ammeter, voltmeter, galvanometer	

Capacitor, general symbol	
Capacitor, polarized	
Amplifier	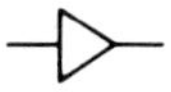

Junction f.e.t. N-type channel Junction f.e.t. P-type channel	
Photodiode	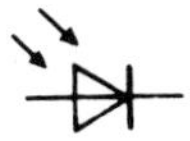
Thyristor	

(BSI) recommend that the symbols shown in BS 3939: 1985 should be used. Some of the main symbols are shown in Table 70.2.

Circuit

Water will not flow around a hydraulic circuit unless the pipework makes a complete path from the supply tank, around the system and back to the tank. This concept also applies to an electrical system - *electricity will not flow unless there is a complete circuit.*

When a circuit incorporating a battery is formed, small electrical charges called *electrons* move around the circuit. These charges pass through the battery, along the cables and back through the battery, the charge movement being initiated by the battery. When electrical flow is taking place, all the electrons in the various parts of the circuit move simultaneously; this shows why a complete 'pathway' must be provided before electron movement can take place.

The action is similar to the flow in a pipe bent to form a loop and filled with water. No flow of water can take place in one part of the pipe unless there is a corresponding water movement in the remainder of the loop.

A simple electrical circuit is formed by connecting a lamp and switch to a battery. Fig. 70.2 shows two ways of illustrating this circuit: pictorial and theoretical (schematic).

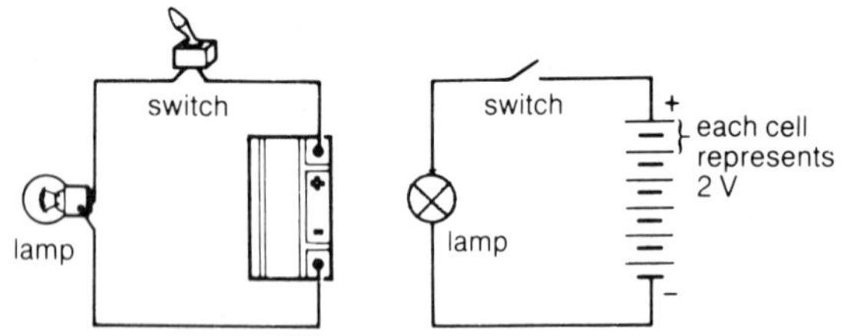

Fig. 70.2 Symbols used to represent electrical components

The flow of electricity is controlled by a switch; this breaks the circuit and stops the flow of current like a tap in a water system.

Closing the switch operates the lamp and causes current to flow through a fine wire filament, this becomes incandescent and gives the illumination.

The terms *closed circuit* and *open circuit* mean that the electrical circuit is complete and incomplete respectively.

Earth return

Using the metal body and frame of the vehicle as a part of the circuit reduces the length of cable and simplifies the layout. The frame is called *earth* (or *ground*) and is connected to one of the battery terminals. Normally the negative terminal is 'earthed' and in this case the polarity of the vehicle is described as 'negative earth'.

Use of the word 'return' is associated with the current flow in a circuit. When a lamp is connected to a battery by two cables, one cable is called the *feed* and the other the *return*. By substituting the vehicle body for the return cable, an *earth return* system is formed (Fig. 70.3). The alternative two-wire arrangement, called *insulated return* is seldom used other than for special-purpose vehicles, e.g. petrol tankers.

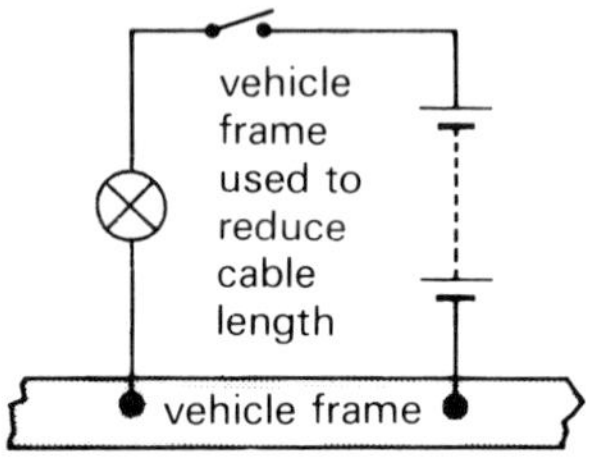

Fig. 70.3 Earth return circuit

It is essential that the battery terminals are connected to give the correct earth polarity. Serious damage to components, especially those containing electronic devices, will occur unless the equipment is connected to give the correct polarity.

Circuit faults

The two faults shown in Fig. 70.4 which occur in a circuit are:

- *Open circuit*
- *Short circuit.*

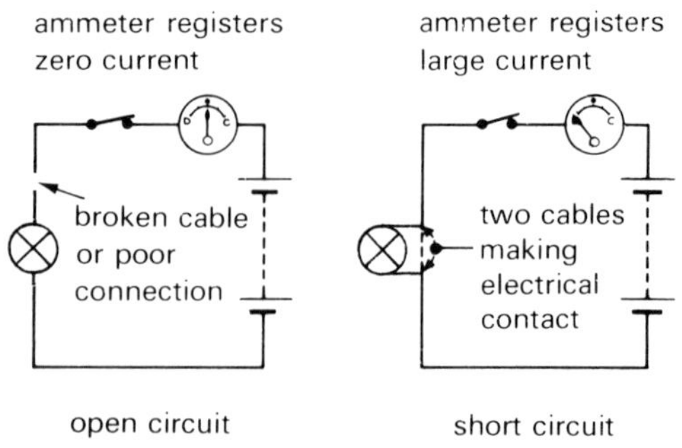

Fig. 70.4 Open and short circuit

Open circuit

This is the term used to describe a break in a circuit. A switch is an intentional means for making an open circuit and is used to control the flow of current. Sometimes an open circuit occurs when it is not required; this fault is caused by a fracture in a cable or a poor connection.

Short circuit

Electricity always takes the shortest route as it flows from source to source, so if a sharp edge of the body cuts through the insulation of the cable, then some of the current will flow to 'earth' rather than travel around the full circuit. The short-circuit path increases the current in the supply cable and if this is excessive, overheating of the cable will cause the cable covering to melt and smoulder. Unless a *fuse* is fitted the short circuit may well cause a fire.

The thickness of the insulation of a cable is dictated by the voltage that the cable has to carry - the higher the voltage, the thicker the insulation.

Fuse

A fuse is a short length of thin wire which is designed to melt and break the circuit if the current exceeds the rated value marked on the fuse. Provided it is of the correct rating, a fuse reduces the risk of fire in the event of a short circuit.

A number of fuses, or other forms of thermal circuit breaker, are fitted to protect the various circuits of a vehicle (Fig. 70.5). Some electrical components require a comparatively large current, so in these cases a fuse having a high 'amperage' is specified.

Main electrical circuits

The battery, which is the energy source when the engine is not running, is connected to a number of separate circuits. The main circuits are:

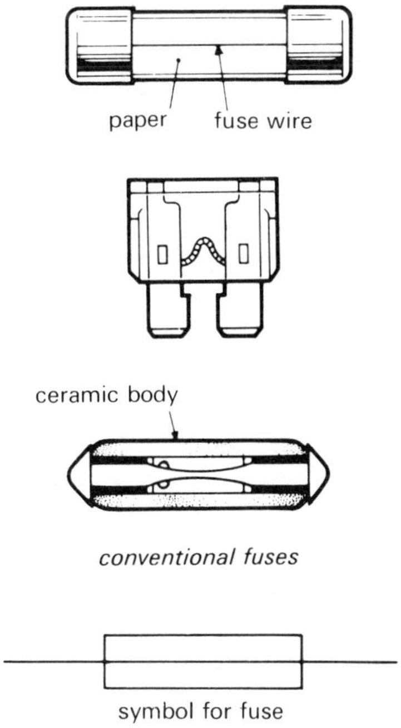

Fig. 70.5 Types of fuse

1 *Charging* - supplies the electrical energy when the engine is running. Also this system maintains the battery in a *fully charged* state.
2 *Ignition* - provides a spark to 'fire' the mixture in the engine cylinders.
3 *Lighting* - gives the necessary exterior and interior illumination.
4 *Starting* - enables the engine to be cranked over at a speed sufficient for it to 'fire' and keep running.
5 *Auxiliary* - includes the various accessories such as windscreen wipers and washers, direction indicators, horn, in-car entertainment, etc.

71 The battery

The purpose of a battery is to 'store' electrical energy which is needed to operate components when the engine is stationary or when the output from the electrical generator is low. It does this by converting the electrical energy supplied to it into chemical energy so that when an electrical current is required the energy flow can be reversed.

Lead–acid battery

Most vehicles use a lead–acid battery. This type of secondary (reversible in charge/discharge) battery has lead plates immersed in an electrolyte (a liquid solution) of sulphuric acid (H_2SO_4) and distilled water.

There are various types of lead–acid battery; these are:

- *Conventional*
- *Low maintenance*
- *Maintenance-free*.

Conventional type

This type shown in Fig. 71.1 consists of a polypropylene or hard rubber container; this houses a number of separate cells of 2 V nominal voltage. A 12 V battery has six cells connected in *series* by lead strips; a series arrangement means that the cells are interconnected with the positive terminal of each cell joined to the negative of the adjacent cell.

An alternative method of connection is to link all positive terminals together and all negative terminals together; an arrangement called *parallel*. This cell layout is not used on vehicles because it gives a total voltage of only 2 V instead of 12 V. On occasions when a parallel arrangement is used, the capacity is increased; this is the period of time that a battery supplies a given voltage before it becomes discharged. When six 2 V cells are connected in parallel, the capacity is about six times as great as that given by the series arrangement.

Each cell is made up of two sets of lead plates, positive and negative, which are placed alternately and separated by an insulating porous material. The plates on the conventional type of battery consist of a lattice-type grid of lead–antimony, which retains the active lead oxide paste material.

A moulded cover seals the cells and either individually screwed caps or a one-piece detachable cover allows the cells to be periodically topped up with distilled water. Vents in the cover permit the gas to escape; this is evident during the charging process and when the battery is used in a discharged state.

Various types of terminal post are used. Some batteries have a post with a hole to enable the connector to be clamped; others use a cylindrical post around which the connector is fitted.

It is essential that the battery is connected to the vehicle in a way that gives the correct earth polarity. Extensive damage to electrical components will occur if the polarity is incorrect. Terminals are marked '+' (positive) and '−' (negative); when cylindrical posts are used, the larger diameter is positive. Also the colours red (or brown) and black (or blue) are used to indicate positive and negative respectively.

Cell action

As the battery discharges (gives out its energy) the lead in the plates undergoes a chemical change and the acid becomes 'weaker', i.e. the 'specific gravity' decreases.

To reverse the action an electrical charge is supplied from an alternator. This changes the lead plates back to their original chemical form and 'strengthens' the electrolyte. A charge current has to pass through the battery in one direction, so a direct current (d.c.) must be used. To produce the chemical reversal of the plates, the charge current flows in the

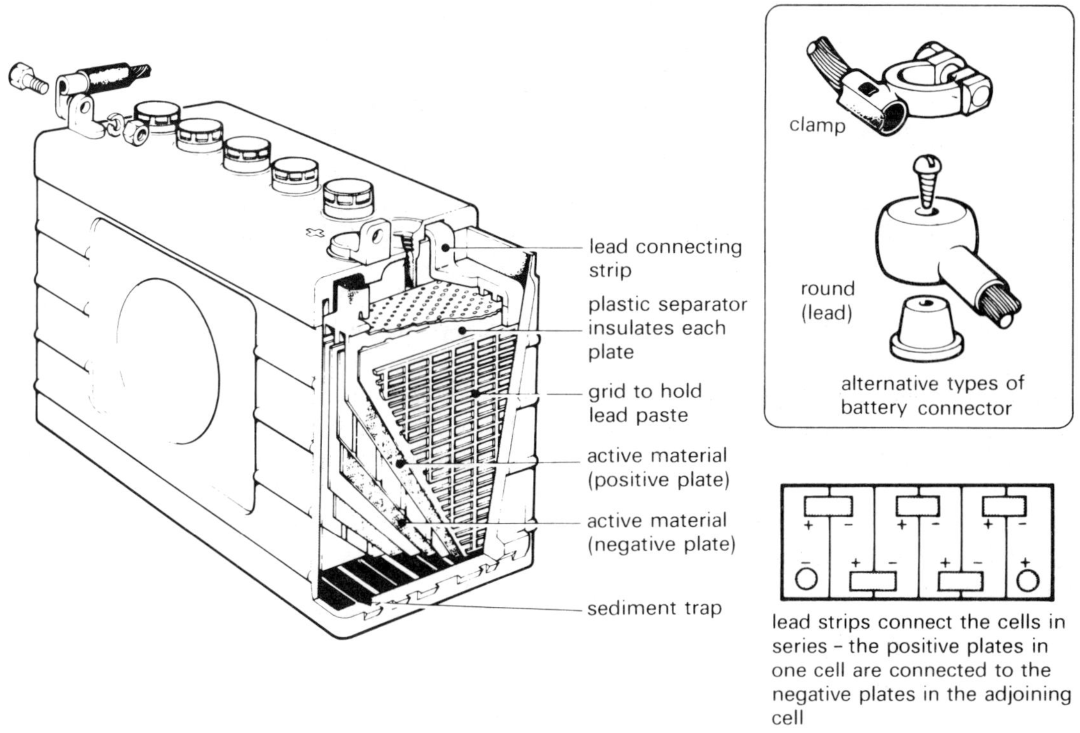

Fig. 71.1 Battery – conventional type

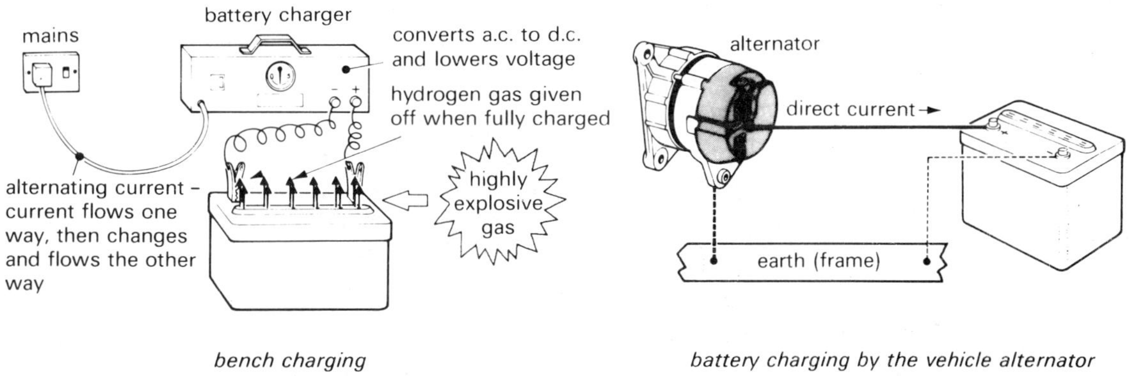

Fig. 71.2 Battery charging methods

opposite direction to the current given out by the battery, so the connection between the battery and alternator (or bench charger) must be positive to positive (Fig. 71.2).

Voltage variation

When the battery is 'taken off charge', the terminal voltage (p.d.) is about 2.1 V, but this quickly drops to 2.0 V, where it remains for the major part of the discharge period. Towards the end of the discharge period, the p.d. falls rapidly until a fully discharged value of 1.8 V is reached. These values represent p.d., so current must be flowing at the normal discharge rate when the voltage reading is taken.

Terminal voltage towards the end of the charging period rises to 2.5–2.7 V, but this quickly falls to about 2.1 V when the charge current ceases.

Capacity

The total area of the plates governs the time for which a battery will supply a discharge current; this represents the capacity of a battery and is often expressed in *ampere-hours* (A h).

A battery rated at 38 A h, based on a 10 hour rate,

should supply a steady current of 3.8 A for 10 hours.

Reserve capacity is an alternative method of expressing the capacity of a battery. This rating indicates the time in minutes for which a battery will deliver a current of 25 A at 25 °C before the cell voltage drops to 1.75 V. A typical 40 A h battery having a reserve capacity of 45 minutes should keep the vehicle in operation for this time after the charging system has failed, provided the electrical load is normal.

Maintenance of conventional-type batteries

Regular maintenance is needed to keep this type of battery in good condition. This work includes the following tasks.

Battery fitting Extensive damage to electrical components results if the earth polarity is accidentally reversed.

Mounting bolts and brackets should hold the battery firmly in place, but the bolts should not be over-tightened.

When removing or refitting a battery,

the earth lead should be removed first and fitted last.

This reduces the risk of the spanner arcing to earth.

For safety reasons, the earth battery terminal should be disconnected before work is carried out on engine parts that are liable to rotate.

Electrolyte level The electrolyte should be maintained at the correct level (e.g. 6 mm above the plates) by topping up with distilled water. Frequent topping up indicates that the battery is being over-charged.

Terminal corrosion Corrosion is reduced by coating the terminal with petroleum jelly or with silicone grease. A corroded terminal is cleaned by immersing the terminal in ammoniated warm water or soda dissolved in water. This solution can also be used to neutralize acid spillage on vehicle body parts.

Use of jump-leads In the event of a discharged battery, an engine can sometimes be started by temporarily connecting, by jump leads, another battery to the car battery. To do this, the two batteries are connected in parallel (positive to positive and negative to negative). After the engine has started and run for a few minutes, the engine must be allowed to slow-run before the jump-leads are disconnected; making sure that the earth lead is removed first.

Battery hazards An acid burn should be treated immediately with sodium bicarbonate solution or, failing this, clean water. Acid splashed into the eye should be bathed immediately in clean water and medical attention should be sought as soon as possible.

Sometimes the acid supplied to a garage needs to be 'broken down' to weaken it before it is used to fill a new battery. For safety reasons, the dilution must be carried out in a glass or earthenware container by adding *acid to water* and NOT water to acid.

Acid splashes on clothes can be neutralized with an alkali, such as ammonia, if holes are to be avoided. Gloves and goggles should be worn when handling or pouring acid; also it is advisable to wear an acid-proof apron.

A large quantity of hydrogen is released from the cells when charging continues beyond the fully charged state. The risk of an explosion from this highly flammable gas is great when connecting the terminals, so care must be taken to avoid a spark. After bench-charging it is wise to let the battery stand for a time before fitting it to the vehicle, especially when the car incorporates a central-locking door system.

Battery tests

These tests should be performed in accordance with the battery manufacturer's recommendations. The following tests relate to a conventional battery and are intended as a guide.

Hydrometer The specific gravity (or relative density) of an electrolyte is measured by a *hydrometer*; the readings indicate the state of charge (Fig. 71.3). Values quoted represent the weight of the electrolyte in relation to the weight of an equal volume of water. Typical values are:

fully charged	1.280
half-charged	1.200
fully discharged	1.150

Electrical load test Also called a *high-rate discharge test* and *drop test*, this equipment indicates the ability of a battery to supply a large current similar to that required to operate the starting motor. The test shows the battery voltage during the time that a large current is being 'drawn' from the battery.

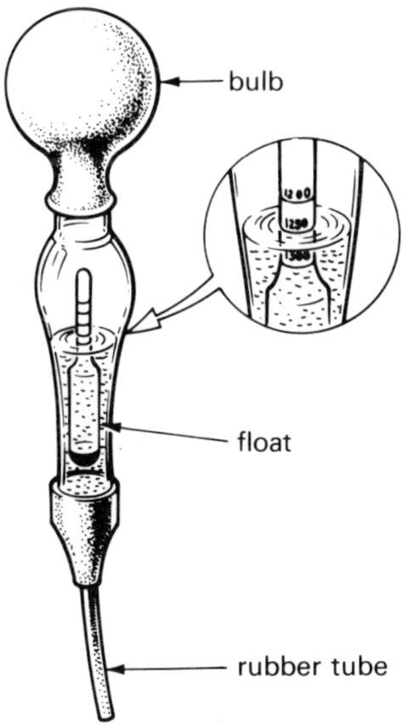

Fig. 71.3 Hydrometer

This is a severe test and must not be prolonged beyond the recommended time. It should only be performed on a fully charged battery.

A 30 A h battery in good condition, tested against a load of about 100 A, should show a reading of 8–10 V for a period of 15 seconds.

Low-maintenance battery

Improved materials and new construction techniques have reduced the need for a battery to be periodically topped up with distilled water to replace losses due to gassing. These new style batteries are fitted to modern vehicles because the reduction in maintenance is an attractive selling feature.

Use of these batteries has been made possible by the improved control of charging rate and voltage obtained from an alternator system as compared with dynamo charging equipment.

Gassing has been reduced by changing the grid material from lead–antimony to an alloy of *lead–calcium*.

When the low-maintenance battery operates with a good charging system and at normal temperature, the electrolyte only has to be checked once per year or at 80 000 km intervals.

Other than in its grid material, the construction of a low-maintenance battery is similar to the conventional type. Since the performance characteristics are based on proven designs, it is normally possible to fit a low-maintenance battery in place of a conventional battery.

Maintenance-free battery

The maintenance-free battery shown in Fig. 71.4 differs in several respects from the conventional battery. The most significant features are that the battery is sealed, except for a very small vent hole, and that it requires no service attention other than to be kept clean.

A test indicator is fitted in the top cover to register

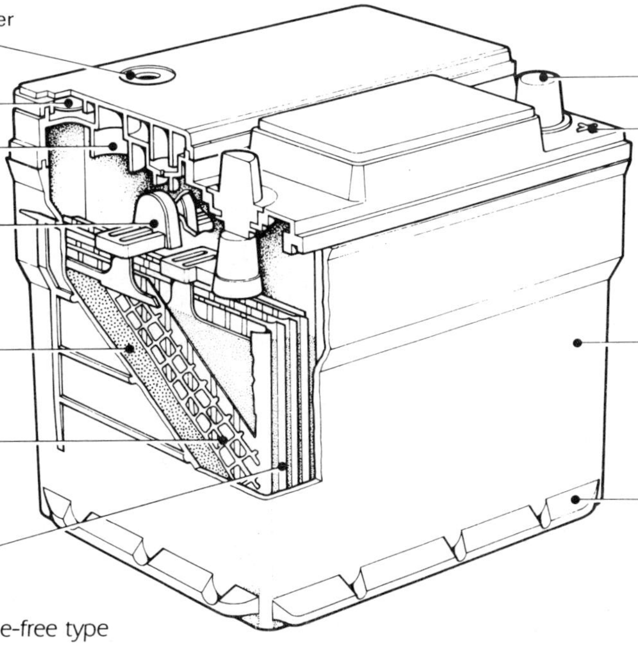

Fig. 71.4 Battery – maintenance-free type

the state of charge. The indicator is a form of hydrometer which signals the battery condition by changing colour. The type illustrated shows green when the battery is charged and serviceable, and green/black or black when recharging is required. When a light yellow is signalled, it indicates that an internal fault is present; in this condition the battery must not be charged or tested. Furthermore, jump leads must not be used when this fault is indicated; instead the battery should be replaced and the alternator output checked.

If the battery is discharged to a point where it cannot crank the engine and in consequence the engine has to be started either by jump-leads or by pushing the car, then it will be impossible for the alternator to recharge the battery. When a battery is in this condition, it must be removed and bench-charged; this is necessary because the voltage needed to recharge it is higher than that given by the charging system of the vehicle.

72 The charging system

When the engine is running, the charging system provides electrical energy for the operation of the vehicle components; also it must maintain the battery in a fully charged state. To fulfil these duties, the generator of the charging system must convert mechanical energy to electrical energy.

There are two types of generator: a *dynamo*, commonly used in the past, and an *alternator*, which is now favoured on most vehicles.

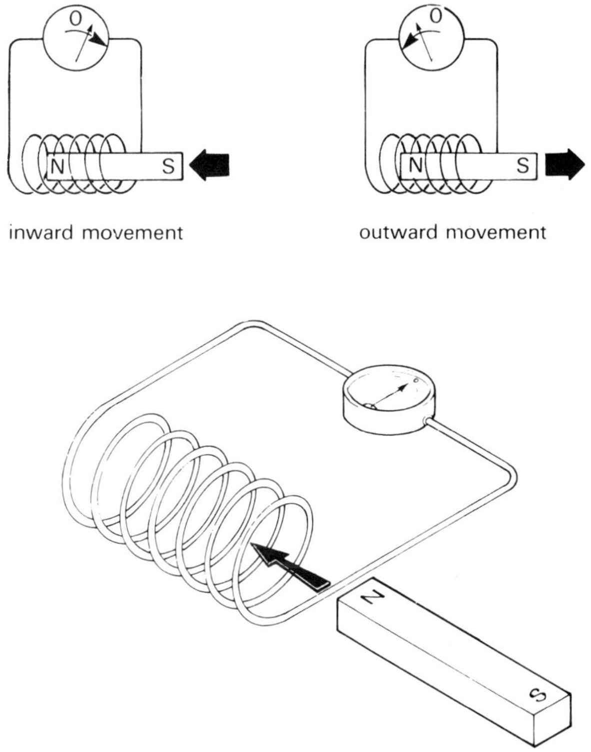

Fig. 72.1 Generation of electric current

Principle of the alternator

The origin of the generator goes back many years. In 1831 Michael Faraday performed an experiment that showed that an electric current was generated when a magnet was moved in a coil of wire (Fig. 72.1). He demonstrated that during the inward movement of the magnet, the needle on the meter moved; this showed that an electric current was flowing in the coil circuit. Withdrawal of the magnet caused the (centre zero) needle to move in the other direction; this indicated that the current was flowing in the opposite direction. At times when there was no relative movement of the magnet and coil, electrical energy was not generated in the coil circuit.

He went on to demonstrate that the electrical energy generated depended on three things:

1 The strength of the magnetic field.
2 The length of the conductor exposed to the magnetic field.
3 The speed at which the magnetic field is cut.

A modern alternator-type generator uses the principle of electromagnetic induction established in Faraday's experiment: a magnet is rotated inside a coil of wire. In this coil an e.m.f. is generated which causes an electric current to be forced around the circuit.

Using the following steps, the function of each part of an alternator can be appreciated.

Generation of an alternating current

Fig. 72.2(a) shows a shaft-driven permanent magnet positioned in a soft iron C-shaped stationary member (stator). This stator provides an easy path for the magnetic flux between the N and S poles of the magnet. Figs. 72.2b–d show how the magnetic lines of force (flux) vary in density and direction as the magnet is rotated.

In Fig. 72.3 a voltmeter is placed across a coil of wire wound around the stator. Rotation of the magnet generates an e.m.f. which varies, as shown by the graph, with the magnet position. The polarity of the magnetic flux in the stator changes as the magnet is

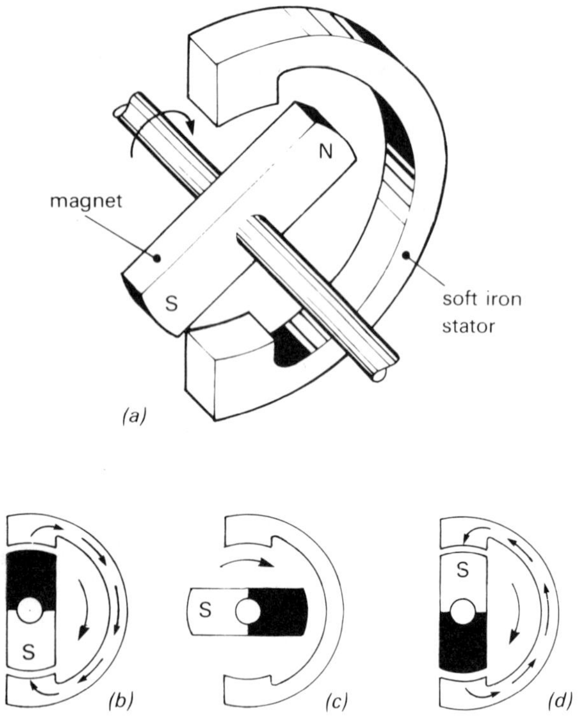

Fig. 72.2 Path of magnetic flux

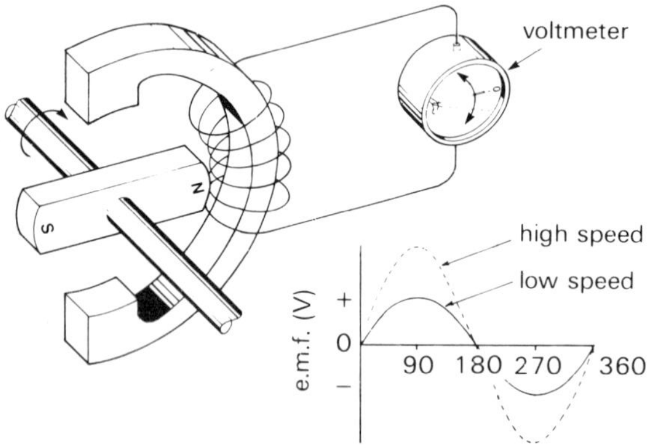

Fig. 72.3 Generation of an a.c. output

rotated, so an alternating current (a.c.) is produced. The amount of current depends on the three factors given by Faraday; since the only variable is the speed, then the faster the magnet is rotated, the higher will be the output.

Space and weight are very important, so the alternator must be designed to give the highest possible output. One step towards improved efficiency is to use a multi-pole magnet arranged in the manner shown in Fig. 72.4. In this case the magnet is placed coaxially with the shaft and, by means of soft iron 'fingers', two extra poles are formed on the rotating

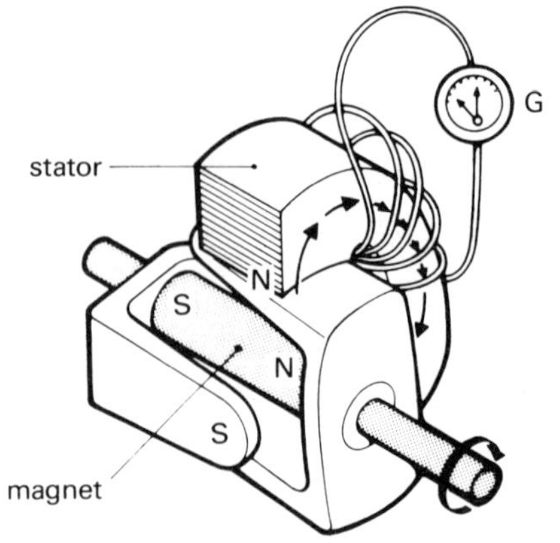

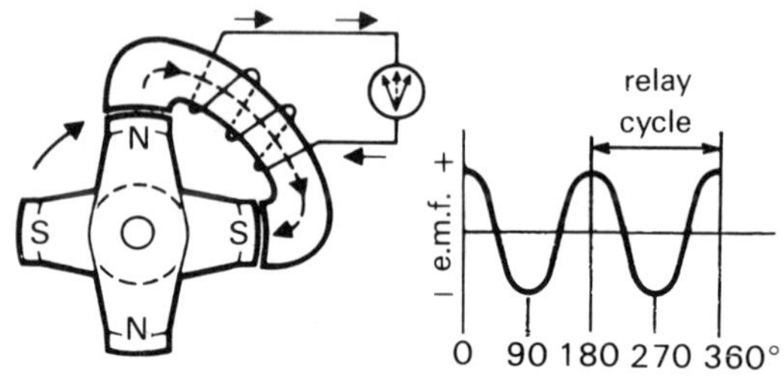

Fig. 72.4 Output from multi-pole magnet

member (rotor). This arrangement doubles the output and makes the unit more efficient.

As before, the output increases as the speed rises up to a point where the rapid changes in current flow prevent any further increase in output. This feature is advantageous because the machine protects itself from being overloaded.

Multi-phase output To utilize more of the rotor poles, another stator winding is added in the position shown in Fig. 72.5. This gives two independent outputs as shown by the graph. The stator winding B gives an output that peaks at a rotor position 45° after the e.m.f. peak of winding A; i.e. the output from the two windings is 45° out of phase and is called *two-phase output*.

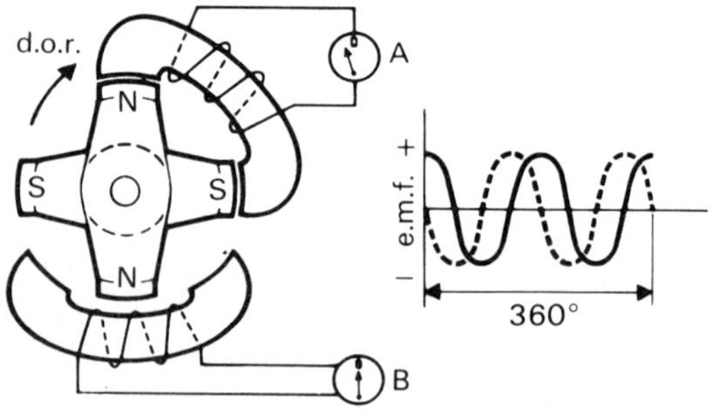

Fig. 72.5 Multi-phase output

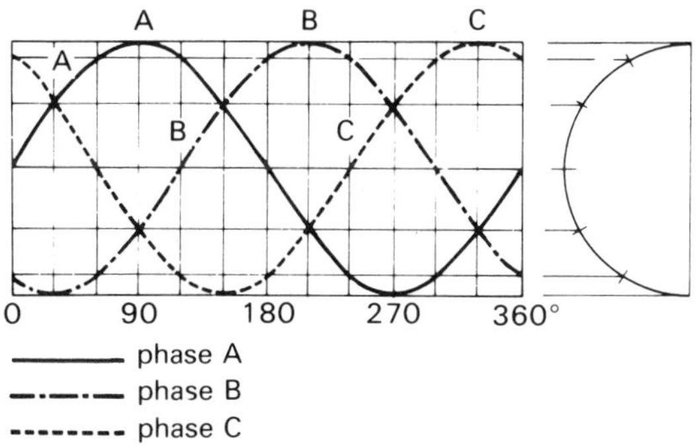

Fig. 72.6 Three-phase output

Similarly if another stator winding is added and all three are spaced out around a multi-pole rotor, then a *three-phase output* is obtained (Fig. 72.6). As the number of magnetic poles is increased, each individual cycle will be shorter; therefore the larger number of a.c. cycles produced in one revolution of the rotor will produce a comparatively smooth peak output of current.

Control of magnet strength When the engine is turning over very fast, the very high e.m.f. output from a permanent magnet type alternator causes both damage to the machine and overcharging of the battery. To overcome these problems, the permanent magnet on the rotor is replaced by an electromagnet. This arrangement allows the strength of the magnetic field to be easily controlled by varying the current supplied to the winding; the greater the current passed to the 'field', the stronger the magnet.

Fig. 72.7 shows the parts of a modern alternator that generate the electrical energy; the type shown is a three-phase, twelve-pole machine. Current to the rotor field is delivered via slip rings; these are two rings of copper on to which rub two carbon brushes.

The three sets of stator windings, which are interconnected together in a *star* or *delta* arrangement, have three output leads, one for each phase.

Rectification of a.c. to d.c.

Alternating current is unsuitable for battery charging, so the output must be rectified (changed) to direct current. Today this is achieved by using a *semi-conductor* device called a *diode*.

Semiconductors Certain materials, such as silicon and germanium, are neither good electrical conductors nor insulators; instead they come in a borderline category. If semiconductor crystals are doped in a

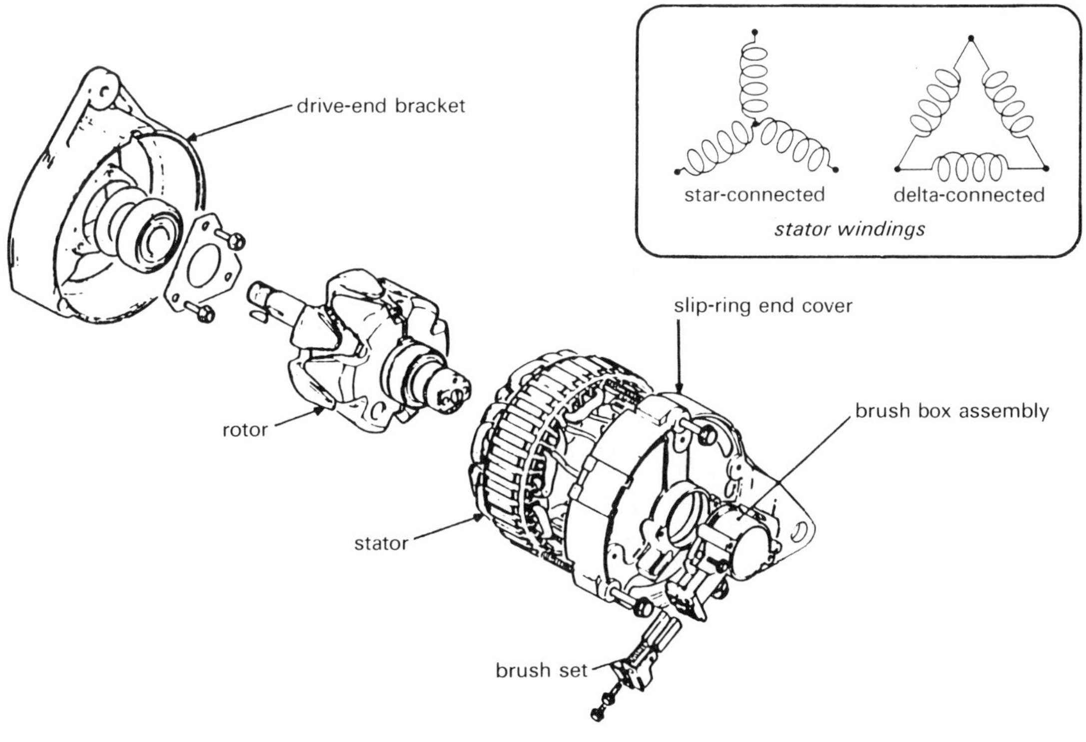

Fig. 72.7 Rotor and stator assembly

special way by adding traces of certain impurities and then arranged with other semiconductors, a range of solid-state electronic devices can be obtained. The most common devices are called *diodes* and *transistors*.

The diode is an electronic 'valve' that allows current to flow in one direction but not the other. It is represented by the symbol shown in Fig. 72.8a. The arrow head indicates the direction in which the diode freely conducts a *conventional-flow* current. (The term 'conventional flow' is used when the current is assumed to flow from positive to negative.)

Diode rectifier Fig. 72.8b shows how a diode can be used in a simple circuit to rectify a.c. to d.c. The one-way action of a single diode blocks the 'reverse' current flow and gives an output, called *half-wave rectification*, as shown by the graph.

Four blocking diodes, set to form a *bridge circuit* (Fig. 72.8c), are needed to ensure that half the output

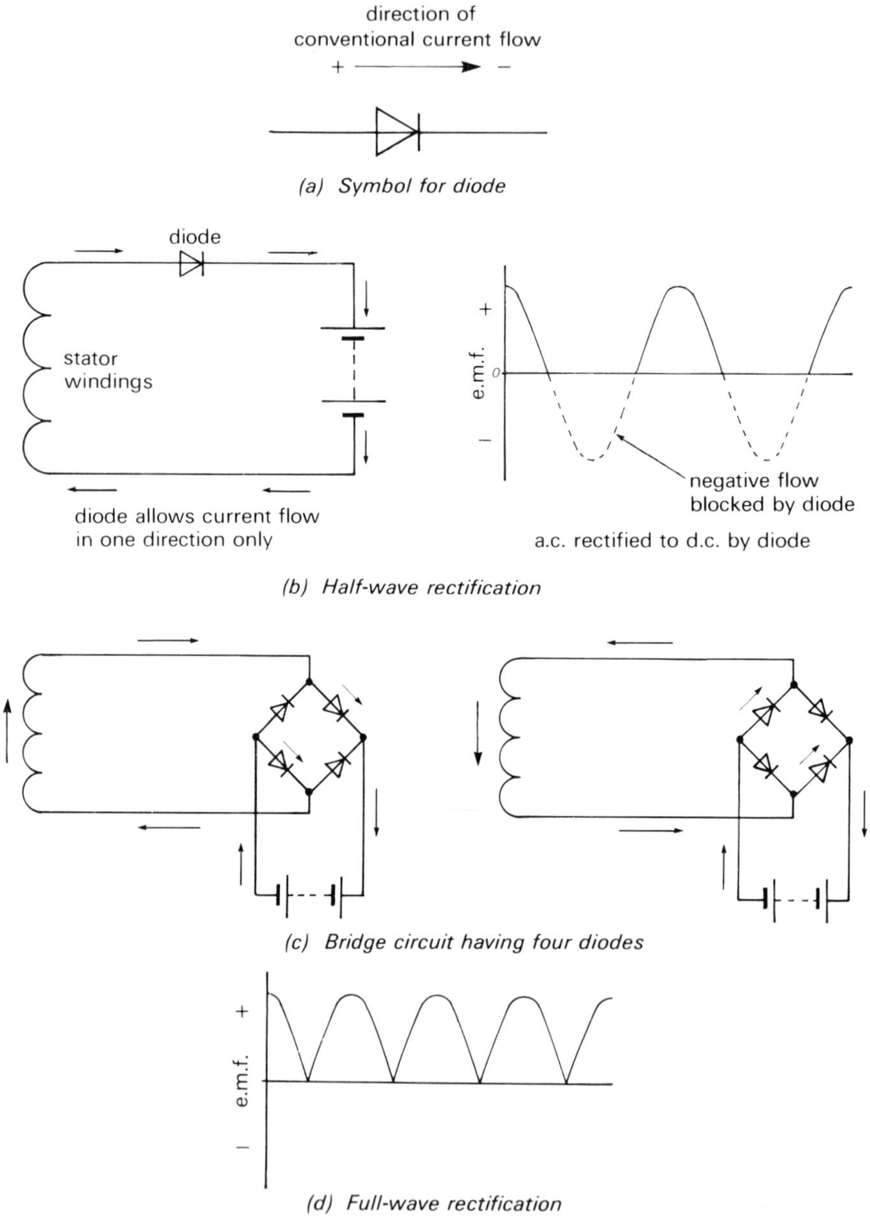

Fig. 72.8 Current rectification

is not lost, i.e. to achieve *full-wave rectification* (Fig. 72.8d). With the diodes arranged in this manner, the battery is always supplied with a d.c. charge, irrespective of the current polarity generated in the stator windings.

A three-phase output from an alternator with three stator windings requires a rectifier having six output diodes arranged in the manner shown in Fig. 72.9. This diode network ensures that the current output is always 'in one direction'. Irrespective of the position and direction of the current-flow arrow on the stator windings, the terminal B+ will always be positive.

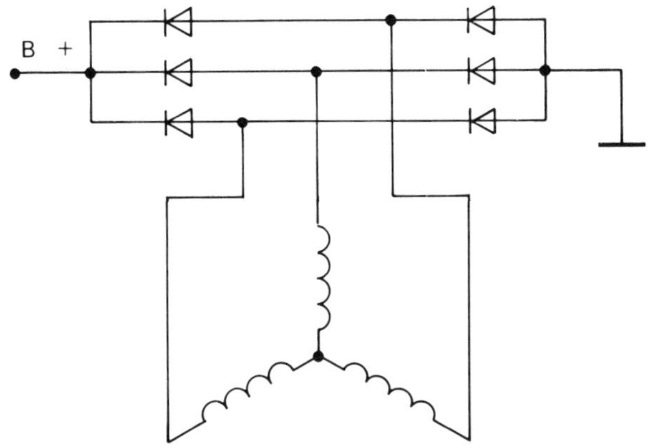

Fig. 72.9 Alternator diode arrangement

Battery reverse flow   When the engine is stationary, and at other times when the alternator output voltage is lower than the battery voltage, current would tend to flow from the battery to the alternator and discharge it. Whereas the old dynamo charging system needed a *cut-out* switch in the main line to prevent this current drain, the one-way characteristic of the diodes in alternator rectifier automatically satisfies this need.

This inherent feature means that the main output terminal on the alternator is subject to full battery voltage, so the fact that this connection is 'live' should be remembered when the alternator is either disconnected or removed.

Voltage output control

The output from an alternator must be limited to a voltage of about 14.2 V; this value equals the p.d. of a battery that is nearing its fully charged state. Setting the alternator to this maximum will allow it to charge at a high rate when the battery is discharged, but as the battery approaches its fully charged state the charging rate will gradually reduce to zero.

Alternator output is controlled by varying the current supplied to the rotor field. The *regulator* performs this current control function by using electronic devices in the form of diodes and transistors. These are retained in a 'micro' metal container, which is mounted under the alternator cover.

Solid-state devices, such as diodes and transistors, are highly sensitive to heat and voltage surges, so during operation and repair, precautions need to be taken to ensure that a regulator (and rectifier pack) is protected from exposure to these conditions.

Alternator operation

The alternator circuit, shown in Fig. 72.10, includes a *charge warning lamp* to signal to the driver when the system is not operating. In addition to this duty, the warning lamp circuit is used initially to excite the field. On switching on the ignition, the small current needed to operate the lamp also passes through the rotor winding and regulator to earth. This creates a weak magnetic field sufficient to commence the charging operation when the alternator starts to rotate. Naturally this operation will not occur if the bulb has 'blown'. Unless 12 V is obtained from the lead which has been disconnected from the 'IND' terminal, the alternator will not charge.

When the unit is operating, three field diodes supply the field with a current as dictated by the regulator. As the engine speed rises from stationary or slow-running, the output voltage at the field diodes also rises. This reduces the current through the warning lamp and causes the lamp to become dim and finally 'go out'.

Surge protection diode   Breakdown of the main transistor in a regulator occurs if the alternator is charging and a poor connection, or similar fault, causes the voltage to increase suddenly. To avoid this damage, a surge protection diode is sometimes fitted between the 'IND' lead and earth. This special diode only conducts when the surge voltage exceeds a given value.

Failure of this diode, following a fault, often causes it to conduct the field supply continuously to earth, so with the field shorted out, the alternator will not charge until the diode is replaced (or disconnected). Replacement of this diode only takes a few minutes.

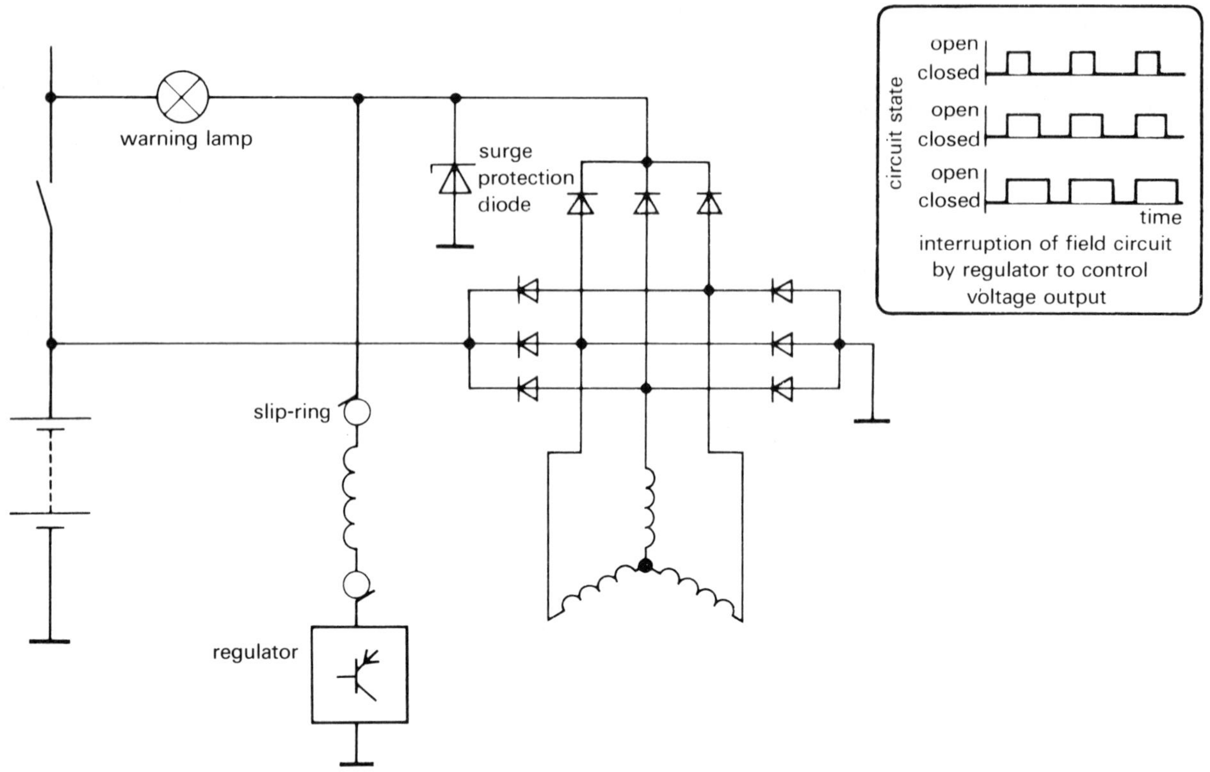

Fig. 72.10 Regulator control of field current and surge protection

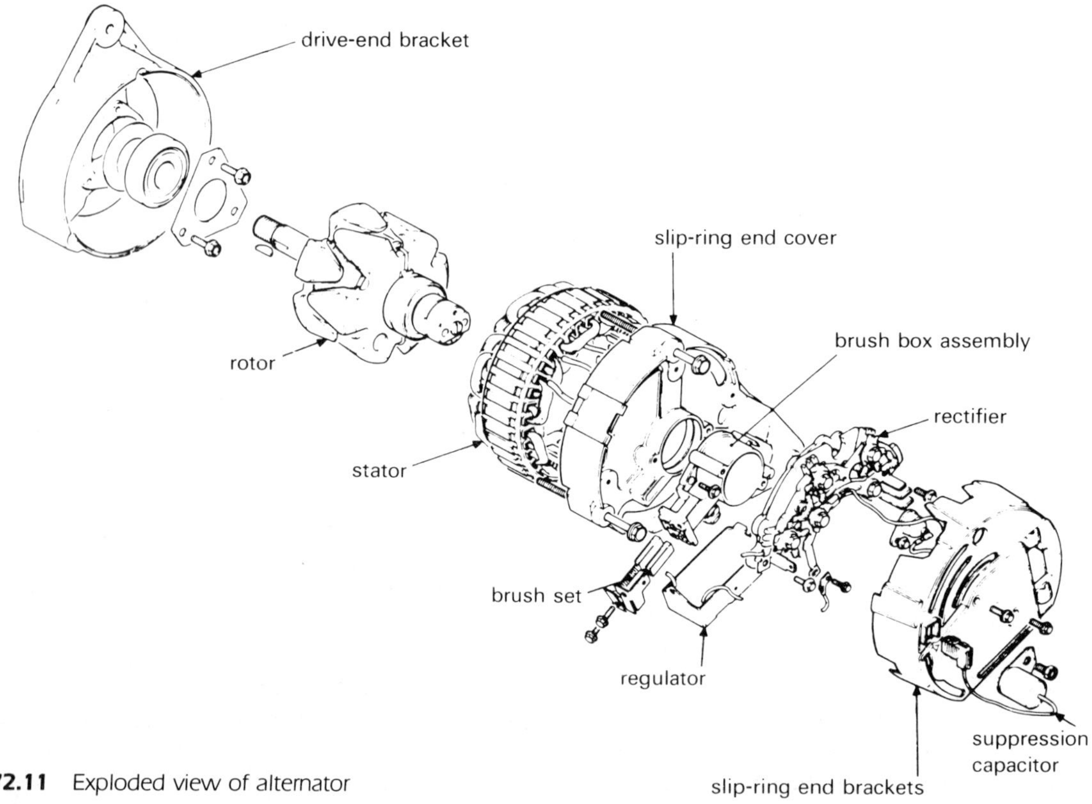

Fig. 72.11 Exploded view of alternator

Alternator construction

The construction of a modern alternator is shown in Fig. 72.11. The unit is driven by a belt at a speed higher than the crankshaft so it produces a good output even when the engine speed is comparatively low.

Alternators can operate at speeds up to 15 000 rev/min and require a high belt tension to prevent slip when a large current is being produced. In view of these operating conditions, ball bearings are used to support the rotor.

Cooling air is pumped through the machine by a cooling fan driven by the rotor shaft.

73 The lighting system

The modern vehicle incorporates many different lights to meet the requirements of the law and the needs of the driver. These lamps are grouped in separate circuits, which include the following:

1 *Side and rear lamps* including lamps for the number plate, glove compartment and instrument panel.
2 *Main driving lamps (headlamps)* including a dip switch to avoid dazzle.
3 *Rear fog lamp(s)* for 'guarding' the rear of the vehicle in conditions of poor visibility.
4 *Auxiliary driving lamps* (front fog lamps).
5 *Reversing lamps*.
6 *Brake lights* to warn a following driver that the vehicle is slowing down.
7 *Directional indicators and hazard warning lights*, which are required to flash at a given rate.
8 *Instrument panel lights* for signalling either the operation of a particular light unit, or the presence of a fault in a section of a lighting circuit.

Circuit arrangement

A very simple circuit consists of a battery, switch and lamp. When more than one lamp is controlled by a single switch, the lamps are arranged in *parallel*; this ensures that full battery voltage is applied to each lamp so as to obtain maximum brilliance.

To save weight, most lighting circuits use an earth return layout, so good 'clean' connections are needed at the numerous earthing points. A 'dirty' connection (i.e. poor earth) causes a high resistance; this reduces the voltage applied to the lamp and results in a poor light.

Circuit diagrams

These are shown in either a *locational* or *compact* theoretical form; the former layout shows each component in its relative position, whereas the latter makes the diagram simpler to follow. Fig. 73.1 shows a simplified lighting circuit represented in both forms.

This parallel circuit has the lamps controlled by three switches:

Switch 1 operates the side and rear lamps. It also supplies:
Switch 2 which operates the headlamps, and supplies:
Switch 3 to distribute the current to either the main beam or the dip-beam headlamp bulbs.

Circuit protection

A single fuse, mounted in the main supply cable, protects a circuit in the event of a severe short. This simple protection system cannot be used in the external lighting supply cable because all lights will go out when the fuse fails, a dangerous situation when the vehicle is travelling at speed along a dark road.

To avoid this danger, manufacturers either fit separate fuses for each light system, or refrain from fusing the headlamp circuit altogether. Fig. 73.2 shows a lighting system which incorporates a series of fuses.

Relay control

When the current is very high, a *relay* is used to relieve the switch of the main current load. The relay acts as a heavy-duty remote switch and is operated when a small current, controlled by the main switch, is supplied to the relay.

Filament lamps

The main details of a *single centre contact* (SCC) side lamp bulb are shown in Fig. 73.3a. Enclosed in a glass container is a helically wound tungsten filament; this is held by two support wires, one attached to an insulated contact in a brass cup and the other

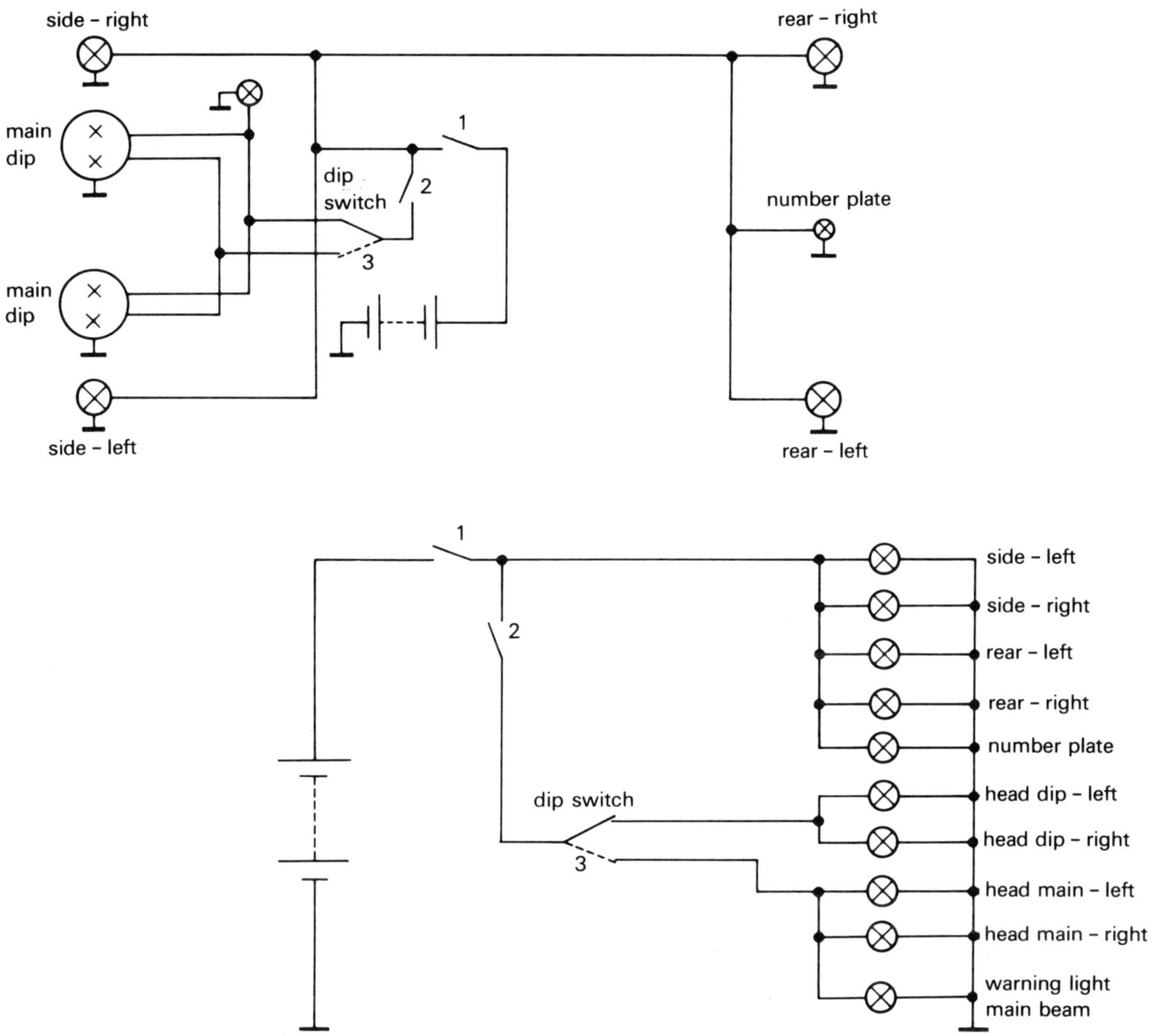

Fig. 73.1 Lighting circuit (simplified)

soldered to the cap to form an earth. A low-wattage vacuum bulb of the type shown, has the air removed to reduce heat loss and prevent oxidation of the filament. The type of bulb shown is secured in its holder by a brass *bayonet cap* (BC); alternative fixing methods are the *miniature screw cap* (MES – Miniature Edison Screw) and the *capless* type.

Filaments of driving lamps are made to operate at a higher temperature by filling the lamp to a slight pressure with an inert gas such as argon. Heat loss from the filament due to gas movement is reduced by winding the filament in a tight helix.

Some driving lamps have two filaments to give a main and dip beam (Fig. 73.3b and c). One end of each filament is connected to the cap and the other ends are soldered to the two lamp contacts. Means must be provided with this type to prevent incorrect fitting of the bulb in the holder.

Tungsten–halogen

Because of its high light output, many modern driving lamps are fitted with this type, which is also called a *quartz-halogen* (Fig. 73.3d). Quartz is used instead of glass and by filling the lamp with halogen gas, vaporization of the filament, which leads to blackening of the glass and filament erosion, is reduced. Another advantage of this type is that its more compact form allows more precise focusing.

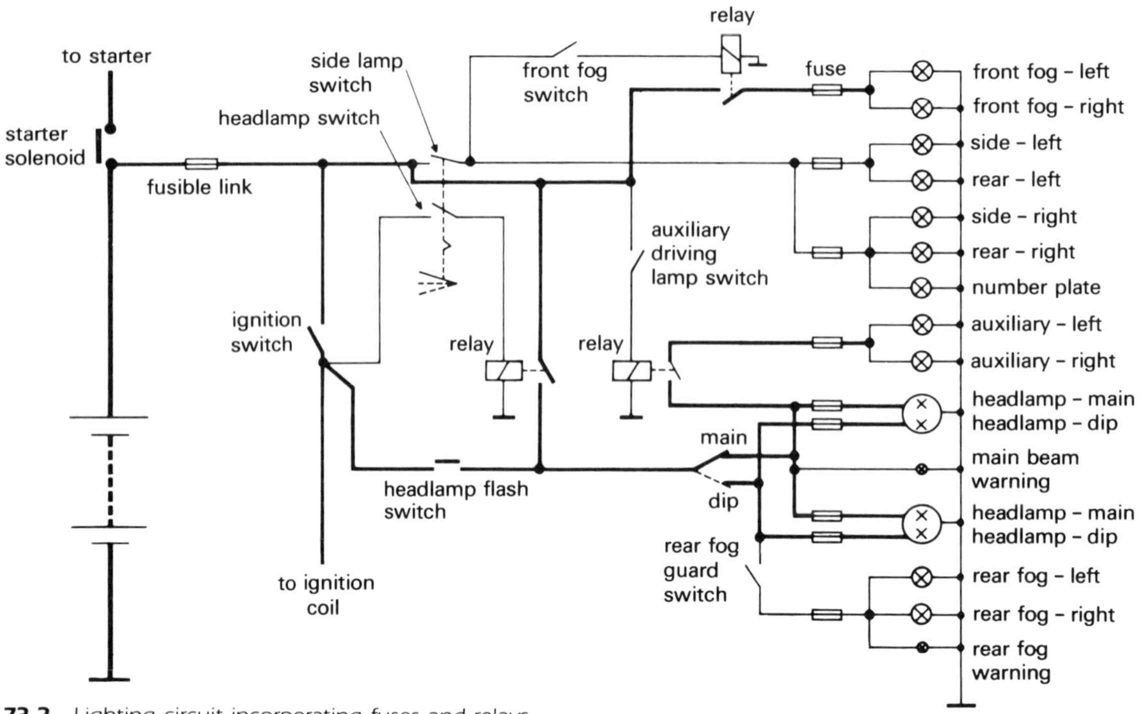

Fig. 73.2 Lighting circuit incorporating fuses and relays

When replacing any bulb, and especially a tungsten-halogen type, a clean cloth should be used to avoid touching and staining the 'glass'.

Sealed beam

Accurate location of the filament at the focal point, and elimination of dirt and moisture on the reflector, are both achieved with a sealed beam lamp unit (Fig. 73.4).

An aluminized glass reflector, fused to the lens, supports the main and dip tungsten filaments; the whole unit is filled with an inert gas. Because the bulb has no independent glass envelope, tungsten deposits are spread over a very large area, so the light efficiency of a sealed beam unit remains high for a long period of time.

The sealed beam unit has two disadvantages; it is costly to replace when a filament fails, and sudden light failure occurs when the lens becomes cracked.

Lamp marking

Regulations state that all bulbs, and other lamp parts, used on a vehicle must be marked with the letter 'E' and a number that identifies the country where approval was given. This mark indicated that the part conforms to the EEC standard specified for a given application.

Reflector

A highly polished aluminized reflector of parabolic shape concentrates and directs the light rays in the required direction. Positioning the main beam filament at the focal point of the reflector gives parallel rays and maximum illumination (Fig. 73.5). This is achieved by a pre-focus bulb to site the filament accurately.

Some vehicles use a multi-segment reflector with several different focal lengths moulded together to form the light unit. These *homofocular* plastic reflectors are very narrow, which allows them to blend in with the body contour.

Lens

A glass lens, consisting of several prismatic block sections, distributes the light to obtain the required illumination. The design of the lens pattern attempts to achieve good illumination for both main and dip positions: a concentrated long-range illumination for the main beam, and a wide spread of light distributed just in front of the vehicle for the dip beam. Regulations insist that to avoid dazzling oncoming drivers the lens must deflect the dip beam downwards and also offset it towards the near-side.

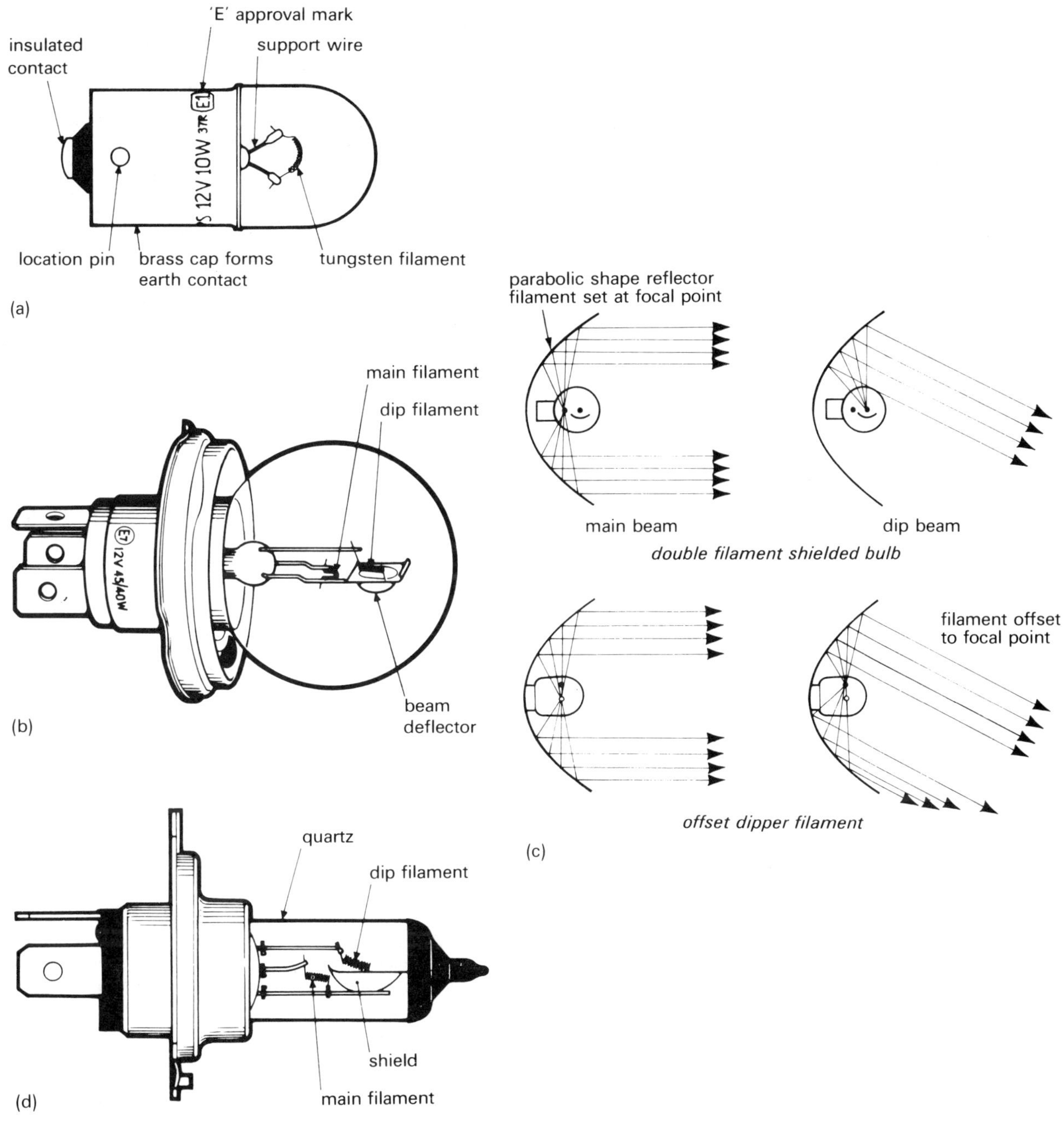

Fig. 73.3 Filament lamps

Four-headlamp system

Optically it is difficult to produce a single lens and reflector unit that gives an illumination that satisfies both main and dip conditions.

To overcome this drawback some vehicles have four headlamps: two for long-distance illumination and two for lighting the area immediately in front of the car. Each one of the outer lamps has two filaments: a dip filament situated at the focal point to give good light distribution, and a second filament positioned slightly away from the focal point to provide near-illumination for main-beam lighting. When the lamps are dipped, the inner lamps that give long-distance illumination are switched off.

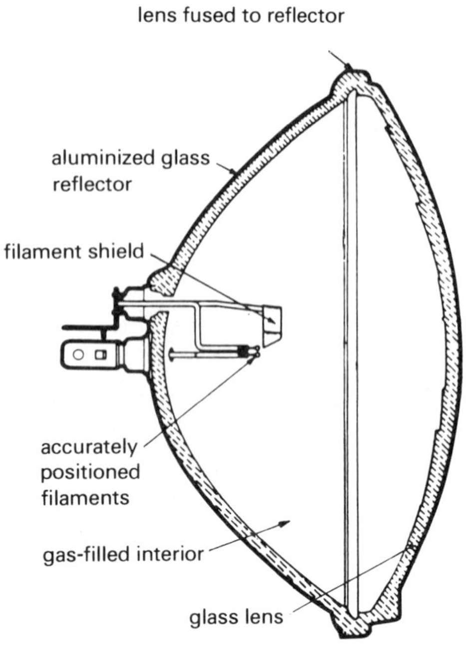

Fig. 73.4 Sealed beam lamp unit

For accommodation reasons the lamps used in this arrangement are smaller than those used on a two-headlamp system.

Headlamp adjustment

Special equipment is available to check and adjust the horizontal and vertical aim of the headlamps. When this is unavailable, the method recommended is to position the vehicle on level ground a given distance in front of a vertical screen set parallel with the headlamps.

After adjusting the tyre pressures and settling the suspension (by bouncing the vehicle), the light pattern given by the dip beam on the screen is observed: this is achieved by covering each lamp in turn. The vertical location of the beam centre (hot spot) of the lamp must be a given distance (recommended by the manufacturer) below the height of the headlamp centre: the horizontal spacing of the hot spots should be equal to the distance between the lamps.

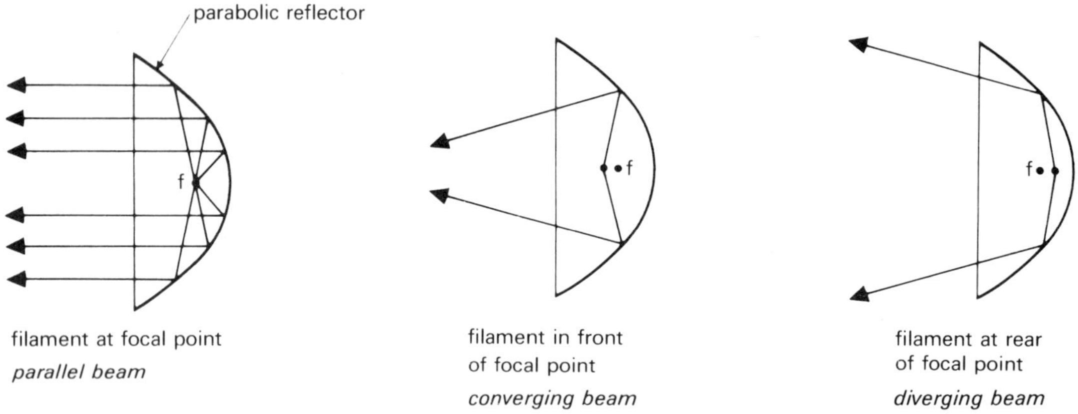

Fig. 73.5 Headlamp filament position

74 The ignition system

Early designs of engine used a *hot tube* type of ignition system. This was an externally heated tube attached to the combustion chamber which was designed to glow red and ignite the fuel–air mixture about t.d.c. at the end of the compression stroke.

As engine speeds increased, the time, relative to crank position, at which the charge was ignited had to be set more accurately, so designers turned to an *electric spark* system invented by Lenoir.

Ignition requirements

Towards the end of the compression stroke of an engine, an electric spark is required to ignite the petrol–air mixture. The spark has to have sufficient energy to start the charge burning and it must be timed to occur at the correct instant.

A system that produces a 'fixed' spark at t.d.c. does not suit modern engines that are designed to operate at various load settings over a wide speed range. Instead the spark must occur well before t.d.c. to suit these conditions and, in addition, the spark timing must be precise and be made to vary when the conditions change. In early systems the spark timing was adjusted by the driver, but modern units perform this duty automatically either by mechanical or electronic means.

Altering the timing to make the spark occur earlier in the operating cycle is described as *advancing the ignition*. Conversely, the term *retard* is used when the timing is changed to make it occur later.

To obtain a high power output from an engine, the maximum cylinder pressure should occur at 12° after t.d.c., irrespective of engine speed and mixture/load conditions. Since it takes a comparatively long time for the burning fuel to build up to its maximum pressure, the spark must be set to allow for this time period. The angle of advance is enlarged when the load is decreased or if either the air/fuel mixture ratio or engine speed is increased. This timing change is necessary because the crankshaft moves through a larger angle during the extra time taken by the gases to build up to the maximum pressure.

Production of high voltage

Although early designs used a comparatively low-voltage trembler arrangement, the introduction of the *high-voltage, timed spark system* considerably improved engine performance. The voltage needed on a high-voltage system to produce a spark depends on the size of the spark plug gap and on the gas pressure in the cylinder. A normal gap is about 0.6 mm (0.024 in), and although only a small charge of about 600 V is required to produce a spark across this gap in the open air, it will take 10–50 times this voltage to fire a plug that is under pressure in an engine. This should be borne in mind when an ignition system is being tested - a spark produced at a plug outside the cylinder does not guarantee that it will spark when it is subjected to cylinder pressure.

A high-voltage spark can be generated by a *magneto* or *coil-ignition* system.

Magneto

This is a small self-contained ignition unit which generates pulses of high-tension current and distributes them to the appropriate cylinders at the correct time.

Motor cycles and small single-cylinder engines use this system, but it is not used on modern cars.

Coil-ignition

This system uses electrical energy produced by a battery or alternator to supply the low-tension ignition current. The ignition coil transforms the voltage to that required to produce a high-voltage spark.

Compared with a magneto, the coil-ignition system makes engine starting much easier. Also it is simpler to control the maximum voltage to suit the conditions.

Coil-ignition systems

The conventional battery-inductive ignition system was introduced by Kettering in 1908, but it was not until the mid 1920s that the system was accepted as a successor to the magneto for use on cars.

Until recently, the main layout of a coil-ignition system, as it is commonly known, did not alter, but the current need to design 'cleaner' engines has demanded ignition systems that produce a higher-energy spark that is timed far more accurately than before. Electronically controlled coil-ignition systems of the *breakerless* type meet these requirements. To avoid confusion, the conventional system will be called the Kettering type.

Kettering type

Fig. 74.1 shows the layout of a basic system that has been in use for many years. The main components are:

Ignition coil This produces the high voltage necessary to cause a spark at the sparking plug. It transforms the battery voltage of 12 V to a low-current, high-voltage charge that is needed to jump the plug gap. The ignition coil has two windings inside its case and these are connected to three external terminals:

1 Low tension (l.t.) to the battery via the ignition switch.
2 Low tension to the contact breaker.
3 High tension (h.t.) to the sparking plug via the distributor. This lead must be highly insulated to prevent loss of the h.t. current.

Contact breaker A high-tension current is produced by the ignition coil when the low-tension circuit from the battery through the coil to the vehicle frame (earth) is interrupted.

It consists of two contact points: a fixed contact, screwed down to a baseplate, and a movable contact that is insulated from the metal parts that surround it. A cam, normally driven by the engine camshaft, operates the movable contact against the reaction of a strip-type spring. The number of lobes on the cam matches the number of cylinders, e.g. a four-cylinder engine has four lobes, so one revolution of the cam will give four sparks. On a two-stroke engine, one spark per revolution per cylinder is required, so the cam is driven at the same speed as the engine instead of half engine speed as required by a four-stroke unit.

A spark at the plug occurs at the instant the contacts separate, so the assembly is set (timed), in relation to the crankshaft, to give a spark at the correct time.

A *capacitor* (also called a *condenser*) is connected in the circuit across the contacts, i.e. in parallel. This reduces arcing of the contacts and in consequence gives a more rapid interruption of the l.t. circuit.

Distributor On a single-cylinder engine the h.t. lead from the coil is taken direct to the sparking plug. A multi-cylinder engine has a number of sparking plugs and each one of these has to be connected to the coil when the high-voltage charge is fed to the plug. The h.t. switch used to select the appropriate plug is called a distributor.

It consists of a hard plastic *distributor cap*, inside which is a *rotor arm*; this is turned at half crankshaft speed by a shaft that drives the contact breaker cam. The h.t. lead from the coil fits in the centre of the cap and a carbon brush rubs on the rotor to transmit the electrical charge.

At the instant the spark occurs, the rotor arm is set to 'point' to the brass 'segment' in the cap that is connected to the plug that requires the charge.

The distributor cap, rotor, contact breaker assembly and automatic advance systems are all incorporated in one unit; this is often called the *distributor unit*.

High-tension leads Highly insulated with PVC plastics or thick rubber, the h.t. leads link the distributor to the sparking plugs and ignition coil.

Radio and TV interference is reduced by using *suppression leads* having a high electrical resistance. The conductor used for the lead is either carbon-impregnated cotton or glass fibre. These leads are generally made up in a set to suit the suppression requirements of the particular vehicle.

Sparking plug The plug consists of a highly insulated centre electrode, and an earth electrode which is welded to a metal body. A typical gap of 0.6 mm (0.024 in) between the electrodes enables a spark to be produced when a high voltage is delivered to the plug (Fig. 74.2).

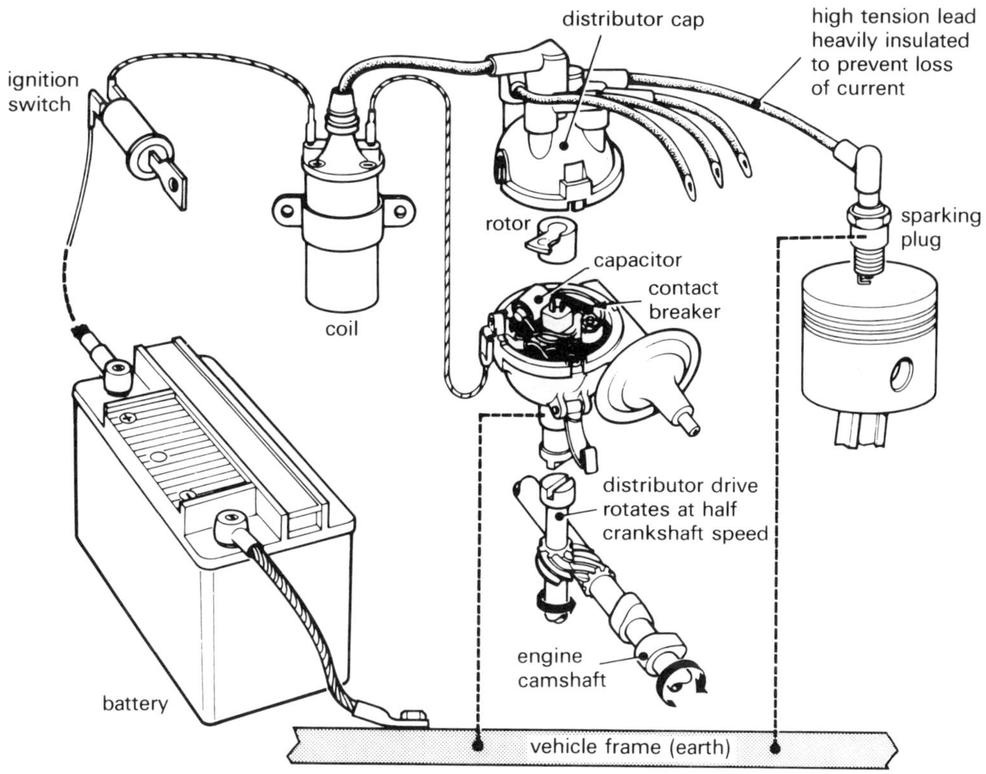

pictorial view of coil-ignition system

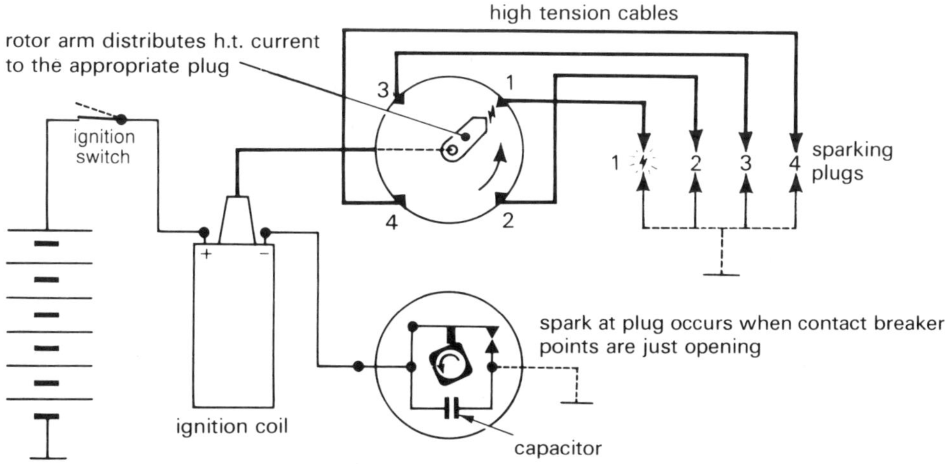

Fig. 74.1 Cool-ignition system – Kettering type

Magnetism and induction

In order to understand the operation of ignition, and other systems, consider the following experiments.

Fig. 74.3a shows a conductor passing through a piece of paper, on to which are scattered some iron filings. When current flows through the conductor, the iron filings arrange themselves in a series of concentric circles. This indicates the presence of a magnetic field. If the process could be slowed down, you would see that on making the circuit, the field moves outwards from the conductor. When the flow of current is

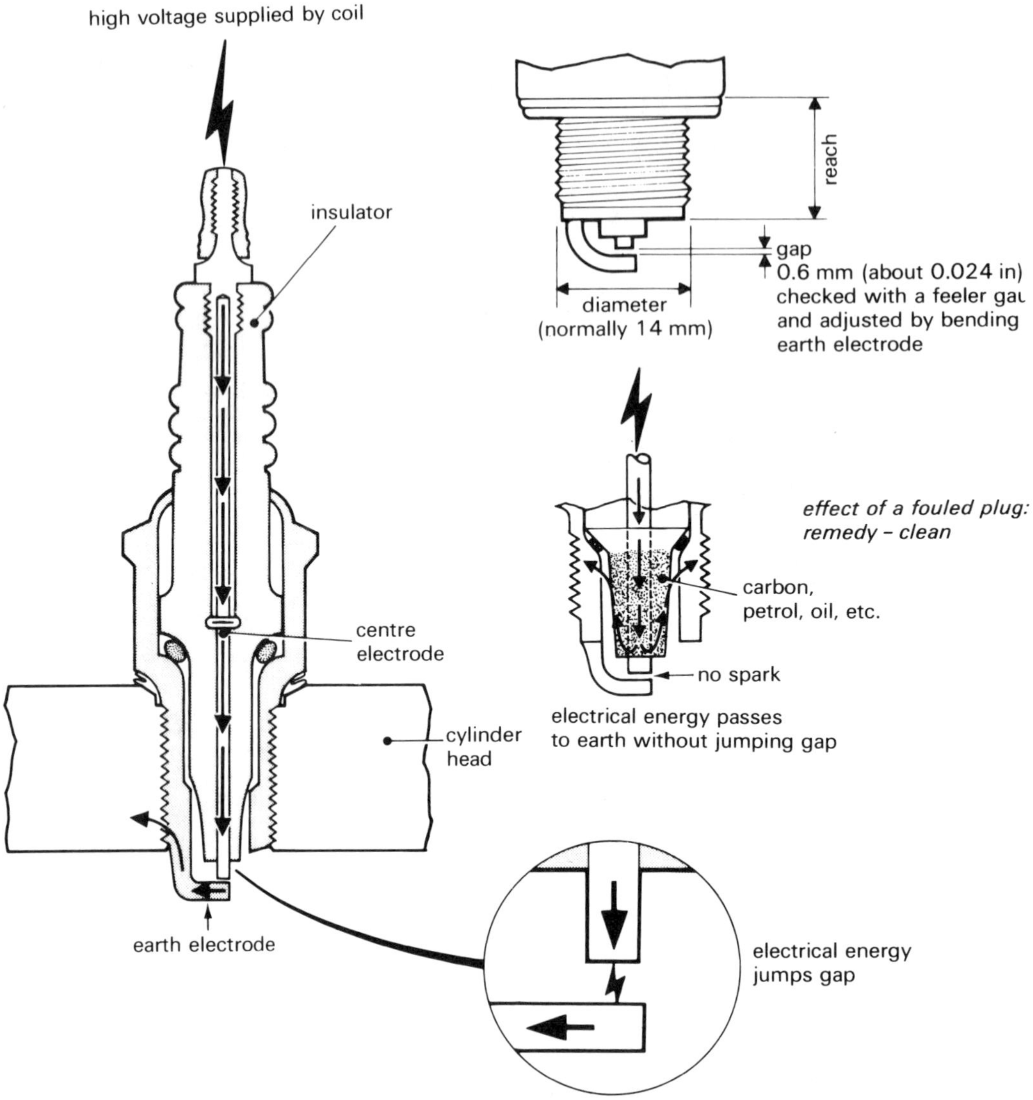

Fig. 74.2 Sparking plug

interrupted, a reverse action occurs - the field collapses.

Fig. 74.3b shows a length of wire wound in the form of a coil. On closing the switch, the magnetic field surrounding each turn of the coil combines with other fields to produce a larger field. A soft iron core, mounted in the centre of the coil, concentrates and intensifies the field. This core becomes a magnet when current is flowing through the coil (Fig. 74.3c); the magnetic strength is governed by the amount of current flowing, and the number of turns on the coil.

Fig. 74.3d shows another coil, termed a secondary winding, which is wound over, or placed near, the original coil or primary winding. A galvanometer, which is an instrument used for detecting a small electrical current, is inserted in the secondary circuit, and the switch is operated. On closing and opening the switch, you will find that the galvanometer needle momentarily flicks one way on switching on, and then the other way on switching off. This was first discovered by Faraday, and from his experiments it was concluded that when a magnetic line of force cuts a conductor (or vice versa), an e.m.f. is induced in that conductor. The magnitude of the e.m.f. depends on:

1 the rate of change of magnetic field
2 the number of turns on the coil.

In the apparatus shown in Fig. 74.3d, the build-up of the field around the primary causes the lines of

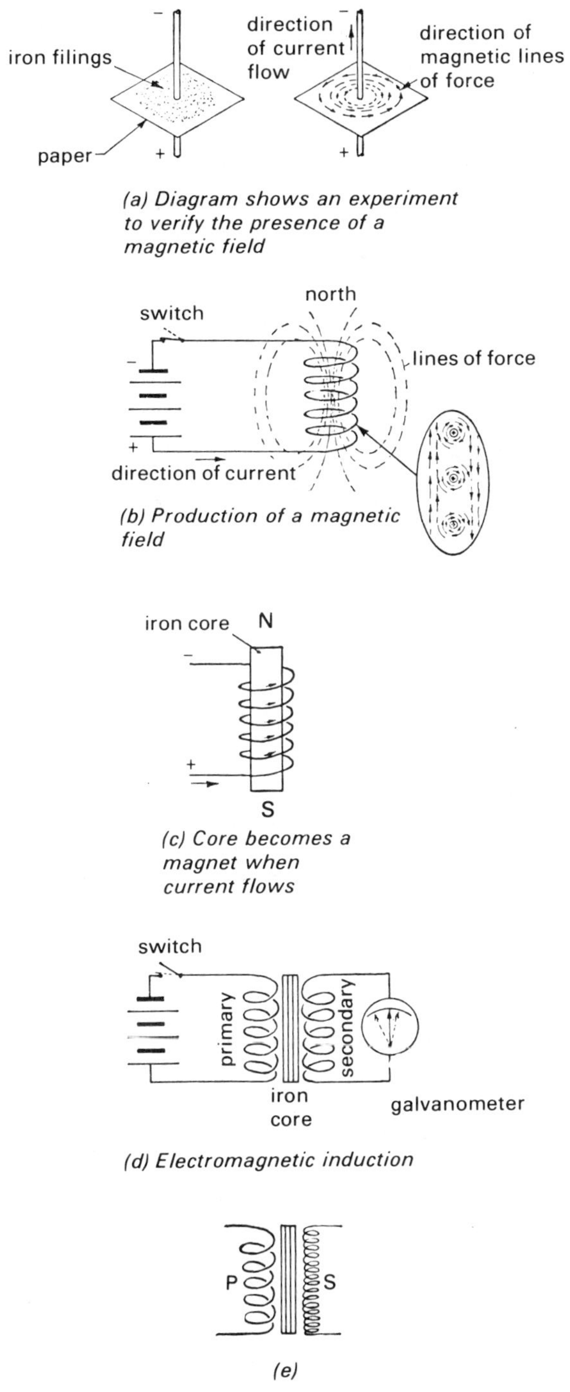

Fig. 74.3 Magnetism and induction

force to cut the secondary, and therefore an e.m.f. is induced in the secondary. Since the build-up is slower than the collapse, a much higher e.m.f. is induced when the circuit is broken.

If the secondary contains more turns than the primary (Fig. 74.3e), it is possible to obtain a higher voltage in the secondary circuit: e.g. if the secondary contains 100 times the number of turns wound on the primary, the e.m.f. will be 100 times greater than the e.m.f. in the primary, assuming the efficiency is 100 per cent. This increase in voltage is balanced by a proportional decrease in the current.

Principle of operation of a coil-ignition system

The layout of an earth-return (ER) ignition system is shown in Fig. 74.4. The system consists of two circuits: primary and secondary.

Primary circuit

Wound around a soft iron core in the coil are several turns of comparatively heavy enamelled wire. This is arranged in series with a battery, ammeter, switch and contact breaker. A capacitor is connected in parallel with the contact breaker.

Secondary circuit

A secondary coil winding, consisting of several thousand turns of fine enamelled wire, is wound under the primary. One end of this winding is joined to the contact breaker terminal, and the other end is connected in series with the distributor and sparking plugs. A 'return path' from the plug, via earth, passes through the battery and primary winding, and so e.m.f. induced in the primary winding is added to the large e.m.f. produced in the secondary winding. This gives a higher coil efficiency.

When the ignition is switched on and the contacts are closed, the current flowing in the primary winding sets up a magnetic field around the iron core of the coil. Opening the contacts, at a time when the spark is required, interrupts the current flowing in the primary circuit, and causes the magnetic field to collapse. During this collapse, the lines of force cut the secondary winding, and induce an e.m.f. in the secondary: a higher e.m.f. than that acting in the primary is obtained, since the secondary contains more turns. The h.t. current is conveyed from the coil to the rotor arm, which, at this time, should be pointing to the correct distributor segment. This is the segment connected to the plug of the cylinder that requires ignition of the charge.

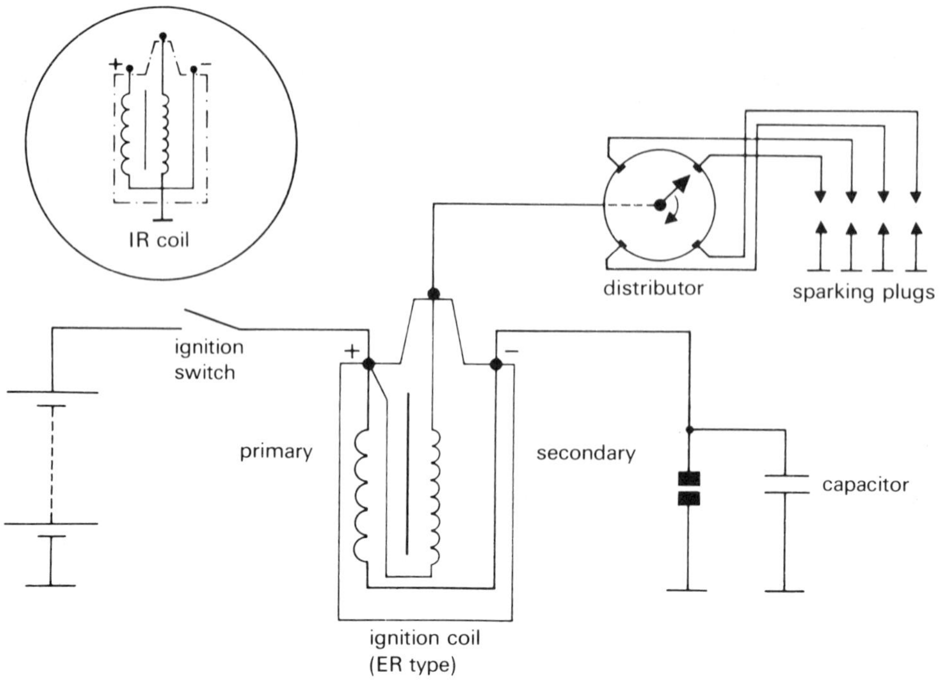

Fig. 74.4 Coil-ignition system

Capacitor

A capacitor is fitted for two reasons:

1 It reduces arcing of the contacts.
2 It ensures a quicker collapse of the magnetic field.

It consists of two sheets of foil or metallized paper, which are separated from each other by at least two sheets of insulating material such as waxed paper. These are wound into a cylindrical shape, and inserted in a metal container. One sheet is joined to the earthed container, and the other sheet is connected by a wire or metal strip to the insulated side of the contact breaker.

When a voltage is applied to the terminals, a current can be made to flow into, but not through, the capacitor. This voltage charges the capacitor to a value equal to the supply voltage, and when this point is reached, the flow of current ceases. If the supply is now disconnected, the charge will be retained for a time, but will gradually 'leak away'. When a charged capacitor is connected to a circuit bridging the terminals, the charge produces a current flow in a direction opposite to the original supply current.

In order to understand the function of a capacitor, consider the operation of an ignition system which has the capacitor disconnected. As the contacts break, the lines of force cut the primary winding as well as the secondary, and therefore an e.m.f. is also induced in the primary winding. This builds up to a pressure sufficient to cause a spark to jump the small contact gap. Arcing maintains current flow in the primary circuit, and thereby prevents the rapid collapse of the magnetic field, as well as causing serious burning of the contacts.

The action is in many ways similar to the hydraulic analogy shown in Fig. 74.5. Water flowing along a pipe at great speed will produce a sudden pressure rise if the tap is shut off quickly. This pressure surge could lead to the discharge of a small amount of water

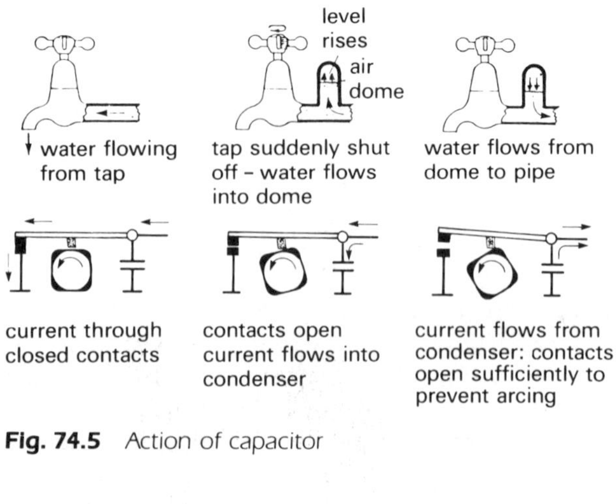

Fig. 74.5 Action of capacitor

through the 'closed' tap, by lifting the tap washer off its seat.

Fitting an air dome to the system allows the water to flow into the dome as the tap is suddently shut off, but after a short time the air forces the water back into the pipe.

The capacitor must perform a duty similar to the buffer action of the air dome. When the contacts part, the capacitor absorbs the self-induced current, and by the time the capacitor is fully charged the contacts have opened sufficiently to prevent a spark occurring at the contact breaker points.

Insulated-return (IR) system

A different coil is required when an IR system is used on a vehicle. One end of the secondary winding coil is connected to a fourth terminal or to the coil case. This terminal (or case) must be earthed to the frame to complete the secondary circuit.

An IR coil may be used on an ER system, but an ER coil cannot be employed on an IR system. The coil type is generally stamped on the base of the coil.

Distributor construction

Fig. 74.6 shows the construction of the complete distributor assembly. This comprises the distributor, contact breaker assembly and automatic advance mechanism.

Distributor

A rotor arm, mounted above the cam and driven at half engine speed, is contacted by a spring-loaded carbon brush. This is fed by an h.t. cable from the coil, and housed at the centre of a moulded Bakelite distributor cap. Cables from the various sparking plugs connect, in firing order, with brass segments held in the cap. These are positioned so that there is a small gap between the rotor and the segment. Ventilation slots, formed at the spigot joint between cap and main body, permit the escape of gases generated by the sparking.

Contact breaker

This consists of two tungsten contacts: a fixed earth contact and a movable insulated contact. This is linked by its return spring to a terminal on the side of the main body (Fig. 74.7). A cam having four lobes (four-cylinder) or six lobes (six-cylinder) operates the contact. Elongated holes, through which the clamping screws pass, allow for adjustment. The contact is set

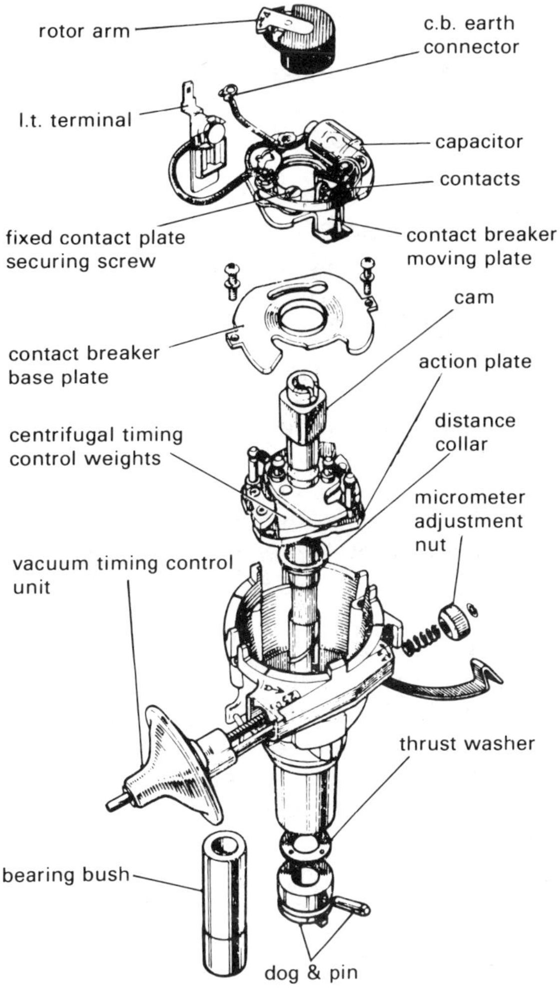

Fig. 74.6 Distributor unit

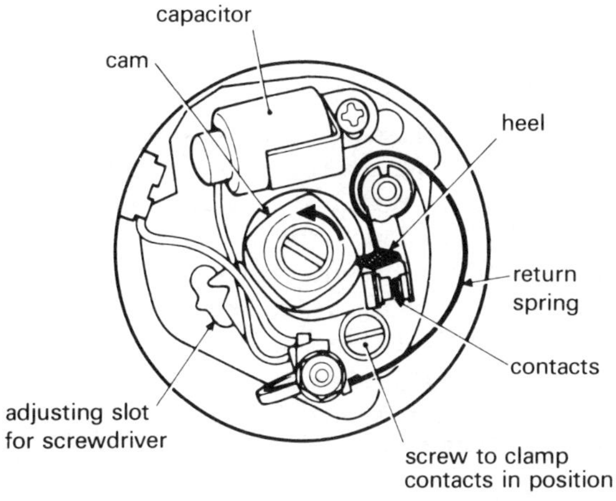

Fig. 74.7 Contact breaker assembly

to provide a cam lift or contact gap of the order of 0.36–0.40 mm (0.014–0.016 in).

A capacitor, earthed by a fixing screw to the contact base plate, connects with the insulated screw on the side of the main body.

Automatic advance

Maximum cylinder pressure should be developed just after t.d.c. This generally means that the spark must be produced before t.d.c., since time is taken for the gas to build up to its maximum pressure. The time factor is fairly constant, but the angle moved by the crankshaft during this time varies in proportion to the engine speed. This means that, as the engine speed is increased, the spark must be advanced.

Alteration of the ignition timing to suit speed and load conditions is performed by two automatic advance–retard mechanisms; these are:

- *Centrifugal* – advances the spark as the speed increases, i.e. speed-sensitive
- *Vacuum* – uses induction manifold depression to advance the spark during light load (cruising) conditions, i.e. load-sensitive.

To advance the spark, either the cam is turned in the direction of rotation (d.o.r.) or the base-plate is moved against the d.o.r.

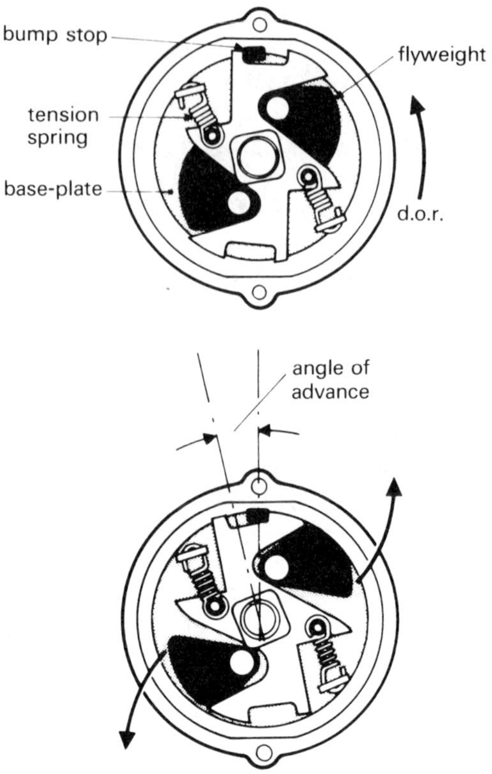

Fig. 74.8 Speed-sensitive centrifugal advance mechanism

Centrifugal advance

The advancement of the cam in the arrangement shown in Fig. 74.8 is produced by two flyweights. These are pivoted on a base-plate which is driven around by the distributor shaft. A contoured face on the driving side of each flyweight acts against a cam-plate; this is integral with the cam so that when the flyweights move outwards, the cam is rotated in the d.o.r.

The flyweight movement produced by a given increase in speed is determined by the strength of the two tension springs fitted between the base-plate and cam-plate. By carefully matching the spring strength and flyweight contour to the ignition timing requirement, a good engine performance can be obtained. A typical advance curve for a 'mechanical system' is shown in Fig. 74.9.

On some engines the rate of advance in the low-speed range has to be greater than at speeds greater than about 1000 rev/min. This can be achieved by using unequal-strength springs with the stronger spring fitted slack on its post.

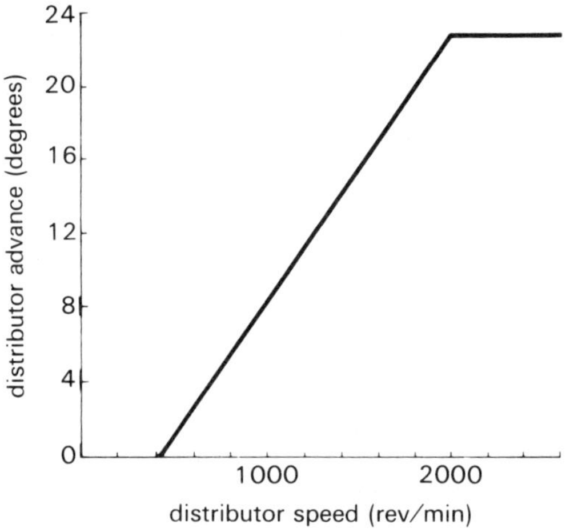

Fig. 74.9 Typical advance given by centrifugal advance mechanism

Vacuum timing control

The normal centrifugal advance mechanism is sensitive to speed, but cannot 'feel' the degree of load on the engine. This means that maximum advance is restricted to avoid engine 'pinking'. The condition is particularly severe when an economy-type carburettor is fitted. Weak mixture and slow burning of the fuel during part-load conditions demand a greater ignition advance than that used for maximum engine load. To overcome this problem, a vacuum control unit is often incorporated in the distributor unit.

The main construction is illustrated in Fig. 74.10. 'Depression' felt at a drilling, generally in the vicinity of the throttle, is communicated to a 'vacuum' chamber. This is fitted with a spring-loaded diaphragm, and is linked to the distributor body.

During cruising conditions, a high depression acts on the diaphragm to give maximum ignition advance. When engine load is increased, the pressure alteration in the 'vacuum' chamber rotates the distributor to a position giving less advance.

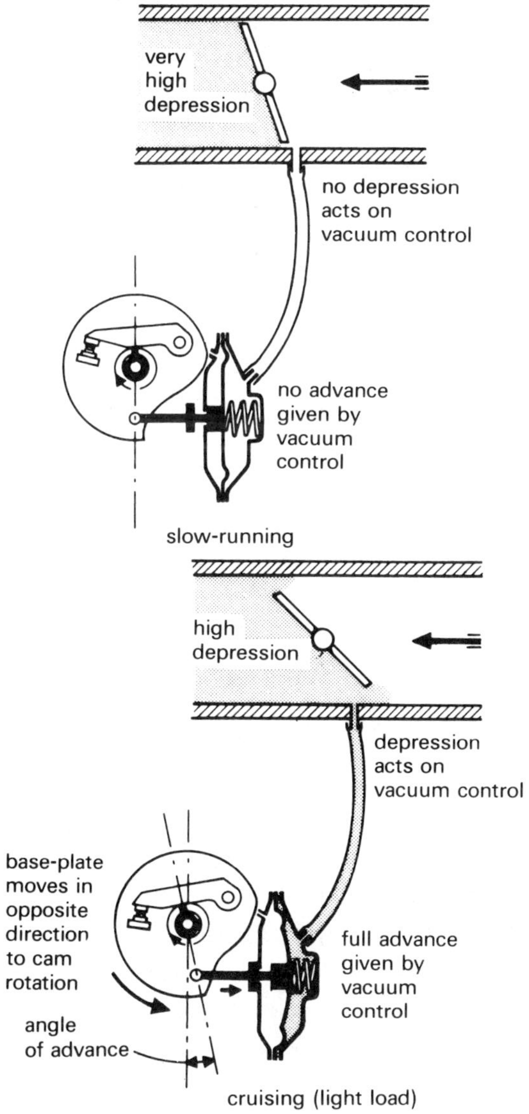

Fig. 74.10 Vacuum advance system

Most modern units are built into the distributor, so only the contact breaker base plate need be rotated by the vacuum control.

Vacuum advance control valves

Precise ignition timing is essential if the emission from an engine is to meet the limits imposed by local regulations. Control of the vacuum advance is often used to meet these requirements and this has meant that additional control valves, such as a *spark delay valve*, have been fitted in the vacuum pipe to control the advance given by the vacuum advance unit. A further development has been the introduction of dual diaphragm chambers; these are more sensitive to the variation in load on an engine.

Sparking plugs

The sparking plug consists of a steel body, which retains a ceramic insulated centre. This electrode connects with the h.t. lead, and is generally made of a nickel alloy. One or more earth electrodes are welded to the body, and adjusted to produce a spark gap suitable for the engine. Most plugs today have a 14 mm dia. thread, but 10 mm and 18 mm are occasionally used. The length of thread, termed the *reach*, is governed by the distance through the cylinder head.

To function successfully, the plug must operate at the correct temperature. A low temperature allows oil and carbon to form on the insulator. This *fouling* causes the electrical charge to short to earth and leads to misfiring. When the plug temperature is too high, the plug electrodes get too hot, and pre-ignition occurs.

The heat range of the plug is governed mainly by the distance between the electrode tip and the heat transfer washer. Fig. 74.11a shows a plug which can disperse its heat quickly. This is known as a *cold* or *hard* plug, and is suitable for an engine that has 'hot' cylinder conditions, e.g. high-performance engines. Moving the copper transfer washer away from the electrode (Fig. 74.11b), raises the plug temperature to give a *hot* or *soft* plug.

Whenever an electric spark occurs, waves of electrical energy are radiated. These cause interference with domestic television receivers and car radios. To limit this interference, the law stipulates that some effective form of suppression must be used. The

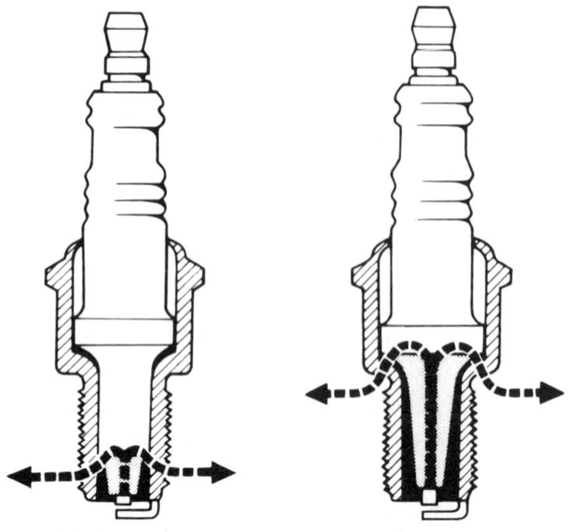

Fig. 74.11 Sparking plug heat range

ignition is the main offender, and so special devices are fitted to the h.t. circuit. These generally take the form of resistors of value between 5000 and 25 000 ohms, which can be incorporated in: (1) a special h.t. lead, (2) a special carbon brush fitted between rotor and centre distributor terminal, or (3) a fixed resistance, which is inserted in the h.t. lead.

Centre electrode polarity

When a metal is heated it emits electrons, which are negative charges of electricity. The hottest part of a sparking plug is the centre electrode, so if this is arranged to be of negative polarity, then the electron flow due to the heat will be in the same direction as the charge given by the ignition coil. A centre electrode having negative polarity gives a more effective spark for a given coil output.

The centre electrode polarity is controlled by the direction of the coil windings in relation to the battery polarity. In order to obtain a 'negative spark' the l.t. connections on the coil are now marked '+' and '−'.

A lead pencil fitted in a disconnected h.t. lead plug terminal with the pencil point held close to the plug terminal shows that the polarity is correct when the flare in the spark appears on the pencil side of the gap.

Contact breaker setting

Dwell angle

In the past the method of setting the contact breaker was to rotate the cam until it was in the full lift position and then measure the gap between the points with a feeler gauge. The method was often inaccurate because it was impossible to take into account the position the cam occupied when the engine was running. Any wear in the driving shaft bearings gave an incorrect gap which led to incorrect timing of the spark and often caused poor ignition performance at high engine speeds because of the comparatively short time that current was flowing in the primary circuit. Furthermore, operation of most contact sets causes a transfer of metal from one contact to the other and this results in a 'pip' building up on one contact and a 'hole' forming in the other. This would defeat the feeler gauge method unless the 'pip' was ground away.

Modern electrical testing equipment generally incorporates a meter which measures the *dwell angle* and this is used as an alternative to the feeler gauge method. As applied to ignition units, *dwell* is the period that the contacts are closed during a cam movement equal to the angle between the cam lobes. (This statement gives the common definition but readers may find cases where the term 'dwell' is used to indicate the 'open' period).

Fig. 74.12 shows a diagram of a cam suitable for a four-cylinder engine. In this case the closed period or 'dwell angle' is 60°. This is typical for a four-lobe cam, whereas a cam used with a six-cylinder engine generally has a dwell angle of 35°; both cases give a tolerance of about ±2°.

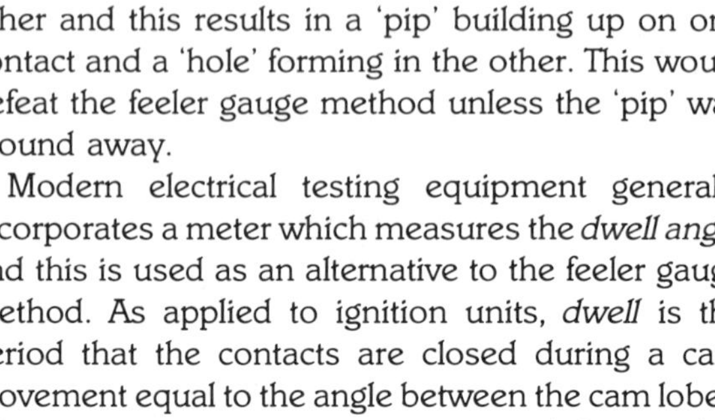

Fig. 74.12 Dwell angle

Some manufacturers state the setting as percentage dwell. This indicates as a percentage the period that the contacts are closed or:

$$\text{Dwell (\%)} = \frac{\text{Dwell angle (closed)}}{\text{Angle between cam lobes}} \times 100$$

The dwell angle for the cam shown in Fig. 74.12 is:

$$\frac{60}{90} \times 100 = 67\%$$

Double contact breakers

Consideration of the dwell angles of units fitted to four and six-cylinder engines show that dwell angle is reduced as the number of cylinders increases. If a single contact breaker were used on an eight-cylinder engine, then the short dwell period would seriously affect the coil's performance at high speeds. To overcome this problem a double contact breaker is often used (Fig. 74.13).

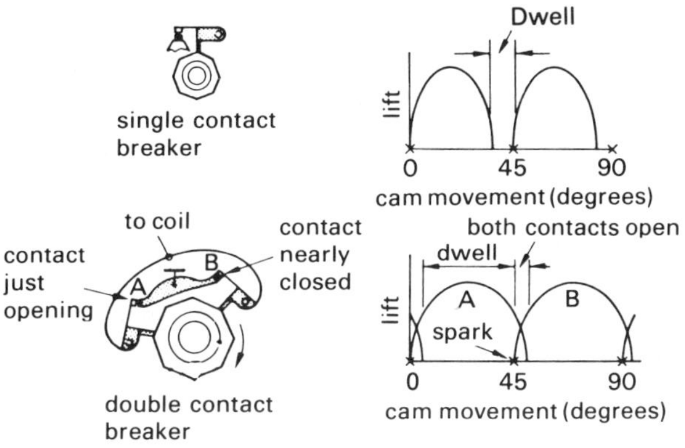

Fig. 74.13 *Advantage of double contact breaker*

Contact set A is connected in parallel with set B so the circuit is only interrupted when both contacts are open simultaneously. As soon as contact A opens to give a spark at the plug, the other contact closes to re-establish the primary circuit.

Ignition coil output

The voltage needed to produce an efficient spark is increased when the engine compression pressure is raised or if the mixture strength is weakened.

To meet modern engine requirements various refinements are made to the ignition system.

Oil-filled coil

Immersing the windings in oil gives the following improvements:

1 Better insulation and greater resistance to moisture.
2 Primary winding operates at a lower temperature, so the lower resistance allows more current to flow.
3 Reduced corona (glow of light) effect.

Ballast resistor

The maximum output from a normal coil decreases as the engine speed is increased, so in order to give a more constant output a ballast resistor is sometimes fitted in the primary circuit. This should not be confused with the resistor fitted to improve cold starting (see page 452).

At low engine speeds the relatively large current flow in the l.t. circuit causes the resistor to get hot, increase its resistance and so limit the current flow. When engine speed is high the resistor runs cooler.

Starting problems

Starting difficulties with modern engines are rare and when they arise the trouble can often be blamed on inadequate servicing. Whenever certain models are prone to a particular fault then attention is focused on possible causes and this generally leads to either a modification or a recommendation of a suitable repair or treatment. Two examples of past trouble are dampness and low battery voltage.

Dampness

May be caused by under-bonnet condensation, climatic conditions or any means which allows water to come into contact with high-tension ignition leads. Moisture on these leads causes the h.t. current to short to earth; consequently this external 'leakage' of electric current prevents a spark occurring at the plug. In the past the manner in which this trouble was avoided was to protect the equipment by preventing moisture from coming into contact with the h.t. system. If this precaution was not followed, then it meant that the distributor cap, with leads attached, had to be removed and taken to a warm place to dry out.

A recent invention has provided a liquid substance which can be applied from an aerosol can to the ignition equipment. This liquid acts as a barrier to the moisture and prevents future trouble, and in cases where the leads are already damp, a spray from the aerosol normally overcomes the problem and allows the engine to start.

Low battery voltage

This is often the cause of an engine failing to start on a cold morning. If the life of the battery is nearing its end, the high starter motor load is sufficient to lower the voltage at the ignition coil to a value less than 9 V. Under these conditions the coil is unable to provide the voltage necessary to produce a spark at the plug even though the starter motor is still operating successfully. This problem is not helped by the driver who operates the starter motor for periods longer than about three seconds at a time, since the longer the motor is operated the lower the battery voltage will become. It is hoped that between each starter operation a period is given for the battery to 'recover'.

Some manufacturers have recognized this problem and have altered the ignition system to a form similar to that shown in Fig. 74.14. This diagram shows that the normal 12 V coil has been replaced by a 7.5 V coil and a ballast resistance inserted in series in the primary circuit. When the engine is running, the resistor drops the voltage from 12 V to 7.5 V. During the starting phase the starter switch shorts out the resistor and applies 'full' battery voltage to the ignition coil.

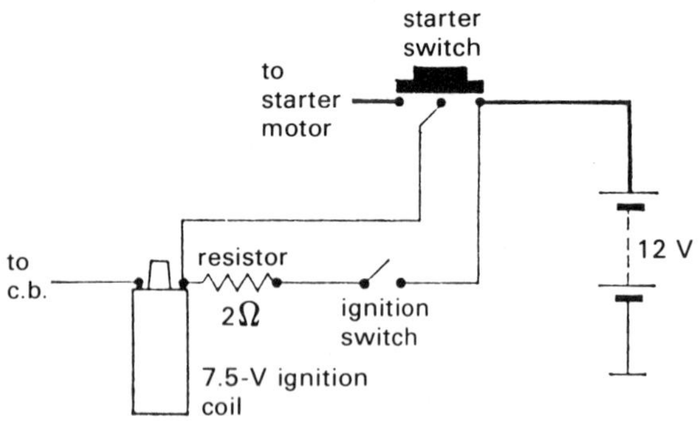

Fig. 74.14 Ballast resistor-ignition system

Ignition timing

Manufacturers normally give detailed instructions for retiming the ignition, but when these are not available, use the following method:

1 Set No. 1 piston to t.d.c. compression.
2 Connect the drive so that contacts are just *opening*, when the rotor is pointing to the distributor segment 'feeding' No. 1 cylinder.
3 Fit h.t. leads to the distributor in the order of firing.
4 Start engine and make final adjustment. If an ignition timing mark is provided on the crankshaft fan pulley or flywheel, a timing light can be used. A vacuum gauge serves as an alternative method when a timing light cannot be employed.
5 Road test.

The use of an offset distributor driving tongue allows the unit to be removed and refitted without disturbing the timing, provided the plate clamped to the distributor has not been moved.

Breakerless ignition

Although the conventional ignition system gave good service, the twin problems of pollution and fuel economy demanded that improvements were made. Modification or elimination of the contact breaker was the main alteration because the traditional contact breaker has the following drawbacks:

1 Ignition timing varies from specification. The timing alters owing to:
   a wear at heel, cam and spindle and errosion of contact faces,
   b contact bounce and inability of the heel to follow the cam at high speed.
2 Dwell angle variation due to change in engine speed or wear at contacts.

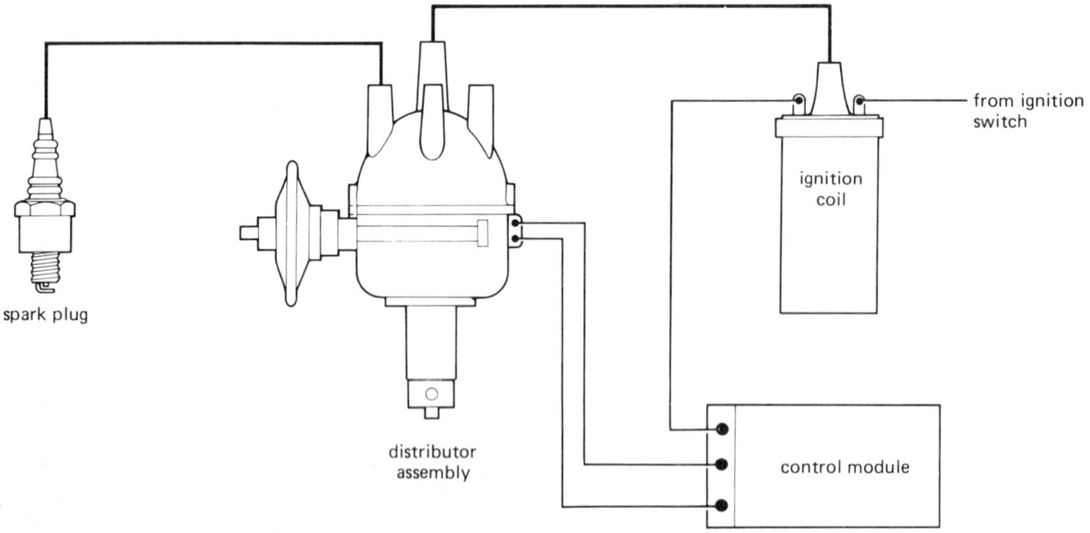

Fig. 74.15 Layout of electronic ignition system (simplified)

3 Servicing requirements are frequent.
4 Cannot effectively control a current greater than about 3 A.

Drawback of conventional coil

The conventional ignition coil primary winding has a large number of turns. During operation of this coil, interruption of the primary circuit *mutually induces* an e.m.f. into the secondary and *self-induces* into the primary winding an e.m.f. of polarity similar to that which existed before the circuit was broken.

Self-inductance slows down the rate of build-up and collapse of the magnetic field and this is one reason why the performance of the standard ignition coil falls as engine speed is increased. The problem becomes acute on engines which demand a large number of sparks per second, e.g. a multi-cylinder two-stroke engine or an eight-cylinder unit.

Low-inductance coil

A low-inductance ignition coil has a smaller number of turns on the primary winding so as to reduce self-inductance. This gives a sharp interruption when the circuit is broken, but the lower primary resistance means that the current flow is much greater; normally it is about three times as great, so contact breaker life is much shorter.

In addition to good high-speed performance the low-inductance coil also gives a good energy release during the spark period, so these advantages make this type of coil popular. Increased use of this coil type is accompanied by the need for a circuit breaker which does not have the drawbacks associated with the conventional contact breaker arrangement.

Electronic systems

The duty of a conventional contact breaker is to:
1 Trigger the system when the spark is required.
2 Interrupt the primary current flow to induce a high voltage in the secondary circuit.

Breakerless ignition systems, sometimes called *transistorized ignition systems*, use solid-state electronic devices to perform the two contact breaker duties. Normally they work in conjunction with a low-inductance ignition coil.

The simplified layout of an electronic system (Fig. 74.15) shows its similarity to a conventional ignition system. The distributor assembly incorporates a common type of automatic advance mechanism and h.t. distributor, but the contact breaker is replaced by a transistorized switch to 'trigger the spark'.

Trigger systems

Fig. 74.15 shows the main layout of an electronic ignition system. The distributor unit is similar to a conventional unit with the exception that the contact breaker is replaced by an electronic switch called a *pulse generator*. This device generates an electrical pulse to signal to the *control module* when a spark at the plug is required. The action is similar to the trigger of a gun – when the trigger is operated the control module breaks the primary circuit; this causes the ignition coil to produce the h.t. charge.

The three main types of pulse generator are:
- *Inductive*
- *Hall*
- *Optical*.

Inductive pulse generator Fig. 74.16 shows one type of inductive pulse generator suitable for a four-cylinder engine. It consists of a permanent magnet, coil winding and iron rotor, which acts as a reluctor to vary the flux flow. Two leads connect this unit with the control module.

When the tooth of the reluctor passes the permanent magnet, the change in magnetic flux induces an e.m.f. in the coil. Since this e.m.f. varies between 0.5 and 100 V (the actual output depending on the speed of movement), the voltage variation, together with the change in frequency, provides a sensing signal that can also be used for purposes other than spark triggering.

The size of the air gap is critical and should not need to be altered during normal servicing of the vehicle.

Some manufacturers do not use the distributor unit, if fitted, as a pulse generator. Instead they use either the flywheel teeth or a toothed ring fitted at the end of the crankshaft. Naturally any damage to the 'teeth' on this ring affects the ignition performance.

Hall generator Fig. 74.17 shows a distributor unit fitted with a Hall generator; it gets this name because the system is based on the *Hall effect*.

The unit consists of a permanent magnet, vane and semiconductor chip, which is supplied with a voltage from the battery. A three-core cable connects the Hall generator with the control module; the leads are signal-input, Hall-output and earth.

As the vane passes between the magnet and chip, a change in magnetic flux causes the chip to switch the sensor off and on; the signal for the spark occurs at the instant when the vane leaves the air gap.

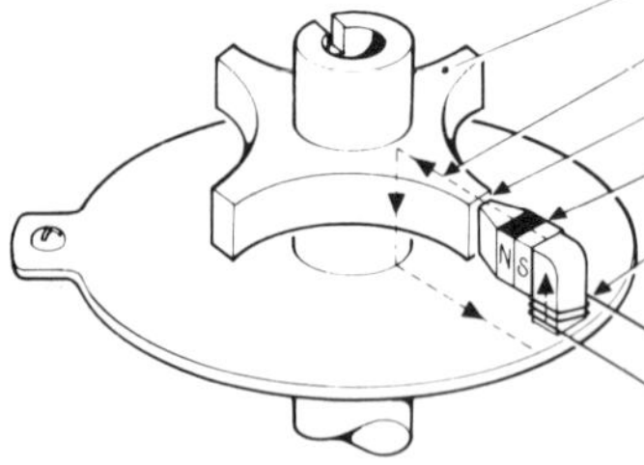

Fig. 74.16 Inductive pulse generator

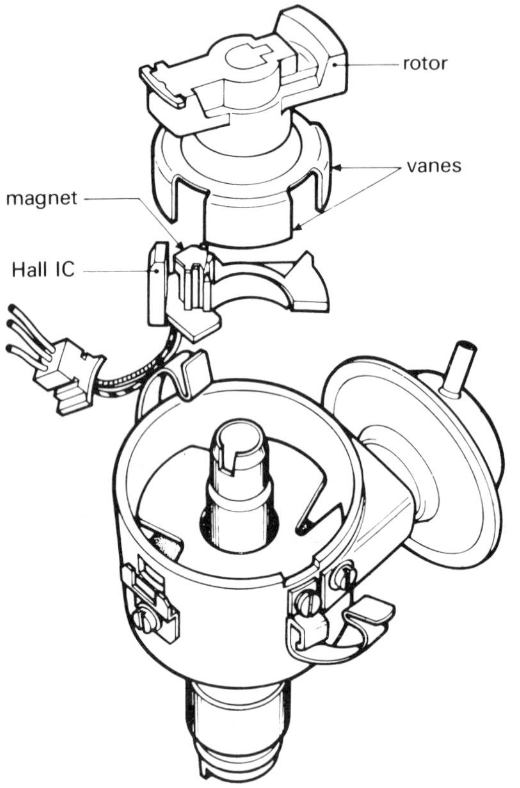

Fig. 74.17 Hall generator (Bosch)

With the ignition switched on, the Hall generator produces its full output to trigger a spark (even when the engine is being turned over by hand), so:

care should be exercised when handling this system owing to the risk of receiving an electric shock.

Optical pulse generator This system senses the spark point by using a shutter to interrupt an infra-red light beam projected by a light-emitting diode (LED) on to a photo-transistor (a semiconductor switch that is sensitive to light intensity). Fig. 74.18 shows a conventional distributor unit modified to take an optical trigger.

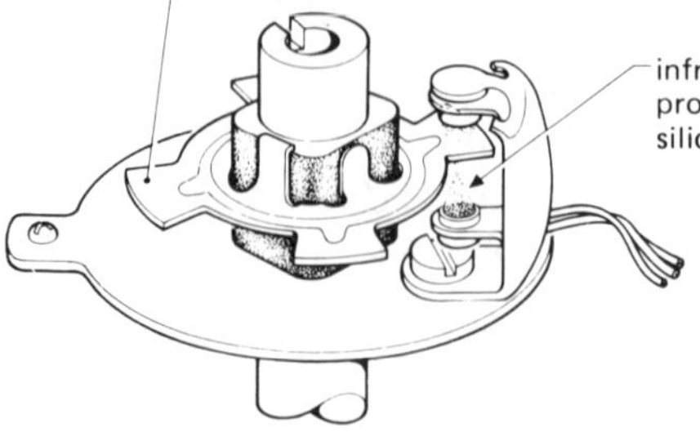

Fig. 74.18 Optical trigger system

Control modules

The duty of the control module, or trigger box, is to switch the primary circuit current on and off in accordance with the signal it receives from the pulse generator. Most cars fitted with electronic ignition layouts use an *inductive storage system*.

Miniaturization of solid-state devices over the past few years has allowed the control module to be accommodated within the distributor; this simplifies the wiring and improves reliability. In addition to its switching duty, the module also varies the dwell so as to control the maximum output voltage applied to the plugs; systems having this feature are often called *constant energy* systems.

The inductive storage system uses a primary circuit layout similar to a Kettering-type except that a robust power transistor (Darlington transistor) in the control module is used to make and break the primary circuit, i.e. to do the same job as a contact breaker.

Principle of a transistor The principle of a transistor is shown in Fig. 74.19. Using the water analogy, it will be seen that water flow in a large pipe 'ce' can be controlled by water pressure in a small pipe 'b'. At times when the pressure of water in the small pipe is insufficient to open the small valve, the large valve will remain closed and the main flow will be interrupted.

A transistor functions in a similar way; it will not allow current flow between 'e' and 'c' unless a small voltage is applied to 'b'. This switching feature of a transistor is of great value, because it enables a large current to be controlled by a very small current. In this application it is used as an alternative to a relay, but whereas a relay is comparatively large and uses moving parts, a transistor is very small and is a solid-state device.

Primary current switching A simplified layout of the circuit is shown in Fig. 74.20. The power transistor is

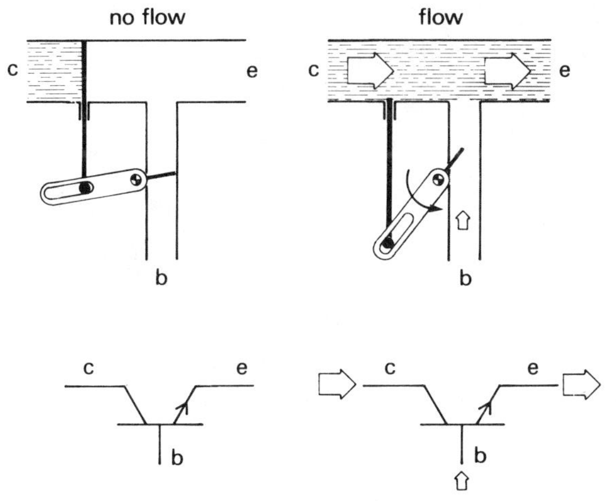

Fig. 74.19 Principle of transistor

connected, by terminals 'c' and 'e', in the primary circuit between the ignition coil and earth, and terminal 'b' is connected to the trigger section of the electronic circuit.

The transistor is switched 'on' when its base receives a small current which is signalled by the trigger. When this takes place, the primary circuit is closed and the ignition coil is able to build up its magnetic flux. At the time when the trigger pulse signals that a spark is required, the current to the transistor base is interrupted and the primary circuit is broken.

The action of the module can be demonstrated by connecting a test lamp between 'earth' and the 'module side of the ignition coil'. Assuming the module and trigger are functioning correctly, a slow rotation of the engine causes the light to go on and off.

Engine management systems

Tighter limits on emissions have forced many manufacturers to use more precise electronic ignition arrangements; many of these incorporate the ignition module with a digital computer. These ignition management systems eliminate the mechanical automatic advance arrangements; instead the timing of the spark, in relation to the trigger signal, is set in accordance with the data held in the pre-programmed memory unit of the computer.

A number of sensors are positioned around the engine to measure operating conditions such as temperature, load and speed. Signal data is fed from these sensors to the computer and after comparing these data with its memory, the computer is able to determine the correct ignition setting.

Some engines are fitted with a *knock sensor* to detect the onset of detonation (Fig. 74.21). When this sensor signals to the ECU that combustion knock is taking place within the cylinder, the computer slightly retards the ignition setting. By continually monitoring the engine, the use of a knock limiter system allows a greater ignition advance than normal to be used.

One great feature of these management systems is that the timing is not affected by component wear, so the timing setting should remain accurate for a very long period of time.

The distributor fitted to a vehicle having an engine management system only needs to fulfil its main duty, namely to distribute the 'spark' to the respective cylinders.

Distributorless system The distributor can be eliminated by connecting the pairs of sparking plugs

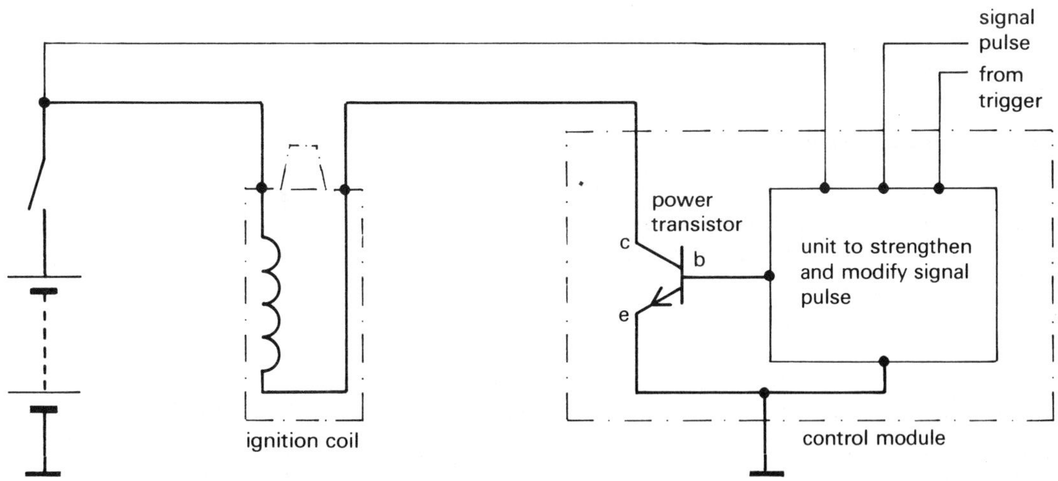

Fig. 74.20 Breakerless ignition circuit

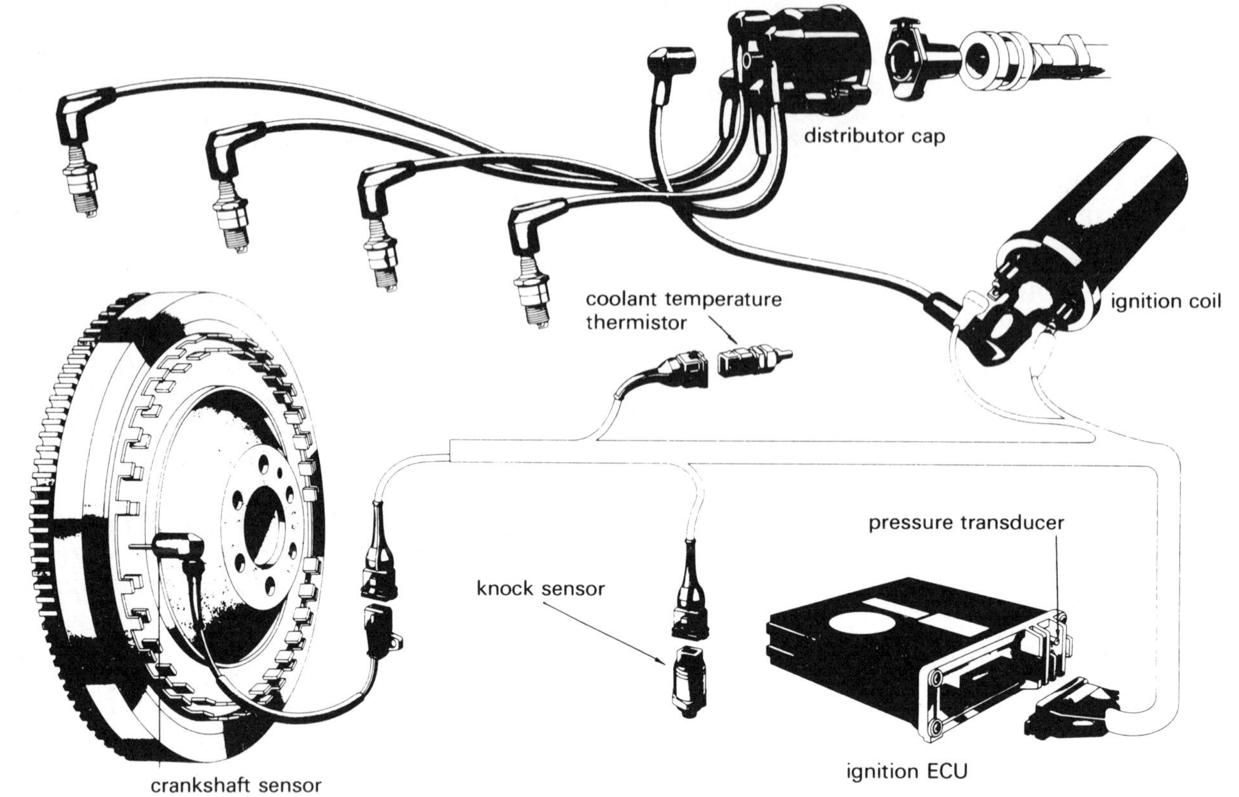

Fig. 74.21 *Programmed ignition system (Rover Group)*

of 'complementary' cylinders in the manner shown in Fig. 74.22. In a four-cylinder engine, the pairs are Nos. 1 and 4, and Nos. 2 and 3. A separate ignition coil is needed for each pair of sparking plugs.

When the primary is interrupted in the normal way, the h.t. voltage produces a spark at both sparking plugs. Since only one cylinder is primed for ignition, the spark in the other cylinder, which is ending its exhaust stroke, will not perform any useful function other than complete the secondary circuit.

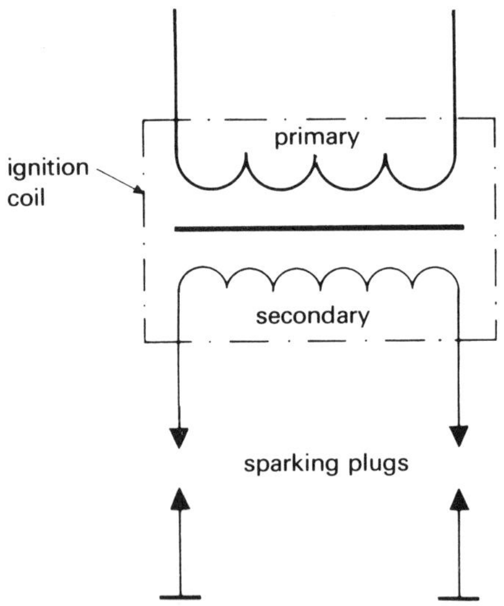

Fig. 74.22 *Distributorless ignition system*

Ignition system maintenance

Lack of maintenance of ignition systems, especially Kettering-type systems, causes more breakdowns than any other engine fault. The following points outline some of the important tasks that should be periodically carried out:

Sparking plugs These should be removed at the recommended times (e.g. every 10 000 km), cleaned, regapped, tested (if equipment is available) and refitted. Carbon can be cleaned off with a wire brush but is best removed by sand-blasting; if this method is used then all traces of sand must be removed from the plug, especially from the thread.

The gap is measured with a feeler gauge and adjustment is carried out by bending the earth electrode.

On refitting, the plug should be tightened to the correct torque. If a torque wrench is unavailable, then the following technique can be used:

Gasket seat Tighten 'finger-tight' and then rotate the plug one quarter-turn with a plug spanner

Taper seat Ensure that the seat is clean, tighten 'finger-tight' and then use a plug spanner to rotate the plug one sixteenth-turn.

Contact breaker About every 10 000 km a new contact set should be fitted. After fitting the new contacts, they must be adjusted to the recommended gap. This can be measured with a feeler gauge after setting the cam in a position where the contacts are fully open.

Use of a dwell-meter gives greater accuracy when setting the contact gap. An increase in the gap reduces the dwell and advances the spark; for example if the dwell angle is reduced by 5°, then the timing will be advanced by 5°.

Ignition timing Marks are normally provided on the crankshaft pulley to enable the ignition timing to be set. An outline procedure is as follows: first the engine is warmed up and then stopped to allow a *stroboscope* to be connected to No. 1 plug lead. After restarting the engine, the slow-running is set to the recommended speed and the flashing light is directed on to the moving pulley. The 'freezing' of the motion enables the position of the timing marks to be observed (N.B. it also gives the impression that other moving parts are stationary, so care must be taken to avoid injury).

Correct timing of the spark is important. If the spark is over-advanced, pinking and perhaps engine damage will result. Retarded timing causes poor power, increased fuel consumption and overheating.

Maintenance of electronic systems

In addition to maintenance of common items such as sparking plugs, the various electronic units are tested for serviceability by using special test sets. After making initial checks on external wiring and multi-pin connectors, individual units are then tested as recommended; if found defective they are either exchanged or replaced.

Extra care should be exercised when handling these extra-high voltage units: *in some cases a shock can be fatal.*

75 The starter system

An electrical *starter motor* is required to crank the engine over at a speed of about 100 rev/min to enable the fuel to be atomized and the mixture to be compressed sufficiently to start the engine. This cranking used to be achieved by the driver turning a starting handle, but today all cars use an electric starting system.

To obtain sufficient cranking torque from the robust electric motor, a large current is needed. On a cold morning this may be up to about 500 A, so thick, short supply cables and a good battery capable of delivering this high current are needed.

The complete system may be considered in three sections:

- Motor
- Starting circuit
- Mechanical engagement systems.

Motor

The motor converts electrical energy supplied from the battery into mechanical power; this is achieved by using an electromagnetic principle similar to that employed for the generation of electrical energy.

Principle of a motor

When two 'like' poles of two magnets are brought together, the magnets push each other apart. From this experiment, the expressions '*like poles repel*' and '*unlike poles attract*' are obtained.

A motor uses the 'repel action' to produce a force as shown in Fig. 75.1. This force produces motion when a current-carrying conductor is placed in a strong magnetic field; the force is caused by the bending of the flux lines. Flux lines may be regarded as elastic bands: when the bands between the poles are stretched, a force is created.

By bending the conductor in the form of a loop, and using carbon–copper brushes to conduct the current to copper commutator segments joined to a conductor situated in a magnetic field, a partial rotation of the loop is achieved. Full rotation can be obtained by using more loops (windings) placed in different angular positions. These loops are wound around a soft iron armature (to concentrate the field) and are joined to commutator segments so as to allow a current to flow to each individual loop as it passes through the densest part of the field.

Formation of the magnetic field Most starter motors have a wound electromagnetic field which is normally connected in series in the main circuit. This arrangement produces a strong field because all of the current that is supplied to the motor must pass through the field windings.

To obtain a high torque from a motor that is required to operate at a low cranking speed demands a large supply of electrical power from the battery. Since the power (watts) is the product of the current (ampere) and voltage, the need for using low-resistance conductors in a motor becomes apparent. Consequently, field windings and armature coils are made from strips of copper or aluminium and high-conductivity copper–carbon brushes are used. In addition to the low-resistance conductors in the motor, the multi-strand copper supply cables, and their connectors, must be of adequate size to carry the high current required by the motor without over-heating or causing the voltage to drop excessively.

Recent developments have seen the introduction of small starter motors which use a permanent magnet field. These lightweight machines achieve their high torque output by using an epicyclic gear between the motor and the driving gear.

Four-brush motors

A reduced resistance and greater current flow can be achieved by using four brushes instead of two. The

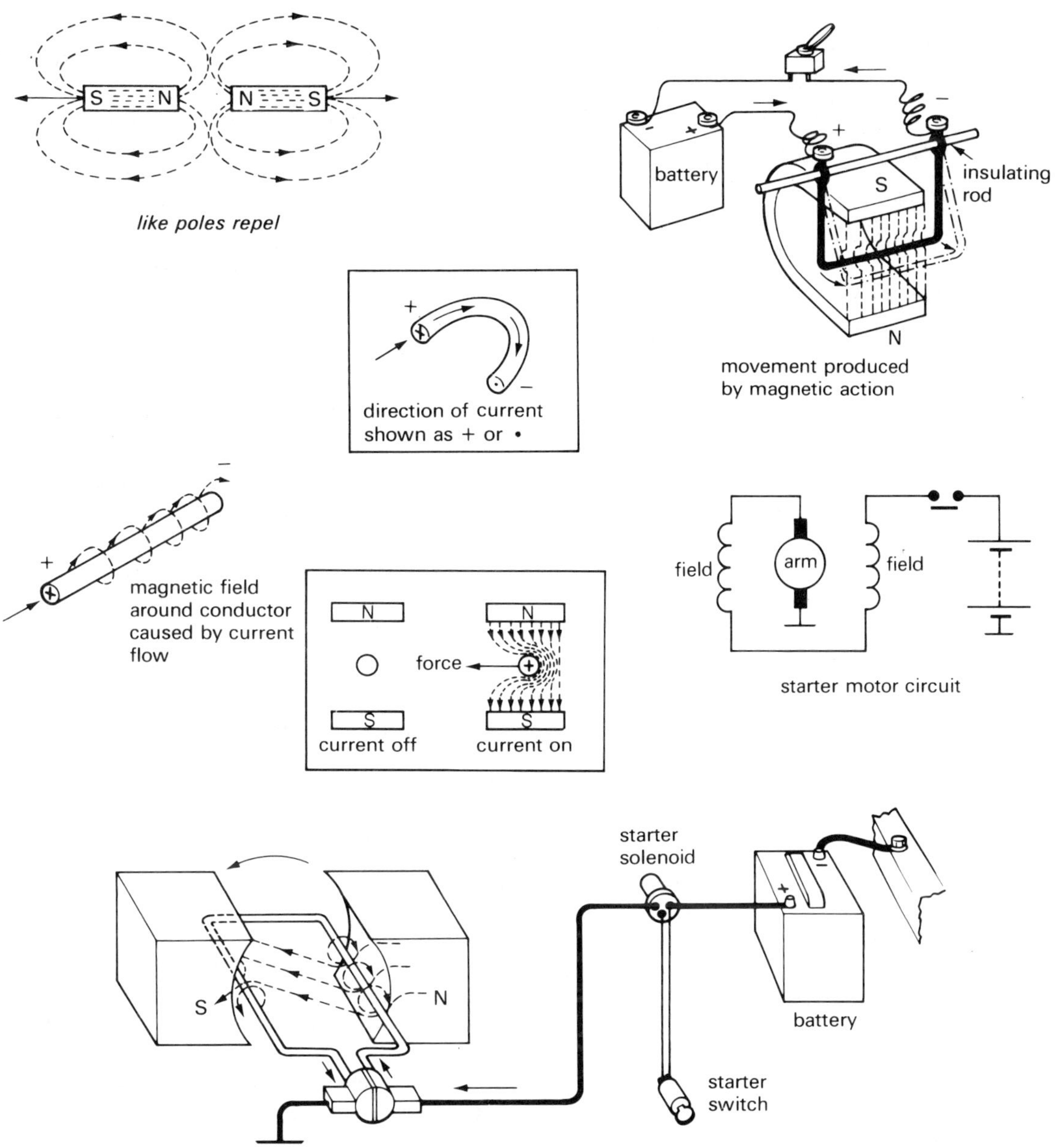

Fig. 75.1 Principle of starter motor

four-brush machines are made in either a *series-wound* or *series-parallel* field arrangement (Fig. 75.2).

Fig. 75.3 shows an exploded view of a four-brush, four-pole motor complete with drive. This type of motor has a field winding bent in a wave form to give four poles in the order NSNS and a face-type commutator; this uses brushes situated on the side of the armature instead of the more common radial position associated with a cylindrically shaped commutator.

Starter motor circuit

Fig. 75.4a shows a pictorial view of a typical cable layout for a starter motor.

The diagram shows two earthing straps; these are used to complete the circuit. Partial failure of either one of the straps will seriously reduce current flow and in consequence will affect motor operation. In addition, the alternative 'earth paths' from the battery to the motor through such parts as engine controls will cause local heating.

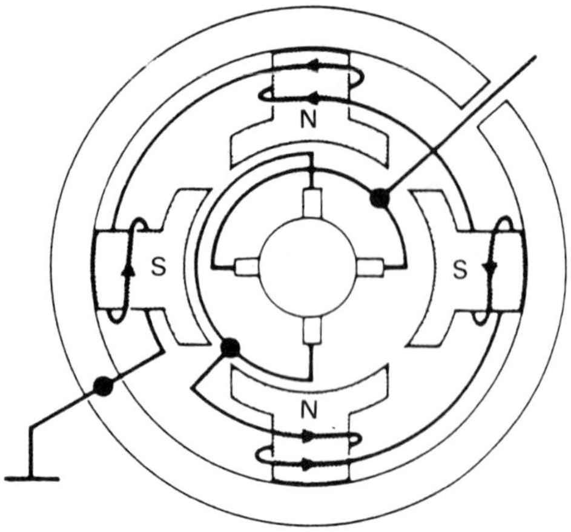

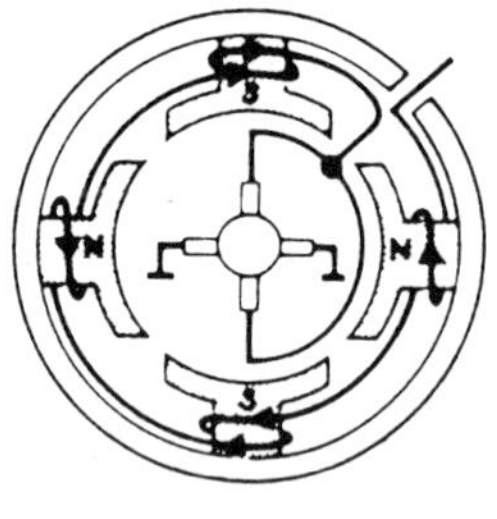

alternative construction

series-wound motor

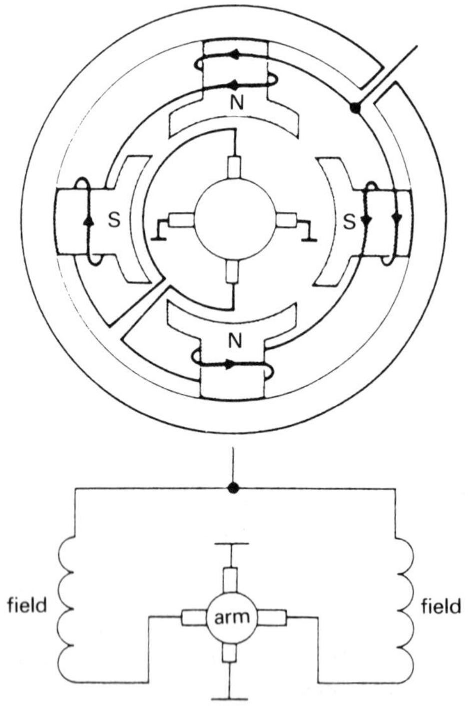

series-parallel motor

Fig. 75.2 Field winding arrangements

Solenoid operation

The inclusion in the circuit of a solenoid allows the length of heavy-duty cable to be kept to a minimum. The voltage drop increases with the length, so when a long cable is used the p.d. applied to the motor is considerably less than that obtained across the battery.

Switching is achieved by a pair of heavy-duty copper contacts in the solenoid. These contacts are closed by a low-current winding in the solenoid. It is energized when a remote switch, normally actuated by the ignition key, is operated (Fig. 75.4b).

The solenoid has two windings: a closing coil and a hold-on coil (Fig. 75.4c). These are connected in parallel with each other so they both operate when the driver's switch is closed. When the motor is in operation only the hold-on coil is in use; this feature reduces the solenoid operating current.

The audible click given from the solenoid when it operates is a useful aid for diagnosis; for example, a repeated clicking sound from the solenoid suggests that the hold-on coil is defective.

Mechanical engagement system

A low gear ratio between the starter motor and the engine enables sufficient torque to be developed to turn the crankshaft. This ratio is obtained by using a pinion on the motor to drive a large gear formed on the flywheel.

Permanent engagement of the pinion causes the armature to overspeed when the engine is running, so a mechanism is used to ensure that the pinion only connects with the flywheel when the engine is being started.

The two engagement systems shown in Fig. 75.5 are used on light vehicles; they are:

- *Inertia engagement*
- *Pre-engagement.*

Inertia engagement

This type relies on the natural reluctance (inertia) of a pinion to change its state of motion. This feature is utilized to move the pinion so that it engages with the teeth of the flywheel when the armature starts to rotate.

The pinion is mounted on a helix which is splined to the starter motor armature. Sudden rotational movement of the armature rotates the helix, but owing to its inertia, the pinion remains stationary; this causes the pinion to slide along the helix and engage

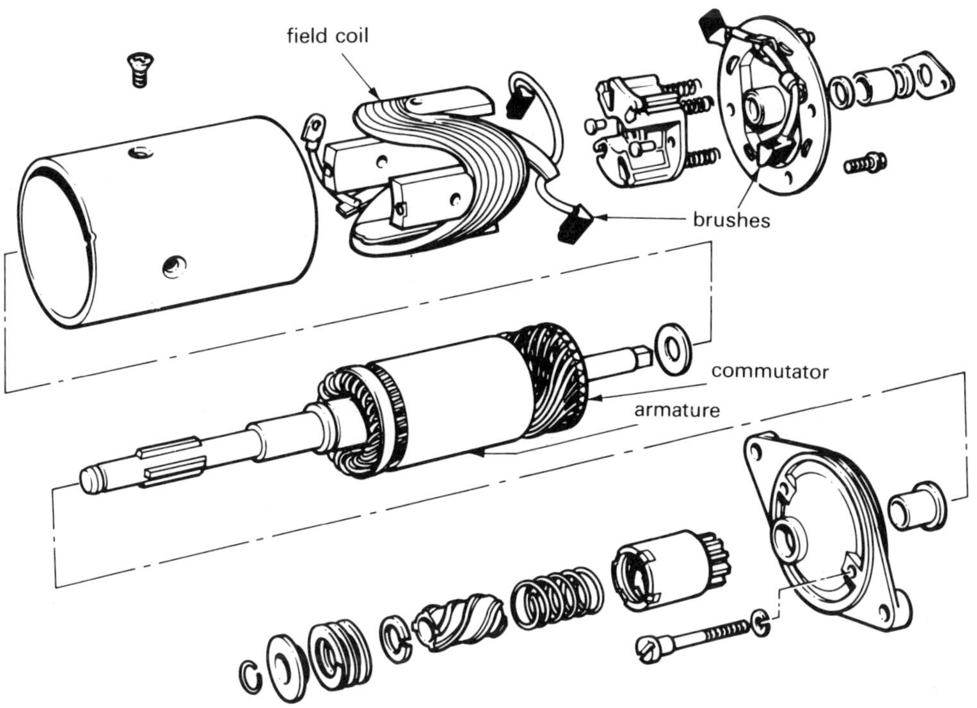

Fig. 75.3 Starter motor and drive (exploded view)

with the flywheel teeth. A 'lead-in' (chamfer) is present on both the pinion and flywheel teeth to aid this initial engagement.

When the engine fires, the increase in engine speed spins the pinion along the helix and causes it to disengage from the flywheel teeth.

The shock at the instant of engagement and disengagement is severe, so a large spring is fitted

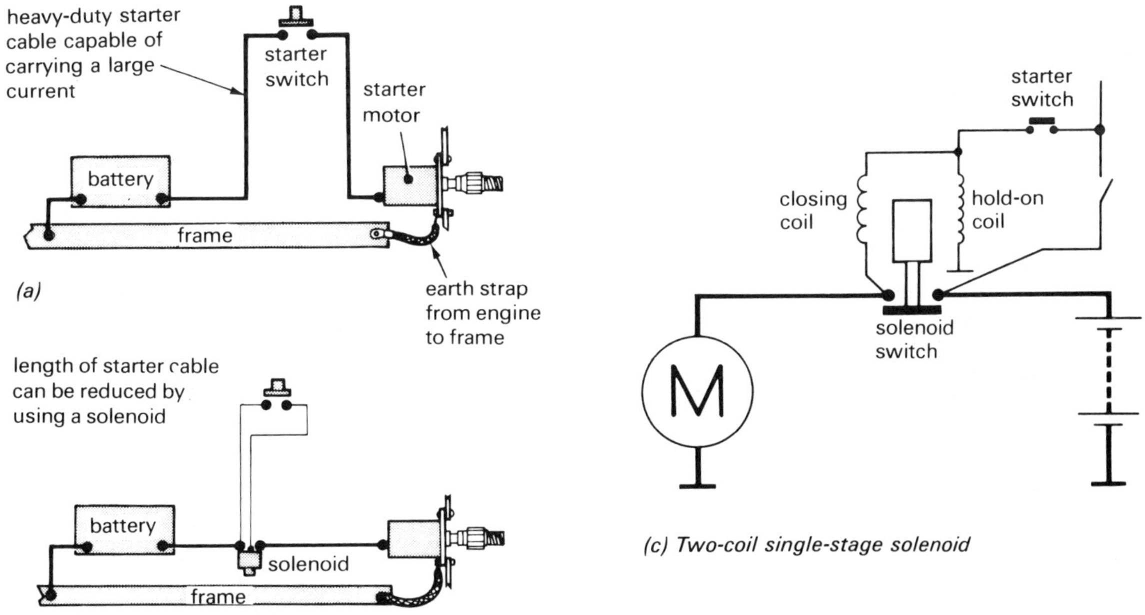

(b) Solenoid – a remote-controlled switch

(c) Two-coil single-stage solenoid

Fig. 75.4 Starter motor circuits

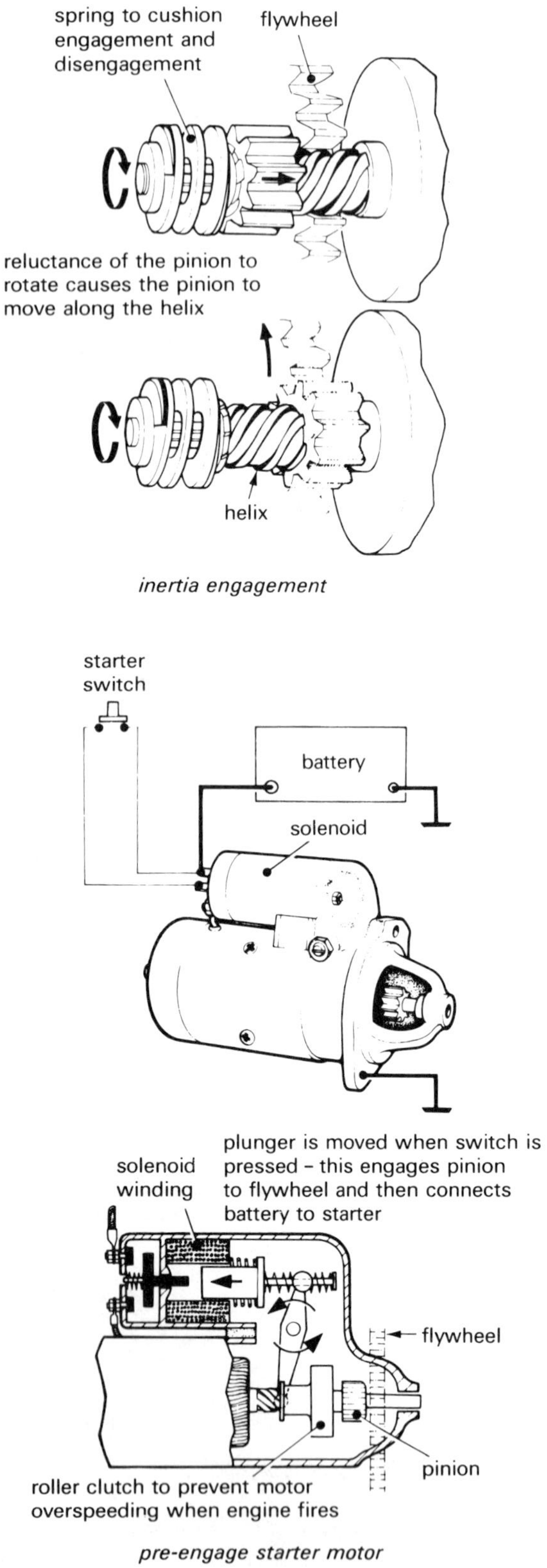

Fig. 75.5 Engagement systems

adjacent to the helix to act as a cushion.

On some systems the vibration of the engine causes the pinion to move along the helix and touch the flywheel. On some models the 'tinkling noise' given from this contact is prevented by using a light spring to hold the pinion away from the flywheel.

Possible faults Two main faults can arise with this system; they are:

- Pinion fails to engage
- Jammed pinion (seized engine).

Pinion fails to engage Assuming the speed of rotation is sufficient, the most frequent cause of a 'sticking pinion' is dirt on the helix. Often this is due to the presence of oil on the splines which collects dust from the clutch.

One recommended lubricant is Molykiron (SAE 5), but if this type of 'dry' lubricant is unavailable then the helix should be left in a dry state after cleaning.

Jammed pinion This occurs after the driver has attempted to start the engine. It is caused by local wear on the flywheel ring gear. On a four-cylinder engine this wear takes place in two regions (six-cylinder, three regions) because the engine normally comes to rest in one of these positions.

When the flywheel tooth wear becomes excessive, the force of pinion impact, combined with shaft flexibility, jams the teeth together and locks the engine.

An emergency method that can be used on a car having a manual gearbox, is to engage top gear and 'rock the car' by pushing it backwards and forwards. The normal method is to remove the motor and examine the flywheel teeth for wear and the starter spindle for alignment. The appropriate repair is then carried out.

A ring gear is normally a 'shrink fit' on the flywheel, so after removing the worn gear, the new gear is heated to the limit recommended to expand the ring; then it is pushed into place, making sure that the chamfer is on the correct side.

Pre-engagement

This type was originally confined to diesel engines, but nowadays it is often used on petrol engines. The change from the inertia type to a pre-engaged drive on petrol engines was due to the difficulty of maintaining pinion engagement with an inertia drive when an engine 'part-fires' during the cranking period.

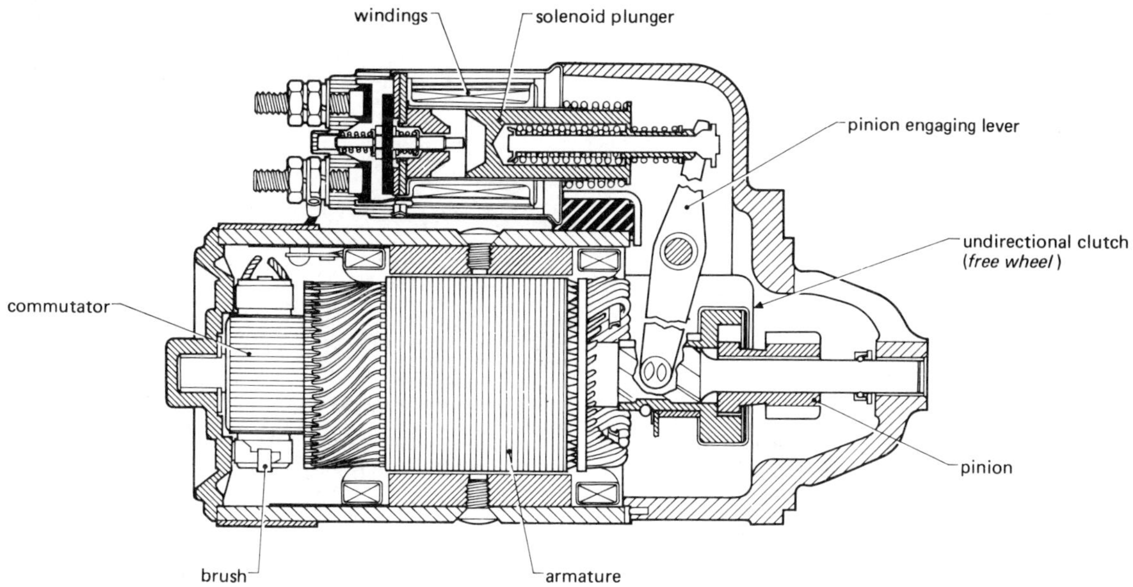

Fig. 75.6 Pre-engaged starter motor

Pre-engaged drives have been used on diesel engines for many years. The use of this type of starter motor drive takes into account the special starting requirements of a CI engine; these include:

1 *Component inertia* Considerable torque is needed to rotate the crankshaft initially and accelerate it to the cranking speed.

2 *Cranking speed* The cranking speed of a diesel engine should be higher than that of a petrol engine to initiate combustion.

3 *High compression pressure* During the cranking period the CI engine draws in a full charge of air and compresses it to a much higher pressure than a petrol engine. As a result the cranking torque is high and the speed is too irregular to maintain engagement of the starter pinion.

Fig. 75.6 shows the constructional details of a pre-engaged type of starter motor. A solenoid plunger is connected to an operating lever; this is pivoted to the casing at its centre and forked at its lower end to engage with a guide ring. This ring acts against a unidirectional roller clutch and pinion. Helical splines, formed on the armature shaft, engage with the driving part of the unidirectional clutch. These splines cause the pinion to rotate slightly when the clutch and pinion are moved axially. A strong return spring in the solenoid holds the lever and pinion in the disengaged position.

When the starter switch is operated, the two-coil solenoid windings become energized and the plunger is drawn into the core. This initial action causes the lower end of the operating lever to move the guide ring and pinion assembly towards the flywheel teeth. This movement, aided by the slight rotation of the pinion, normally gives full engagement of the gears. After this initial action, extra travel of the solenoid plunger causes the main contacts to close: this connects the battery to the motor.

Drive from the armature is transmitted to the unidirectional clutch and pinion by helical splines.

Sometimes the initial movement causes the pinion teeth to butt against the flywheel teeth; as a result full engagement is prevented. When this occurs, a spring in the linkage flexes and allows the solenoid plunger sufficient movement to close the main contacts. As soon as the armature and pinion start to rotate, the teeth slip into mesh and a spring pushes the pinion to its driving position.

After the engine has started, the pinion speed will exceed the armature speed. If the motor is still in use, the rollers in the unidirectional clutch will be unlocked and the clutch will slip to protect the motor.

Release of the starter switch de-energizes the solenoid and allows the return spring to open the switch contacts. This occurs well before the pinion disengages and so avoids overspeeding of the motor. Further movement of the plunger causes the operating lever to withdraw the pinion fully from the flywheel.

76 Auxiliary equipment

The modern car contains many items of electrical equipment in addition to the components already described in this book. These additional systems and components are classed as auxiliary equipment.

As years go by, more and more electrical equipment is added to the vehicle; some electrical items may be regarded as desirable but non-essential, whereas other items have to be fitted to meet statutory requirements. History shows that a non-essential of yesteryear becomes a statutory requirement of today, e.g. horns, windscreen wipers and washers, directional indicators, brake lamp. Nowadays all new vehicles must be fitted with these systems, and statutory regulations, nowadays framed by the EEC, govern the construction and method of operation.

Vehicle equipment

Auxiliary equipment fitted on a modern vehicle includes:

- *Horn* The law requires an audible warning device which emits a continuous note that is neither too loud nor harsh in sound.
- *Windscreen wipers* The law requires a wiper on the driver's side that operates effectively.
- *Windscreen washer* The law requires a screen washer to be fitted to clean the driver's side of the windscreen.
- *Directional indicators* The law requires that three coloured signal lamps are fitted in approved positions on each side of the vehicle, together with a tell-tale indicator. The rating of each lamp is specified together with the flashing rate (60–120 flashes per minute).
- *Brake lamp* The law requires the fitting of a lamp which must automatically light up when the brake is applied.
- *Reversing lamp* The law states that no more than two white lamps may be used, the wattage of each lamp not being greater than 24 W; also regulations cover the switching requirements.
- *Instrumentation system* This includes: fuel level, coolant temperature, and vehicle condition monitoring arrangements for detecting system faults.
- *Clock*
- *Interior and courtesy lights*
- *Fan and heater motor*
- *Rear window demister*
- *Window winding*
- *Central door locking*
- *In-car entertainment system*
- *Door mirrors*
- *Sun roof*
- *Etc, etc.*

Lighting regulations, which are revised periodically, vary from one country to another, so when a bulb, or lamp part, is renewed, it is necessary to fit the type recommended by the manufacturer.

Most auxiliary systems are fused, either in groups or individually. In the past, often only two auxiliary fuses were fitted, but the modern car uses a large number to give maximum protection and minimum loss of components when a single fuse 'blows'. Normally the fuses, or other forms of circuit breaker, are mounted together on one or more fuse panels situated in various places on the vehicle.

A circuit layout of the main items of the auxiliary equipment is shown, in simplified form, in Fig. 76.1 together with typical fuse ratings. It will be seen that each item of auxiliary equipment has its own individual circuit. In view of the large amount of equipment fitted on modern vehicles, most manufacturers now publish a number of separate circuit diagrams instead of showing the complete electrical system on one diagram; for example, one diagram is used to show the wiring layout of the ignition and

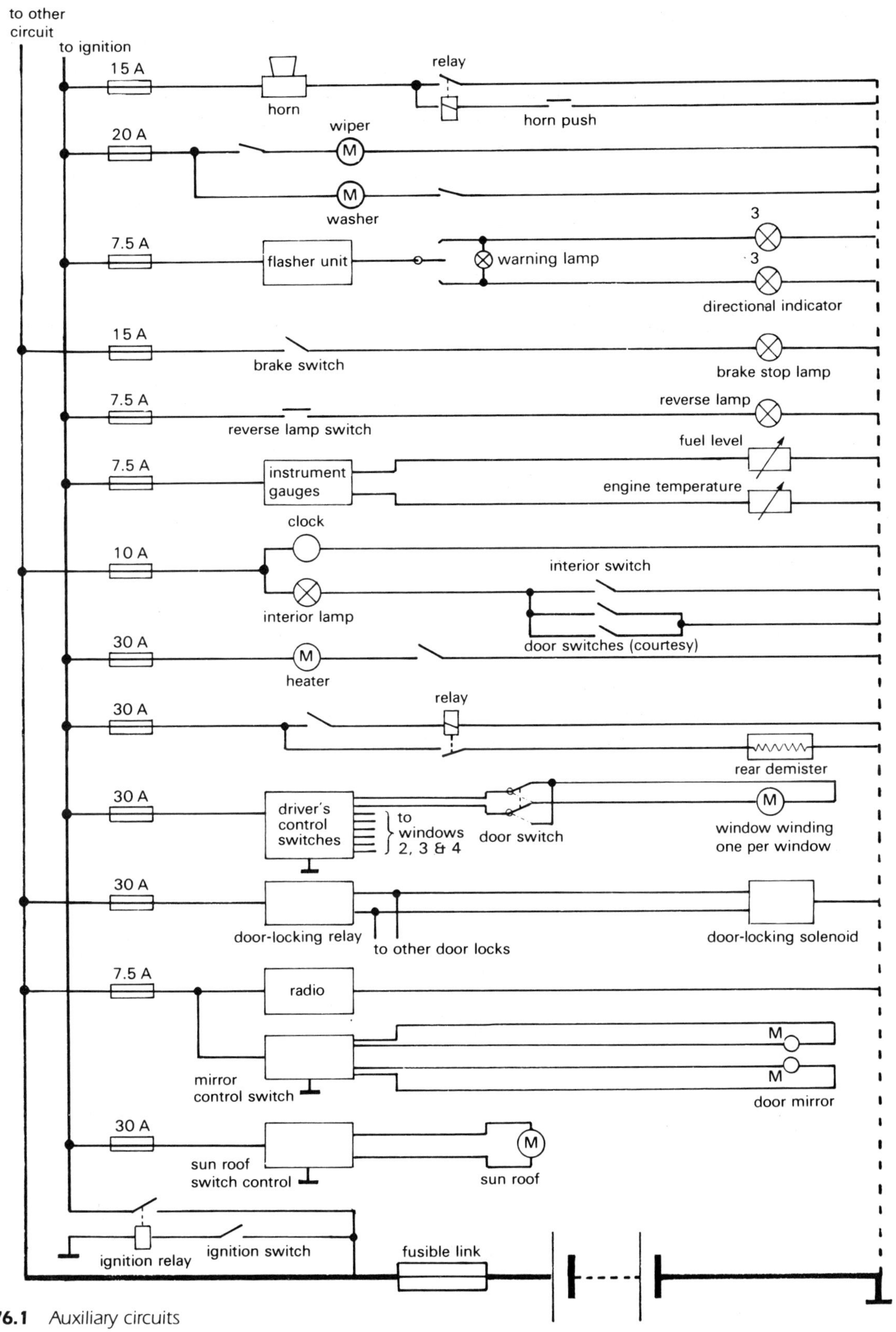

Fig. 76.1 Auxiliary circuits

fuelling systems and another the charging and starting systems.

Whenever a fault has to be pin-pointed, the wiring diagram shows the interconnection, and location, of the various circuit elements. This information is necessary if the cause of a particular problem has to be diagnosed.

Space in this book limits the study of auxiliary systems, so only the directional indicating system and simple instrumentation are covered. If more information on other systems is required, the reader is referred to the companion book published by the author entitled *Fundamentals of Automotive Electronics*.

Directional indicators

Fig. 76.2 shows the layout of a typical directional indicator circuit having three lamps on each side of the vehicle. When the switch is moved to the left or right, current is supplied to the appropriate lamps. Regular interruption of the current to give a flashing light is performed by the flasher unit; this is situated on the battery side of the main switch.

If a vehicle breaks-down on the highway the driver should be able to warn other drivers by arranging all the directional indicator lamps to flash simultaneously. This hazard-warning feature is activated by an additional switch controlled by the driver.

Today, various types of flasher unit are in use: thermal, capacitor and electronic. These units all have three electrical connections: input, output and earth. A flasher unit has to meet international standards which include the provision of an audible and visual system to signal the failure of a lamp. Behaviour of the 'tell-tale' lamp varies with the type of system; a thermal vane-type flasher causes the lamp to stay on if one bulb fails, whereas an electronic type causes the lamp to flash at twice the normal rate.

Thermal-type flasher

The *vane-type* flasher unit, shown in Fig. 76.3 is a simple, low-cost unit which consists of a rectangular, snap-action, spring-steel vane that is supported at a point mid-way along the longer side. A thin metal ribbon, diagonally connected to the corners of the vane, pulls the vane towards the base. A pair of contacts, one at the centre of the ribbon and the other fixed to the base, make the electrical circuit. This flasher unit is fitted in series with the lamps, and current to supply these lamps is taken through the metal ribbon, vane and contacts.

Operation of the indicator switch instantly activates the signal lamps. The heating effect of this current on the metal ribbon extends its length and allows the vane to click upwards to its natural position. This action opens the contacts, breaks the circuit and extinguishes the signal lights. A short time after this action the lack of heating current causes the ribbon to cool and contract; this clicks the vane downwards and once again closes the contacts to repeat the cycle of events.

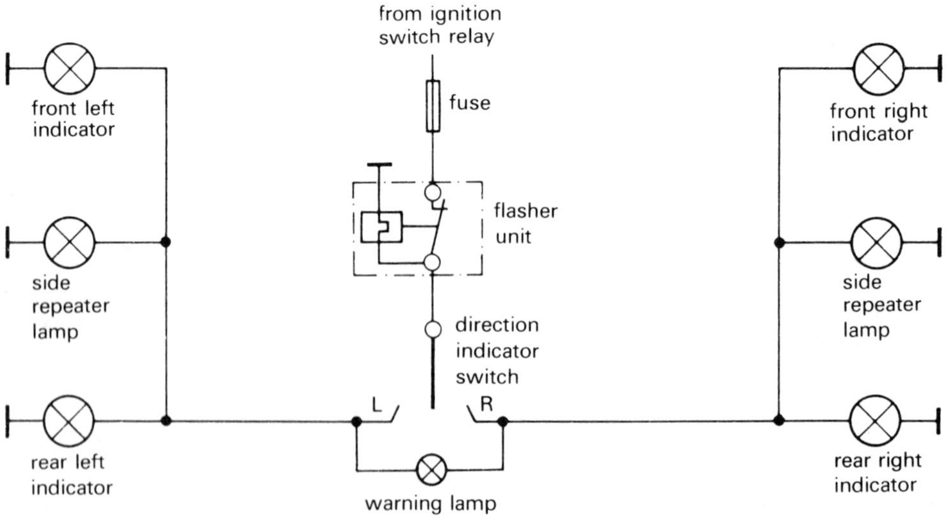

Fig. 76.2 Directional indicator circuit

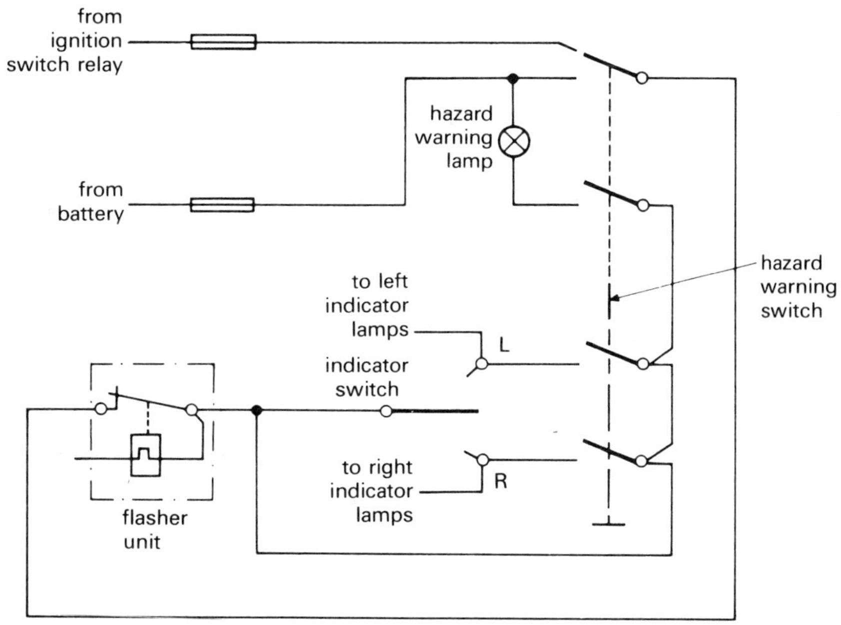

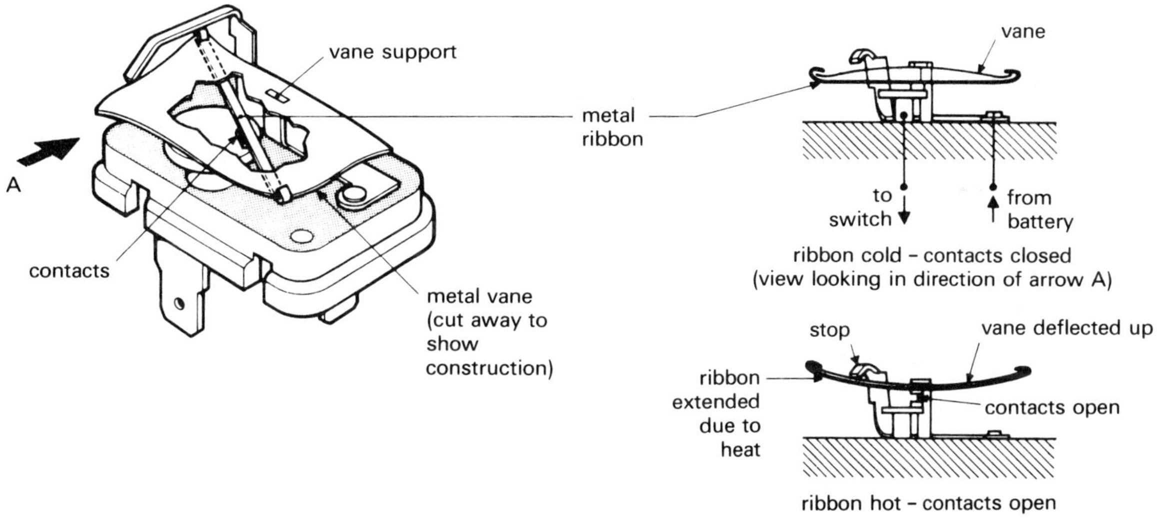

Fig. 76.3 Vane-type flasher

Electronic type

This type normally controls the lamp current with an electromagnetic relay; this is energized by an electronic circuit which uses a capacitor as a timer to control the rate of flashing.

Thermal-type fuel gauge

This type of fuel contents indicator uses the bending action which occurs when a brass-steel bimetallic strip is heated. In the case of the fuel gauge, the strip is wound with a fine wire and heated by an electric current. The amount that the strip bends depends on the current that is passing through the heating coil; the greater the current, the greater the bending action. By interlinking the strip with a fuel gauge needle, as shown in Fig. 76.4, an indication of the current that is flowing in the circuit is given.

The fuel tank unit (sender unit) is a float-operated variable resistance. As fuel is poured into the tank, movement of the float reduces the resistance of the sensor. This causes a greater current to flow in the

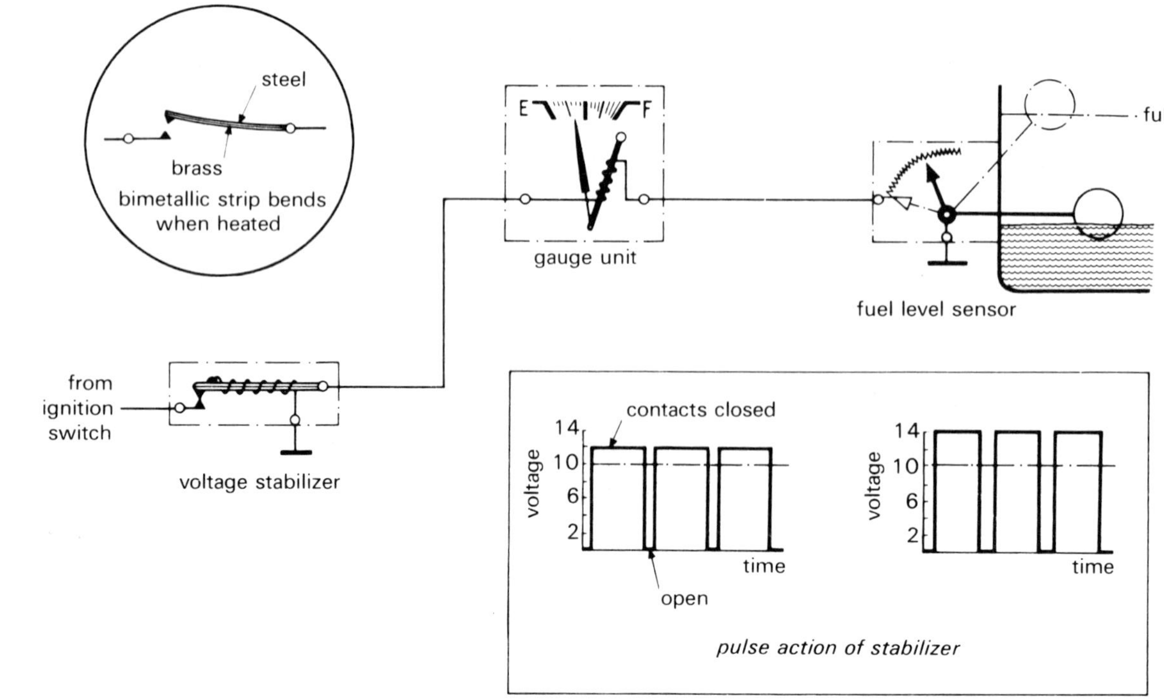

Fig. 76.4 Thermal-type fuel gauge

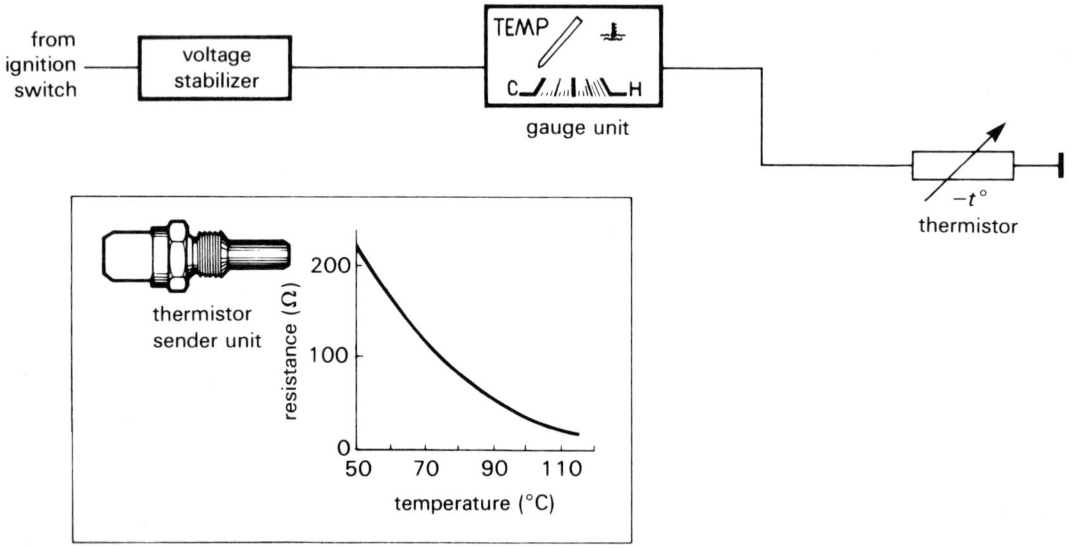

Fig. 76.5 Thermal-type engine temperature gauge

circuit which results in a larger deflection of the gauge needle.

A false reading will occur unless the voltage applied to the circuit is kept constant; this is the duty of the *voltage stabilizer*. By continually opening and closing a pair of contacts to give a series of pulses, the stabilizer maintains a mean voltage of 10 V (or 7 V depending on the system) across the circuit. Many modern vehicles use an electronic voltage stabilizer instead of the bimetallic type shown in Fig. 76.4.

Care must be taken to avoid an explosion when testing the operation of the system with the sensor unit removed and positioned close to the tank. Generating sparks within a petrol-saturated tank is safe, but it is highly dangerous to generate them in an atmosphere where both oxygen and petrol vapour are present.

Thermal-type engine temperature gauge

The conventional thermal-type gauge used for this instrument is similar to that used to indicate the fuel contents. The gauge reading is affected by supply voltage, so the system is fed via a voltage stabilizer, the same unit as that used to supply other thermally operated gauges.

The engine temperature sensor is generally situated in the coolant system on the engine side of the thermostat. Normally the brass bulb that is in contact with the coolant contains a sensing capsule called a *thermistor*. This semiconductor resistor pellet is thermally sensitive, but varies its resistance in the opposite way to most metals: when the temperature is increased, the resistance is decreased. The thermistor's high *negative temperature coefficient* (n.t.c.) of resistance causes the resistance to vary from about 220 Ω at 50 °C to 20 Ω at 115 °C.

The circuit for a bimetallic indicator gauge and thermistor is shown in Fig. 76.5. In operation, a rise in engine temperature lowers the thermistor resistance and increases the current passing through the gauge unit; this causes the bimetallic strip to bend more than before.

Damage will occur if the thermistor is connected directly to a battery.

77 Fault diagnosis

Although obvious electrical faults can be easily spotted, knowledge of electrical principles is essential if the cause of a more involved fault is to be pinpointed.

In view of the many items of electrical and electronic equipment fitted on a modern vehicle, it is recommended that electrical tests are restricted to systems in which the tester is competent. Whereas in the past haphazard use of simple test equipment perhaps resulted in damage to a meter, or a flash of an electric spark, modern electronic control units can easily be damaged beyond repair - a costly way to obtain experience.

Types of electrical fault

Failure of a particular system, or part of a system, is the cry of the driver when some item of electrical equipment does not function in the accepted manner. Often this claim does not name a particular item, so it is the duty of the diagnostician, by driver interrogation and practical tests, to identify the circuit, and sub-circuit, in which the defective part is situated in order to locate the defect.

Statements made by the driver such as 'The engine will not start' and 'The windows will not close' are vague; as a result the repair is likely to take much longer than necessary. Additional information obtained from the driver not only saves time: it also prevents unnecessary replacement of one or more costly components.

The terms *symptom*, *fault* and *cause* are sometimes misunderstood. In this chapter these terms are used to mean the following:

Symptom - change in behaviour of a given part or system that is normally witnessed by the driver.
Fault - general cause of failure which gives the symptom.
Cause - that which produces the effect or fault.

For example,

Symptom:	Engine will not start when cold.
Fault:	Ignition system defective.
Cause:	High resistance across contact breaker (dirty contacts).

(The *remedy* is to fit a new contact set.)

Intermittent faults

One great problem with electrical equipment is that certain faults, often caused by a high resistance at a connection point, only show up on odd occasions. These intermittent faults are difficult to trace because the high resistance is not always present. Sometimes they can be linked to a particular operating or ambient condition, at other times the system operates perfectly under all conditions. When the system is functioning normally, even the use of sophisticated electrical test equipment fails to show that the system is not in perfect order. This is little consolation to a driver when the fault recurs, especially if the fault is associated with one of the control systems of the vehicle.

In an attempt to satisfy the vehicle owner some repair establishments attempt to solve the problem by using 'swamp tactics'; they keep fitting different parts until the fault disappears. This is a costly exercise which is not always appreciated by the person who has to pay the bills.

Many potential faults of an intermittent nature could be avoided if the manufacturer was prepared to use higher-quality connectors, but cost factors often preclude this. In consequence, it is anticipated that this type of fault will increase over the next few years. This means that either a diagnostician must be trained to be competent to undertake this type of work or that repair personnel are made aware of their limitations.

Main circuit faults

In addition to obvious defects, the three main circuit faults are:

1 *Open circuit* - a physical break in the circuit that prevents current flow
2 *Resistance in circuit* - a defect that reduces current flow
3 *Short circuit* - breakdown of insulation that causes the current to by-pass a section of the circuit.

These faults can be detected by various checks and tests, the depth of diagnosis being governed by the skill of the tester and the electrical test equipment available.

Circuit testing

Visual check

Having identified the precise symptom, especially the conditions in which it occurs, a visual check of the suspected system is the first action. In the majority of electrical failures, the cause of a particular fault is a 'blown' bulb, defective fuse, loose connection or broken wire in the circuit.

In any battery-powered system, the initial checks must cover the condition of the battery and its connectors. Switching on the headlights and noting their brilliance when the defective system is operated is a useful guide to the state of charge of a battery, especially with symptoms associated with engine starting. A system that is dependent on the battery will not respond correctly if the driver leaves the lights on, or if the alternator drive belt is slack.

The diagnostician should not be content with just clearing the fault that caused the stated symptom; the reason why the fault developed should be determined. For example, poor starting motor operation may be caused by a discharged battery: recharging is part of the remedy; the other part is finding the reason why the battery became discharged, i.e. establishing whether the initial fault was caused by a defective battery, poor charging rate or driver negligence.

Test lamp

A simple test lamp can easily be made by soldering two cables to a 12 V, 6 W bulb as shown in Fig. 77.1. Connecting a crocodile clip and pointed probe to the cables enables the lamp to be used as a *continuity tester to locate the position of an open circuit*.

Broken cables, points of very high resistance and

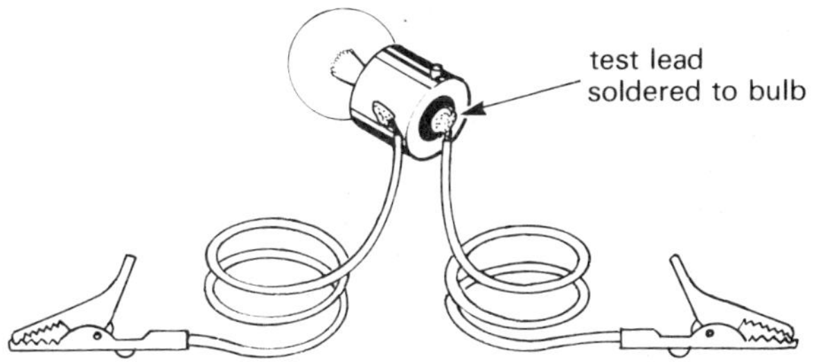

Fig. 77.1 Test lamp

poor earths in basic systems can be quickly diagnosed by a test lamp, provided the circuit path is known. However, the test lamp should not be used on electronically controlled circuits unless actual test points are specified by the manufacturer.

Where possible the first test point should be made at the power source; after that further tests are made at connection points along the current flow path.

Some manufactured test lamp sets are made to suit a particular vehicle system; the 'lamp' used in these units is often a voltage-sensitive *light emitting diode* (l.e.d.). This semiconductor device will not light up until a given voltage is applied, so by selecting a single diode, or range of diodes, to suit the circuit voltage, a simple test kit is obtained. This test set not only tests continuity: it also verifies that a pre-determined minimum voltage is available at some specific point of a given system.

Use of test lamp Knowledge of the circuit and its 'behaviour' is needed before a test is carried out.

The amount of information obtained from the test depends on the type of circuit. In the test of the primary circuit of an ignition system shown in Fig. 77.2 it is possible to check the continuity of the circuit and the general condition of the contact breaker. Since the majority of ignition faults are caused by defects in the primary circuit, the test lamp provides a quick and simple means of diagnosing the cause of many of these faults.

With the contact breaker (c.b.) set open and the ignition switched on, no current will flow in the circuit. The lamp will light at points 1-4, which shows that there is good continuity between 1 and 4.

Connecting the lamp to either of points 3 or 4 and closing the contact breaker should extinguish the light; this is because the potential difference (p.d.) between the lamp connections is zero. If the lamp remains lit when the c.b. is closed, then the test suggests that the c.b. is dirty.

Use of voltmeter This meter has a very high resistance and is used to measure the difference in electrical potential between two points of a circuit.

An important fact to remember when making a voltmeter test is that the voltage in a circuit only falls after a current has passed through a resistor. Unless current is flowing in a circuit, the voltage obtained at all points on the 'live' side will be equal to the electro motive force (e.m.f.) of the power source. Since the main purpose of a voltmeter test is to locate any 'unwanted' resistance, it is essential that the appropriate switches are set in the 'closed' position.

Consider a suitably scaled voltmeter substituted for the test lamp shown in Fig. 77.2. With the c.b. open, the readings at points 1–4 will be about 13 V; this is the e.m.f. of a 12 V battery.

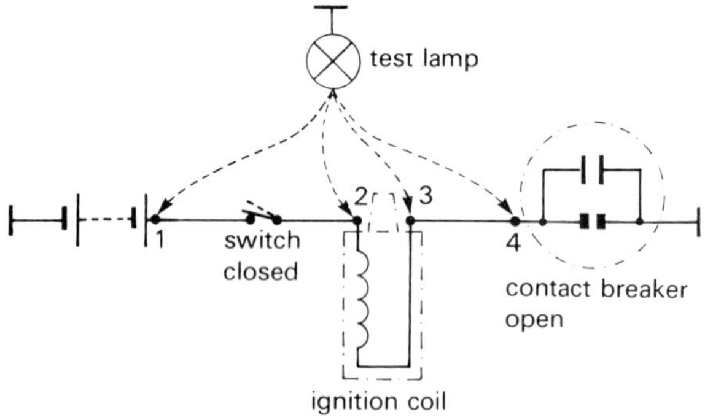

Fig. 77.2 Circuit check with test lamp

Voltmeter readings with the c.b. closed indicate the p.d. across the section of the circuit spanned by the voltmeter. A circuit in good condition should give readings of about 12 V at points 1 and 2 and 0 V at points 3 and 4.

Any unwanted resistance in the battery coil section will give a drop in p.d. at the point where the resistance is located. Similarly, if the c.b. is dirty, the p.d. at points 3 and 4 will be higher than 0 V. The extent of this resistance can also be measured by placing the voltmeter 'across' the contacts, i.e. connecting one lead to the fixed contact and the other to the moving contact. Normally the *voltage drop* across the c.b. should not be greater than 0.2 V, so to measure this small value a voltmeter having a 0–2 V scale is generally used.

Use of ammeter This meter has a very low resistance, so it will be damaged if it is connected incorrectly; its purpose is to measure the current that is flowing in a circuit. Normally the meter is scaled in amperes, but if an ammeter is to be used on electronic circuits, the very small current flow requires a very accurate meter capable of reading in milliamperes (1 mA = 0.001 A).

Measurement of current in the circuit shown in Fig. 77.2 will require the ammeter to be connected in series with the circuit (Fig. 77.3). A typical reading for this type of system with the c.b. closed is about 4 A, but this will be reduced if an unwanted resistance is present in any part of the circuit, especially the c.b. Conversely, if a *short circuit* exists, such as when the switch-coil cable *shorts to earth*, then a higher ammeter reading is registered: the greater the short, the higher the reading.

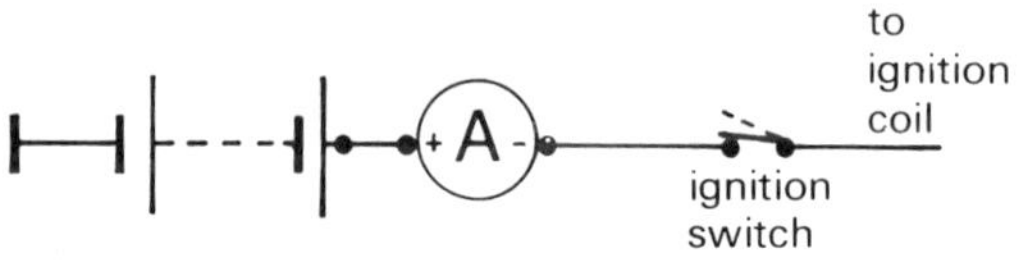

Fig. 77.3 Ammeter position to measure current in circuit

With all meters it is necessary that the measurement being taken is within the range of the meter. In cases of multi-scale meters, it is advisable initially to select a high scale and then reduce the range accordingly.

Use of ohmmeter This meter applies a low voltage across the circuit between the test probes. From the current flow produced, the meter is able to compute the resistance in ohms. An adjustment is provided on the meter to enable the scale to be zeroed to suit the voltage of the internal battery.

Fig. 77.4 shows how an ohmmeter is connected to measure the resistance of the primary winding of an ignition coil. It should be noted that the coil is disconnected from the circuit during this test because when the coil is left in the circuit the ohmmeter will measure the resistance of both the coil winding and the external circuit including the earth path.

Since the ohmmeter uses an internal battery to measure resistance, it should not be applied to any circuit that includes an electronic control unit, unless specified by the manufacturer.

Use of Ohm's Law to calculate resistance Ohm's Law states that:

$$\text{Resistance} = \frac{\text{Voltage}}{\text{Current}}$$

When an ohmmeter is not available, this expression can be used to measure the resistance of a com-

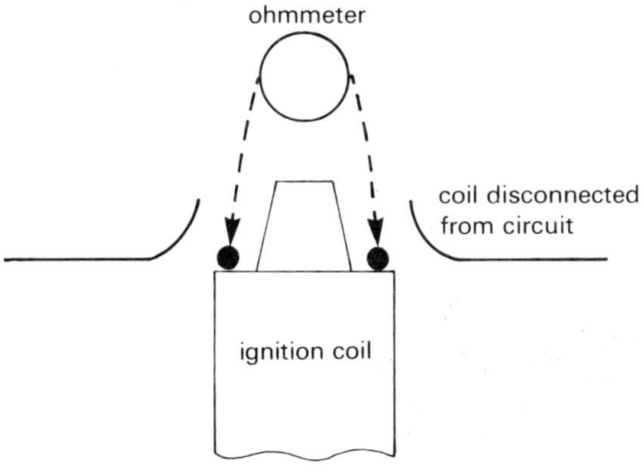

Fig. 77.4 Ohmmeter connected to measure resistance

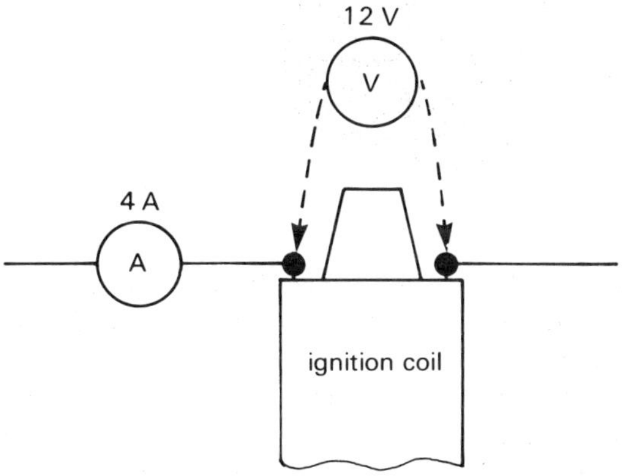

Fig. 77.5 Voltmeter–ammeter method for measurement of resistance

ponent. Arranging a voltmeter and ammeter in the manner shown in Fig. 77.5 and noting the reading given by each meter, the resistance can be calculated. In the example given the resistance is $\frac{12}{4} = 3\,\Omega$.

Use of multi-meters This type of meter combines an ammeter, voltmeter and ohmmeter in one unit and provides a number of different scale ranges to suit most applications.

A good multi-meter is expensive, so it must be treated with respect. Also care must be exercised to ensure that the meter is set to the correct scale BEFORE attempting to take a reading.

Fuses and overload protection

A circuit breaker, or fuse similar to the types shown in Fig. 77.6, is designed to break the circuit in the event of a short circuit or overload. This arrangement reduces the risk of fire and protects the circuit components from damage.

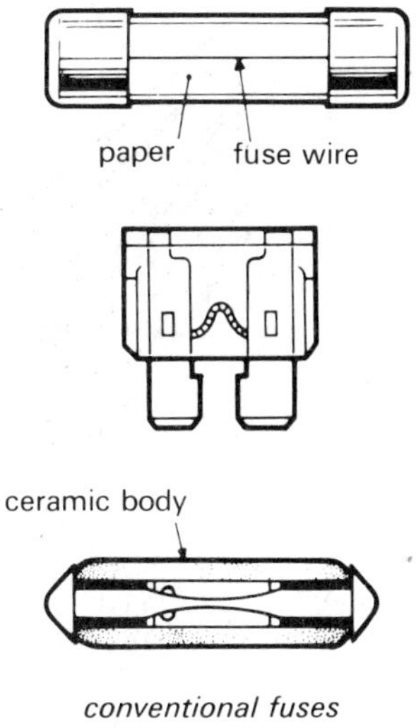

Fig. 77.6 Types of fuse

When a fuse melts, a new fuse of the correct rating is inserted; the system is then operated to see if it functions correctly. If a short circuit exists, the fuse will blow immediately; in this case the circuit must be checked out to pinpoint the cause.

A simple check is made by using a test lamp in place of the fuse. The test is started by switching off all systems that are controlled by the fuse, and then each separate circuit is switched on in turn: the lamp lights when the defective circuit is brought into action. When the faulty circuit has been identified, a visual check of the cables should be made. If this examination fails to establish the cause, then each section of the circuit should be disconnected in turn to locate the defective part.

Cables

The various electrical components are connected to the supply by low-resistance cables. Copper cables, stranded to give good flexibility, are generally used to connect components remotely situated. Where several cables follow a common path they are taped together to form a loom or wiring harness. This construction reduces both the risk of chafing against the metal frame and breakage of the cable due to vibration. To obtain maximum protection, the cables leave the loom at a point where the component is situated.

The quoted size of a cable refers to the wire diameter and the number of strands. If the cable is too small for its length, or for the current it has to carry,

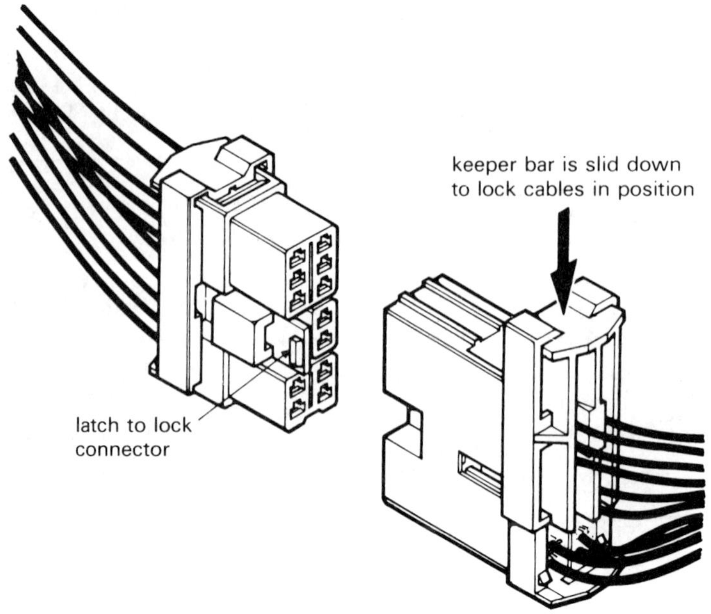

Fig. 77.7 Cable connector

then it will overheat and produce a voltage drop which affects the performance of the component, e.g. a light will not give its full brilliance and a starting motor will not revolve at the correct speed.

The current to be carried by a cable depends on the circuit to which it is fitted: this means that different-sized cables are used to save weight. Most cables, other than thick starter cables, have a wire diameter of 0.30 mm. Table 77.1 gives a general guide for cables of average length.

When a defective cable is renewed, it should be replaced with a cable of similar size. If a new circuit is to be installed, the maximum current load should be estimated in order to ascertain the cable size.

Table 77.1 Cable ratings and applications

Conductor size (No. of strands/diam., mm)	*Maximum current rating (ampere)*	*Application*
9/0.30	5.75	Lightly loaded circuits
14/0.30	8.75	Ignition circuits, sidelamps and tail-lamps, general body wiring
28/0.30	17.5	Headlamps, horns, heated rear windows
65/0.30	35	Ammeter circuit
120/0.30	60	Alternator charging circuit (heavy-duty)

Table 77.2 Colours for circuits

Circuit	*BSI*	*Letter code British*	*German*
Earth connections	black	B	SW
Ignition circuits	white	W	WS
Main battery feed	brown	N	BR
Sidelamps	red	R	RT
Auxiliaries controlled by ignition switch	green	G	GN
Auxiliaries not controlled by ignition switch	purple	P	VI
Headlamps	blue	U	BL

Assuming the length is not exceptional, a maximum current of 0.5 A per 0.30 mm strand is suggested.

Cable covering and coding

Today most cables are insulated with PVC plastics. This material has good resistance to fire, petrol and oil, but it should be noted that PVC gives off dangerous fumes when it is heated.

To aid identification the cable covering is coloured. The colour code used on a vehicle depends on the parent country of the manufacturer. Table 77.2 shows the colours used for the principal circuits.

In addition to the base colour, some cables have a thin tracer line running along the cable. The colour of this tracer identifies the part of the circuit formed by this cable; for example, a cable shown as UW has a blue base colour and a white tracer.

Connectors

Although bullet connectors of the soldered and crimped forms are still used to join cables, the need for greater terminal security and improved protection against the ingress of salt and moisture has demanded the use of more efficient connectors.

Fig. 77.7 shows a typical plug-and-socket connector used to join a number of cables. At the ends of the connector, the terminals are exposed; this provides a test point for meter checks.

A suspected unwanted resistance in a connector can be tested by carrying out a voltage-drop test across the pair of terminals. The connector must not be parted prior to this test because evidence of the fault will be lost. In the case of an intermittent fault, the simulation of engine heat can be produced by applying heat to the connector by a hair dryer. If a high resistance is detected, then the connector should be replaced with a new one of the approved design.

The so-called remedy achieved by 'waggling' the connector to make a good contact only provides a temporary solution; the effective repair is to fit a new connector.

Index

acceleration pumps, fitted to carburettors 93
accumulators *see* batteries
acid preparation for batteries 426
Ackermann steering layout 313–15
Ackroyd-Stuart, Herbert 175
acting, direct; servo 410
adhesive force 309, 385
air conditioning 347–56
  construction and operation 348
  refrigerant for 348
  systems 349, 354
air-cooling systems
  cylinders 138
  water-cooling compared with 138
air filters, in combination with silencers 126
air–fuel mixture, ratio 80
  defined 80
  function of inlet manifold in distribution of 507
  in starting from cold 100, 104
  methods of achieving correct 87
  provision for slow running 90, 100
  significance of 80, 92, 131
air suspension systems 362
all-speed governors, defined 186
alternator charging system 429–35
aluminium alloys, properties for engine parts 30, 51, 170
ampere 419
anti-dive suspension 368
anti-foam agents, in lubricants 156
anti-freeze 149
anti-knock rating *see* fuel, octane number
anti-lock braking system (ABS) 412–17
aquaplane 379
Arnott vane-compressor superchargers 202
articulated lorries, layout 13
asbestos-free friction material 386, 397
atomization 83
automatic advance, in coil-ignition system 448
automatic transmission systems and epicyclic gearing 250–75
axle casings 305
axle shafts 5, 306
  types 306
axles
  and types of drive 293–304
  four-wheel drive layout 308–12
  front 5, 367
  multi-drive 12
  rear, and final drive by bevel gears 5
  relation to Ackermann steering unit 313
  stub, reasons for adoption 313

balance, problems of, in single-cylinder engines 64, 69
balance masses, in structure, functions of crankshafts 67
ball bearings 171
banjo casings for axles 305
batteries 2, 6, 424–8
  functions 424
  lead–acid type 424
  maintenance of 426
baulk ring synchromesh gearing 238
bearings 167–73
  big-end 26
  functions 167
  in compression-ignition engines 170, 181
  lubrication of 173
  materials 169
  of crankshaft 22, 167–70
  types
    ball 171
    bushes 167
    roller 171
    split bearings 167
  use of thrust washers 170
Belleville disc-spring washers 301
bellows-type thermostats in cooling systems 146
bevel gears
  adjustments 296
  pinion positioning 293
  pre-loading 296
  to control noise 293
  lubricants for 294
  need for 293
  types
    hypoid 294
    spiral 293
    straight 293
big-end bearing 26
bleed screws, in compression-ignition fuel system 184, 189
bleeding of brakes 388, 406
bodywork 9, 343
bore of cylinder, defined 17
Borg and Beck clutches 215
Borg Warner type automatic gearbox 250, 256, 258
Bosch, Robert 183
Bosch D-Jetronic 116
  K-Jetronic 113–16
  L-Jetronic 119

Mono-Jetronic
VE-type fuel pump for CI engine 190–96
bottom dead centre, defined 15
bounce, problems of valve 38
Bourdon gauge 165
box chassis frame sections 338
brake fluid 388, 405
brake servos, in automatic transmission systems 245
brake shoes
in drum brakes 6, 386, 391–6
use of automatic adjusters 394
brakes 6, 385–417
air-powered 390, 411
anti-lock 412–17
diagnosis, correction of faults 406
disc type 6, 387, 396
factors affecting road–tyre adhesion 385
fluid properties for 388, 405
legal requirements 390
operating systems
compressed air 390, 411
hydraulic 388, 399–417
hydraulic servo 410
mechanical 387
servo assistance 392, 407–11
vacuum assistance 390, 407
Burt–McCullum sleeve valves 36
bushes 167
by-pass oil filters 162

cables, colour coding 474
cam and peg steering box 325
cam followers, in valve-operating gear 40, 44
camber angle, in steering systems 318
cams
in fuel injection pumps 185, 188, 193
in valve-operating gear 41, 46
camshaft
drives for 41–4
materials 41
carburettor 75–105
acceleration pumps fitted to 93
adjustment
for efficient starting from cold 94
for slow running 90, 101
choke 85, 95, 97, 100
constant-choke 87–99
economizers fitted to 92
electronic 104
engine running on problem 92
functions 80
methods of mixture correction 87–9
air bleed systems 88
basic principles 84–7
constant-depression types 87, 100–5
over-correction 88
mixture compensation 87–9
power system 92
strangler cold-starting 95
transfer flat spot 90
see also multi-carburettors, multi-choke carburettors
carden joint 279
carrier-type rear axle 305
cast-iron, properties for engine parts 30, 34, 49, 51
casting, in manufacture of crankshafts 24
catalyst 133
catalytic converter 133
c.d. type carburettor 100–5
cellular (film-type) radiators 145
centre-point steering 318
centric vane compressor 202
centrifugal blower superchargers 202
cetane ratings, in fuels for compression-ignition engines 178
chain drive, for operation of camshafts 42
channel chassis frame sections 338
charging systems 429–35
chassis
in vehicle development 6, 337–40
sections 338
choke tubes, in structure of carburettors 86, 97, 99, 100
choke valve 95
Citroën, hydro-pneumatic suspension system 362
clearance volume, defined 17
Clerk, Dugald 19
Clerk two-stroke engine 19, 199
closed-loop fuel control system 134
clutch 213–21
adjustment of 217, 218
development and purpose of 3, 5, 213
diaphragm-spring 215
faults 219
in automatic gearing 245, 247, 250, 257
'judder', causes, control 218, 220
multi-disc, in automatic transmission systems 260
multi-plate 214, 221
single-plate 214–20
construction, functions 214–18
crimping 217
use of strap drive 215
torque transmitted by 221
unidirectional 248
clutch linings 214, 221
coaches 14
combustion 58, 80, 178
avoidance of detonation 58
calorific value of fuel 80
factors determining rate of 81
in compression-ignition engines 178
optimum rate 58
process defined 58
significance of mixture strength 80
combustion chamber
for compression-ignition engines 179–81
direct injection type 179
indirect injection type 179
thermal efficiency 58
volumetric efficiency 58, 60
types
'bath-tub' 60
four valve 63
hemispherical 61
L-head 60

commutator 458
compensators for mechanically operated brakes 388
compression-ignition engines 12, 175–98
advantages, disadvantages 5, 182
combustion chambers 179–81
four-stroke operation 175
fuel systems 183–98
expulsion of air from 183, 186
layout 184, 190, 192
governors 186–9, 194, 196
history 175
injection pumps 183–96
injection systems 179, 183
lubrication 182, 187
necessary properties in fuels for 178
problems of control at idling, maximum speeds 186
starters 463
inertia 460
pre-engaged 460
starting from cold 180, 188, 195, 451
supercharging of 177, 200, 203
two-stroke operation 176
valve timing 181
compression ratio
defined 17
in compression-ignition engines 175
in petrol engines
relation to engine efficiency 199
compression rings *see* piston rings
condensers (capacitors), in coil-ignition systems 443, 446
conduction in cooling system 142
conductor, electrical 420
connecting rods 15, 26–8
general design 26
materials 26, 28
constant-choke carburettors 87
constant-load synchromesh gearing 237
constant-mesh gearing 237
constant-velocity universal joint 282–5, 291, 310
contact breakers in coil-ignition systems 442, 447, 457
continuously variable transmission (CVT) 228, 250
convection in cooling system 142
convertible (drop-head) bodies 9
coolants 138, 141, 149–51
advantages, disadvantages of water 139, 141
freezing, in cold weather 139, 149
cooling systems 137–54
air 138, 140
in Wankel engine 211
liquid 141–53
air compared with 138
circulation systems 142, 151
coolants 141, 149–51
heat transfer in 142
methods for control of temperature 146, 152
pumps for 143
radiator types 144–6
copper-lead, for engine bearings 170
correction in carburation, defined 88
corrosion, in cooling systems, causes, control 151
countermasses, in structure of crankshaft 67
crankcase, casting in unit with cylinder block 48
crank pin, in petrol engine structure 15
crankshaft 21–5
balance weights 21, 70
bearings 22, 167–70
causes of unbalance 21, 68–74
construction in compression-ignition engines 181
flywheel attachment 22
for multi-cylinder engines 22, 66–8, 70–2
front end of shaft 24
hardening 25
in petrol engine structure, functions 15
main features 15, 21
oil retainers 23
crimping, in single-plate clutch adjustments 217
cross flow 63
cross-ply tyres 379
CVT transmission 228, 250
cylinder 15
bore 17
side force applied by piston to wall of 29, 35
cylinder block 48
casting in unit with crankcase 48
materials 48
monobloc casting 48
cylinder head 50–2
advantages of fixed 48
construction, functions 50
fixing, tightening 51
materials 51
problems of detached 49
cylinder liners 49–50
dry 49
slip 50
wet 50

dampers 67, 101, 218, 364
functions 67, 218
torsional oscillation 68
types 364
dampness, problem of, in starting 451
Day, Joseph 19
Day two-stroke engine 19
dead centres in petrol engine function, defined 15
de-Dion drive 288
delivery valves, in fuel injection pumps 185, 193
Derv, volatility of 182
detergents, as additives in lubricants 156, 182
detonation, causes, control of 58
diaphragm-spring clutches 215
Diesel, Dr Rudolf 175
Diesel engines *see* compression-ignition engines
Diesel knock, causes, correction, 178
differential gears 299–304
limited slip differential 301–4
locking arrangement 301
purposes, principles 5, 299
visco type 302, 310
differential, pressure; brakes 404
diodes, in alternator charging systems 432
direction indicators 464, 466

disc brakes 386, 396–8
dished wheels 319
distributor, in coil-ignition systems 442, 447
distributorless ignition system 455
distributor pumps, in compression-ignition engines 188–96
domed piston crowns 30
double contact breakers 451
double-declutching 234
double-helical gears 227, 237
double link and radius rod independent front suspension 368
drag links, in steering systems 316
dribble, elimination in fuel systems of compression-ignition engines 185
driveability 135
drive arrangements 7, 288–92, 308–12
drive shaft 5, 287
drop arms, in steering systems 316
drop-head (convertible) bodies 9
drum brakes 386, 391–6
dry cylinder liners 49
Dunlop, J. B. 378
duo-servo (self-energizing) brakes 394
dwell-angle period, in function of ignition units 450
dynamic wheel unbalance 383

earth circuits, in electrical system 422, 461
eccentric-rotor oil pumps 159
eccentric-vane oil pumps 160
econimizers, fitted to carburettors 92
eight-wheeled vehicles 12
electric fuel pumps 77, 113, 122
electric-powered vehicles 207
electrical systems, equipment 6, 424–75
  battery 424–8
  charging 429–35
  earth return 422, 447
  for accessories 464–5
  ignition 441–57
  lighting 436–40
  starter system 458–63
  testing and maintenance 422, 470–5
  see also individual items
electrical terms 419–23
electrolytes, substances, levels 424, 426
electromotive force (e.m.f.) 419
electronic control unit (ECU) 104, 112, 115, 116–125, 266, 274, 335, 413
electronic ignition 453
emission, exhaust 59, 81, 91, 112, 131–6, 197, 205, 206, 207, 211, 212
engine balance 68–74
  component 71
  dynamic 69
  firing interval 71
  primary 69
  secondary 71
  static 71
engine management system 455
engine noise *see* silencers
engine size, relation to swept volume 17
engine, square and oversquare 17
engines *see under different types of engine*
enrichment devices in carburettors 92
epicyclic gearing 227, 244–7, 250–75, 277, 311
epitrochoid principle in Wankel engine 210
estate cars *see* station wagons
ethylene glycol, use as anti-freeze 151
exhaust gases
  composition of 81, 131–6
  harnessing to superchargers 199, 202–6
  recirculation of 60
  relation of air/fuel ratio to composition of 81
exhaust-heated inlet manifolds 107
exhaust manifolds 128
  design, performance 128
  fixing of inlet manifold to (hot spot) 107
exhaust pipe, in design of exhaust manifold 128
exhaust systems, component parts 128–30
expansion tanks, for sealed cooling systems 153

fade, brake 387, 396
fail safe 327, 407
fan belt 143, 148
fan pulley, mounting on crankshaft 24
fans 148
  electric motor 148
Ferguson formula for front-wheel drive 309
field, magnetic 429, 444, 458
fillet rolling 25
filters
  combination with silencers 126
  in fuel supply systems
    of compression-ignition engines 183, 184
    of petrol engines 77, 114–25
  in lubrication systems 158, 161–3
final drive gears
  in front-wheel drive layout 291
  in structure of rear axle 5, 293–304
firing order in multi-cylinder engines 65–8
five-cylinder engine 67
fixed-choke carburettors 87–99
flasher 466
float chamber, in simple carburettor 84
float indicators, for checking oil levels 166
fluid coupling flywheel 250
flywheel
  attachment, functions, in structure of crankshaft 22
  in hydraulic transmission systems 250–75
Ford
  CVT transmission 275
  gearbox 241, 259, 275
Fottinger, Dr 250
four-stroke engines
  compression-ignition system 175
  cycle of operations in petrol engine 16–17, 53–5
  see also individual aspects, parts
four-valve combustion chamber 63
four-wheel
  drive mechanisms 8, 308–12
  steering 331–6
    Honda 332

Mazda 332
frames 6, 337–45
alignment, checking, straightening 339
construction 337–9, 341
integral types 341–5
energy-absorbing 339
materials 338, 341
repairs 339
free-wheel unit 248
freezing, dangers in cooling systems 149
Freon refrigerant 348
friction 155
causes 155
factors determining force necessary to overcome 155, 385
front axles 5, 313
front hub construction 326
front-wheel drive 8, 291, 310
fuel
calorific value 80
combustion defined 58
in compression-ignition engines 175, 178, 182
need for correct vaporization 81
octane, cetane ratings 58, 178
provision of mixture for slow-running 89–91
significance of mixture strength 80
types 80, 178
fuel consumption
function, operation of carburettor economizers 92
in compression-ignition engines 180, 182, 200
in gas turbine engines 209
result of use of turbochargers 199, 204
fuel feed pumps, in compression-ignition engines 184
fuel filters 184
fuel governors 186–9, 194, 196
fuel injection 82, 112–25
CI engine 175–98
electronic 112, 115, 117
flap metering 117
hot-wire metering 117, 119
mechanical 113–16
multi-point 112–24
single point 112
spark-ignition engine 112–25
throttle body 124
vane metering 117
fuel spray
factors affecting behaviour of 106, 197
fuel supply systems
in petrol engines 75–136, 112–25
filters 75, 77, 114–25
filters combined with silencer 126
pipe-lines, connections 79
pumps 76–8, 113–24
tank 75
types 12, 76, 82
in compression-ignition engines 183–98
full-flow oil filters 162
fully-floating
axle shafts 306
gudgeon pins 34
fusing, causes, cure 423, 436, 473

gas turbine engines 207–9
gaskets, between cylinder head and block 51
gearbox 35, 223–76
automatic, construction and operation 250
electronically controlled 274
faults and tests 254, 274
lubrication of 269, 274
manual
constant-mesh 227, 237
sliding-mesh 227, 231–6, 237
synchro-mesh 227, 237
gear jumping, causes 236
gear whine, causes 236
gearing
calculation of ratio 224, 232, 246, 295
development, purpose of 223–30
differential in rear axles 299–304
limited slip 301–4
purpose, principles 5, 299
faults with 236, 237
front- and rear-wheel drive 243, 291
final-drive 293–304
planetary 244–7, 256–60
types
double helical 227, 237
epicyclic 227, 244–7, 256–60, 311
helical 227, 237, 293
spur 227, 231
gear-type oil pumps 158–61, 264, 328, 410, 413
generators 429–33
Girling brake mechanism 402
glow plugs 180
governors
in automatic transmission systems 266
in compression-ignition engines
construction, functions 186–9, 194, 196
gravity-feed fuel supply system 75
gudgeon pin 27, 34–5

Hall generator in ignition system 453
harmonic balancer 73
hatchback 9
header tanks, for liquid cooling systems 141
headlamps
adjustment 440
cable colour coding 474
dipping 436
two, four-lamp systems 439
heater plug starting of CI engine 180
heaters 347
helical spring suspension 358
high-rate discharge battery tests 426
Hooke, Robert 279
hooke-type universal joints 279
horn 464
hot spot, in exhaust-heated manifolds 107, 111
Hotchkiss drive 288
Hydragas suspension 362
hydraulic automatic transmission systems 250–75

hydraulic brakes 388, 399–417
hydraulic governors for compression-ignition engines 188, 189, 191
hydraulic power-assisted steering 327–9
hydraulic servo brakes 392, 407–11
hydraulic transmission systems 250–75
Hydrolastic suspension 360
hydrometer battery tests 426, 428
hydropneumatic suspension 362
hygroscopic substances 406
hypoid bevel gears 294
hysteresis, in rubber suspension systems 360

idling and maximum speed governors, defined 186
ignition systems 6, 441–57
  breakerless 452–6
  cable colour coding 474
  condensers 442, 446
  contact-breakers 442, 447, 457
  distributorless 455
  distributors 442, 447
  double contact-breakers 451
  electronic 452–7
  functions of electric circuit 442
  optical 454
  principles of coil-ignition system 441–7
  sparking plugs 442, 449, 456
  timing 57, 135, 442, 448, 452, 457
  transistorized 452
independent
  front suspension 5, 367–70
  rear suspension 5, 371–4
indium, use in copper-lead bearings 170
induction hardening of crankshafts 25
induction manifolds 106–111, 181
induction robbery 108–11
induction turbulence 107, 134, 179
inductive storage 454
inertia force 69
inertia fuel switch 77, 123
inertia valve; brake 402
injection pumps in compression-ignition engines
  in layout of fuel system 184, 190
  types
    distributor pumps 188–96
    in-line pumps 184–8
injectors 196
  servicing 197
inlet manifolds 106–11
  essential features 106, 181
  for four-cylinder engines 108
  for six-cylinder engines 110
  functions in distribution of air–fuel mixture 108
  types
    exhaust-heated 107
    water-heated 107
in-line cylinder engines 65–7
  firing order 65–7
  numbering methods 65–7
in-line injector pumps in compression-ignition engines 184–8
insulator, electrical 420
integral frame construction 341–5
Invar steel, for piston skirts 32

jacking points 10
jet engines, principles of 207–9
'judder', causes, control 218, 220
jump leads, use of 426

Kettering ignition system 442
kick-down in automatic gearbox 269
king pin, in steering systems and independent front suspension 317
  inclination 318
Knight sleeve valve 36
knock sensor 455
knocking (ticking) in gearbox, causes 236
  see also detonation

Lambda closed-loop fuel control 116
laminated (leaf) spring suspension 357
lamps 436–40, 464
Lanchester, Frederick 73, 158
layout of vehicles 1–14
layshaft (countershaft) 231
lead-acid type batteries 424–8
  capacity 425
  care and maintenance, topping up 426
  construction 424, 427
  electrolyte 424, 426
  fitting 426
  preparation of acid for 426
  safety precautions in use of 426
  tests 426
  voltage variation 425
lead-bronze, use in bearings 170
lead in petrol 135
leading and trailing shoe brakes 391
lean-burn engine 134
Lenoir, Etienne 18
lights *see* lamps *and under individual lights*
lighting system 436–40
limited-slip differential mechanism 301
lip-type seals, for retaining oil in crankshaft 23
liquid cooling systems
  air-cooling systems compared with 138
  circulation system 142, 151
  coolants 141, 150
  pumps for 143
  radiator types 144–6
load-apportioning valve for brakes 404
Lockheed hydraulic brakes 399
lorries *see* trucks
lozenging of chassis frames 337
lubricants
  extreme pressure, for bevel gears 294
  primary function 155
  viscosity 155
lubrication 155–66
  boundary 157
  hydrodynamic 157
  of axle differential mechanism 301

of bearings 157
of compression-ignition engines 182, 187
of gearboxes 236, 238
of fuel system in compression-ignition engines 31
of parts of governors 187
of universal joints 280
Lucas-Girling anti-lock brake 416
Luft 119

MacPherson independent front suspension 365, 369
magnetism and induction 443
main filter, in lubrication system 161-3
main journals, in structure, functions of crankshafts 21
mainshaft (third motion shaft) 231
manifold absolute pressure (MAP) 116
manifold 106-11, 128, 181
heated 107, 111
heating 107, 180
plenum chamber in 110
resonating chamber 110
siamesed port 108
manifold pressure
acceleration pump controlled by 94
enrichment jet controlled by 93
MAP sensor 116
mean effective pressure
defined 80
determining factors 80
effect of variation of air/fuel ratio 81
mean torque in single-cylinder engines 64
mechanical efficiency 59
mechanical fuel pumps 76
mechanical governors, for compression-ignition engines 186, 189, 192
mechanically operated brakes 387
mid-engine 7
mineral oils as lubricants 156
mini-bus 4
Mitsubishi
engine balance system 73
Silent Shaft balancer 73
mixing chamber of carburettor 86
mixture correction 87, 88
air bleed systems 88
over-correction 88
principles of 89
monoxide, carbon 91, 131-4
muffler *see* silencer
multi-carburettors, purpose, design of 97-9, 108
multi-choke carburettors 97-9, 108
multi-cylinder engines 64-74
arrangement 65
in line 66
vee 67
cylinder numbering methods 65
multi-hole nozzles, in fuel injectors in compression-ignition engines 179, 197
multimeter, use of 473
multi-plate clutch 214, 221, 257, 260
multi-stage torque converter 255

natural circulation cooling systems defined 141
negative offset 319
negative scrub radius 319
negative wheel castor 320
nitriding of crankshafts 25
nodular iron 24
nozzles, in fuel injectors in compression-ignition engines 179, 197
NSU, use of Wankel-type engine 209

octane rating 58
offset gudgeon pin 35
offset, negative; steering 319
ohm 420
ohmmeter, use of 420, 472
Ohm's law 420, 472
oil
additives 156
control (scraper) piston rings 33
cooling of 157
engines *see* compression-ignition engines
functions in lubrication system 158
pressure
indication 164
relief valves 161
properties 156
pumps 158-61
types of 158-61
oiliness 156
oilways, in lubrication systems 159
open circuit, testing for 422, 471
optical trigger 454
oscillation, torsional 67
Otto, Dr N. A. 18
Otto cycle 18
overcooling, troubles resulting from 138
overdrive
principles 277
use of epicyclic gearing 260, 277
overgear 225
overhead valve 40
overlay, bearing 170
oversteering, causes, control 314-16
oxides of nitrogen, NO_x 59, 131-6, 199

Panhard rod 289
parallel link independent rear suspension 371
Perkins Prima CI engine 180
petrol 16
process of vaporization mixing with air 81
unleaded 135
see also air-fuel mixture, carburettor
petrol engines 5
compression ratio, defined 17
cycle of operations 18-20
definition, indication of size 17
main parts 15-17
method of working 15-20
petrol injection 112-25
phases of combustion 178

phasing in fuel injection of compression-ignition engines 185
pinion positioning in bevel gears 293-5, 297
Pintaux nozzles, in fuel injectors in compression-ignition engines 180, 197
pintle nozzles, in fuel injectors in compression-ignition engines 180, 197
piston 15, 29-32
  assembly of split-skirt 31
  assembly together with connecting rod 27
  clearance 32
  cooling of, by lubricating oil 157
  details 30
  function in four-stroke cycle 18
  in petrol engine structure and functions 15
  in two-stroke cycle 19
  main features 29
  materials 30
  structure in compression-ignition engines 181
piston crown 30
piston rings 32-4
  compression, pressure, gas, defined 32-3
  fit of 33
  main features 32
  materials 34
  oil-control, scraper rings 33
piston skirt, construction functions 31
piston slap, causes 35
planetary gearing 227, 244-8
plunge-type CV joints 284
pneumatic governors, for compression-ignition engines 187
polarity plug electrode 450
pollutants, discharge of 59, 81, 91, 112, 131-6, 164, 197, 205, 206, 207, 211, 212
pollution 112, 131-6, 164, 197, 207
poppet valves, structure, functions 36-7
Porsche engine balance system 73
positve crankcase ventilation (PCV) 133, 164
positive wheel castor 320
potential difference 419
power, electrical 420
power-assisted steering 327
power jets in carburettors 92
power take-off 236
power unit 5, 207
pre-engaged starter drive 462
pre-ignition
  causes, effects 137, 175
  defined 37, 137
pressure cooling systems, avoidance of scalds 152
pressure plate 214-16
  disphragm spring 215
  in clutch construction of single plate 214
pressure-regulating valve
  fuel injection 114, 121, 123, 190, 193
  hydraulic brake 402
  lubrication 161
primary (clutch, first motion) shaft 231
propeller shaft 6, 285-7
  composite type 286
  two-piece 286
  whirling speed of 286
PTFE bearings 169, 368
pumps
  for liquid cooling systems 143
  for power-assisted steering 327
  in hydraulic converters 251-5
  in lubrication systems 158-61
  in automatic transmission systems 264
  see also injection pumps
pump-feed fuel supply systems 76
  see also electric fuel pump, mechanical fuel pump
push-rods, in valve-operating gear 46

rack and pinion steering box 325, 329
radial-ply tyres 379
radiation in cooling system 143
radiators 144-6
  adaptation for pressure cooling systems 151
  blinds 147
  construction 144
    cellular or film-type 144
    tube and fin 144
  cross-flow type 145
  design for liquid cooling systems 144
  shutters 148
ram cylinders, for power-assisted steering 327
rating, speed; tyres 381
rating, ply; tyres 381
rear axles 293-312
  and final drive by bevel gears 293-8
  differential 299-304
structure, functions 299
rear lamps 436
rear-wheel drive, reasons for general adoption of 7
recirculating ball steering box 325
rectifier, electrical 431
regulators
  in alternator charging systems 433
  in automatic transmission systems 263
remote control, gearbox 241
resistance, air; rolling and gradient 224, 343
resistance in circuits, testing for 471
resistor, ballast 451
reticular tin 170
reversed Elliot stub axles 317
Ricardo, Sir Harry 178
Ricardo Comet combustion chambers 179
ridge-dodger ring 33
ring belt of piston, structure, functions 29
Rochas, M. Beau de 18
rockers, in valve-operating gear 46-7
rocking couple 70
roller followers, in fuel injection pumps 185, 195
roller tappets, in valve-operating gear 44
Roots-type blowers 201
  in compression–ignition engines 177
rotary valve 36
rotors in Roots blowers 201
rubber suspension 360
Rzeppa CV joints 282

SAE 155, 405
SCD chain adjuster 42
Schrader valve 351, 380
scraper rings *see* piston rings, oil-control
screw and nut steering box 325
scroll seal for crankshaft oil retainer 23
seat belts 10
seating, seats 10
self-aligning torque 316
self-servo action of brakes 392
semi-drop centre wheel rims 375
semi-floating axle shafts 306
semi-floating gudgeon pins 34
sequential carburettor 97
servo assistance 407–11
shock absorbers *see* suspension systems
shooting-brakes *see* station wagons
Shorrocks vane-compressor superchargers 202
short circuits, testing for 422, 471
side (L-head) valve 60
  see also valves
side-valve engines 60
sidelamps 436
silencers
  exhaust 129
  in combination with air cleaners 126
silica gel dryer 352
simple harmonic motion (SHM) 71–4
Sinclair, H. 250
single-cylinder engines
  functioning, limitations 64
  problem of balance 64, 69
single-hole nozzles, in fuel injectors for compression-ignition engines 180, 197
six-cylinder engines 66
  exhaust manifolds 129
  inlet manifolds 110
size of petrol engines, definition 17
sleeve valves 36
slide valves 36
sliding-mesh gearbox 231–6
  disadvantages 237
  four-speed 231–4
slip cylinder liners 50
slow running
  adaptation of carburettor for 90, 100
  adjustment of carburettor for 90
  appropriate mixture strength for 90
  volume-controlled system 90
sludge 164
small-end of connecting rods 27
solenoid switches in starter motors 460
solid piston skirt 31
sparking plugs 442, 449, 456
speedometer drive 243
spherical roller bearings 171
spiral bevel gears 293
split bearings 167
split casings for axles 305
split line brake 402
springs
  in clutch construction, functions of single-plate 214
  diaphragm 215
  suspension 357–63
  valve 38
square engine 17
squish effect 60, 179
stabilizer, voltage 468
starter systems 458–63
  construction 458
  drive 460
  faults, causes, correction 462
  inertia 460
  layout 459
  pre-engaged drive 460
  solenoids 460, 461
starting from cold 95, 115, 451
  in compression-ignition engines 180, 188, 195
static wheel unbalance 383
station wagons, layout 10
stators
  in alternator charging systems 431
  in construction, function of hydraulic converters 250
steam engine 211
steering systems 313–36
  Ackermann layout 313–15
  components
    front hub construction 326
    power assistance 327–9
    steering gearbox 317, 324
    steering joints 317, 325
  four-wheel 331–6
  oversteer and understeer 314
  swinging beam 313
  wheel alignment 322
stepped transmission system 228
stretch bolts 51
Stirling engine 211
strangler valve 95
strap drive, in single-plate clutches 215
stroke, defined 15
stub axle, in steering layouts 2, 313, 327
supercharging, superchargers 199–206
  principles 199
  types
    centrifugal blower 202
    Roots blower 177, 199, 201
    turbocharger 199–206
    vane compressor 201
surge protection diode in alternator 433
surging of valve springs 38
suspended vacuum servo 407
suspension systems
  dampers for 364–6
  independent front 5, 367–70
  independent rear 5, 371–4
  purposes 5, 357
  types
    air 362
    Hydrolastic 360
    hydropneumatic 362
    metal spring 357–9

rubber 360
swash plate 212, 351
swept (displacement) volume
defined 17
in calculation of compression ratio 17
swinging arm, in independent suspension systems 369, 371
swinging beam steering 313
swinging half-axles 371
swirl (air flow), in function of compression-ignition engines 179
symptom, fault and cause, definition of 470
synchro-mesh gearing
need for 237
types
baulk ring 238
constant load 237

tandem master cylinder, in hydraulic brakes 399
tank *see* fuel supply system
taper roller bearings 171, 296, 326
tappets
barrel, roller, in valve-operating gear 44
clearance 40
hydraulic 44-5
in fuel injection pumps 185
materials 44
temperature, control in cooling systems 146
temperature gauges 469
thermal efficiency 58, 175, 182, 199
thermal slot piston 31
thermostats in cooling systems 146
thermosyphon cooling systems 141
Thomson, R. W. 378
three-cylinder engine 70
three-element torque converters 254
three-quarter floating axle shafts 306
throttle 86, 90, 268
throw, in function of crankshaft 21
toe-out on turns 314
top dead centre 15
marking of position for valve timing 56
torque converters, in construction, function of automatic gearbox 250-5
torque fluctuations 64
torque reaction 288
torque-to-yield method of bolt tightening 51
torque tube drive 288
torsional oscillation 67
torsion bar suspension 359
total volume
defined 17
in calculation of compression ratio 17
Toyo Kogyo, use of Wankel-type engine 210
Tracta CV joint 282
track arm 314
track rod
in steering layouts 314
linkages, in independent front suspension 370
traction control system (TCS) 417
tractive effort 226, 229, 288, 309
transaxle 292, 310
transfer ports, in two-stroke cycle 19
transistor, principle of 454
transmission systems 5
automatic 250-75, 308-12
transverse link independent front suspension 367, 371
trigger, ignition 453
Tripode CV joint 282, 284
tube- and fin-type radiators 144
tubeless tyres 379
tubular chassis frame sections 338
tungsten-halogen lamp 437
turbocharger 199, 202-6
boost limitation of 205
construction and operation 202
function 203
intercooler 205
turbulence 58, 59, 135, 178
twin overhead camshaft layout 42, 46
twin-rotor Wankel engines 211
two leading shoe brakes 392
two-stroke engines
cycle of operations 19
pressure charging 177, 199
tyre slip angles 314
tyre wear, causes, control 323
tyres 378-82
construction 378
cross-ply (diagonal ply) 379
inflation pressure 380
interchanging 381
marking of 381
legal requirements 382
radial-ply 379
tubeless 379
wear patterns of 323, 382
valves 380

unbalance, in operation of crankshaft 21, 68-74
undergear 225
understeer, causes, control 314-16
unitary type of construction 6, 341-5
universal joints 279-87
in structure, functions of 5, 6
front-wheel drive mechanism 291, 327
rear axles 5, 373
lubrication of 280
need for 5, 279
types 279-87
constant velocity 282-5, 373
flexible fabric 279
hooke-type 279
layrub couplings 280
rubber trunnion mountings 280

vacuum exhauster system 410
vacuum timing controls, in coil-ignition system 448
valve guide 37, 163
valve timing 44, 56
in compression-ignition engines 181
in connecting up of camshaft drives 44
in relation to piston and crankshaft position 56

valves
  aluminized 39
  clearance 40
  functions 36
  in hydraulic, automatic transmission systems 263–75
  in pressure cooling systems 146, 151–3
  in thermostats 146
  materials 38
  operating gear 40
  pressure regulating, in hydraulic brakes 402
  problems of bounce, surging 38
  springs 38
  timing 44, 53–7, 181
  types 36, 263
vane compressor-type superchargers 202
vans, layout of 11
vaporization, function of 81
variable-depression carburettors 87–99
vee-cylinder engines 65, 67, 111
vegetable oils as lubricants 156
vehicle
  development 1–3
  heavy 11
  light 4
  light commercial 11
  structure layout 327–45
venturi control unit
  in structure of carburettor 85, 97, 100
  relation to pneumatic governors 187
venturi, variable 100–5
vertical overhead valves 61
viscous couplings
  cooling system 148
  transmission 302, 310
viscosity of lubricants 155
  index improvers 156
  necessary properties 155
  relation to temperature (viscosity index) 155
Volkswagen
  Beetle 7
  gearbox 248
volt 419
voltage regulators, in charging 433
voltage stabilizer 468
volumetric efficiency 58, 108

Wankel engines 209–11
water, as coolant; advantages, disadvantages 139, 141
water cooling systems *see* liquid cooling systems, water as coolant
water-heated inlet manifolds 108
water jacket, design 142
water pumps, in cooling systems 143
watt 420, 458
wax-element-type thermostats 146
Weiss CV joint 282
well-based wheel rims 375
wet cylinder lines 50
wheel castor 320
wheel hubs, lubrication of bearings in 307, 326
wheel shimmy 367
wheels 4, 375–7
  alignment 322, 374, 382
  balancing 383
  checking and adjusting alignment 322
  space-saver 377
  types of rim 375
  see also braking systems, suspension systems, tyres
whirling speed of shafts 286
white metal, use for bearings 169
windscreen wipers 464
wishbone-type independent front suspension 367
worm and roller steering box 325
worm and sector steering box 324
worm gears 295, 324

Yield point 51

Zenith carburettors 101